ATLAS OF THE BREEDING BIRDS OF NEVADA

DAS 2004

BREEDING BIRDS
OF NEVADA

TED FLOYD

CHRIS S. ELPHICK

GRAHAM CHISHOLM

KEVIN MACK

ROBERT G. ELSTON

ELISABETH M. AMMON

AND JOHN D. BOONE

ILLUSTRATIONS BY RAY NELSON
MAPS BY ROBERT G. ELSTON

WITH FOREWORDS BY
SENATOR HARRY REID
AND C. RICHARD TRACY

UNIVERSITY OF NEVADA PRESS
RENO AND LAS VEGAS

University of Nevada Press, Reno, Nevada 89557 USA

Copyright © 2007 by University of Nevada Press

Species drawings copyright © 2007 by Ray Nelson

All rights reserved

Manufactured in China

Library of Congress Cataloging-in-Publication Data

Atlas of the breeding birds of Nevada / Ted Floyd . . . [et al.] ; with forewords

by Senator Harry Reid and C. Richard Tracy , illustrations by Ray Nelson ;

maps by Robert G. Elston.

p.   cm.

Includes bibliographical references and index.

ISBN-13: 978-0-87417-695-7 (alk. paper)

1. Birds—Nevada. 2. Birds—Nevada—Geographical distribution. I. Floyd,

Ted, 1968–

QL684.N3A85 2007

598.09793—dc22   2006034347

The paper used in this book meets the requirements of American

National Standard for Information Sciences—Permanence of Paper

for Printed Library Materials, ANSI Z.48-1984. Binding

materials were selected for strength and durability.

FIRST PRINTING

16   15   14   13   12   11   10   09   08   07

5   4   3   2   1

Frontispiece: Illustration of Black-throated Sparrow copyright © 2004 by
David Allen Sibley

CONTENTS

ILLUSTRATIONS

FOREWORD BY SENATOR HARRY REID

Nevada is a land of superlatives. It has been America's fastest-growing state for years, and its ever-expanding population is among the nation's most urbanized. Away from the cities, Nevada is also unique, with more public land, by percentage, than any other state save Alaska. And while Nevada is the nation's most arid state, it also has more mountain ranges than any other—314! Yet, paradoxically, the richness and diversity of Nevada's landscapes, and the wildlife they support, are often overlooked and underappreciated. Among those who have not visited our state, misperceptions abound, perhaps the most common being that Nevada is a vast, flat desert, home to rattlesnakes, coyotes, and tumbleweed, and, of course, a place to detonate nuclear devices. Nevadans know better, but still there is mystery and uncertainty surrounding the true extent—and the condition—of our biological resources. Systematic treatments of our flora and fauna are far more limited than is the case within more densely populated states.

It is my belief that our rapid growth, coupled with our great plenitude of deserts, rangelands, forests, and mountains, and our scarcer, but biologically important, streams and wetlands, provide us with both an opportunity and an obligation to grow responsibly—to learn all that we can about our natural resources so we can intelligently integrate the needs of the economy and the environment. The book you now hold is an important step toward developing and promoting that understanding. It is one of the relatively few statewide breeding bird atlases in the West, and one of the first anywhere to incorporate predictive modeling or provide habitat-based inventories. The *Atlas of the Breeding Birds of Nevada* is all the more remarkable in that it was generated in large measure through the efforts of scores of dedicated volunteers, citizen-scientists following in the great tradition of America's natural historians. The *Atlas* also provides a positive model for what we can accomplish when resource mangers, academic researchers, private landowners, and volunteers cooperate.

A work of this scale and scope can be expected to expand our knowledge, even in unexpected ways. For example, I note with pleasure that two of the *Atlas*'s biggest surprises—breeding Gilded Flickers and Rufous-crowned Sparrows—were found very near my home in Searchlight. I am sure you too will find some interesting surprises in the regions of the state with which you are most familiar.

It is a given that the *Atlas* will be widely consulted by federal, state, and local resource managers and academics. Those entrusted with overseeing our public lands, wildlife, and natural resources have labored for far too long under the limitations imposed by a shortage of reliable data and general resources, and this book will help to close that gap. But I also hope the *Atlas* will help to deepen the interest of a wider cross-section of Nevadans in the natural heritage of our state. I hope many will familiarize themselves with the birds that share our landscapes, and learn what they have to teach us about the health and integrity of our environment. Only then will we be able to pass this knowledge on to our children, families, and friends. From such understanding comes appreciation, and from appreciation comes stewardship.

FOREWORD BY C. RICHARD TRACY

University of Nevada, Reno, Department of Biology

This book is an important legacy of Graham Chisholm whose ambitious vision included founding the Great Basin Bird Observatory (GBBO) and creating the Nevada Breeding Bird Atlas project. The concept of a breeding bird atlas for Nevada seemed to be a daunting task, because there was scant infrastructure and a seemingly insufficient number of skilled birdwatchers in the mid 1990s to handle a major census of birds in such a large and diverse state. Chisholm came to the University of Nevada, Reno's Biological Resources Research Center for help, because he did not see any way that GBBO could afford to census the density of sampling sites typical of breeding bird atlases in eastern states. I suggested to him that by using digital spatial modeling, we could interpolate the distributions of Nevada's breeding birds based on data from a smaller subset of potential sampling sites. The approach required collecting data from approximately 800 spatially random plots distributed among several ecosystems in the state. Even though this approach would require many fewer sample sites than other atlases, it was not an easy one. For example, approximately 15 samples fell randomly within the Nevada Test Site, and one of those random sites fell very close to Area 51. We had meetings to explain how important it would be for us to stick to the scientific sampling design in order to infer the geographic range of each bird species, but imagine trying to explain the need to use binoculars to identify birds near Area 51!

In spite of formidable obstacles, GBBO helped develop the infrastructure, rallied the birding community, and raised the funds required to complete the Nevada Breeding Bird Atlas project. The result is a book that is unquestionably an invaluable resource to biologists, state and federal agencies, conservationists, birdwatchers, and many others. Nevada needs this book! Nevada has exceedingly sensitive species and ecosystems at a time when change to its ecosystems occurs at rates greater than at any time in history. Nevada is visited by bird species that travel thousands of miles to breed in tiny riparian strips in a sea of desert vegetation. In some cases (such as the Yellow-billed Cuckoo and the

Southwestern Willow Flycatcher), these visiting populations have very small numbers of individuals, and the year-to-year variation in their population numbers in Nevada is so huge that it would be impossible to track population trends over time. Fires are now occurring in the Mojave and Great Basin Deserts at rates that could cause some Nevada shrublands to be completely replaced by alien annual grasses and mustards. As urban development in Clark County and agricultural development in northern counties require more-and-more water, lakes become playas and wetlands disappear into deserts. The result is that dozens of species of snails, amphibians, and fish are going extinct—and birds could follow. When Walker Lake no longer has fish (a likely outcome within the next decade), there will be no loon festival in the town of Hawthorne, American White Pelicans will not travel there from Anaho Island to feed, and none of Nevada's other piscivorous birds will be able to consider Walker Lake their home. The atlas will help us manage Nevada's ecosystems so that "likely outcomes" do not mean that we have to lose our state's precious biological resources.

We should not forget that this *Atlas* was built partially from statistical models, the efficacy of which depends upon tacit assumptions. In fact, it is likely that the atlas project has missed finding at least a few species that are Nevada residents, and we know that some uncertainty remains over the distributional details of many other species. We should also be aware that these kinds of uncertainties can be attributed, to some extent, to the methods used by the atlas project. What all of this means is that the atlas project should be considered a beginning, not the ultimate answer, for developing our understanding of Nevada's bird species.

This excellent *Atlas* provides an enormously valuable benchmark, and it also provides solid hypotheses about the biology and ecology of bird species in Nevada. Now we need to pursue those hypotheses so that we can facilitate the needed science that will help us protect our state's biological resources. Because Graham Chisholm lived in, and continues to love, Nevada, we now have a maturing infrastructure to continue the science and management of its precious biological resources so that our children and grandchildren, hopefully, will be able to experience no fewer species than are present in Nevada today.

PREFACE

The *Atlas of the Breeding Birds of Nevada* is the result of a great deal of careful planning and a monumental amount of work. It seems appropriate to start it with the personal perspectives of those who played key roles in guiding the project from inception to publication. The origins of the atlas project are perhaps best described in the words of John Swett, who was on the initial atlas project planning team:

> The genesis of the *Atlas* can be traced back to an August 1996 Western Working Group meeting of Partners in Flight held in Sierra Vista, Arizona. The Western Working Group struggled with the enormous task of taking national Partners in Flight priorities and evaluating them at the regional and state levels so that state, regional, and continental bird conservation plans could be written. The group held wide-ranging discussions on species prioritization, habitat associations, and population and habitat objectives. The available data varied from state to state and from physiographic area to physiographic area, but one fact was abundantly clear to the four members of Nevada Partners in Flight present at this meeting: the paucity of data for Nevada was unmatched. During the evening social hour, Jim Ramakka, Bob Flores, Joe Kahl, and I discussed potential solutions to this problem and decided to initiate the Nevada breeding bird atlas based on the ongoing *Arizona Breeding Bird Atlas* that Joe Kahl and I had worked with as volunteers. Jim Ramakka, the Nevada Partners in Flight state chair at that time, asked Graham Chisholm in the summer of 1996 to take the lead for this important project. In December 1996, Nevada Partners in Flight voted to endorse the newly formed Great Basin Bird Observatory's role in conducting the Nevada breeding bird atlas, thus setting the project in motion.
>
> Prior to the atlas project, Nevada's young birding community had no experience with large-scale volunteer-based data collection efforts. The project had a galvanizing

effect on birdwatchers in Nevada. They came together as a community to collect data that only people with their experience know how to collect, and to contribute to a better scientific understanding of bird distributions in the state. The project was fueled by an extraordinary partnership between government agencies, conservation initiatives, academic researchers, and private citizens, whose alliance in pursuit of the atlas still reverberates through the community. Among its results are the statewide land bird monitoring program, the Nevada Bird Count, and a formal framework for all bird monitoring in Nevada, the Nevada Coordinated Bird Monitoring Plan. These efforts represent an unusually strong convergence of groups who share a common interest: better information on bird populations in Nevada.

GRAHAM CHISHOLM: FOUNDER AND FORMER BOARD CHAIR, GREAT BASIN BIRD OBSERVATORY

With Nevada Partners in Flight's support for a Nevada breeding bird atlas secured, Jim Eidel, Larry Neel, and I explored the idea of forming the Great Basin Bird Observatory (GBBO) to become the coordinating entity for the project. GBBO was incorporated in December 1996, when the decision to take on the atlas project was finalized. The atlas project was to be GBBO's central focus for the next five years. GBBO played a key role in coordinating the efforts of the academic community (most prominently the University of Nevada, Reno's Biological Resources Research Center), public agencies, corporate and foundation supporters, and the volunteer base, and made it possible to implement an ambitious project that no single preexisting entity could have undertaken alone. GBBO has since enjoyed the role of seeing the project through to publication and has taken the lead on several statewide bird population monitoring programs as the next logical step to follow the comprehensive inventory of bird distributions in Nevada.

TED FLOYD: ATLAS COORDINATOR

Many books about Nevada natural history open with an apology for the "Sagebrush Sea" or the "Big Nothing." I will take a different approach by recounting my earliest impressions of Nevada, which I visited for the first time in November 1998. The descent into the Reno airport was spectacular. A feisty snowstorm was blowing down off the Sierra Nevada, and a labyrinth of valley streams stretched out before me. My hotel room was in downtown Reno, within a two-minute walk of dazzling Hooded Mergansers (uncommon elsewhere in the region), loud-mouthed Western Scrub-Jays (of the distinctive *californica* subspecies), and comical American Dippers (which breed right in Reno, I would later learn).

Early the next morning, I traveled to Pyramid Lake with Graham Chisholm and Jim Eidel, two of the founders of the Great Basin Bird Observatory. Swirling gray clouds and rugged mountain foothills imparted a stark beauty to the desert landscape just north of Sparks. No amount of perspective or experience prepared me for the spectacle of Pyramid Lake—immense, dazzling blue, with whitecaps glinting like crystal in the wan winter light. We spent most of the morning in the shelter of Hardscrabble Creek Canyon, a secluded drainage on the west side of Pyramid Lake. Here we found resident birds such as Bewick's Wrens and Black-billed Magpies, early winter visitors such as Townsend's Solitaires, and even a couple of strays—a pert little Winter Wren and a close-up White-throated Sparrow.

I was hooked on Nevada birding, and I longed for a return visit. In fact, I moved to Reno, all the way from Philadelphia, just two months later in January 1999 to take over the reins of the *Atlas of the Breeding Birds of Nevada* as the project coordinator. Although spring comes slowly to northern Nevada, I couldn't help but notice signs of breeding bird activity all around me: American Kestrels poking about a nest site at an old brick building on South Virginia Street, a Cooper's Hawk carrying sticks to a tall cottonwood in urban Oxbow Park, and American Dippers singing loudly above the din of traffic on the Arlington Street Bridge.

In the years that followed I was fortunate to get to visit much of Nevada—a land that encompasses major chunks of the Great Basin and the Mojave Desert as well as thin slivers of the distinctive Sierra Nevada and Columbia Plateau ecoregions.

In the far southern reaches of Nevada, one may find Gilded Flickers and Bendire's Thrashers breeding in the lowland Joshua tree forests; Gray Vireos, Rufous-crowned Sparrows, and Black-chinned Sparrows in the pinyon-juniper-covered foothills; and Grace's Warblers and Flammulated Owls in the tall pine forests of the mountains. In lowland riparian situations, Blue Grosbeaks, Phainopeplas, and Lucy's Warblers are fairly common, and threatened or endangered breeders such as Yuma Clapper Rails, Western Yellow-billed Cuckoos, and Southwestern Willow Flycatchers can be found in remnant patches of suitable habitat.

Western Nevada, where the Sierra Nevada slices across the boundary with California, is home to a highly distinctive avifauna. White-headed Woodpeckers and Red-breasted Sapsuckers breed in the pine forests of the Carson Range, and Pileated and Black-backed Woodpeckers show up from time to time. Cassin's Vireos sing from aspens, and Nashville Warblers are frequently heard in montane shrub assemblages of manzanita and mountain mahogany. Vaux's Swifts and Hermit Warblers are occasional summer visitors, although breeding has not been confirmed for either species.

Northeastern Nevada offers the birder an intriguing mix of lush agricultural valleys and jagged mountain ranges that harbor a distinctive "eastern" avifauna. Bobolinks can be fairly common breeders in hayfields and wet meadows, and Eastern Kingbirds are infrequent but present as breeders along hedgerows and woodland edges. Species such as Grasshopper Sparrows and Gray Catbirds are reported from time to time, although breeding has not been confirmed in recent years. Up in the hills, Swainson's Thrushes and MacGillivray's Warblers sing loudly from the leafy aspen forests. And higher still—above tree line—there are Himalayan Snowcocks, American Pipits, and Black Rosy-Finches.

Finally, no description of Nevada's birds would be complete without mention of the great "middle"—the "sagebrush sea" that leaps instantly to mind when most people think of Nevada. It is a land of austere beauty with its own complement of unique bird life. Black-throated Sparrows sing weakly in the greasewood, Gray Flycatchers make their homes in the tall sagebrush, and introduced Chukars flush along the dusty backcountry roads. In the astonishingly fertile marshes and desert oases, small colonies of Black Terns and Franklin's Gulls return to breed each year, garnet-eyed Eared Grebes build their flimsy floating nests, and stunningly attired Black-necked Stilts and American Avocets nest by the thousands.

RAY NELSON: ATLAS ILLUSTRATOR

The first time Ted Floyd and I discussed the artwork for the atlas, it was clear to me that we shared a common vision for the aesthetic appeal of the document. We both wanted to do something just a little bit different. Consequently, the illustrations are not the standard, predictable poses ordinarily depicted in avian identification guides. That type of illustration serves its purpose well, but this book cannot hope to substitute for a field guide, which readers of the *Atlas* no doubt possess already.

Instead, the drawings are primarily composed to tell the story of the birds that breed in Nevada. Most of the images refer to features of breeding birds: courtship displays, breeding-season plumages, paired males and females, birds carrying food or fecal sacs, birds singing, parents on the nest, chicks or eggs in the nest, adults with their brood, and so on. Some refer to an anecdote in the text, demonstrate a landscape common to the state, or illustrate an identifiable Nevada landmark. On rare occasions I just had an itch that needed scratching, and the result is the artist's preference of the moment.

The development time for each image was split fairly evenly between research and actual drawing. I read accounts of nesting and breeding behavior from many sources and also considered my personal observations to interpret some aspect that I hoped would be of interest to the viewer. Although I regularly examined video footage and photos of a species prior to drawing, I always set aside any reference material while I sketched the motion or position of the bird. I used reference material again during the final inking to ensure, for instance, that the Gadwall hadn't picked up a physical characteristic unique to the Bushtit.

Our wish is that the illustrations in this book will enlighten you, or prompt you to confirm or question your recollections about species behavior, or will in a few cases simply amuse you.

Overleaf: map of prominent locations, p. xvi; map of sample block locations, p. xvii

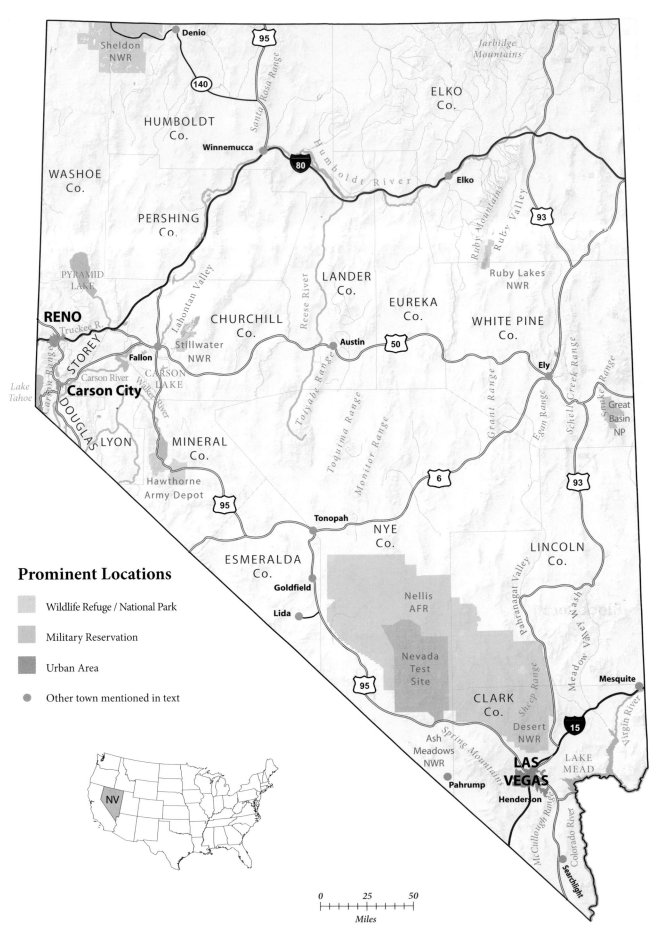

Prominent Locations

Wildlife Refuge / National Park

Military Reservation

Urban Area

● Other town mentioned in text

Denio

Sheldon NWR

HUMBOLDT Co.

Santa Rosa Range

140

95

Winnemucca

WASHOE Co.

PERSHING Co.

PYRAMID LAKE

RENO

Truckee R.

STOREY

Carson River

Carson City

Lake Tahoe

DOUGLAS

Carson Range

LYON

Fallon

CARSON LAKE

Stillwater NWR

Lahontan Valley

CHURCHILL Co.

Reese River

Walker River

MINERAL Co.

Hawthorne Army Depot

95

Toiyabe Range

LANDER Co.

Austin

50

Toquima Range

Monitor Range

EUREKA Co.

WHITE PINE Co.

Ely

Grant Range

Egan Range

Schell Creek Range

Snake Range

Great Basin NP

6

93

Tonopah

NYE Co.

ESMERALDA Co.

Goldfield

Lida

95

Nellis AFR

Nevada Test Site

LINCOLN Co.

Pahranagat Valley

Meadow Valley Wash

93

Mesquite

Virgin River

15

ELKO Co.

Jarbidge Mountains

80

Humboldt River

Elko

Ruby Mountains

Ruby Valley

Ruby Lakes NWR

Desert NWR

Ash Meadows NWR

Spring Mountains

CLARK Co.

Sheep Range

LAS VEGAS

LAKE MEAD

Pahrump

Henderson

McCullough Range

Colorado River

Searchlight

NV

0   25   50

Miles

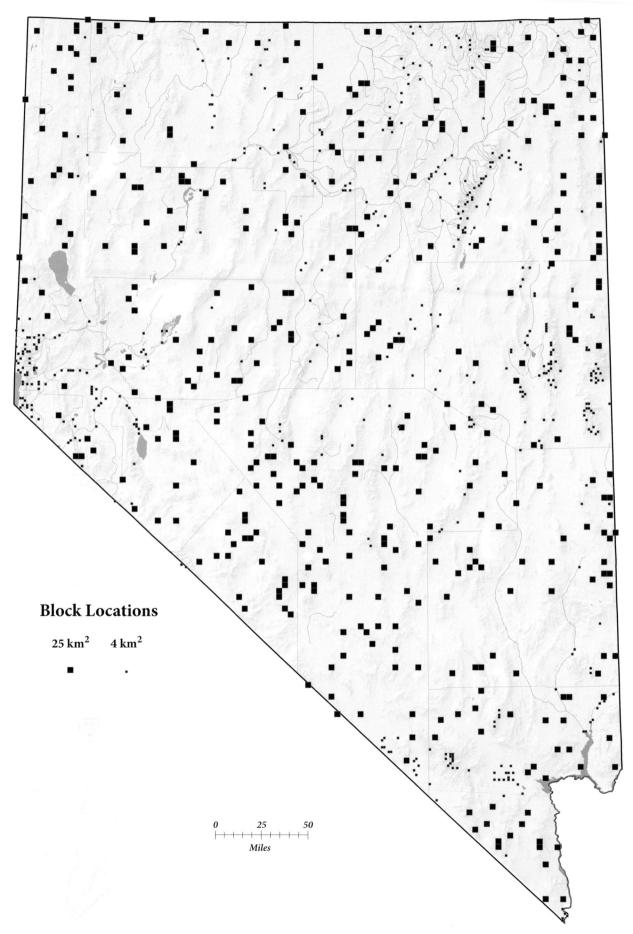

Block Locations

25 km²   4 km²

0   25   50

Miles

PART I

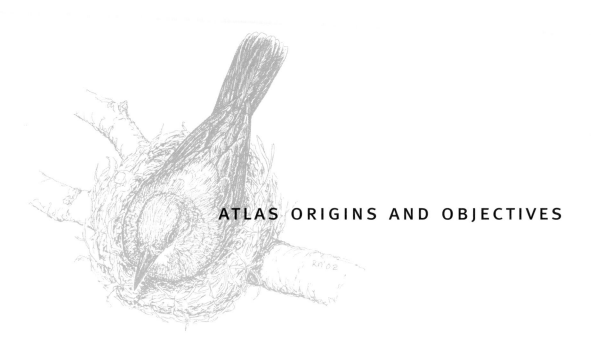

ATLAS ORIGINS AND OBJECTIVES

This book is the culmination of a great deal of work by a great many people. Like most breeding bird atlases, the primary objective of the Nevada project was to describe the geographic distribution of the state's breeding birds. From the outset, however, it was clear that achieving even this simple goal would be a major challenge. Nevada is larger than the states of Maryland, Delaware, New Jersey, New York, Connecticut, Rhode Island, Massachusetts, Vermont, and New Hampshire combined. But those states also have a human population that is about thirty-five times larger than Nevada's. Needless to say, the task of recruiting enough birders to conduct a volunteer-based surveying project throughout the entire state was far more formidable than it would have been elsewhere. To make matters worse, Nevada is not the most accessible of states. Apart from the rugged landscape, with more than three hundred mountain ranges, vast expanses of sagebrush and creosotebush, and elevations ranging from about 300 feet (100 meters) at the Colorado River to over 13,100 feet (4,000 meters) at Boundary Peak, there are relatively few roads in the state, and there are large areas of military land where access is greatly restricted. The restricted areas associated with Nellis Air Force Range alone cover more land than the state of Connecticut. These constraints had made Nevada's avifauna one of the most poorly known in North America when the atlas project began.

THE HISTORY

With Graham Chisholm of the Great Basin Bird Observatory willing to act as a catalyst, the Nevada Working Group of Partners in Flight agreed in December 1996 to support the atlas project. One of the first steps was the creation of the Atlas Steering Committee, which oversaw and guided the project from its inception. This committee comprised a mixture of people from government agencies, nonprofit conservation groups, academic institutions, supporting businesses, and the birding community. The project was coordinated by, and funded through, the Great Basin Bird Observatory (GBBO), a nonprofit organization that was formed to take on the atlas project. Although volunteers conducted much of the fieldwork, there were still considerable direct costs. The basic budget for the atlas, approximately $590,000, was provided by a range of grants and gifts from agencies, foundations, businesses, and private individuals. In addition, considerable in-kind support was contributed in many forms and from many sources. These included expert interviews and consultancy, time allocated for data analysis, provision of promotional materials, volunteers' donation of travel expenses, and many others.

Academic researchers were a part of the atlas project from the beginning, and GBBO and the Biological Resources Research Center (BRRC) at the University of Nevada, Reno worked closely together throughout. BRRC in particular helped with the initial study design and the components of the study that used Geographic Information Systems (GIS) technology. Peter Brussard and Richard Tracy served on the Atlas Steering Committee. Tim Wade, formerly of BRRC, helped design the atlas sampling scheme, and Robert Elston subsequently provided GIS guidance through the data collection, analysis, and map production phases of the study. All of these sources of support were crucial, and the project could not have been completed without them.

Data collection began in 1997, when 87 observers collectively visited 111 of the projected 800 atlas blocks. The success of this first year convinced everyone that the project was feasible, and three full-scale field seasons followed in 1998–2000. Volunteer birders, including many from other states, conducted much of the work; by the end of the project more than four hundred birders had contributed to the atlas database. Given the scope of the task, the GBBO also hired full-time fieldworkers each year to supplement the data collected by volunteers.

Initially the project was coordinated by Graham Chisholm, GBBO's executive director at the time. At the end of the initial field season, in fall 1997, Chris Elphick joined the atlas team to help with data analysis and the predictive modeling project, and Kevin Mack was hired to assist with volunteer and field crew coordination. By the end of the second field season it was apparent that the atlas needed a full-time manager to run the project. Ted Floyd was hired in January 1999 to oversee the last two years of fieldwork, to organize and analyze the data, and to coordinate the atlas book's production. During the last years of the project, Ted was assisted especially by Kevin Mack, who managed the database, and by Robert Elston at BRRC, who oversaw the GIS work relating to data analysis, predictive modeling, and map production.

After writing the initial draft of the manuscript, Ted Floyd left GBBO to become the editor of *Birding* magazine for the American Birding Association. Elisabeth Ammon, John Boone, and Chris Elphick, along with Ted, were responsible for revising and completing the manuscript. Robert Elston produced the maps, and Kevin Mack proofed all of the data included in the species accounts. Finally, Ray Nelson was commissioned to illustrate the species accounts.

PROJECT GOALS

From the outset, a main objective of the Nevada breeding bird atlas project was to quantify the diversity and grandeur of the state's bird life. Although several accounts of Nevada's birds and many detailed local studies had been published, there had never been a concerted effort to systematically describe the precise distribution patterns of all bird species across the entire state. With both the resource management and birding communities showing growing interest in Nevada's birds, the time seemed right to seek answers to fundamental questions about Nevada's avifauna.

The atlas project was designed to do more than just satisfy the curiosity of birders and ornithologists, though. It was also intended to gather data that would help to guide major bird conservation initiatives in the state such as the Clark County Desert Conservation Program's Multiple Species Habitat Conservation Plan (Clark County 2000). By building into the project an attempt to understand and predict species distributions, the atlas project was also intended to guide efforts to identify additional, as yet unsurveyed places where bird species might occur.

We hope that researchers and resource managers can build on the foundation that this book provides and proceed to ask more complex questions about the management and conservation of bird populations in Nevada. How important is sagebrush for Sage Sparrows? What demographic processes account for changes in population size and range for declining species? Can Bell's Vireos be found where our models predict a high likelihood of occurrence? Can we design detailed ecological studies on individual species that allow those models to be extended and improved?

The atlas results have largely confirmed preexisting ideas about the status and distribution of some species in the state; Gambel's Quail, Greater Roadrunner, and White-headed Woodpecker are good examples. For many more species the atlas has provided details on distribution that were not previously known; examples include Long-billed Curlew, Cassin's Kingbird, and Pine Siskin. The atlas has significantly redefined our thinking about some species—about their presence and range in the state; Gilded Flicker, Black-billed Magpie, and Gray Vireo come to mind in that regard.

In a real sense, the results of any breeding bird atlas are obsolete the moment they are published, simply because bird populations are naturally dynamic. The species accounts discuss many species whose ranges are in flux. Declining populations often catch our attention, but there are also species that are increasing in numbers and expanding their Nevada ranges; a short list includes Double-crested Cormorant, Red-shouldered Hawk, Eurasian Collared-Dove, Anna's Hummingbird, American Crow, Cliff Swallow, Bewick's Wren, Blue-gray Gnatcatcher, Great-tailed Grackle, and Lesser Goldfinch.

This book thus cannot be the final word on the distribution and abundance of our breeding avifauna. Instead, the real point of the *Atlas of the Breeding Birds of Nevada* is to galvanize future research and to inspire new questions. It is our hope that you will *use* the atlas: Use it to guide your research on bird populations in Nevada. Use it as a birding reference. Use it as an inspiration to go out and find new breeding populations in previously understudied parts of the state. Use it to clarify where our predictions are right—and where they are wrong. And then come back in twenty years to help repeat the exercise so that we can see how things have changed.

CONSERVATION ISSUES AND INITIATIVES
AFFECTING NEVADA'S BIRDS

While a large number of Nevada bird species are declining, many others are increasing in numbers or expanding their ranges. This highly dynamic avian fauna is presumed to be at least partly governed by changes that have happened on the Nevada landscape since European settlement. A current summary of conservation and management issues for birds in Nevada and Utah can be found in Floyd (2005). This chapter summarizes these issues and describes the current government programs and other projects that address birds and their conservation in the state. Resource management plans initiated and implemented by federal land managers are particularly important to the fate of Nevada's birds, because 87% of the land in Nevada is managed by the federal government. However, some of the lands that are richest in bird life, such as many desert springs and floodplains of streams and rivers, are privately owned, making stewardship by private landowners a critical component of bird conservation in the state.

GROWTH AND WATER

Nevada has for many years had the fastest-growing population in the nation. It is also the most arid state in the country. Nevada gets most of its water from winter precipitation and is largely excluded from the generous midsummer monsoons that inundate parts of Arizona and Utah. The rapid development has put unprecedented pressure on Nevada's water resources, including both surface and groundwater reserves. For instance, water rights to the Walker River's flows are allocated at over 140%; in other words, only in years with well above average water flows does any appreciable amount flow into the ecologically important Walker Lake. Walker Lake is one of the best-known terminal lakes and wetlands of the Great Basin. Like Mono Lake, Pyramid Lake, and the Lahontan Valley, it has no outlet and, historically, was the end of the line for all water that entered the drainage. All of Nevada's terminal lakes and wetlands were—and still are—major resources for aquatic birds, both for fish eaters such as loons and pelicans, and for invertebrate feeders such as shorebirds and many ducks.

Recent years have seen a number of successful efforts to acquire water rights to restore in-stream flows to some of Nevada's rivers. For example the Truckee River and Pyramid Lake have benefited from the purchase of water rights by Great Basin Land and Water and by the U.S. Bureau of Reclamation's Truckee River Operating Agreement. Similar acquisitions by The Nature Conservancy, the U.S. Fish and Wildlife Service, and the state of Nevada have established water rights for conservation for the Lahontan Valley wetlands. These water rights are critical to maintain the long-term ecological health of these important habitats.

Nevada's existing cities are seeing explosive population growth today. The greater Las Vegas area is leading the trend, but many lesser-known communities like Pahrump, Fernley, and Mesquite are also seeing their populations surge. Native habitats are being converted to urban and suburban settings at an increasing rate, including the mesquite and acacia-covered expanses in the Las Vegas Valley and much of the historic wetlands of the Truckee Meadows and Carson Valley. Increased pressure on available water resources is among our greatest concerns. Most municipal water currently comes from the major rivers, using water rights that once were put to agricultural use. Currently, the available water is largely allocated, and further growth in the cities will require increased use of groundwater aquifers or trans-basin water imports, an ecologically thorny issue. The largest currently proposed project of this nature is a plan to pump groundwater in eastern Nevada's Lincoln and White Pine counties and import it by pipelines into southern Nevada. Developers in the Reno area are currently looking for ways to import water from the rural areas north of Reno and Sparks. Understandably, major concerns exist about the possible ecological impacts of these projects: unsustainable rates of groundwater pumping can dry up local springs and wetlands supplied by groundwater. In a desert state like Nevada, these small "oases" are often the primary source of water, cover, and invertebrate prey populations for birds amid large expanses of desert uplands. Nevada will have to join other desert states in a search for ways to manage this finite resource in the context of explosive population growth and the need to maintain and improve prospects for the state's avian and other wildlife populations.

PUBLIC LAND AND LAND BIRD HABITATS

Nevada's remaining riparian woodlands are a continuing concern for bird conservationists. Channelization, conversion of lands for agricultural use, livestock grazing, erosion and downcutting of streambeds, water diversion, and impoundments are major causes of habitat loss or degradation. Most of these impacts began well before monitoring programs were in place, making it particularly difficult to estimate the overall extent of the habitat change or its effects. There is little doubt, however, that just about *any* management measure that leads to the recovery of riparian woodlands along our streams and rivers should be considered a net gain for riparian birds. In Northern Nevada, The Nature Conservancy and the Bureau of Land Management have begun to acquire large sections of the lower Truckee River corridor for open space and habitat restoration. The Pyramid Lake Paiute Tribe has led efforts to restore flows to the lower Truckee River suitable for the endangered cui-ui, a fish endemic to the Pyramid Lake system, and for recovery of riparian woodlands.

The options for riparian recovery are much more restricted in southern Nevada. Most of the native riparian woodlands have been invaded—and often displaced—by the exotic tamarisk, and most areas along the Colorado River and its tributaries are so intensively developed that a reversal to a natural riparian setting may not be possible. Complicating matters, the endangered Southwestern Willow Flycatcher and Yuma Clapper Rail occur more and more often in tamarisk-dominated or otherwise heavily modified habitats, and the most radical habitat restoration efforts would have impacts on these and other rare riparian bird species. These circumstances produce massive challenges to all projects designed to maintain and improve conditions for birds in riparian habitats. The Bureau of Reclamation's Lower Colorado River Multi-Species Conservation Program, designed to increase the amount of suitable habitat for native birds in this region, is the largest-scale and probably most ambitious of these programs, but other agencies, such as the Lake Mead National Recreational Area (National Park Service), are also engaged in habitat improvement and species conservation projects.

Other habitat types that support unique bird communities are also subject to change in Nevada. For instance, land managers have long been concerned about the decline and degradation of the state's limited aspen groves. The damage has been attributed to a wide array of causes, including livestock grazing, timber cutting, and climate change. To better understand this problem, the USDA Forest Service's Humboldt-Toiyabe District, which includes all of Nevada's national forests, recently mapped all aspen stands in the district. The Nevada Bird Count project, a bird-monitoring program led by the Great Basin Bird Observatory, now supplies standardized bird population data for aspen stands so that stand condition can be statistically related to bird composition and abundance.

The Humboldt-Toiyabe District of the USDA Forest Service is also currently completing its twenty-five-year Forest Plan revision, which details how wildlife habitats will be managed on national forest lands. The Forest Service, which manages approximately 8% of the federal lands in Nevada, has responsibility for much of the higher-elevation bird habitats, including areas of high species diversity such as the Carson Range in far western Nevada, the Spring Mountains of southern Nevada, and the Jarbidge Range in far northern Nevada. Historically somewhat immune to many of the issues facing other national forests, like timber production and oil and gas development, the Humboldt-Toiyabe District is now seeing increased interest from energy developers, and applications for oil and gas development in White Pine County's National Forests are currently under consideration. The Forest Service is also under increasing pressure to implement fuel reduction projects that involve substantial removal of vegetation on national forests that are encroached by settlements, and many of these areas are also seeing an unprecedented surge in motorized outdoor recreation.

The Bureau of Land Management (BLM) is the largest land manager in the state, overseeing approximately 67% of Nevada's federal public land. Three of the six BLM districts in the state are currently revising their Resource Management Plans. In addition, the agency is confronting issues related to water pipeline rights-of-way and the potential impacts from geothermal and wind farm development. A major focus of the BLM's land use planning involves management of pinyon-juniper and sagebrush habitats, which are the subject of significant conservation concerns. Cheatgrass invasion, ungulate grazing, and the extensive wildfires that burned millions of acres around the turn of the millennium have so decimated and degraded sagebrush-dominated shrublands that many researchers feel they are in immediate danger of being lost (e.g., Knick et al. 2003). Since European settlement, pinyon-juniper woodlands have expanded into lower elevations than those they were historically known to occupy. The causes of this process are not entirely clear, but changes in fire regimes seem a likely factor. Other pinyon-juniper stands have become so dense that the shrub understory, which is important to many birds, is crowded out; usually, fires burn the entire stand. A Nevada research group of the Forest Service's Rocky Mountain Research Station has been studying the changes in pinyon-juniper woodlands and is working on management guidelines for maintaining a healthy mosaic of stand types (e.g., Miller and Rose 1999, Tausch 1999). Recently planned large-scale energy development projects for central and eastern Nevada threaten to fragment the habitats of sagebrush and pinyon-juniper birds that are not particularly tolerant of this process (e.g., the Greater Sage-Grouse and the Sage Sparrow).

CONSERVATION OPPORTUNITIES

In southern Nevada, many recent wildlife conservation efforts have been initiated under the auspices of Clark County's Desert Conservation Program, specifically the Clark County Multiple Species Habitat Conservation Plan (MSHCP), which went into effect in 2001 (Clark County 2000). The MSHCP started out as a Desert Tortoise Habitat Conservation Plan designed to prevent a net loss of the habitat of the threatened desert tortoise, but it soon became obvious that development in Clark County would affect other species as well. The new multiple species plan considers those other species; it currently includes eight bird species, and another seven bird species are under evaluation for inclusion. The accomplishments of this program, funded by the sale of public lands for development, include a variety of land acquisition, public education, conservation, and restoration projects, as well as the funding of research and monitoring projects designed to determine the status, population trends, and ecological requirements of the species affected by development in the county.

A large portion of the atlas project that led to this book was funded by the Clark County MSHCP and by the U.S. Bureau of Reclamation's Lower Colorado Region office. Specifically, both the MSHCP and the Bureau of Reclamation needed a better understanding of the distribution and numbers of the birds within Clark County, and of the habitats used by each species. In order to determine net losses to habitats or populations, the first step is to catalog exactly where birds occur, what habitats they use, and approximately how common they are. The determination of habitat preferences improves the ability to predict impacts on populations. The predictive models generated during the atlas data analysis may be of particular interest to large programs such as the Clark County MSHCP. Such models allow accurate estimation of the chances that particular bird species will occur in particular areas. The reliability of these models was tested in Clark County in order to statistically determine how well the model could predict the occurrence of different species. Now that these model tests have been completed (their results will be published elsewhere), the next step is to evaluate the need for additional information on species' habitat use and natural history in order to better predict where they occur.

The concept of species-based habitat conservation plans is taking a particularly strong hold in Nevada and some surrounding states. After the initial successes of the Clark County MSHCP, the Bureau of Reclamation completed and implemented a Multi-Species Conservation Program for the entire lower Colorado River region in 2004. It is hoped that large-scale efforts like these will open new doors for regional conservation planning and

bird-monitoring programs. Habitat conservation plans similar to the Clark County plan are currently being developed for the Virgin River and for Lincoln County, and others are likely to be produced in the future.

Several organizations are actively working to improve our understanding of bird populations and management needs in Nevada. The Nevada Partners in Flight chapter functions as a multi-agency umbrella initiative for statewide land bird conservation planning. The group's Bird Conservation Plan (Neel 1999a) identifies the state's major bird habitat types and priority bird species for conservation. The atlas, which Nevada Partners in Flight was largely responsible for initiating, adopted this approach and used the same habitat types for block selection. In 2002, Nevada Partners in Flight was also responsible for initiating the Nevada Bird Count, an ongoing statewide land bird monitoring program that is being implemented under the guidance of the Great Basin Bird Observatory. Such a standardized, annual bird-monitoring program is the key to detecting and understanding changes in Nevada's bird populations.

Several other large-scale planning efforts with ramifications for bird populations are also under way in Nevada. For instance, the BLM leads the Great Basin Restoration Initiative, which was developed to address conservation concerns brought on by the wildfires that recently burned millions of acres of Great Basin desert uplands in Nevada. The program brings a wide variety of stakeholders together to identify and address the issues involved in managing sagebrush and other upland habitats. The National Audubon Society's Important Bird Area Program recently identified thirty-seven Important Bird Areas in Nevada. This program is currently preparing conservation plans for maintaining and enhancing these hotspots for bird diversity in the state. Also, the Nevada Department of Wildlife recently completed the Nevada Comprehensive Wildlife Conservation Strategy, which will be among the state's most ambitious efforts toward wildlife conservation planning. All of the federal and state agencies and several nongovernmental organizations that have responsibilities for wildlife habitat contributed to this plan, which delineates conservation priorities and strategies for achieving conservation goals for wildlife populations and the habitats on which they depend in Nevada.

Because much of the land in Nevada is managed by one federal agency or another, these agencies and Congress are ultimately responsible for addressing many of the issues that affect the state's landscapes. The political debates and difficult decisions about the use of public lands in the United States will have a tremendous impact on Nevada. The challenge lies in ensuring that the qualities that make Nevada a unique place for both birds and people are preserved as the human population grows and public use and management of these lands change in the coming decades.

ATLAS METHODS

In the preface to his magnificent *Marin County Breeding Bird Atlas,* Dave Shuford wrote: "Beyond the grid-based distribution maps, there is no set formula (nor should there be) as to what warrants inclusion in a breeding bird atlas, or as to whether it even need be a book." The Nevada Breeding Bird Atlas Steering Committee agreed with this sentiment, but the logistical constraints imposed by Nevada's geography required us to depart somewhat from the traditional grid-based distribution maps. The maps in this atlas are based on a grid of contiguous blocks, but our data collection methods and maps differ from those of other breeding bird atlases.

Nevada is a big state with few people, a combination that is not well suited to traditional atlas methods, which rely on the twin assumptions of a large volunteer base of birders working in an accessible and manageably sized land area. Most breeding bird atlas projects subdivide the study area into a series of contiguous "blocks," or grid cells, and then attempt to compile a list of the species breeding within each cell on the grid. As the number of birders per land area declines, compromises have to be made. For example, block size can be increased; this makes the number of locations that need to be visited more manageable, but it also reduces the resolution of the data. Another option is to restrict surveys to a subset of blocks, which has the drawback of leaving unvisited "holes" in the resulting maps. Yet another approach is simply to record the point location of all records, regardless of any underlying grid.

In western North America, where large areas with few birders are the rule, atlas steering committees have confronted the problem in various ways. The *Colorado Breeding Bird Atlas* (Kingery 1998) is a successful example of an atlas survey that used a grid system considerably more coarse grained than those usually used in the East. Other atlas projects have increased the time span of the study, devoting up to ten years to data collection; the recently completed *Arizona Breeding Bird Atlas* and the ambitious *New Mexico Breeding Bird Atlas,* now under way, have taken this route. Yet another approach is to conduct atlas projects at the county level, as has been done in California, where many excellent county atlases have been published. Finally, there are atlases that have sampled only a subset of the land area covered; for example, by selecting a subset of blocks randomly or according to some systematic scheme. The *South Dakota Breeding Bird Atlas* (Peterson 1995) combined several of these sampling methods.

Geographic Information Systems (GIS) technology has added a new twist to atlas production. The "Washington Breeding Bird Atlas" (Smith et al. 1997) resolved the problem of limited sampling by combining remote sensing data describing patterns of land cover with information on bird habitat use to produce predicted breeding distributions that supplement the field data. More recently, the *Oregon Breeding Bird Atlas* (Adamus et al. 2001) combined sampling at several different scales to gain the advantages of both large, contiguous grid cells and smaller, more intensively surveyed blocks.

Our approach in Nevada was to use a habitat-stratified random-sampling design to select a subset of all possible blocks to sample. Despite the lengthy name, the idea is actually quite simple. In essence, we divided the state into sixteen major habitat types (see map on p. 12, which shows open water and barren zones) and then randomly selected a subset of blocks corresponding to each habitat type where we would collect data on breeding birds. The actual fieldwork then proceeded in a manner familiar to anyone who has done atlas work elsewhere. By limiting our fieldwork to a subset of blocks, we ensured that we would obtain detailed information about a set of precise locations while still maintaining representative coverage across the state.

ADVANTAGES OF STRATIFIED RANDOM SAMPLING

The project used stratified random sampling because it provided a data set suitable for predicting statewide distributions. Picking sample sites randomly within habitat types ensured that the data collected were representative of the entire state and minimized the chance of bias.

Numerous biases can arise if site selection is not randomized. For example, it is likely that people would focus on places that are already known to be good for birds, or that appear to have good habitat, or that are close to home. Although this approach might make for more interesting birding, it would not produce a data set representative of the state as a whole, and consequently would be useless for making extrapolations about which species are likely to be found in unvisited areas. When sites are picked randomly, such biases are diminished. Although atlas workers might have had to drive farther or spend more time in areas where there were few species, the resulting data are more useful because they allow statisticians to extrapolate more accurately about what birds are found in areas that were not visited during fieldwork.

The decision to stratify the sampling by habitat types was also a way to improve predictions about birds that are habitat specialists; this is especially important for rare habitats. Stratification ensured that each major habitat type received at least a minimum amount of coverage by atlas workers. If sites had been selected completely at random, without regard to habitat, then an enormous number of blocks would have had to be surveyed to ensure that the rarest habitats, and the birds associated with them, were represented. Obviously, this would defeat the purpose of using a sampling scheme to make the amount of fieldwork more manageable. In addition, there is always a strong temptation—even for the most neutral and unbiased of observers—to spend more time in productive riparian strips and less time exploring desolate expanses of greasewood or creosotebush. And some habitats, while not necessarily uninteresting, are scarce or remote in Nevada; ash woodlands in southern Nevada and alpine meadows in central and northern Nevada are good examples. Stratification ensured that all habitats, even the very rare and inaccessible ones, were included in the atlas sampling.

Finally, stratified random sampling enabled us to minimize our efforts in habitat types that occupy very large areas in the state. A good example is the salt desert scrub habitat that is widespread in west-central Nevada. This habitat is home to Common Ravens, Horned Larks, Black-throated Sparrows, and relatively little else. It would have been nice, of course, if we could have covered every square mile of salt desert scrub; but that was neither possible nor necessary. It was not necessary because this habitat is fairly homogeneous: one salt desert scrub block looks much like another. Consequently, we sampled only enough salt desert scrub habitat to detect large-scale geographic patterns in order to be able to make inferences about what birds would be found in the areas that were not sampled.

DISADVANTAGES OF STRATIFIED RANDOM SAMPLING

Stratified random sampling has disadvantages, too, of course. In our project these fell into three categories: problems caused by surveying only a sample of all possible sites, problems with using a stratified random scheme as the sampling method, and problems associated with predicting bird distributions from the data that were collected.

Certainly, some people will consider random sampling far inferior to the regular, rectangular grid system that so many other atlas projects have employed. But if it is possible to make reasonably accurate inferences about species distributions from surveys of a subset of sites, then this concern becomes unwarranted, and the extra work of visiting all sites unnecessary. Moreover, as noted above, exhaustive surveys are simply not possible in some regions of North America.

Although stratified random sampling has distinct advantages from the statistical perspective of reducing biases in the resulting data, it also introduces some problems. From a logistical standpoint, randomization can be troublesome because some of the sites selected are difficult—occasionally impossible—to visit. In these cases, alternative random sites are usually selected as replacements, but with the recognition that the randomization process becomes less perfect. Another problem is that random selection often does not pick sites that are known to have important bird populations or that are often visited by birders. The omission of these well-known birding sites may seem odd. Typically, however, these are not sites that atlas projects are designed to investigate (because their bird life is already well known), and information about breeding birds at these sites is usually collected anyway (see "Incidental Observations," below).

Perhaps the biggest problem with stratifying the sampling design is that it requires sites to be classified into strata. This would be fine if it were possible to view the state as neatly divisible into little squares that can be assigned to *either* habitat A *or* habitat B *or* habitat C. In parts of Nevada this is actually possible. Think of the "sagebrush sea" or the carpet of creosotebush-dominated scrublands in southern Nevada, where each block is a fairly uniform tract. In many areas, however, the blocks include a mixture of several habitats—often separated by indistinct boundaries where one habitat merges into another. Even the sagebrush sea has many small patches of other habitats—a small stream or a ranch house surrounded by shade trees—embedded within the sagebrush habitat. Consequently, rules have to be devised that define how much of a particular habitat a block must minimally contain in order to be classified as that habitat type.

Even deciding how to subdivide the continuous variation of nature into a discrete set of habitats is fraught with difficult, and often arbitrary, decisions, because habitats can constantly be subdivided into narrower and narrower categories. For instance, our atlas scheme combines all "Mojave" habitat into a single class, even though some bird species generally occur only in the subset of major blocks that have Joshua trees. There are probably additional differences associated with the abundance of Joshua trees and a host of other factors, too. And this problem undoubtedly repeats itself for every other defined habitat class. Thus, the set of categories we used was a compromise between capturing the range of variation in the bird assemblages that exist in the state and the number of "habitat types" that were manageable in terms of efficient data collection and analysis.

A final set of problems with our approach is that it assumes that we can infer statewide distributions from the subset of sites selected. Although we are reasonably confident that our sampling scheme produced better predictions than alternative methods would have done, this does not necessarily mean that the predictions are as good as we would like them to be. Many issues impinge on the accuracy of statistical predictions, not the least of which is the "noisiness" of a complex natural world. This topic is discussed in more detail below in the "Predicting Species Distributions" chapter (pp. 25–28).

ATLAS HABITAT TYPES AND BLOCK CLASSIFICATION

Habitat types rarely come in neat, uniform blocks. We had to develop a system for assigning habitat types to blocks that could then be sampled to ensure adequate coverage of each habitat. We first created a map describing the distribution of these habitats across the state using data collected as part of the national Gap Analysis Program (GAP). The Nevada GAP vegetation map used for the atlas project (produced by the second iteration of the GAP project) was based on satellite imagery data collected during the summer months between 1988 and 1991 (Homer 1997). GAP vegetation data are classified using the National Vegetation Classification System, a standardized system created by the Federal Geographic Data Committee. The GAP project recognized more than sixty different community types in Nevada. To simplify things, we aggregated these different categories into sixteen broad habitat types that seemed likely to be relevant when describing bird distributions (see map of habitat types on p. 12).

The sixteen habitat classes, which were used for selecting atlas blocks, were the following (scientific names of plants are listed in the Appendix):

or trees that grow along borders and ditches, and are an important habitat component for many birds.

As exemplified in the photo of Ruby Valley, the agricultural areas of Nevada provide important habitats for such species as the Bobolink, Long-billed Curlew, and Sandhill Crane. Photograph by Bob Goodman.

ALPINE—This largely herbaceous vegetation type is found above timberline on the tallest mountaintops (above 10,000 feet, or 3,000 meters). It consists of a variety of forbs, sedges, and grasses, along with low-growing shrubs and trees such as willows, limber pines, and bristlecone pines. An example of this habitat type is the summit of Wheeler Peak in the Snake Range, as illustrated here. When the snow is gone, birds such as rosy-finches, American Pipits, and White-crowned Sparrows breed in these habitats. Photograph by Bob Goodman.

AGRICULTURAL—In Nevada, this habitat type is rather restricted and occurs in valley bottoms, frequently near riparian areas or former wetlands. It consists of irrigated, cultivated fields or pastures. Often, agricultural areas are lined by broadleaved shrubs

ASH—This riparian woodland is found only in the Mojave Desert portion of Nevada. It is dominated by velvet ash and screwbean mesquite and is found along small spring outflows. In Nevada, this habitat type is largely restricted to Ash Meadows National Wildlife Refuge and the Amargosa Valley in southern Nye County, where it provides important habitat "oases" for many breeding and migrant songbirds. Photograph by Bob Goodman.

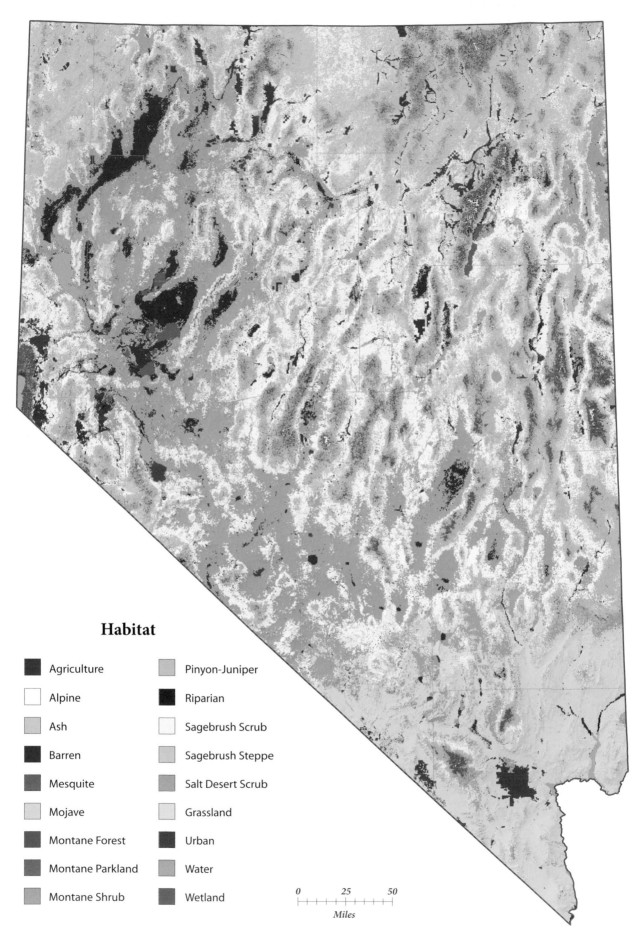

Habitat

- Agriculture
- Alpine
- Ash
- Barren
- Mesquite
- Mojave
- Montane Forest
- Montane Parkland
- Montane Shrub
- Pinyon-Juniper
- Riparian
- Sagebrush Scrub
- Sagebrush Steppe
- Salt Desert Scrub
- Grassland
- Urban
- Water
- Wetland

0   25   50
Miles

GRASSLAND—This habitat type is broadly defined to include all vegetation covers that are dominated by perennial and annual grasses. It is neither widespread nor common, and it is found primarily in the northern and central portions of Nevada.

MESQUITE—Mesquite habitat is found in scattered clumps, mostly in the western Mojave Desert of Nevada, which includes southern Nye and western Clark Counties. It is dominated by mesquite, but may also have tamarisk and Torrey saltbush. Mesquite also grows as part of riparian woodlands, in which case it is classified as "riparian" habitat. Mesquite woodlands support several bird species that are primarily associated with this habitat type. For instance, the Phainopepla requires mistletoe-infected trees such as those shown in the photo of a mesquite woodland near the town of Pahrump. The Spring Mountains are in the background. Photograph by Lisa Crampton.

MOJAVE—The Mojave habitat type is broadly defined to include most lower-elevation desert scrublands of the Mojave Desert. It is most often dominated by creosotebush, but also includes the biologically important yucca woodlands and cholla cactuses, as seen here near Mormon Mesa. Photograph by Bob Goodman.

MONTANE FOREST—This habitat type accounts for much of the coniferous (and occasionally broadleaf) woodlands of the higher elevations of the Great Basin, the Sierra Nevada, and the Mojave Desert. While it excludes the "dwarf conifers" (pinyon pine and junipers), it does consist most often of the taller pines (e.g., ponderosa or Jeffrey pines, lodgepole pine, or limber pine), firs, or spruces. Aspen stands, such as the one shown here on Peavine Peak, may also be categorized as "montane forest" when they occur away from riparian zones. Patches of the biologically important aspen woodlands are sometimes interspersed with conifers. Photograph by Bob Goodman.

MONTANE PARKLAND—Montane meadows and clearings within surrounding montane forest comprise the montane parkland habitat type. In Nevada, montane parkland is relatively rare, often consisting of small seeps and wet meadow habitats, as photographed here in the Little Valley of the Carson Range. Photograph by Elisabeth Ammon.

MONTANE SHRUBLAND Montane shrublands, found at middle to high elevations, consist of patchy stands of mostly broad-leaved shrubs or shrubby trees, such as manzanita, oak, currant, or snowberry in the Sierra Nevada region; and oak, mountain mahogany, sagebrush, and rabbitbrush in other parts of the state. Mountain mahogany woodlands, as shown here in the foothills around Peavine Peak of far western Nevada, are a commonly encountered example of this habitat type. Photograph by Bob Goodman.

PINYON-JUNIPER—One of the most widespread habitat types in Nevada, pinyon-juniper woodlands are savannah-like dwarf conifer forests that occur at middle elevations throughout the state. They are usually dominated by pinyon pine and/or several species of juniper, and they often have sagebrush-dominated shrublands as an understory. Along the state's northern tier of counties, approximately north of the Interstate 80 corridor, junipers predominate and pinyon pines are largely absent (Charlet 1996). The photo shows a typical pinyon-juniper landscape near the town of Ely in eastern Nevada. Photograph by Bob Goodman.

RIPARIAN—Riparian areas are restricted in this desert state, but they provide critical habitat for many birds. Riparian habitats include a wide variety of woodlands that grow in the floodplains of Nevada's larger rivers and along streams of all sizes and elevation ranges. Dominant trees and shrubs may include cottonwood, willow, alder, and aspen in the Great Basin, as well as exotic species such as Russian olive or tamarisk. In the Mojave Desert, riparian areas may be dominated by mesquite, tamarisk, willow, and, at higher elevations, aspen and willow. The top photo depicts a typical lowland Great Basin river floodplain along the Walker River, which features highly altered riparian woodlands. Higher-elevation riparian woodlands of Nevada are often dominated by aspen and willows, as seen here, immediately above, at a stream in the northern Ruby Mountains. Photographs by Bob Goodman.

SAGEBRUSH SCRUB—This habitat type includes all sagebrush-dominated shrublands that have relatively little forb or grass understory. These shrublands, with plants that may grow as tall as 6 feet (2 meters), function as important habitat for birds that nest in shrubs. Sagebrush scrublands are widespread throughout the central Great Basin and are particularly widespread in the central and eastern portions of Nevada. The photo here illustrates a typical sagebrush shrubland setting in the foothills of the Lake Range, just east of Pyramid Lake in western Nevada. Photograph by Bob Goodman.

SALT DESERT SCRUB—Salt desert scrub includes a variety of lowland desert vegetation that thrives in the (often-salty) valley bottoms of the Great Basin and Mojave Desert. Shadscale, Bailey's greasewood, Torrey saltbush, and fourwing saltbush are usually the dominant shrubs in these barren-looking habitats. The photo shows the Fernley Sink near the town of Fernley. Photograph by Bob Goodman.

SAGEBRUSH STEPPE—The steppe appears to be an almost even mixture of sagebrush-dominated shrublands and grassy areas that also feature a significant forb component. This habitat type is most abundant in the northernmost portions of the state where the Columbia Plateau extends into Nevada. It is an important habitat type for birds such as the Greater Sage-Grouse and Vesper Sparrow. The photo shows sagebrush steppe in the vicinity of scenic High Rock Canyon in northwestern Nevada. Photograph by Bob Goodman.

URBAN—In Nevada, urban habitats may take the form of inner-city landscapes that feature palm trees or other ornamental shade trees, heavily modified riparian areas, or suburban settings that offer bird feeders and a variety of ornamental and native woodland covers. The habitat cover map used for the atlas also included major commercial operations in the "urban" classification, which caused counterintuitive classifications in some cases. For example, the Hawthorne Army Depot south of Walker Lake in Mineral County was classified as "urban," but it is primarily a facility for ammunition storage and other military functions. Vegetation at the depot largely consists of salt desert species, and thus the site lacks most of the habitat features typically used by urban birds. More typical urban settings, however, provide important stopover habitats for migrating songbirds, and they often harbor distinct breeding bird communities as well. This photo shows an urban habitat at the Truckee River near downtown Reno. Photograph by Bob Goodman.

WETLAND—In Nevada, wetland marshes are largely restricted to valley bottoms where springs and larger watercourses provide terminal wetlands, such as the Lahontan Valley in western Nevada and the Ruby Valley in northeastern Nevada. Many wetland areas have been modified for agriculture, but several biologically important areas are protected as Wildlife Management Areas and National Wildlife Refuges. Permanent wetlands typically feature emergent vegetation such as bulrushes, sedges, cattails, and rushes. Wetlands are important habitats for a variety of aquatic birds for breeding, migration stopovers, and wintering. This photo shows a playa wetland area, Swan Lake, north of Reno. Photograph by Bob Goodman.

In addition to these sixteen habitat types used for classifying blocks, observers also reported other habitats that were not used during block selection. The following habitats were used to classify observer reports:

OPEN WATER—Many lakes and reservoirs, such as Lake Mead along the Colorado River and Rye Patch Reservoir along the lower Humboldt River, consist of open water with little wetland vegetation. Some, notably Pyramid and Walker lakes, are important waterbird sites. Shown here is a view of Pyramid Lake, with the "Pyramid" and Anaho Island visible in the background. Photograph by Bob Goodman.

The sixteen habitat types used for block selection were subdivided into two groups: those that exist in extensive patches that could be sampled at a relatively large scale, and those that occur in smaller patches that required finer-scale sampling. We then used our habitat map to determine the location of each of our sample blocks. We used 5 kilometer (3.1 mile) × 5 kilometer (25 square kilometers, or 9.6 square miles) sampling blocks for the extensive habitats, such as sagebrush steppe and Mojave. To select blocks for these habitat types, we generated a 5 kilometer × 5 kilometer grid and overlaid it on the habitat map. If a cell contained at least 80% of a particular habitat type, it was assigned to that type. From this grid of relatively uniform blocks we selected a subset to serve as atlas blocks. The number of blocks per habitat type was proportional to the area of each habitat type in the state, as calculated from the habitat map. In other words: the more common the habitat type, the more blocks of that habitat type we sampled.

We used 2 kilometer (1.2 mile) × 2 kilometer (4 square kilometers or 1.5 square miles) sampling blocks for the less common and patchier habitats such as montane forest and wetland. As with the 25 square kilometer blocks, the number of blocks sampled per habitat type was proportional to the area each habitat type covered in the state. These blocks were selected in a manner similar to the larger blocks. The rarity and patchiness of the restricted habitats made it impossible to select only blocks with at least 80% of a single habitat, though. Even at this smaller scale, blocks that are almost entirely mesquite, for example, are rare. Hence, blocks were assigned to particular habitats based on a smaller proportion of their area covered by that habitat. For instance, montane forests were defined as blocks containing at least 60% montane forest habitat, and montane parkland blocks contained at least 40% parkland. Table 1 summarizes the rules we used to assign blocks to a habitat type.

BARREN—Although not as widespread as some would expect in a desert state, barren areas are an important component of the Nevada landscape. They include all unvegetated land covers, such as inhospitable salt flats, or playas; sand dunes; and the biologically important cliffs, talus slopes, and rock outcroppings that several bird species use as nesting habitat. Barren habitat in the Goshute Range is depicted here. Photograph by Bob Goodman.

TABLE 1. ATLAS BLOCK HABITAT TYPES, SIZES, AND SAMPLING

HABITAT TYPE	BLOCK SIZE	PERCENTAGE OCCURRENCE ON BLOCK	ORIGINAL NUMBER OF BLOCKS	NUMBER OF SUBSTITUTE BLOCKS	NUMBER OF BLOCKS SAMPLED
Agricultural	4 km²	80%	25	0	24
Alpine	4 km²	30%	12	0	12
Ash	4 km²	30%	5	0	3
Grassland	4 km²	80%	20	2	21
Mesquite	4 km²	30%	4	0	4
Mojave	25 km²	80%	49	0	45
Montane Forest	4 km²	60%	81	41	111
Montane Parkland	4 km²	40%	20	1	19
Montane Shrubland	4 km²	80%	39	6	40
Pinyon-Juniper	25 km²	80%	48	3	48
Riparian	4 km²	200 meters	118	3	111
Sagebrush Scrub	25 km²	80%	106	2	100
Sagebrush Steppe	25 km²	80%	82	2	77
Salt Desert Scrub	25 km²	80%	108	9	106
Urban	4 km²	80%	45	0	39
Wetland	4 km²	80%	9	0	9

Note: Blocks were subdivided into sixteen habitat types and, depending on habitat type, were either 25 km² or 4 km² in size. Blocks were selected randomly from the pool of potential blocks that contained a minimum amount of the given habitat type; these amounts varied among habitats and are given in the table. Blocks used for testing the predictive models are excluded.

For the very rarest habitat types, even our site selection procedure for restricted habitats identified only a tiny number of atlas blocks. Many of these rare habitats are considered especially important for birds, either because they harbor unusual species (such as mesquite, ash, and alpine) or because they support an especially wide variety of species (such as wetlands). The Atlas Steering Committee considered the atlas's two main site selection strategies, described above, inadequate for these important habitats and augmented the number of blocks by identifying all known sites containing these habitat types and then randomly selecting blocks from this pool.

Riparian habitats presented yet another problem. Since riparian zones occur in long, narrow strips and are typically separated by large expanses of other habitats, they rarely constituted a large percentage of a block's area. Hence, riparian blocks were chosen on the basis of the total length of riparian habitat in a block, rather than the area. A stream present in a block had to be at least 656 feet (200 meters) long in order for it to be assigned as "riparian."

ORIGINAL BLOCK SELECTION

The Atlas Steering Committee determined at the outset that approximately 800 breeding bird atlas blocks should be visited during the course of the project. This total was considered sufficiently large to adequately sample all major habitats and logistically feasible given budget constraints and the number of birders in the state. We actually managed to visit 769 blocks.

SUBSTITUTE BLOCKS

By the end of the third year of fieldwork it was clear that it would be impossible to visit 71 of the 771 originally selected blocks. Some of the blocks were simply inaccessible, usually for reasons of physics; sheer cliffs, for example, are hard to survey. In a few cases, private landowners refused to allow access. Likewise, al-

though we received considerable help from the Departments of Defense and Energy, some of the selected blocks on their lands were deemed off limits to atlas workers. Sixty-nine of the 71 inaccessible blocks were replaced with substitute blocks selected using the same techniques used to choose the initial set of blocks and chosen to create a one-to-one correspondence with the habitat types of inaccessible blocks. The locations of all 769 surveyed blocks are depicted on the map on page xvii. Other prominent locations often mentioned in the species accounts are shown in the map on page xvi.

INCIDENTAL OBSERVATIONS

The main goal of our fieldwork was to maximize survey coverage for atlas blocks. Blocks were often separated by long distances, though, and many observers spent time birding in areas that did not fall within a block. Consequently, we encouraged people to submit "incidental observations" of breeding birds from outside the blocks. Such records were especially important for uncommon species encountered in few blocks. For the more common species, we encouraged people to submit incidental observations primarily in cases where breeding could be confirmed.

While they were useful, the incidental observations were not habitat-stratified data, and their locations were not randomly selected. They tended to come from places and habitats that presented some combination of easy access, nice scenery, and good birding. Pahranagat National Wildlife Refuge, the Ruby Mountains, and the Interstate 80 corridor show up frequently in the incidental observations database, for example, while lesser-known and remoter places like the Plutonium Valley, Chloride City, and Roach do not. Moreover, our request that observers prioritize their incidental observation reports toward less common species, a high certainty of actual breeding, or both means that these data are largely unsuitable for direct comparisons with the data set from atlas blocks.

HOW THE FIELDWORK WAS CONDUCTED

While habitat-stratified random sampling may be unfamiliar to veterans of breeding bird atlas projects in other areas, our techniques for actually finding birds were the tried-and-true methods used everywhere. The basic strategy was to send people to atlas blocks with instructions to find as many species of breeding birds as possible in each block. They were to try to answer the questions: Does species A occur here? And if so, does it breed here?

The answer to the first question is a simple "yes" or "no," which has obvious advantages for interpretation. The answer to the second question ought in principle to be straightforward as well, but unfortunately, the evidence for actual breeding is often circumstantial, and a definite answer is not always possible. We discuss this matter in greater detail later in this chapter.

Observers were not told to spend a fixed amount of time in any given block. The more remote or inaccessible blocks could be visited only once during the four-year project. In contrast, some blocks that overlapped with popular birding sites near urban centers were visited dozens of times by multiple parties each year of the project. Similarly, no actual limits were set for an "atlas season," mostly because we knew very little about the timing of breeding for many birds in Nevada. The database indicates that most atlas work was done from early spring to early summer (approximately March through early June) in the Mojave region of Nevada, and from late spring to midsummer (May through mid-July) in the Great Basin region.

We were tremendously pleased with the ability of atlas workers to gain access to blocks. Lengthy hikes and overnight camping were required in many instances, and along with bird records we received many reports of encounters with rattlesnakes, black bears, and mountain lions. Sometimes atlas workers had to put in extra effort to secure permission for entry onto private or restricted government lands. The vast majority of landowners and government officials were extraordinarily generous in giving access to their lands; Nellis Air Force Range is a good example. Dedicated atlas workers and helpful military personnel worked together to get good coverage on the totally unknown northern reaches of this vast site, and were rewarded by finding previously unknown populations of Cactus Wrens and Ladder-backed Woodpeckers.

For each field trip, atlas workers filled out field cards (or incidental observation forms in the case of sightings outside a block) and submitted them to the Great Basin Bird Observatory for archiving, data entry, and analysis (see sample card on facing page). Before cards were accepted, they were reviewed to ensure that the information was complete, and they underwent a three-stage quality control check. First, all field cards were reviewed for anomalous records (e.g., very rare breeders, species far outside their known range, etc.). Second, data were entered and thoroughly reviewed by comparing the entered data with the original field cards. This review involved checking for data entry errors, for appropriate use of breeding codes (see below), and for records that seemed odd in any way. Third, the data were mapped and summarized by habitat type to highlight any remaining odd-ities (that is, records that did not fit the general distribution pattern for the species or reported the species in an unusual habitat). Two people were involved in each stage of the review. When questions arose, the atlas coordinator made every effort to contact the original observer to supply the missing information or verify the observation. All of the original field cards remain on file at the Great Basin Bird Observatory.

BREEDING CODES

Summer is the breeding season for most Nevada birds, of course, but the mere sighting of a bird during the warmer months does not necessarily indicate local breeding. An example familiar to many Nevadans is the sight of a flock of American White Pelicans spiraling high above the desert—often nowhere near appropriate breeding habitat. In other cases, local breeding is suspected but difficult to prove. Think of the Turkey Vulture—eminently observable but fiendishly difficult to find on the nest. Therefore, each species seen in a block was assigned a code describing the type of breeding evidence found. These codes were grouped into four categories corresponding to different levels of proof for breeding and forming a hierarchy of certainty: confirmed breeders at the top, probable breeders one notch below, possible breeders lower still, and presumed nonbreeders at the bottom. These codes are based on a set of standards created by the North American Ornithological Atlas Committee (NORAC 1990) and are similar to those used in most breeding bird atlases.

Presumed Nonbreeders (atlas code Ø) were usually late migrants (e.g., a White-crowned Sparrow in the Mojave Desert), strays and vagrants that do not breed in Nevada (e.g., an American Redstart singing at an oasis), individuals that spend the summer here but do not breed (e.g., many, but not all, Ring-billed Gulls), and birds that do breed in the state but are sometimes seen far from their breeding grounds (e.g., pelicans flying over the desert).

Possible Breeders (atlas code PO) were birds recorded in a block that could well have been breeding there but for which there was no direct evidence of actual nesting. Two types of possible breeder were recognized:

Bird seen or heard in potential breeding habitat (code **PO:#**). Examples include a Ruby-crowned Kinglet heard calling (but not singing) in a fir forest, a Red-naped Sapsucker pecking at an aspen, or a dead Flammulated Owl on a forest road.

Singing male (code **PO: ✕**). Singing was broadly defined. Examples include a Ruby-crowned Kinglet heard singing in a fir forest, a Red-naped Sapsucker observed drumming on an aspen, or a Flammulated Owl heard hooting in the woods.

Probable Breeders (code **PR**) were species exhibiting behaviors strongly suggesting that they were breeding nearby, but for which it was not absolutely certain that breeding was occurring within the block's boundaries. Probable breeders could be sorted into several types:

BLOCK WORK THIS YEAR

Visit #	Date	Field Work Party Hours	Number of Observers	Field Work *Total Hours of all Observers	Travel to and from Home to Block Total Hours of all Observers	Travel to and from Home to Block Total Miles All Vehicles
1	5/8	1 hr	1			
2	5/19	driving past headquarters				
3	6/12	3 hrs	1			
4	6/16	only at headquarters 10 min also 7/10				
5	6/28	5 hrs				
6	7/6	4 hrs				
7	7/24	only at headquarters 20 min	1			
8	7/27	1 hr	1			
Total	xxxx	8/12 1 hr xxxxxxxxxx			○	

*Party Hours time Number of Observers = Total Hours   ⟨14⟩ ⟨1⟩ ⟨14⟩   ≈150 round trip *6 = mi

900

Other Observers (Names)	Street Address, City, Zip	Phone/email

DATA ENTRY CHECKLIST
(Please complete before returning card)

Is there a breeding code for every species?
Is there a habitat code for every species?
Is there an abundance code for every species?
Did you calculate total party hours?
Did you calculate mileage?
Did you give habitat information on front page?

BLOCK COMPLETION CHECKLIST
(Please evaluate each criterion for this block)

Visit all habitats? YES/NO
Spend at least 25 hours? YES/NO
At least 50% confirmed? YES/NO
At least 1 nocturnal visit? YES/NO
What % species do you think have been missed?

25%

NEVADA BREEDING BIRD ATLAS
FIELD CARD

ATLAS BLOCK INFORMATION

Block worked in previous year?	Atlas Year 1998	Do you want block again next year?
Yes No		Yes No

Map(s) Name OVERTON VALLEY OF FIRE EAST

Atlas Block Number 6715 LOW RIPARIAN

Atlas Region ____

Habitat types in block (estimate % for each):

R — Riparian 15% ____%
OW — lake 40% ____%
W — wetlands, fields 30% ____%
U/S — urban, housing 5% ____%
    ranches
Mo — Mohave 10%   100%

ATLASER

Name Dorothy Crowe
Address 245 Jamieson Circle
City Henderson State NV Zip 89014

Telephone (H) (___) _____
          (W)(___) _____

E-mail _____

Printing Courtesy of Nevada Power Company

BREEDING CODES
(See handbook for detailed descriptions)

OBSERVED (Ob)

O Non-breeder or migrant observed during breeding season

POSSIBLE (Po)

Possible breeder observed or heard
X Singing male

PROBABLE (Pr)

P Pair
T Territory (singing at same place on 2 visits or other evidence of defense)
C Courtship or display (including copulation)
N Visiting nest site
A Agitated behavior
I Incubating (brood patch on bird in hand)
B Nest building (wrens, verdin, or woodpeckers)

CONFIRMED (Cf)

CN Carrying nesting material
NB Nest building (not to be used for wrens, verdin, or woodpeckers)
DD Distraction behavior
UN Used nest
FL Fledged young
ON Occupied nest
CF Carrying food
FY Feeding fledglings
FS Fecal sac
NE Nest with eggs
NY Nest with young

ABUNDANCE CODES

Total Pairs in Block

A1 1 breeding pair
A2 2-10 breeding pairs
A3 11-100 breeding pairs
A4 101+ breeding pairs

HABITAT CODES

R Riparian
W Wetland
OW Open Water/Lake
G Grassland
SDS Salt Desert Scrub
SS Sagebrush Steppe
SB Sagebrush Scrub
Mo Mojave
Ash Ash
Me Mesquite
MF Montane Forest
MS Montane Shrub
MP Montane Parkland
J Juniper
PJ Pinyon Juniper
Alp Alpine
Ag Agriculture
U/S Urban/Suburban/Ranchstead
B Barren

* Species marked with an asterisk are poorly known breeders in the state and additional information is required. Please complete verification forms for the marked species and any species not on the field card. Verification forms for species found breeding outside of their typical range are also requested.

† When known, please give information on cowbird hosts.

Hab	Abu	Species	Ob visit	Ob code	Po visit	Po code	Pr visit	Pr code	Cf visit	Cf code
		Pied-billed Grebe								
OW	A2	Eared Grebe			3 6/12	#				
		Western Grebe								
OW	A3	Clark's Grebe							6/12	FL 3
W		American White Pelican	6/12	O	7/27	#?				
OW	A2	Double-crested Cormorant							5/8	NY 1
		American Bittern								
		Least Bittern								
R OW	A2	Great Blue Heron							5/8	NY 1
R OW	A1	Great Egret			6/12	#				3
R OW	A2	Snowy Egret							6/12	FL 3
		Cattle Egret								
		Green Heron								
R OW	A2	Black-crowned Night Heron			6/12	#				3
OW		White-faced Ibis			5/19	#?				2
		Trumpeter Swan								
		Canada Goose								
		Wood Duck								
		Green-winged Teal								
		Mallard								
		Northern Pintail								
		Blue-winged Teal								
		Cinnamon Teal								
		Northern Shoveler								
		Gadwall								
		American Wigeon								
		Canvasback								
		Redhead								
		Ring-necked Duck								
		Lesser Scaup								

Seven singing males (code **PR:S**). This code was used when at least seven singing males were heard within the block boundaries. Typically, it was used when a block could not be visited for long enough to obtain a higher level of proof, and in situations such as a forest full of singing Hermit Thrushes or a city neighborhood alive with calling Inca Doves.

Pair (code **PR:P**). This code refers to the observation of a presumed mated pair (e.g., drake and hen Gadwalls swimming together in a marsh or male and female tanagers sitting close to each other in a pine tree). It does *not* refer simply to two birds of the same species in close proximity to each other (e.g., two Common Ravens at the same road kill). This code was widely misused (as in the example of the ravens), and many **PR:P** records were downgraded to **PO:#** or even **Ø** during the data review process.

Territorial behavior (code **PR:T**). This code was used when there was evidence of a bird holding territory. Most commonly it referred to a bird observed singing at the same place on two different dates. An example might be observations of a MacGillivray's Warbler singing from a certain perch on a chokecherry several days apart. Other forms of territorial defense, such as a male robin chasing another male robin, were also coded as territorial behavior.

Courtship or display (code **PR:C**). Any form of courtship behavior or breeding display would warrant the use of this code. Examples include food exchange by Cedar Waxwings or Greater Roadrunners, and aerial flights of male Calliope Hummingbirds or Long-billed Curlews. The assumption (usually verifiable in the field) was that the displaying bird was in the presence of a prospective partner. This code was also used for copulation, which was frequently observed in species such as Mallards and House Finches.

Nest visit (code **PR:N**). This code was used for nest visits that could not be linked unequivocally to breeding. The presence of a bird at a nest is not necessarily indicative of actual breeding. Early in the season, many raptors inspect old nests but do not actually reuse them. Other species, such as corvids, are insatiably curious and attend nests throughout the breeding season for various reasons (often of the mischievous sort). Compare this code with **CO:ON**.

Agitated behavior (code **PR:A**). Birds behaving in an agitated manner that indicated the presence of a nearby nest were given this code. An example might be a Lincoln's Sparrow drooping its wings, puffing up its crown, calling frequently, and flying in short spurts around a wet mountain meadow. Such behavior usually indicates breeding, but since it can sometimes also be observed in nonbreeders, it does not actually confirm breeding. Compare this code with **CO:DD**.

Incubating (code **PR:I**). This rarely used code referred to captured birds (usually at banding stations) that were found to have brood patches indicating that they were actively attending a nest with eggs or young. Birds with brood patches sometimes travel significant distances, though, and the capture of an individual with a brood patch does not necessarily indicate breeding in the immediate vicinity or in the block where the bird was caught.

Nest building (code **PR:B**). In most cases, nest building is obvious evidence of breeding (see codes **CO:CN** and **CO:NB**). Certain species (e.g., Verdins, woodpeckers, and some wrens), though, build nests that are not actually used for eggs and young. These may be used as part of the courtship process, to distract predators from the real nests, or as shelter from the elements. Stronger evidence of breeding is required for these species.

Confirmed Breeders (atlas code **CO**) were birds whose behavior provided incontrovertible evidence of nesting in the block.

Carrying nest material (code **CO:CN**). In late spring or early summer, Double-crested Cormorants carrying sticks or American Robins with their faces stuffed with straw are common sights. Invariably, such sights mean that the birds are working on a nest, and usually the nest is nearby.

Nest building (code **CO:NB**). This code could refer to any form of nest-building activity—from a Calliope Hummingbird weaving gossamer into its tiny nest to a Golden Eagle ramming parts of trees into its cliff dwellings. As noted above, however, nest building is not always a sign of confirmed breeding.

Distraction display (code **CO:DD**). Some species perform stereotyped distraction displays only around their nest or young. The classic example is the broken-wing ruse of the Killdeer. Many waterfowl also engage in this sort of behavior. Compare this code with atlas code **PR:A**.

Used nest (code **CO:UN**). This code was applied when a previously used nest could be assigned to a particular species. Old bird nests, however, are often difficult to identify, limiting the frequency with which the code could be used. The "Dutch ovens" of Black-billed Magpies, the pendant pouches of Bullock's Orioles, and the monstrous productions of Bushtits are obvious exceptions. Note that old Cactus Wren and Verdin nests—even though they can often be identified—did not warrant use of this atlas code (see also **PR:B**).

Fledged young (code **CO:FL**). Along with the code for pairs (**PR:P**), this was perhaps the most misused atlas code. It was intended to be used only for barely fledged, probably dependent, young birds presumed to be in the immediate vicinity of their nests. While reviewing the atlas data, all records using this code were scrutinized to ensure that this was indeed the case. When there was doubt as to whether the record applied to recently fledged young, it was downgraded to a more conservative code.

Occupied nest (code **CO:ON**). This code was used for a bird actually sitting on a nest: a Swainson's Hawk poking up out of a platform of sticks, an Eared Grebe hunkered down on a floating mat of marsh vegetation, or a Costa's Hummingbird wedged into its miniature teacup of a nest. Compare this code with **PR:N**.

Carrying food (code **CO:CF**). As the breeding season progresses, birders often see birds bringing food to their nests. Examples might include a Yellow-billed Cuckoo carrying a caterpillar or a Western Meadowlark transporting a juicy grub. Certain birds—especially raptors and corvids—sometimes carry food for personal gain, however, either to cache it for later consumption, as part of a breeding display, or simply until they find a good place to sit and eat. Observers were asked not to use this code for such species unless there was compelling evidence that food was actually being transported to young.

Feeding fledglings (code **CO:FY**). Young birds of many species are unable to forage for themselves for some time after they leave the nest. During this period it is possible to observe parents feeding young birds, as in the case of juvenile European Starlings begging for food in suburban yards.

Fecal sac removal (code **CO:FS**). Many birds carry the fecal sacs that contain their young's excrement away from the nest immediately after they are produced. This behavior is especially common in passerines, but it was rarely reported—in part, perhaps, because it is difficult to distinguish fecal sacs from food items at a distance (code **CO:CF**).

Nest with eggs (code **CO:NE**). One of the thrills of atlas work was chancing upon a nest with eggs in it: tiny eggs in the tidy nest of a Broad-tailed Hummingbird, large oblong eggs in the downy nest of a Cinnamon Teal, or the shorter oval eggs of a Snowy Plover lying in the middle of a playa. It was difficult to know exactly what species the eggs belonged to, of course, but many atlas workers were willing to wait around for the parents to return to the nest to clinch the identification.

Nest with young (code **CO:NY**). Even more thrilling was coming across a nest crowded with nestlings: dark, dumpy Broad-tailed Hummingbirds, olive-brown Cinnamon Teal fluff balls, or wide-eyed Snowy Plover chicks. Young birds are particularly vulnerable at this stage, and many observers chose to keep their distance rather than risk harm to the nestlings; consequently, this code was not frequently reported.

Only the highest-ranking breeding code was reported for each species in any given block. Thus, if an observer first heard a Gray Vireo singing (**PO:✕**) and later observed it engaged in agi-

tated behavior (**PR:A**), only the latter code was applied. Likewise, in situations involving multiple observers or multiple years of study, only the highest-ranking breeding code was applied. The following scenario illustrates the point: in 1998, observer A sees a Hooded Oriole in appropriate breeding habitat (**PO:✕**); in 2000, observer B finds a Hooded Oriole building a nest (**CO:NB**) in the same block; and later in 2000, observer A discovers a pair of Hooded Orioles (**PR:P**) in that block. The correct code in this case would be **CO:NB**, because it is the strongest piece of evidence for actual breeding in the block.

HABITAT INFORMATION

In addition to collecting information on breeding bird activity, we asked observers to report the habitat types that each species used. Although our sampling design involved assigning habitat types to each block before doing any fieldwork, these assignments could not be used to determine which specific habitats particular species were using, for at least two reasons.

First, blocks were always assigned to a single, predominant habitat type even though it would be truly anomalous for a block to contain only one habitat. There is always a good chance that small patches of less common habitats will occur within a 25 square kilometer (9.6 square mile) area. These patches undoubtedly influenced the birds that were found in a block, especially when they involved available water (e.g., small wetlands and riparian strips, or irrigated farmland and human habitation). Thus, simply associating bird observations with the habitat assigned to a particular block would give a very misleading picture of the habitats each species was actually using, especially at the small spatial scales that may be important for understanding nest placement and foraging behavior.

A second problem with using the habitat assigned to each block rather than on-site observations is that those assignments were based on maps created in 1996. Although the satellite images used for producing habitat maps provide a wealth of information that is available nowhere else, such data are not perfect, and erroneous habitat classifications undoubtedly exist in the habitat map. In addition, conditions can change rapidly, and the habitat type present when the images were obtained might not have been there by the time the block was surveyed, especially near the state's major urban centers.

To overcome these problems, we asked observers to estimate the amount of each habitat in their blocks and to report the habitat type in which they actually observed each species. The first form of habitat information provided a way to "ground-truth" the GAP-based block selection process. The second provided more precise information on the habitats in which birds were found.

The observer-supplied habitat codes can best be understood by way of the following example. Observer A visits a southern Nevada block that we had classified as pinyon-juniper. On entering the block, the observer soon finds herself in a mixed-age pinyon-juniper woodland and quickly ticks off the expected species: Northern Flicker, Gray Vireo, Western Scrub-Jay, Spotted Towhee, Black-chinned Sparrow, and so on. Working her way downslope, she comes across a small riparian patch where she finds more Spotted Towhees plus a Blue-gray Gnatcatcher and two Ash-throated Flycatchers. Finally, she determines that one corner of the block actually cuts into some lower-elevation Mojave habitat, which supports Black-tailed Gnatcatcher, Black-throated Sparrow, Verdin, Common Raven, and many more Ash-throated Flycatchers. On finishing her fieldwork, Observer A estimates that her block was 90% pinyon-juniper, 9% Mojave, and 1% riparian. Note that the GAP-based habitat description was not "wrong"; on the contrary, it correctly characterized the major habitat type as pinyon-juniper. The observer's information was crucial, though, in determining the actual habitat composition of the block.

In the preceding example, an observer saw several birds in multiple habitats. Spotted Towhees were found in pinyon-juniper and riparian, and Ash-throated Flycatchers were found in riparian and Mojave habitats. This has an important consequence for interpreting the habitat tables reported in each species account in this book: **Sometimes the total number of habitat observations exceeds the total number of blocks in which a particular species was found.** The Whip-poor-will is a case in point: it was found in only one block, but four habitat codes were reported for the species.

On the whole, observer-supplied habitat codes were fairly crude because we always emphasized the importance of identifying as many birds as possible in each block and confirming breeding in as many species as possible. Obtaining information on habitat use was icing on the cake, to be obtained as time permitted. Researchers wishing to perform follow-up analyses on the atlas data should thus use the observer-supplied habitat codes with caution. A few examples of potential pitfalls in the interpretation of the habitat codes follow:

- *Definition of habitat types.* Observers differed greatly in what they considered, say, "riparian" habitat. A babbling brook surrounded by cottonwoods out in the middle of the desert is clearly riparian habitat. But what about a dry wash with taller desert shrubs but no riparian-obligate plant species? Or a tamarisk grove at the edge of Lake Mead? On the whole, observers often equated "riparian" with "good habitat." Also, "riparian" frequently seems to have been used as a catch-all for "mesic" or "wet."
- *The problem of patch size.* A related problem has to do with small habitat chunks nested within large expanses of another habitat type. Consider a little farm pond with enough cattails to support breeding Song Sparrows or Red-winged Blackbirds. Some observers would have called this a "wetland" habitat. Others would have considered it part of the larger farmland setting and would have called it "agricultural."
- *The problem of microhabitats.* Birds seldom depend on just one major habitat type during the breeding season. Often, the actual breeding site (typically difficult to observe) and the foraging grounds (where most observations are actually made) are quite distinct. For example, in Nevada,

Long-eared Owls are essentially riparian obligates with regard to nest placement, but nearly all of their foraging takes place in wide-open desert or grassland habitats. Even more extreme is the example of the White-throated Swift, which breeds only on remote rock faces but can be seen flying over any habitat. Finally, there is the example of the American White Pelican: almost all of the Nevada nests are restricted to a single island in the middle of Pyramid Lake, but foraging pelicans can be seen a hundred miles away. Assigning meaningful habitats to such birds clearly can be tricky.

- *"Weird" habitats.* The term "sagebrush breeders" does not usually call to mind Cinnamon Teal, Dusky Grouse, Northern Saw-whet Owls, and Bullock's Orioles. In fact, when we received field cards for these species in sagebrush habitat, we initially assumed the records were in error. But they were not. Cinnamon Teal sometimes build their nests well out in the desert, and Bullock's Orioles occasionally hang their pendant nests from tall sagebrush. And while Dusky Grouse and Northern Saw-whet Owls may not actually nest in the sagebrush, they certainly occur in this habitat during the breeding season. The basic point is that most birds use a much wider array of habitat types than is often imagined.
- *Bias toward "good" habitats.* The potential for observers to give disproportionate attention to good habitat took two forms. First, observers tended to gravitate toward "good" habitats within any particular block: the plantings around a ranch house way out in the desert, a montane meadow in an otherwise monotonous stand of ponderosa pine, or a riparian corridor through an urban center. In general, though, observers seem to have tried to spend time in each of the habitat types within a particular block.

The second problem involved incidental observations, which were not related to habitat coverage; observers simply went wherever they wanted to go. A good example of the result is the Rock Wren: 6 out of 431 (1%) block observations were assigned to riparian habitats, while 15 out of 99 (15%) incidental observations were made in riparian areas. Conversely, 80 (19%) of the block habitat observations were in barren areas, while only 10 (10%) of the incidental observations were made in this habitat. On the whole, birders prefer riparian habitats over barren habitats, and this preference is reflected in the incidental Rock Wren observations. During block observations, though, birders did not have the option of simply ignoring areas with barren habitat; consequently, the block data are likely to be more representative of habitat use by Rock Wrens.

ABUNDANCE CODES

Although assessing the abundance of different species was not a major goal of the atlas project, we did ask atlas workers to estimate the total number of breeding pairs of each species in the blocks they surveyed. The different levels of estimated abundance were as follows:

A1—just one pair in the entire block ("uncommon")
A2—two to ten pairs in the block ("fairly common")
A3—eleven to one hundred pairs in the block ("common")
A4—more than one hundred pairs in the block ("abundant")

The species accounts in this book sometimes note abundance using the terms given in parentheses in the list above. Although we have standardized the use of these terms when they refer to data from the atlas project, similar standardization of these terms used when citing other works is impossible, and we caution readers accordingly. **Also, remember that two different block sizes were used for atlas fieldwork, and take that into account when comparing abundance scores across habitats.**

Finally, note that these estimates were intended to refer to the total number of breeding pairs per block, not just the number of pairs actually seen during fieldwork. We feel that abundance estimates were often too low, for (at least) the following three reasons:

- In some cases, observers seem to have counted only the number of pairs actually seen. For scarce and conspicuous species (e.g., Golden Eagle), the number of pairs seen may be close to the total number of pairs present. But for most other species (e.g., Brewer's Sparrow), the total number of pairs seen is almost always much lower than the total number of pairs present—even in blocks with thorough coverage.
- In workshops and in the field, we noticed that observers usually underestimated (often by a wide margin) the number of birds in any given area. A case in point is the Vermilion Flycatcher. This bird is exceptionally conspicuous and very easy to identify. Based on demographic research conducted at the Warm Springs Ranch in northern Clark County (USGS 1998, 2000), we know that scores of pairs breed in an area of about 1,200 acres (485 ha; much smaller than a 25 square kilometer atlas block). Yet, every visitor to this site during the atlas years came back with estimates much lower than the actual number of breeding pairs present.

- A related problem is that observers sometimes underestimated the size of the area spanned by an atlas block. Walking the perimeter and crisscrossing each of the diagonals of a Mojave block, for example, would require a hike of more than 21 miles (34 kilometers). And even this effort would only scrape the tip of the iceberg of Black-throated Sparrow territories that are usually present in this habitat type.

As with the habitat codes described above, the summary statistics for abundance codes given in the species accounts refer to *all* reported abundance codes. **Since some blocks were visited more than once, the total number of abundance estimates sometimes exceeds the total number of blocks in which a particular species was found.** As with the habitat codes, the Whip-poor-will is a good example: the species was found in only one block, but it was twice reported as being fairly common because of multiple visits to this block.

SUMMARY

At the most basic level, the goal of a breeding bird atlas is to provide detailed range maps for all species that breed in a given region. Consequently, we focused our resources and efforts on achieving this goal as rigorously and thoroughly as possible. In contrast, and as noted above, the additional data provided in the form of habitat codes and abundance estimates were collected less rigorously and are thus somewhat anecdotal in nature. Not all observers provided this supplemental information, and the methods for collecting it were less formally defined. Detailed analysis of habitat use and abundance patterns is also beyond the scope of this book. Still, the database we created represents the first comprehensive description of habitat selection and abundance patterns for most bird species throughout the state of Nevada. We would therefore welcome and encourage additional analysis of the data collected during the project.

PREDICTING SPECIES DISTRIBUTIONS

The atlas sampling scheme was designed to allow the prediction of species distributions in portions of the state that fieldworkers did not actually visit. The process of making the statistical models from which these predictions are derived is continuing, with successive rounds of model refinement and testing. To date, we have completed an initial round of model building, and the results are presented in this book as probabilistic occurrence maps for each species. Future rounds of model testing and refinement are anticipated as additional distributional data are generated. Given our desire to get the atlas data published in a timely fashion, however, those results must await future publications. In this section we explain the model that we used to generate the predictive maps.

THE PREDICTIVE MODEL

The many approaches scientists have developed for predicting species distributions (Scott et al. 2002) range from the very simple to the mathematically overwhelming, with the amount of detail increasing as models are refined for individual species. Constructing individualized models for each of Nevada's breeding bird species would have been a gargantuan undertaking, especially given the lack of detailed information on many species, so we took the simple route. The model that we created incorporates only latitude, longitude, elevation, and broad habitat types and can be applied to any species. It proceeds with just two steps, which we repeated separately for every species.

Step one: In the first step, we used all of the block and incidental observations of possible, probable, and confirmed breeders to determine the range of values for latitude, longitude, and elevation within which the species was found. This process created a three-dimensional "box" within which the atlas data suggest that occurrence during the breeding season is plausible. To make the model somewhat more realistic, we extended this "plausible range" to include the entirety of any mountain range or basin that lay partially within the zone identified by the latitudinal and longitudinal limits. For all locations that lay outside this region we set the probability of finding the species to be zero.

Step two: In the second step, we estimated the likelihood of finding a species at each location within the plausible breeding range, based on the habitat type present. This part of the model used only the block data from the region identified in step one. Incidental observations were not used because they did not sample each habitat in proportion to its occurrence, and could therefore have produced a highly biased estimate of a bird's habitat use. For each of the atlas-defined habitats (see Table 1 for a list) we calculated the proportion of blocks in which evidence of breeding (possible, probable, or confirmed) was obtained for a species. We then used these proportions to assign probabilities to sites that had not already been given a probability of zero during step one. For example, if a species was seen in half of all the montane forest blocks within the region identified in step one, then the model assumes that there is a 50% chance that the species occurs during the breeding season wherever montane forest occurs, as long as it lies within that same region. This process was repeated for each habitat, and the resulting probabilities were used to produce a map that indicates the likelihood of finding the species at each point in the state.

To explain this process further we offer a simplified example for an imaginary species: the Greater Slothopper. Let us pretend that slothoppers occur from downtown Winnemucca south to the Las Vegas Strip, and from Reno in the west to Elko in the east. Let us also assume that they are found only between elevations of 2,000 and 6,000 feet. Step one creates the prediction that there is no chance of finding the species during the breeding season anywhere outside this defined area—that is, north of Winnemucca, east of Elko, south of Las Vegas, or west of Reno. Furthermore, even within the "box," the model predicts that the species will not occur below 2,000 feet or above 6,000 feet.

An examination of the block data shows that this imaginary species occurs in thirty of the forty urban blocks that lie within the geographic region of plausible occurrence, and in two of the ten agricultural blocks. Consequently, step two of the model predicts that there is a 75% (30/40) chance of the species being found in urban areas, and a 20% (2/10) chance of it occurring in agricultural areas. Since the species was never seen in any of the blocks assigned to other habitats, the model will assume that the chance of occurrence in all other habitats is zero.

Following the mapping conventions used in our species accounts, and given all of the results described above, the predictive map for this species would show white space (equivalent to a 0% chance of occurrence) for all areas outside the geographic and elevation limits identified in step one. Inside this region, all urban habitat would show up red, which indicates that there is at least a 50% chance of finding the species if a block-sized area is searched in the highlighted zones; agricultural areas would appear yellow (indicating an 11–25% chance of occurrence). All other habitats within the plausible region would be white because the species was not recorded in blocks assigned to the other habitats.

HOW WELL DOES THE MODEL REFLECT REALITY?

The advantage of this model is that it is relatively easy to understand and simple to implement. On the other hand, its simplicity undoubtedly limits its accuracy. In particular, it is unlikely that the model will accurately capture the role of habitat features that occur at small spatial scales, or that it will adequately describe complex habitat relationships. Habitats that occur in small patches pose a particular problem because the model reflects use of these habitats only indirectly. For example, Canyon Wrens are usually found in rock outcrops. Rock outcrops tend to be small relative to the size of an atlas block, and consequently were not one of the major habitat classes used in the atlas study design. Rock outcrops are embedded within many of the habitat classes that were used, however, and the rate at which Canyon Wrens are found in each of the defined habitats probably reflects how often these various habitats contain rock outcrops. So, to predict how often Canyon Wrens occur in a given habitat *on average,* the model probably works fine. Because predictions are not based directly on the presence of rock outcrops, however, the model will not account for subtle variation in the abundance of rock outcrops within habitat types. For example, if outcrops are more common in some areas of Mojave habitat than others, and this in turn affects the occurrence pattern of Canyon Wrens, there would be no way for the model to show this variation.

Similar problems exist for habitats for which there were atlas blocks, but that also occurred as small fragments embedded within other habitats. For example, small wetlands or riparian patches frequently lie within other habitats, as do houses, shelterbelts, and other features that might be classified as urban or agricultural if they occurred in larger patches. The occurrence of these "embedded" habitat fragments sometimes created what appeared to be anomalous patterns. For instance, many wetland species were predicted to occur—at low levels—in terrestrial habitats such as sagebrush scrub and salt desert scrub. In most cases, this was not because wetland species actually occurred out in the desert far from water. Rather, it reflects the fact that blocks dominated by a particular terrestrial habitat may also contain a small wetland. In other words, a prediction that there is a 2% chance of finding a Western Grebe in Mojave habitat actually means that there is a 2% chance that a block-sized patch of Mojave habitat will contain a patch of wetland habitat occupied by Western Grebes.

Another limitation of the model is that it incorporates only a very narrow subset of the factors that influence a bird's distribution. Basically, the model assumes that only gross habitat differences are important, yet any birder knows that a myriad of little details can make one spot slightly more or less likely to contain a particular species. When averaged out across many different sites, these details probably do not mask a general pattern, but at any specific location they might lead to misleading predictions.

INTERPRETING THE PREDICTIVE MAPS

Given these caveats, it is reasonable to ask how a user should interpret the model's predictions—and indeed, whether the predictions can be trusted at all. The model's results are presented as a series of maps that show the estimated chances of finding each species at sites throughout the state. As a guide to broad distributional patterns, these maps can be expected to be quite accurate. For instance, the predictive map for Say's Phoebe indicates that the species occurs throughout the state but is more likely to be encountered in the Mojave habitat of southern Nevada than in the desert habitats of the Great Basin. The predictive maps are less likely to be accurate for specific locations, however, and the more specific the location, the lower the accuracy is likely to be. This is not to say that the maps cannot be used to identify specific locations that are worth searching for a given species. Indeed, if a user focuses more on the relative probability values than on the specific numbers, the maps should be helpful most of the time. In general, the best places to find a species will be shaded red, places where there is a reasonable chance of finding a species will be shaded orange or yellow, and places where there is only a very slim chance of finding a species will be blue or white.

Even when viewed with the appropriate caveats, some predictive maps will still show unexpected results that require explanation. In most cases, we address these issues in the text of the species accounts. Some of these apparent problems recur frequently, however, so we provide some general explanations here:

1. Predicted occurrence outside expected range. The predictive maps frequently suggest that there is some chance of finding a species breeding outside its known elevation limits or habitat associations. In some cases, especially for wetland and riparian species, these predictions reflect the occurrence of "embedded" habitats, as described above. A second explanation for these apparent overpredictions is that they result from migrants being recorded as potential breeders. During migration, birds often occur at lower elevations, or in different habitats, than those used for breeding. Songbirds often sing repeatedly during spring migration, and in some species certain individuals may even appear to be paired. Consequently, determining whether such birds are actually potential nesters, especially late in the spring, is often a difficult judgment call. During their review of the data, the atlas coordinators attempted to eliminate records of probable migrants, but inevitably some will have slipped through this vetting process. These records likely account for situations where species that are known to nest primarily at high elevations in Nevada, such as the Ruby-crowned Kinglet, are predicted to have some chance of occurring at low elevations. Similarly, they might explain why a species is predicted to occur farther south than would be expected based on its known breeding range. Although the problems described here may cause the model to overestimate the chance of a species occurring in a particular habitat or region, in most cases such overestimates are not severe. For instance, sometimes a large area where a species simply does not breed is dark blue on the predictive map (≤5% chance of occurrence) rather than white (0% chance), but rarely is it orange (26–50%) or red (≥51%).

2. Overestimated probabilities near a species' geographical range limits. A second problem, more serious but less common, involves high probabilities estimated in areas where a species is not very likely to occur. Generally, this problem was observed within relatively small areas near the geographical limits of a species' range, and it was most common along the transition zone between the Mojave Desert and the Great Basin. Several species that are largely restricted to the Mojave Desert, for example, were predicted to have an unrealistically high chance of occurring in the northernmost part of the Mojave. These patterns require a detailed explanation, and some readers may prefer to skip ahead and simply accept our assurance that things sometimes look "odd" in this region because of the model's simplicity.

For those interested in the details, recall that probabilities are assigned to habitats during the model's second step. As described previously, these probabilities are calculated using only the data from the atlas blocks that lie within the geographic and elevation limits determined in the model's first step. When blocks of a particular habitat type are rare within a species' range, it becomes harder to get an accurate estimate of a species' actual likelihood of occurrence in these habitats. This is an example of a more general "sample size" effect. For instance, if you flip a coin twice, you can easily get "heads" 100% of the time—that is, on both tosses. If, however, you flip the coin one hundred times, the chance of getting 100% heads is miniscule; instead, you will likely get a result that closely approximates the true probability (50%). This sample size problem as it relates to atlas blocks and the predictive model is best illustrated with an example.

Consider Bendire's Thrasher, a species that rarely if ever breeds north of the Mojave. Great Basin habitats (such as salt desert scrub) occur primarily north of the region in which the model considers occurrence of this species to be plausible. The boundary between the Mojave and the Great Basin, however, is not a clearly defined straight line, and small areas of salt desert scrub do lie within this plausible region for Bendire's Thrashers. The number of surveyed salt desert scrub blocks within this plausible region was very small, though, making it difficult to estimate accurately the probability of occurrence in this habitat. If Bendire's Thrashers happened to be present in a large proportion of the salt desert scrub blocks that were sampled—perhaps because those blocks also contained small areas of habitat more typical in the Mojave Desert—the model will extrapolate this result to all salt desert patches within the plausible region. Thus, if two of three salt desert scrub blocks within the plausible region contained sufficient habitat for Bendire's Thrashers to be present, an expected likelihood of 67% would be applied to all salt desert scrub in the region. Just as two coin tosses cannot be relied upon to give an accurate estimate of the probability of "heads," very limited sampling of a habitat can produce misleading results about the occurrence of a particular bird species.

Sometimes, late migrants exacerbated the problem of overprediction near a species' range limit. For instance, the Ruby-crowned Kinglet map shows a very high probability of finding the species at lower elevations in the mountains of southern Nevada. This result is clearly wrong and probably arose because a very small number of Mojave blocks were present within the elevation range for the species, and late migrant Ruby-crowned Kinglets were seen in a few of them.

3. Predicted occurrence in central mountain ranges. Several species that breed in Nevada nest primarily in mountain ranges near the state's periphery, such as the Carson Range in far western Nevada, the Santa Rosa and Jarbidge ranges in far northern Nevada, the Snake Range in far eastern Nevada, and the Spring Mountains in southern Nevada. Many of the state's central mountain ranges support habitats similar to those of the peripheral ranges—montane forests, riparian areas, and alpine zones, for example—and it is thus not surprising that the predictive maps highlight any central mountain ranges that are similar in elevation and habitat to those ranges to the east and west where the species was documented as a breeder. In fact, however, these central Nevada ranges frequently lack species that occur near the state's borders, perhaps because they are more isolated or smaller than those near the periphery. These biogeographic factors make it less likely that species will get to the central ranges in sufficient numbers to support persistent populations.

While the predicted occurrence of some species in central mountain ranges might be incorrect, the maps do indicate areas where populations, if they should exist, are most likely to be discovered, and consequently provide a valuable guide for anyone interested in advancing our knowledge of bird distributions in the state. We therefore encourage readers of this book to actively use the predictive maps in searching for new locations for breeding birds.

4. Systematic habitat classification problems. Although the currently available habitat maps have improved substantially over their predecessors, mapping habitat types is still an imprecise exercise. Sometimes patches or even regions are classified incorrectly; on some other occasions, a patch or region does not fit well into the existing set of simplified categories. In terms of our predictive maps, the most common and visually apparent example of this problem involves the Hawthorne Army Depot in Mineral County, at the south end of Walker Lake. A large portion of this reserve consists of ammunition bunkers scattered regularly within an expanse of highly altered salt desert scrub habitat. The habitat map classifies this region as "urban," although it has almost none of the features (houses, food sources, exotic plantings, trees, easy access to water, bird feeders, etc.) that we would normally associate with an urban area. Although the predicted occurrence of "urban" species is typically very high at the depot for this reason, most birds do not seem to regard Hawthorne Army Depot as an urban resource.

Additionally, our habitat map sometimes failed to distinguish true native grasslands from agricultural fields or irrigated pastures. This resulted in some predictive map errors as well, although they are not as striking as those associated with Hawthorne Army Depot.

5. Visualizing predicted occurrences in smaller habitat patches. Most of the atlas-defined habitat types—such as sagebrush steppe, Mojave, and pinyon-juniper woodland—commonly occupy large, relatively contiguous swaths of the landscape. The probabilities of occurrence associated with these habitats are easy to visualize on the predictive maps. Other habitats, however, such as alpine tundra, ash woodland, and riparian zones, tend to occur in much smaller or narrower patches. Even when these habitat patches are correctly classified in the habitat map and handled correctly by the model, the results may be hard to see on the predictive maps. For instance, a quick glance at the predictive map for the Black Rosy-Finch suggests that the species does not occur in the state at all. Closer examination, however, reveals a high probability of occurrence, albeit in alpine areas that have a very limited extent. The same phenomenon occurs with many riparian-associated birds. A significant portion of the riparian zones where the model considers them likely to occur are simply too narrow to be visible when plotted on predictive maps at the scale possible in this book.

Ultimately, the accuracy of the predictive maps can be evaluated only by systematically testing how often they are correct. Researchers working in collaboration with the Great Basin Bird Observatory are currently conducting such tests, focusing at first on Clark County, but hopefully extending their research to the rest of the state. Information about these and any subsequent tests will be posted on the Great Basin Bird Observatory's Web site (www.gbbo.org) as they are completed.

OVERVIEW OF THE RESULTS

I f there is one thing Nevada's birders agree on, it is that most of the published range maps currently available are woefully inadequate in their portrayal of the distribution and abundance of the state's birds. The inexactitude of maps is to some extent inevitable, of course; most field guides simply cannot show more than the broad outline of a species' range. In Nevada, the extremes of geography make the consequences of these simplifications seem especially acute, because large swaths of inappropriate habitat are often marked as occupied, and areas of suitable habitat are unrecognized because they are poorly known. The paucity of published information on Nevada's birds and their habitats means that some maps are just wrong, even in their portrayal of range boundaries.

The *Atlas of the Breeding Birds of Nevada* presents up-to-date and thorough distribution data on all of the bird species that breed in Nevada. The authors of earlier books on Nevada's birds were often forced to estimate or guess at range limits. There was a surprising degree of uncertainty about the ranges of even many widely distributed and common species before the atlas fieldwork began: How far south do Brewer's Sparrows breed? What is the northern range limit of the Verdin? How extensive is the range of the Northern Shoveler? In other cases there was virtually no information on a species' basic status and distribution in Nevada. This was true both for species with rather limited ranges, such as the Gilded Flicker and the Bobolink, and for fairly widespread species with disjunct breeding ranges in the state, such as all three of the nuthatches. The atlas results provide new occurrence records for these and many other species, and the predictive maps often suggest areas for additional fieldwork and discovery.

The main results of the atlas are chronicled in the species accounts that form the bulk of this volume. This chapter provides an overview of the data collected during the atlas project and a summary of the general patterns that emerged from the study.

THE ATLAS YEARS: 1997–2000

It is always tempting to look at the range maps in a breeding bird atlas and think of them as the "last word" on the status and distribution of the breeding avifauna of a particular region. It is more appropriate, though, to view atlas range maps as a snapshot from the finite period of time during which the fieldwork was completed. Ten years from now, the breeding ranges of Anna's Hummingbird and the Great-tailed Grackle will probably differ from those shown in this book. It is difficult even to imagine Nevada's avifauna a hundred years from now: Ferruginous Hawks, Calliope Hummingbirds, and Abert's Towhees could be replaced by California Condors, Barred Owls, and Northern Cardinals!

Even without significant large-scale and long-term environmental changes (such as sagebrush loss and degradation or global climate change), *any* four-year period is distinctive, unique, and irreproducible. Consider the period 1997–2000: it started out with wetter years than normal and finished up with drier years than average. One major consequence of this rapid transition from wet to dry was the creation of a high potential fuel load in the form of above-average rangeland productivity in the mid-1990s, which led to spectacular burns during the fire seasons of 1999 and 2000. Many habitats that atlas workers visited in 1997 and 1998 were almost completely burned in 1999 and 2000. Users of the atlas data should therefore be aware that local occurrences of sagebrush species such as Greater Sage-Grouse and Brewer's Sparrow may be appreciably different in the first decade of the twenty-first century than they were during the last four years of the twentieth.

In some cases, such as the example of the burned areas cited above, the explanation for local presence or absence of birds is fairly evident. In other cases, there is no obvious explanation for patterns observed during the atlas years. For example, longtime birders were surprised not to encounter Painted Redstarts during the atlas project. In years past, the species was recorded during summer in the mountains of southern Nevada; and it may one day be found breeding in the state. But not one of the atlas workers saw a Painted Redstart during the atlas years. Other species that may well be part of the breeding avifauna of Nevada were not confirmed as breeders during the atlas project; examples include Spotted Owl, Vaux's Swift, Hermit Warbler, Grasshopper Sparrow, and Gray-crowned Rosy-Finch. Conversely, atlas workers found unexpected birds whose occurrences may turn out to be mere anomalies in the long term; possible examples include Pileated Woodpecker, Gilded Flicker, and Rufous-crowned Sparrow.

BLOCK COVERAGE

Of the 771 atlas blocks initially selected, 71 were found to be inaccessible due to restricted access or impassibility and had to be replaced (see pp. 16–17 for a description of block selection methods). By the end of the final field season, fieldworkers had visited 769 atlas blocks. An additional 78 blocks in Clark County were visited by a special field team in 2000. Data from these 78 blocks are being used to test the accuracy of the predictive models created from the main atlas data set, and records from these blocks are shown on the atlas range maps as incidental observations.

At least 418 people contributed to the data gathered during the atlas project. Over the course of the study, these individuals spent more than 8,500 hours surveying blocks, for a total of more than 14,600 observer-hours. Fieldworkers spent at least 5,200 hours traveling more than 155,000 miles (ca. 250,000 km) to distant field sites. Table 2 breaks down these figures for each year of the project.

Other atlas projects have assessed how adequately individual blocks were covered by determining how closely the species lists from each block approach some expected number of species or by examining the proportion of all species for which breeding was confirmed. Such assessments are inherently subjective and require considerable prior knowledge. Identifying the number of expected species in any of our blocks in any objective way was deemed very difficult, and we felt it would have presupposed knowledge about the distributions we were trying to determine. With the wide variation in habitat types found in Nevada (e.g., compare a sagebrush block with no water source with one with extensive shallow wetlands), it was also clear that different standards would have had to be set for each habitat. Breeding confirmation rates also proved to be difficult to use. Setting some target confirmation rate to be met in all blocks would have assumed that a constant proportion of all species seen in a block actually bred there. Given the wide variation in bird species composition found across our blocks, this assumption did not seem reasonable. Hence, we made no attempt to assess whether blocks had achieved "adequate" coverage.

We did, however, collect data on the amount of time observers spent in each block (see map on facing page). Since individual observers varied in how intensively they searched an area, and blocks varied in how easy they were to search, these data provide only an approximate sense of how much effort was expended searching each site. Nonetheless, they do provide some insight into potential biases that might exist in the data due to the way in which different areas were surveyed.

BIRD OBSERVATIONS

Over the four years of the study, fieldworkers obtained evidence of breeding for 265 species of birds. Of these, 243 were confirmed as breeders in Nevada. After reviewing the species list and examining new information gathered since the atlas fieldwork was completed, we considered it likely that at least 257 species currently breed in Nevada.

Among the most exciting finds were the Gilded Flicker and Rufous-crowned Sparrow, which were found in the extensive Joshua tree woodlands near the town of Searchlight in far southern Nevada. Atlas workers also found several Mojave bird species in a spot well north of the ecoregional boundary of the Mojave Desert, on the Nellis Air Force Range near the town of Tonopah. Cactus Wren, Ladder-backed Woodpecker, and Scott's Oriole were reported at this small island of Mojave Desert located in what are otherwise Great Basin landscapes. Other interesting discoveries included the unexpectedly widespread distributions of the Black-chinned Sparrow and Gray Vireo, which are commonly found in pinyon-juniper woodlands of the Mojave Desert. The atlas confirmed that they actually occur as far north as Lincoln and White Pine counties, well beyond the Mojave Desert's northern limits. Also, Bobolinks and Swainson's Thrushes

TABLE 2. OBSERVER EFFORT DURING THE NEVADA BREEDING BIRD ATLAS PROJECT

YEAR	NUMBER OF OBSERVERS	NUMBER OF BLOCKS VISITED	PARTY-HOURS OF FIELDWORK[‡]	OBSERVER-HOURS OF FIELDWORK[‡]	HOURS SPENT TRAVELING[‡]	MILES TRAVELED TO VISIT BLOCKS[‡]
1997	87	111	>1,339	>1,918	>691	>27,309
1998	117	248	>2,240	>3,794	>1,763	>37,189
1999	121	434	>2,832	>5,750	>1,733	>53,543
2000	137	266	>2,086	>3,144	>1,012	>36,870
Total	418[†]	769[†]	>8,497	>14,606	>5,199	>154,911

[†]The annual figures do not sum up to the total because many observers participated in the project in more than one year, and some blocks received coverage during more than one year.
[‡]These figures underestimate actual time and mileage because some observers did not report effort and many observers clearly underestimated travel time and distances.

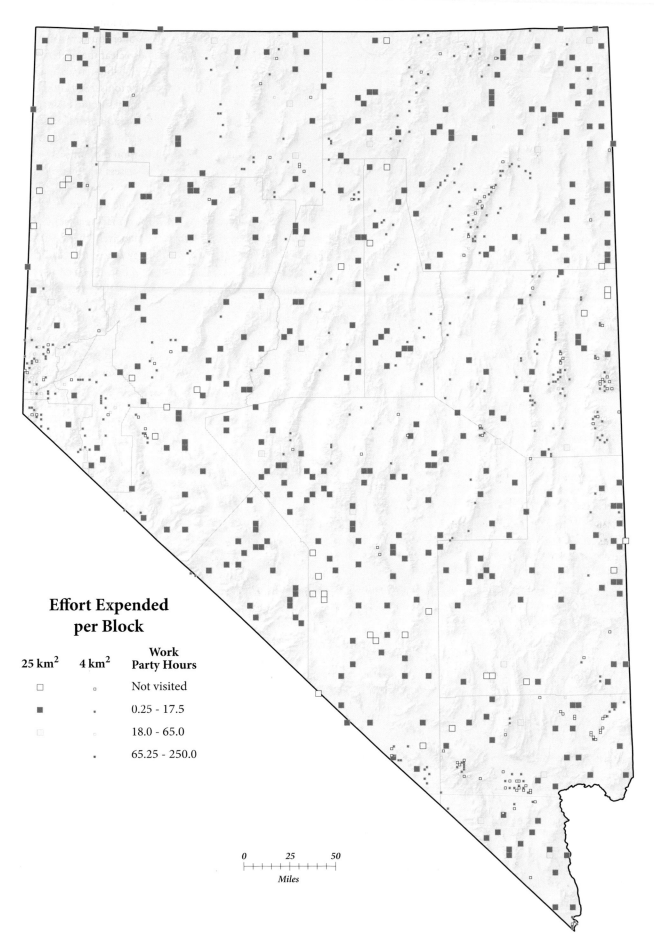

Effort Expended
per Block

25 km²	4 km²	Work Party Hours
☐	▫	Not visited
■	▪	0.25 - 17.5
☐	▫	18.0 - 65.0
▪	▪	65.25 - 250.0

0   25   50

Miles

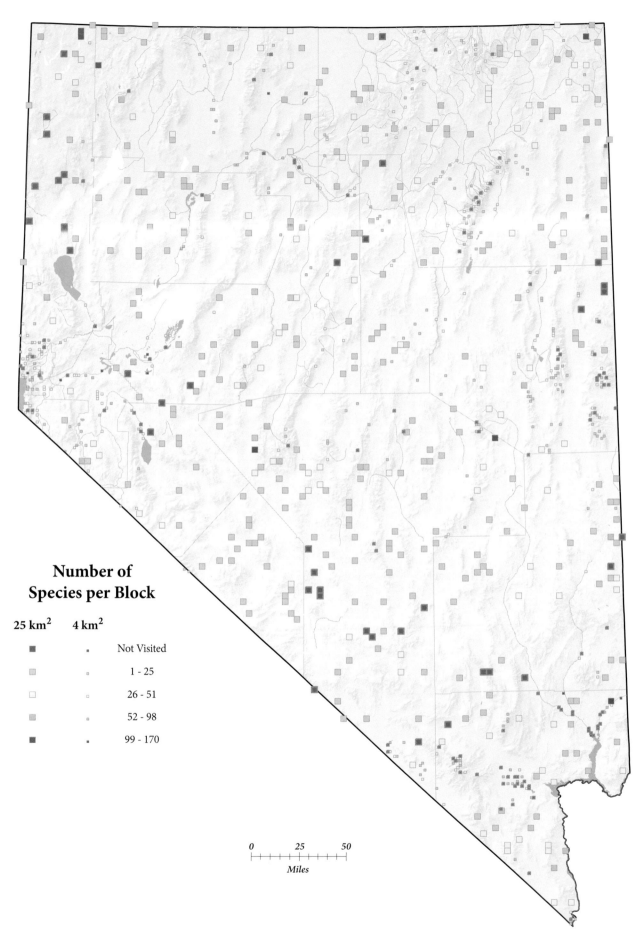

Number of
Species per Block

25 km²   4 km²

Not Visited

1 - 25

26 - 51

52 - 98

99 - 170

0   25   50

Miles

TABLE 3. THE TWENTY-FIVE MOST FREQUENTLY ENCOUNTERED SPECIES DURING THE ATLAS PROJECT

SPECIES	NUMBER OF BLOCKS IN WHICH SPECIES WAS RECORDED[†]
1. Brewer's Sparrow	577
2. Common Raven	541
3. Mourning Dove	457
4. Horned Lark	369
5. Northern Flicker	340
6. Western Meadowlark	338
7. Rock Wren	318
8. American Robin	315
9. Black-throated Sparrow	314
10. Brown-headed Cowbird	298
11. Red-tailed Hawk	296
12. American Kestrel	277
13. Sage Thrasher	269
14. Sage Sparrow	253
15. Brewer's Blackbird	247
16. Loggerhead Shrike	246
17. Spotted Towhee	235
18. Turkey Vulture	223
19. House Finch	222
20. Mountain Chickadee	215
21. Green-tailed Towhee	213
22. Chipping Sparrow	212
23. Killdeer	205
24. Common Nighthawk	200
25. Northern Harrier	191

Note: Data are from blocks only.

[†]This refers only to blocks in which possible, probable, or confirmed breeders were documented. Atlas blocks with presumed nonbreeders are excluded from the totals. For an explanation of this distinction, see pp. 18–21.

were found in more locations than expected during the atlas survey, mostly in northern Nevada.

Unfortunately, it is impossible to determine beyond a doubt *all* the species that currently breed in Nevada. The older literature generally includes little information about the breeding status of recorded birds. Some species are difficult to detect during surveys that use standard atlas methods. For example, there are historic records of Elf Owls breeding along the lower Colorado River in Nevada (Rosenberg et al. 1991), but no recent breeding has been confirmed for this species. Also, Spotted Owls breed close to the Nevada border in the Sierra Nevada, and may at least occasionally breed in Nevada as well (see p. 540–41). Additional surveys will be necessary to determine the current status of both species in Nevada.

Among the most frustrating species were several that were suspected to breed in Nevada but were either not found at all (e.g., the Northern Waterthrush in the far northern Jarbidge Range) or could not be confirmed as breeders (e.g., the Hermit Warbler in the far western Carson Range or the recently introduced Sharp-tailed Grouse in northern Nevada mountain ranges). Another species "missing" from the atlas database is the Eurasian Collared-Dove, which apparently did not colonize the state until after fieldwork was completed (see p. 540).

Table 3 lists the twenty-five species that were found in the greatest number of atlas blocks. The most frequently encountered species was the Brewer's Sparrow, which is found almost exclusively in sagebrush-dominated habitats during the breeding season. Other sagebrush-associated species, such as Sage Thrasher and Sage Sparrow, also ranked high in encounter frequency. Mountain Chickadee and Chipping Sparrow are typical of pinyon-juniper habitats in Nevada. The high encounter frequencies of these species reflect the widespread occurrence of these two habitat types in the state of Nevada. The remainder of the list of most frequently encountered species comprises birds with a fairly generalized habitat use, such as Common Raven Mourning Dove, and Western Meadowlark.

The Nevada atlas database contains 32,396 records; 29% of these records involved confirmed breeders, and an additional 23% involved probable breeding. It comes as no surprise that there was considerable variation in the ease with which different species could be found, and breeding confirmed. Most owls are hard enough to catch a glimpse of, let alone to see on the nest; thus, most owl records were of possible breeders. Pairs of waterfowl (presumed mates) are easy to view, and many species of ducks, geese, and swans thus had high rates of probable breeding. And species such as the Black-billed Magpie and American Robin, which build conspicuous nests and tend to their young in plain view of human observers, had very high confirmed breeding rates. The median confirmation rate, calculated across all species using only data from atlas blocks, was 27%, and ranged from 0% to 100% for individual species. Confirmation rates were typically higher for incidental observations, but this simply reflects the biases inherent in that subset of the data.

Species richness patterns across the state (see map on facing page) show many areas with relatively low richness—fewer than twenty-six species. Hotspots with ninety-nine or more species of birds occurred along river corridors, such as the Virgin, Carson, and Truckee rivers, and in a number of scattered locations throughout the central and northern parts of the state. Species richness varied greatly among habitats, with the highest richness found in wetlands, followed by ash, agricultural, and riparian habitats, and the lowest number of species found in salt desert scrub and in grasslands (Table 4).

TABLE 4. BIRD SPECIES RICHNESS AMONG HABITAT TYPES IN NEVADA BASED ON ATLAS BLOCK DATA

HABITAT TYPE	AVERAGE NUMBER OF SPECIES
Agriculture	43.4
Alpine	30.2
Ash	46.7
Grassland	14.1
Mesquite	22.0
Mojave	24.9
Montane Forest	34.9
Montane Parkland	32.8
Montane Shrubland	33.1
Pinyon-Juniper	35.7
Riparian	41.3
Sagebrush Scrub	19.8
Sagebrush Steppe	36.8
Salt Desert Scrub	13.5
Urban	29.1
Wetland	56.6

One important lesson the atlas highlights is that Nevada has a wildly diverse and dynamic landscape. Mountain ranges along the state's periphery stand out as sometimes surprising strongholds for several Nevada birds. The Carson Range in the far west, which is an extension of the northern Sierra Nevada, has many birds that are rarely, if ever, found elsewhere in the state. The far northern mountains, such as the Jarbidge, Santa Rosa, and Montana ranges, support populations of birds that are characteristic of the sagebrush steppe and other vegetation covers typical of the Columbia Plateau. The Spring Mountains and Sheep Range in southern Nevada support disjunct populations of several montane birds, some of which may prove to be taxonomically distinct from populations found in other parts of the state. Mention of the McCullough and Virgin mountains of Clark County conjures fond memories for many atlas workers, who were treated to rare sightings of birds typically found south of Nevada.

And the bird life of central Nevada is by no means uninspiring. The Ruby Mountains and surrounding valleys have several rare species of birds; for example, the Himalayan Snowcock, which is found nowhere else in North America, and the Trumpeter Swan, which does not breed elsewhere in the state. Other ranges, such as the Toiyabe, Monitor, Snake, and Schell Creek ranges of central and eastern Nevada, are strongholds for some species of conservation concern such as the Pinyon Jay.

PART II

SPECIES ACCOUNTS

It goes without saying that the species accounts are the meat and potatoes of any breeding bird atlas. After all, it is in these accounts that the reader will find significant new information about the status and distribution of breeding birds in an area. In many breeding bird atlases, each species account also provides an overview of the bird's natural history (what the nest is made of, clutch size, etc.) along with specific details on occurrence and distribution.

The *Atlas of the Breeding Birds of Nevada* departs somewhat from this model by focusing almost entirely on information with direct relevance to Nevada and offering little about basic breeding biology. The reasons for taking this approach are twofold. First, there are widely available general references that provide such information. Foremost among these is probably the *Birds of North America* series, available on-line at http://bna.birds.cornell.edu/BNA/. *The Birder's Handbook* (Ehrlich et al. 1988) and *Lives of North American Birds* (Kaufman 1996) also include summary information for individual species, and the *Sibley Guide to Bird Life and Behavior* (Elphick et al. 2001) provides an overview for each family of North American birds. Second, omitting this general information allowed us to devote more space to data from Nevada. Much of the material in the species accounts consists of new data, or a new look at older data with the goal of providing a new synthesis. Finally, because this atlas project was originally designed to provide information for bird conservation and management in Nevada, we summarize the information available on these topics.

TEXT

The text of each species account consists of three parts: an overview, an analysis of the bird's breeding distribution in Nevada and adjoining regions, and a discussion of its conservation status and management issues in the state and surrounding regions. The purpose of the overview is to give the reader a feel for salient features of the bird's behavior, breeding biology, or distribution in Nevada. Species identification is discussed in some accounts, breeding biology in others, and conservation imperatives in yet others. There is no fixed formula for this section; the main point is to emphasize memorable features of the species in question.

The "Distribution" section of the species accounts discusses atlas survey results in relation to earlier literature from Nevada and from nearby states. The older Nevada literature is described in order to place the atlas results into a historical context. Only a few major reviews of Nevada's breeding avifauna are available, and none emphasizes quantitative study of breeding birds. Major pre-atlas resources with direct relevance to Nevada include Ridgway 1877; Fisher 1893; Linsdale 1936, 1951; Ryser 1985; Alcorn 1988; and Titus 2003. All of these sources were consulted extensively during the process of drafting the atlas species accounts. In addition to major historical and current references on Nevada's avifauna, we consulted various books that deal more narrowly with aspects of species' biology with direct relevance to the *Atlas of the Breeding Birds of Nevada*. These fall into three categories: (1) books that deal with only a geographically restricted portion of Nevada; for example, Chisholm and Neel's (2002) thorough treatment of the birds of the Lahontan Valley; (2) books that deal only with particular taxa in Nevada; for example, Herron et al.'s (1985) detailed treatise on the hawks and owls of Nevada; and (3) studies on nonavian taxa that are relevant to birds or employ methods similar to those of the atlas project; for example, Charlet's (1996) Nevada conifer atlas.

The atlas fieldwork coincided with a renaissance of ornithological studies in Nevada and the greater region. Several important works became available in time to consult while preparing the atlas manuscript; for example, Paige and Ritter 1999, Neel 1999a, Clark County 2000, Chisholm and Neel 2002, Dobkin and Sauder 2004, and Walters 2004. Several periodical publications with direct relevance to the breeding avifauna of Nevada also came into existence during the atlas years, particularly the journal *Great Basin Birds* (published by the Great Basin Bird Observatory) and the Great Basin Regional Report in the journal *North American Birds*. The Nevada Birds list server, archived since its inception in June 2000 at <list.audubon.org/archives/ nvbirds.html>, is another important resource for the student of bird distributions in Nevada. We also consulted many experts on the state's avifauna to obtain more detailed information on the distributions of certain species than could be obtained using atlas methods. **For game birds especially, the Nevada Department of Wildlife and the National Wildlife Refuge system have much more comprehensive data on breeding occurrences and recent population trends than are outlined in the *Atlas of the Breeding Birds of Nevada*.** We encourage readers to consult the many reports and publications released by these agencies.

Along with reviewing the relevant Nevada literature we consulted major works on the status and distribution of birds in the states surrounding Nevada and elsewhere in the West. This was done because neither the major ecoregions (in Nevada, these include the Great Basin, Mojave Desert, Columbia Plateau, and Sierra Nevada) nor the birds observe state boundaries. For instance, in order to understand the restricted range of the Black-capped Chickadee in Nevada, it is necessary to consult literature covering nearby areas in Utah, Idaho, and Oregon.

Works from adjoining states that we consulted extensively include the following: for Oregon, Littlefield 1990, Contreras and Kindschy 1996, Contreras 1999, Adamus et al. 2001, and Marshall et al. 2003; for Idaho, Stephens and Sturts 1998; for Utah, Behle 1985, Behle et al. 1985, and UOSBRC 2005; for Arizona, Monson and Phillips 1981, Rosenberg et al. 1991, and Corman and Wise-Gervais 2005; and for California, Garrett and Dunn 1981, McCaskie et al. 1988, Gaines 1992, Small 1994, and various California county breeding bird atlases. Farther afield, we found that some atlases from more distant areas of the North American West—notably those for Colorado (Kingery 1998), Washington (Smith et al. 1997), and Saskatchewan (Smith 1996)—and from other parts of the world were sometimes relevant to interpreting atlas data from Nevada.

Beyond the realm of regional ornithological treatises for the West and elsewhere, we also consulted the broader literature. Several sources that we referred to repeatedly were from the 1992–2002 *Birds of North America* series, the North American Breeding Bird Survey (Sauer et al. 2005), and the primary journal literature (especially *Condor* and *Wilson Bulletin*). Throughout, we follow the nomenclature and taxonomic sequence in the seventh edition (1998) and supplements thereto of the *AOU Check-list of North American Birds*.

To provide full context for atlas findings, we attempted a fairly comprehensive review of all these sources as they relate to each species. The "Distribution" section of each account also contains relevant information on atlas habitat data and abundance estimates. (See pp. 21–23 for a discussion of some limitations on interpreting these data.) Habitat data and abundance estimates are also often discussed in the context of other literature, and the reader should be advised that definitions for habitat types and, more important, for bird abundance categories (e.g., the terms "common" and "uncommon") vary widely in their use from author to author. Discussions that strictly involve Nevada atlas data use these terms according to the definitions provided on pp. 22–23.

The immediate goal of any breeding bird atlas is, of course, the production and interpretation of range maps. The longer-term objective for projects of this scale is to provide data that will be analyzed and interpreted for additional purposes. To this end, the text of every species account ends with a "Conservation and Management" section. For this section, we consulted recent information on bird population trends, the most important being data from the national Breeding Bird Survey, which has been conducted nationwide since the 1960s (Saver et al. 2005). Details about this program, which is administered by the U.S. Geological Survey, are available online at http://www.mbr-pwrc.usgs.gov/ bbs/. We also included information on the conservation status of species from several bird conservation initiatives and species conservation plans that have ranked species according to conservation concern. The *North American Landbird Conservation Plan* of Partners in Flight (Rich et al. 2004) ranks the land bird species of highest concern as Watch List species and species of moderate concern as Stewardship species. The *Bird Conservation Plan* of Nevada's Partners in Flight chapter recognizes the land birds of highest concern in the state as Priority species (Neel 1999a). The waterbird conservation plans for the United States and for the Intermountain West rank species according to High, Moderate, and Low Concern (Kushlan et al. 2002, Ivey and Herziger 2005), and the *U.S. Shorebird Conservation Plan* uses these and the highest ranking of Highly Imperiled (Oring et al. 2005). Finally, we reported rankings of species recognized in the two largest multi-species conservation plans in Nevada: the Clark County Multiple Species Habitat Conservation Plan (Clark County 2000) and the Lower Colorado River Multi-species Conservation Program (BOR-LCR 2004). Both of these programs rank the species of highest concern as Covered species, and species that merit further investigation as Evaluation species.

Since conservation issues relevant to many species have not been studied in Nevada, we have often drawn on findings from nearby or more distant regions, accepting that there is uncertainty when extrapolating those results into a Nevada context. Despite the limited extent of state-specific information, it is a very exciting time to be involved in the planning and monitoring of bird conservation and management projects in Nevada, and we hope that the "Conservation and Management" sections of each species account will inspire many new questions and approaches.

DISTRIBUTION MAPS

Many devotees of breeding bird atlases consider the range maps to be the single most useful component of the published book. As explained on pp. 9–10, the maps for the Nevada atlas were generated a little differently from those in other breeding bird atlases. Readers are strongly encouraged to study the methods used to produce the maps of this atlas before plunging headfirst into the species accounts!

Each account includes two maps: one showing the actual data collected during fieldwork, and a second, smaller one showing the likelihood of finding a species in different parts of the state. This "Probability of Occurrence" map, or predictive map, was generated based on a statistical model that used the field data to predict a general distribution pattern and the relative chance of finding a species at locations throughout the state. See the chapter on predictive modeling (pp. 25–28) for details.

The data maps show all locations where potentially breeding individuals were found during the atlas project. Different symbols differentiate between data from atlas blocks and incidental observations. This distinction is important for any formal analysis of the data, but is less relevant when simply describing the known distribution limits of a species. We used circles (○, ●, etc.) to denote records from atlas blocks, and triangles (△, ▲, etc.) to denote incidental observations. The shading of the symbols indicates the breeding status: confirmed (black), probable (gray), or possible (white). Circles, by default, are centered on the southwest corner of their associated block. For aesthetic reasons, on some blocks that were situated partly or mostly outside Nevada's borders, the default symbol location was shifted slightly so that at least half of the symbol fell within Nevada's borders.

For most species, we did not map data that referred to presumed nonbreeders (atlas code Ø; see p. 18). We made this decision because many of these records represented migrants or other nonbreeding individuals, and plotting them would have complicated the interpretation of maps that focus on breeding distributions. Reports of birds that were not actually nesting nearby nevertheless provided meaningful information about some species. This is especially true for colonial breeders that travel far from their colonies to feed. For example, almost all of Nevada's American White Pelicans nest at a single location in the state and travel long distances to forage. Hence, most pelican records probably do relate to breeding birds, although they were not breeding in the vicinity of the sighting. Moreover, these observations provide useful information about the distances pelicans travel from their colony to feed, and concentrations of observations quite likely identify important feeding areas. For the case of the pelican, where observations of nonbreeders were thought to provide valuable information, we generated a map that uses a cross (✕) to denote each sighting.

In addition to the bird data, each map shows county boundaries, in order to orient readers, and selected land cover types that shed light on the distribution pattern. Since species respond to different geographic features and habitats, we chose not to overlay the bird data on a uniform map. Instead, we decided which land cover types to show based on our understanding of the bird's biology and on what the atlas data revealed about their habitat selection. For example, the maps for montane species typically show all of the state's montane forests, while those for species associated with the southern deserts show the distribution of Mojave habitat. In deciding what features to include on the maps, we tried to balance our desire to illustrate the patterns discussed in the text with the need to keep maps simple and understandable. These choices were *not* based on formal statistical analyses, which a full understanding of the species' distributional patterns and habitat use would require.

The smaller predictive map included in each species account shows the output from a statistical model that used atlas records, geographical information, and land cover data to predict species distributions throughout the state. It is with these maps that we diverge furthest from the typical breeding bird atlas. The chapter entitled "Predicting Species Distributions" (see pp. 25–28) discusses the rationale behind the predictive maps, the methods used to create them, and their reliability. We explain below how these maps should be interpreted.

Every birdwatcher knows that it is impossible to be certain that a particular species will occur in a particular place. Consequently, our predictive model does not attempt to forecast precisely where each species will and will not occur. Instead, the model uses the atlas data to estimate the probability that a species will be found in an area. The predictive maps thus show how the chance of finding a species changes from one region of the state to another. Each map includes up to six colors, each indicating a different chance of finding the species. White areas indicate where the species is expected to be absent as a breeder. Dark blue shading is used for areas where the model suggests that there is a very low (<5%) chance of finding the species breeding, and dark red shading identifies areas with the highest chance (>50%) of finding the species breeding.

Because of their size, the predictive maps for some species, especially those that are relatively uncommon, show little or no variation in occurrence probabilities. Indeed, for some rare species, potentially occupied areas may be so small that the maps will appear to be entirely white. Atlas users interested in the finer-scale variation that the maps published in this volume cannot show can consult the Great Basin Observatory's Web site (www.gbbo.org), which shows higher resolution maps that clearly identify where the model predicts each species is most likely to be found. This Web site will also report any future research that uses, tests, or refines the predictive models produced using the atlas database.

SUMMARY STATISTICS IN TABLES

In addition to the maps and interpretive text, each species account includes tables that provide breeding confirmation statistics (separated into atlas blocks and incidental observations), a breakdown of habitat use (also separated into atlas blocks and incidental observations), and abundance estimates (only for atlas blocks; abundance was not estimated for incidental observations). In the following, we summarize the key points to consider when reviewing the data tables (for more details, see pp. 18–23).

1. The totals reported in the three tables often do not match. There are several reasons for this discordance. First, some atlas blocks were visited multiple times, resulting in different abundance estimates. Second, in some cases a species was seen in multiple habitats within a block. Third, observers sometimes did not report habitat or abundance data at all.

2. The habitat tables show observer-reported habitats, not the habitat types originally assigned to atlas blocks by the GAP vegetation map.

3. Rounding errors occurred in the percentages reported, sometimes resulting in totals slightly greater or less than 100%.

4. All quantitative data reported in the tables are relatively coarse. More focused studies are needed to better determine bird densities and descriptions of bird-habitat relationships. The species account tables merely attempt to provide an overview of these relationships to the extent that atlas data can provide one.

Additional analyses of the atlas data are encouraged, and we will be glad to discuss potential uses with investigators interested in using the data.

SUPPLEMENTAL SPECIES ACCOUNTS

Finally, twenty-eight species are treated in short species accounts placed after the full species accounts. These species, for which breeding was not confirmed during the atlas project, were included for a variety of reasons. Some were confirmed after the atlas surveys were completed or were confirmed earlier by other means. Some have bred here in the past. Yet others were reported only as probable or possible breeders during the atlas surveys. This section also includes introduced species that are not yet well established, and species that have potential to spread into the state in the near future.

CANADA GOOSE
Branta canadensis

Once a symbol of wilderness, the Canada Goose is now generally thought of in connection with golf courses and urban parks. In Reno, the species nests not only in city parks but also atop high-rise buildings. Away from human habitation, the status of the species is less clear. Numbers appear to be increasing in Nevada, as they are across most of the continent. There is a suspicion, however, that the migratory instincts of the Canada Goose have been eroded by the creation of permanent food supplies in agricultural and suburban areas.

DISTRIBUTION

Canada Geese breed across nearly all of Canada and most of the United States except in the most southerly regions (Mowbray et al. 2002). Alcorn (1988) reported breeding in western and northern Nevada, but not in the south. Atlas fieldwork also revealed concentrations of breeders along the drainages of west-central Nevada and scattered breeding records across the northern half of the state. Atlas results further suggested that the species is beginning a range expansion into southern Nevada. Breeding was confirmed at several sites in Lincoln County, where Alcorn's (1988) account does not mention the species as occurring. There were also three unconfirmed atlas breeding records in Clark County, and nesting was confirmed in Las Vegas's Lorenzi Park in 1998 and 2000 (C. Titus and D. Blake, pers. comm.) and in Hidden Valley near the town of Moapa in 2001 (R. Saval, pers. comm.).

Canada Geese are easily approached and conspicuous on their nests, resulting in a relatively high confirmation rate (48%) in occupied blocks. The species was often rated as fairly common, with multiple pairs estimated to be present. Most records were from aquatic habitats, with a few records from urban and agricultural habitats. Not surprisingly, our predictions suggest that Canada Geese are most likely to be found in areas with extensive wetlands and along major river drainages. The species' ability to use small wetland and riparian habitat patches is reflected in the model's prediction of a small chance of occurrence throughout a large swath of northern Nevada.

CONSERVATION AND MANAGEMENT

Around cities and farms, the increase in Canada Goose populations has been staggering, to the point that geese are sometimes considered pests. Management efforts now tend to focus on preventing overpopulation rather than on conservation (Mowbray et al. 2002). Population growth has been documented throughout North America, despite the fact that Canada Geese are heavily hunted (Mowbray et al. 2002). Local increases have been attributed to various causes, but human activities are always a central theme (Shuford 1993, Rudesill in Burridge 1995:35, Smith et al. 1997, Winn in Kingery 1998:68–69). Populations are also growing in Europe, where the species was introduced around 1650 (Kirby and Sjöberg in Hagemeijer and Blair 1997:75).

Despite the burgeoning population numbers overall, concern for the Canada Goose has been raised on several fronts. First, the status of "traditional" migratory populations in native habitats is uncertain. Though Ryser (1985) considered Canada Goose populations in Nevada to be largely unchanged since the nineteenth century, the atlas data show only scattered records of confirmed breeding away from the Reno and Lahontan Valley areas. At Ruby Lake National Wildlife Refuge, the average breeding population of about 130 pairs produced young at a rate far below the refuge's management objectives during the atlas years (USFWS 2001), and according to longtime residents of the area, goose numbers have dropped considerably in the last few decades (J. Mackay, pers. comm.). Second, many Canada Geese have become sedentary and are now summering in areas where previously they were known only to winter. Year-round food and artificially enhanced nest sites, both provided courtesy of humans, are among the suspected incentives for this "lifestyle" change (Small 1994).

HABITAT USE

HABITAT	ATLAS BLOCKS	INCIDENTAL OBSERVATIONS
Agricultural	7 (7%)	1 (3%)
Grassland	0 (0%)	1 (3%)
Open Water	21 (22%)	11 (30%)
Riparian	32 (33%)	4 (11%)
Urban	6 (6%)	3 (8%)
Wetland	30 (31%)	16 (44%)
TOTAL	96	36

Total numbers reported in the Habitat Use, Breeding Status, and Abundance tables may differ from each other (see pp. 22–23 for details). Percentage sums may differ slightly from 100% due to rounding.

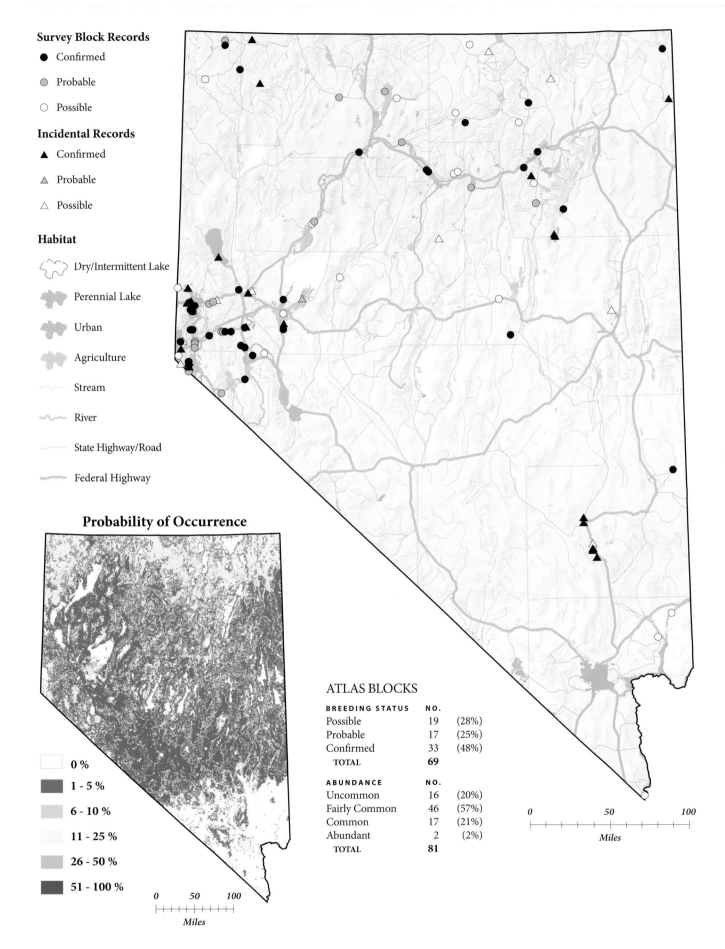

Survey Block Records

● Confirmed

● Probable

○ Possible

Incidental Records

▲ Confirmed

▲ Probable

△ Possible

Habitat

⬭ Dry/Intermittent Lake

⬭ Perennial Lake

⬭ Urban

⬭ Agriculture

〜 Stream

〜 River

— State Highway/Road

— Federal Highway

Probability of Occurrence

☐ 0 %

▨ 1 - 5 %

▨ 6 - 10 %

▨ 11 - 25 %

▨ 26 - 50 %

▨ 51 - 100 %

0   50   100
Miles

ATLAS BLOCKS

BREEDING STATUS	NO.	
Possible	19	(28%)
Probable	17	(25%)
Confirmed	33	(48%)
TOTAL	69	

ABUNDANCE	NO.	
Uncommon	16	(20%)
Fairly Common	46	(57%)
Common	17	(21%)
Abundant	2	(2%)
TOTAL	81	

0   50   100
Miles

TRUMPETER SWAN
Cygnus buccinator

When one thinks of conservation success stories, the Trumpeter Swan often comes to mind. The bird was brought to the brink of extinction by habitat destruction and hunting. In the latter half of the twentieth century, however, the Trumpeter Swan staged a remarkable comeback—the result of direct intervention and active management by wildlife biologists. In Nevada, the Trumpeter Swan was reintroduced into the Ruby Valley beginning in 1949; the population gradually increased, and several pairs are now resident in the valley.

DISTRIBUTION

The Trumpeter Swan was widespread in western and central North America prior to its dramatic range retraction and population decline. The species was apparently native to the eastern half of the Great Basin, including eastern Nevada (Ryser 1985), and John Muir's reference to swans at Mono Lake in June 1875 (Wolfe 1979) suggests that they may have occurred in the western Great Basin as well. By the early twentieth century, however, the species had been completely extirpated from the state (see Linsdale 1936). Trumpeter Swans were reintroduced during the 1940s and 1950s, when birds were transplanted from Montana to the Ruby Valley (Banko 1960). Both Banko (1960) and Ryser (1985) provided substantial detail on this reintroduction project, and Alcorn (1988) gave a fragmentary record of known occurrences (of migrants or winterers) elsewhere in the state.

Trumpeter Swans continue to breed at Ruby Lake National Wildlife Refuge but have not spread elsewhere in the state. Two incidental records from the Ruby Valley came in during the atlas project. Both records were of confirmed breeders in freshwater marshes. The Ruby Valley breeders are part of a growing, but still highly disjunct, population of Trumpeter Swans in western North America. Because the species is so rare in Nevada, a predictive map could not be generated. Beyond the state borders, the nearest breeding populations are in southern Oregon and Idaho (Mitchell 1994, Contreras 1999, Adamus et al. 2001).

CONSERVATION AND MANAGEMENT

Ryser (1985:125) called the Trumpeter Swan recovery "heartwarming," and on the whole this is a reasonable assessment: almost all populations of the species are increasing, and it has recently expanded its range eastward (Mitchell 1994). The Trumpeter Swan's population growth rate has been slow, however, and individual populations may still require special attention.

Because of the fragmented nature of the Trumpeter Swan's distribution, flocks must be managed on an individual basis (Mitchell 1994). In Nevada, there were six to seven pairs at Ruby Lake during the atlas years 1997–2000 that produced an average of five cygnets each year (USFWS 2001). Since 2000, the population has increased to eight pairs, but no cygnets were produced from 2001 to 2003 (J. Mackay, pers. comm.). Disturbances around the nest are a problem for this species, and low water levels on the refuge during this period may have made nests more accessible to coyotes and other predators. A key management question concerns the fate of the cygnets produced on the refuge.

Young birds join the wintering flock of about forty swans and tend to leave with these winterers (J. Mackay, pers. comm.). Neither the origin of the wintering birds nor the destination of the departing young is known. It is also not clear why more birds do not stay to breed at the refuge. Nonetheless, the persistence of this population for more than fifty years is encouraging and suggests that attempts to translocate birds to protected wetlands elsewhere in the state might also be successful.

HABITAT USE

HABITAT	ATLAS BLOCKS	INCIDENTAL OBSERVATIONS
Wetland	0 (0%)	2 (100%)
TOTAL	0	2

Total numbers reported in the Habitat Use, Breeding Status, and Abundance tables may differ from each other (see pp. 22–23 for details). Percentage sums may differ slightly from 100% due to rounding.

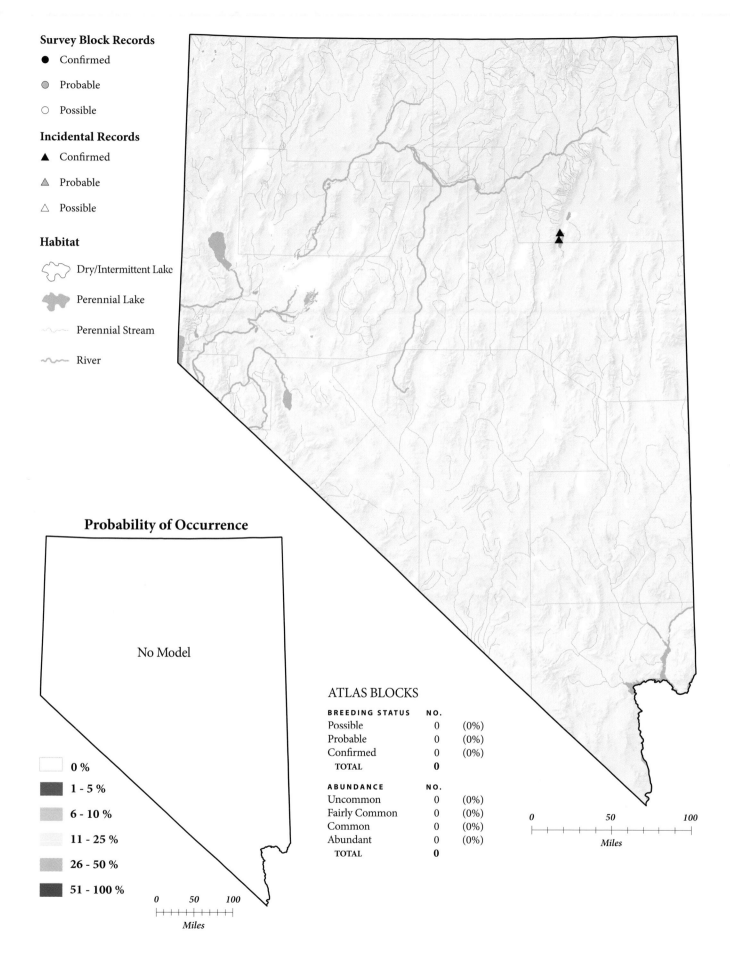

Survey Block Records
● Confirmed
◐ Probable
○ Possible

Incidental Records
▲ Confirmed
△ Probable
△ Possible

Habitat
Dry/Intermittent Lake
Perennial Lake
Perennial Stream
River

Probability of Occurrence

No Model

0 %
1 - 5 %
6 - 10 %
11 - 25 %
26 - 50 %
51 - 100 %

0   50   100
Miles

ATLAS BLOCKS

BREEDING STATUS	NO.	
Possible	0	(0%)
Probable	0	(0%)
Confirmed	0	(0%)
TOTAL	0	

ABUNDANCE	NO.	
Uncommon	0	(0%)
Fairly Common	0	(0%)
Common	0	(0%)
Abundant	0	(0%)
TOTAL	0	

0   50   100
Miles

WOOD DUCK
Aix sponsa

"Extravagant" and "over the top" are terms that come to mind when one thinks of the Wood Duck. The bird is stunningly—even garishly—attired, but it is no coquette. Wood Ducks favor wooded waterways and spook at the slightest disturbance. Their exquisite plumage and reclusive habits have proved an irresistible challenge to sportsmen, and the combination of hunting and habitat loss reduced populations during the early twentieth century. The species might even have been extirpated from Nevada during this time, and its recovery probably resulted from improved protection from overhunting, the provision of nest boxes, and regeneration of riparian vegetation.

DISTRIBUTION

Wood Ducks were confirmed as breeders along the Carson and Truckee rivers of western Nevada, and additional records of probable breeding were reported from the Humboldt and Walker rivers. Atlas data suggest that they are not especially numerous; half of the abundance estimates involved only one pair per occupied block. Wood Ducks are probably more common than the atlas data suggest, however, as several hundred birds are known to occur within the Lahontan Valley (C. Nicolai, pers. comm.).

Wood Ducks have a disjunct range in North America. Ryser (1985) described the species as a peripheral element of the Great Basin avifauna, and today its Nevada distribution remains limited to the major drainages of the west-central portion of the state. Within the small geographic area from which records came, how-

ever, our model predicts that Wood Ducks have a moderately high chance of occurring. Linsdale (1936) described the species as formerly uncommon and cited no Nevada records from the early twentieth century, but Alcorn (1988) noted several reports of pairs and breeding records during the 1940s and 1950s.

All but one of the nineteen atlas habitat records describe Wood Ducks as being in riparian or wetland habitats. The species is a cavity nester and requires trees old enough to support nest holes or the presence of nest boxes (Hepp and Bellrose 1995). At a larger spatial scale, the occurrence of the Wood Duck is determined by the distribution of streams and rivers (Bellrose and Holm 1994). The absence of birds elsewhere in the state, and from much of the rest of the Intermountain West (Hepp and Bellrose 1995), is presumably due to the long distances between patches of suitable habitat and the small size of these patches, which reduce the chance that birds will colonize them and that populations will persist.

CONSERVATION AND MANAGEMENT

Wood Ducks require wooded wetlands and riparian zones for breeding, and the species benefits directly from riparian habitat enhancement or restoration. Nest site availability may be limiting in areas where there are few large trees with natural cavities, and providing nest boxes is often beneficial (Hepp and Bellrose 1995). It is important, however, to space boxes correctly and to place them in inconspicuous locations to minimize the amount of intraspecific brood parasitism. When boxes are clustered in the open, the rate at which females lay eggs in the nests of other females can be high enough to result in significant egg losses (Eadie et al. 1998).

The overall prognosis for the Wood Duck is good, and it is encouraging that the species has steadily increased in numbers throughout most of its breeding range, particularly in the West (Sauer et al. 2005). In Nevada, the prospects for the Wood Duck also seem auspicious. The species is likely one of the prime beneficiaries of the ongoing protection and restoration efforts along the Truckee, Carson, and Walker rivers, and with sufficient riparian habitat recovery it could well spread into other riparian areas in the region.

HABITAT USE

HABITAT	ATLAS BLOCKS	INCIDENTAL OBSERVATIONS
Agricultural	1 (7%)	0 (0%)
Riparian	10 (71%)	2 (40%)
Wetland	3 (21%)	3 (60%)
TOTAL	14	5

Total numbers reported in the Habitat Use, Breeding Status, and Abundance tables may differ from each other (see pp. 22–23 for details). Percentage sums may differ slightly from 100% due to rounding.

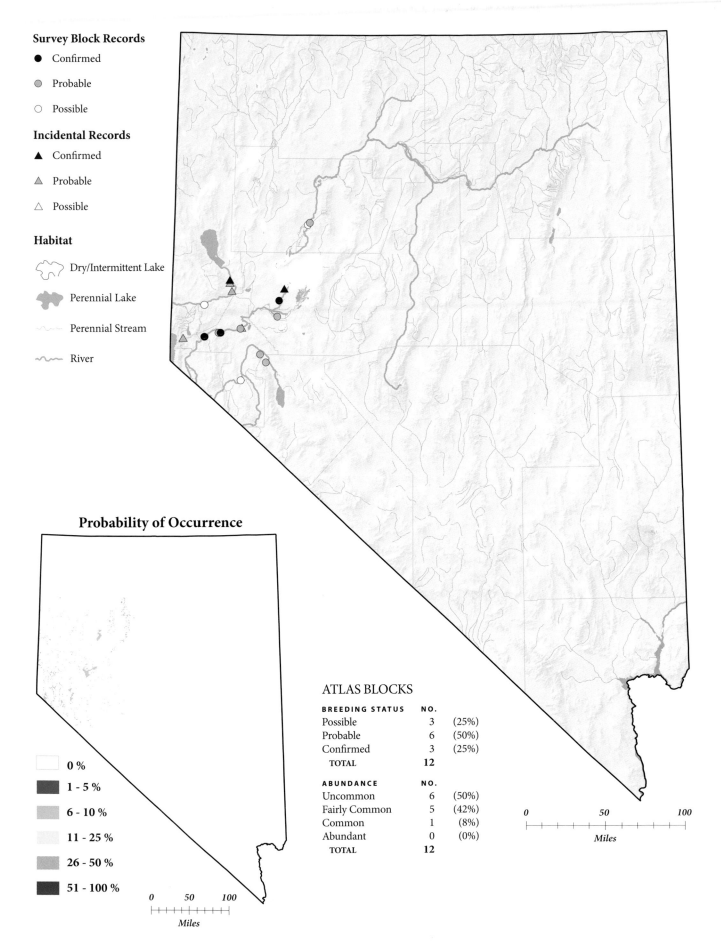

Survey Block Records

● Confirmed

● Probable

○ Possible

Incidental Records

▲ Confirmed

▲ Probable

△ Possible

Habitat

Dry/Intermittent Lake

Perennial Lake

Perennial Stream

River

Probability of Occurrence

☐ 0 %

■ 1 - 5 %

■ 6 - 10 %

☐ 11 - 25 %

■ 26 - 50 %

■ 51 - 100 %

0   50   100

Miles

ATLAS BLOCKS

BREEDING STATUS	NO.	
Possible	3	(25%)
Probable	6	(50%)
Confirmed	3	(25%)
TOTAL	12	

ABUNDANCE	NO.	
Uncommon	6	(50%)
Fairly Common	5	(42%)
Common	1	(8%)
Abundant	0	(0%)
TOTAL	12	

0   50   100

Miles

GADWALL
Anas strepera

What the Gadwall lacks in showiness it makes up for in subtle beauty. Its plumage has a certain understated dapperness. It is also one of the most widespread and common waterfowl species in Nevada. Breeders can be found around the state in large marshes, on small farm ponds, and even in urban settings. The Gadwall appears to have been restricted to the northern part of the state at one time, but it has made recent inroads into southern Nevada.

DISTRIBUTION

The Nevada range of the Gadwall seems to be in the midst of an expansion. As recently as the late twentieth century the species apparently did not breed in southern Nevada (Alcorn 1988), and during the early 1900s only a few Gadwalls were present on Nevada lakes and ponds during the summer (Linsdale 1936). This view is consistent with the general observation that the Gadwall was once absent from most of the western United States (LeSchack et al. 1997).

During the atlas years, Gadwalls were found throughout much of Nevada. Only in Esmeralda County and in similarly dry adjoining regions of Mineral and Nye counties was the species completely absent. The occurrence of Gadwalls in northern Nevada was not surprising; earlier reports describe them as common to abundant in this part of the state (Ryser 1985). More surprising were the scattered confirmations of breeding from eastern and southern Nye County, the Pahranagat Valley, and the Las Vegas area.

In general, breeding Gadwalls require wetlands with emergent grassy vegetation and protected islands (LeSchack et al. 1997), and most of the reported habitat associations were consistent with this requirement. Gadwalls' use of small wetlands embedded within other habitats accounts for the prediction that there is at least a low chance of their occurrence throughout even the drier areas in the state. In Washington, the species reportedly shows a marked preference for urban habitats (Smith et al. 1997), so further expansion into the urban centers of western and southern Nevada might be expected.

More than ten Gadwall pairs were reported for about a quarter of the blocks in which abundance was estimated, and in two blocks more than one hundred pairs were estimated to be present. At Ruby Lake National Wildlife Refuge, Gadwalls are the most abundant dabbling duck, with more than five hundred pairs nesting annually (USFWS 2001). Overall, breeding was confirmed in about a third of the blocks where the species was seen. Atlas surveys conducted elsewhere have suggested that breeding Gadwalls may be difficult to detect (Fingerhood in Brauning 1992:427–428, Gallagher 1997)—perhaps because females look similar to female Mallards—and their true abundance and frequency of breeding may have been underestimated.

CONSERVATION AND MANAGEMENT

Gadwall populations in North America have grown steadily since at least the 1960s, and the species has expanded from its traditional western range, especially to the east (LeSchack et al. 1997, Sauer et al. 2005). Overall, populations still seem to be faring well, and the apparent range expansion into southern Nevada accords with the broader continental pattern.

In Nevada, conservation concerns center on maintaining suitable habitat in large wetlands, which in this desert state can be subject to significant annual variations in water levels. At the local level, Gadwall populations can be encouraged in several ways. Gadwalls nest relatively late in the season (LeSchack et al. 1997), and this should be considered when manipulating water levels in managed wetlands. They also frequently nest on, and may prefer, well-vegetated islands that are protected from mammalian predators by surrounding deep water (LeSchack et al. 1997). In general, Gadwalls seem to prosper at protected sites that are managed for breeding waterfowl, and they readily use a variety of artificial impoundments (LeSchack et al. 1997).

HABITAT USE

HABITAT	ATLAS BLOCKS	INCIDENTAL OBSERVATIONS
Agricultural	3 (4%)	0 (0%)
Open Water	20 (24%)	12 (33%)
Riparian	21 (26%)	6 (17%)
Sagebrush Steppe	1 (1%)	0 (0%)
Urban	1 (1%)	0 (0%)
Wetland	36 (44%)	18 (50%)
TOTAL	82	36

Total numbers reported in the Habitat Use, Breeding Status, and Abundance tables may differ from each other (see pp. 22–23 for details). Percentage sums may differ slightly from 100% due to rounding.

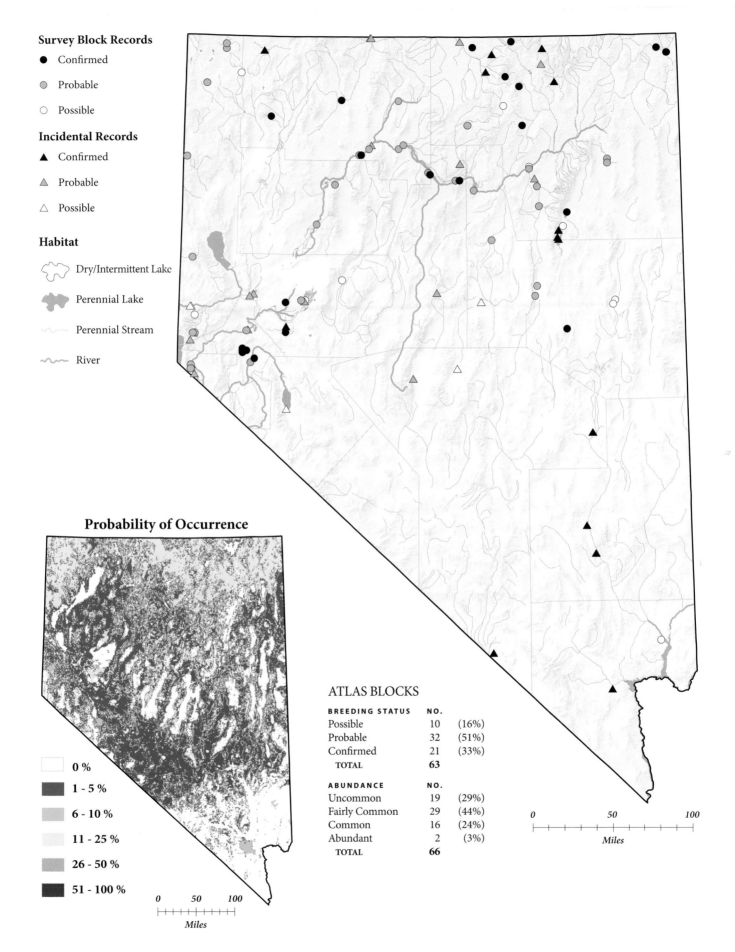

Survey Block Records

● Confirmed

● Probable

○ Possible

Incidental Records

▲ Confirmed

▲ Probable

△ Possible

Habitat

Dry/Intermittent Lake

Perennial Lake

Perennial Stream

River

Probability of Occurrence

0 %

1 - 5 %

6 - 10 %

11 - 25 %

26 - 50 %

51 - 100 %

0   50   100
Miles

ATLAS BLOCKS

BREEDING STATUS	NO.	
Possible	10	(16%)
Probable	32	(51%)
Confirmed	21	(33%)
TOTAL	63	
ABUNDANCE	NO.	
Uncommon	19	(29%)
Fairly Common	29	(44%)
Common	16	(24%)
Abundant	2	(3%)
TOTAL	66	

0   50   100
Miles

AMERICAN WIGEON

Anas americana

The handsome American Wigeon is among the least likely of Nevada's dabbling ducks to be encountered throughout much of the state. Nevada is at the periphery of the species' range, which is centered on the Canadian prairies, and breeding is largely restricted to the state's northeastern corner. The American Wigeon breeds in the Ruby Valley, but—perhaps surprisingly—does not nest regularly in the extensive wetland complexes of the Lahontan Valley.

DISTRIBUTION

Nevada lies toward the southern limit of the American Wigeon's breeding range, and earlier accounts describe its distribution as being restricted to the northern part of the state (Linsdale 1936, Alcorn 1988). The atlas project data are consistent with this assessment. The American Wigeon was reported from ten blocks, and there were incidental observations from ten additional sites. All records were from the northern half of Nevada, and most were from the northeastern corner. Breeding was confirmed only in Eureka and Elko counties.

American Wigeons were not very numerous, with fewer than ten pairs recorded in all but one of the occupied blocks. This result, too, is consistent with earlier accounts. Ryser (1985) called the species occasional to uncommon on its Nevada breeding grounds, and Alcorn (1988) termed it uncommon.

Continentally, the geographic range of the American Wigeon has increased in recent years as habitat has increased in the eastern states and provinces (Mowbray 1999). Closer to home, the American Wigeon has recently begun to breed at desert lakes (e.g., Bridgeport and Crowley reservoirs) just across the Nevada border in east-central California (Gaines 1992). Nonetheless, the species remains a rare summer visitor to the extensive wetlands of the Lahontan Valley (Chisholm and Neel 2002), and no breeding was reported from this area during the atlas years.

Wherever they occur, American Wigeons seem to favor open freshwater wetlands with adjoining cover (Mowbray 1999). Wigeon hens are more likely to take their broods out onto open water than are females of other dabbling ducks (Palmer 1976), suggesting that the species was unlikely to be missed when present in a surveyed area. Confirmed breeding was not reported from the Ruby Valley during the atlas project, although U.S. Fish and Wildlife Service surveys show that at least forty to seventy-five pairs nest at Ruby Lake annually (USFWS 2001).

CONSERVATION AND MANAGEMENT

In recent decades the American Wigeon has undergone a range expansion, especially in eastern North America, but this expansion has been coupled with substantial population losses in the heart of its range. Favorable conditions in the mid-1990s allowed populations to increase, but they still remain well below the levels of the early 1980s (Mowbray 1999).

There is little quantitative information on the population status of the American Wigeon in Nevada, and there is no specific information on management practices that favor the species. In the northern prairies, managed grazing and burning have been used to create dense vegetation in terrestrial habitats adjacent to the open aquatic habitats where adults and ducklings feed (Kruse and Bowen 1996). The applicability of these methods to nesting sites in Nevada, where the effects of grazing and fire are different from those in the prairies, remains to be tested.

HABITAT USE

HABITAT	ATLAS BLOCKS	INCIDENTAL OBSERVATIONS
Open Water	3 (25%)	4 (36%)
Riparian	2 (17%)	1 (9%)
Wetland	7 (58%)	6 (55%)
TOTAL	12	11

Total numbers reported in the Habitat Use, Breeding Status, and Abundance tables may differ from each other (see pp. 22–23 for details). Percentage sums may differ slightly from 100% due to rounding.

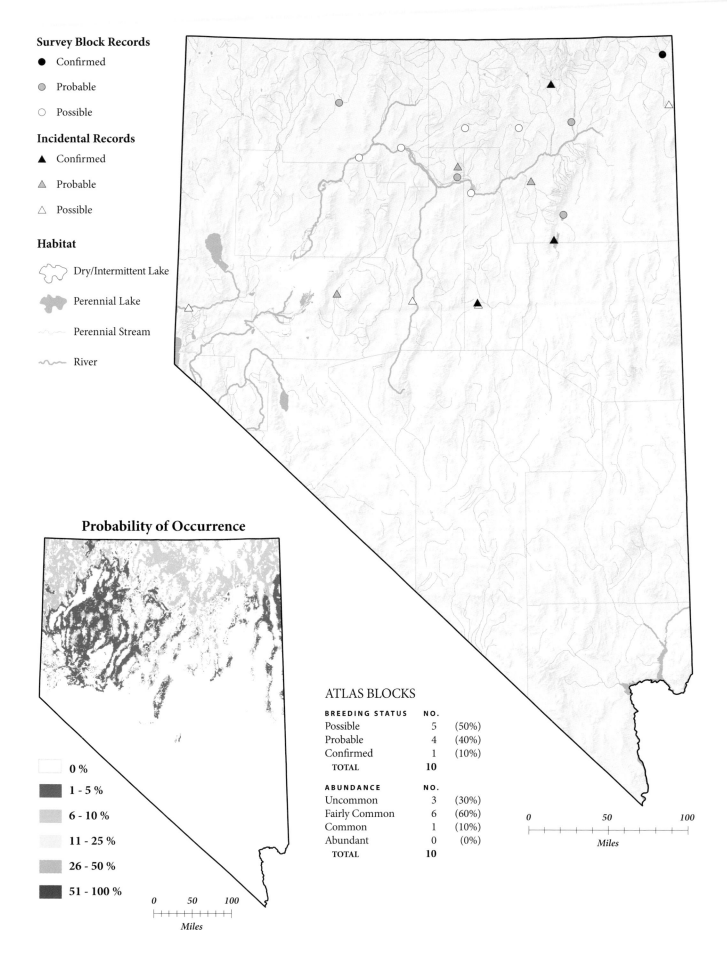

Survey Block Records
● Confirmed
◐ Probable
○ Possible

Incidental Records
▲ Confirmed
▲ Probable
△ Possible

Habitat
Dry/Intermittent Lake
Perennial Lake
Perennial Stream
River

Probability of Occurrence

- 0 %
- 1 - 5 %
- 6 - 10 %
- 11 - 25 %
- 26 - 50 %
- 51 - 100 %

0   50   100
Miles

ATLAS BLOCKS

BREEDING STATUS	NO.	
Possible	5	(50%)
Probable	4	(40%)
Confirmed	1	(10%)
TOTAL	10	

ABUNDANCE	NO.	
Uncommon	3	(30%)
Fairly Common	6	(60%)
Common	1	(10%)
Abundant	0	(0%)
TOTAL	10	

0   50   100
Miles

MALLARD
Anas platyrhynchos

Thoroughly at home in city parks and on farm ponds, the adaptable Mallard has fared well in the agricultural and urbanized regions of Nevada. But Mallards are common and widespread in natural habitats, too. The extent to which Nevada's Mallards are derived from feral stock is not clear, but population growth in urban areas may be at least in part attributable to birds brought in from elsewhere. Investigating the extent to which feral and wild populations interact is potentially a fruitful area for research, although at this stage it is possible that populations are too intermixed for us to ever fully understand the relationship between them.

DISTRIBUTION

The Mallard is generally regarded as the most widely distributed and abundant duck in the Northern Hemisphere (Drilling et al. 2002). It is likewise widely distributed and fairly common in Nevada, with a breeding distribution similar to that of the less common Gadwall. Mallards are present everywhere except in the dry desert landscape of Esmeralda County and adjoining portions of Mineral and Nye counties, and our predictions make it clear that the species has a high chance of occurring in many areas. Earlier accounts attest to the species' presence as a breeder in northern Nevada, but the breeding populations in Clark, Lincoln, and southern Nye counties may be more recently established. Alcorn (1988) listed records from all Nevada counties but implied that the Mallard withdraws from southern Nevada during the breeding season.

Mallards are well known for their catholic choice of aquatic habitats. In the West, they are most numerous in valley wetlands, but they also range into the mountains, where they occur on small streams and lakes (Behle et al. 1985, Smith et al. 1997). During the atlas project, breeding was confirmed even at eleva-

tions above 7,200 feet (2,200 meters). Most records came from wetland and riparian habitats, as expected, but reports from urban and agricultural habitats were not uncommon. Areas featuring these habitat types also figure prominently in our predicted distribution for this species. A few sightings from upland habitats presumably reflect the willingness of Mallards to nest some distance from water and to use even the tiniest of wet spots for breeding. Although Mallards were widespread in the state, the records clustered in the counties around Reno and Carson City will not come as a surprise to those familiar with the avifauna of city parks and golf course ponds. Similarly, in the south, there were several records from around Las Vegas. Elsewhere, the species' distribution appears closely tied to the major river systems.

CONSERVATION AND MANAGEMENT

Nevada's Mallard population appears to be in good health, and the species has benefited from human activity. Mallards make good use of waste grain in agricultural areas and are successful breeders in highly developed areas; they also respond well to wetland protection and enhancement.

Concerns have been raised, especially for areas near urban centers, that surging populations of mostly urban or introduced Mallards may mask declines in wild populations. For instance, Garrett and Dunn (1981) suggested that wild populations have declined in California. In Nevada, many Mallard populations nest far from urban centers, so this problem seems less likely to apply here. Nonetheless, as with Canada Geese, it is important not to let the successes of urban populations draw attention away from the value of maintaining good habitat conditions and high waterfowl production at other wetlands around the state.

HABITAT USE

HABITAT	ATLAS BLOCKS	INCIDENTAL OBSERVATIONS
Agricultural	12 (6%)	2 (3%)
Montane Forest	1 (<1%)	0 (0%)
Montane Shrub	0 (0%)	1 (1%)
Open Water	35 (16%)	13 (18%)
Riparian	93 (43%)	20 (28%)
Sagebrush Scrub	1 (<1%)	0 (0%)
Urban	11 (5%)	5 (7%)
Wetland	63 (29%)	30 (42%)
TOTAL	216	71

Total numbers reported in the Habitat Use, Breeding Status, and Abundance tables may differ from each other (see pp. 22–23 for details). Percentage sums may differ slightly from 100% due to rounding.

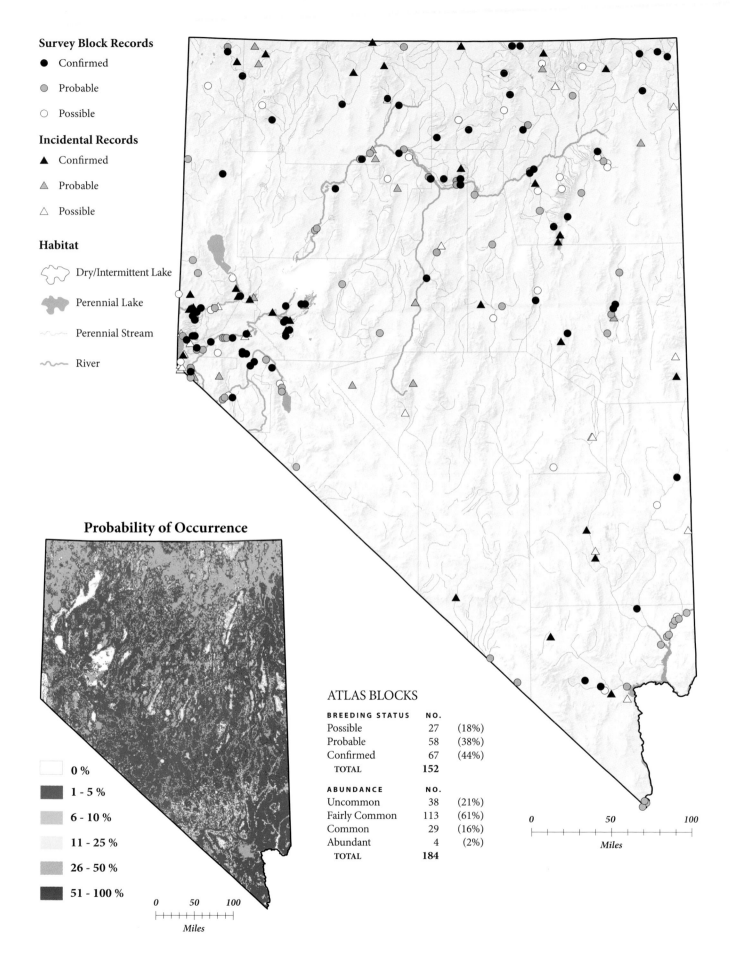

Survey Block Records

● Confirmed

● Probable

○ Possible

Incidental Records

▲ Confirmed

▲ Probable

△ Possible

Habitat

Dry/Intermittent Lake

Perennial Lake

Perennial Stream

River

Probability of Occurrence

☐ 0 %

■ 1 - 5 %

■ 6 - 10 %

☐ 11 - 25 %

■ 26 - 50 %

■ 51 - 100 %

0   50   100

Miles

ATLAS BLOCKS

BREEDING STATUS	NO.	
Possible	27	(18%)
Probable	58	(38%)
Confirmed	67	(44%)
TOTAL	152	

ABUNDANCE	NO.	
Uncommon	38	(21%)
Fairly Common	113	(61%)
Common	29	(16%)
Abundant	4	(2%)
TOTAL	184	

0   50   100

Miles

BLUE-WINGED TEAL
Anas discors

Finding a pair of Blue-winged Teal in Nevada is thrilling for two reasons. The first is obvious: the male, with his bold facial pattern of a gleaming crescent moon superimposed on a lead gray background, is striking to behold. The second reason is that the Blue-winged Teal is decidedly uncommon in Nevada. The causes of its scarcity are not immediately obvious, for the species is common throughout much of its extensive North American range. It was formerly common in Nevada, and appropriate habitat seems to be available in various parts of the state.

DISTRIBUTION

Atlas workers found Blue-winged Teal in ten blocks that were widely scattered across northern Nevada. There were also four incidental observations, all from the northern half of the state. The only confirmed breeding records came from western Nevada—one from near the Carson River and three from the wetlands of the far northwestern corner of the state. No more than ten pairs were thought to be present in any of the blocks for which abundance estimates were provided.

The atlas results are consistent with a recent assessment by Titus (2003), who considered the species to be rare in summer in northwestern Nevada, uncommon in summer in the northeast, and rare in winter in the south. In the past, however, the Blue-winged Teal seems to have been more common. Ryser (1985) analyzed the earlier literature and concluded that the species formerly was abundant in Nevada, but that its numbers had dwindled considerably. Alcorn (1988) called the Blue-winged

Teal uncommon in summer but cited widespread breeding at locations such as Sheldon and Pahranagat national wildlife refuges, the Ruby and Lahontan valleys, and Las Vegas Wash. Linsdale (1936) called the species fairly common and even provided records from Esmeralda County—an area that produced few records of any duck species during the atlas years.

Of course, it is possible that some Blue-winged Teal were simply missed in some places due to the limitations of the atlas sampling protocol and the challenges associated with identifying females (Garrett and Dunn 1981, Littlefield 1990). The widespread absence of the species from vast swaths of the state and the low predicted chance of occurrence in all habitats, however, suggest that the scarcity is real. Intensive atlas coverage in the Lahontan Valley and annual surveys of breeding waterfowl at Ruby Lakes National Wildlife Refuge (USFWS 2001) corroborate this overall pattern.

Currently, the Blue-winged Teal is not common in surrounding states either. It breeds locally in scattered locations across Oregon (Adamus et al. 2001) and is considered to be an uncommon and sparsely distributed summer resident of northern and central Utah (Behle et al. 1985). In California, it is present mainly in the northeast, and in Arizona, in the east-central part of the state (Small 1994, Rowher et al. 2002, Wise-Gervais in Corman and Wise-Gervais 2005:60–62).

CONSERVATION AND MANAGEMENT

The reasons for the apparent decline of the Blue-winged Teal in Nevada are unclear. Ryser (1985) suggested that the present landscape of northern Nevada is not really suitable for a species that is more at home in the extensive grasslands and marshes of north-central North America. Interestingly, Linsdale (1936) specifically cited wet meadows in Nevada as a habitat in which the Blue-winged Teal was rather common during the breeding season. It may thus be the case that large-scale agricultural development in some parts of Nevada contributed to the decline. In Saskatchewan, however, the Blue-winged Teal readily accepts artificial habitats such as canals and ditches, and is thought to have suffered comparatively little from agricultural development (Smith 1996).

HABITAT USE

HABITAT	ATLAS BLOCKS	INCIDENTAL OBSERVATIONS
Open Water	4 (33%)	0 (0%)
Riparian	1 (8%)	1 (20%)
Wetland	7 (58%)	4 (80%)
TOTAL	12	5

Total numbers reported in the Habitat Use, Breeding Status, and Abundance tables may differ from each other (see pp. 22–23 for details). Percentage sums may differ slightly from 100% due to rounding.

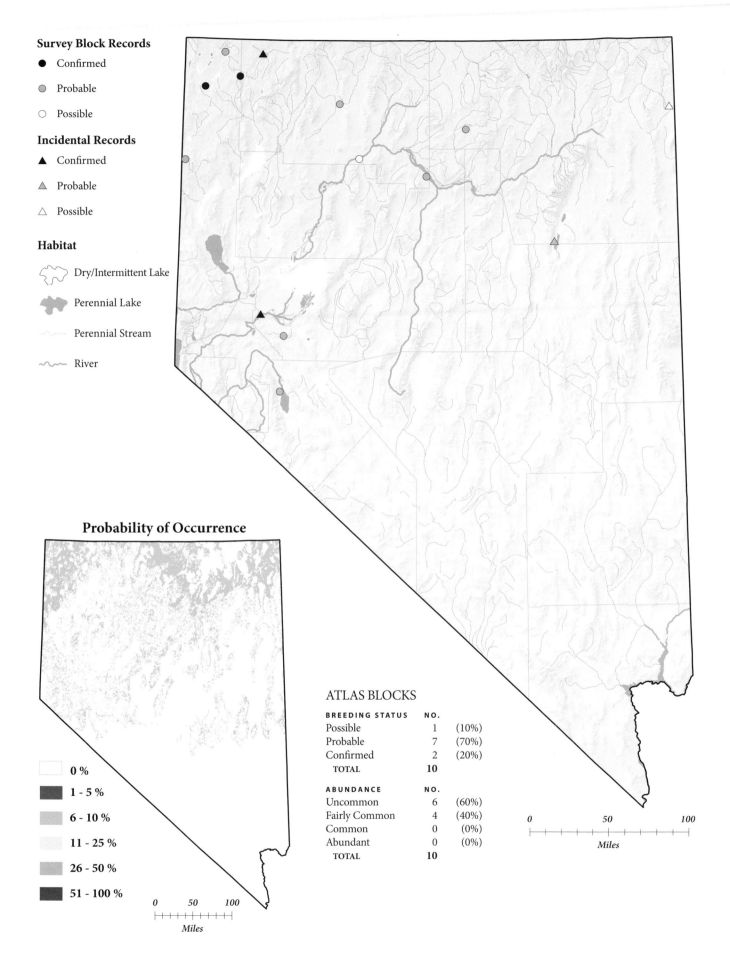

Survey Block Records

● Confirmed

● Probable

○ Possible

Incidental Records

▲ Confirmed

▲ Probable

△ Possible

Habitat

Dry/Intermittent Lake

Perennial Lake

Perennial Stream

River

Probability of Occurrence

	0 %
	1 - 5 %
	6 - 10 %
	11 - 25 %
	26 - 50 %
	51 - 100 %

0   50   100
Miles

ATLAS BLOCKS

BREEDING STATUS	NO.	
Possible	1	(10%)
Probable	7	(70%)
Confirmed	2	(20%)
TOTAL	10	

ABUNDANCE	NO.	
Uncommon	6	(60%)
Fairly Common	4	(40%)
Common	0	(0%)
Abundant	0	(0%)
TOTAL	10	

0   50   100
Miles

CINNAMON TEAL
Anas cyanoptera

In physical appearance the Cinnamon Teal is a metaphor for the desert landscape. The fiery red eyes and sandstone-colored plumage dominate the first impression. On closer inspection, however, its rich blue speculum gleams like an oasis. The Cinnamon Teal is fairly common in the lowland wetlands of northern Nevada, but farther south its status is less clear and there are signs of a range contraction.

DISTRIBUTION

Atlas workers confirmed breeding Cinnamon Teals throughout the northern half of Nevada, but not in the south. Mineral, Esmeralda, Nye, Lincoln, and Clark counties produced no confirmed breeding records, although each county did have records of possible and probable breeders. The earlier literature, in contrast, attests to the presence of the species throughout southern Nevada (Alcorn 1988). For instance, at the turn of the nineteenth century, Fisher (1893) considered the Cinnamon Teal a common breeder in the Pahranagat Valley.

In the areas surrounding southern Nevada, the Cinnamon Teal is a sporadic breeder in the lower Colorado River drainage (Rosenberg et al. 1991), a fairly common breeder in southern California (Small 1994), and a widespread but local breeder in Arizona (Wise-Gervais in Corman and Wise-Gervais 2005:62–63).

Although Cinnamon Teal were widespread in northern Nevada—which lies at the heart of the species' range (Gammonley 1996)—atlas workers rarely estimated them to be very numerous. This result is not entirely consistent with the earlier literature. For instance, Alcorn (1988) considered the Cinnamon Teal to be one of Nevada's most common breeding ducks, and Ryser (1985) called the species abundant in the Great Basin and provided density estimates for several sites that would have met the atlas definition of "abundant." As many as 200–350 pairs have bred annually at Ruby Lakes National Wildlife Refuge (USFWS 2001), suggesting that there are large populations in at least some parts of the state. The Cinnamon Teal is also rated as a common or abundant breeder throughout Utah (Behle et al. 1985) and in southeastern Oregon (Littlefield 1990, Contreras and Kindschy 1996, Adamus et al. 2001).

Cinnamon Teal were reported from all aquatic habitats and from other habitats that presumably contained small wetlands suitable for breeding. Like other waterbirds that use small wetlands, Cinnamon Teal are predicted to occur over a large area, albeit with a low likelihood in most places. In general, they use a range of wetland situations—Linsdale (1936:37) described them as "present wherever water occurs"—including highly alkaline habitats. However, breeding sites tend to have dense emergent vegetation (Gammonley 1996).

CONSERVATION AND MANAGEMENT

The Cinnamon Teal is among the least common and least studied waterfowl species in North America (Gammonley 1996), and conservation strategies need to recognize the limitations of our current knowledge. One limitation comes from the lack of detailed population trend data. This situation is unusual for a North American dabbling duck and exists because standardized waterfowl breeding surveys focus on the northern prairies, where most other species are most common.

Given the paucity of knowledge about the species and the importance of the Intermountain West to its maintenance, the Cinnamon Teal should be a high conservation priority in Nevada. Basic habitat and demographic variables remain to be studied in the state, and it seems likely that habitat preservation will figure prominently in conservation strategies. Habitat enhancement is likely to be valuable, too, and will probably involve activities similar to those that benefit other waterfowl, such as the creation of island nesting areas protected from mammalian predators.

HABITAT USE

HABITAT	ATLAS BLOCKS	INCIDENTAL OBSERVATIONS
Agricultural	5 (5%)	0 (0%)
Open Water	19 (20%)	6 (15%)
Riparian	25 (26%)	7 (18%)
Salt Desert Scrub	1 (1%)	0 (0%)
Sagebrush Scrub	1 (1%)	0 (0%)
Wetland	44 (46%)	27 (68%)
TOTAL	95	40

Total numbers reported in the Habitat Use, Breeding Status, and Abundance tables may differ from each other (see pp. 22–23 for details). Percentage sums may differ slightly from 100% due to rounding.

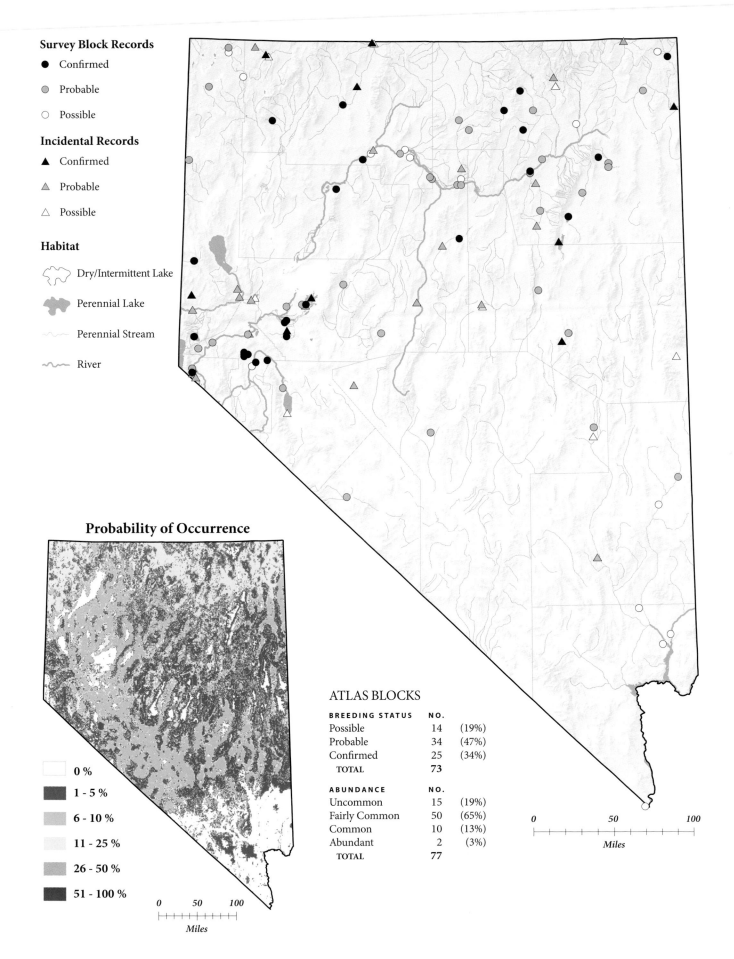

Survey Block Records

- ● Confirmed
- ● Probable
- ○ Possible

Incidental Records

- ▲ Confirmed
- △ Probable
- △ Possible

Habitat

- Dry/Intermittent Lake
- Perennial Lake
- Perennial Stream
- River

Probability of Occurrence

- ☐ 0 %
- 1 - 5 %
- 6 - 10 %
- 11 - 25 %
- 26 - 50 %
- 51 - 100 %

0   50   100
Miles

ATLAS BLOCKS

BREEDING STATUS	NO.	
Possible	14	(19%)
Probable	34	(47%)
Confirmed	25	(34%)
TOTAL	73	

ABUNDANCE	NO.	
Uncommon	15	(19%)
Fairly Common	50	(65%)
Common	10	(13%)
Abundant	2	(3%)
TOTAL	77	

0   50   100
Miles

NORTHERN SHOVELER
Anas clypeata

With its distinctive hunchbacked flight profile, oversized bill, and frenetic feeding behavior, the Northern Shoveler comes across as rather awkward and inelegant. And there is something improbable about the male's splashily patterned plumage—gray-brown across the back, with bright rufous flanks, a gleaming white breast, and a dark green head. In Nevada, Shovelers are found in the major drainages and wetland systems across the northern half of the state.

DISTRIBUTION

The atlas results portray the Northern Shoveler as occurring spottily throughout northern Nevada, with most records coming from within or near major refuges and management areas, and from larger rivers. Breeding was confirmed at only six sites, and there were only two records from the southern counties, neither of confirmed breeding. Atlas workers estimated multiple pairs in over two-thirds of the occupied blocks.

The atlas findings are largely consistent with recent studies from elsewhere in the region and with earlier Nevada studies. Alcorn (1988) reported breeding in northern Nevada but not in the south, and Ryser (1985) considered the species to be common throughout the Great Basin.

The Northern Shoveler is a common summer resident in Utah (especially toward the north) and eastern Oregon (Behle et al. 1985, Adamus et al. 2001). In the lower Colorado River Valley, the species is uncommon (Rosenberg et al. 1991), and modern-day breeding remains unconfirmed in Arizona (Corman in Corman and Wise-Gervais 2005:596).

Nearly all atlas records were from aquatic habitats, as expected. There were single records from agricultural and urban areas, but these were presumably associated with small ponds or wetlands within larger swaths of developed land. Although widespread in a variety of wetland types, the Northern Shoveler is not indifferent to local conditions. It typically occurs in open, shallow water where it can feed unhindered by emergent vegetation, and sewage treatment ponds seem especially suitable in this regard (DuBowy 1996).

CONSERVATION AND MANAGEMENT

Overall, North American populations of the Northern Shoveler are stable or increasing (DuBowy 1996, Sauer et al. 2005). The species' status is not especially well known in Nevada, where few waterfowl surveys are conducted away from the major refuges and management areas.

Presumed mechanisms of population regulation include habitat availability, predation, and weather (especially drought) (DuBowy 1996). During the breeding season the Northern Shoveler feeds mainly on invertebrates, and arthropod availability may influence population health. For example, Littlefield (1990) speculated that Northern Shoveler numbers at Malheur National Wildlife Refuge in Oregon are affected by aquatic arthropod densities. More work is needed, as the basic dietary requirements of the Northern Shoveler remain to be fully determined. It is interesting to note that even though Northern Shovelers' consumption of arthropods is considered a fairly recent discovery (DuBowy 1996), Fisher (1893) observed them to be voracious eaters of flies in the Owens Valley of California.

Northern Shovelers respond well to habitat improvements on the breeding grounds. For instance, they have benefited both from wetland restoration efforts conducted through the Prairie Pothole Joint Venture and from improvements to grasslands achieved through the Conservation Reserve Program (Ehresman in Jackson et al. 1996:82–83). In general, Northern Shovelers prosper in wetlands that are managed for large expanses of open water (Kaminski and Prince 1981). Sewage treatment ponds may have played a role in the species' recent expansion into the eastern United States (DuBowy 1996), and this "resource" should not be overlooked in Nevada.

HABITAT USE

HABITAT	ATLAS BLOCKS	INCIDENTAL OBSERVATIONS
Agricultural	1 (3%)	0 (0%)
Open Water	8 (26%)	2 (22%)
Riparian	6 (19%)	0 (0%)
Urban	1 (3%)	0 (0%)
Wetland	15 (48%)	7 (78%)
TOTAL	31	9

Total numbers reported in the Habitat Use, Breeding Status, and Abundance tables may differ from each other (see pp. 22–23 for details). Percentage sums may differ slightly from 100% due to rounding.

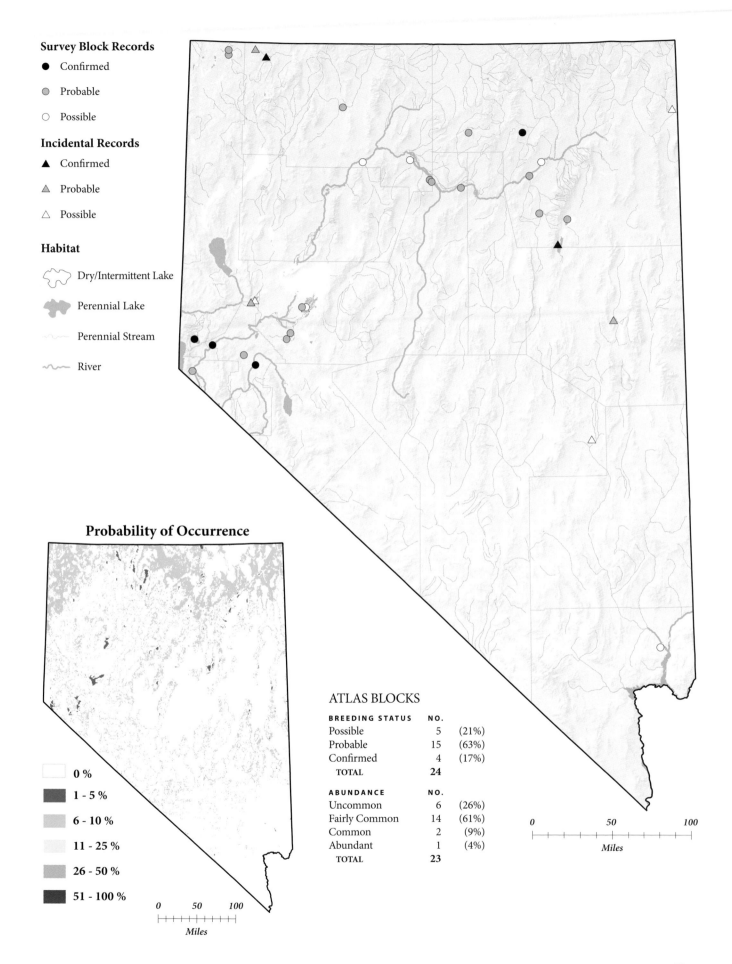

Survey Block Records

● Confirmed

● Probable

○ Possible

Incidental Records

▲ Confirmed

▲ Probable

△ Possible

Habitat

Dry/Intermittent Lake

Perennial Lake

Perennial Stream

River

Probability of Occurrence

☐ 0 %

■ 1 - 5 %

☐ 6 - 10 %

☐ 11 - 25 %

☐ 26 - 50 %

■ 51 - 100 %

0   50   100
Miles

ATLAS BLOCKS

BREEDING STATUS	NO.	
Possible	5	(21%)
Probable	15	(63%)
Confirmed	4	(17%)
TOTAL	**24**	

ABUNDANCE	NO.	
Uncommon	6	(26%)
Fairly Common	14	(61%)
Common	2	(9%)
Abundant	1	(4%)
TOTAL	**23**	

0   50   100
Miles

NORTHERN PINTAIL
Anas acuta

The smart-looking Northern Pintail is one of the more wide-spread breeding ducks in Nevada, yet it is also a species of conservation concern. Populations are declining across the continent, and the Northern Pintail is one of the few waterfowl species yet to show clear signs of recovery in response to the extensive wetland conservation efforts of the past two decades. In Nevada—at the southern edge of their range—pintails vary from locally abundant to absent at presumably suitable locations.

DISTRIBUTION

Atlas data show Northern Pintails occurring throughout northern Nevada, with records from most of the major refuges, management areas, and river systems, as expected. A handful of records came from southern Nevada, but breeding was confirmed only in the northern counties. At a statewide level, this distribution pattern is consistent with that noted by Alcorn (1988). At a regional level, the absence of Northern Pintails from southern Nevada is consistent with the species' absence from the lower Colorado River Valley (Rosenberg et al. 1991) and decreasing abundance toward southern Utah (Behle et al. 1985). Elsewhere in the region, Northern Pintails breed spottily along the eastern slope of the Sierra Nevada in California (Small 1994) and through eastern Oregon and southern Idaho (Stephens and Sturts 1998, Adamus et al. 2001).

Pintails were usually not very numerous where encountered, with approximately 90% of the occupied blocks estimated to have fewer than 10 pairs. Throughout the Great Basin, Northern Pintails range from uncommon to abundant (Ryser 1985), and the species does occur in high numbers in a few areas of Nevada. For instance, approximately 150–200 pairs were reported annually at Ruby Lakes National Wildlife Refuge during the atlas years (USFWS 2001).

Most records came from the three major aquatic habitat types, but there were also a few records from agricultural habitats. Nearly all of the earlier accounts emphasize the ephemeral nature of the habitats in which Northern Pintails are found. Austin and Miller (1995), for example, cited semipermanent

wetlands, seasonal ponds, and intermittent lakes as suitable for breeding by Northern Pintails. Compared with other ducks, pintails are not quite as tightly associated with wetland habitats for breeding, and females can sometimes be found in upland habitats, such as sagebrush, more than a mile from water (e.g., Gaines 1992). The predictive map shows a moderate chance of finding pintails in the northern part of the state, with decreasing likelihood as one moves south. As is the case for all aquatic species, though, the actual likelihood of occurrence depends on the presence of suitable ponds or wetlands within the terrestrial habitat matrix.

CONSERVATION AND MANAGEMENT

Although the Northern Pintail is one of the most abundant ducks in North America, populations have declined considerably since the 1970s, and numbers are well below targets set by the North American Waterfowl Conservation Plan (Austin and Miller 1995). Pintails are quick to abandon undesirable habitats, and populations often decline during droughts. But populations can recover quickly when suitable flooded habitat becomes available.

The Northern Pintail was heavily hunted in the past, and a moderate harvest still occurs. The overall effect of hunting is not well known, although the production of young seems to have a greater influence on populations than do survival rates during the nonbreeding season (Austin and Miller 1995). Female Northern Pintails that nest in the uplands, far from water, are perhaps more vulnerable to mammalian predators than other species of ducks that nest exclusively on islands or on floating vegetation. Ducklings hatched from such nests are also presumably vulnerable during their long trek to the shore.

HABITAT USE

HABITAT	ATLAS BLOCKS	INCIDENTAL OBSERVATIONS
Agricultural	3 (4%)	0 (0%)
Open Water	16 (24%)	1 (7%)
Riparian	14 (21%)	1 (7%)
Wetland	34 (51%)	12 (86%)
TOTAL	67	14

Total numbers reported in the Habitat Use, Breeding Status, and Abundance tables may differ from each other (see pp. 22–23 for details). Percentage sums may differ slightly from 100% due to rounding.

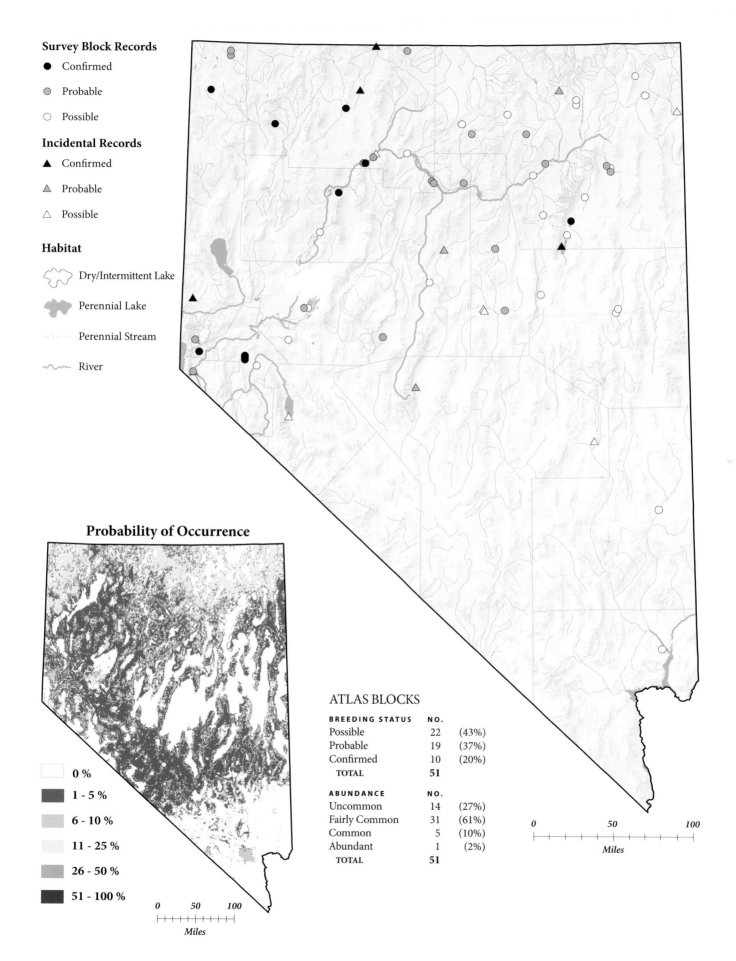

Survey Block Records

- ● Confirmed
- ● Probable
- ○ Possible

Incidental Records

- ▲ Confirmed
- ▲ Probable
- △ Possible

Habitat

Dry/Intermittent Lake

Perennial Lake

Perennial Stream

River

Probability of Occurrence

- 0 %
- 1 - 5 %
- 6 - 10 %
- 11 - 25 %
- 26 - 50 %
- 51 - 100 %

0   50   100

Miles

ATLAS BLOCKS

BREEDING STATUS	NO.	
Possible	22	(43%)
Probable	19	(37%)
Confirmed	10	(20%)
TOTAL	**51**	

ABUNDANCE	NO.	
Uncommon	14	(27%)
Fairly Common	31	(61%)
Common	5	(10%)
Abundant	1	(2%)
TOTAL	**51**	

0   50   100

Miles

GREEN-WINGED TEAL

Anas crecca

The dumpy, yet dainty, Green-winged Teal breeds widely throughout north-central and northeastern Nevada. It can be found in small numbers in the valley marshes, but it is also a bird of wooded rivers and even montane riparian areas. The species' abundance in Nevada has always been relatively low, but populations here appear to be fairly stable.

DISTRIBUTION

There is a decidedly northern aspect to the Green-winged Teal's distribution in North America, with most breeders occurring north of the border between the United States and Canada. Nevada is at the southern limit of the breeding range, so the species has never been common here. Ryser (1985) considered it to be uncommon in the Great Basin, and Alcorn (1988:46) cited "limited numbers" in northern Nevada. The species appears never to have bred in southern Nevada (Linsdale 1936, Alcorn 1988). In adjacent states, the Green-winged Teal also occurs rather sporadically and in modest numbers (Behle et al. 1985, Small 1994, Piest in Corman and Wise-Gervais 2005:66–67), although Adamus et al. (2001) considered them to be more common and widespread in Oregon than previously thought.

Atlas workers found Green-winged Teal in thirty-nine blocks, and there were incidental observations from another fourteen locations. Abundance estimates indicated that fewer than ten pairs were present in most of the occupied blocks, although teal were more numerous in a few locations. The bulk of the records came from northern Nevada, with most confirmed records coming from Elko County. This northern distribution is also reflected in the predictive map, which does not show a high probability of occurrence anywhere but suggests that finding Green-winged Teal is more likely in the northernmost latitudes of the state. Within northern Nevada, however, the Green-winged Teal was essentially absent from the productive terminal wetlands of the Sierra Nevada's drainages. This result is surprising given the large number of ducks these wetlands support, but

it reinforces a recent assessment of the Green-winged Teal's status in the Lahontan Valley (Chisholm and Neel 2002).

The breeding habitat of the Green-winged Teal is more reminiscent of the Wood Duck's than that of other ducks in the genus *Anas*. While it can be found in marshlands on valley floors, it will also nest along wooded streams or on beaver ponds (Gaines 1992). Atlas records were fairly evenly divided among wetland, riparian, and open-water habitats. In Washington, Green-winged Teal occur in wetlands with shrubby vegetation or trees (Smith et al. 1997), and in Utah, they breed at elevations as high as 10,000 feet (3,000 meters) (Behle et al. 1985).

CONSERVATION AND MANAGEMENT

Unlike many North American waterfowl, the Green-winged Teal did not suffer dramatic population declines during the twentieth century (Sauer et al. 2005). This species breeds primarily to the north of the prairies, in areas that remain largely uninhabited by humans, and its nesting habitats have consequently been less affected by human activity than have those of other dabbling ducks (Johnson 1995). The species also appears to have responded well to land management projects such as the U.S. Department of Agriculture's Conservation Reserve Program (Ehresman in Jackson et al. 1996:72–73).

These ducks require dense cover, which should be considered when managing habitats at a local level. Moreover, habitat degradation, although not a concern throughout much of the Green-winged Teal's distribution, is considered to be more serious in the southern portion of its range than elsewhere (Johnson 1995). Given the absence of any obvious declines in Green-winged Teal populations in Nevada, however, and the species' widespread abundance elsewhere, it is not a high conservation priority in the state at present.

HABITAT USE

HABITAT	ATLAS BLOCKS	INCIDENTAL OBSERVATIONS
Agricultural	1 (2%)	1 (8%)
Open Water	13 (27%)	2 (15%)
Riparian	14 (29%)	2 (15%)
Wetland	20 (42%)	8 (62%)
TOTAL	48	13

Total numbers reported in the Habitat Use, Breeding Status, and Abundance tables may differ from each other (see pp. 22–23 for details). Percentage sums may differ slightly from 100% due to rounding.

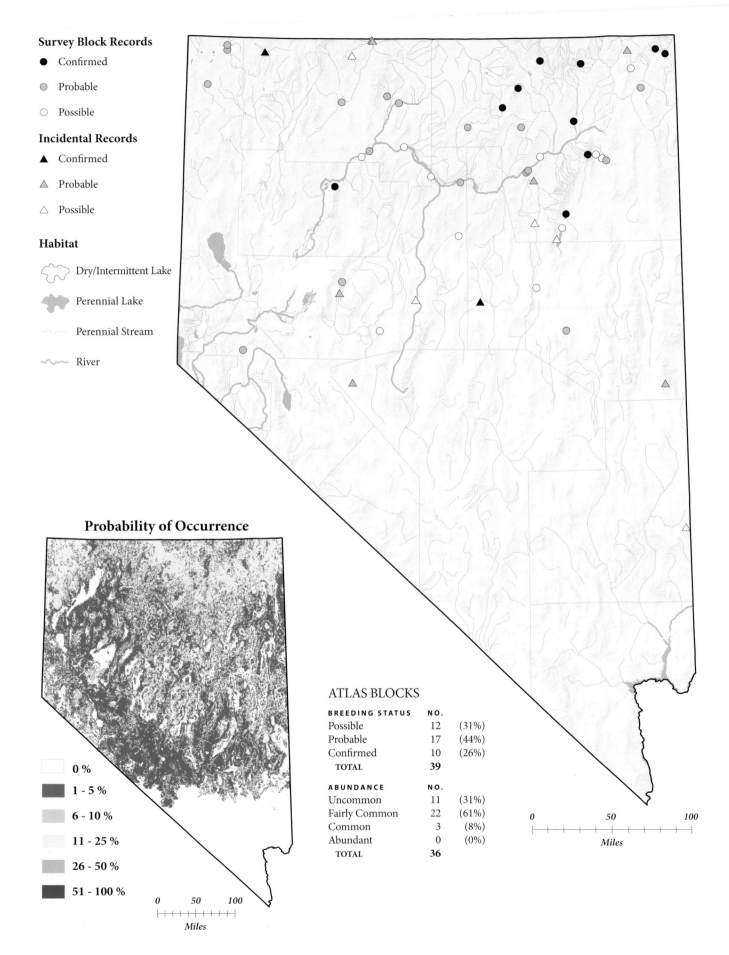

Survey Block Records
- ● Confirmed
- ● Probable
- ○ Possible

Incidental Records
- ▲ Confirmed
- △ Probable
- △ Possible

Habitat
- Dry/Intermittent Lake
- Perennial Lake
- Perennial Stream
- River

Probability of Occurrence

- ☐ 0 %
- ■ 1 - 5 %
- ▨ 6 - 10 %
- ☐ 11 - 25 %
- ▨ 26 - 50 %
- ■ 51 - 100 %

0   50   100
Miles

ATLAS BLOCKS

BREEDING STATUS	NO.	
Possible	12	(31%)
Probable	17	(44%)
Confirmed	10	(26%)
TOTAL	39	

ABUNDANCE	NO.	
Uncommon	11	(31%)
Fairly Common	22	(61%)
Common	3	(8%)
Abundant	0	(0%)
TOTAL	36	

0   50   100
Miles

CANVASBACK
Aythya valisineria

On migration and during the winter, the Canvasback is a common sight on large bodies of open water throughout Nevada. During the breeding season, however, these regal ducks take on a different demeanor. Although widespread in the state, they nest primarily at a few large wetlands in northern Nevada. They tend to favor wetlands with dense emergent vegetation, and their well-concealed nests can be difficult to locate.

DISTRIBUTION

All but one atlas record came from northern Nevada counties, and the lone record from the south was from northern Nye County. Canvasbacks were found in only eight atlas blocks and were not particularly common in any of them—no blocks were estimated to have more than ten breeding pairs. There were also five observations of confirmed breeders at other sites. One of these sites, at Ruby Lake National Wildlife Refuge, was an exception to the general pattern of limited abundance. The refuge regularly supports several hundred nesting Canvasbacks and represents the species' most southerly large breeding population (Kruse et al. 2003).

The scattered, sparse, and northerly distribution that atlas workers found for Canvasbacks in Nevada is largely consistent with Ryser's (1985) determination that the species is rare to uncommon throughout the Great Basin except in the Ruby Valley. This rarity is reflected in our predictive map, which suggests that the likelihood of seeing breeding Canvasbacks is low throughout the state. Alcorn (1988) cited breeding south at least to Overton Wildlife Management Area in Clark County, but there were no atlas records from that far south.

The Canvasback is not common in adjacent states, either. There are no breeding records from the lower Colorado River Valley (Rosenberg et al. 1991), and the species is scarce as a breeder in California (Small 1994), Arizona (Corman in Corman

and Wise-Gervais 2005:596), and Utah (Behle et al. 1985). In Oregon, breeding is limited to scattered locations in the southeast (Littlefield 1990, Contreras and Kindschy 1996, Adamus et al. 2001), and in Idaho also to portions of the southeast (Stephens and Sturts 1998).

All reported habitats for this species were aquatic, as expected. Details on the Canvasback's ecological requirements in Nevada can be found in Noyes and Jarvis (1985) and Kruse et al. (2003). Canvasbacks can be hard to find in the marshes and grassy wetlands they favor, and it is possible that they were overlooked in some places.

CONSERVATION AND MANAGEMENT

Although there have been fluctuations, Canvasback populations within the United States have remained mostly stable since the 1950s (USFWS 2003, Sauer et al. 2005). This favorable finding should be tempered, however, by local and regional concerns relevant to Nevada populations. For instance, at Ruby Lake National Wildlife Refuge, the stronghold for the species in the state, annual reproduction has fallen well below management objectives (USFWS 2001).

A long-term study by Kruse et al. (2003) found that reproductive success at Ruby Lake between 1970 and 2000 was influenced mainly by water levels, which in turn were related to the amount of snowpack in the mountains. In contrast to the situation in the northern prairies, Canvasback productivity at Ruby Lake declines in years with high water levels. Nests that hatched ducklings were found in shallower water than unsuccessful nests, and were farther from land and in less dense, but wider, bands of emergent vegetation. High rates of nest loss during deep-water events were attributed largely to avian predators, which can find nests more easily when emergent vegetation is largely inundated. Almost three-quarters of all Canvasback nests in this study also contained eggs laid by other ducks—primarily Redheads (see Sorenson 1998). But there was little evidence that this nest parasitism had any negative effects, presumably because ducklings are largely capable of feeding themselves, and looking after one or two adoptees requires little extra work from the parents.

HABITAT USE

HABITAT	ATLAS BLOCKS	INCIDENTAL OBSERVATIONS
Open Water	3 (37%)	1 (25%)
Riparian	1 (13%)	0 (0%)
Wetland	4 (50%)	3 (75%)
TOTAL	8	4

Total numbers reported in the Habitat Use, Breeding Status, and Abundance tables may differ from each other (see pp. 22–23 for details). Percentage sums may differ slightly from 100% due to rounding.

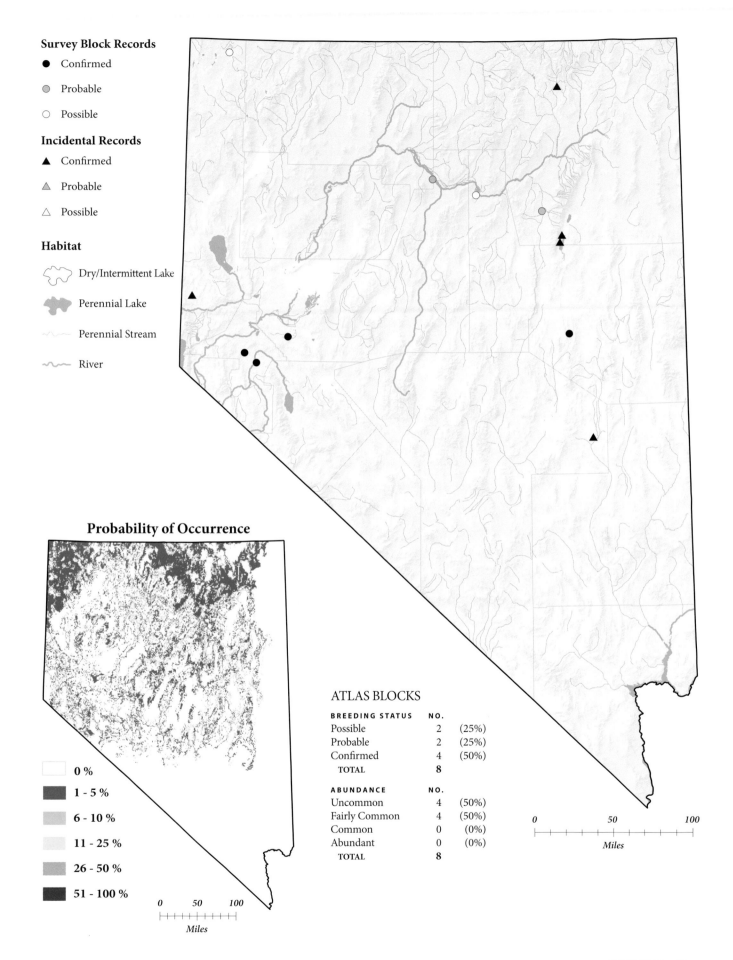

Survey Block Records

● Confirmed

● Probable

○ Possible

Incidental Records

▲ Confirmed

▲ Probable

△ Possible

Habitat

Dry/Intermittent Lake

Perennial Lake

Perennial Stream

River

Probability of Occurrence

☐ 0 %

■ 1 - 5 %

■ 6 - 10 %

■ 11 - 25 %

■ 26 - 50 %

■ 51 - 100 %

0   50   100
Miles

ATLAS BLOCKS

BREEDING STATUS	NO.	
Possible	2	(25%)
Probable	2	(25%)
Confirmed	4	(50%)
TOTAL	8	

ABUNDANCE	NO.	
Uncommon	4	(50%)
Fairly Common	4	(50%)
Common	0	(0%)
Abundant	0	(0%)
TOTAL	8	

0   50   100
Miles

REDHEAD
Aythya americana

The Redhead is the most widespread of the bay ducks in Nevada, and except for the Ruddy Duck it is the only diver that breeds regularly in the southern part of the state. Superficially similar in appearance to Canvasbacks, Redheads are found at a wider variety of sites and are more likely to be encountered almost anywhere in the state except in the Ruby Valley.

DISTRIBUTION

The Redhead occurs only in North America, where it breeds in the prairie potholes of the upper Midwest and in the arid interior of the western United States. Ryser (1985) considered the species to be abundant in the Great Basin, and Alcorn (1988) listed it as a summer resident in northern Nevada with especially high concentrations in the Ruby Valley. Linsdale (1936) likewise listed the Redhead as common, especially in western Nevada.

The atlas results suggest that the species is somewhat more widespread than the earlier literature indicated, with confirmed breeding throughout northern Nevada and south to the Pahranagat Valley. The only region that produced no records was the south-central portion of the state—in and around Esmeralda County, where few waterfowl of any species were found. Most abundance estimates (80%) from occupied blocks indicated that fewer than ten pairs were present. Very high concentrations can occur where conditions are right, however—for example, in the Lahontan Valley, where there was an average of more than 1,300 nests annually between 1967 and 1995 (Chisholm and Neel 2002), and at Ruby Lake National Wildlife Refuge, where more than three hundred pairs nested each year from 1998 to 2001 (USFWS 2001).

To the north and east of Nevada, the Redhead is a common breeder, occurring through much of the southeastern half of Oregon (Littlefield 1990, Contreras and Kindschy 1996, Adamus et al. 2001), and especially around the Great Salt Lake in Utah

(Behle et al. 1985). To the south, its distribution is less certain. Garrett and Dunn (1981) considered its status in southern California to be complex, and its status as a breeder in the Owens Valley of eastern California was not clear to Small (1994). There seem to be a few breeding records from the Mono Basin of California (Gaines 1992). The species formerly bred along the lower Colorado River (Garrett and Dunn 1981).

All of the atlas records referred to aquatic habitats, as expected. Redheads are predicted to occur widely in Nevada, although generally not with a very high probability. They are omnivorous and are often found in the shallow areas near the edges of pools, where a variety of foods are available. For a good summary of the species' foraging biology, based on fieldwork in Nevada, see Noyes and Jarvis (1985).

CONSERVATION AND MANAGEMENT

The Redhead population in North America has remained relatively stable since the 1950s (USFWS 2003). In Nevada, population trends are not well known away from the major refuges and management areas where systematic surveys are conducted. But there is some basis for local concern; at Ruby Lake, for instance, annual production of young is only about a quarter of the refuge's target number (USFWS 2001).

The Redhead is a facultative brood parasite, with females sometimes laying some of their eggs in the nests of other species. The incidence of parasitism is quite high in some regions (Joyner 1983, Ryser 1985, Sorenson 1998); for example, about 70% of 811 Canvasback nests studied at Ruby Lake contained Redhead eggs (Kruse et al. 2003). Unlike the most infamous North American avian brood parasite, the Brown-headed Cowbird, the Redhead is not considered a threat to other species because Redhead chicks are precocial and do not require the same level of care and attention as a cowbird chick. The host parents are usually able to raise their own young in addition to any adopted ducklings.

HABITAT USE

HABITAT	ATLAS BLOCKS	INCIDENTAL OBSERVATIONS
Open Water	14 (31%)	5 (31%)
Riparian	8 (18%)	0 (0%)
Wetland	23 (51%)	11 (69%)
TOTAL	45	16

Total numbers reported in the Habitat Use, Breeding Status, and Abundance tables may differ from each other (see pp. 22–23 for details). Percentage sums may differ slightly from 100% due to rounding.

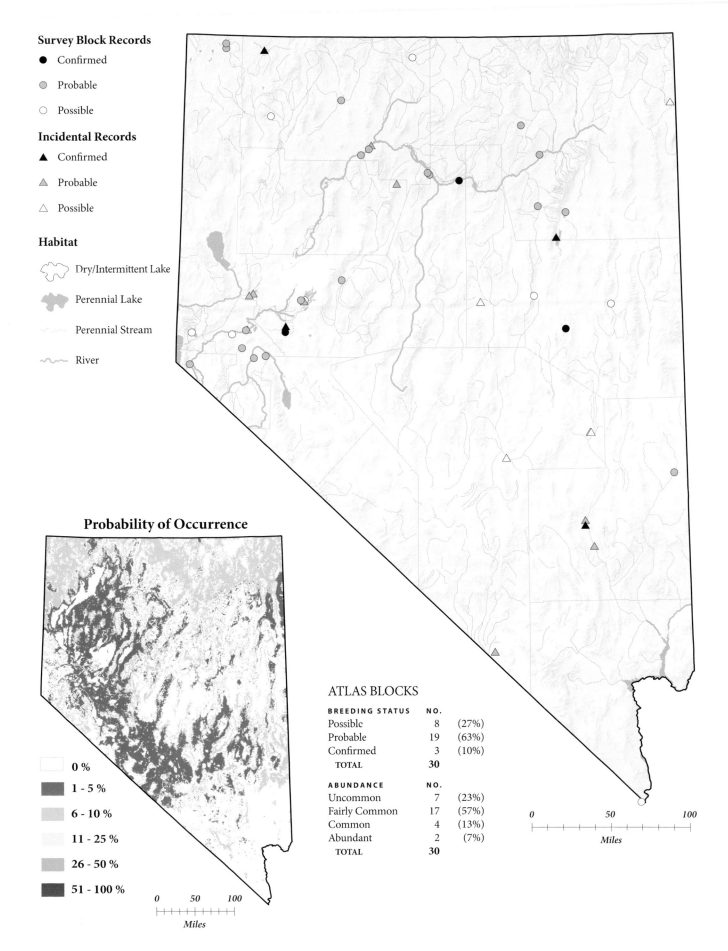

Survey Block Records

● Confirmed

● Probable

○ Possible

Incidental Records

▲ Confirmed

▲ Probable

△ Possible

Habitat

Dry/Intermittent Lake

Perennial Lake

Perennial Stream

River

Probability of Occurrence

☐ 0 %

■ 1 - 5 %

■ 6 - 10 %

■ 11 - 25 %

■ 26 - 50 %

■ 51 - 100 %

0   50   100
Miles

ATLAS BLOCKS

BREEDING STATUS	NO.	
Possible	8	(27%)
Probable	19	(63%)
Confirmed	3	(10%)
TOTAL	30	

ABUNDANCE	NO.	
Uncommon	7	(23%)
Fairly Common	17	(57%)
Common	4	(13%)
Abundant	2	(7%)
TOTAL	30	

0   50   100
Miles

RING-NECKED DUCK

Aythya collaris

Like those of the Lesser Scaup and Trumpeter Swan, the Ring-necked Duck's breeding range in Nevada is largely limited to the extensive wetlands of the Ruby Valley. The population that breeds in the valley appears to be stable, but the number of breeding pairs is relatively low. Moreover, the Ring-necked Ducks that breed in the Ruby Valley are some distance from other permanent breeding populations. The possibility of smaller populations elsewhere in northern Nevada remains, and provides a future challenge for birders.

DISTRIBUTION

The Ring-necked Duck reaches the southern limit of its range in Nevada, and is not especially common in the state. The small population that breeds in the Ruby Valley is estimated to contain forty to seventy pairs, with an annual production of fifty to one hundred young (USFWS 2001).

The species is largely absent from surrounding regions. Breeding had not been confirmed in Utah at the time Behle et al. (1985) published their checklist of Utah birds, and there appear to be no breeding records from near the Nevada border in Idaho (Stephens and Sturts 1998). In Oregon, however, local breeding occurs through much of the state except the region west of the Cascades (Adamus et al. 2001). In California, known breeding areas are relatively few and are limited to the northern part of the state (McCaskie et al. 1988, Gaines 1992, Small 1994).

Although the only confirmed breeding records for the atlas came from the Ruby Valley, it is possible that the species breeds elsewhere in the state as well, as suggested by the predictive map.

Probable or possible breeding was documented at several other northern locations. Since Ring-necked Ducks readily colonize new breeding sites (Hohman and Eberhardt 1998), these areas should be considered for follow-up study. Two records from southern Nevada were considered to involve nonbreeders and were not mapped; however, observers should be on the lookout for breeders throughout the state.

All habitats reported for this species were either open water or wetlands. Emergent vegetation seems to be a key element for lowland-nesting populations in regions such as southeastern Oregon (Contreras and Kindschy 1996) and the Ruby Valley (J. Mackay, pers. comm.). Ring-necked Ducks in the western United States sometimes nest at high elevations, and their habitat requirements are somewhat different in such situations. Montane beaver ponds are used in Colorado (Versaw in Kingery 1998:92–93), and forested wetlands in Washington (Smith et al. 1997). Ring-necked Ducks have not yet been discovered breeding in montane forests in Nevada, but the possibility should not be dismissed.

CONSERVATION AND MANAGEMENT

Through much of North America, the Ring-necked Duck appears to be expanding its range and increasing in numbers (Hohman and Eberhardt 1998). The species may be a recent addition to the breeding avifauna of Nevada (see Linsdale 1936), and the number of summer records from the southern Sierra Nevada has increased as well (Gaines 1992). Ring-necked Ducks have also become increasingly common in Utah (Behle et al. 1985) and Colorado (Versaw in Kingery 1998:92–93), and as winterers in southern California (Garrett and Dunn 1981). Unlike several other duck species, ring-neckeds continued to thrive through the 1980s and 1990s (Hohman and Eberhardt 1998).

Management in Nevada should emphasize the conservation of the Ruby Valley population and a search for possible breeding grounds elsewhere. Ring-necked Ducks readily use constructed wetlands, which should be considered as a part of any conservation strategy in potential breeding areas. Factors with negative impacts on the Ring-necked Duck are not well known, and the direct effects of hunting and most environmental toxins are reported to be either minor or unsubstantiated (Hohman and Eberhardt 1998). Lead shot, however, has long been considered to be especially harmful to the species and remains a potential problem today (Bellrose 1959, Hohman and Eberhardt 1998). The extent to which these factors influence Ring-necked Ducks in Nevada is unclear.

HABITAT USE

HABITAT	ATLAS BLOCKS	INCIDENTAL OBSERVATIONS
Open Water	3 (50%)	1 (50%)
Wetland	3 (50%)	1 (50%)
TOTAL	6	2

Total numbers reported in the Habitat Use, Breeding Status, and Abundance tables may differ from each other (see pp. 22–23 for details). Percentage sums may differ slightly from 100% due to rounding.

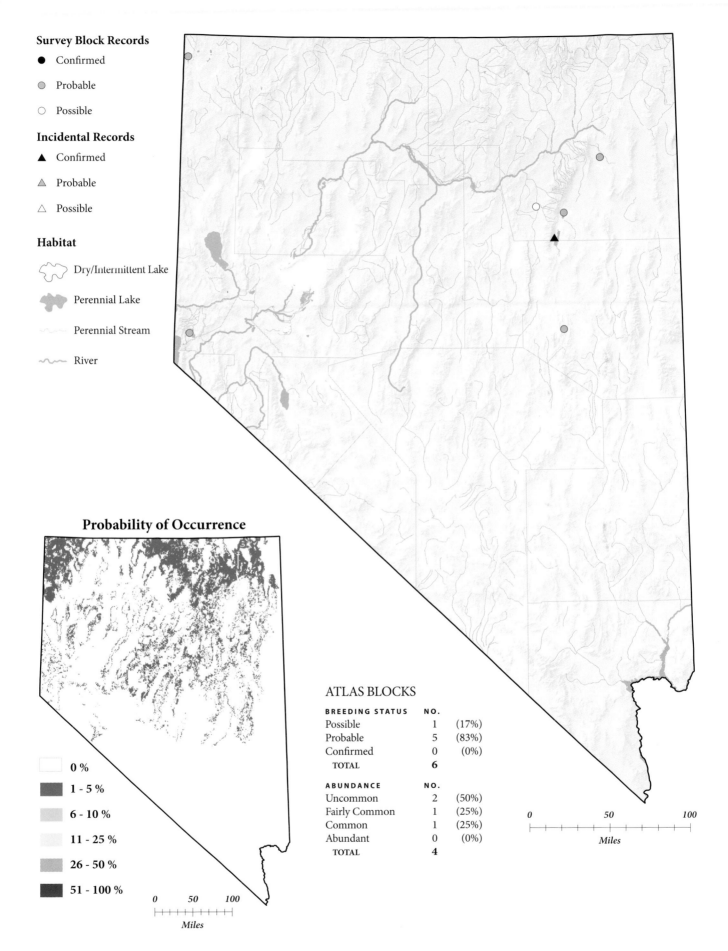

Survey Block Records

● Confirmed

◐ Probable

○ Possible

Incidental Records

▲ Confirmed

△ Probable

△ Possible

Habitat

Dry/Intermittent Lake

Perennial Lake

Perennial Stream

River

Probability of Occurrence

0 %

1 - 5 %

6 - 10 %

11 - 25 %

26 - 50 %

51 - 100 %

0   50   100
Miles

ATLAS BLOCKS

BREEDING STATUS	NO.	
Possible	1	(17%)
Probable	5	(83%)
Confirmed	0	(0%)
TOTAL	6	

ABUNDANCE	NO.	
Uncommon	2	(50%)
Fairly Common	1	(25%)
Common	1	(25%)
Abundant	0	(0%)
TOTAL	4	

0   50   100
Miles

LESSER SCAUP
Aythya affinis

Like many other ducks, the smartly attired Lesser Scaup is near the southern limit of its breeding range in Nevada. Most of the state's population nests in the marshes of the Ruby Valley, although smaller numbers occur elsewhere in the northern regions. And like most other diving ducks that breed in Nevada, the Lesser Scaup is neither particularly abundant nor greatly threatened. Habitat preservation at a few critical sites is the key to preventing its numbers from dipping to dangerously low levels.

DISTRIBUTION

Most of Nevada's breeding Lesser Scaup are found in the Ruby Valley. The annual breeding population was estimated at just over two hundred pairs there, with annual production of between 190 and 240 young during the atlas years (USFWS 2001). Other confirmed records came from two Elko County locations farther to the north, and there were scattered atlas records of unconfirmed breeders from elsewhere in northern Nevada. Lesser Scaup have also bred sparingly in the Mason Valley, at Wildhorse Reservoir, and possibly at Sheldon National Wildlife Refuge (Alcorn 1988), though they were not confirmed at these locations during atlas fieldwork. Our predictive model suggests the potential for breeding through most of northern Nevada, though the probability of occurrence is likely underestimated in the far northeast. Breeding occurs only where suitable wetlands are present within the habitats highlighted by the model, however, and even there the chance of finding this species is relatively low. For instance, there are no breeding records from the extensive wetlands of the Lahontan Valley (Chisholm and Neel 2002).

The population in the Ruby Valley is probably the largest in Nevada or surrounding areas. There do not appear to be any breeding records from adjoining portions of Idaho, though Lesser Scaup occur farther east (Stephens and Sturts 1998) and the species is a rare breeder in northern Utah (Behle et al. 1985). It is not clear if there are small numbers of breeders in Arizona (Corman in Corman and Wise-Gervais 2005:597), but the species' breeding presence is more pronounced in northern California and Oregon (Small 1994, Adamus et al. 2001).

Most of the reported habitats were open water, which is not surprising for a diving duck such as the Lesser Scaup. The actual nest, however, is usually placed on dry land rather than in vegetation over water (Austin et al. 1998). At Malheur National Wildlife Refuge, for example, Lesser Scaup tend to nest on islands, in moist meadows, and even in dry upland areas (Littlefield 1990).

CONSERVATION AND MANAGEMENT

Even though its numbers have been declining since the early 1980s, the Lesser Scaup is still among the most numerous of North America's diving ducks (Austin et al. 1998). Its status in Nevada is somewhat uncertain. The population in the Ruby Valley has been relatively stable in recent years, but there is little definitive information about the smaller populations elsewhere in northern Nevada. Linsdale (1936) gave no breeding records for Nevada, and it is possible that the species is a recent arrival in the state.

Austin et al. (1998, 2000) summarized the major determinants of population health for the Lesser Scaup, which appear to include nest predation, water quality, and water availability. The species is adaptable, can be found in a variety of wetland situations, and seems to be relatively unaffected by hunting at the present time. An interesting testimony to its adaptability comes from Lake Erie, where wintering populations flourish on the recently introduced zebra mussel (Custer and Custer 1996). Even at the margins of its range in Nevada the Lesser Scaup does not seem greatly threatened. Because of its generally low numbers, however, continued monitoring seems prudent.

HABITAT USE

HABITAT	ATLAS BLOCKS	INCIDENTAL OBSERVATIONS
Open Water	8 (67%)	1 (50%)
Wetland	4 (33%)	1 (50%)
TOTAL	12	2

Total numbers reported in the Habitat Use, Breeding Status, and Abundance tables may differ from each other (see pp. 22–23 for details). Percentage sums may differ slightly from 100% due to rounding.

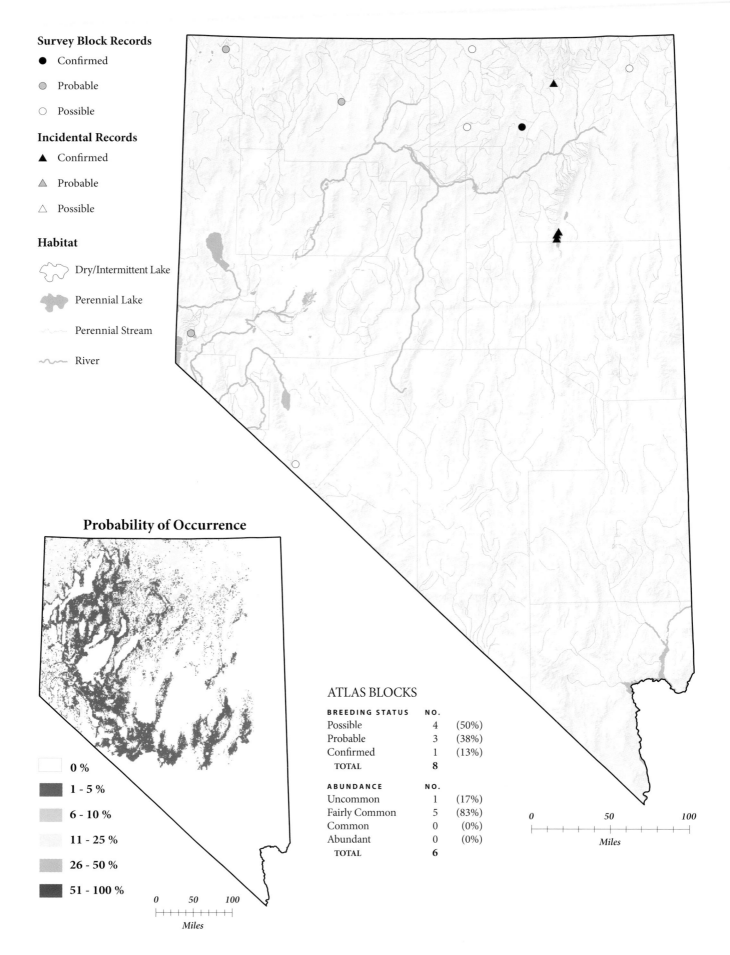

Survey Block Records

● Confirmed

● Probable

○ Possible

Incidental Records

▲ Confirmed

▲ Probable

△ Possible

Habitat

Dry/Intermittent Lake

Perennial Lake

Perennial Stream

River

Probability of Occurrence

☐ 0 %

■ 1 - 5 %

■ 6 - 10 %

☐ 11 - 25 %

■ 26 - 50 %

■ 51 - 100 %

0   50   100
Miles

ATLAS BLOCKS

BREEDING STATUS	NO.	
Possible	4	(50%)
Probable	3	(38%)
Confirmed	1	(13%)
TOTAL	8	

ABUNDANCE	NO.	
Uncommon	1	(17%)
Fairly Common	5	(83%)
Common	0	(0%)
Abundant	0	(0%)
TOTAL	6	

0   50   100
Miles

COMMON MERGANSER
Mergus merganser

Muscular looking and submarine shaped, the Common Merganser is a distinctive inhabitant of the forested lakes and fast-flowing streams of northern Nevada. Adults eat fish while the newly fledged young forage primarily on aquatic insect larvae. Family groups are vulnerable to disturbance by humans (Gaines 1992), and individuals are susceptible to water pollution (Mallory and Metz 1999). The presence of breeding Common Mergansers usually signifies a healthy water body.

DISTRIBUTION

Breeding records for the Common Merganser came from the terminal drainages of the Sierra Nevada in the west and from the Humboldt River drainage in the north of the state. There were also records from the Colorado River in the south, but these were assumed to have been of nonbreeding birds and do not appear on the map. Most abundance estimates suggested the presence of multiple pairs, but, not surprising for a solitary breeder, there were no blocks in which more than ten pairs were thought to occur. Common Mergansers were reported from various aquatic habitats, with most records coming from riparian areas.

The atlas results do not differ greatly from earlier studies in the state, although they imply a somewhat wider distribution than was previously described. Ryser (1985) considered the Common Merganser to be peripheral in the Great Basin, with limited breeding in western Nevada. Alcorn (1988) gave records from scattered locations in northern Nevada, including Goose Creek and the Owyhee River in Elko County, Pyramid Lake, and Lake Tahoe. Chisholm and Neel (2002) reported breeding along the upper Humboldt River. There are no records of breeding from anywhere in southern Nevada. Our model predicts low to moderate probabilities of encountering Common Mergansers roughly within the latitudinal and longitudinal limits defined by the actual atlas observations; the riparian patches with the highest odds are generally too small to show up on the printed map.

It seems likely that the Common Merganser's true Nevada breeding range extends to the east of the predicted range, especially into northeastern Elko County.

The Common Merganser breeds in northern California (Small 1994), widely but locally in Oregon and Idaho (Stephens and Sturts 1998, Adamus et al. 2001), and rarely in Utah (Behle et al. 1985). For breeding habitat, Common Mergansers require mountainous or forested waterways, and they generally favor larger streams and rivers (Small 1994, Contreras and Kindschy 1996, Mallory and Metz 1999). They use a wide variety of nest sites. Tree cavities are typical, but the species also nests among boulders and in cliff crevices (Littlefield 1990, Smith et al. 1997). There are even records of nests on lighthouses, covered bridges, and hay bales (Shuford 1993).

CONSERVATION AND MANAGEMENT

North American populations of the Common Merganser are stable or increasing (Mallory and Metz 1999), and this pattern seems to be borne out by anecdotal observations in Nevada. The species is not imperiled in the state, but it is of interest as an indicator of good water quality. Common Mergansers require water with good visibility for hunting underwater, and they therefore avoid rivers and lakes with high levels of turbidity or plant growth (Shuford 1993, Burridge in Burridge 1995:42).

As top predators in aquatic ecosystems, Common Mergansers are also vulnerable to bioaccumulation of pesticides and toxic metals (Mallory and Metz 1999). Their predatory proclivities have made Common Mergansers subjects of persecution in some areas because of perceived conflicts with commercial fishing interests (Mallory and Metz 1999). Nesting Common Mergansers seem to favor areas that are undisturbed by humans (Gaines 1992), as can also be inferred from a review of the atlas records along the Truckee River: Common Mergansers were scarce in and near the cities of Reno and Sparks but nested both upstream and downstream of the metropolitan area.

HABITAT USE

HABITAT	ATLAS BLOCKS	INCIDENTAL OBSERVATIONS
Open Water	5 (19%)	0 (0%)
Riparian	18 (69%)	2 (100%)
Urban	2 (8%)	0 (0%)
Wetland	1 (4%)	0 (0%)
TOTAL	26	2

Total numbers reported in the Habitat Use, Breeding Status, and Abundance tables may differ from each other (see pp. 22–23 for details). Percentage sums may differ slightly from 100% due to rounding.

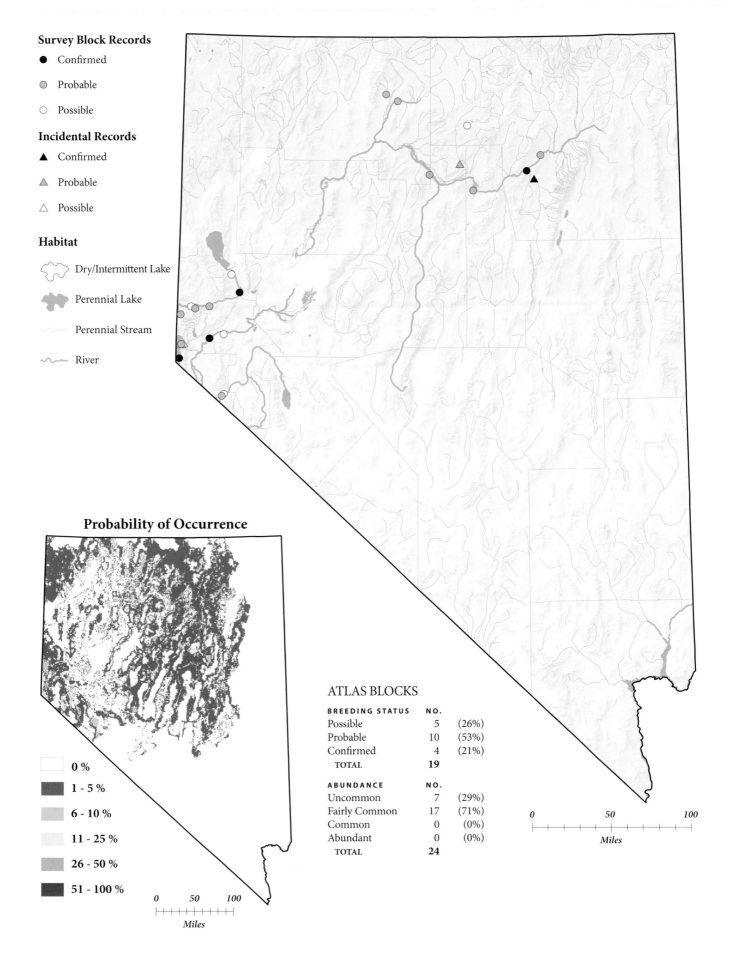

Survey Block Records

● Confirmed

● Probable

○ Possible

Incidental Records

▲ Confirmed

▲ Probable

△ Possible

Habitat

Dry/Intermittent Lake

Perennial Lake

Perennial Stream

River

Probability of Occurrence

☐ 0 %

■ 1 - 5 %

■ 6 - 10 %

■ 11 - 25 %

■ 26 - 50 %

■ 51 - 100 %

0   50   100
Miles

ATLAS BLOCKS

BREEDING STATUS	NO.	
Possible	5	(26%)
Probable	10	(53%)
Confirmed	4	(21%)
TOTAL	19	

ABUNDANCE	NO.	
Uncommon	7	(29%)
Fairly Common	17	(71%)
Common	0	(0%)
Abundant	0	(0%)
TOTAL	24	

0   50   100
Miles

RUDDY DUCK
Oxyura jamaicensis

Proportioned like the proverbial "rubber ducky," the male Ruddy Duck in breeding plumage is a case study of clashing parts and colors. His oversized bill glistens as if with a fresh coat of bright blue paint, and his intense white cheek patch stands out against the black and chestnut hues of the rest of his body. Fair-sized populations of these comical ducks can be found in valley wetlands throughout much of Nevada. The Ruddy Duck is especially likely to be found in vegetation-choked waters with a rich invertebrate prey base.

DISTRIBUTION

Ruddy Ducks were confirmed as breeders in most of the major wetland complexes of Nevada: in the valleys of the west, at scattered locations in Elko County, and at several southern locales. Unconfirmed breeding records were widespread, too. Indeed, the species was found everywhere except in Esmeralda County and adjoining regions of Mineral and Nye counties, where the dry bottomlands and rocky hills provide insufficient habitat even for the resourceful Ruddy Duck.

Multiple pairs of Ruddy Ducks were estimated to occur in most of the blocks for which abundance estimates were made, with at least ten pairs in almost one-third. The species was a confirmed breeder in 38% of the blocks in which it was found. These results match earlier accounts, which describe the Ruddy Duck as a common breeder in the Great Basin at scattered locales throughout the state (Ryser 1985, Alcorn 1988). Linsdale (1936) also described the Ruddy Duck as present statewide, so it appears that its overall distribution in Nevada has not changed greatly during the past century.

In nearby states, the Ruddy Duck nests occasionally in the lower Colorado River Valley (Rosenberg et al. 1991) and in much of western California, but not around Mono Lake and farther south along the Nevada-California border (Gaines 1992, Brua 2001). In Oregon, Ruddy Ducks breed most abundantly in the southeast, with more scattered breeding in the northern parts of the state (Littlefield 1990, Contreras and Kindschy 1996, Adamus et al. 2001). Confirmed breeding is spotty throughout southwestern Idaho (Stephens and Sturts 1998). The species is considered to be a common summer resident in much of Utah, however, with especially good numbers near the Great Salt Lake (Behle et al. 1985).

For breeding habitat, Ruddy Ducks require wetlands with dense emergent vegetation. In the Lahontan Valley, they are typically found in deeper water surrounded by cattails and hardstem bulrush (Chisholm and Neel 2002). Ruddy Ducks are evidently tolerant of human activity; in southern California, for example, most confirmed breeding records came from urban parks with wetlands that had abundant cattails (Gallagher 1997). Although Ruddy Ducks are most commonly found in valley wetlands, they also range up to 7,000 feet (2,100 meters) in California (Small 1994).

CONSERVATION AND MANAGEMENT

Although the Ruddy Duck is widespread and fairly common in much of our region, it is susceptible to range contractions. These have occurred in southern California, where the species was formerly more widespread as a nester (Garrett and Dunn 1981). Even where Ruddy Ducks are common during summer, many individuals may not actually be breeders (Shuford 1993).

The Ruddy Duck is a brood parasite that lays its eggs both in the nests of other Ruddy Ducks and in those of other duck species (Joyner 1983). Other parasitic ducks, such as Redheads, lay their eggs in Ruddy Duck nests. Both parasitism of Ruddy Ducks and parasitism by Ruddy Ducks occur at relatively low rates (Brua 2001), however, and the impact on populations is probably small.

Ruddy Ducks respond well to habitat enhancement. At Fish Creek National Wildlife Refuge in Utah, their population increased following creation of new habitat with a rich invertebrate fauna (Ryser 1985). Even in urban environments, such as Orange County, California, the Ruddy Duck can be attracted by planting or preserving cattail stands in wetlands (Gallagher 1997).

HABITAT USE

HABITAT	ATLAS BLOCKS	INCIDENTAL OBSERVATIONS
Open Water	15 (45%)	10 (42%)
Riparian	2 (6%)	0 (0%)
Salt Desert Scrub	0 (0%)	1 (4%)
Wetland	16 (49%)	13 (54%)
TOTAL	33	24

Total numbers reported in the Habitat Use, Breeding Status, and Abundance tables may differ from each other (see pp. 22–23 for details). Percentage sums may differ slightly from 100% due to rounding.

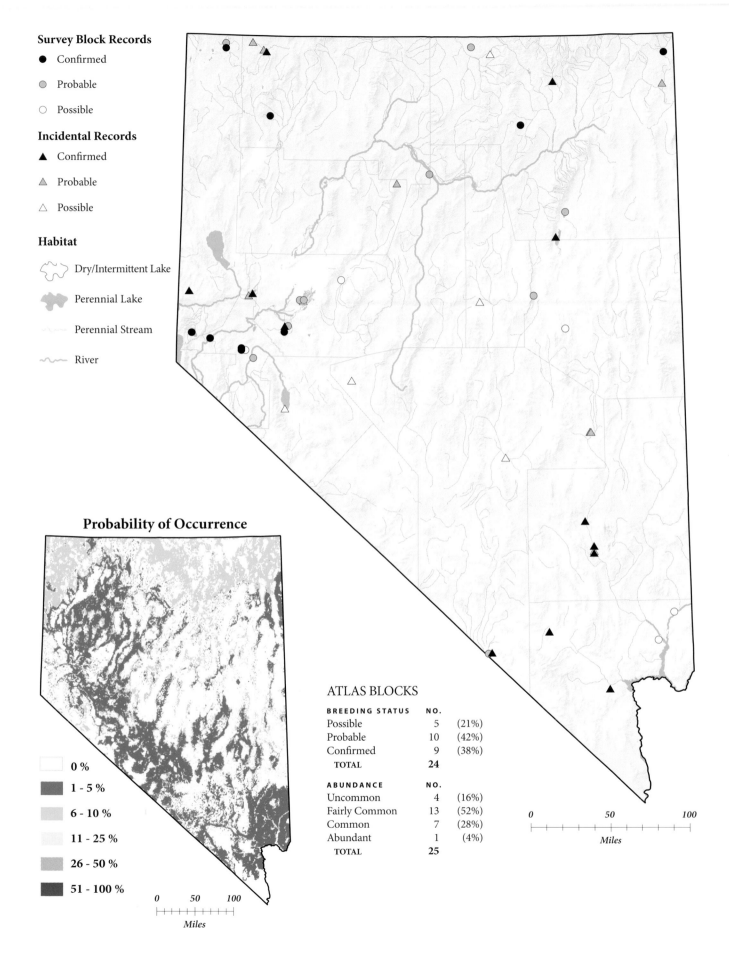

Survey Block Records

● Confirmed

● Probable

○ Possible

Incidental Records

▲ Confirmed

▲ Probable

△ Possible

Habitat

Dry/Intermittent Lake

Perennial Lake

Perennial Stream

River

Probability of Occurrence

0 %

1 - 5 %

6 - 10 %

11 - 25 %

26 - 50 %

51 - 100 %

0   50   100
Miles

ATLAS BLOCKS

BREEDING STATUS	NO.	
Possible	5	(21%)
Probable	10	(42%)
Confirmed	9	(38%)
TOTAL	24	

ABUNDANCE	NO.	
Uncommon	4	(16%)
Fairly Common	13	(52%)
Common	7	(28%)
Abundant	1	(4%)
TOTAL	25	

0   50   100
Miles

CHUKAR
Alectoris chukar

The Chukar, a native of Eurasia and the Middle East, was first released in Nevada in 1935 (Christensen 1954). These birds flourished in the climate and terrain of the Great Basin and soon began feeding on invasive cheatgrass, another nonnative. Chukars became established so quickly that a hunting season was opened in 1947, and the Chukar is now Nevada's primary upland game bird (Christensen 1996). The species occurs statewide, and in places like the grassy foothills of Humboldt County it is nearly impossible to drive through the backcountry without flushing Chukars.

DISTRIBUTION
Atlas records came from most of the state, but were sparser toward the east and south, a pattern first noted fifty years ago by Christensen (1954). The only large area seemingly devoid of Chukars was in and around Lincoln County. Chukars have been reported from this remote and rugged area (Alcorn 1988), though, and they may simply have been missed during atlas surveys. Chukars were confirmed as breeders in only 18% of the occupied blocks, but because they are not generally migratory, many of the unconfirmed observations may indicate areas with breeding birds. This cannot always be assumed, of course; Chukars move about after breeding, and long dispersal movements between breeding sites have been reported (Christensen 1954, Walter 2002). For instance, Chisholm and Neel (2002) reported sightings in the Lahontan Valley but noted that breeding there was unlikely.

Nevada constitutes a large portion of the Chukar's North American breeding range (Christensen 1996), which extends into far eastern California and northern Arizona, throughout Utah, and north into eastern Oregon, southwestern Idaho, and limited regions beyond (Behle et al. 1985, Small 1994, Chris-

tensen 1996, Stephens and Sturts 1998, Adamus et al. 2001, Corman in Corman and Wise-Gervais 2005:74–75). Throughout this range Chukars thrive on rocky, arid slopes in various types of shrub, scrub, and grassland vegetation. At present, Chukars are thought to be fully established in most or all of the regions where they are capable of long-term persistence (Christensen 1996), and populations are no longer routinely supplemented by additional releases. In Nevada, they are most often seen in low- and mid-elevation sagebrush habitats, but there were also a number of atlas records from higher-elevation pinyon-juniper and montane shrub habitats. The predictive map indicates a moderate chance of finding Chukars throughout most of Nevada, with lower probabilities in salt desert and Mojave habitats. The moderate probabilities predicted for the northern tier of counties may underestimate true occurrences.

CONSERVATION AND MANAGEMENT
The introduction of the Chukar into the Great Basin has been highly successful (Ryser 1985); limited population trend data indicate many more Chukars in Nevada now than in the 1960s (Sauer et al. 2005). Hunting-related population monitoring suggests that Chukar numbers vary substantially in response to precipitation and range productivity (Christensen 1996; see also Ryser 1985). A primary limiting factor is drinkable water, which game management programs often supplement with artificial watering holes known as "guzzlers" (Small 1994, Christensen 1996). Given that Chukars feed liberally on cheatgrass and other invasive plants, they may benefit from heavy grazing and range fires (Christensen 1996). Hunting itself is managed and poses no population-level threats (Christensen 1996).

HABITAT USE

HABITAT	ATLAS BLOCKS	INCIDENTAL OBSERVATIONS
Barren	6 (7%)	0 (0%)
Grassland	3 (4%)	2 (5%)
Mojave	3 (4%)	0 (0%)
Montane Forest	0 (0%)	1 (3%)
Montane Parkland	0 (0%)	1 (3%)
Montane Shrub	10 (12%)	6 (16%)
Pinyon-Juniper	9 (11%)	5 (14%)
Riparian	5 (6%)	7 (19%)
Salt Desert Scrub	4 (5%)	0 (0%)
Sagebrush Scrub	22 (27%)	10 (27%)
Sagebrush Steppe	18 (22%)	4 (11%)
Urban	1 (1%)	1 (3%)
Wetland	1 (1%)	0 (0%)
TOTAL	82	37

Total numbers reported in the Habitat Use, Breeding Status, and Abundance tables may differ from each other (see pp. 22–23 for details). Percentage sums may differ slightly from 100% due to rounding.

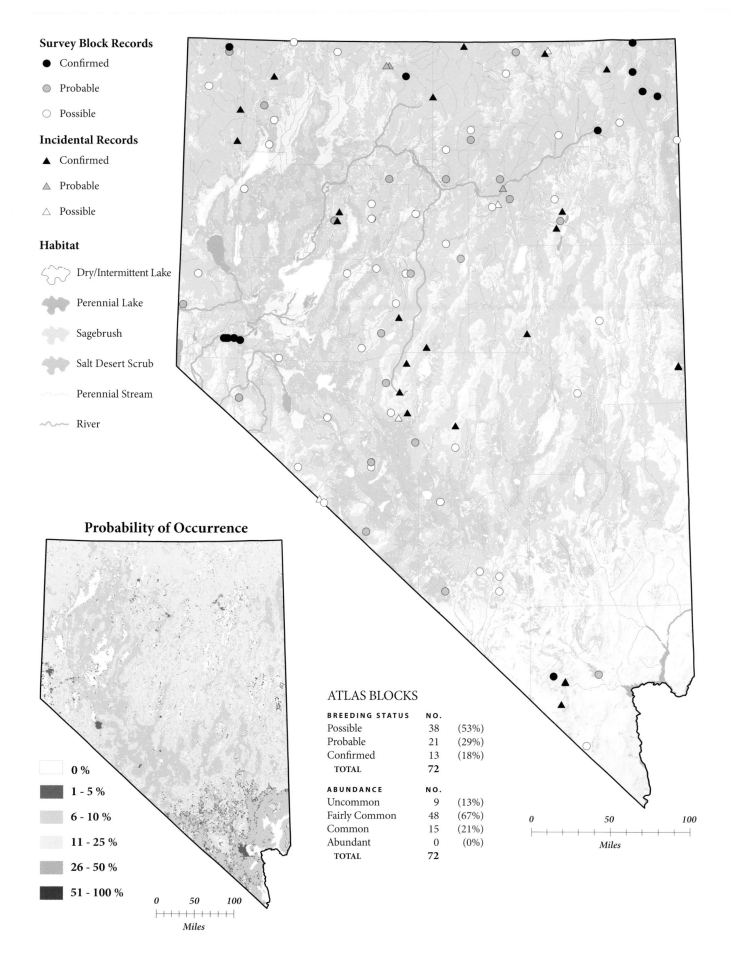

Survey Block Records

● Confirmed

● Probable

○ Possible

Incidental Records

▲ Confirmed

▲ Probable

△ Possible

Habitat

Dry/Intermittent Lake

Perennial Lake

Sagebrush

Salt Desert Scrub

Perennial Stream

River

Probability of Occurrence

☐ 0 %

■ 1 - 5 %

☐ 6 - 10 %

☐ 11 - 25 %

☐ 26 - 50 %

■ 51 - 100 %

0   50   100
Miles

ATLAS BLOCKS

BREEDING STATUS	NO.	
Possible	38	(53%)
Probable	21	(29%)
Confirmed	13	(18%)
TOTAL	72	

ABUNDANCE	NO.	
Uncommon	9	(13%)
Fairly Common	48	(67%)
Common	15	(21%)
Abundant	0	(0%)
TOTAL	72	

0   50   100
Miles

HIMALAYAN SNOWCOCK
Tetraogallus himalayensis

The Himalayan Snowcock provides an irresistible inducement for birding fanatics from all across North America to visit Nevada. The Ruby Mountains of Elko County are the only place in the New World where an introduced population of this central Asian native has taken hold. These large game birds are thought to be widespread and fairly common here in the alpine zone, but they can be difficult to find. They are most likely to occur on the steep, rocky slopes of the tallest peaks, in the company of pikas, Rock Wrens, and Golden Eagles. Sometimes they venture out into alpine meadows and stunted thickets, where they can be seen alongside two other high-elevation specialists, the Black Rosy-Finch and the American Pipit.

DISTRIBUTION

In 1963, in what turned out to be one of the last successful introductions of an exotic game bird into North America, the Nevada Department of Wildlife released Himalayan Snowcocks into the highlands of the Ruby Mountains (Christensen 1998). Supplemental releases continued for a number of years, and eventually snowcocks established a self-sustaining breeding population. Introductions were also attempted in several other Nevada locations, but those populations never became established. Other states and provinces (Oregon, Wyoming, Washington, and Alberta) received birds via Nevada's captive-breeding program but never released them because of concerns about disease and second thoughts about releasing nonnative species into ecosystems (Christensen 1998). The Ruby Mountains population was sufficiently stable by 1980 to open a hunting season (Alcorn 1988), and the birds are now prized as trophies. Hunting pressure remains modest, however, because of the snowcock's wariness and the rugged terrain it inhabits (Christensen 1998).

Within its very limited Nevada range, the Himalayan Snowcock can be found on steep, rocky, sparsely vegetated slopes and in adjacent alpine meadows (Christensen 1998). Steep slopes appear to be a key element for the snowcock, which flies downhill to escape hunters and predation-minded Golden Eagles (Morse et al. 1995, Christensen 1998). Most birders who have experience with this species in Nevada report it from the upper Lamoille Canyon in Elko County. Its typical habitat consists of boulder-strewn snowfields with a few stunted willow thickets, surrounded on all sides by nearly barren talus slopes.

The atlas fieldwork shed little new light on the distribution and abundance of the Himalayan Snowcock in Nevada. The species was reported from two atlas blocks in the Ruby Mountains, and was confirmed as a breeder at three other Ruby Mountains sites that fell outside atlas blocks. Snowcocks are also established in the East Humboldt Range, which is contiguous with the northern Ruby Mountains (Christensen 1998), but none were recorded there during atlas work. Himalayan Snowcocks have not spread beyond these two ranges, and are not expected to do so.

CONSERVATION AND MANAGEMENT

The usual conservation issues are probably not relevant for an introduced game bird occupying a tiny, nearly inaccessible sliver of Nevada. The snowcock's current population size and trends are not known, but it has been estimated—based on extrapolation of population densities in the species' native range—that about a thousand Himalayan Snowcocks may be present in the Ruby Mountains in spring (Christensen 1998). From a wildlife enthusiast's perspective, the Himalayan Snowcock certainly has attracted attention to a truly spectacular mountain range with a rich ensemble of alpine bird life. Visiting birders often take advantage of "snowcock tours." Some might not agree that the weird, wailing calls and sociable clucking of the snowcock enrich the Ruby Mountains experience, but a snowcock seeker cannot help but be impressed by this tough bird's elusiveness and the fragile alpine ecosystem it inhabits.

HABITAT USE

HABITAT	ATLAS BLOCKS	INCIDENTAL OBSERVATIONS
Alpine	4 (80%)	3 (100%)
Barren	1 (20%)	0 (0%)
TOTAL	5	3

Total numbers reported in the Habitat Use, Breeding Status, and Abundance tables may differ from each other (see pp. 22–23 for details). Percentage sums may differ slightly from 100% due to rounding.

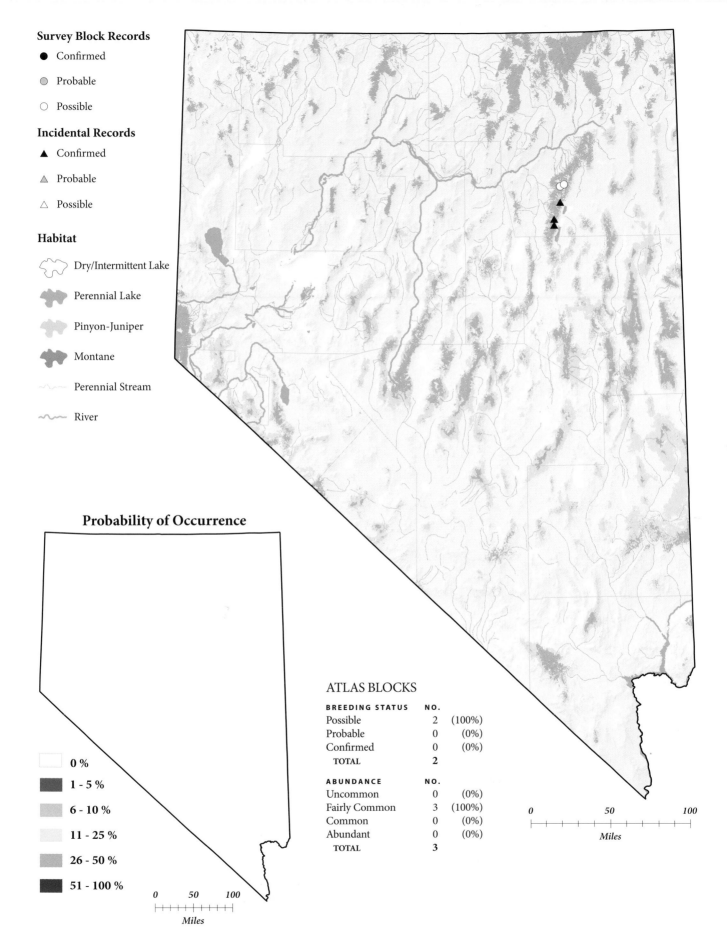

Survey Block Records

● Confirmed

◉ Probable

○ Possible

Incidental Records

▲ Confirmed

◬ Probable

△ Possible

Habitat

Dry/Intermittent Lake

Perennial Lake

Pinyon-Juniper

Montane

Perennial Stream

River

Probability of Occurrence

☐ 0 %

■ 1 - 5 %

■ 6 - 10 %

☐ 11 - 25 %

■ 26 - 50 %

■ 51 - 100 %

0   50   100
Miles

ATLAS BLOCKS

BREEDING STATUS	NO.	
Possible	2	(100%)
Probable	0	(0%)
Confirmed	0	(0%)
TOTAL	2	

ABUNDANCE	NO.	
Uncommon	0	(0%)
Fairly Common	3	(100%)
Common	0	(0%)
Abundant	0	(0%)
TOTAL	3	

0   50   100
Miles

GRAY PARTRIDGE
Perdix perdix

In North America, the introduced Gray Partridge is most abundant in the croplands and grasslands of the northern Great Plains. Smaller populations have also become established in the West, the most southerly of these in northern Nevada. Conditions in the Great Basin are somewhat marginal for the species, though, and Gray Partridges are relatively uncommon here. Perhaps for this reason they receive only modest attention from Nevada hunters.

DISTRIBUTION

The Gray Partridge, a broadly distributed Eurasian species, was introduced into North America in the early 1900s (Carroll 1993), and into Nevada around 1923 (Ryser 1985). Populations were also established elsewhere in the West, including in northern Utah, Idaho, eastern Oregon, and southeastern Washington (Behle et al. 1985, Carroll 1993, Stephens and Sturts 1998, Adamus et al. 2001). Attempts to establish the species in California were unsuccessful (Small 1994).

Alcorn (1988) described this bird as locally common in northern Nevada. His account suggests a wider distribution and greater abundance than our atlas workers and Titus (2003) found. Breeding was confirmed by atlas workers at only two sites: in northeastern Elko County and in central White Pine County. There were additional records from Humboldt and Lander counties, where breeding populations are known to be established (W. Molini, pers. comm.), though breeding was not confirmed in these locations by atlas fieldwork.

The Gray Partridge seems to prefer farmland as its main habitat, both in its native range and in North America (Potts 1986, Carroll 1993). In the northern Great Plains and the Midwest, it is most apt to thrive in regions dominated by traditional agricul-

tural practices that offer fields of cereal grains, row crops, shelterbelts, and hedgerows (Carroll 1993). Northern Nevada has little to offer in this regard, and Gray Partridges certainly show no great affinity for Nevada's limited agricultural areas. Instead they seem to occur in shrubby or grassy uplands, which are perhaps reminiscent of their Old World steppe and grassland habitats (Aebischer and Kavanagh in Hagemeijer and Blair 1997:212–213). Even within these habitats and within their northeastern Nevada range, the probability of occurrence is fairly low.

CONSERVATION AND MANAGEMENT

In the case of the Gray Partridge, several factors militate against the birder's common knee-jerk indifference to the conservation concerns of exotic game birds. Monitoring data are somewhat limited, but populations in the Great Plains region appear to be declining—substantially in some areas (Sauer et al. 2005). Interestingly, these declines in overall abundance have occurred in concert with some southward range expansions (Carroll 1993). Hunting pressure in North America is modest at best, and does not seem to be a factor in the declines (Carroll 1993). In Europe, populations declined substantially during the second half of the twentieth century, primarily as the result of changes in agricultural practices, such as increased pesticide application, hedgerow removal, and automated irrigation (Potts 1986, Carroll 1993). Similar problems might also apply in North America, suggesting that the species could act as a bellwether for broader conservation concerns (Carroll 1993).

The ecology of the Gray Partridge in Nevada also may be of general interest because the species appears to rely less on agricultural lands here than it does elsewhere in its range. Western populations in general appear to be more stable than those farther east, though trend data are limited and often not very reliable (Sauer et al. 2005). Within Nevada, populations are known to fluctuate greatly (Alcorn 1988), but the general impression of most observers is one of reasonable stability over the long term (W. Molini, pers. comm.).

HABITAT USE

HABITAT	ATLAS BLOCKS	INCIDENTAL OBSERVATIONS
Grassland	1 (25%)	0 (0%)
Riparian	1 (25%)	0 (0%)
Sagebrush Scrub	1 (25%)	1 (50%)
Sagebrush Steppe	1 (25%)	1 (50%)
TOTAL	4	2

Total numbers reported in the Habitat Use, Breeding Status, and Abundance tables may differ from each other (see pp. 22–23 for details). Percentage sums may differ slightly from 100% due to rounding.

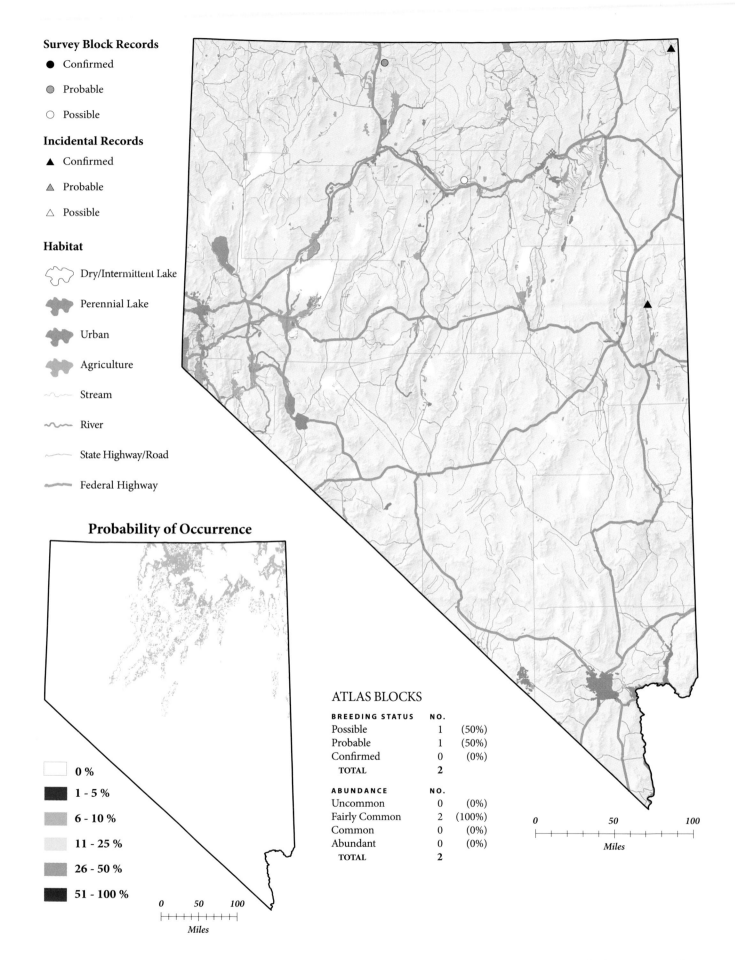

Survey Block Records

● Confirmed

◉ Probable

○ Possible

Incidental Records

▲ Confirmed

△ Probable

△ Possible

Habitat

⬡ Dry/Intermittent Lake

⬡ Perennial Lake

⬡ Urban

⬡ Agriculture

〜 Stream

〜 River

— State Highway/Road

━ Federal Highway

Probability of Occurrence

☐ 0 %

■ 1 - 5 %

■ 6 - 10 %

☐ 11 - 25 %

■ 26 - 50 %

■ 51 - 100 %

0   50   100
Miles

ATLAS BLOCKS

BREEDING STATUS	NO.	
Possible	1	(50%)
Probable	1	(50%)
Confirmed	0	(0%)
TOTAL	2	

ABUNDANCE	NO.	
Uncommon	0	(0%)
Fairly Common	2	(100%)
Common	0	(0%)
Abundant	0	(0%)
TOTAL	2	

0   50   100
Miles

RING-NECKED PHEASANT
Phasianus colchicus

Fine looking and reportedly fine tasting, the Ring-necked Pheasant has long been among the most popular of the exotic North American game birds. Pheasants have been widely introduced in Nevada, and they are sometimes common in lowland locations—especially around farms, towns, and riparian corridors. The species is prone to local extirpation in our region, however, and most populations in Nevada are relatively small.

DISTRIBUTION

Ring-necked Pheasants were first introduced into the West in the late nineteenth century (Giudice and Ratti 2001). Pheasants became best established in the extensive cultivated areas of the Great Plains and Midwest, but they occur nearly coast-to-coast (Giudice and Ratti 2001). The date of the first Nevada introductions is unclear, but pheasants are not mentioned in Linsdale's (1936) book on the birds of Nevada. By the late 1980s pheasants were established in some of the state's agricultural lowlands (Alcorn 1988).

The atlas records were divided between a tight cluster of observations along the lower Colorado and Virgin river drainages and scattered observations across the lowlands of the northern and western counties. Most of the northern records were from areas near rivers. Of twenty-two locations where pheasants occurred, however, breeding was confirmed in only two. Ongoing releases of captive-raised birds, often conducted by private hunting clubs in Nevada and surrounding regions, make it difficult to determine whether sufficient wild breeding occurs for populations to remain viable without supplementation (Adamus et al. 2001, Giudice and Ratti 2001). At least some breeding populations have become established in nearby states, though, including California, Utah, and Idaho (Behle et al. 1985, Small 1994, Stephens and Sturts 1998).

Throughout their North American range, pheasants are most partial to farmlands that include other habitat features, such as forested borders, hedgerows, ditches, or marshes. They are flexible, though, and also occur in native habitats (Giudice and Ratti 2001). Atlas fieldworkers reported pheasants most often in riparian zones, followed by agricultural and urban areas, and a few other habitats. This preference for relatively low elevations and moist habitats is also reflected on the predictive map, although the model probably overestimates the chance of finding this species in south-central Nevada.

CONSERVATION AND MANAGEMENT

Management of the Ring-necked Pheasant clearly revolves around its long history as an introduced game bird, which began when pheasants were first introduced into Europe from central Asia (Bijlsma and Hill in Hagemeijer and Blair 1997:218–219). As is the case with many game birds, which frequently have high reproductive rates, population dynamics tend to exhibit periodic booms and busts (Giudice and Ratti 2001). This variability, in conjunction with restocking and hunting activities, greatly complicates any assessment of trends. Pheasants appear to be declining over much of North America (Sauer et al. 2005), however, and the industrialization of agricultural operations is a likely causative factor (Giudice and Ratti 2001). Pheasant management often focuses on retaining or restoring the habitat features that characterized traditional agricultural landscapes, and on adjusting the timing of mechanical crop harvesting to minimize nest disturbances (Giudice and Ratti 2001). In Nevada, the overall impression is that populations are not stable. In the Lahontan Valley, for instance, unfavorable agricultural practices and diseases have apparently caused declines (Chisholm and Neel 2002).

HABITAT USE

HABITAT	ATLAS BLOCKS	INCIDENTAL OBSERVATIONS
Agricultural	6 (23%)	1 (25%)
Grassland	3 (12%)	0 (0%)
Riparian	9 (35%)	1 (25%)
Sagebrush Scrub	1 (4%)	0 (0%)
Sagebrush Steppe	1 (4%)	0 (0%)
Urban	4 (15%)	1 (25%)
Wetland	2 (8%)	1 (25%)
TOTAL	26	4

Total numbers reported in the Habitat Use, Breeding Status, and Abundance tables may differ from each other (see pp. 22–23 for details). Percentage sums may differ slightly from 100% due to rounding.

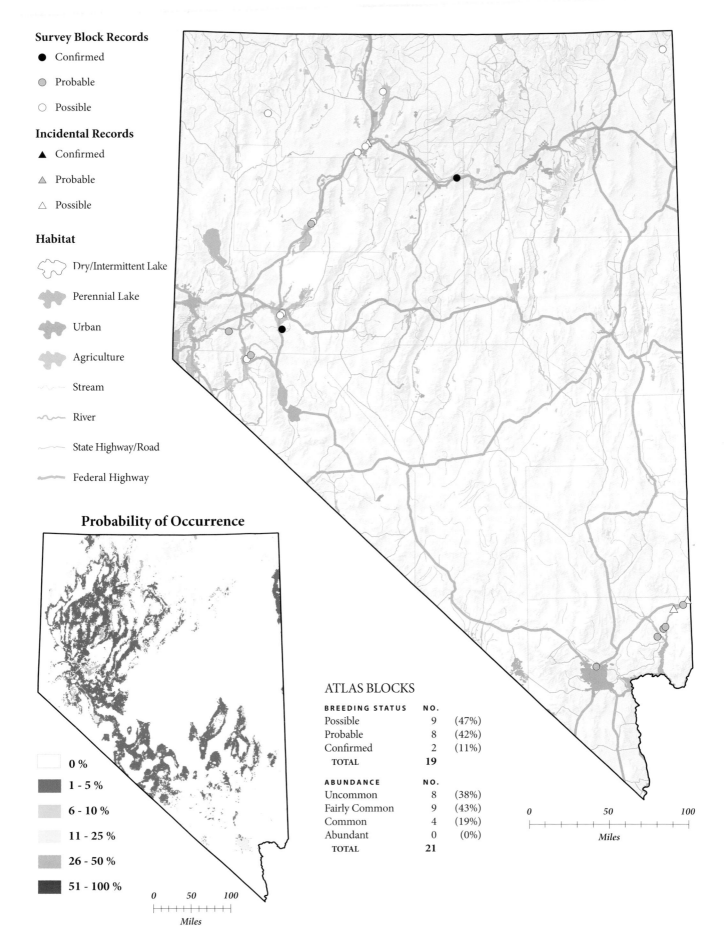

Survey Block Records

● Confirmed

◉ Probable

○ Possible

Incidental Records

▲ Confirmed

△ Probable

△ Possible

Habitat

Dry/Intermittent Lake

Perennial Lake

Urban

Agriculture

Stream

River

State Highway/Road

Federal Highway

Probability of Occurrence

☐ 0 %

■ 1 - 5 %

■ 6 - 10 %

☐ 11 - 25 %

■ 26 - 50 %

■ 51 - 100 %

0   50   100
Miles

ATLAS BLOCKS

BREEDING STATUS	NO.	
Possible	9	(47%)
Probable	8	(42%)
Confirmed	2	(11%)
TOTAL	19	

ABUNDANCE	NO.	
Uncommon	8	(38%)
Fairly Common	9	(43%)
Common	4	(19%)
Abundant	0	(0%)
TOTAL	21	

0   50   100
Miles

RUFFED GROUSE

Bonasa umbellus

A bird of broadleaved forests of the northern United States and Canada, the Ruffed Grouse barely enters Nevada, and only then with a little help. There appear to be no verifiable records for the species in Nevada prior to the release of transplanted birds in 1963, and it is assumed that the bird is not native to the state. Suitable habitat exists in and around Elko County, though, where the species now appears to persist. Ruffed Grouse are valued as game birds, and are managed accordingly. Birders and hikers in the aspen stands of the state's northeastern mountains can expect occasionally to flush Ruffed Grouse and be startled by their characteristically explosive takeoff.

DISTRIBUTION

Ruffed Grouse were introduced into Elko County in 1963 (Ryser 1985, Alcorn 1988) and have been present in the area ever since. There were few atlas records, however: three in atlas blocks and four incidental observations. Breeding was confirmed in the Jarbidge Mountains, Ruby Mountains, and Schell Creek Range, and an unconfirmed breeding record was reported from the far northeast. The species has also been introduced into the Santa Rosa Range (W. Molini, pers. comm.), although that area produced no atlas records.

The current Nevada range of the Ruffed Grouse is quite restricted, and the degree to which birds are self-sustaining is unclear. To the east and north, however, in Utah, Idaho, and Oregon, native populations of this grouse are fairly widespread (Behle et al. 1985, Stephens and Sturts 1998, Adamus et al. 2001). California has a small native population in the far northwestern corner of the state (Small 1994) and an introduced population near Lassen Peak in Shasta County (McCaskie et al. 1988).

The Ruffed Grouse occurs widely in the forests of northern North America and seems to have a strong inclination toward more mesic deciduous woodlands (Rusch et al. 2000). Atlas habitat records were too few to allow generalizations, but in the Intermountain West, aspen stands and riparian areas are the primary habitats (Rusch et al. 2000). Alcorn (1988) considered any aspen forests in Nevada potentially suitable for colonization by this species. The predictive map indicates such areas in northeastern Nevada, but the probability of occurrence is fairly low throughout.

CONSERVATION AND MANAGEMENT

The population dynamics of the Ruffed Grouse have been well studied, and in the more northerly parts of its range the species is known for its regular population cycles (Rusch et al. 2000). Farther south, this regularity does not seem to exist, but Ruffed Grouse numbers still vary substantially from year to year. Combined with hunting pressure and logging of forests, these population dynamics led to the implementation of hunting limits and active management of the bird, starting in the late nineteenth century. Eastern grouse populations benefited from the forest regrowth that followed the abandonment of agriculture in many areas, although as forests mature, the habitat becomes less suitable again (Rusch et al. 2000). In Washington, timber harvesting appears to have created the second-growth hardwood and mixed forest types in which the Ruffed Grouse flourishes (Smith et al. 1997). Rusch et al. (2000) considered Ruffed Grouse numbers to be declining somewhat in the eastern United States and relatively stable in the West. Breeding Bird Survey data, however, suggest the opposite: declines in the West but not in the East (Sauer et al. 2005).

Nevada populations are hunted, but they are periodically restocked within their limited range. Thus, population trends and their contributing factors are hard to decipher. The support of the hunting community means that the prognosis for continued presence in Nevada is good. Ruffed Grouse harvests are carefully regulated throughout the species' range, and in many areas habitat is managed to create conditions favorable for them (Rusch et al. 2000).

HABITAT USE

HABITAT	ATLAS BLOCKS	INCIDENTAL OBSERVATIONS
Montane Forest	1 (25%)	1 (20%)
Montane Shrub	2 (50%)	0 (0%)
Riparian	1 (25%)	3 (60%)
Sagebrush Steppe	0 (0%)	1 (20%)
TOTAL	4	5

Total numbers reported in the Habitat Use, Breeding Status, and Abundance tables may differ from each other (see pp. 22–23 for details). Percentage sums may differ slightly from 100% due to rounding.

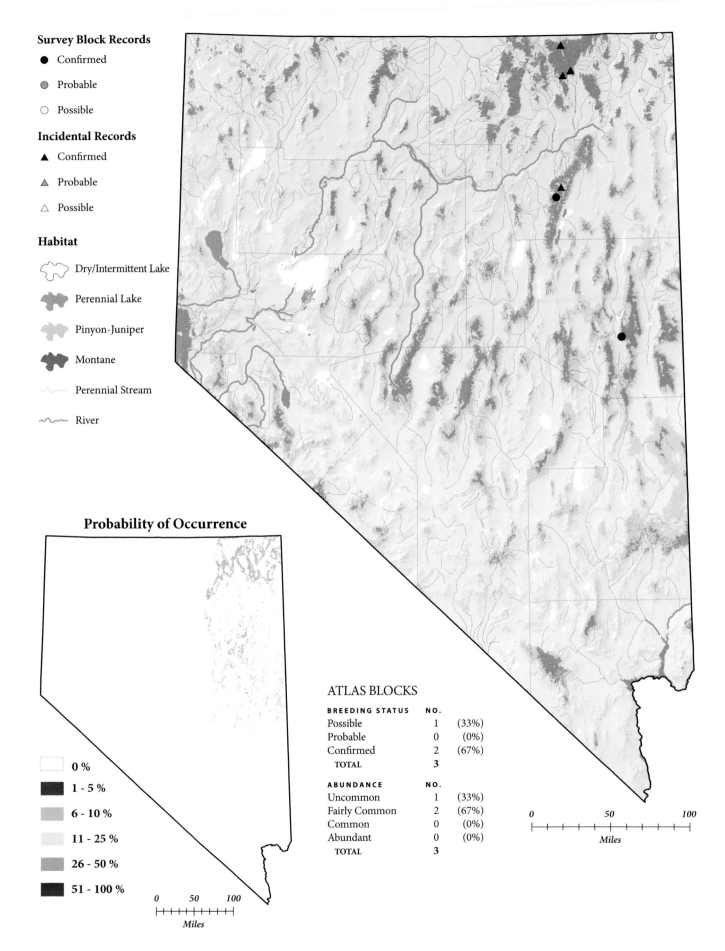

Survey Block Records

● Confirmed

◐ Probable

○ Possible

Incidental Records

▲ Confirmed

▲ Probable

△ Possible

Habitat

Dry/Intermittent Lake

Perennial Lake

Pinyon-Juniper

Montane

Perennial Stream

River

Probability of Occurrence

☐ 0 %

■ 1 - 5 %

▨ 6 - 10 %

☐ 11 - 25 %

▨ 26 - 50 %

■ 51 - 100 %

0   50   100
Miles

ATLAS BLOCKS

BREEDING STATUS	NO.	
Possible	1	(33%)
Probable	0	(0%)
Confirmed	2	(67%)
TOTAL	3	

ABUNDANCE	NO.	
Uncommon	1	(33%)
Fairly Common	2	(67%)
Common	0	(0%)
Abundant	0	(0%)
TOTAL	3	

0   50   100
Miles

GREATER SAGE-GROUSE

Centrocercus urophasianus

The Greater Sage-Grouse, an icon of the "sagebrush sea," has come to symbolize many of the environmental changes in our region, from degradation of sagebrush to loss of riparian areas. Conservationists, managers, and landowners are working hard to stabilize and recover populations of this magnificent lekking bird in Nevada, but many challenges remain.

DISTRIBUTION

Greater Sage-Grouse were reported throughout northern and east-central Nevada but were found only sporadically in west-central Nevada, where breeding was not confirmed. More detailed studies by the Nevada Department of Wildlife have identified additional mountain ranges with breeding populations. The atlas distribution is largely consistent with earlier accounts (Linsdale 1936, Alcorn 1988); however, Nevada's sage-grouse populations have declined in abundance and have become increasingly fragmented (Stiver 2001), matching trends elsewhere (Paige and Ritter 1999, Schroeder et al. 1999, Dobkin and Sauder 2004). Greater Sage-Grouse also occur in eastern California, southeastern Oregon, southern Idaho, parts of Utah, and in the northern Rocky Mountain states, but they have disappeared from many parts of their historical range (Schroeder et al. 1999).

Greater Sage-Grouse use sagebrush steppe habitats with significant bunchgrass and forb components. As the predictive map suggests, these diverse plant communities are most common along Nevada's northern edge. Sparsely vegetated lek sites within the sagebrush matrix—where birds gather to select mates—are

also required, and riparian areas, wet meadows, springs, and seeps offer foraging opportunities for broods as the range dries out in late summer. Sage-grouse often move substantial distances in search of seasonally appropriate microhabitats (Connelly et al. 2004), and impediments to these movements can have dire consequences for local populations (Paige and Ritter 1999, Schroeder et al. 1999, Stiver 2001). NWF (2002) and Connelly et al. (2004) discuss habitat use in Nevada and beyond in greater detail.

CONSERVATION AND MANAGEMENT

Sage-grouse declines were noted as early as the 1930s (Linsdale 1936, Paige and Ritter 1999), but serious conservation concerns came to the forefront only during the 1960s–1980s (Connelly et al. 2004). Declines have been attributed to destruction, degradation, and fragmentation of sagebrush shrublands; heavy grazing, especially in the wet areas the birds use in late summer; changes in fire frequency favoring establishment of exotic annual grasses or sagebrush monocultures; regulated hunting; and poaching (Paige and Ritter 1999, Schroeder et al. 1999, Dobkin and Sauder 2004). While there is little doubt about the historical declines, present trends are harder to pinpoint (Dobkin and Sauder 2004). Some authors have reported continued rangewide declines (Dobkin and Sauder 2004), while others have suggested that numbers are stabilizing in some areas, including Nevada (Paige and Ritter 1999, Connelly et al. 2004). Without active management, however, continued declines are expected (Connelly et al. 2004).

In response to proposals to list the Greater Sage-Grouse under the U.S. Endangered Species Act, the Governor's Sage-Grouse Team—a coalition of conservationists, managers, local and state governments and agencies, and land users—was formed to recommend conservation actions in Nevada (Espinosa 2004). On the federal front, the U.S. Fish and Wildlife Service has now determined that the species does not warrant listing, but concerns about the bird's prospects remain. Immediate action to reverse population losses, as called for by Partners in Flight (Rich et al. 2004), will no doubt be challenging to achieve. Proposed strategies include establishing habitat preserves; restoring native plant communities; maintaining dense patches of tall sagebrush with a mosaic of age and size classes; protecting springs, seeps, and wet meadows; and minimizing pesticide use and livestock overgrazing (Paige and Ritter 1999, Schroeder et al. 1999).

HABITAT USE

HABITAT	ATLAS BLOCKS	INCIDENTAL OBSERVATIONS
Agricultural	1 (2%)	0 (0%)
Alpine	1 (2%)	0 (0%)
Grassland	2 (3%)	2 (6%)
Montane Forest	0 (0%)	1 (3%)
Montane Shrub	5 (8%)	2 (6%)
Riparian	4 (7%)	2 (6%)
Sagebrush Scrub	16 (27%)	12 (39%)
Sagebrush Steppe	31 (52%)	11 (35%)
TOTAL	60	30

Total numbers reported in the Habitat Use, Breeding Status, and Abundance tables may differ from each other (see pp. 22–23 for details). Percentage sums may differ slightly from 100% due to rounding.

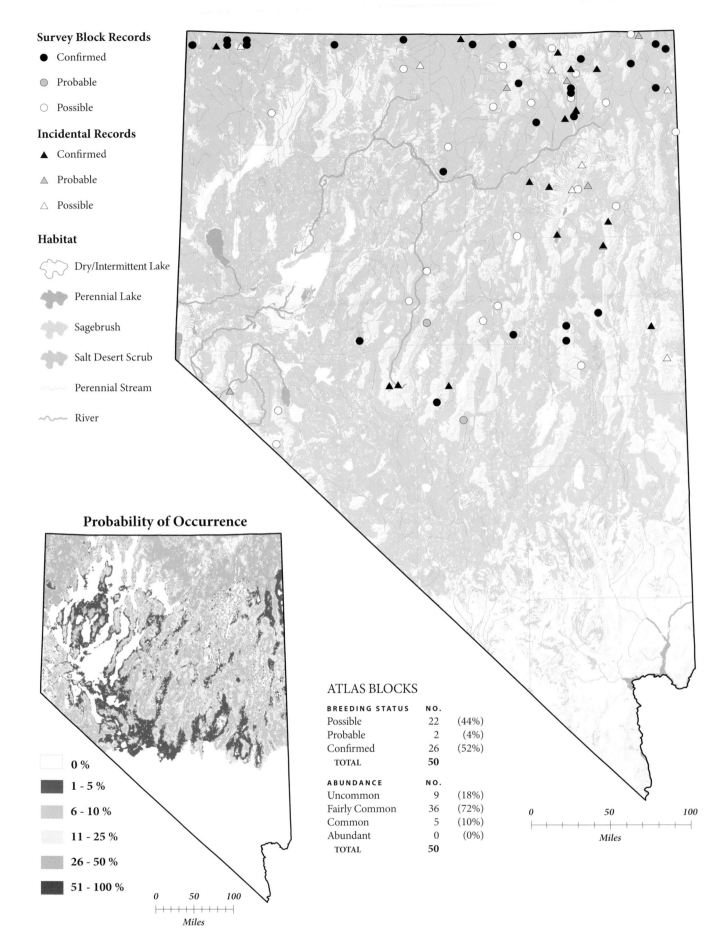

Survey Block Records

- ● Confirmed
- ● Probable
- ○ Possible

Incidental Records

- ▲ Confirmed
- ▲ Probable
- △ Possible

Habitat

Dry/Intermittent Lake

Perennial Lake

Sagebrush

Salt Desert Scrub

Perennial Stream

River

Probability of Occurrence

☐	0 %
■	1 - 5 %
☐	6 - 10 %
☐	11 - 25 %
☐	26 - 50 %
■	51 - 100 %

0   50   100

Miles

ATLAS BLOCKS

BREEDING STATUS	NO.	
Possible	22	(44%)
Probable	2	(4%)
Confirmed	26	(52%)
TOTAL	**50**	

ABUNDANCE	NO.	
Uncommon	9	(18%)
Fairly Common	36	(72%)
Common	5	(10%)
Abundant	0	(0%)
TOTAL	**50**	

0   50   100

Miles

SOOTY/DUSKY GROUSE

Dendragapus fuliginosus and *D. obscurus*

For the duration of the atlas project, Sooty Grouse and Dusky Grouse were considered a single species—the "Blue Grouse." As this book was in the final stages of production, however, this species was split into two, both of which breed in Nevada. The Sooty Grouse is the Pacific coastal form, and inhabits the dense forests of the Carson Range, while the Dusky Grouse occur farther east, occupying not only montane forests but also sometimes adjacent sagebrush habitat. Here, we treat these species collectively, but an important goal for future researchers will be to identify clearly the boundary between the ranges of the two forms.

DISTRIBUTION

Atlas records came from many of Nevada's major mountain ranges. Clustered sightings occurred in the Carson Range in the far west, the Jarbidge Mountains in the northeast, and several other ranges in the east. Records were sparse in the central ranges and absent from the south. The species may have been missed in some areas, as "Blue Grouse" can be hard to find (Gaines 1992, Contreras and Kindschy 1996). Other accounts agree that Dusky Grouse are most common in the northeastern part of the state but give less consistent reports of its status in the south (Linsdale 1936, Ryser 1985, Alcorn 1988, Titus 2003). Linsdale (1936) noted also Dusky Grouse in the central mountains, just east of the single atlas record from the Toiyabe Range, and Alcorn (1988) also reported the species in the Diamond Mountains, which lie just south of the Ruby Mountains, and in the Grant Range of northeastern Nye County.

In surrounding states, "Blue Grouse" occur throughout the Sierra Nevada and in far northern California, western Oregon, and much of Idaho (Gaines 1992, Small 1994, Stephens and Sturts 1998, Adamus et al. 2001). The breeding distribution of Dusky Grouse in Utah and northern Arizona is rather patchy and restricted, much as it is in Nevada (Behle et al. 1985, Zwickel 1992, Bradley in Corman and Wise-Gervais 2005:78–79).

Both species are mountain woodland birds, although interior populations may also use sagebrush and other rangeland habitats more than a mile from forest edges (Zwickel 1992). Alcorn (1988) cited several occurrences in the valleys of Nevada, and atlas fieldworkers also reported occasional use of lower-elevation shrub habitats. The predictive map clearly indicates a largely montane orientation and shows that Nevada's far western and far eastern ranges offer the best chance of finding each species. The map also highlights several mountain ranges where breeding birds may be awaiting future confirmation.

CONSERVATION AND MANAGEMENT

Nevada populations probably have declined during the past century (Alcorn 1988). Elsewhere, declines have been reported in central and southern California (Small 1994) and Utah (Behle et al. 1985). Breeding Bird Survey data also suggest a decline in the global population since the 1960s (Sauer et al. 2005). Causes of declines are not well known, although forestry practices may play a role. Both grouse are commonly hunted, but there is little indication of overharvesting (Zwickel 1992). On top of long-term trends, population numbers can vary substantially from year to year in response to weather conditions and other factors (Zwickel 1992). The "Blue Grouse" is on the Partners in Flight Watch List, with a management goal of doubling the Sierra Nevada populations (Rich et al. 2004). The challenge in Nevada is to get better information on each species' breeding distribution, especially in mountain ranges without recent confirmed breeding, and on population trends.

HABITAT USE

HABITAT	ATLAS BLOCKS	INCIDENTAL OBSERVATIONS
Montane Forest	25 (53%)	7 (37%)
Montane Parkland	3 (6%)	1 (5%)
Montane Shrub	8 (17%)	3 (16%)
Pinyon-Juniper	1 (2%)	0 (0%)
Riparian	7 (15%)	6 (32%)
Sagebrush Scrub	1 (2%)	0 (0%)
Sagebrush Steppe	2 (4%)	2 (11%)
TOTAL	47	19

Total numbers reported in the Habitat Use, Breeding Status, and Abundance tables may differ from each other (see pp. 22–23 for details). Percentage sums may differ slightly from 100% due to rounding.

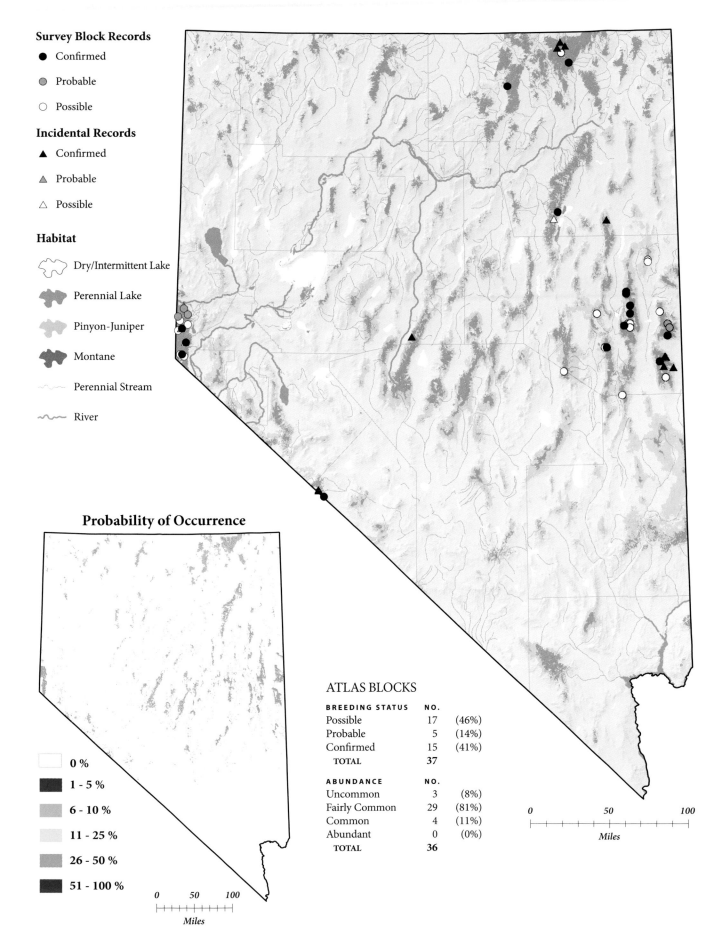

Survey Block Records

● Confirmed

● Probable

○ Possible

Incidental Records

▲ Confirmed

▲ Probable

△ Possible

Habitat

Dry/Intermittent Lake

Perennial Lake

Pinyon-Juniper

Montane

Perennial Stream

River

Probability of Occurrence

☐ 0 %

■ 1 - 5 %

■ 6 - 10 %

11 - 25 %

26 - 50 %

■ 51 - 100 %

0   50   100
Miles

ATLAS BLOCKS

BREEDING STATUS	NO.	
Possible	17	(46%)
Probable	5	(14%)
Confirmed	15	(41%)
TOTAL	37	

ABUNDANCE	NO.	
Uncommon	3	(8%)
Fairly Common	29	(81%)
Common	4	(11%)
Abundant	0	(0%)
TOTAL	36	

0   50   100
Miles

WILD TURKEY
Meleagris gallopavo

Wily and resourceful, the introduced Wild Turkey occurs in scattered locales in the lowlands of far southern and west-central Nevada. It is most common in riparian woodlands, but small flocks are also seen in agricultural fields. Although Wild Turkeys are shy and difficult to approach, they are not birds of the wilderness. They tend to be found close to small towns and in actively farmed rural areas. The present status of the Wild Turkey in Nevada is uncertain, and is complicated by a long history of official and unofficial releases.

DISTRIBUTION

Originally from the East, lower Midwest, and parts of the Four Corners region of the southwestern United States, the Wild Turkey has a long history of introductions and domestication. For instance, it was introduced into Europe as early as the sixteenth century (Eaton 1992, Hudec in Hagemeijr and Blair 1997:221). Turkeys are not native to Nevada, and Ryser (1985) and Alcorn (1988) reported several introductions beginning in 1960. It is not entirely clear whether earlier releases occurred as well. Today, Wild Turkeys are spottily distributed in Nevada but absent from most areas of the state. Atlas records came from the Carson River drainage, from two widely separated areas along the Humboldt River, and from two sites in southern Nevada.

Given the small number of sites where Wild Turkeys were seen, it is difficult to generalize about their habitat in Nevada; however, agricultural and riparian areas appear to be important. These two habitats account for most of the records provided by the atlas observers and are the main habitats present in the parts of the state where Wild Turkeys are predicted to occur. Whether these predictions hold up, however, will likely depend on whether releases have been conducted in a particular area. In general, Wild Turkeys seem to like having access to larger trees for evening roosting (Eaton 1992), and they also require access to drinking water (Nish 1973). During daylight hours, Wild Turkeys forage on the ground in various lowland habitats in the vicinity of the roosting site.

The status of the Wild Turkey in Nevada is uncertain, and it is not clear whether any of the Nevada populations are self-sustaining. Similar situations exist in most of the surrounding states, where scattered introduced populations are found (Behle et al. 1985, Small 1994, Stephens and Sturts 1998, Adamus et al. 2001). The exception is Arizona, where Wild Turkeys are native, although even there, numbers have been supplemented by re-introductions (Moors in Corman and Wise-Gervais 2005:80–81).

CONSERVATION AND MANAGEMENT

Intense hunting pressure on native Wild Turkeys began soon after Europeans settled North America. Birds were often shot off their roosts, resulting in drastic population declines that bottomed out by the early twentieth century (Eaton 1992). Regulation of hunting and targeted transplants reversed this trend, and Wild Turkey numbers increased steadily within the species' historical range from the mid-twentieth century onward (Eaton 1992, Sauer et al. 2005). At present, the Wild Turkey's popularity as a game bird effectively ensures conservation attention, and annual monitoring of harvests provides a mechanism to track the status of populations.

At a global level, Wild Turkey introductions have often been successful, with populations established in places as diverse as the Czech Republic, Hawaii, Tasmania, and New Zealand (Eaton 1992). Introduced populations are inherently vulnerable to extirpation, however, and many introduction attempts fail to result in established populations. At present, Wild Turkeys seem to be only tenuously established in Nevada.

HABITAT USE

HABITAT	ATLAS BLOCKS	INCIDENTAL OBSERVATIONS
Agricultural	4 (31%)	0 (0%)
Grassland	1 (8%)	0 (0%)
Riparian	5 (38%)	2 (100%)
Sagebrush Scrub	1 (8%)	0 (0%)
Urban	1 (8%)	0 (0%)
Wetland	1 (8%)	0 (0%)
TOTAL	13	2

Total numbers reported in the Habitat Use, Breeding Status, and Abundance tables may differ from each other (see pp. 22–23 for details). Percentage sums may differ slightly from 100% due to rounding.

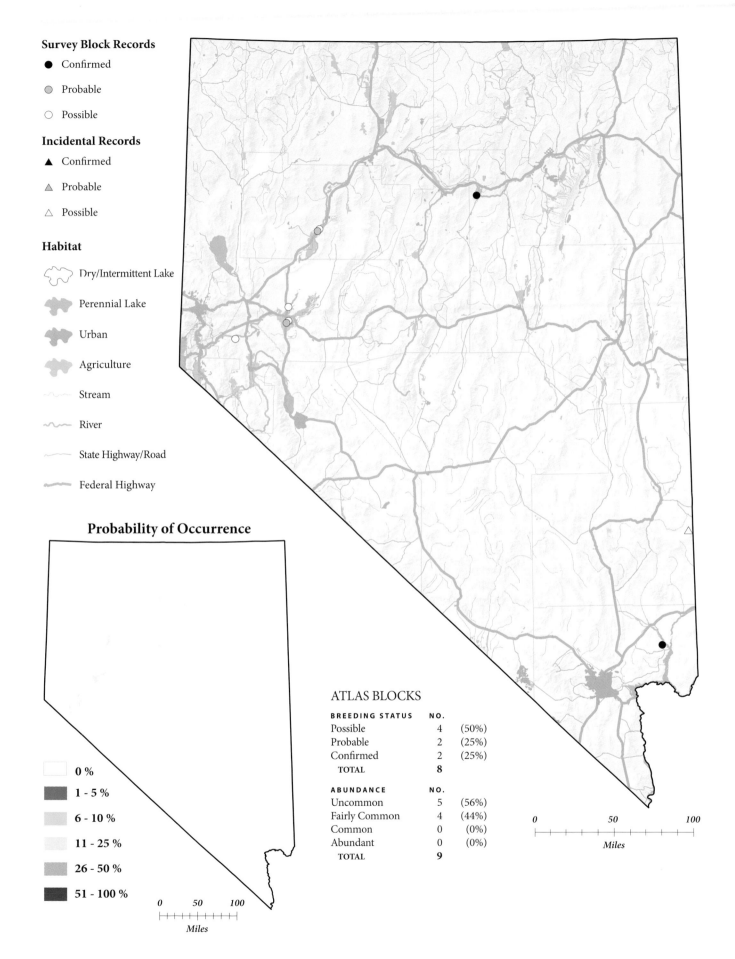

Survey Block Records

● Confirmed

◐ Probable

○ Possible

Incidental Records

▲ Confirmed

△ Probable

△ Possible

Habitat

Dry/Intermittent Lake

Perennial Lake

Urban

Agriculture

Stream

River

State Highway/Road

Federal Highway

Probability of Occurrence

☐ 0 %

■ 1 - 5 %

■ 6 - 10 %

■ 11 - 25 %

■ 26 - 50 %

■ 51 - 100 %

0   50   100
Miles

ATLAS BLOCKS

BREEDING STATUS	NO.	
Possible	4	(50%)
Probable	2	(25%)
Confirmed	2	(25%)
TOTAL	8	

ABUNDANCE	NO.	
Uncommon	5	(56%)
Fairly Common	4	(44%)
Common	0	(0%)
Abundant	0	(0%)
TOTAL	9	

0   50   100
Miles

MOUNTAIN QUAIL
Oreortyx pictus

The penetrating *kuwoooop!* of the Mountain Quail imparts an otherworldly quality to the manzanita-covered slopes of the Carson Range. Though frequently heard, this quail is hard to see. It is reluctant to flush and seems always to be underneath the most impenetrable ground cover. Even when performing their annual elevational migrations, Mountain Quail actually walk up and down the mountainsides! Decades of introductions and a dearth of historical data make it difficult to assess the extent of this quail's original distribution.

DISTRIBUTION
Atlas workers found Mountain Quail almost exclusively in or near the Carson Range. The outlying record came from a pinyon-juniper block in northwestern Nye County. Experimental releases of Mountain Quail in the nearby Desatoya Range during the atlas period (D. Delehanty, pers. comm.) may have been the source of this bird. Breeding was confirmed only in far western Nevada during atlas work, but Gutiérrez and Delehanty (1999) reported that breeding has also occurred in central and northern Nevada.

Earlier studies in Nevada indicate a more extensive distribution than the atlas workers documented. Ryser (1985) reported Mountain Quail in the Toiyabe Mountains, and Alcorn (1988) gave records north to the Santa Rosas in Humboldt County, east to Elko and Eureka counties, and south all the way to Goldfield and Lida in Esmeralda County. At least some of these outlying populations—which may have been introduced—may persist

(W. Molini, pers. comm.), but Mountain Quail are secretive and well concealed, and atlas workers could easily have missed them.

The Mountain Quail's distribution extends through the Cascade Range, Sierra Nevada, California's coastal ranges, and southern California's desert ranges, with a small population also present in western Idaho (Small 1994, Stephens and Sturts 1998, Gutiérrez and Delehanty 1999, Adamus et al. 2001). These quail occupy a variety of habitat types across a broad elevation gradient, but in all cases they prefer dense cover and shun open areas (Gutiérrez and Delehanty 1999). Nevada populations are strongly associated with mountain forests and shrublands, but farther south in the Mojave Desert of California, Mountain Quail also inhabit lower elevations. They are predicted to have a relatively high chance of occurring in the mountains of west-central Nevada, but the likelihood of actually finding them undoubtedly diminishes as one moves away from the Sierra Nevada.

CONSERVATION AND MANAGEMENT
Although Mountain Quail are almost certainly native to Nevada's small chunk of the Sierra Nevada, their historical status farther east is poorly known and was muddied by introductions well before the first systematic observations were made (e.g., Linsdale 1936). Even now the distribution of Mountain Quail in Nevada's interior ranges is uncertain, and it is further complicated by substantial year-to-year population fluctuations (Alcorn 1988) and continued introductions.

Breeding Bird Survey data do not indicate significant long-term population changes rangewide or in any of the three Pacific Coast states; insufficient data are available from Nevada for analyses (Sauer et al. 2005). Local declines and range contractions have been noted in eastern Oregon, Idaho, Nevada, and Washington (Gutiérrez and Delehanty 1999, Adamus et al. 2001). Some of these declining populations may have been introduced, but others, including those in Nevada, may be native. Historical and current threats are not well known, however, and active management is limited to setting harvest limits and conducting reintroductions in areas where populations have declined (Gutiérrez and Delehanty 1999). Because of its restricted geographical distribution and possible declines, Partners in Flight considers the Mountain Quail to be a Watch List species in the Sierra Nevada (Rich et al. 2004).

HABITAT USE

HABITAT	ATLAS BLOCKS	INCIDENTAL OBSERVATIONS
Montane Forest	3 (30%)	0 (0%)
Montane Parkland	2 (20%)	0 (0%)
Montane Shrub	1 (10%)	0 (0%)
Pinyon-Juniper	3 (30%)	2 (100%)
Urban	1 (10%)	0 (0%)
TOTAL	10	2

Total numbers reported in the Habitat Use, Breeding Status, and Abundance tables may differ from each other (see pp. 22–23 for details). Percentage sums may differ slightly from 100% due to rounding.

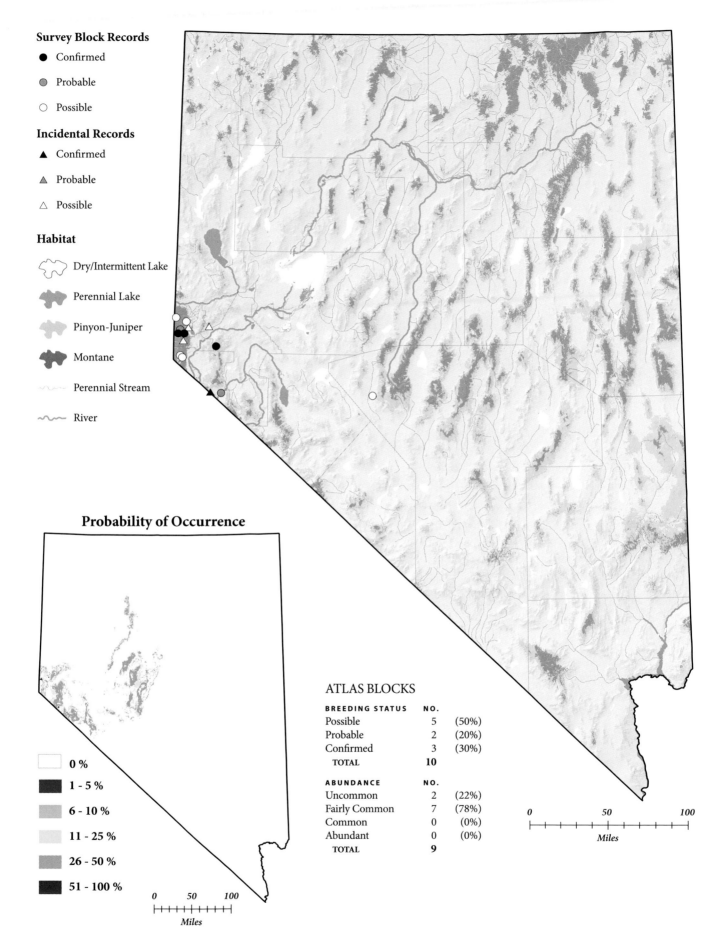

Survey Block Records
● Confirmed

◐ Probable

○ Possible

Incidental Records
▲ Confirmed

▲ Probable

△ Possible

Habitat

Dry/Intermittent Lake

Perennial Lake

Pinyon-Juniper

Montane

Perennial Stream

River

Probability of Occurrence

□	0 %
■	1 - 5 %
	6 - 10 %
	11 - 25 %
	26 - 50 %
■	51 - 100 %

0   50   100
Miles

ATLAS BLOCKS

BREEDING STATUS	NO.	
Possible	5	(50%)
Probable	2	(20%)
Confirmed	3	(30%)
TOTAL	10	

ABUNDANCE	NO.	
Uncommon	2	(22%)
Fairly Common	7	(78%)
Common	0	(0%)
Abundant	0	(0%)
TOTAL	9	

0   50   100
Miles

CALIFORNIA QUAIL
Callipepla californica

If you live in or near Reno, you are probably well acquainted with the California Quail. New residents of the city and surrounding areas are invariably enamored of this exotic-looking—though surprisingly ubiquitous—bird and startled by its loud, maniacal wailing. But California Quail are not just townies; they are habitat generalists that occur from montane riparian woodlands to sagebrush deserts. California Quail are now distributed throughout much of northern Nevada, but this was not always the case. The more eastern populations are not native, and even the birds in far western Nevada are of uncertain origin.

DISTRIBUTION

The geographical range of the California Quail is considerably more extensive today than it was during the nineteenth century, and may still be expanding. The bird's native range is thought to have included California, Baja California, southern Oregon, and far western Nevada (Calkins et al. 1999). Introductions extended this range to previously unoccupied or depopulated parts of these states, and also into Utah, Washington, Idaho, far eastern Arizona, and even as far afield as Hawaii and New Zealand (Behle et al. 1985, Stephens and Sturts 1998, Calkins et al. 1999, Adamus et al. 2001, Corman in Corman and Wise-Gervais 2005:84–85). Introductions in Nevada began as early as 1862 (Alcorn 1988), and as a result, it is quite difficult to reconstruct the quail's historic range here.

Atlas workers often found the California Quail in moderate to large numbers, with at least ten pairs estimated to occur in more than 40% of the occupied blocks. Most records came from the Truckee, Carson, and Walker river drainages of far western

Nevada. To the north and east, occurrences were much more scattered.

California Quail tolerate a broad array of habitats and withstand arid conditions well. As is the case elsewhere in their range (Calkins et al. 1999), California Quail in Nevada commonly occur in sagebrush, and they also have a significant presence in agricultural areas such as the Lahontan Valley (Chisholm and Neel 2002). Atlas habitat data also indicate that riparian habitat is heavily used and confirm that a significant portion of the overall population inhabits residential areas. The predictive map highlights the high chance of finding California Quail in urban, agricultural, and riparian habitats. It also predicts a moderate chance of occurrence in the valleys of northern Nevada, which is probably accurate in the west, but less so as one moves east. This prediction, however, may indicate the potential for further spread of this species into new parts of the state.

CONSERVATION AND MANAGEMENT

California Quail declined in their native range from the early settlement period through the nineteenth century, probably because of hunting pressure and habitat destruction (Calkins et al. 1999). In recent decades, however, reintroductions, new introductions, hunting restrictions, and the quail's increased use of human-altered habitats have reversed this trend, and California Quail are now increasing in numbers in Nevada and throughout much of their range (Sauer et al. 2005). Introductions can sometimes create problems, however. For example, California Quail may have displaced Mountain Quail in Idaho (Calkins et al. 1999). In other cases, introductions may have blurred the lines between geographically distinct subspecies (Calkins et al. 1999).

The prognosis for the California Quail in Nevada is excellent. It is popular with hunters and bird lovers alike, adapts well to habitats that have been greatly modified by humans, and can apparently be hunted quite heavily without detriment to the population. It responds well to habitat enhancement such as "guzzler" projects that provide watering holes, and many introductions have been wildly successful (Calkins et al. 1999).

HABITAT USE

HABITAT	ATLAS BLOCKS	INCIDENTAL OBSERVATIONS
Agricultural	12 (8%)	1 (3%)
Grassland	1 (<1%)	0 (0%)
Montane Forest	3 (2%)	2 (7%)
Montane Parkland	2 (1%)	1 (3%)
Montane Shrub	6 (4%)	0 (0%)
Pinyon-Juniper	8 (5%)	0 (0%)
Riparian	40 (26%)	12 (40%)
Salt Desert Scrub	4 (3%)	0 (0%)
Sagebrush Scrub	31 (20%)	5 (17%)
Sagebrush Steppe	17 (11%)	0 (0%)
Urban	26 (17%)	9 (30%)
Wetland	2 (1%)	0 (0%)
TOTAL	152	30

Total numbers reported in the Habitat Use, Breeding Status, and Abundance tables may differ from each other (see pp. 22–23 for details). Percentage sums may differ slightly from 100% due to rounding.

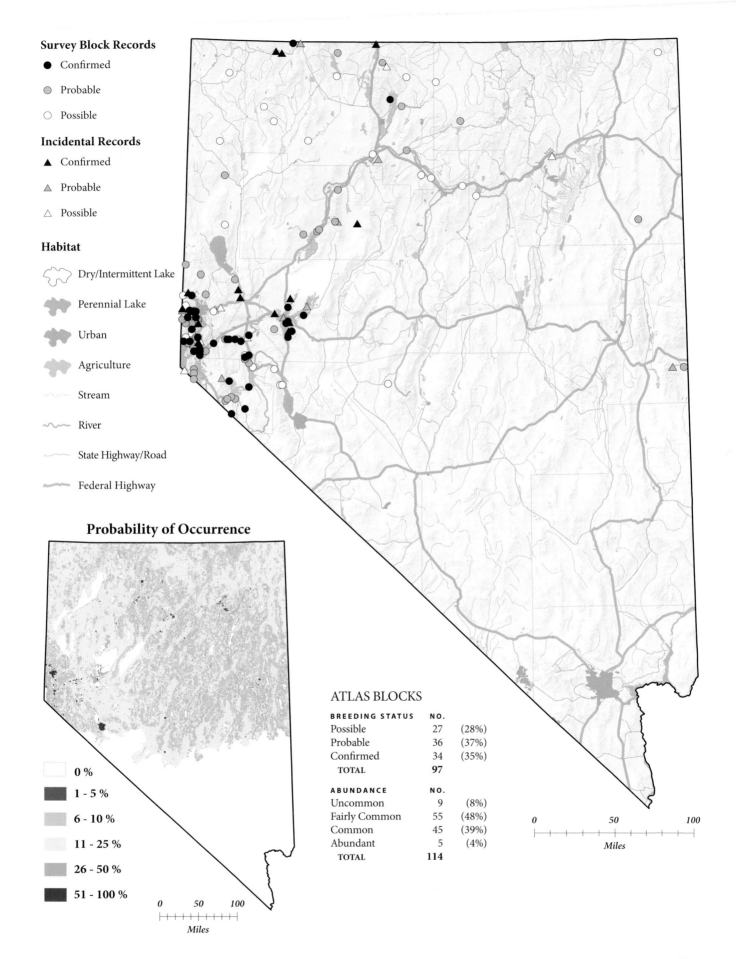

Survey Block Records

● Confirmed

● Probable

○ Possible

Incidental Records

▲ Confirmed

▲ Probable

△ Possible

Habitat

Dry/Intermittent Lake

Perennial Lake

Urban

Agriculture

Stream

River

State Highway/Road

Federal Highway

Probability of Occurrence

0 %

1 - 5 %

6 - 10 %

11 - 25 %

26 - 50 %

51 - 100 %

0   50   100
Miles

ATLAS BLOCKS

BREEDING STATUS	NO.	
Possible	27	(28%)
Probable	36	(37%)
Confirmed	34	(35%)
TOTAL	97	

ABUNDANCE	NO.	
Uncommon	9	(8%)
Fairly Common	55	(48%)
Common	45	(39%)
Abundant	5	(4%)
TOTAL	114	

0   50   100
Miles

GAMBEL'S QUAIL
Callipepla gambelii

The colorful Gambel's Quail brings an element of comic relief to the austere deserts of southern Nevada. In a typical encounter, a birder gets a glimpse of this avian spheroid as it waddles through dense ground cover, cackling and muttering to itself. The bird is striking to behold: wildly patterned in blacks and whites and lovely desert pastels, and capped off with a bizarre black topknot that sticks straight out of its forehead. But the clownish antics of this Nevada native belie the fact that it is resourceful, hardy, and well suited to life in the hot deserts of southern Nevada.

DISTRIBUTION
Atlas fieldworkers found Gambel's Quail in most of the Mojave Desert region of Nevada, usually in small to medium-sized coveys. Atlas data and the predictive map were consistent with earlier accounts of the bird's Nevada range, which characterize it as a common resident of southern deserts and absent farther north (Fisher 1893, Linsdale 1936, Ryser 1985, Alcorn 1988).

Gambel's Quail is also common in southern Utah (Behle et al. 1985) and southeastern California (Small 1994), but the heart of its range is the Sonoran Desert of southwestern Arizona and northwestern Mexico, where it is the only native gallinaceous bird (Brown et al. 1998, Wise-Gervais in Corman and Wise-Gervais 2005:86–87). These quail have been introduced with varying success into Colorado, portions of Utah, and elsewhere (Behle et al. 1985, Levad in Kingery 1998:152–153).

Gambel's Quail occupy various types of brushy lowland vegetation, but they gravitate toward riparian areas and associated habitats, including tamarisk stands, mesquite-filled washes, and Nevada's limited ash habitats (Brown et al. 1998). Riparian areas

may be attractive because they provide aboveground roost sites (Brown et al. 1998). Whether or not these quail require drinking water is a matter of dispute. Populations do congregate around water sources, including "guzzlers," but survival and productivity apparently do not suffer when water sources are withdrawn (Brown et al. 1998). Probably a more important determinant of year-to-year abundance, which can vary greatly depending on winter precipitation, is the amount of succulent green vegetation in the spring (Brown et al. 1998). Like the California Quail farther north, some Gambel's Quail populations have "adopted" suburban and agricultural habitats and are fixtures in many residential neighborhoods. The association with humans is indicated by the high predicted chance of finding them in southern towns and cities. The highest predicted probabilities at mid-elevation, however, may be overestimates due to small sample sizes in foothills habitats.

CONSERVATION AND MANAGEMENT
Gambel's Quail appears to be relatively secure throughout its range. Some conservation concern exists because the world population of the species resides within a limited geographical area. Determining whether the loss and degradation of desert riparian areas in the southwestern United States has an effect on the species, therefore, would be worthwhile. Additionally, the close link between plant production and quail abundance generates concerns about the possible effects of invasive annual grasses, though such impacts have not been stringently evaluated (Brown et al. 1998). Livestock grazing is a concern in other parts of the species' range. Because of its potential vulnerability, Gambel's Quail is a Partners in Flight Stewardship species for the southwestern region (Rich et al. 2004).

Determining recent population trends is problematic because yearly fluctuations in abundance produce "noisy" Breeding Bird Survey data and mask strong trends (Sauer et al. 2005). At present, however, Gambel's Quail remains abundant and appears to be faring well in the southern deserts.

HABITAT USE

HABITAT	ATLAS BLOCKS	INCIDENTAL OBSERVATIONS
Agricultural	4 (5%)	1 (2%)
Ash	4 (5%)	0 (0%)
Grassland	3 (4%)	0 (0%)
Mesquite	15 (19%)	4 (8%)
Mojave	15 (19%)	12 (23%)
Montane Shrub	2 (3%)	0 (0%)
Open Water	0 (0%)	1 (2%)
Pinyon-Juniper	3 (4%)	3 (6%)
Riparian	18 (23%)	17 (33%)
Salt Desert Scrub	5 (6%)	2 (4%)
Sagebrush Scrub	0 (0%)	3 (6%)
Sagebrush Steppe	1 (1%)	1 (2%)
Urban	8 (10%)	6 (12%)
Wetland	0 (0%)	2 (4%)
TOTAL	78	52

Total numbers reported in the Habitat Use, Breeding Status, and Abundance tables may differ from each other (see pp. 22–23 for details). Percentage sums may differ slightly from 100% due to rounding.

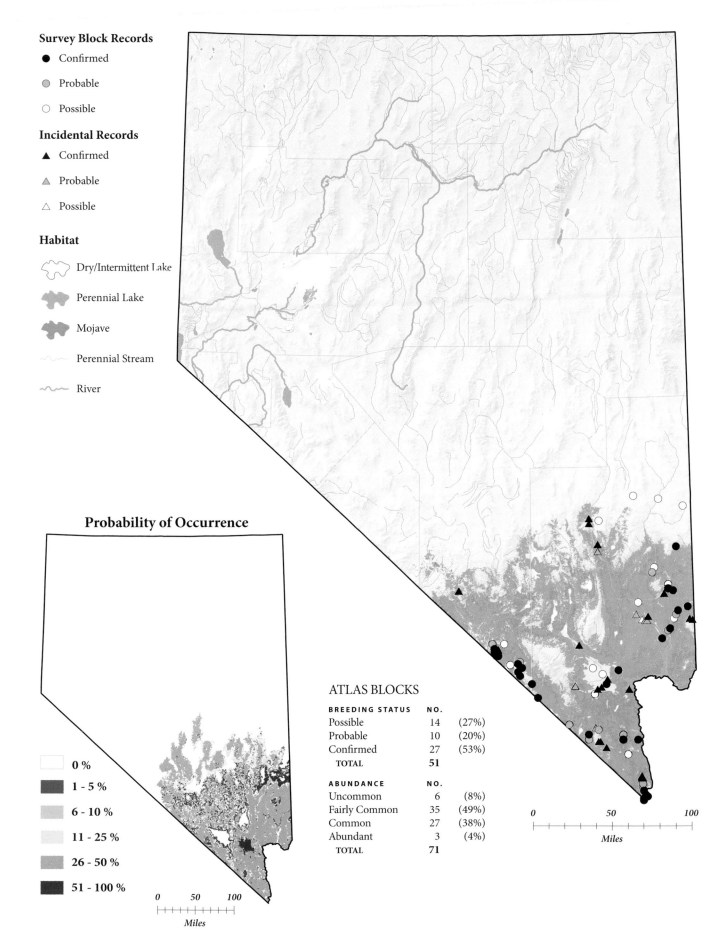

Survey Block Records

● Confirmed

◑ Probable

○ Possible

Incidental Records

▲ Confirmed

◬ Probable

△ Possible

Habitat

Dry/Intermittent Lake

Perennial Lake

Mojave

Perennial Stream

River

Probability of Occurrence

☐ 0 %

■ 1 - 5 %

■ 6 - 10 %

☐ 11 - 25 %

■ 26 - 50 %

■ 51 - 100 %

0   50   100
Miles

ATLAS BLOCKS

BREEDING STATUS	NO.	
Possible	14	(27%)
Probable	10	(20%)
Confirmed	27	(53%)
TOTAL	51	

ABUNDANCE	NO.	
Uncommon	6	(8%)
Fairly Common	35	(49%)
Common	27	(38%)
Abundant	3	(4%)
TOTAL	71	

0   50   100
Miles

PIED-BILLED GREBE
Podilymbus podiceps

The Pied-billed Grebe is a rather reclusive bird during the summer months, but it is a fairly widespread breeder throughout much of northern Nevada and also occurs in a few places farther south in the state. Grebes are naturally attracted to large wetlands associated with major river drainages, but this species can turn up just about anywhere there is open water and emergent vegetation. Farm ponds, irrigation ditches, and sewage treatment facilities are all acceptable.

DISTRIBUTION
Pied-billed Grebes breed across most of the United States and Mexico (Muller and Storer 1999). They are widely distributed in the West, though there are pockets of regional scarcity in parts of the Southwest and Pacific Northwest (Muller and Storer 1999, Wise-Gervais in Corman and Wise-Gervais 2005:92–93). In Nevada, Pied-billed Grebes were recorded primarily from the northern half of the state, with clusters in the Lahontan and Ruby valleys and scattered reports from isolated wetlands elsewhere in the north.

Reports from southern Nevada came primarily from sites along the Colorado River and its tributaries. Pied-billed Grebes were found in thirty blocks, which seems rather low for a species that both Ryser (1985) and Alcorn (1988) considered common in Nevada. These grebes can be difficult to detect (Muller and Storer 1999), however, and in some blocks they may have gone unnoticed in the dense cattail marshes they call home. The bird's solitary lifestyle and understated courtship displays probably also contributed to the lack of records and appearance of a sparse distribution in the state. Observers who encountered Pied-billed Grebes generally estimated that there were ten or fewer pairs present, but there were four blocks where the birds were more numerous.

Open water, wetlands, and riparian habitats accounted for nearly all of the records. The single record from Mojave habitat was an anomaly, but it is worth noting that Pied-billed Grebes have fairly flexible habitat requirements and will use small wetlands embedded within other habitats. Gallagher (1997) and Muller and Storer (1999) reported breeding on farm ponds and in urban areas. The species is predicted to occur over a wide area because of its habitat flexibility, albeit with a low chance of being encountered in most places.

CONSERVATION AND MANAGEMENT
Pied-billed Grebe populations have increased in the United States since the 1980s, but this pattern is influenced greatly by increases in the central portion of the continent (Sauer et al. 2005). In the West as a whole, Breeding Bird Survey data suggest a decline over this same time span, although there seems to be a lot of local variation (Sauer et al. 2005). Reported trends are questionable, however, because the Breeding Bird Survey is not designed to detect secretive marsh birds such as Pied-billed Grebes (Muller and Storer 1999).

Recent increases in Colorado have been attributed to the proliferation of farm ponds and irrigation reservoirs (Potter in Kingery 1998:40–41). Increases in Marin County, California, have likewise been connected with habitat alteration by humans (Shuford 1993). On the other hand, reported declines in the East have been tentatively linked to eutrophication of wetlands caused by nutrient inputs from human sources (Muller and Storer 1999). The overall picture, however, is that of an adaptable species that is likely to respond well to wetland restoration. Forbes et al. (1989) outlined the specific requirements for successful nesting. In Nevada, a two-pronged approach to management seems appropriate. First, it is advisable to protect existing marshes from degradation and development. Second, wetland restoration and mitigation projects, even in Nevada's urban centers, could benefit this species.

HABITAT USE

HABITAT	ATLAS BLOCKS	INCIDENTAL OBSERVATIONS
Mojave	0 (0%)	1 (4%)
Open Water	15 (39%)	12 (52%)
Riparian	4 (11%)	3 (13%)
Wetland	19 (50%)	7 (30%)
TOTAL	38	23

Total numbers reported in the Habitat Use, Breeding Status, and Abundance tables may differ from each other (see pp. 22–23 for details). Percentage sums may differ slightly from 100% due to rounding.

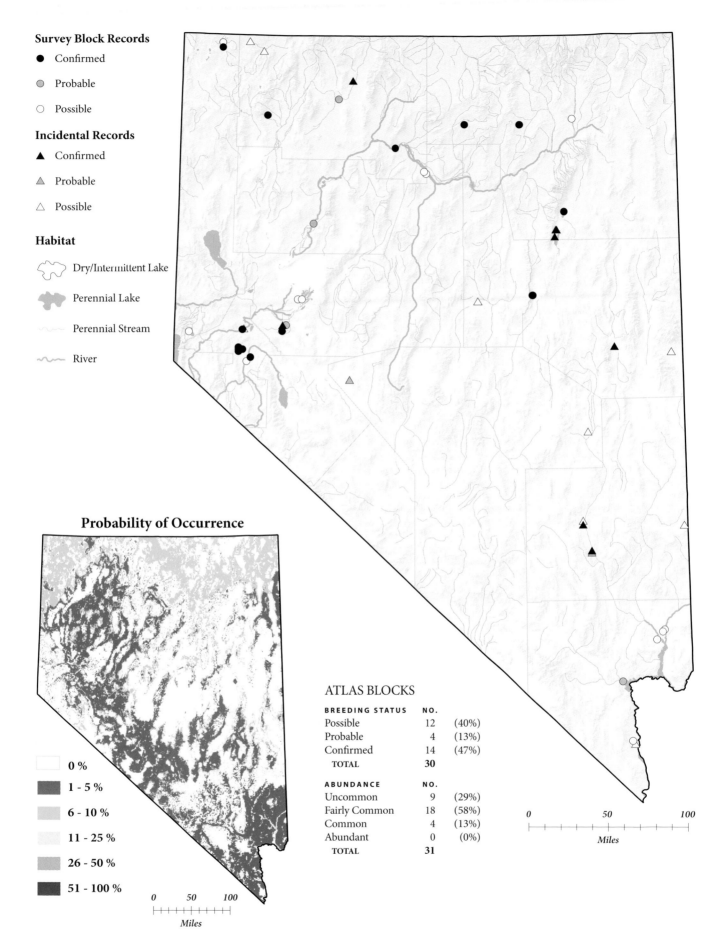

Survey Block Records

- ● Confirmed
- ● Probable
- ○ Possible

Incidental Records

- ▲ Confirmed
- ▲ Probable
- △ Possible

Habitat

- Dry/Intermittent Lake
- Perennial Lake
- Perennial Stream
- River

Probability of Occurrence

- 0 %
- 1 - 5 %
- 6 - 10 %
- 11 - 25 %
- 26 - 50 %
- 51 - 100 %

0   50   100
Miles

ATLAS BLOCKS

BREEDING STATUS	NO.	
Possible	12	(40%)
Probable	4	(13%)
Confirmed	14	(47%)
TOTAL	30	

ABUNDANCE	NO.	
Uncommon	9	(29%)
Fairly Common	18	(58%)
Common	4	(13%)
Abundant	0	(0%)
TOTAL	31	

0   50   100
Miles

EARED GREBE
Podiceps nigricollis

Unlike the reclusive and solitary Pied-billed Grebe, Eared Grebes are colonial breeders that are typically quite conspicuous on their breeding grounds. They build floating nests in shallow water, sometimes right out in the open, and their courtship displays are elaborate and visually striking. Eared Grebes can be very numerous in the bigger marshes of northern Nevada, but they occur in small wetlands as well. Although the number of Eared Grebes breeding in Nevada is a small fraction of the continent's population, the Great Basin is extremely important for the species during migration, especially in fall, when the bulk of the American population uses the region's lakes as a staging area.

DISTRIBUTION

Eared Grebes were found in twenty-three blocks, and block data combined with incidental observations confirmed breeding at seventeen locations. Records were most common in the larger wetlands of northern Nevada, although there were scattered records across the northern counties. Farther south, breeding was confirmed at several sites in Clark and Lincoln counties, primarily along the White and Colorado rivers. The species is predicted to range throughout the lower-elevation portions of the state, but the probability of finding it in any particular location is low, simply because the chance of finding suitable wetland habitat in Nevada is low.

In adjacent states, Eared Grebes are common breeders, especially in the larger marshes of the Great Basin. Important nesting concentrations are found at Malheur National Wildlife Refuge in Oregon and in the marshes around the Great Salt Lake in Utah (Ryser 1985). In southeastern Oregon, there is also a pattern of scattered occurrence (Adamus et al. 2001), similar to that in

Nevada, further reflecting the sparse distribution of wetlands throughout the Great Basin.

Eared Grebes are usually colonial breeders and can occur in high densities; however, they also occur in loose, semicolonial assemblages, or even as single pairs on small bodies of water (Cullen et al. 1999). The atlas did not uncover any especially large colonies, and observers estimated the presence of more than ten pairs in only about one-fifth of the blocks in which the species was found. Other authors reported estimates of several hundred nests at both Carson Lake and at Stillwater National Wildlife Refuge in the mid-1990s (Chisholm and Neel 2002).

CONSERVATION AND MANAGEMENT

Data from the Breeding Bird Survey suggest a sustained increase in Eared Grebe numbers since the 1960s (Sauer et al. 2005). However, this survey is not designed to count colonial wetland birds, and its results should be viewed with caution (Cullen et al. 1999). The most serious threats to the species probably occur outside the breeding season, especially during migration.

Migrating Eared Grebes are susceptible to spectacular groundings, or "wrecks," during stormy weather (Ryser 1985, Jehl 1993, 1996). Unexplained die-offs in the Great Basin date back over a century (Fisher 1893), suggesting that large-scale mortality among Eared Grebes migrating through Nevada cannot be attributed to human causes alone.

In the fall, more than one million Eared Grebes gather to molt and fatten up on brine shrimp at Mono Lake, just to the west of Nevada in California; even larger numbers can be found at the Great Salt Lake in Utah (Boyd and Jehl 1998). Together, these two sites harbor most of the continent's Eared Grebes on migration (Cullen et al. 1999). Although the former threat of increasing water salinity at Mono Lake due to water diversions appears to have been alleviated (Hart 1996), concerns always remain when a population is highly concentrated within a small area. Thus, the Intermountain West Waterbird Conservation Plan considers the Eared Grebe a species of High Concern in the Intermountain West, and the North American Waterbird Conservation Plan considers it a species of Moderate Concern at the continental level (Kushlan et al. 2002, Ivey and Herziger 2005).

HABITAT USE

HABITAT	ATLAS BLOCKS	INCIDENTAL OBSERVATIONS
Open Water	21 (60%)	15 (58%)
Riparian	1 (3%)	0 (0%)
Wetland	13 (37%)	11 (42%)
TOTAL	35	26

Total numbers reported in the Habitat Use, Breeding Status, and Abundance tables may differ from each other (see pp. 22–23 for details). Percentage sums may differ slightly from 100% due to rounding.

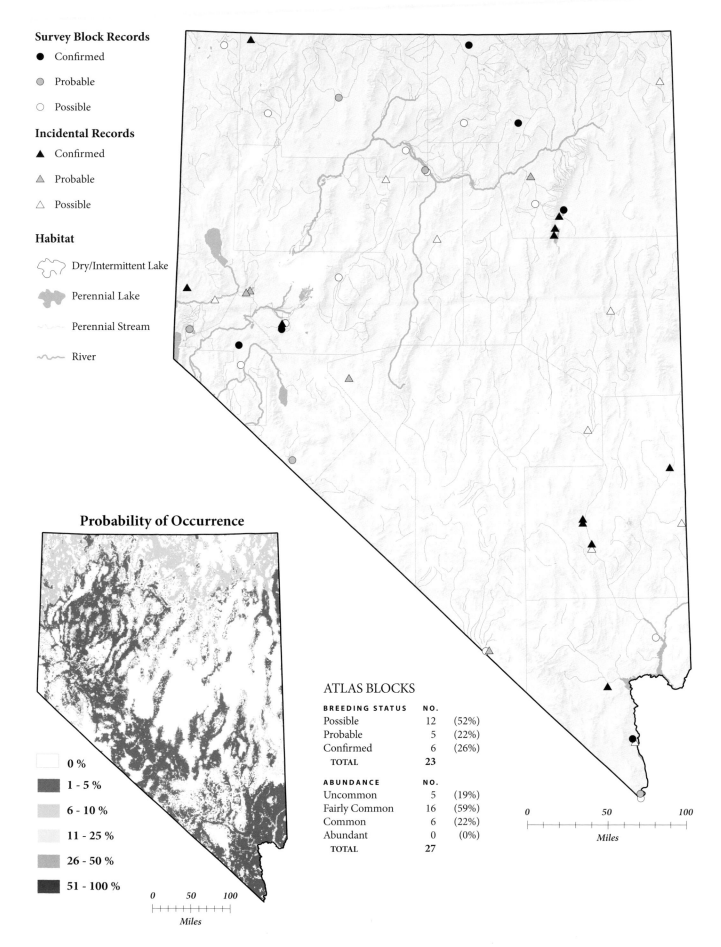

Survey Block Records

● Confirmed

● Probable

○ Possible

Incidental Records

▲ Confirmed

▲ Probable

△ Possible

Habitat

Dry/Intermittent Lake

Perennial Lake

Perennial Stream

River

Probability of Occurrence

	0 %
	1 - 5 %
	6 - 10 %
	11 - 25 %
	26 - 50 %
	51 - 100 %

0   50   100
Miles

ATLAS BLOCKS

BREEDING STATUS	NO.	
Possible	12	(52%)
Probable	5	(22%)
Confirmed	6	(26%)
TOTAL	23	

ABUNDANCE	NO.	
Uncommon	5	(19%)
Fairly Common	16	(59%)
Common	6	(22%)
Abundant	0	(0%)
TOTAL	27	

0   50   100
Miles

WESTERN GREBE
Aechmophorus occidentalis

Few sights are more spectacular than the courtship displays of the Western Grebe. Mated pairs engage in a complex ritual of behaviors with names as bizarre—"ratchet-pointing," "dip-shaking," "barge-trilling"—as the postures to which they refer. The elaborate "weed ceremony," in which the birds approach each other in the water and dance, each with a bill-full of aquatic plants, is especially enchanting. Although Western Grebes are conspicuous inhabitants of lakes and marshes throughout western North America, much remains unknown about their distribution. In part this is because most studies have not distinguished between this species and the similar Clark's Grebe, which until recently was considered a color morph of the Western Grebe (AOU 1985).

DISTRIBUTION

The Western Grebe is widespread but very sparsely distributed as a Nevada breeder, with records from only eleven blocks and breeding confirmed in ten locations. Breeding sites were clustered in the wetlands of the west-central portion of the state and in Elko County, with scattered records elsewhere. In general, Western Grebes select breeding sites with extensive areas of open water for feeding and with tracts of emergent marsh plants to which they attach their floating nests (Storer and Nuechterlein 1992).

Ryser (1985) described colonies of *Aechmophorus* grebes in the Great Basin that often numbered in the hundreds or even thousands of pairs. Such high densities appear to be rare in

Nevada; most abundance estimates indicated far fewer birds. Alcorn (1988) viewed Western Grebes as most numerous in western and southern Nevada. The atlas data are not entirely consistent with this assessment, and our predictive map suggests that although the chance of finding the species is low throughout Nevada, it is higher in the northern two-thirds of the state than in the south. The area over which the model predicts a small chance of finding this species is determined by the distribution of terrestrial habitats that occasionally contain small wetlands.

CONSERVATION AND MANAGEMENT

Breeding Bird Survey data indicate a recent rangewide increase in *Aechmophorus* grebe abundance (Sauer et al. 2005). These roadside surveys are of uncertain value for monitoring colonial waterbirds such as the Western Grebe, however, because they do not always sample wetlands well. Additionally, most of the past surveys combined numbers for Western Grebes and Clark's Grebes. In several states, though, the general pattern does seem to be one of recent range expansion (Janssen 1987, Dinsmore in Jackson et al. 1996:44–45, Righter in Kingery 1998:44–45).

Still, the Western Grebe has been beset by recent and historic threats. The Intermountain West Waterbird Conservation Plan considers it a species of High Concern in the Intermountain West, and the North American Waterbird Conservation Plan has designated it a species of Moderate Concern continentally (Kushlan et al. 2002, Ivey and Herziger 2005). A century ago, these birds were hunted for their feathers, which were used to adorn women's clothing. Since then, numbers have recovered, and Ryser (1985) credited federal protection in the early twentieth century with benefiting Great Basin populations. Mid-century use of pesticides may have harmed California populations, and there has been recent concern throughout the species' range about disturbances created by boats and other recreational activities (Storer and Nuechterlein 1992).

The major present-day threat to Western Grebes in Nevada, however, is probably the loss of suitable nonbreeding habitat. Increasing salinity in Walker Lake due to water diversions, for instance, is likely to diminish fish populations, making the site increasingly unsuitable for the thousands of grebes that stage there (McIvor 2003). Maintaining healthy fish populations in Pyramid Lake, where several thousand postbreeding Western Grebes gather annually (Mack 2000, 2001), is also important to the future management of the species.

HABITAT USE

HABITAT	ATLAS BLOCKS	INCIDENTAL OBSERVATIONS
Open Water	11 (61%)	8 (57%)
Wetland	7 (39%)	6 (43%)
TOTAL	18	14

Total numbers reported in the Habitat Use, Breeding Status, and Abundance tables may differ from each other (see pp. 22–23 for details). Percentage sums may differ slightly from 100% due to rounding.

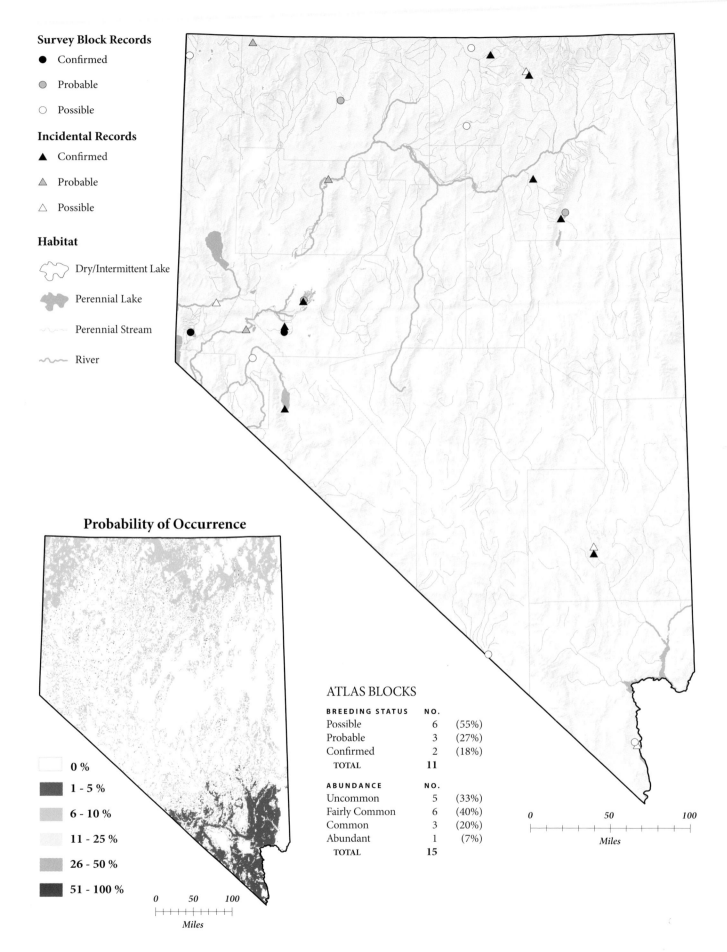

Survey Block Records

● Confirmed

● Probable

○ Possible

Incidental Records

▲ Confirmed

▲ Probable

△ Possible

Habitat

Dry/Intermittent Lake

Perennial Lake

Perennial Stream

River

Probability of Occurrence

☐	0 %
■	1 - 5 %
■	6 - 10 %
☐	11 - 25 %
■	26 - 50 %
■	51 - 100 %

0   50   100
Miles

ATLAS BLOCKS

BREEDING STATUS	NO.	
Possible	6	(55%)
Probable	3	(27%)
Confirmed	2	(18%)
TOTAL	11	

ABUNDANCE	NO.	
Uncommon	5	(33%)
Fairly Common	6	(40%)
Common	3	(20%)
Abundant	1	(7%)
TOTAL	15	

0   50   100
Miles

CLARK'S GREBE
Aechmophorus clarkii

The stately Clark's Grebe and its look-alike close cousin, the Western Grebe, together constitute one of the more vexing identification challenges for Nevada birders. The two species differ consistently in bill color and call note, and there are general, but variable, differences in facial pattern and overall coloration. Otherwise, these birds are nearly identical in most respects. Although the characteristics that serve to distinguish them by sight and sound have received a fair amount of study, their respective distributions and ecology are still not well understood because the older literature usually does not distinguish between the two species.

DISTRIBUTION

Breeding Clark's Grebes were confirmed in five blocks and at nine other locations, and the species was reported from more than twenty distinct sites. Generally, the distribution of Clark's Grebes was similar to that of Western Grebes, with clusters of records in west-central Nevada, Elko County, and the river drainages of the southeast. The predictive maps of the two species show some differences, however, with Clark's Grebes perhaps more likely than Western Grebes to occur in parts of the south, and somewhat less likely to be in the center of the state. As with Western Grebes, however, the chance of finding Clark's Grebes anywhere is low and depends on the presence of appropriate wetland habitat, which often occurs at a resolution too fine for our model to capture with precision.

Earlier studies point to Nevada, and more generally to the Great Basin, as a hotspot for Clark's Grebes, and concentrations of hundreds have been reported (Neel 1999a). During atlas fieldwork, however, the species was ranked as abundant only at Carson Lake in Churchill County. Just outside Nevada, Goose Lake (Modoc County, California) supports a Clark's Grebe breeding colony that may be the largest in the world (Small 1994).

The relative abundances of Clark's and Western Grebes are difficult to gauge. Breeding Clark's Grebes substantially outnumber Western Grebes in western Nevada and in the Lahontan Valley (Neel 1999a, Chisholm and Neel 2002). In contrast, annual fall surveys of Pyramid Lake found Western Grebes to greatly outnumber Clark's Grebes (e.g., Mack 2000), presumably due to a large influx of postbreeding Western Grebes from the north. Looking beyond Nevada, the general pattern seems to be that breeding Western Grebes outnumber Clark's Grebes in the Great Basin parts of California, Oregon, and Utah (Ratti 1981).

In broader geographical terms, the distributions of the two species overlap extensively, although Clark's Grebe has a smaller, more southerly breeding range than the Western Grebe (Storer and Nuechterlein 1992). Patterns of local distribution for the two species are not well known, however, and it is unclear whether they occupy different habitats or compete with each other for resources. Righter (in Kingery 1998:46–47) suggested that the two species occupy different microhabitats, but they will also nest in close proximity to each other (Neel 1999a).

CONSERVATION AND MANAGEMENT

The fact that many earlier studies did not distinguish Clark's Grebes from Western Grebes presents a challenge when it comes to designing conservation strategies. Although the taxonomic split with the Western Grebe was proposed early by Storer (1965) and received additional support from Nuechterlein's (1981) study on reproductive isolating mechanisms, it was not formally recognized until 1985.

Clark's Grebe is a Nevada Partners in Flight Priority species because its habitat and management needs are poorly understood and because the Great Basin is an important stronghold for the species (Rich et al. 2004). For now, it seems prudent to manage the species in the manner laid out generally for *Aechmophorus* grebes, which is summarized in the Western Grebe species account (pp. 102–3) and in Neel 1999a.

HABITAT USE

HABITAT	ATLAS BLOCKS	INCIDENTAL OBSERVATIONS
Open Water	9 (45%)	7 (47%)
Riparian	3 (15%)	0 (0%)
Urban	0 (0%)	1 (7%)
Wetland	8 (40%)	7 (47%)
TOTAL	20	15

Total numbers reported in the Habitat Use, Breeding Status, and Abundance tables may differ from each other (see pp. 22–23 for details). Percentage sums may differ slightly from 100% due to rounding.

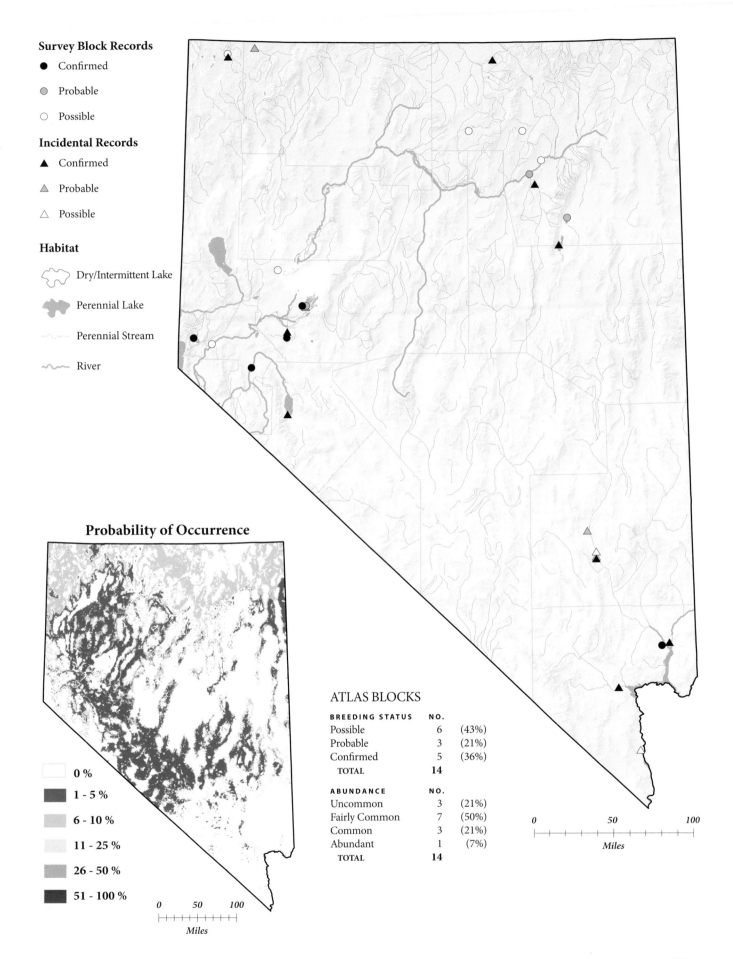

Survey Block Records

● Confirmed

● Probable

○ Possible

Incidental Records

▲ Confirmed

▲ Probable

△ Possible

Habitat

Dry/Intermittent Lake

Perennial Lake

Perennial Stream

River

Probability of Occurrence

☐ 0 %

1 - 5 %

6 - 10 %

11 - 25 %

26 - 50 %

51 - 100 %

0   50   100
Miles

ATLAS BLOCKS

BREEDING STATUS	NO.	
Possible	6	(43%)
Probable	3	(21%)
Confirmed	5	(36%)
TOTAL	14	

ABUNDANCE	NO.	
Uncommon	3	(21%)
Fairly Common	7	(50%)
Common	3	(21%)
Abundant	1	(7%)
TOTAL	14	

0   50   100
Miles

AMERICAN WHITE PELICAN
Pelecanus erythrorhynchos

Anaho Island, in the middle of Pyramid Lake, is home to Nevada's only permanent colony of American White Pelicans. It is one of the largest breeding colonies in the United States, and visitors typically return with comic tales about thousands of hulking, clumsy pelicans shuffling around their nest sites and squabbling amongst themselves. The sight of a pelican flock gracefully soaring high above the desert—twisting synchronously on invisible air currents and shimmering like crystals in the sky—evokes a more reverent response.

DISTRIBUTION

The American White Pelican is a familiar sight throughout Nevada, but breeding is largely confined to Anaho Island, which averaged about 4,500 pairs during the 1980s and 1990s, with a peak of 6,500 pairs (Neel 1999a). Nesting was also confirmed at Franklin Lake in Elko County in 1999, and small numbers of pelicans sometimes breed in the Carson Sink in Humboldt County when the water is deep enough to create islands protected from predators (Neel 1999a, Chisholm and Neel 2002).

No colonies fell within atlas blocks, but pelicans were seen in twenty-nine blocks and at five other locations. All observations save those at the breeding sites mentioned previously involved nonbreeding birds or birds away from their colony; many were seen flying overhead. For example, a pair reported near Denio in Humboldt County was presumed to be two unmated foragers away from a colony. Given the species' highly specialized breeding habitat, it is unlikely that important breeding went unnoticed during the atlas project.

American White Pelicans forage far from their colonies (Evans and Knopf 1993). Birds fitted with satellite transmitters in western Nevada, for instance, made repeated round trips to sites as far away as California's Central Valley (Yates 1999). Consequently, sightings away from colonies can help to identify critical foraging habitat. Most records of foraging individuals came from the Lahontan Valley wetlands. Daily counts of up to one thousand pelicans during the nesting season, and up to eight thousand in late summer, demonstrate the value of these wetlands to the species (Chisholm and Neel 2002). In addition, a string of sightings along the Humboldt River suggests that this waterway is an important travel corridor.

Because there was no evidence of breeding in any of the atlas blocks, it was impossible to generate a predictive map for breeding pelicans. The occurrence map for this species includes all sightings (not just those involving breeding evidence) in order to show that the area used by foraging pelicans is much larger than might be assumed based on the species' highly localized breeding.

CONSERVATION AND MANAGEMENT

Conservation of the American White Pelican revolves around its status as a highly colonial breeder. Only four or five pelican colonies are extant in the Great Basin (Ryser 1985, Neel 1999a), and there are just a few more in all of the western United States (Sidle et al. 1985). Pelicans readily desert their colonies in the face of predation and human disturbance (Evans and Knopf 1993), and numbers can vary considerably from year to year (Neel 1999a). For these reasons the American White Pelican is designated a species of Moderate Concern in the North American Waterbird Conservation Plan (Kushlan et al. 2002), and it is a species of High Concern in the Intermountain West Waterbird Conservation Plan (Ivey and Herziger 2005).

In addition to protecting colonies, it is important to be vigilant about pelican foraging areas. Feeding pelicans need shallow water filled with fish (Neel 1999a), and it seems clear that the Lahontan Valley wetlands such as Carson Lake and Stillwater National Wildlife Refuge are of great importance to the birds breeding at Pyramid Lake. If nesting and feeding areas are managed together, we can be cautiously optimistic about the American White Pelican's future. The species' range and numbers have expanded in recent decades (Evans and Knopf 1993), and colonies have even become established in artificial habitats such as irrigation reservoirs in Colorado (Potter in Kingery 1998:48–49).

HABITAT USE

HABITAT	ATLAS BLOCKS	INCIDENTAL OBSERVATIONS
Barren	0 (0%)	1 (33%)
Wetland	0 (0%)	2 (67%)
TOTAL	0	3

Total numbers reported in the Habitat Use, Breeding Status, and Abundance tables may differ from each other (see pp. 22–23 for details). Percentage sums may differ slightly from 100% due to rounding.

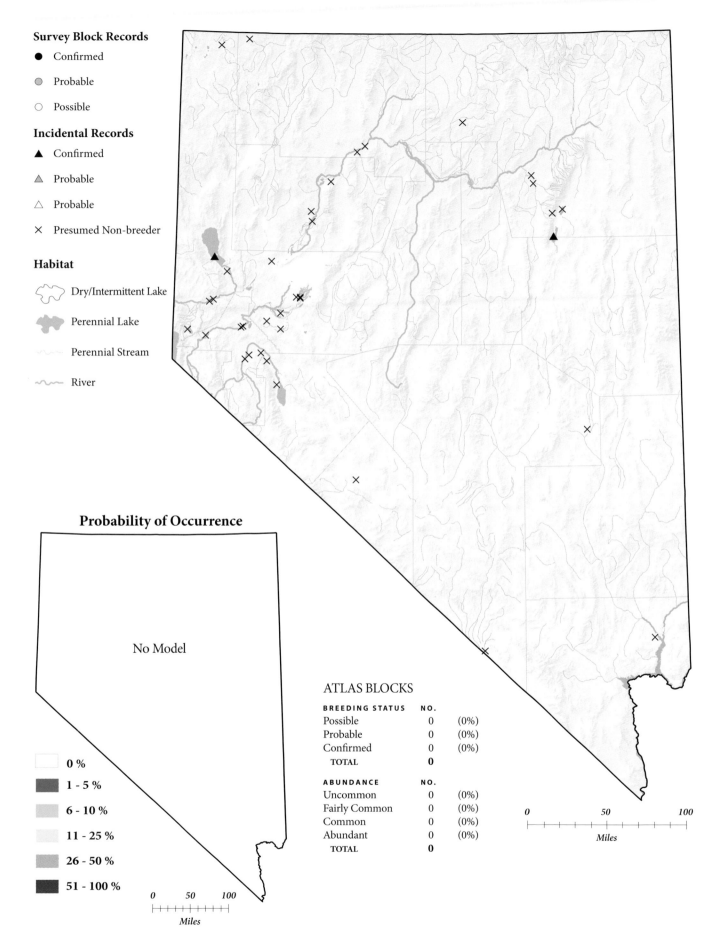

Survey Block Records

● Confirmed

◐ Probable

○ Possible

Incidental Records

▲ Confirmed

▲ Probable

△ Probable

✕ Presumed Non-breeder

Habitat

⬭ Dry/Intermittent Lake

▨ Perennial Lake

〜 Perennial Stream

〜 River

Probability of Occurrence

No Model

☐ 0 %

▨ 1 - 5 %

▨ 6 - 10 %

▨ 11 - 25 %

▨ 26 - 50 %

▨ 51 - 100 %

0   50   100
Miles

ATLAS BLOCKS

BREEDING STATUS	NO.	
Possible	0	(0%)
Probable	0	(0%)
Confirmed	0	(0%)
TOTAL	**0**	

ABUNDANCE	NO.	
Uncommon	0	(0%)
Fairly Common	0	(0%)
Common	0	(0%)
Abundant	0	(0%)
TOTAL	**0**	

0   50   100
Miles

DOUBLE-CRESTED CORMORANT
Phalacrocorax auritus

The Double-crested Cormorant is an increasingly common sight in Nevada. This jet black fish eater can now be found on most of our major lakes and rivers, but many sightings involve non-breeding summer visitors or breeders foraging far from their nesting colonies. Cormorants prosper wherever there is clean water filled with fish. Their piscivorous habits have created conflicts with the commercial fish-farming industry, and cormorant persecution is on the upswing in many areas, often without good evidence that the perceived conflict is real.

DISTRIBUTION

During much of the twentieth century the Double-crested Cormorant seemed to be quite scarce in Nevada. Linsdale (1936) cited Anaho Island in Pyramid Lake as the only breeding colony in the state, and Ryser's (1985) survey of the Great Basin provided breeding records only from the western edge of Nevada. Alcorn (1988) gave brief details on a few additional breeding colonies in southern Nevada.

Double-crested Cormorants were widely noted during the atlas survey period, however, which seems to indicate a recent and perhaps ongoing expansion in the state. They were observed in twenty-four blocks, although most of the observations involved presumed nonbreeders. The species was confirmed as a breeder in only one block—near Overton Wildlife Management Area at the north end of Lake Mead. There were six additional reports of confirmed breeding from widely scattered locations across the state, including the Ruby Valley in Elko County, the Pahranagat Valley in Lincoln County, and Pyramid Lake. The Double-crested Cormorant was also confirmed as a breeder in an urban setting, with a well-established colony at Virginia Lake in Reno and suspected breeding in many areas just downstream

from Reno along the Truckee River drainage. As the predictive map suggests, this is a species that could turn up in many parts of the state, but there is a low chance of it occurring in any particular spot.

Assessing the extent to which summer sightings refer to non-breeding birds is difficult, and is further complicated by the Double-crested Cormorant's rather fractured breeding distribution in the states surrounding Nevada (see Hatch and Weseloh 1999). Nonbreeders occur throughout the species' range, but some breeders also forage many miles from their colonies (Hatch and Weseloh 1999). And while large gatherings at breeding colonies are usually easy to detect, an occasional single pair is much less obvious. For example, the only confirmed Double-crested Cormorant record during the Marin County atlas project involved a single nest in a Great Blue Heron colony (Shuford 1993).

CONSERVATION AND MANAGEMENT

The mechanisms of population regulation in cormorants are poorly known (Hatch and Weseloh 1999) but seem to involve a complex array of natural and anthropogenic factors. Heavy population losses in Washington were precipitated by a recent El Niño event (Wilson 1991), and some losses in Nevada have been attributed to nest predation by California Gulls (Ryser 1985). Local population declines in California (Small 1994) and Utah (Mitchell 1975) have been attributed to various human-related causes.

Despite some local losses, the overall trend for the species in recent years has been rapid population increases rangewide (Hatch 1995). The general increase is due in part to the creation of new habitats, such as irrigation reservoirs in Colorado (Andrews and Righter 1992), but much of it probably involves recolonization of areas from which cormorants were extirpated by persecution in the nineteenth and early twentieth centuries, and by pesticides in the 1940s and 1950s (Hatch 1995). The recent good fortunes of the Double-crested Cormorant have not come without a cost; the species is increasingly seen as a threat by the commercial fishing industry in some areas. In Nevada, however, this is not an issue, and the cormorant's prognosis seems good.

HABITAT USE

HABITAT	ATLAS BLOCKS	INCIDENTAL OBSERVATIONS
Barren	0 (0%)	1 (6%)
Open Water	5 (50%)	6 (35%)
Riparian	2 (20%)	1 (6%)
Urban	0 (0%)	1 (6%)
Wetland	3 (30%)	8 (47%)
TOTAL	10	17

Total numbers reported in the Habitat Use, Breeding Status, and Abundance tables may differ from each other (see pp. 22–23 for details). Percentage sums may differ slightly from 100% due to rounding.

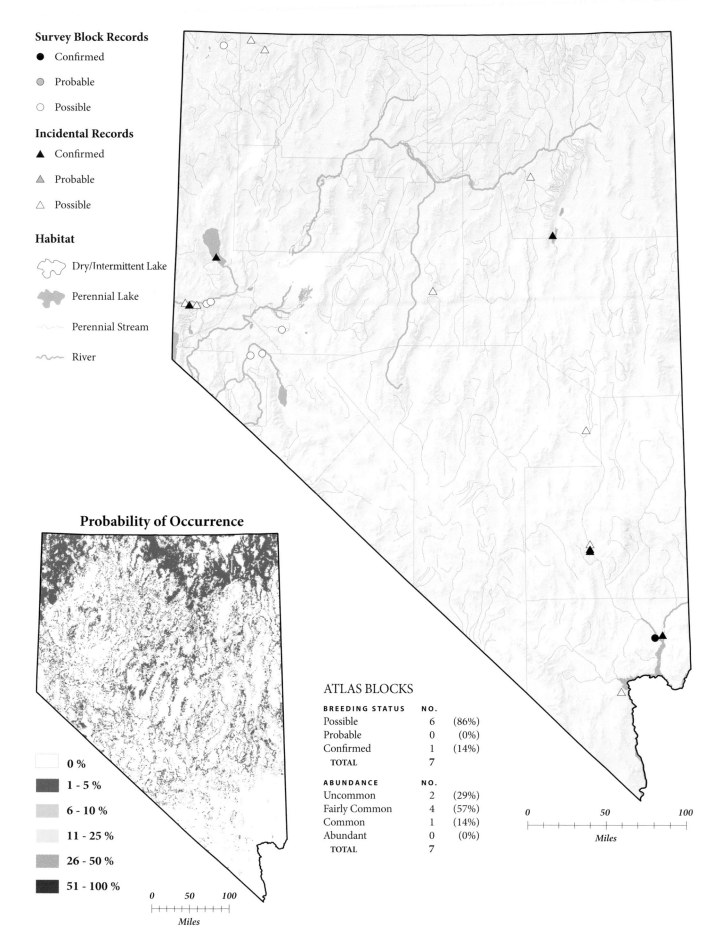

Survey Block Records

● Confirmed

◉ Probable

○ Possible

Incidental Records

▲ Confirmed

▲ Probable

△ Possible

Habitat

Dry/Intermittent Lake

Perennial Lake

Perennial Stream

River

Probability of Occurrence

☐ 0 %

■ 1 - 5 %

■ 6 - 10 %

☐ 11 - 25 %

■ 26 - 50 %

■ 51 - 100 %

0   50   100

Miles

ATLAS BLOCKS

BREEDING STATUS	NO.	
Possible	6	(86%)
Probable	0	(0%)
Confirmed	1	(14%)
TOTAL	7	

ABUNDANCE	NO.	
Uncommon	2	(29%)
Fairly Common	4	(57%)
Common	1	(14%)
Abundant	0	(0%)
TOTAL	7	

0   50   100

Miles

AMERICAN BITTERN
Botaurus lentiginosus

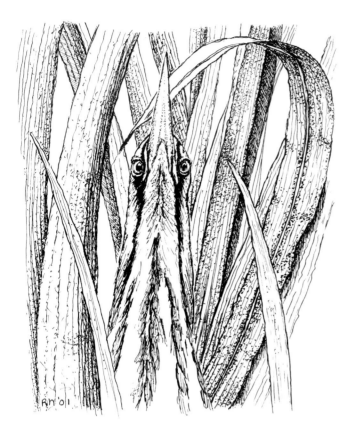

A master of camouflage and deception, the American Bittern is a widespread but very sparsely distributed species in Nevada. Everything about its lifestyle makes it hard to find. Unlike most herons, the American Bittern is a solitary nester loath to stand out in the open. It is crepuscular and tends to be most common in dense cattail marshes where birders seldom go. When alarmed or threatened, American Bitterns "freeze" in an upright pose, their long lines blending in perfectly against the backdrop of marsh vegetation. The bizarre call of the American Bittern is easy enough to hear and is instantly recognizable, but it has a frustratingly ventriloquial quality that has thwarted many a birder intent on actually finding the elusive "thunder pumper."

DISTRIBUTION

American Bitterns have long been considered to be widespread in the state. Linsdale (1936:32) wrote that they were resident "wherever marshes occur," and Ryser (1985) classified them as common in marshes throughout the Great Basin. During the atlas project, American Bitterns were found in ten widely scattered blocks and at nine other locations. Most records came from the large marshes and major river systems in the northern parts of the state, but a few reports of possible breeding came from farther south. Breeding was confirmed only in the Ruby Valley, but given the difficulty of finding the nests or young of this species,

it seems likely that dedicated marshland surveys would show that breeding also occurs elsewhere.

Most atlas sightings came from wetland habitats, as expected. A few records were assigned to riparian habitats, perhaps reflecting a misuse of the term "riparian" as a catch-all for all things aquatic in this generally arid environment. Only a single pair was thought to be present in six of the eight blocks for which abundance estimates were provided, and no blocks were estimated to contain more than ten pairs. The densely vegetated marshlands that bitterns inhabit are notoriously difficult to census, however, and undercounting seems likely, as previous atlas projects have also suggested (e.g., Smith et al. 1997). High densities are sometimes detected; Alcorn (1988) cited a report of up to two hundred nesting pairs in the Ruby Valley marshes of northeastern Nevada.

Overall, the atlas results suggest that the American Bittern is less ubiquitous than earlier accounts indicate, but the lack of quantitative data and the bittern's proclivity for hiding in dense vegetation make comparisons very difficult. Moreover, the predictive map demonstrates the potential, albeit small and dependent on the presence of requisite marsh habitat, for finding this species in many areas of the state.

CONSERVATION AND MANAGEMENT

Basic natural history information on the American Bittern is scanty, but the general consensus is that the species is declining (Gibbs et al. 1992a, Sauer et al. 2005), most acutely in the Midwest and in western Canada. The pattern in and around Nevada, however, is complex. The notion that the species has declined in the state would be consistent with a pattern described for parts of neighboring California (Gaines 1992, Shuford 1993, Small 1994), but Breeding Bird Survey data suggest that numbers in California have actually increased since the 1960s (Sauer et al. 2005). These discrepancies highlight the difficulty of monitoring this cryptic marshland bird.

Conservation of the American Bittern in Nevada clearly revolves around protection of wetlands with well-established tracts of dense marsh vegetation. Careful monitoring of this elusive species will be important, especially in the many wetlands where water depth and other wetland conditions fluctuate substantially.

HABITAT USE

HABITAT	ATLAS BLOCKS	INCIDENTAL OBSERVATIONS
Riparian	2 (22%)	2 (29%)
Wetland	7 (78%)	5 (71%)
TOTAL	9	7

Total numbers reported in the Habitat Use, Breeding Status, and Abundance tables may differ from each other (see pp. 22–23 for details). Percentage sums may differ slightly from 100% due to rounding.

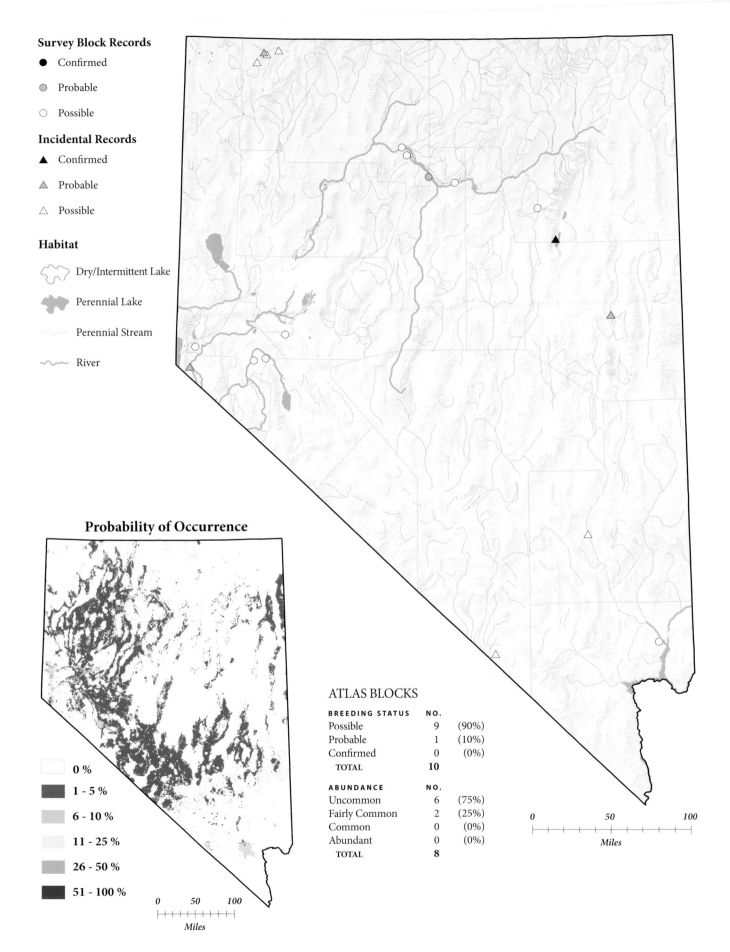

Survey Block Records

● Confirmed

◉ Probable

○ Possible

Incidental Records

▲ Confirmed

◮ Probable

△ Possible

Habitat

Dry/Intermittent Lake

Perennial Lake

Perennial Stream

River

Probability of Occurrence

☐ 0 %

1 - 5 %

6 - 10 %

11 - 25 %

26 - 50 %

51 - 100 %

0   50   100
Miles

ATLAS BLOCKS

BREEDING STATUS	NO.	
Possible	9	(90%)
Probable	1	(10%)
Confirmed	0	(0%)
TOTAL	10	

ABUNDANCE	NO.	
Uncommon	6	(75%)
Fairly Common	2	(25%)
Common	0	(0%)
Abundant	0	(0%)
TOTAL	8	

0   50   100
Miles

LEAST BITTERN
Ixobrychus exilis

The Least Bittern, one of the most secretive birds in the state, is even harder to locate than the American Bittern. Least Bitterns rarely venture into the open, and their varied vocalizations are not well known among birders. In Nevada, it is possible to catch a glimpse of this diminutive and dapper heron in the large freshwater marshes of the Lahontan Valley and the lowlands of the Colorado River drainage. Since atlas and other records were few, however, it is difficult to gauge the species' status in the state.

DISTRIBUTION

Prior to the atlas project, Least Bitterns were seldom reported in Nevada, and neither Ryser (1985) nor Alcorn (1988) reported breeding in the state. Titus (2003) considered the species to be rare in northern Nevada and local in the south. Atlas data support the notion that Least Bitterns are scarce in Nevada, and the predictive map reflects a likelihood of encountering the species only in a very few areas. Breeding was confirmed at Carson Lake in Churchill County, where an immature was seen in 1997 and an unfledged juvenile in 1998 (Chisholm and Neel 2002). Least Bitterns have been sighted in the nearby Stillwater marshes over the years, but atlas workers saw none in that area. Two of the three atlas records came from southern Nevada: a probable breeding record at Overton Wildlife Management Area in Clark County and a possible breeder at Pahranagat National Wildlife Refuge in Lincoln County. After the atlas project, a family group with dependent young was repeatedly observed at the Henderson Bird Viewing Preserve in Las Vegas Valley during the summer of 2005 (J. Branca, pers. comm.). All known records in the state were from freshwater marshes with dense emergent vegetation.

Some uncertainty exists over the distribution and status of Least Bitterns in the states adjacent to Nevada. They have reportedly bred at Honey Lake and in the Central Valley of California, in the Klamath and Malheur basins in Oregon, and in the marshes around the Great Salt Lake (Ryser 1985, Gibbs et al. 1992b). Garrett and Dunn (1981) considered the species to be common along the lower Colorado River, but its overall status in California is less certain (Small 1994). The recent *Oregon Breeding Bird Atlas* produced only one record of possible breeding in the state (Adamus et al. 2001). Confirmed breeding was more common, but quite scattered, in southern Arizona (Corman in Corman and Wise-Gervais 2005:102–103).

CONSERVATION AND MANAGEMENT

Because the Least Bittern is difficult to study, its conservation needs are not well known. Although quantitative data are lacking, the species is widely thought to be declining based on anecdotal evidence (Gibbs et al. 1992b). The Least Bittern is considered a species of Moderate Concern in the Intermountain West Waterbird Conservation Plan (Ivey and Herziger 2005) and is a Covered species under the Lower Colorado River Multi-Species Conservation Program (BOR-LCR 2004). Wetland destruction clearly poses a threat. It has been suggested that Least Bitterns need relatively large patches of wetland habitat with dense emergent vegetation, and they may be susceptible to water pollution (Gibbs et al. 1992b). Proximity to human habitation, however, is not necessarily a deterrent.

Generalized monitoring programs such as the Breeding Bird Survey are widely recognized as inadequate for species like the Least Bittern because they undersample patchily distributed wetlands and are ill-suited for detecting very secretive birds (Conway and Gibbs 2005, Sauer et al. 2005). Atlas techniques also may be insufficient for accurate assessments, and it is possible that Least Bitterns occur more frequently in Nevada (and elsewhere) than the data suggest. More thorough surveys in the areas from which atlas records came, as well as in other areas with a history of sightings (e.g., the Ruby Valley marshes), will help to clarify the status of the species in the state.

HABITAT USE

HABITAT	ATLAS BLOCKS	INCIDENTAL OBSERVATIONS
Wetland	2 (100%)	1 (100%)
TOTAL	2	1

Total numbers reported in the Habitat Use, Breeding Status, and Abundance tables may differ from each other (see pp. 22–23 for details). Percentage sums may differ slightly from 100% due to rounding.

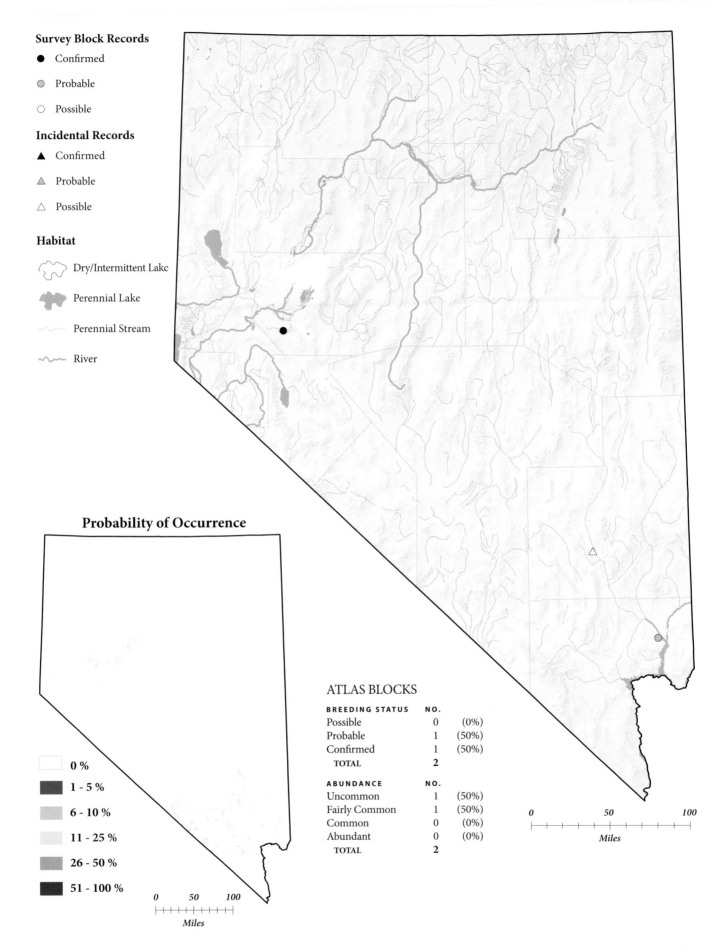

Survey Block Records

● Confirmed

◉ Probable

○ Possible

Incidental Records

▲ Confirmed

△ Probable

△ Possible

Habitat

⬡ Dry/Intermittent Lake

⬤ Perennial Lake

〜 Perennial Stream

〜 River

Probability of Occurrence

☐ 0 %

■ 1 - 5 %

▨ 6 - 10 %

▧ 11 - 25 %

▨ 26 - 50 %

■ 51 - 100 %

0   50   100
Miles

ATLAS BLOCKS

BREEDING STATUS	NO.	
Possible	0	(0%)
Probable	1	(50%)
Confirmed	1	(50%)
TOTAL	2	

ABUNDANCE	NO.	
Uncommon	1	(50%)
Fairly Common	1	(50%)
Common	0	(0%)
Abundant	0	(0%)
TOTAL	2	

0   50   100
Miles

GREAT BLUE HERON
Ardea herodias

The elegant profile of the Great Blue Heron is a characteristic sight in wetlands throughout much of North America. The species is widely cited for its success and adaptability, and it seems to have incurred relatively low losses from the environmental problems that have put many of America's other colonial waterbirds at risk during the past century. Great Blue Herons can be found in various aquatic habitats in Nevada: large marshes and river systems are often used, but so are farm ponds, fishing holes, and golf courses.

DISTRIBUTION

The Great Blue Heron is generally considered a common and widespread breeder throughout Nevada (Ryser 1985, Alcorn 1988). Atlas results show a distribution shaped like a great arc, sweeping northeast from Reno to Elko, and then south to Laughlin and beyond. There were no records from Esmeralda County and few from Nye, Mineral, Lander, and Eureka counties. Not surprisingly, the distribution neatly corresponds to the major waterways in the state, with birds found throughout the drainages emanating from the Sierra Nevada, along the Humboldt River and its tributaries, and along the major rivers of southern Nevada.

Great Blue Herons were confirmed as breeders at twenty-two locations. Abundance estimates were typically low, with only six reports of more than ten pairs in a block. This is surprising for a species that often occurs in colonies of several hundred pairs (Butler 1992), and it suggests that many Nevada colonies are rather small. However, a recent report raises the possibility that Nevada has additional colonies that atlas fieldwork did not detect (see Neel 2001).

The majority of Great Blue Herons recorded by atlas workers were classified as possible breeders. Great blues travel widely in search of food, and many of these possible breeders—assuming they were breeding at all—may have been observed far from their actual breeding location. The wide range of locales in which Great Blue Herons were seen accounts for the wide area of potential occurrence shown on the map. Although the model results might accurately depict the probability of seeing Great Blue Herons, it seems likely that most breeding is restricted to the major drainages from which most of our records came.

CONSERVATION AND MANAGEMENT

Great Blue Heron populations seem to be doing well in most of their geographical range (Sauer et al. 2005) and are not considered at risk at the continental level (Kushlan et al. 2002). The size of Nevada's breeding population is subject to large fluctuations, however, and there is some evidence of a decline in recent decades. The number of nests in the Lahontan Valley fluctuated from more than 600 to fewer than 20 during the 1980s (Chisholm and Neel 2002). During the 1990s, there were fewer pairs overall in this region, with totals ranging from about 40 to 180 (Neel 1998, 1999b, 2001). Presumably these fluctuations are linked to the availability of suitable habitat, but the specific factors influencing population size are unknown. These factors resulted in a designation of Moderate Concern in the Waterbird Conservation Plan for the Intermountain West (Ivey and Herziger 2005).

Great Blue Herons seem to have two conservation needs. First, they require protection at colony sites, where they are vulnerable to deliberate harassment as well as unintended disturbances (Butler 1992). As it did for other herons, the atlas identified a number of colony locations that can be used to design a statewide monitoring program for colonial waterbirds. Additional sites might also exist, especially along the drainages identified earlier in this account. Second, it is possible that Great Blue Herons are limited by the availability of suitable foraging conditions. Threats to foraging areas include the loss of wetland habitat to deliberate draining or periodic drought, and the accumulation of pesticides and heavy metals and increases in salinity as a result of water diversions and contaminated return flows.

HABITAT USE

HABITAT	ATLAS BLOCKS	INCIDENTAL OBSERVATIONS
Agricultural	7 (10%)	1 (2%)
Grassland	1 (1%)	0 (0%)
Mojave	0 (0%)	1 (2%)
Open Water	4 (6%)	4 (10%)
Riparian	34 (47%)	22 (55%)
Salt Desert Scrub	1 (1%)	0 (0%)
Urban	1 (1%)	0 (0%)
Wetland	25 (34%)	12 (30%)
TOTAL	73	40

Total numbers reported in the Habitat Use, Breeding Status, and Abundance tables may differ from each other (see pp. 22–23 for details). Percentage sums may differ slightly from 100% due to rounding.

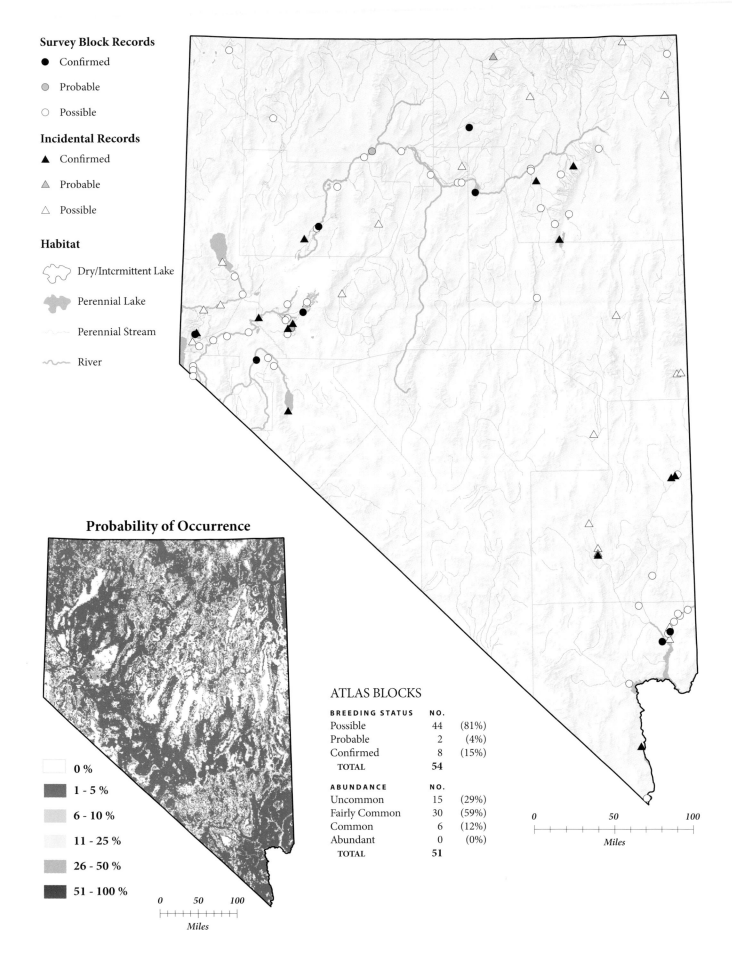

Survey Block Records

● Confirmed

● Probable

○ Possible

Incidental Records

▲ Confirmed

▲ Probable

△ Possible

Habitat

Dry/Intermittent Lake

Perennial Lake

Perennial Stream

River

Probability of Occurrence

	0 %
	1 - 5 %
	6 - 10 %
	11 - 25 %
	26 - 50 %
	51 - 100 %

0   50   100

Miles

ATLAS BLOCKS

BREEDING STATUS	NO.	
Possible	44	(81%)
Probable	2	(4%)
Confirmed	8	(15%)
TOTAL	54	

ABUNDANCE	NO.	
Uncommon	15	(29%)
Fairly Common	30	(59%)
Common	6	(12%)
Abundant	0	(0%)
TOTAL	51	

0   50   100

Miles

GREAT EGRET
Ardea alba

The Great Egret can be found in aquatic habitats as diverse as tiny urban ponds, lush permanent marshes, and ephemeral desert sinks. Whatever the venue, this stately heron commands the instant attention of birders and nonbirders alike. The Great Egret is resourceful and adaptable, but it is not immune to damage from human activities. Populations declined when they were hunted in the late nineteenth century, and although they have recovered, they remain susceptible to habitat degradation, pollution, and colony disturbance.

DISTRIBUTION

Great Egrets were reported from twenty-three blocks, although only three blocks produced confirmed breeding records. Breeding also was confirmed at three of eight sites for which incidental observations were submitted. All of the confirmed breeding records came from northern Nevada, although there were also reports of probable breeding from the Pahranagat Valley in Lincoln County and from southern Nye County. The species was typically rated as uncommon or fairly common, with more than ten pairs estimated in only two blocks. Most reports came from wetland or riparian habitat, as would be expected.

As is true for several aquatic species with clumped breeding distributions, the atlas results paint a somewhat misleading impression of the Great Egret's abundance in Nevada. Although reported at relatively few sites, the state's population likely numbers a few hundred pairs. During the 1980s and 1990s, for instance, breeding occurred at five sites in the Lahontan Valley. The total population size at these colonies reached almost five hundred pairs in the 1980s (Chisholm and Neel 2002), and estimates from the 1990s fluctuated between about fifty and two

hundred pairs (Neel 1998, 1999b, 2001). Although the Lahontan Valley is clearly a focal area for Great Egrets in the state, breeding is also well established in the Ruby Valley and along the Humboldt River in Elko and Eureka counties.

Great Egret numbers appear to have increased in Nevada over the past century. Linsdale (1936), for instance, considered Great Egrets to be infrequent visitors in the state. Elsewhere in the region, the species is listed as rare statewide in Utah (UOSBRC 2005), but in California it is a regular breeder in the Klamath Basin, throughout the Central Valley, and along the Colorado River (Small 1994, McCrimmon et al. 2001). In the northern Great Basin, Great Egrets breed at several colonies in southeastern Oregon (Adamus et al. 2001).

CONSERVATION AND MANAGEMENT

Evidence from the Breeding Bird Survey suggests that Great Egret numbers have steadily increased throughout the continent since the 1960s, and especially in the West (Sauer et al. 2005). The species appears to be a recent arrival in Washington (Smith et al. 1997), and seems to have increased its breeding range in California, too (see Burridge in Burridge 1995:31, Gallagher 1997).

While the recent population increases are encouraging, it should be noted that they are part of a long-term recovery following devastating losses during the past two centuries (McCrimmon et al. 2001). In the nineteenth and early twentieth centuries, hunters killed enormous numbers of Great Egrets to harvest their graceful plumes for the millinery trade. The impacts of hunting were exacerbated by the draining of wetlands and water pollution, which are considered the major causes of declines since the 1930s (McCrimmon et al. 2001). Subsequently, the banning of the pesticide DDT in particular and stricter pesticide regulations in general have helped Great Egret populations to recover (Pratt and Winkler 1985).

The continued recovery of the Great Egret will depend on various factors. Water quality remains a significant issue, as recent population declines at the pollution-plagued Salton Sea in California indicate (Small 1994). Worse is the widespread and largely unregulated use of organochlorine pesticides on the Great Egret's extensive Latin American wintering grounds. At the local level, disturbances around nesting colonies can cause nest abandonment (Shuford 1993).

HABITAT USE

HABITAT	ATLAS BLOCKS	INCIDENTAL OBSERVATIONS
Agricultural	2 (6%)	0 (0%)
Open Water	1 (3%)	0 (0%)
Riparian	9 (27%)	2 (33%)
Urban	1 (3%)	0 (0%)
Wetland	20 (61%)	4 (67%)
TOTAL	33	6

Total numbers reported in the Habitat Use, Breeding Status, and Abundance tables may differ from each other (see pp. 22–23 for details). Percentage sums may differ slightly from 100% due to rounding.

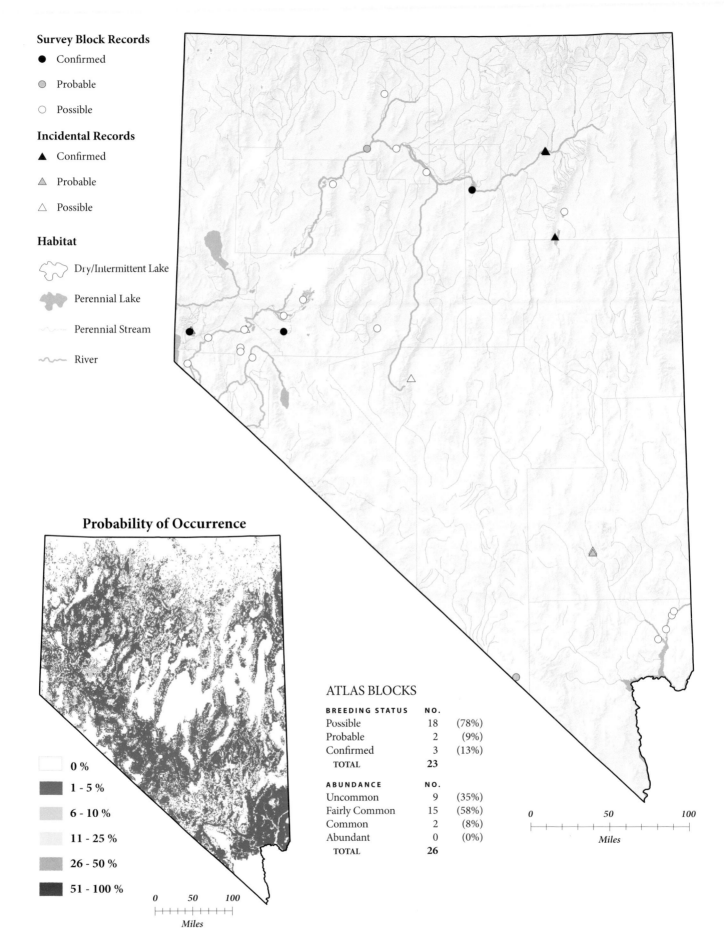

Survey Block Records

● Confirmed

● Probable

○ Possible

Incidental Records

▲ Confirmed

▲ Probable

△ Possible

Habitat

Dry/Intermittent Lake

Perennial Lake

Perennial Stream

River

Probability of Occurrence

☐ 0 %

1 - 5 %

6 - 10 %

11 - 25 %

26 - 50 %

51 - 100 %

0   50   100

Miles

ATLAS BLOCKS

BREEDING STATUS	NO.	
Possible	18	(78%)
Probable	2	(9%)
Confirmed	3	(13%)
TOTAL	**23**	
ABUNDANCE	NO.	
Uncommon	9	(35%)
Fairly Common	15	(58%)
Common	2	(8%)
Abundant	0	(0%)
TOTAL	**26**	

0   50   100

Miles

SNOWY EGRET
Egretta thula

It is hard to say whether the Snowy Egret is more appreciated for its visual splendor or for its comical foraging behavior. It seems to combine the grace of the Great Egret and the charm of the Cattle Egret, and it surpasses both in the diversity of its hunting tactics. Darting to and fro in shallow water, the Snowy Egret stirs up the water with its bright yellow toes and then captures its prey with a quick thrust of its slender black bill.

DISTRIBUTION

The Nevada distribution of the Snowy Egret is similar to the Great Egret's, although the latter was found at fewer sites during atlas fieldwork. Snowy Egret breeding records were scattered across the northern tier of counties in a general association with the state's major rivers and agricultural areas. Overall, Snowy Egrets were found in thirty-seven blocks and at an additional twenty-one sites. Despite the relatively large number of Snowy Egret records, breeding was confirmed at only seven locations— a pattern similar to those we found for most colonial waterbirds in Nevada. It is also likely that colonies were missed in some areas. For instance, Snowy Egret breeding was not confirmed in the Ruby Valley, despite the presence of up to 150 pairs at Ruby Lake in the 1980s (Alcorn 1988).

The statewide distribution of the Snowy Egret seems somewhat more extensive than the older literature indicates (e.g., Linsdale 1936), although this pattern might simply reflect the recolonization of areas that were emptied of egrets by depredations during the plume trade era. The Snowy Egret's overall abundance, however, does not seem to have increased greatly in recent years. Ryser (1985) considered Snowy Egrets to be common in northern Nevada, and Alcorn (1988) called them abundant in some places.

In contrast, the bulk of the atlas abundance estimates were for fewer than 10 pairs in a block, and no colonies of more than 100 pairs were identified. Some Nevada colonies can be quite large, with more than 300 pairs counted at the Lahontan Valley colonies in years prior to the atlas (Alcorn 1988, Neel 1998, Chisholm and Neel 2002). During the atlas years, however, the population in the Lahontan Valley declined from 182 pairs in 1997 to only 20 pairs in 2000 (Neel 1998, 1999b, 2001).

CONSERVATION AND MANAGEMENT

Perhaps even more so than the Great Egret, the Snowy Egret has benefited from protection from persecution. Rangewide, populations increased considerably during the latter half of the twentieth century (Sauer et al. 2005). Coupled with this increase in numbers were range expansions, detailed for various parts of the West by Shuford (1993), Burridge (in Burridge 1995:32), Gallagher (1997), and Ryder (in Kingery 1998:56–57). There have also been declines in some areas, however, including the Salton Sea and nearby areas of southern California and Arizona (Parsons and Master 2000, Sauer et al. 2005).

A recent population increase is difficult to trace quantitatively in Nevada, but there is little doubt that the Snowy Egret's fortunes have improved during the past century. Fisher (1893) did not mention this species in southern Nevada, and Ryser's (1985) analysis of Ridgway's expedition produced no records for northern Nevada. By Linsdale's (1936) time Snowy Egrets occurred in small numbers, but even by Alcorn's time (1988) they seem not to have been breeding in southern Nevada.

Concerns about the conservation of Snowy Egrets remain, especially given the large population fluctuations observed in the Lahontan Valley. The species is susceptible to water pollution, which may be the cause of declines at the Salton Sea. In some respects, however, Snowy Egrets appear to be more resilient than other herons. They seem to cope with human disturbance relatively well (Parsons and Master 2000), as evidenced, for instance, by the occurrence of breeding records from Virginia Lake in downtown Reno.

HABITAT USE

HABITAT	ATLAS BLOCKS	INCIDENTAL OBSERVATIONS
Agricultural	1 (2%)	0 (0%)
Grassland	0 (0%)	1 (5%)
Open Water	2 (4%)	2 (9%)
Riparian	23 (49%)	10 (48%)
Urban	0 (0%)	1 (5%)
Wetland	21 (45%)	7 (33%)
TOTAL	47	21

Total numbers reported in the Habitat Use, Breeding Status, and Abundance tables may differ from each other (see pp. 22–23 for details). Percentage sums may differ slightly from 100% due to rounding.

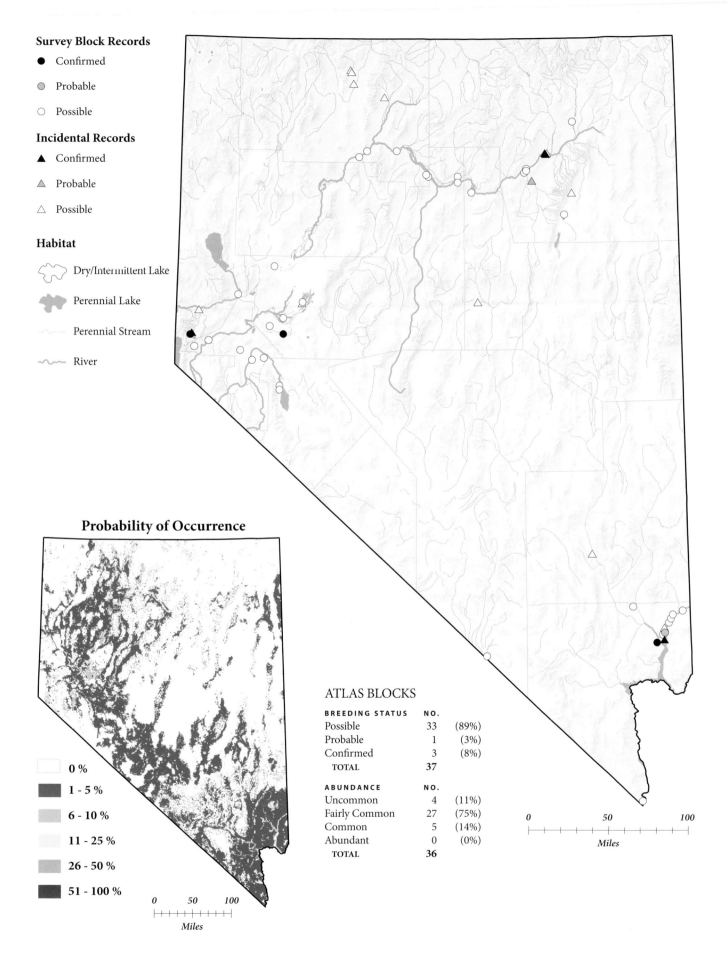

Survey Block Records

● Confirmed

● Probable

○ Possible

Incidental Records

▲ Confirmed

▲ Probable

△ Possible

Habitat

Dry/Intermittent Lake

Perennial Lake

Perennial Stream

River

Probability of Occurrence

- [ ] 0 %
- 1 - 5 %
- 6 - 10 %
- 11 - 25 %
- 26 - 50 %
- 51 - 100 %

0   50   100
Miles

ATLAS BLOCKS

BREEDING STATUS	NO.	
Possible	33	(89%)
Probable	1	(3%)
Confirmed	3	(8%)
TOTAL	37	

ABUNDANCE	NO.	
Uncommon	4	(11%)
Fairly Common	27	(75%)
Common	5	(14%)
Abundant	0	(0%)
TOTAL	36	

0   50   100
Miles

CATTLE EGRET
Bubulcus ibis

The Cattle Egret is an Old World species that colonized the Americas in the early twentieth century. Spreading north from the Atlantic Coast of South America, it has been breeding in North America since the 1950s. Cattle Egrets established a toehold in Nevada in the mid-1970s, and population numbers have apparently stabilized. The birds occur primarily in areas such as Carson Lake where wetlands lie in close proximity to pastures. Elsewhere, Cattle Egrets are seen annually along the larger river drainages of northern Nevada and along the Muddy and Virgin rivers of southern Nevada.

DISTRIBUTION

The first reports of Cattle Egrets were from Ruby Lake National Wildlife Refuge, Carson Lake, and the Nevada Test Site in southern Nevada (Alcorn 1988). By 1980, breeding had been documented (Ryser 1985). Atlas fieldwork produced records in seven widely scattered locations, with breeding confirmed at two sites. Observers estimated multiple pairs—but never more than 10—to be present wherever the species was encountered, and all encounters were in riparian or wetland habitats.

The confirmed breeding sites included Carson Lake in Churchill County, where there has been a colony of 40–100 breeding pairs since the 1980s (L. Neel, pers. comm.). Breeding has also been reported elsewhere in the Lahontan Valley, where up to 225 pairs have nested (Chisholm and Neel 2002). During recent years, the number of pairs in this region fluctuated from as few as 17 pairs in 1994 to as many as 160 pairs in 2000 (Neel 1998,

1999b, 2001). The second site where the atlas fieldwork confirmed breeding was in Elko County, with a colony at Ryndon along the Humboldt River. Possible breeders were also documented in three blocks along the Virgin and Muddy rivers and in two blocks along the Humboldt River. Both drainages offer Cattle Egrets their basic habitat requirements: water and agriculture. The likelihood of finding this bird is predicted to be moderate in the limited areas where both of these requirements are met, in both northern and southern Nevada.

CONSERVATION AND MANAGEMENT

Like the Great-tailed Grackle and Eurasian Collared-Dove, the Cattle Egret invaded Nevada under its own power, but with considerable indirect help from humans. The Cattle Egret's early success in the Western Hemisphere was likely related to deforestation and the creation of pastures in South America during the late nineteenth century. This habitat conversion allowed vagrants from Africa to become established and set the scene for the subsequent population spread northward across the Caribbean and through Central America into the United States.

It is somewhat surprising that the Cattle Egret has not spread more rapidly in the Great Basin region. Ryser (1985) predicted continued population expansion in Nevada, but this seems not to have happened. This result is consistent with the suggestion that the Cattle Egret's invasion of North America has slowed its pace in many areas (Telfair 1994). Garrett and Dunn (1981) noted that the speed with which Cattle Egrets invaded southern California was nearly unparalleled, but several years later, Small (1994) reported subsequent declines in that region.

The future prospects for the Cattle Egret in Nevada are uncertain. Cattle Egrets are highly migratory and prone to wandering (Telfair 1994), so colonies could show up wherever favorable spots exist. Our predictive map suggests a more limited potential range than for other egrets, but it is based on so few records that it probably underrepresents the full range of conditions where the species could conceivably occur.

HABITAT USE

HABITAT	ATLAS BLOCKS	INCIDENTAL OBSERVATIONS
Riparian	3 (60%)	1 (100%)
Wetland	2 (40%)	0 (0%)
TOTAL	5	1

Total numbers reported in the Habitat Use, Breeding Status, and Abundance tables may differ from each other (see pp. 22–23 for details). Percentage sums may differ slightly from 100% due to rounding.

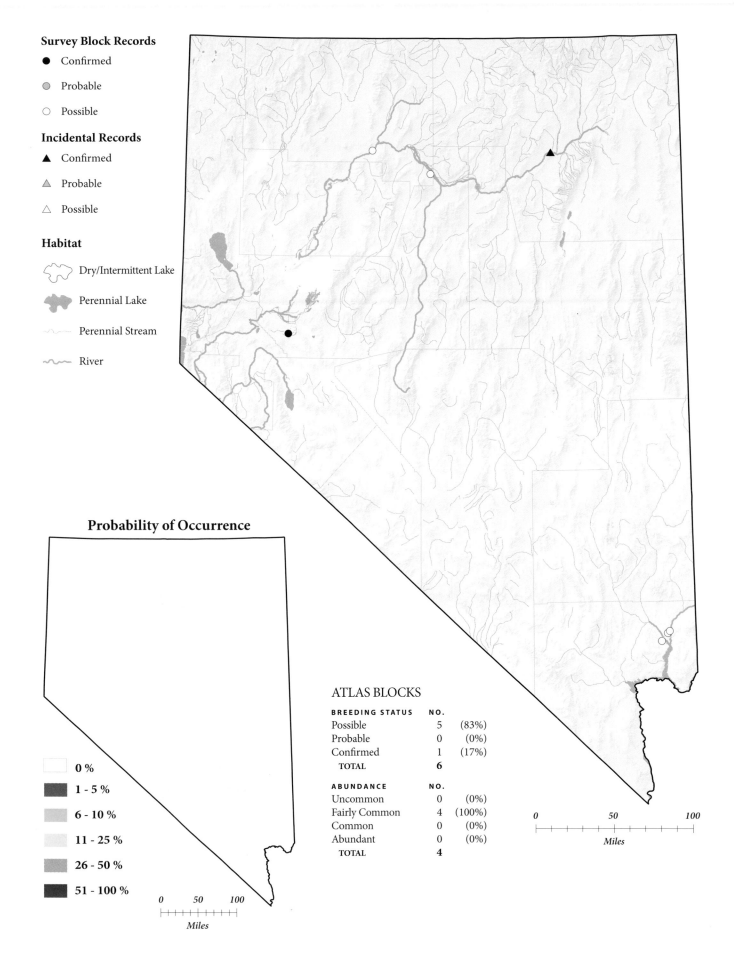

Survey Block Records
● Confirmed
◐ Probable
○ Possible

Incidental Records
▲ Confirmed
△ Probable
△ Possible

Habitat
Dry/Intermittent Lake

Perennial Lake

Perennial Stream

River

Probability of Occurrence

☐ 0 %
▨ 1 - 5 %
▨ 6 - 10 %
▨ 11 - 25 %
▨ 26 - 50 %
▨ 51 - 100 %

0   50   100
Miles

ATLAS BLOCKS

BREEDING STATUS	NO.	
Possible	5	(83%)
Probable	0	(0%)
Confirmed	1	(17%)
TOTAL	6	

ABUNDANCE	NO.	
Uncommon	0	(0%)
Fairly Common	4	(100%)
Common	0	(0%)
Abundant	0	(0%)
TOTAL	4	

0   50   100
Miles

GREEN HERON
Butorides virescens

Bitterns aside, the Green Heron is North America's smallest heron, and Nevada's rarest. This charismatic fisher, somehow reminiscent of an aquatic Lewis's Woodpecker with its dark purplish reds, greens, and slaty blues, was not confirmed as a breeder in the state during the atlas project. Nonetheless, the many records from various locations suggest that the Green Heron is poised to become a regular breeder here.

DISTRIBUTION
Atlas records for Green Herons came from three parts of Nevada. The major concentration was in the south, where sightings occurred mostly along the Colorado and Virgin rivers and their tributaries. Within this region, the area immediately north of Lake Mead produced the best evidence of nesting. Green Herons were also strongly suspected to be breeding at Lorenzi Park in Las Vegas, where birds were often seen between 1997 and 2000, including fledged immatures in both 1999 and 2000 (C. Titus, pers. comm.). To the north, there was also a small cluster of records in the vicinity of Reno and Carson City. One of these records, from along the Carson River, merited probable breeding status, while the rest simply involved birds seen in suitable breeding habitat. There was also a possible breeding record at Ruby Lake in northeastern Nevada. Elsewhere in the region, the species is reported to breed in the northeastern Great Basin near the Bear River in Utah (Ryser 1985), and to the south and west in Arizona and Cal-

ifornia (Davis and Kushlan 1994, Small 1994, Wise-Gervais in Corman and Wise-Gervais 2005:112–113).

The Green Heron's secretive habits and solitary nesting behavior make it rather difficult to reconstruct historical patterns of occurrence (Ryser 1985). Atlas data appear to be consistent with previous reports of regular sightings within the major wetlands and drainages of the state, although few of them involved confirmed breeding (Alcorn 1988). Linsdale (1936) reported collecting two birds from a location opposite Fort Mojave on the Colorado River in Clark County, but it is not clear how common or widespread Green Herons were as breeders at that time. More recently, a nest with four young was found at Overton Wildlife Management Area (Clark County) in 1993 (D. Parmelee, unpubl. data). While we cannot be certain whether atlas data refer to nonbreeding or breeding birds, these ancillary reports suggest that breeding may be likely in at least some of these cases. Additionally, most of the atlas abundance estimates exceeded one pair per block, a result more consistent with small breeding populations than with isolated wanderers.

CONSERVATION AND MANAGEMENT
According to Breeding Bird Survey data, Green Herons are declining overall in North America but are increasing in the West (Sauer et al. 2005). Monitoring this species requires a different approach from that appropriate for most herons, because Green Herons are not very colonial. Consequently, there is little current information on their status in Nevada, and they generally go unnoticed by monitoring efforts designed for colonial species. Broad-based monitoring programs such as the Breeding Bird Survey are also unlikely to provide much information about Green Herons in Nevada, because the species occurs so sporadically that it is often not detected by generalized monitoring protocols.

As a relatively uncommon species that appears to be secure in the surrounding region, the Green Heron is not a high conservation priority in Nevada. Birders should nevertheless be on the lookout everywhere for evidence of breeding, because such observations will likely provide our best information about the status of Green Herons in the near future.

HABITAT USE

HABITAT	ATLAS BLOCKS	INCIDENTAL OBSERVATIONS
Mojave	0 (0%)	1 (10%)
Riparian	6 (60%)	7 (70%)
Urban	0 (0%)	1 (10%)
Wetland	4 (40%)	1 (10%)
TOTAL	10	10

Total numbers reported in the Habitat Use, Breeding Status, and Abundance tables may differ from each other (see pp. 22–23 for details). Percentage sums may differ slightly from 100% due to rounding.

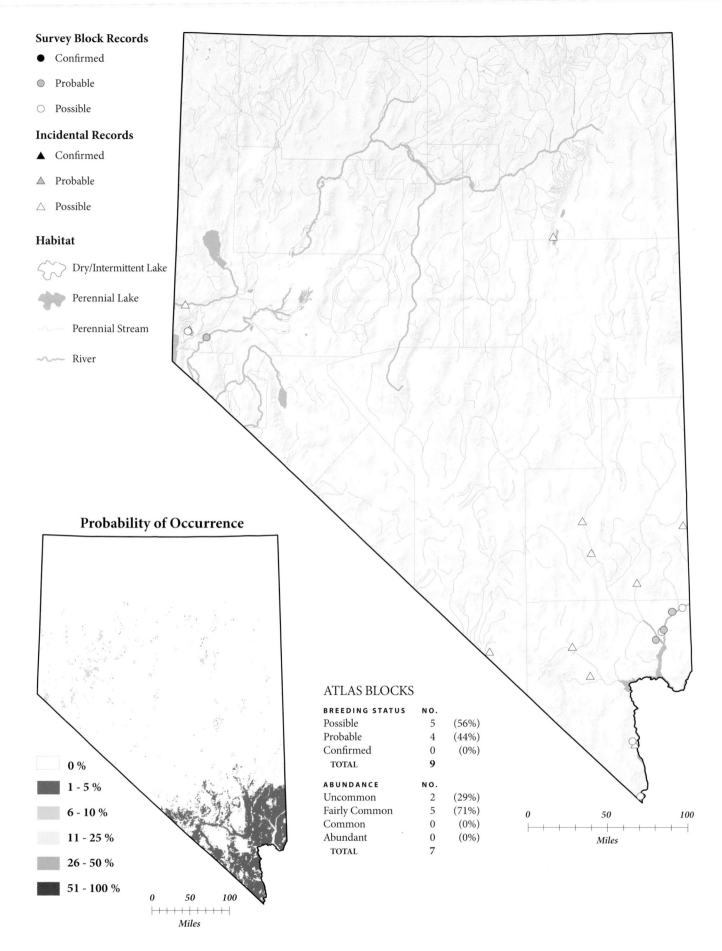

Survey Block Records
● Confirmed
● Probable
○ Possible

Incidental Records
▲ Confirmed
△ Probable
△ Possible

Habitat
Dry/Intermittent Lake

Perennial Lake

Perennial Stream

River

Probability of Occurrence

☐ 0 %

■ 1 - 5 %

■ 6 - 10 %

☐ 11 - 25 %

■ 26 - 50 %

■ 51 - 100 %

0   50   100
Miles

ATLAS BLOCKS

BREEDING STATUS	NO.	
Possible	5	(56%)
Probable	4	(44%)
Confirmed	0	(0%)
TOTAL	9	

ABUNDANCE	NO.	
Uncommon	2	(29%)
Fairly Common	5	(71%)
Common	0	(0%)
Abundant	0	(0%)
TOTAL	7	

0   50   100
Miles

BLACK-CROWNED NIGHT-HERON
Nycticorax nycticorax

By day, Black-crowned Night-Herons are usually seen slumbering under cover of a dense riparian tangle or standing forlornly in a marsh. But by night they are active in the wetlands and riparian zones of Nevada. Their explosive *quok!* is one of the characteristic sounds of the nighttime hours. Despite their mostly nocturnal habits and the relative scarcity of aquatic habitats in Nevada, Black-crowned Night-Herons were widespread and frequently noted during atlas surveys.

DISTRIBUTION

The Black-crowned Night-Heron's distribution in Nevada is similar to those of the other colonial herons. Breeding was confirmed at eighteen sites, including most of the state's major wetlands. In northwestern Nevada, breeding definitely occurred at Pyramid, Washoe, and Carson lakes; at Lahontan Reservoir and Lemmon Valley Marsh; and along the Carson River. Farther east, confirmed breeders were reported from several points along the Humboldt River as well as at Ruby Lake. And in southern Nevada, nesting was confirmed at several sites along the Virgin River, and even in Las Vegas.

The Black-crowned Night-Heron's range in Nevada as revealed by atlas data is consistent with earlier descriptions (Ryser 1985, Alcorn 1988). Similarly, the habitat associations reported were largely unsurprising. The species is generally aquatic and accepts a variety of aquatic habitats, including small ponds, urban lakes, and agricultural ditches (Davis 1993). This use of small wetland habitats, which often occur embedded within other habitat types, accounts for the very wide area over which our model predicts some small chance of finding the species. For instance, Black-crowned Night-Herons will nest in sagebrush habitat close to wetlands, and even in backyards near city lakes.

Abundance was estimated to exceed 10 pairs in only 13% of the blocks for which abundances were estimated, and no blocks were thought to have more than 100 pairs. Previous surveys estimated that 40–331 pairs of Black-crowned Night-Herons nested in the Lahontan Valley between 1992 and 2000 (Neel 1998, 1999b, 2001); however, these estimates are far below the 1,800 pairs noted in 1987 (Chisholm and Neel 2002). It is likely that some currently active breeding sites were not recorded during the atlas; for instance, several colonies present in the Lahontan Valley in the 1980s (Chisholm and Neel 2002) were not reported by atlas fieldworkers.

CONSERVATION AND MANAGEMENT

The Black-crowned Night-Heron is one of the world's most widespread herons, occurring on all continents except Australia and Antarctica. Population trends probably are not especially well monitored by generalized programs such as the Breeding Bird Survey, although these data do suggest an overall increase in numbers since the 1960s, especially in the West (Sauer et al. 2005). Historical accounts from Nevada (Fisher 1893, Linsdale 1936) suggest that numbers were lower in the past than they were during the atlas period. In contrast to this presumed statewide increase, populations at Ruby Lake and in the Lahontan Valley are reported to have declined substantially since the 1980s (Alcorn 1988, Chisholm and Neel 2002).

Black-crowned Night-Herons are ecologically versatile and take a wide variety of foods. Their colonies vary greatly in size, and they frequently nest alongside other wading birds (Davis 1993). They appear to be more tolerant of human disturbance near their colonies than other herons, often occurring in developed areas, but they are vulnerable to loss of their wetland habitats. They also are susceptible to environmental toxins, suffering from eggshell thinning when exposed to high pesticide loads (Custer et al. 1983, Davis 1993). For these reasons, both the North American Waterbird Conservation Plan and the Intermountain West Waterbird Conservation Plan classify the Black-crowned Night-Heron as a species of Moderate Concern (Kushlan et al. 2002, Ivey and Herziger 2005)

HABITAT USE

HABITAT	ATLAS BLOCKS	INCIDENTAL OBSERVATIONS
Agricultural	3 (5%)	0 (0%)
Open Water	4 (6%)	2 (8%)
Riparian	29 (45%)	10 (39%)
Sagebrush Steppe	0 (0%)	1 (4%)
Urban	0 (0%)	2 (8%)
Wetland	28 (44%)	11 (42%)
TOTAL	64	26

Total numbers reported in the Habitat Use, Breeding Status, and Abundance tables may differ from each other (see pp. 22–23 for details). Percentage sums may differ slightly from 100% due to rounding.

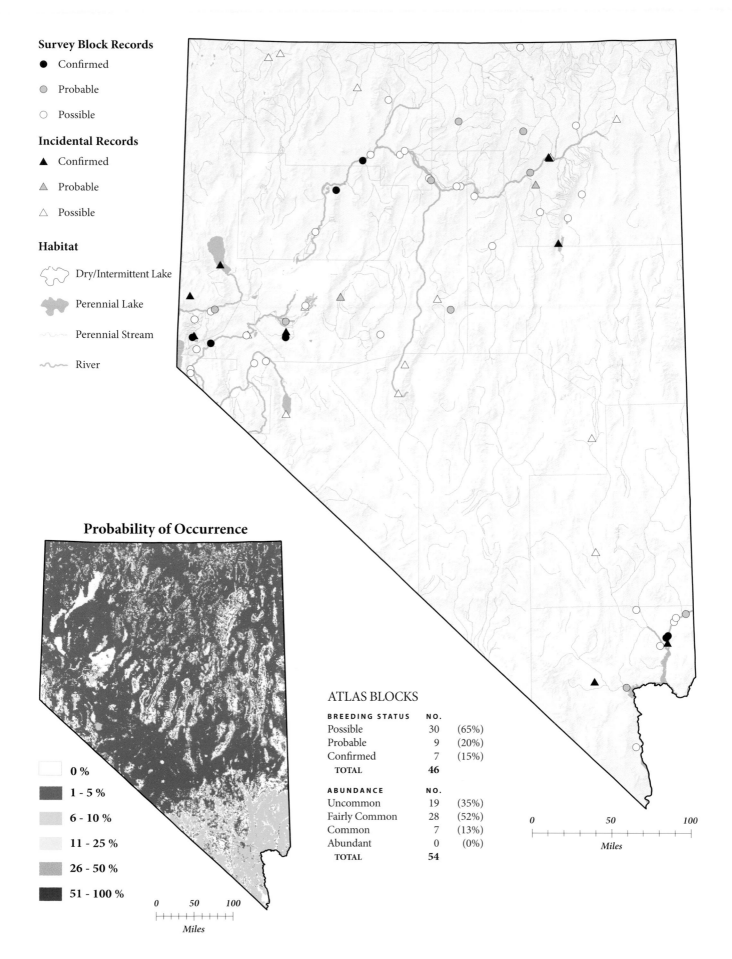

Survey Block Records

● Confirmed

● Probable

○ Possible

Incidental Records

▲ Confirmed

▲ Probable

△ Possible

Habitat

Dry/Intermittent Lake

Perennial Lake

Perennial Stream

River

Probability of Occurrence

0 %

1 - 5 %

6 - 10 %

11 - 25 %

26 - 50 %

51 - 100 %

0   50   100
Miles

ATLAS BLOCKS

BREEDING STATUS	NO.	
Possible	30	(65%)
Probable	9	(20%)
Confirmed	7	(15%)
TOTAL	46	

ABUNDANCE	NO.	
Uncommon	19	(35%)
Fairly Common	28	(52%)
Common	7	(13%)
Abundant	0	(0%)
TOTAL	54	

0   50   100
Miles

WHITE-FACED IBIS
Plegadis chihi

The attire of the White-faced Ibis is exquisite: metallic green, rich chestnut, or shimmering pink, depending on how the light catches the feathers. The species is resident along the coast of the Gulf of Mexico, in central Mexico, and in southern California, and is a summer visitor throughout much of the interior West. It occurs in abundance in the Great Basin and more sparingly in southern Nevada, and it seems to prosper wherever there are extensive wetlands, seasonally flooded grasslands, or irrigated agricultural fields.

DISTRIBUTION

White-faced Ibises have been common in Nevada at least since Robert Ridgway's visit in the 1860s (Ridgway 1877). Linsdale (1936) reported the species from six, and Alcorn (1988) from eleven, of the state's seventeen counties. During the atlas project, White-faced Ibises were reported in all counties except Esmeralda and Mineral. Records came from sixty-seven locations, and there was a decidedly northern aspect to their distribution. Breeding was confirmed in only three blocks and at six additional sites, consistent with a highly colonial species that concentrates in a small number of nesting areas. Compared with the herons and egrets, the White-faced Ibis appears to have a widespread distribution that is less tied to major wetlands and rivers. Confirmed breeding and probable breeding, however, remain concentrated in these established strongholds for colonial waterbirds, and it is likely that many reports of possible breeders involved birds that were foraging on irrigated land or in small wetlands but not nesting nearby.

Breeding was confirmed at several sites in western Nevada, and the Lahontan Valley remains a key area for the species. With the exception of a large dip in the early 1990s, 2,400–6,500 pairs nested there annually between 1986 and 1999 (Chisholm and Neel 2002); overall estimates for northwestern Nevada were above 8,000 pairs during the atlas years (Neel 1998). Colonies

have been found in at least ten locations in this region, but many are not occupied every year (Neel 1998, 1999b). Since the mid-1990s, the Carson Lake colony in Churchill County has evidently held the most birds (Neel 1998, 1999b). Elsewhere in the state, two confirmed breeding records in eastern Nevada were of particular interest: fledglings were observed at a playa in White Pine County, and nest building was documented at Key Pittman Wildlife Management Area in Lincoln County.

Ibis colonies tend to occur in large wetlands, and beds of hardstem bulrush seem to be the favored nesting habitat (Neel 1999a). Almost all confirmed breeding records came from wetland habitats. Breeding ibises are well known for commuting tens of miles to search for food in various flooded habitats, however (Ryder and Manry 1994), which probably accounts for the range of other habitats reported and the correspondingly wide area over which the map predicts a low probability of finding the species during the breeding season.

CONSERVATION AND MANAGEMENT

The Great Basin contains a substantial proportion of the world's breeding White-faced Ibises. North America populations have increased in recent years, following substantial declines in the 1960s and 1970s due to habitat loss and pesticide impacts (Ryder and Manry 1994, Earnst et al. 1998). In Nevada, the White-faced Ibis is a Nevada Partners in Flight priority species. The species has been a monitoring priority for more than two decades, and its numbers have been relatively stable (Neel 1999a, Chisholm and Neel 2002). Management strategies were summarized by Ryder and Manry (1994) in general, and for Nevada in particular by Neel (1999a). Protection of the Carson Lake colony has the highest priority, but the maintenance of other sites is also important because of the species' propensity to shift among colony sites in response to changing conditions. In addition to maintaining flooded areas to provide foraging habitat, it will also be important to limit disturbance and to monitor the effects of mammalian predators.

HABITAT USE

HABITAT	ATLAS BLOCKS	INCIDENTAL OBSERVATIONS
Agricultural	6 (13%)	0 (0%)
Grassland	1 (2%)	1 (3%)
Open Water	6 (13%)	6 (17%)
Riparian	8 (18%)	4 (11%)
Salt Desert Scrub	0 (0%)	1 (3%)
Wetland	24 (53%)	23 (66%)
TOTAL	45	35

Total numbers reported in the Habitat Use, Breeding Status, and Abundance tables may differ from each other (see pp. 22–23 for details). Percentage sums may differ slightly from 100% due to rounding.

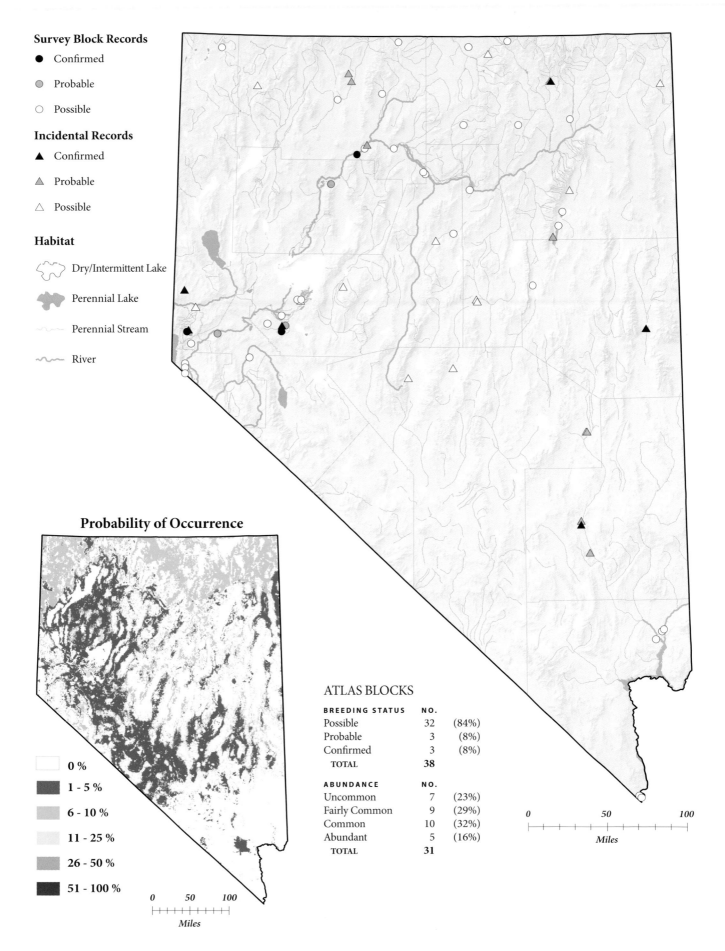

Survey Block Records

● Confirmed

● Probable

○ Possible

Incidental Records

▲ Confirmed

▲ Probable

△ Possible

Habitat

Dry/Intermittent Lake

Perennial Lake

Perennial Stream

River

Probability of Occurrence

☐ 0 %

■ 1 - 5 %

■ 6 - 10 %

☐ 11 - 25 %

■ 26 - 50 %

■ 51 - 100 %

0   50   100
Miles

ATLAS BLOCKS

BREEDING STATUS	NO.	
Possible	32	(84%)
Probable	3	(8%)
Confirmed	3	(8%)
TOTAL	38	

ABUNDANCE	NO.	
Uncommon	7	(23%)
Fairly Common	9	(29%)
Common	10	(32%)
Abundant	5	(16%)
TOTAL	31	

0   50   100
Miles

TURKEY VULTURE
Cathartes aura

What could be more emblematic of a lazy summer afternoon than the sight of a Turkey Vulture rocking unsteadily on upturned wings as it soars high overhead? Turkey Vultures feed entirely on carcasses and have unusual adaptations for this diet, including a strong resistance to pathogens and a highly developed sense of smell. They can be found wherever there is carrion, and roadsides and agricultural areas seem to be especially favored. Turkey Vultures are easy to spot but notoriously difficult to confirm as breeders because they usually place their nests in inaccessible caves and on rocky cliffs.

DISTRIBUTION

Turkey Vultures were recorded in every Nevada county, but breeding was confirmed in only 2 of the 223 blocks where the species was seen. Relative to its abundance, no species was more difficult to confirm as a breeder. Confirmed breeding records came from Elko County, where a bird was discovered on a nest in the Ruby Mountains, and from Churchill County. A third confirmation involved an incidental observation of a nest with young in the Carson City area. Even evidence for probable breeding was hard to come by.

Other atlas projects have reported similar difficulties describing the breeding distribution of Turkey Vultures. For example, few breeders were confirmed in Marin County, California, even though the Turkey Vulture is perhaps the most wide-ranging bird in that county (Shuford 1993). Since it is essentially impossible to distinguish between breeding and nonbreeding individuals (Kirk and Mossman 1998), most Nevada atlas reports that were submitted as "observed" birds (i.e., thirty-nine cases using code Ø) were upgraded to "possible" breeders, assuming that suitable nesting habitat for this species was potentially present in most blocks.

Turkey Vultures are also common and widespread summer residents in all of the states surrounding Nevada (Kirk and Moss-

mann 1998), and it is tempting to characterize them as habitat generalists given the many habitats in which they are observed. Closer inspection, however, shows that the likelihood of finding Turkey Vultures is not the same for all habitats. The predictive map shows a greater chance of finding vultures in the Mojave Desert and in montane habitats of the northern half of the state than at lower elevations in the Great Basin or in urban areas. Within montane areas, vultures seem most likely to occur along the lower slopes of mountains.

A possible explanation for the observed distribution is that vultures favor habitats with rocky cliffs and caves suitable for nesting. Interpreting the distribution data for this species must be done cautiously, however, because so few records refer to known breeders. An alternative explanation for the distribution patterns is that mountain ranges, and especially their lower slopes, generate updrafts that foraging vultures use to get and remain aloft (e.g., see Kirk and Mossman 1998).

CONSERVATION AND MANAGEMENT

Ryser (1985:212) described "an alarming decrease in abundance [of Turkey Vultures] in Nevada during recent years." According to Breeding Bird Survey data and other sources, however, declines are typically local, and Turkey Vultures are increasing in numbers across their North American range (Kirk and Mossman 1998, Sauer et al. 2005).

Locally, Turkey Vultures may be adversely affected by various human activities (Kirk and Mossman 1998). They were once persecuted because of the false belief that they regularly prey on farm animals and spread diseases. Pesticides and heavy metals pose additional threats, especially because vultures are prone to feed on animals that have been killed by toxins or that contain lead shot. Turkey Vultures also are viewed as a major hazard to military aircraft, having caused several serious accidents.

HABITAT USE

HABITAT	ATLAS BLOCKS	INCIDENTAL OBSERVATIONS
Agricultural	14 (5%)	0 (0%)
Alpine	1 (<1%)	1 (1%)
Barren	2 (<1%)	3 (4%)
Grassland	8 (3%)	6 (7%)
Mesquite	1 (<1%)	2 (2%)
Mojave	27 (9%)	9 (11%)
Montane Forest	21 (7%)	3 (4%)
Montane Parkland	2 (<1%)	0 (0%)
Montane Shrub	16 (6%)	7 (8%)
Open Water	1 (<1%)	0 (0%)
Pinyon-Juniper	42 (14%)	6 (7%)
Riparian	40 (14%)	24 (29%)
Salt Desert Scrub	15 (5%)	1 (1%)
Sagebrush Scrub	44 (15%)	12 (15%)
Sagebrush Steppe	41 (14%)	4 (5%)
Urban	8 (3%)	1 (1%)
Wetland	5 (2%)	4 (5%)
TOTAL	288	83

Total numbers reported in the Habitat Use, Breeding Status, and Abundance tables may differ from each other (see pp. 22–23 for details). Percentage sums may differ slightly from 100% due to rounding.

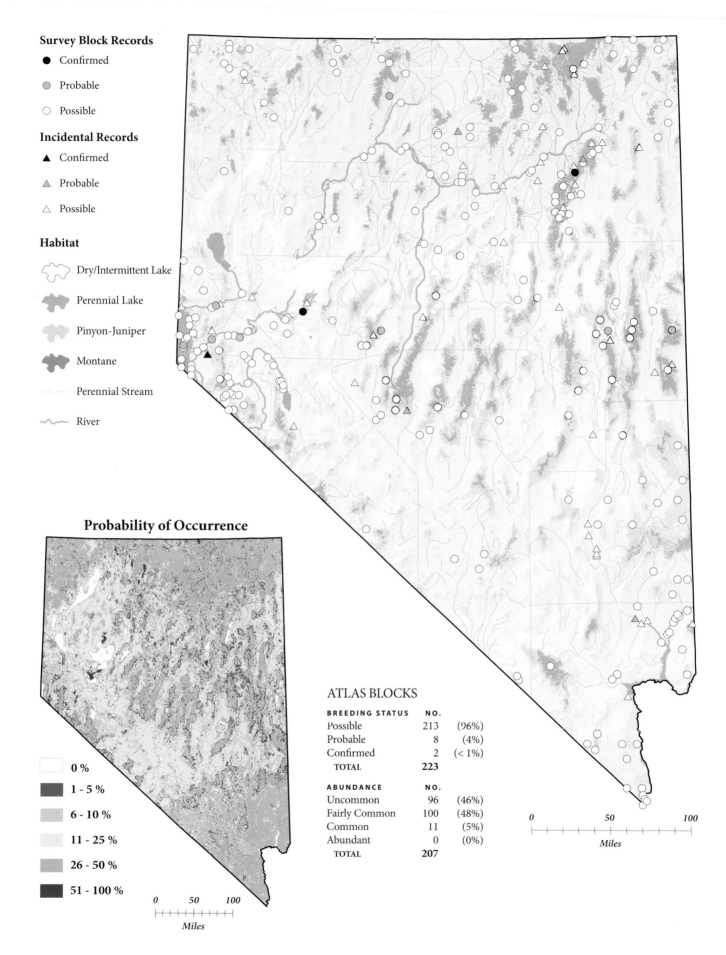

Survey Block Records

● Confirmed

● Probable

○ Possible

Incidental Records

▲ Confirmed

▲ Probable

△ Possible

Habitat

Dry/Intermittent Lake

Perennial Lake

Pinyon-Juniper

Montane

Perennial Stream

River

Probability of Occurrence

	0 %
	1 - 5 %
	6 - 10 %
	11 - 25 %
	26 - 50 %
	51 - 100 %

0   50   100
Miles

ATLAS BLOCKS

BREEDING STATUS	NO.	
Possible	213	(96%)
Probable	8	(4%)
Confirmed	2	(< 1%)
TOTAL	**223**	

ABUNDANCE	NO.	
Uncommon	96	(46%)
Fairly Common	100	(48%)
Common	11	(5%)
Abundant	0	(0%)
TOTAL	**207**	

0   50   100
Miles

OSPREY
Pandion haliaetus

A thoroughly distinctive raptor, the majestic Osprey is an increasingly common sight over the major rivers and large lakes of Nevada. Most birds are wanderers or nonbreeding residents, but breeding has been documented at several sites in western Nevada. Future prospects for the Osprey, in Nevada and beyond, are probably good: it readily accepts artificial nest sites, and poisoning and persecution are no longer the threats they once were. Area birders can play an important role in documenting possible future breeding range expansions of the species in Nevada.

DISTRIBUTION

The Osprey is highly migratory and prone to wandering even during the breeding season (Poole et al. 2002). It has been widely observed during the summer months in Nevada, but for many years the only well-documented breeding records came from the east shore of Lake Tahoe (Herron et al. 1985, Ryser 1985). Recently, Chisholm and Neel (2002) reported breeding at Lahontan Reservoir, confirming Alcorn's (1988) older records from this site. Atlas fieldworkers added several new confirmed and possible breeding locations in western Nevada, and also identified possible breeders in Clark, Elko, Lincoln, and Nye counties. Nests are typically quite conspicuous, so it seems likely that Ospreys have not yet begun breeding in southern Nevada; however, the clustering of summer records suggests that they may do so in the near future.

In neighboring states, Ospreys breed most widely in Idaho and Oregon (Stephens and Sturts 1998, Adamus et al. 2001). California has a fairly substantial breeding population (Small 1994), mostly in the north, whereas breeding in Utah and Arizona is more restricted (Behle et al. 1985, Driscoll in Corman and Wise-Gervais 2005:120–121). Throughout their range Ospreys feed only on fish and usually occur near large bodies of water. Most atlas records came from aquatic habitats, as expected, but there were also a few records from other habitat types. This was doubtless due to some combination of wandering nonbreeders and breeders foraging some distance away from the nest. The predictive map accurately suggests a very low chance of finding Ospreys throughout most of the state. The predictions, however, could be greatly improved by including the actual distribution of suitable Osprey fishing habitat in the model.

CONSERVATION AND MANAGEMENT

The Osprey is migratory and cosmopolitan in distribution, so its conservation status must be viewed in a global context. Widespread and precipitous declines in numbers from the 1950s to the 1970s are attributable largely to pesticides, and subsequent recoveries are the result of pesticide restrictions (Poole et al. 2002). Local declines, however, can occur in response to other factors. For example, Shuford (1993) pointed out that the Osprey's decline in California preceded the DDT era and was probably caused by habitat loss and persecution. The widespread adoption of artificial nesting platforms also suggests that nest site availability can sometimes limit populations.

Osprey populations have increased substantially throughout North America since the 1970s, with some populations doubling in numbers (Poole et al. 2002, Sauer et al. 2005). The species is protected in the United States under the Migratory Bird Treaty Act, and it has benefited from relocation projects in parts of its range (Buckelew and Hall 1994, Walsh et al. 1999). The species has also benefited from the provision—intentional or otherwise—of artificial nest platforms in several regions (Poole et al. 2002). The Osprey's tolerance of people is yet another factor that seems to be working in its favor (Gallagher 1997). Herron et al. (1985) anticipated a range expansion in Nevada, and recent developments seem to have borne out this prediction.

HABITAT USE

HABITAT	ATLAS BLOCKS	INCIDENTAL OBSERVATIONS
Agricultural	1 (10%)	0 (0%)
Montane Forest	3 (30%)	1 (13%)
Open Water	2 (20%)	2 (25%)
Riparian	3 (30%)	2 (25%)
Wetland	1 (10%)	3 (38%)
TOTAL	10	8

Total numbers reported in the Habitat Use, Breeding Status, and Abundance tables may differ from each other (see pp. 22–23 for details). Percentage sums may differ slightly from 100% due to rounding.

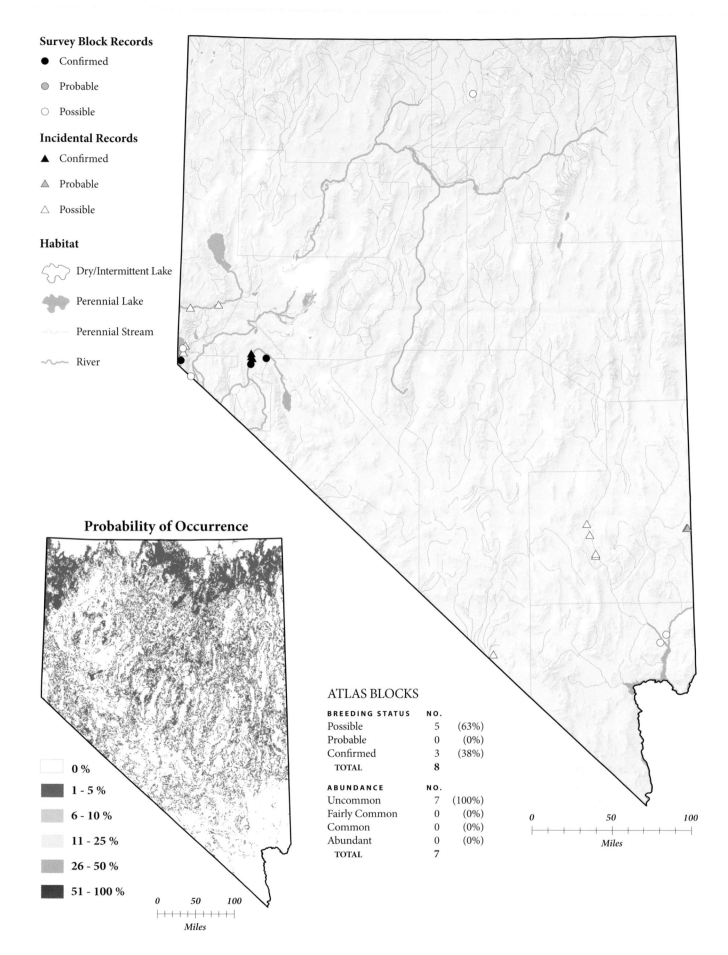

Survey Block Records

● Confirmed

● Probable

○ Possible

Incidental Records

▲ Confirmed

▲ Probable

△ Possible

Habitat

Dry/Intermittent Lake

Perennial Lake

Perennial Stream

River

Probability of Occurrence

☐ 0 %

■ 1 - 5 %

■ 6 - 10 %

☐ 11 - 25 %

■ 26 - 50 %

■ 51 - 100 %

0   50   100
Miles

ATLAS BLOCKS

BREEDING STATUS	NO.	
Possible	5	(63%)
Probable	0	(0%)
Confirmed	3	(38%)
TOTAL	8	

ABUNDANCE	NO.	
Uncommon	7	(100%)
Fairly Common	0	(0%)
Common	0	(0%)
Abundant	0	(0%)
TOTAL	7	

0   50   100
Miles

BALD EAGLE
Haliaeetus leucocephalus

Bald Eagles are on the upswing practically everywhere, and they are an increasingly common sight in Nevada during the winter months. Their summer presence, however, remains very limited. Wandering birds can be seen on occasion through much of the state, but nesting was confirmed only once by atlas fieldworkers. Bald Eagles might increase their breeding activity here, but nesting will be limited by the relative scarcity of their prime habitat—tall trees near large bodies of water.

DISTRIBUTION

Breeding Bald Eagles were nearly absent from Nevada during the twentieth century, and were rather uncommon in the nineteenth. The species nested on Pyramid Lake's Anaho Island in the 1860s, and probably elsewhere as well (Ryser 1985, Alcorn 1988). In the modern era, there have been several nesting attempts at widely scattered locations, but most did not produce fledged young (K. Kritz, pers. comm.). In 1998, a pair of Bald Eagles nested in the Lahontan Valley (Chisholm and Neel 2002), and this nest was the only confirmed atlas breeding record. Eaglets were fledged from this site in every subsequent year through 2003 (K. Kritz, pers. comm.). Atlas workers noted possible breeding in six other widely spaced locations, three of which were in atlas blocks. Most of these records offered fairly meager evidence of breeding and probably referred to nonbreeding summer wanderers. An abandoned nest site in Eureka County was reported to have been built by Bald Eagles, but this could not be verified. K. Kritz (pers. comm.) reported a successfully fledged young in 2000 from a nest at Marlette Lake (Washoe County), a nesting attempt near Carson City in 2001, and hatching of a chick (which subsequently died) in the Ruby Valley in 2005.

In adjacent states, Utah shares Nevada's paucity of breeding records (Behle et al. 1985, Buehler 2000). Arizona does a bit better, with a geographically limited but relatively numerous breeding population (Driscoll in Corman and Wise-Gervais 2005:126–127). The species breeds still more widely along lakes and rivers in northern California, western Oregon, and much of Idaho (Small 1994, Stephens and Sturts 1998, Buehler 2000, Adamus et al. 2001). Bald Eagles are primarily fish eaters with a taste for carrion, and they usually nest in forests or tall trees near large water bodies (Buehler 2000). Such situations are few in Nevada, but one notable exception is the east shore of Lake Tahoe, where breeding has not yet been recorded. Hopeful observers should be aware, however, that along developed shorelines, eagles may nest farther in the uplands than they would in more pristine settings (Buehler 2000). The predictive map hints at where this species might nest, but it is based on too few sightings to be very reliable. An alternative model that identifies forested lands in close proximity to good fishing areas would probably narrow the range of potential Bald Eagle nesting sites.

CONSERVATION AND MANAGEMENT

The Bald Eagle's decline and subsequent recovery constitute one of the best-known conservation case studies of the twentieth century. DDT, and the egg-shell thinning it causes, was primarily responsible for the rangewide decline (Buehler 2000), but direct persecution was perhaps a more important cause of historical declines in Nevada (Herron et al. 1985).

Bald Eagles have been increasing steadily rangewide over the last thirty-five to forty years (Sauer et al. 2005), and the total population is now thought to number around 100,000 birds (Buehler 2000). After decades of federal Endangered status, the Bald Eagle was downgraded to Threatened in 1995, and it may soon be removed from the federal endangered species list altogether. Documenting breeding seems like a straightforward endeavor given the visibility of nests, but many nesting attempts fail, so careful nest monitoring is required (Buehler 2000). Some conservationists have proposed introductions and the provision of artificial perches in regions with very few Bald Eagle nests, like Nevada, but these methods are expensive and not always effective (Buehler 2000). In any case, with Bald Eagles recovering well in most areas, conservation funds might be more profitably directed elsewhere.

HABITAT USE

HABITAT	ATLAS BLOCKS	INCIDENTAL OBSERVATIONS
Agricultural	1 (33%)	0 (0%)
Riparian	1 (33%)	2 (67%)
Sagebrush Steppe	1 (33%)	0 (0%)
Wetland	0 (0%)	1 (33%)
TOTAL	3	3

Total numbers reported in the Habitat Use, Breeding Status, and Abundance tables may differ from each other (see pp. 22–23 for details). Percentage sums may differ slightly from 100% due to rounding.

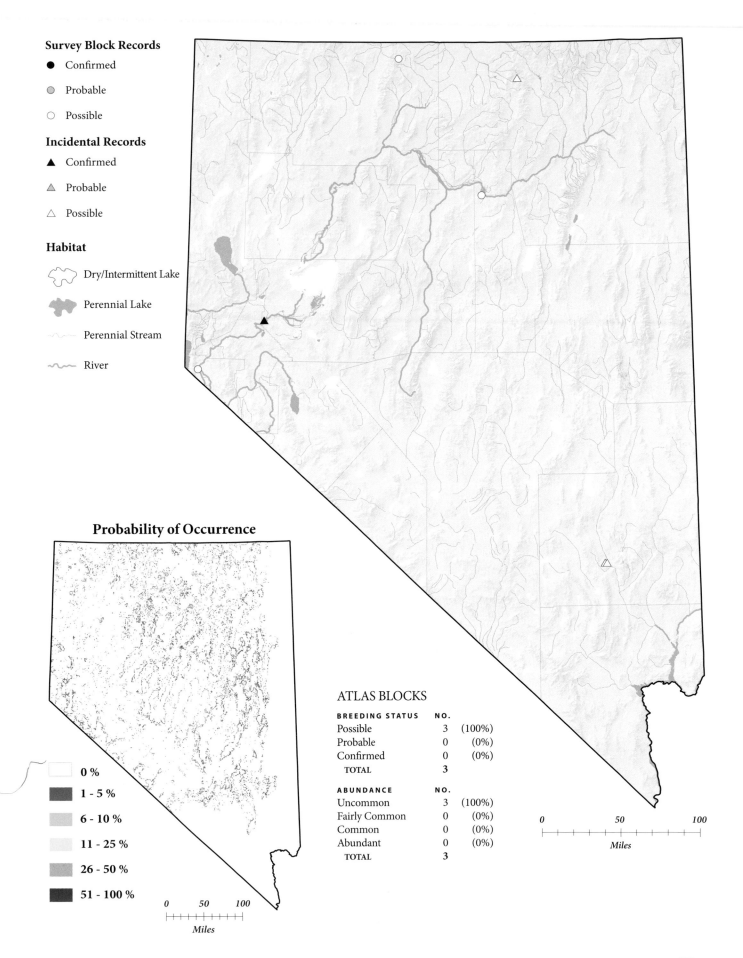

Survey Block Records

● Confirmed

◉ Probable

○ Possible

Incidental Records

▲ Confirmed

△ Probable

△ Possible

Habitat

Dry/Intermittent Lake

Perennial Lake

Perennial Stream

River

Probability of Occurrence

☐ 0 %

■ 1 - 5 %

■ 6 - 10 %

☐ 11 - 25 %

■ 26 - 50 %

■ 51 - 100 %

0   50   100
Miles

ATLAS BLOCKS

BREEDING STATUS	NO.	
Possible	3	(100%)
Probable	0	(0%)
Confirmed	0	(0%)
TOTAL	3	

ABUNDANCE	NO.	
Uncommon	3	(100%)
Fairly Common	0	(0%)
Common	0	(0%)
Abundant	0	(0%)
TOTAL	3	

0   50   100
Miles

NORTHERN HARRIER
Circus cyaneus

There is no shortage of open country in Nevada; nor does there seem to be a shortage of Northern Harriers to patrol it in their stately fashion, looking for prey. These raptors are found in all sorts of treeless expanses over the length and breadth of the state, but they are especially fond of marshes and agricultural areas. Harriers are most often seen in flight, scrutinizing the grass below for voles and mice, and they tend to keep their distance from observers. Around the nest, the harrier is just as apt to be wary, warding off intruders with piercing yelps and menacing dives.

DISTRIBUTION

Northern Harriers were recorded in all of Nevada's counties, but, as Alcorn (1988) also noted earlier, more frequently in the northern than in the southern half of the state. Nesting was difficult to document, with confirmed breeders in only 15% of the occupied blocks.

To our east and north, harriers are widespread breeders in Utah, Idaho, and eastern Oregon (Behle et al. 1985, Stephens and Sturts 1998, Adamus et al. 2001). The harrier's breeding range in California, now somewhat reduced from historical levels, covers the Central Valley, some coastal regions, and the northeast (Small 1994, MacWhirter and Bildstein 1996). Harriers also breed in a small number of Arizona locations (Wise-Gervais in Corman and Wise-Gervais 2005:128–129).

Harriers nest on the ground in open country—even in places as inhospitable as shadscale deserts—and often forage some distance from the nest. The species usually reaches its highest densities in lowland marshes. Linsdale (1936), Herron et al. (1985), and Alcorn (1988) all stressed the harrier's association with wetlands, and marshes and riparian areas accounted for more than one-quarter of the atlas records. As MacWhirter and Bildstein (1996) noted, however, western populations also have a substantial presence in dry, treeless uplands, and these types of locations accounted for many of the atlas records, which were distributed among a large number of habitat categories. The predictive map shows a clear north–south trend, and also highlights

the importance of agricultural areas, which are predicted to be the likeliest spots to harbor this species.

CONSERVATION AND MANAGEMENT

Although the Northern Harrier occurs widely in North America, it is a species of concern in some regions. Harriers declined across their range during the twentieth century due to loss of wetlands, agricultural practices, and eggshell thinning attributed to DDT. These declines may have followed earlier increases in the East during the period of rapid forest clearing (MacWhirter and Bildstein 1996). The declines seem to have slowed since the 1960s, but populations have not fully stabilized (Sauer et al. 2005) and the species has protected status in several regions (MacWhirter and Bildstein 1996). Ongoing loss of wetland habitats and native grasslands remains a concern, and full recovery of populations will probably require habitat conservation measures. MacWhirter and Bildstein (1996) noted that continued or accelerating declines would indicate "widespread disfunction" of the wetlands and uplands on which this species depends.

Harriers seem to have been relatively unaffected by the initial settlement of Nevada (Ryser 1985), perhaps in part because of their extensive use of uplands. Nevada populations probably fluctuate from year to year in response to prey abundance and weather, as they do elsewhere (MacWhirter and Bildstein 1996), but they do not appear to be in any short-term danger (Herron et al. 1985).

HABITAT USE

HABITAT	ATLAS BLOCKS	INCIDENTAL OBSERVATIONS
Agricultural	25 (10%)	3 (4%)
Alpine	1 (<1%)	0 (0%)
Barren	1 (<1%)	0 (0%)
Grassland	16 (6%)	4 (5%)
Mesquite	2 (<1%)	0 (0%)
Mojave	1 (<1%)	2 (3%)
Montane Forest	4 (2%)	1 (1%)
Montane Parkland	2 (<1%)	0 (0%)
Montane Shrub	25 (10%)	2 (3%)
Open Water	0 (0%)	2 (3%)
Pinyon-Juniper	5 (2%)	3 (4%)
Riparian	35 (14%)	22 (29%)
Salt Desert Scrub	33 (13%)	3 (4%)
Sagebrush Scrub	18 (7%)	10 (13%)
Sagebrush Steppe	57 (22%)	6 (8%)
Urban	1 (<1%)	1 (1%)
Wetland	31 (12%)	17 (22%)
TOTAL	257	76

Total numbers reported in the Habitat Use, Breeding Status, and Abundance tables may differ from each other (see pp. 22–23 for details). Percentage sums may differ slightly from 100% due to rounding.

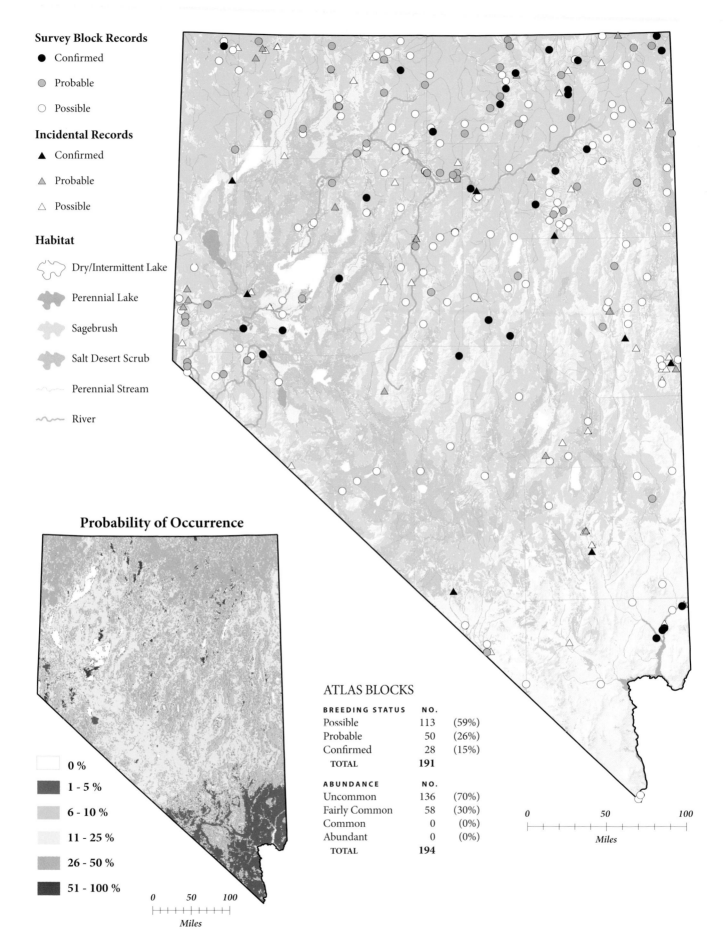

Survey Block Records

● Confirmed

● Probable

○ Possible

Incidental Records

▲ Confirmed

▲ Probable

△ Possible

Habitat

Dry/Intermittent Lake

Perennial Lake

Sagebrush

Salt Desert Scrub

Perennial Stream

River

Probability of Occurrence

☐ 0 %

1 - 5 %

6 - 10 %

11 - 25 %

26 - 50 %

51 - 100 %

0   50   100

Miles

ATLAS BLOCKS

BREEDING STATUS	NO.	
Possible	113	(59%)
Probable	50	(26%)
Confirmed	28	(15%)
TOTAL	191	
ABUNDANCE	NO.	
Uncommon	136	(70%)
Fairly Common	58	(30%)
Common	0	(0%)
Abundant	0	(0%)
TOTAL	194	

0   50   100

Miles

SHARP-SHINNED HAWK

Accipiter striatus

Rn'02

Its savage behavior at winter feeders can easily give one a bad impression of the Sharp-shinned Hawk. The "sharpie" is one of our tiniest raptors, with slender proportions, delicate plumage, a secretive demeanor, and shrill call notes. The diminutive males of this highly dimorphic species are not much bigger than some of their passerine prey. Because they are reclusive and difficult to distinguish from the similar Cooper's Hawk, a great deal remains unknown about this raptor's status in Nevada.

DISTRIBUTION

Atlas workers found Sharp-shinned Hawks in Nevada's major mountain ranges, but the hawks were absent from the northwest and south-central parts of the state, and were nearly absent from the far south. Breeding was difficult to confirm, with only a 16% confirmation rate in atlas blocks. Interestingly, confirmed breeding records came from only four mountain ranges arrayed along a very narrow latitudinal band across the center of the state, where Nevada's largest contiguous forests occur.

Elsewhere in the region, Sharp-shinned Hawks breed widely in Utah (Behle et al. 1985) and in much of Idaho and Oregon (Stephens and Sturts 1998, Adamus et al. 2001). Their breeding range in Arizona is mostly along the Mogollon Rim and in upper-elevation areas in the southeast (Wise-Gervais in Corman and Wise-Gervais 2005:130–131). The exact status of the Sharp-shinned Hawk in California is uncertain, but breeding is prob-

ably most frequent in the northern forests (Small 1994, Bildstein and Meyer 2000).

Sharp-shinned Hawks are birds of forested regions. They accept a large variety of forest types, although they generally use those with some conifers (Bildstein and Meyer 2000). They also may be somewhat restricted to larger tracts of forest. Within the Great Basin, they show an affinity for higher-elevation riparian woodlands (Ryser 1985), and most of the atlas records came from montane forest habitat. Some records also came from pinyon-juniper habitat, and the species was earlier reported from mountain mahogany woodlands in Nevada (Herron et al. 1985). Sharp-shinned Hawks, like Nevada's other two accipiters, are predicted to have a moderate probability of occurrence in mid-elevation woodlands. They are the least likely of the three species to be found at very high elevations, however, and are also less likely than Cooper's Hawks to be seen at low elevations.

CONSERVATION AND MANAGEMENT

The Sharp-shinned Hawk is considered to be one of the more difficult birds to survey in North America (Bildstein and Meyer 2000), with the result that relatively little is known about its status and management needs. Declines were noted during the 1950s–early 1970s, when the pesticide DDT was affecting many predatory birds, but populations began to bounce back after DDT was banned (Bildstein and Meyer 2000). Apparent recent declines in the East seem to be caused by an increase in the number of birds staying in the north for the winter rather than migrating south (Viverette et al. 1996). Breeding Bird Survey data are uniformly deficient in coverage for this species (Sauer et al. 2005), and thus are of little help in clarifying its status.

In Nevada, Linsdale (1936) characterized Sharp-shinned Hawks as common, while Alcorn (1988) regarded them as uncommon. Whether this difference reflects an actual decline in the interim is merely conjectural; current threats to the species in Nevada are not well known. Herron et al. (1985) considered pesticide impacts to be a problem in the 1980s, but little direct evidence points to that as a current problem in Nevada. Like all hawks, this species was persecuted throughout much of the twentieth century, but this threat has abated in recent decades. There is no reason to believe that Nevada's Sharp-shinned Hawk population is currently at risk, but much better monitoring is required to verify this tentative conclusion.

HABITAT USE

HABITAT	ATLAS BLOCKS	INCIDENTAL OBSERVATIONS
Montane Forest	19 (53%)	1 (10%)
Montane Parkland	2 (6%)	0 (0%)
Montane Shrub	2 (6%)	0 (0%)
Pinyon-Juniper	7 (19%)	2 (20%)
Riparian	5 (14%)	6 (60%)
Sagebrush Steppe	1 (3%)	1 (10%)
TOTAL	36	10

Total numbers reported in the Habitat Use, Breeding Status, and Abundance tables may differ from each other (see pp. 22–23 for details). Percentage sums may differ slightly from 100% due to rounding.

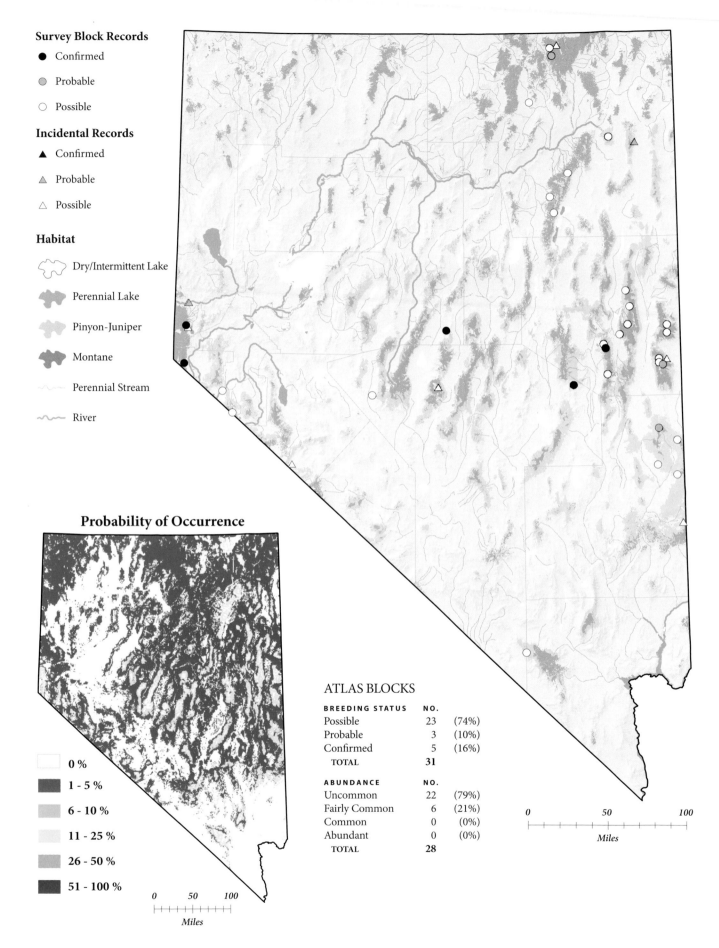

Survey Block Records

● Confirmed

◐ Probable

○ Possible

Incidental Records

▲ Confirmed

△ Probable

△ Possible

Habitat

Dry/Intermittent Lake

Perennial Lake

Pinyon-Juniper

Montane

Perennial Stream

River

Probability of Occurrence

0 %

1 - 5 %

6 - 10 %

11 - 25 %

26 - 50 %

51 - 100 %

0   50   100
Miles

ATLAS BLOCKS

BREEDING STATUS	NO.	
Possible	23	(74%)
Probable	3	(10%)
Confirmed	5	(16%)
TOTAL	31	

ABUNDANCE	NO.	
Uncommon	22	(79%)
Fairly Common	6	(21%)
Common	0	(0%)
Abundant	0	(0%)
TOTAL	28	

0   50   100
Miles

COOPER'S HAWK
Accipiter cooperii

RN'02

Like Nevada's other accipiters, Cooper's Hawk is a sexually dimorphic species that is usually seen in montane woodlands. Unlike Sharp-shinned Hawks and Northern Goshawks, though, Cooper's Hawks range down into low-elevation riparian zones, most notably along the Virgin River, and sometimes breed in urban settings. For example, many observers reported an active nest in Reno every spring during and after the atlas years.

DISTRIBUTION

Cooper's Hawks were widely noted in the mountains that form a broad band across the midsection of the state, from Reno through Ely to the Utah border. There were also multiple records in southern Nevada, mainly from the Spring Mountains, the Sheep Range, and the Virgin River. A few scattered records came from the northern and south-central parts of the state. Cooper's Hawks were frequently sighted in montane forests and pinyon-juniper woodlands, but riparian habitat was recorded for more than a quarter of all block sightings and for more than half of the incidental observations. Breeding in lowland riparian areas had not been noted prior to the atlas fieldwork, and earlier accounts portray Cooper's Hawks as primarily montane breeders in the state (Linsdale 1936, Herron et al. 1985, Alcorn 1988).

Cooper's Hawks were found in about twice as many blocks as Sharp-shinned Hawks or Northern Goshawks, and they are probably the most common of Nevada's accipiters (see Herron et al. 1985). They breed widely in all of the states surrounding Nevada, although they are largely absent as breeders from the lower Colorado River region of southeastern California and southwestern Arizona (Behle et al. 1985, Rosenberg et al. 1991, Rosenfield and Bielefeldt 1993, Small 1994, Stephens and Sturts

1998, Adamus et al. 2001, Wise-Gervais in Corman and Wise-Gervais 2005:132–133).

Cooper's Hawks are forest birds, and they inhabit diverse forest types. Riparian habitat may be particularly important in the southwestern parts of their range (e.g., Rosenberg et al. 1991, Small 1994, Wise-Gervais in Corman and Wise-Gervais 2005:132–133). These hawks appear to be more tolerant of habitat fragmentation and human proximity than Nevada's other accipiters, and they tend to gravitate toward forest edges (Rosenfield and Bielefeldt 1993). The predictive map resembles those for the other two accipiters, but Cooper's Hawks appear to be most likely to occur at lower elevations, while Northern Goshawks are most likely to occur in the higher-elevation forest habitats.

CONSERVATION AND MANAGEMENT

Cooper's Hawks are secretive and difficult to study (Rosenfield and Bielefeldt 1993), although less so than the other accipiters, and good population trend information is lacking. Declines likely occurred in the twentieth century due to habitat destruction and pesticides, but the situation during the last few decades is harder to characterize. Recent population increases are suspected in some parts of the East, although the species retains special protection in many eastern states (Rosenfield and Bielefeldt 1993). Western populations are thought to have been more stable, but local declines stemming from riparian habitat degradation have been noted in relatively recent times in California and the lower Colorado River region (Rosenberg et al. 1991, Rosenfield and Bielefeldt 1993, Small 1994).

Cooper's Hawk is a Priority species for Nevada Partners in Flight (Neel 1999a), although the true status of our populations is not well known. Potential threats include habitat fragmentation, fires, logging, grazing, and pesticides (Neel 1999a; but see Herron et al. 1985, and Rosenfield and Bielefeldt 1993). Twenty years ago, both Herron et al. (1985) and Ryser (1985) considered Nevada populations to be healthy, but the atlas results should be followed up with additional monitoring to evaluate current trends.

HABITAT USE

HABITAT	ATLAS BLOCKS	INCIDENTAL OBSERVATIONS
Agricultural	1 (1%)	0 (0%)
Mojave	1 (1%)	0 (0%)
Montane Forest	24 (29%)	11 (24%)
Montane Parkland	3 (4%)	0 (0%)
Montane Shrub	5 (6%)	1 (2%)
Open Water	0 (0%)	1 (2%)
Pinyon-Juniper	19 (23%)	3 (7%)
Riparian	23 (27%)	25 (56%)
Salt Desert Scrub	1 (1%)	0 (0%)
Sagebrush Scrub	4 (5%)	2 (4%)
Sagebrush Steppe	2 (2%)	1 (2%)
Urban	0 (0%)	1 (2%)
Wetland	1 (1%)	0 (0%)
TOTAL	84	45

Total numbers reported in the Habitat Use, Breeding Status, and Abundance tables may differ from each other (see pp. 22–23 for details). Percentage sums may differ slightly from 100% due to rounding.

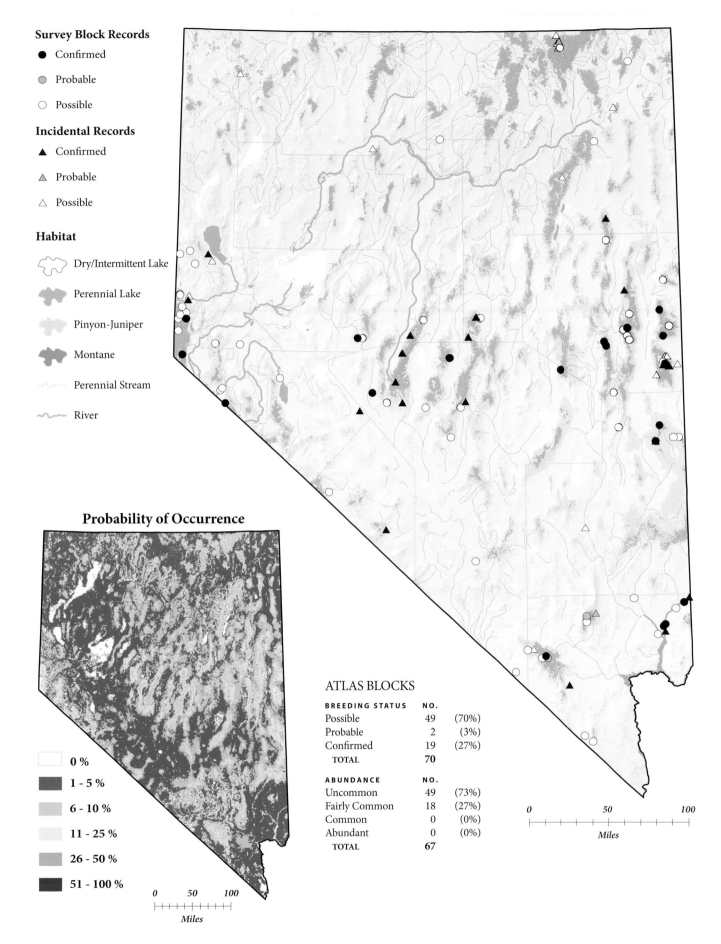

Survey Block Records

● Confirmed

◉ Probable

○ Possible

Incidental Records

▲ Confirmed

▲ Probable

△ Possible

Habitat

Dry/Intermittent Lake

Perennial Lake

Pinyon-Juniper

Montane

Perennial Stream

River

Probability of Occurrence

☐ 0 %

1 - 5 %

6 - 10 %

11 - 25 %

26 - 50 %

51 - 100 %

0   50   100
Miles

ATLAS BLOCKS

BREEDING STATUS	NO.	
Possible	49	(70%)
Probable	2	(3%)
Confirmed	19	(27%)
TOTAL	70	

ABUNDANCE	NO.	
Uncommon	49	(73%)
Fairly Common	18	(27%)
Common	0	(0%)
Abundant	0	(0%)
TOTAL	67	

0   50   100
Miles

NORTHERN GOSHAWK
Accipiter gentilis

One of the best ways to get to know Northern Goshawks is to accidentally wander into a breeding territory. These raptors are aggressively defensive during breeding and will attempt (often successfully) to drive away any birder or hiker who gets too close to a nest. Away from their nests, though, goshawks can be hard to find. In a typical encounter, a birder catches a brief glimpse of this big, bluish accipiter as it darts through mid-story branches or zips across a forest clearing.

DISTRIBUTION

Northern Goshawks were found in all of Nevada's mid-latitude mountain ranges and in the mountains of northern Elko County. Unconfirmed breeding records were also reported from the Ruby Mountains and from the Spring Mountains in the far south. Linsdale (1936) mentioned only one breeding location for the species in Nevada, but Herron et al. (1985), Alcorn (1988), and Neel (1999a) reported a much broader distribution more consistent with the atlas findings. Additional populations have been reported from the Santa Rosa and other northwestern ranges (W. Molini, pers. comm.). The increase in known breeding sites over time likely represents a growing number of knowledgeable observers rather than an actual population increase (Herron 1999).

Nesting Goshawks are territorial and widely spaced, and thus are not particularly common anywhere (Squires and Reynolds 1997, Herron 1999). They breed throughout most forested areas of the states surrounding Nevada (Behle et al. 1985, Squires and Reynolds 1997, Stephens and Sturts 1998, Adamus et al. 2001,

Wise-Gervais in Corman and Wise-Gervais 2005:134–135), except in southern California, where they are quite rare (Small 1994). Northern Goshawks nest in various forest types throughout their range, with a preference for taller, mature stands with significant canopy closure. They forage in their nesting habitat or in a diverse set of adjacent habitats (Squires and Reynolds 1997). In Nevada and Utah, most Northern Goshawks nest in aspen "stringers," the bands of aspen that trace mountain streams and ephemeral drainages (Behle et al. 1985, Herron et al. 1985, Ryser 1985, Younk and Bechard 1994, Herron 1999). Largely consistent with this observation, most habitat records from the atlas fieldwork were from montane forests and riparian habitats (aspen was not recorded as a separate habitat type). Other records came from shrub-dominated habitats that were likely used for foraging or travel. More than the other accipiters, goshawks are tightly associated with Nevada's montane forest "islands," where the predictive map indicates there is a fair chance of finding them. More detailed information on microhabitat and landscape-level determinants of goshawk nesting in Nevada can be found in Herron et al. 1985 and Neel 1999a.

CONSERVATION AND MANAGEMENT

The Northern Goshawk is a Nevada Partners in Flight Priority species (Neel 1999a) due to its small population size and to widespread concern that it may be declining in the West (Reynolds et al. 1992). Nest disturbance and some types of logging practices are cited as threats to goshawks rangewide (Squires and Reynolds 1997). In Nevada, loss and degradation of suitable aspen stands is probably the most pressing concern (Neel 1999a).

Goshawk populations are difficult to survey (Squires and Reynolds 1997), and Breeding Bird Survey data are probably not very informative for this species. Herron et al. (1985) estimated three hundred territories in the state twenty years ago, but the current figure may be higher (Herron 1999). Most goshawks in Nevada nest on national forest lands, where they have Sensitive Species status and are managed under a detailed set of guidelines (summarized in Reynolds et al. 1992). In addition, the Humboldt-Toiyabe District of the USDA Forest Service and the Nevada Department of Wildlife are currently devising a cooperative statewide monitoring plan for Nevada goshawk populations.

HABITAT USE

HABITAT	ATLAS BLOCKS	INCIDENTAL OBSERVATIONS
Montane Forest	9 (36%)	8 (40%)
Montane Parkland	1 (4%)	0 (0%)
Montane Shrub	4 (16%)	1 (5%)
Pinyon-Juniper	4 (16%)	2 (10%)
Riparian	5 (20%)	8 (40%)
Sagebrush Scrub	1 (4%)	1 (5%)
Sagebrush Steppe	1 (4%)	0 (0%)
TOTAL	25	20

Total numbers reported in the Habitat Use, Breeding Status, and Abundance tables may differ from each other (see pp. 22–23 for details). Percentage sums may differ slightly from 100% due to rounding.

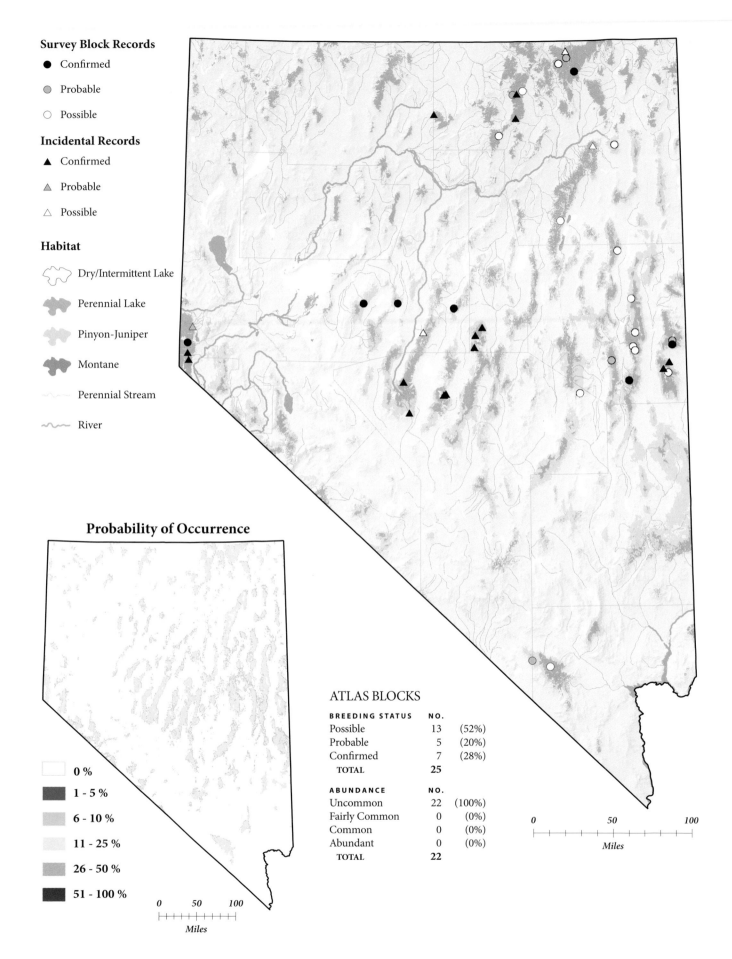

Survey Block Records

● Confirmed

● Probable

○ Possible

Incidental Records

▲ Confirmed

▲ Probable

△ Possible

Habitat

Dry/Intermittent Lake

Perennial Lake

Pinyon-Juniper

Montane

Perennial Stream

River

Probability of Occurrence

0 %

1 - 5 %

6 - 10 %

11 - 25 %

26 - 50 %

51 - 100 %

0   50   100
Miles

ATLAS BLOCKS

BREEDING STATUS	NO.	
Possible	13	(52%)
Probable	5	(20%)
Confirmed	7	(28%)
TOTAL	25	

ABUNDANCE	NO.	
Uncommon	22	(100%)
Fairly Common	0	(0%)
Common	0	(0%)
Abundant	0	(0%)
TOTAL	22	

0   50   100
Miles

RED-SHOULDERED HAWK
Buteo lineatus

Prior to the atlas survey, the Red-shouldered Hawk was considered an occasional wanderer into Nevada. The conventional wisdom was that in the West, the species was restricted to central and coastal California. Several records during the atlas period, however, including one confirmed breeding record, seem to indicate an ongoing range expansion from California. The extent to which increased observer effort contributed to this result is not completely clear, and for now it is difficult to prognosticate about the future of the Red-shouldered Hawk in Nevada.

DISTRIBUTION

Until the late 1990s, the Red-shouldered Hawk was rarely seen in Nevada. Linsdale (1936) reported no records of the species, nor was it included in Herron et al.'s (1985) summary of Nevada raptors. Ryser (1985) considered the Red-shouldered Hawk accidental in the Great Basin, while Alcorn (1988) cited only a handful of records throughout the state. Chisholm and Neel (2002) discussed recent increases in Red-shouldered Hawk sightings in the Lahontan Valley and speculated on the possibility of breeding in western Nevada. In the last year of the atlas, fieldworkers observed Red-shouldered Hawks at several widely scattered Nevada locations. Evidence of breeding was scant for most of these records. At Pahranagat National Wildlife Refuge, however, Mike San Miguel and Tom Wurster saw a juvenile begging for food; this constituted the first known breeding record for Nevada. In the fall of 2000, just after the atlas fieldwork was completed, Red-shouldered Hawks were reported from at least fourteen widely scattered locales in Nevada (Floyd and Stackhouse 2001)—an unprecedented showing by the species in the state.

The Red-shouldered Hawk is primarily an eastern species. Its only established breeding stronghold in the West is in western and central California (Small 1994), home to the *elegans* subspecies (Crocoll 1994). The recent increase in Red-shouldered Hawk sightings in Nevada appears to be part of a larger pattern, however. The species was first found nesting on the eastern slope of the Sierra Nevada in California in 1992 (Small 1994). Breeding was recently confirmed for the first time in southwestern Oregon (Adamus et al. 2001), and numbers also appear to have increased in the lower Colorado River Valley, where breeding is rare (Rosenberg et al. 1991). In Idaho, Arizona, and Utah, Red-shouldered Hawks are still considered accidental visitors or rare wanderers, although they nest in Arizona on rare occasions (Behle et al. 1985, Stephens and Sturts 1998, Corman in Corman and Wise-Gervais 2005:591, UOSBRC 2005).

Since Red-shouldered Hawks were not found in any atlas blocks, it was not possible to generate a predictive map, and records were too few overall to draw general conclusions about habitat use. In California, the *elegans* subspecies generally occurs in riparian and oak woodlands, but the species as a whole is quite adaptable and breeds in many different forest types throughout its range (Crocoll 1994).

CONSERVATION AND MANAGEMENT

Unlike the late-twentieth-century recoveries of the Bald Eagle, Osprey, and Peregrine Falcon in Nevada, the apparent influx of Red-shouldered Hawks into the state was largely unanticipated. Significant declines occurred in the eastern part of the range in the twentieth century, presumably due to pesticides, habitat loss, and human disturbances around nest sites. More recently, populations have apparently been increasing rangewide (Sauer et al. 2005). Breeding Bird Survey data from California, where Nevada's birds are presumed to have originated, suggest a significant increase in numbers in recent decades (Sauer et al. 2005), which may be driving the greater frequency of sightings in neighboring states. Nevadans occasionally grouse about California imports, but the Red-shouldered Hawk is a welcome addition to the state's avifauna. Nevada birders throughout the state should be on the lookout for additional breeding attempts by this elegant, sleek raptor.

HABITAT USE

HABITAT	ATLAS BLOCKS	INCIDENTAL OBSERVATIONS
Agricultural	0 (0%)	1 (20%)
Pinyon-Juniper	0 (0%)	1 (20%)
Riparian	0 (0%)	2 (40%)
Wetland	0 (0%)	1 (20%)
TOTAL	0	5

Total numbers reported in the Habitat Use, Breeding Status, and Abundance tables may differ from each other (see pp. 22–23 for details). Percentage sums may differ slightly from 100% due to rounding.

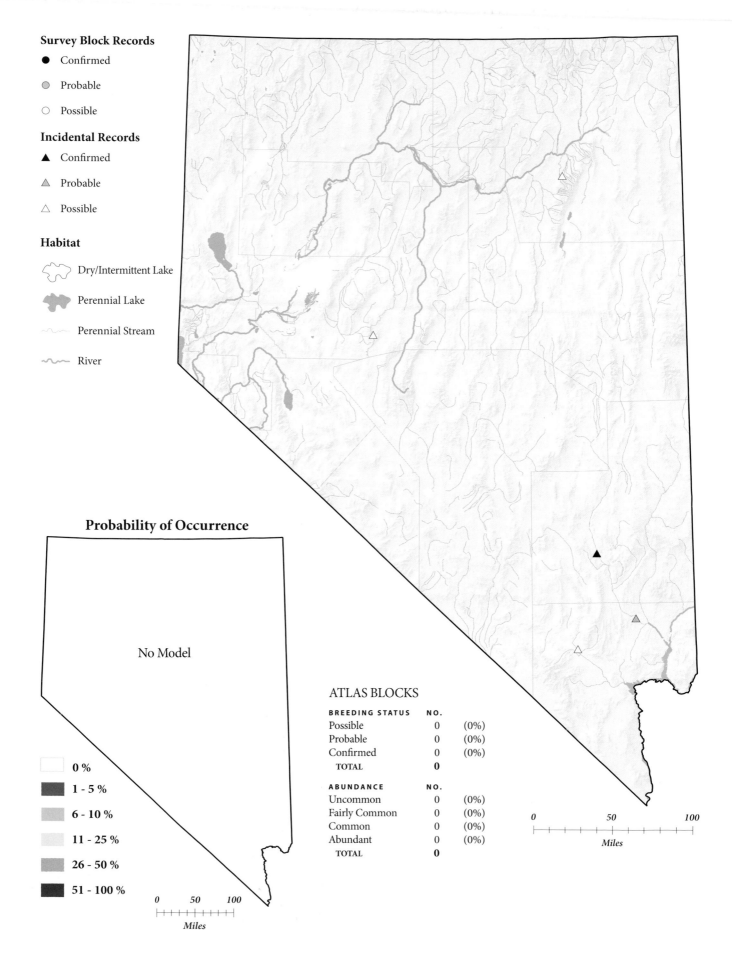

Survey Block Records

● Confirmed

◉ Probable

○ Possible

Incidental Records

▲ Confirmed

▲ Probable

△ Possible

Habitat

Dry/Intermittent Lake

Perennial Lake

Perennial Stream

River

Probability of Occurrence

No Model

☐ 0 %

■ 1 - 5 %

■ 6 - 10 %

☐ 11 - 25 %

■ 26 - 50 %

■ 51 - 100 %

0   50   100
Miles

ATLAS BLOCKS

BREEDING STATUS	NO.	
Possible	0	(0%)
Probable	0	(0%)
Confirmed	0	(0%)
TOTAL	0	

ABUNDANCE	NO.	
Uncommon	0	(0%)
Fairly Common	0	(0%)
Common	0	(0%)
Abundant	0	(0%)
TOTAL	0	

0   50   100
Miles

SWAINSON'S HAWK
Buteo swainsoni

In early April, when the mountains are still snowcapped and a chill hangs in the air, Swainson's Hawks return to the open agricultural valleys of Nevada, having just completed their amazing migration from the South American pampas. Although most raptor populations are recovering from earlier declines, Swainson's Hawks appear still to be in trouble. Land use changes on their breeding grounds may play some role, but the hawks also face numerous perils during migration and on their wintering grounds. Swainson's Hawk is therefore a species of great conservation interest in Nevada.

DISTRIBUTION

Most atlas records came from northern Nevada, running northeastward from western Nevada through the Lahontan Valley and the Humboldt River drainage. The atlas records largely match earlier accounts by Linsdale (1936) and Herron et al. (1985), who reported summering Swainson's Hawks only in the northern two-thirds of the state. Neel (1999a), however, also noted breeding records as far south as Ash Meadows National Wildlife Refuge, and indeed, there were two confirmed atlas records from that region in southern Nye County.

The distribution of Swainson's Hawks centers on the valleys of the Intermountain West and the Great Plains. They breed in eastern Oregon, southern Idaho, southeastern Arizona, and most of Utah (Behle et al. 1985, Stephens and Sturts 1998, Adamus et al. 2001, Wise-Gervais in Corman and Wise-Gervais 2005:142–143). In California, most breeding occurs in the Central Valley and near the northern Nevada border, but populations have declined in historical times (Small 1994). The relatively large number of atlas records in Nevada seems to contradict Herron et al. (1985), who estimated fewer than 150 pairs in the state.

Swainson's Hawk is a bird of open grasslands and shrublands from the Great Plains westward. It has adapted well to agricultural areas, especially where native habitat and farmland are interspersed (England et al. 1997). Herron et al. (1985) found most Nevada nests in the agricultural valleys, and about a fourth of the atlas records came from agricultural habitat. Swainson's Hawks typically nest in scattered deciduous trees near open areas where they forage for insects and other small prey (England et al. 1997). Their use of farmland is reflected in the predictive map, which shows a moderately high chance of occurrence in agricultural areas. The chance of occurrence is quite low everywhere else, although it increases slightly in the far northern reaches of the state. The somewhat higher probabilities in the northern Mojave Desert probably overrepresent the species' occurrence in that part of Nevada.

CONSERVATION AND MANAGEMENT

Swainson's Hawks were once among the West's most abundant raptors, but for some time after European settlement they were persecuted as "varmints"—which is ironic because these hawks prey on agricultural pests (England et al. 1997). Accounts suggesting precipitous declines through the early 1900s (see England et al. 1997) probably include Nevada populations (Herron et al. 1985, Ryser 1985), and numbers have likely continued to decline (England et al. 1997, Paige and Ritter 1999). Contributing causes may include pesticide-induced winter mortality in South America, declines in breeding habitat quality, loss of grasslands and suitable nesting trees within agricultural settings, and other habitat changes (England et al. 1997, Paige and Ritter 1999).

There is considerable concern for Swainson's Hawk, which is a Partners in Flight Watch List species rangewide (Rich et al. 2004) and a Priority species in Nevada (Neel 1999a). Implementation of existing conservation strategies (e.g., Neel 1999a) that address the ecological requirements of this species is probably the most important management strategy.

HABITAT USE

HABITAT	ATLAS BLOCKS	INCIDENTAL OBSERVATIONS
Agricultural	12 (23%)	7 (26%)
Grassland	7 (13%)	3 (11%)
Mojave	0 (0%)	2 (7%)
Montane Shrub	1 (2%)	1 (4%)
Pinyon-Juniper	1 (2%)	0 (0%)
Riparian	13 (25%)	3 (11%)
Salt Desert Scrub	2 (4%)	0 (0%)
Sagebrush Scrub	3 (6%)	6 (22%)
Sagebrush Steppe	8 (15%)	2 (7%)
Urban	1 (2%)	2 (7%)
Wetland	4 (8%)	1 (4%)
TOTAL	52	27

Total numbers reported in the Habitat Use, Breeding Status, and Abundance tables may differ from each other (see pp. 22–23 for details). Percentage sums may differ slightly from 100% due to rounding.

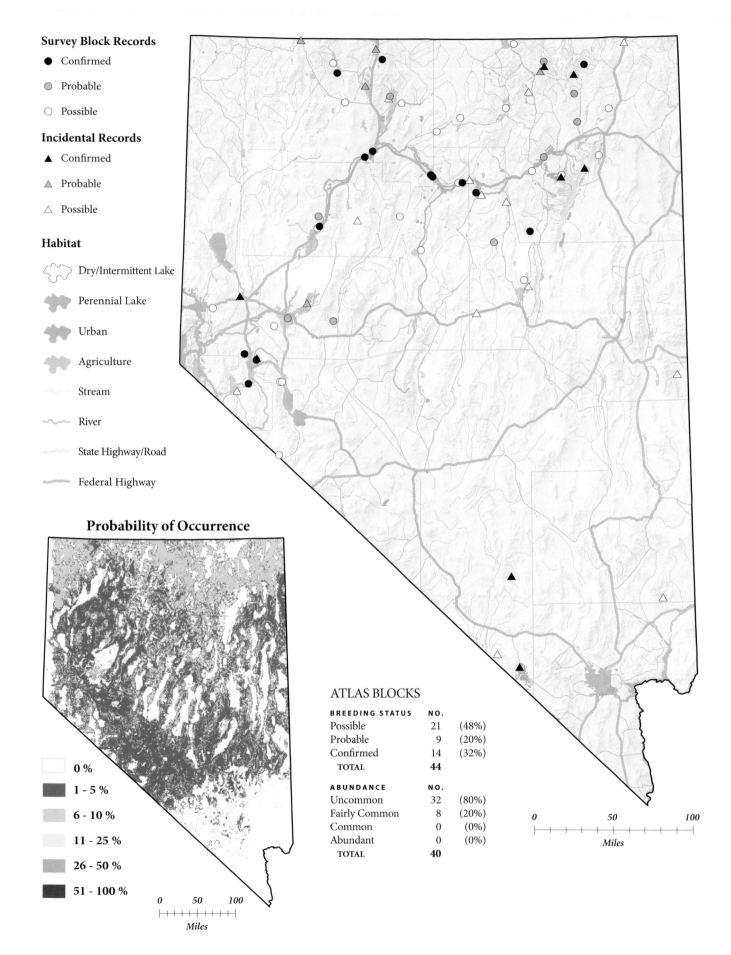

Survey Block Records

● Confirmed

● Probable

○ Possible

Incidental Records

▲ Confirmed

▲ Probable

△ Possible

Habitat

Dry/Intermittent Lake

Perennial Lake

Urban

Agriculture

Stream

River

State Highway/Road

Federal Highway

Probability of Occurrence

0 %

1 - 5 %

6 - 10 %

11 - 25 %

26 - 50 %

51 - 100 %

0   50   100

Miles

ATLAS BLOCKS

BREEDING STATUS	NO.	
Possible	21	(48%)
Probable	9	(20%)
Confirmed	14	(32%)
TOTAL	44	

ABUNDANCE	NO.	
Uncommon	32	(80%)
Fairly Common	8	(20%)
Common	0	(0%)
Abundant	0	(0%)
TOTAL	40	

0   50   100

Miles

RED-TAILED HAWK

Buteo jamaicensis

In Nevada, when someone says they have seen "a hawk," they usually mean a Red-tailed Hawk, for there are few Nevada habitats that do not support this widespread and versatile buteo. They can be found in wilderness areas in the high mountains, in the suburbs, in the low desert, and virtually everywhere in between. The Red-tailed Hawk requires only three basic elements for successful breeding: a mature tree set in an open space for hunting, a prey base consisting primarily of small mammals, and some protection from human disturbance.

DISTRIBUTION

Red-tailed Hawks were found in all Nevada counties, and breeding was confirmed throughout the state. Dense clusters of records occurred throughout the state as well, but this pattern may be partly due to greater survey efforts in the more accessible areas. Red-tailed Hawks are not only ubiquitous now; they seem to have been common ever since ornithologists started observing Nevada's avifauna (Ridgway 1877, Linsdale 1936, Herron et al. 1985, Alcorn 1988).

The Red-tailed Hawk is widespread in the western parts of its U.S. range (Preston and Beane 1993), and, for a raptor, is considered fairly common to common in the states surrounding Nevada (Behle et al. 1985, Small 1994, Stephens and Sturts 1998, Adamus et al. 2001, Wise-Gervais in Corman and Wise-Gervais 2005:146–147). The only sizable areas where the species tends to be scarce are large expanses of dense forest and completely treeless deserts; for instance, Red-tailed Hawks are considered uncommon breeders in the deserts of the lower Colorado River region (Rosenberg et al. 1991).

Red-tailed Hawks are habitat generalists, but they do require elevated nest and perch sites (Preston and Beane 1993). Within Nevada, they breed in settings as diverse as large cities, isolated agricultural lands, and remote mountain areas—on all but the very highest, treeless mountaintops (Herron et al. 1985). Not surprisingly, atlas workers observed Red-tailed Hawks in virtually all habitats, but most often in riparian areas, montane forests, sagebrush, and Mojave habitat. These patterns are also apparent on the predictive map; but even in areas of salt desert scrub, the probability of occurrence is estimated to be fairly high.

CONSERVATION AND MANAGEMENT

Red-tailed Hawk populations are increasing in much of North America (Preston and Beane 1993, Sauer et al. 2005). In the West, the species has benefited from human actions such as the planting of exotic trees (Shuford 1993) and fire suppression that has allowed patches of woodland to expand into open rangelands (Preston and Beane 1993). Although some persecution may still occur, it appears that Red-tailed Hawk populations are regulated primarily by other factors (Preston and Beane 1993).

Breeding Bird Survey data suggest a stable population in Nevada (Sauer et al. 2005), and the species appears to have adapted to local changes within the state (Herron et al. 1985). For example, Red-tailed Hawks have fared well in the Lahontan Valley (Chisholm and Neel 2002), a region greatly altered by humans in the twentieth century. The species enjoys good publicity in Nevada and is generally applauded for its habit of eating rodents (Herron et al. 1985). Given its current and projected population health, it might seem that the Red-tailed Hawk should be a low conservation priority in our area. Western Nevada, however, has high concentrations of wintering birds (Preston and Beane 1993) and may also be important to populations that breed outside our region.

HABITAT USE

HABITAT	ATLAS BLOCKS	INCIDENTAL OBSERVATIONS
Agricultural	25 (6%)	5 (3%)
Alpine	4 (1%)	0 (0%)
Barren	8 (2%)	1 (<1%)
Grassland	3 (<1%)	7 (5%)
Mesquite	2 (<1%)	1 (<1%)
Mojave	39 (10%)	18 (12%)
Montane Forest	51 (13%)	14 (9%)
Montane Parkland	5 (1%)	1 (<1%)
Montane Shrub	30 (7%)	4 (3%)
Open Water	2 (<1%)	0 (0%)
Pinyon-Juniper	48 (11%)	11 (7%)
Riparian	56 (14%)	45 (29%)
Salt Desert Scrub	20 (5%)	1 (<1%)
Sagebrush Scrub	39 (10%)	26 (17%)
Sagebrush Steppe	49 (12%)	7 (5%)
Urban	18 (4%)	10 (7%)
Wetland	5 (1%)	2 (1%)
TOTAL	404	153

Total numbers reported in the Habitat Use, Breeding Status, and Abundance tables may differ from each other (see pp. 22–23 for details). Percentage sums may differ slightly from 100% due to rounding.

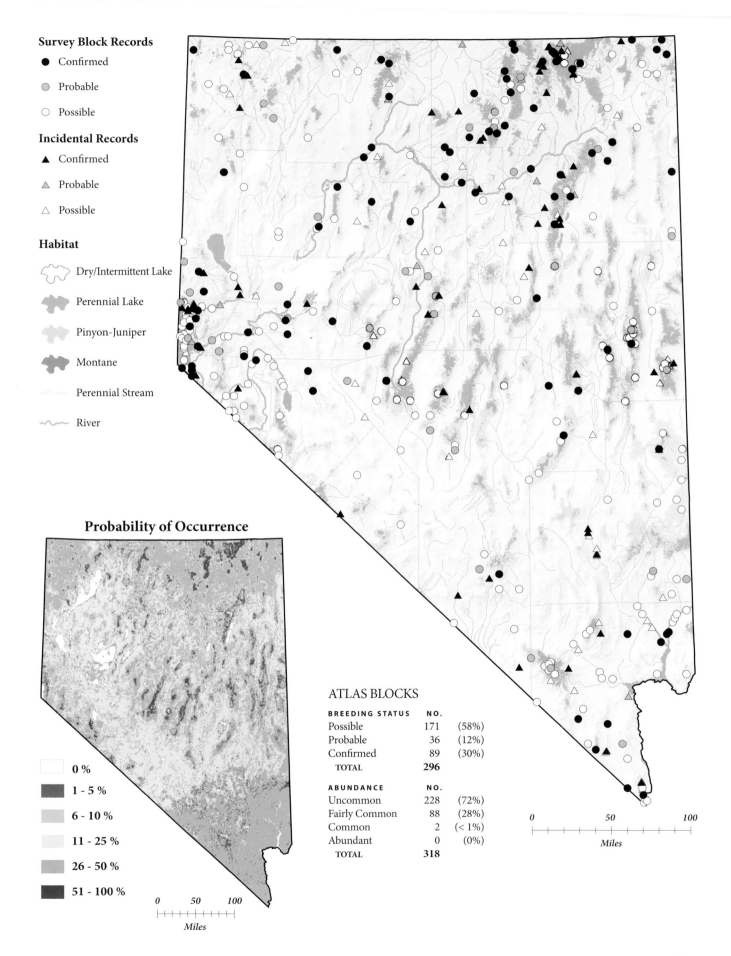

Survey Block Records

● Confirmed

● Probable

○ Possible

Incidental Records

▲ Confirmed

▲ Probable

△ Possible

Habitat

Dry/Intermittent Lake

Perennial Lake

Pinyon-Juniper

Montane

Perennial Stream

River

Probability of Occurrence

- 0 %
- 1 - 5 %
- 6 - 10 %
- 11 - 25 %
- 26 - 50 %
- 51 - 100 %

0   50   100
Miles

ATLAS BLOCKS

BREEDING STATUS	NO.	
Possible	171	(58%)
Probable	36	(12%)
Confirmed	89	(30%)
TOTAL	**296**	

ABUNDANCE	NO.	
Uncommon	228	(72%)
Fairly Common	88	(28%)
Common	2	(< 1%)
Abundant	0	(0%)
TOTAL	**318**	

0   50   100
Miles

FERRUGINOUS HAWK
Buteo regalis

With its ghostly plumage and melancholy call, the Ferruginous Hawk seems right at home in the blanched sagebrush desert of northeastern Nevada. Nowhere common, Ferruginous Hawks are most likely to be found where the sagebrush is interspersed with occasional junipers. According to some estimates, the range-wide population size may be only about ten thousand birds (Bechard and Schmutz 1995). Conservation in Nevada and elsewhere is therefore a high priority.

DISTRIBUTION

Atlas records for the Ferruginous Hawk were concentrated in the east-central part of Nevada—in Lander, Eureka, White Pine, and southern Elko counties—with breeding confirmed as far south as central Nye County. Although Herron et al. (1985) cited isolated breeding records from throughout the state, atlas workers found Ferruginous Hawks to be rare in Nevada's far northern tier, and essentially absent from the far west and far south (see also Ryser 1985).

Nevada lies along the southwestern edge of the Ferruginous Hawk's breeding range, which extends from the northwestern Great Plains to the Great Basin. There are scarcely any California nesting records (Small 1994), and the species does not breed along the lower Colorado River Valley (Rosenberg et al. 1991). Scattered breeding occurs in northern Arizona and eastern Oregon, and a bit more commonly in southern Idaho, western Utah, and northeast into Wyoming and the northern plains

(Behle et al. 1985, Stephens and Sturts 1998, Adamus et al. 2001, Dobkin and Sauder 2004, Corman in Corman and Wise-Gervais 2005:148–149).

Like the Northern Harrier and Prairie Falcon, the Ferruginous Hawk is a bird of open country. It inhabits grasslands and shrublands, and avoids forests, steep terrain, and high elevations (Bechard and Schmutz 1995, Dobkin and Sauder 2004). As the predictive map indicates, Ferruginous Hawks are most likely to be found in sagebrush shrublands, with lower probabilities of occurrence in salt desert scrub and sagebrush steppe. Several writers (Ryser 1985, Alcorn 1988) have emphasized the importance of juniper trees as nest sites, but atlas surveys produced few records from pinyon-juniper blocks. In general, Ferruginous Hawks seem to use landscapes with scattered trees present among the sagebrush for nest sites and open foraging grounds all around (Bechard and Schmutz 1995). More complex models that describe this combination of habitat features will likely improve predictions about the species' range in Nevada.

CONSERVATION AND MANAGEMENT

The Ferruginous Hawk is uncommon throughout its range, and there is some concern that its numbers have been declining. These hawks target fairly large mammalian prey such as rabbits and ground squirrels, and the greatest historical impacts may have occurred with the near extirpation of prairie dogs from the Great Plains (Paige and Ritter 1999). Current population trends are unclear. The general impression is one of decline (Bechard and Schmutz 1995), but the limited Breeding Bird Survey data do not show a reliable trend (Sauer et al. 2005). Total numbers are low, and populations are thought to be sensitive to heavy grazing, conversion of rangeland, and nest disturbance (Bechard and Schmutz 1995, Paige and Ritter 1999, Dobkin and Sauder 2004).

Herron et al. (1985) reported Nevada populations to be healthy and stable, but quantitative data to support this conclusion are limited. Statistics provided by Herron et al. (1985) and Bechard and Schmutz (1995) indicate that 10% of the world's population of Ferruginous Hawks may breed in Nevada. Because of its rarity and uncertain status, Nevada Partners in Flight has designated the Ferruginous Hawk a Priority species (Neel 1999a). Better information on population status and current trends is needed for future conservation planning.

HABITAT USE

HABITAT	ATLAS BLOCKS	INCIDENTAL OBSERVATIONS
Agricultural	3 (8%)	1 (4%)
Grassland	4 (11%)	4 (14%)
Pinyon-Juniper	8 (22%)	6 (21%)
Riparian	2 (6%)	0 (0%)
Salt Desert Scrub	5 (14%)	4 (14%)
Sagebrush Scrub	11 (31%)	12 (43%)
Sagebrush Steppe	3 (8%)	1 (4%)
TOTAL	36	28

Total numbers reported in the Habitat Use, Breeding Status, and Abundance tables may differ from each other (see pp. 22–23 for details). Percentage sums may differ slightly from 100% due to rounding.

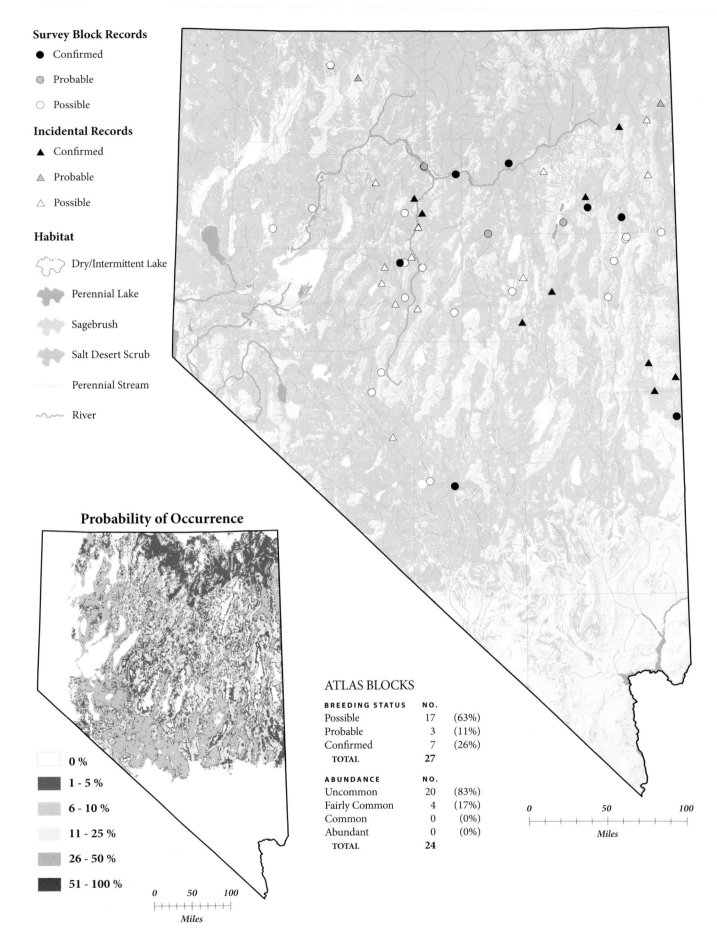

Survey Block Records

● Confirmed

● Probable

○ Possible

Incidental Records

▲ Confirmed

▲ Probable

△ Possible

Habitat

Dry/Intermittent Lake

Perennial Lake

Sagebrush

Salt Desert Scrub

Perennial Stream

River

Probability of Occurrence

0 %

1 - 5 %

6 - 10 %

11 - 25 %

26 - 50 %

51 - 100 %

0   50   100

Miles

ATLAS BLOCKS

BREEDING STATUS	NO.	
Possible	17	(63%)
Probable	3	(11%)
Confirmed	7	(26%)
TOTAL	27	

ABUNDANCE	NO.	
Uncommon	20	(83%)
Fairly Common	4	(17%)
Common	0	(0%)
Abundant	0	(0%)
TOTAL	24	

0   50   100

Miles

GOLDEN EAGLE

Aquila chrysaetos

Solitary, unapproachable, and always imposing, the Golden Eagle is one of our most diagnostic avian indicators of wilderness and unspoiled solitude. This impressive hunter, named for its distinctively colored nape and crown plumage, is never especially abundant, but it is widespread in the rugged canyons, sagebrush foothills, and high mountains of Nevada. In the past, the Golden Eagle had a rocky relationship with humans, but the species is now viewed with tolerance and even appreciation. With persecution significantly reduced, Nevada's Golden Eagle population is considered stable and may even be increasing.

DISTRIBUTION

Atlas workers found Golden Eagles throughout the state, although less frequently toward the south. The species' Nevada distribution has probably changed little during the past century (Herron et al. 1985, Ryser 1985), although detailed assessments are not available. In surrounding states, the only large areas without well-established breeding populations are California's Central Valley, the lower Colorado River region, northwestern Oregon, and northern Idaho (Behle et al. 1985, Small 1994, Stephens and Sturts 1998, Adamus et al. 2001, Kochert et al. 2002, Driscoll in Corman and Wise-Gervais 2005:150–151).

Golden Eagles maintain large home ranges around the nest and are thus spread thinly throughout their breeding range. In Nevada, they are year-round residents (perhaps supplemented in the winter by some visitors from the north), but they range so widely that they can be seen in virtually any habitat. For example, the two most commonly reported habitats for the species during the atlas period were sagebrush scrub and sagebrush steppe, both of which are unsuitable for breeding per se. Instead, Golden Eagles nest in rugged crags, canyons, cliffs, and mountains, coming down into the valleys to hunt (Ryser 1985). Golden Eagles may move upslope, too, after the snow has melted (Gaines 1992).

Probability of occurrence is estimated to be moderate to high throughout all of Nevada except the extreme south, and reaches its peak in the far north.

CONSERVATION AND MANAGEMENT

It is difficult to fathom that a bird as majestic as the Golden Eagle was a victim of persecution in relatively recent times, but such was the case. Herron et al. (1985), Ryser (1985), and Alcorn (1988) discussed the extensive hunting and poisoning that occurred in Nevada. The basic problem was ranchers' traditional antagonism toward any potential livestock predator. Less appreciated was the fact that Golden Eagles' preferred prey items, large rodents and rabbits, actually feed on the same forage as livestock (Herron et al. 1985).

Persecution of Golden Eagles has dropped considerably in recent years, and the species is protected under the federal Bald and Golden Eagle Protection Act. Even now, though, it is estimated that a large proportion of Golden Eagle deaths are the direct or indirect results of humans (Kochert et al. 2002). In addition to shooting and poisoning, modern threats include collisions with vehicles or other objects and electrocution by power lines. Eagles are also relatively intolerant of close human presence, and urbanization has rendered many former breeding areas unsuitable (Small 1994, Rudesill in Burridge 1995:52, Kochert et al. 2002). As is true for all raptors, determining population trends is difficult without monitoring that is directed toward individual species. Some sources report ongoing declines in the West (Kochert et al. 2002), while others suggest relative stability or small increases (Sauer et al. 2005).

Nevada populations are thought to be stable (Herron et al. 1985), and the atlas data clearly indicate that Golden Eagles are still widespread and frequently encountered. Certainly, a bird that favors rocky canyons for nesting, open shrublands for hunting, and a muted human presence should be right at home in Nevada, and it seems that Golden Eagles are.

HABITAT USE

HABITAT	ATLAS BLOCKS	INCIDENTAL OBSERVATIONS
Agricultural	2 (1%)	1 (2%)
Alpine	5 (3%)	1 (2%)
Barren	14 (8%)	10 (15%)
Grassland	5 (3%)	3 (5%)
Mojave	3 (2%)	1 (2%)
Montane Forest	13 (7%)	3 (5%)
Montane Parkland	4 (2%)	2 (3%)
Montane Shrub	18 (10%)	7 (11%)
Open Water	1 (<1%)	0 (0%)
Pinyon-Juniper	24 (13%)	5 (8%)
Riparian	8 (4%)	3 (5%)
Salt Desert Scrub	17 (9%)	3 (5%)
Sagebrush Scrub	34 (19%)	17 (26%)
Sagebrush Steppe	30 (17%)	9 (14%)
Wetland	1 (<1%)	1 (2%)
TOTAL	179	66

Total numbers reported in the Habitat Use, Breeding Status, and Abundance tables may differ from each other (see pp. 22–23 for details). Percentage sums may differ slightly from 100% due to rounding.

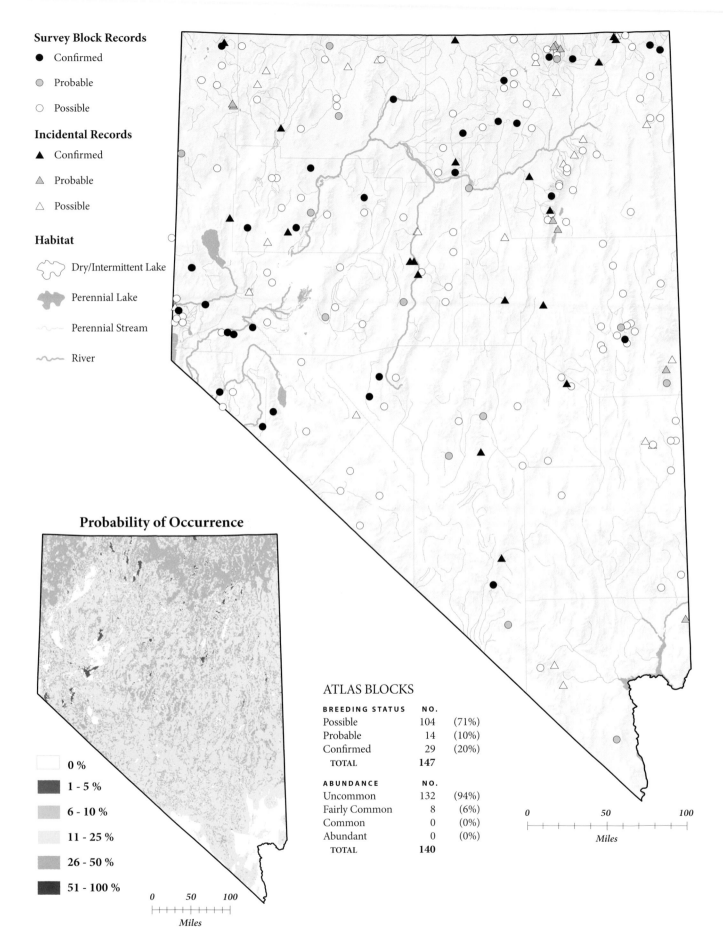

Survey Block Records

● Confirmed

● Probable

○ Possible

Incidental Records

▲ Confirmed

▲ Probable

△ Possible

Habitat

Dry/Intermittent Lake

Perennial Lake

Perennial Stream

River

Probability of Occurrence

0 %

1 - 5 %

6 - 10 %

11 - 25 %

26 - 50 %

51 - 100 %

0   50   100

Miles

ATLAS BLOCKS

BREEDING STATUS	NO.	
Possible	104	(71%)
Probable	14	(10%)
Confirmed	29	(20%)
TOTAL	147	

ABUNDANCE	NO.	
Uncommon	132	(94%)
Fairly Common	8	(6%)
Common	0	(0%)
Abundant	0	(0%)
TOTAL	140	

0   50   100

Miles

AMERICAN KESTREL
Falco sparverius

Most North American raptors are clad in somber browns, grays, and buffs, but not the American Kestrel. This petite falcon is downright flashy: males are orange and blue, with bold black barring and spots, and an ornate facial pattern to rival the Wood Duck's. Undoubtedly Nevada's most abundant bird of prey, the American Kestrel is also one of the most easily observed and entertaining to watch as it hovers in search of prey. Kestrels are tolerant of human presence and often nest in towns, farms, and even major urban areas.

DISTRIBUTION

American Kestrels have historically been numerous and widespread in Nevada (Linsdale 1936, Ryser 1985). They were reported from about a third of all atlas blocks and from many additional sites. Kestrel records came from all of Nevada's counties, and breeding was confirmed throughout the state. As the predictive map indicates, atlas records were more densely distributed toward the north, confirming a previous assessment by Herron et al. (1985). Multiple pairs of American Kestrels were noted in nearly half of the occupied atlas blocks, a pattern that differs from those for the larger hawks.

The American Kestrel occurs throughout the continent in a broad range of open and semiopen habitats. Kestrels breed widely in all of the states surrounding Nevada (Behle et al. 1985, Small 1994, Stephens and Sturts 1998, Adamus et al. 2001, Smallwood and Bird 2002, Wise-Gervais in Corman and Wise-Gervais 2005:154–155). Large expanses of dense forest and the lowest, hottest, most barren desert areas are about the only places where they may be absent or scarce (Rosenberg et al. 1991, Small 1994).

The American Kestrel is our only cavity-nesting diurnal raptor, requiring either a previously excavated tree cavity or a suitable substitute such as a crevice in a building or a nest box (Smallwood and Bird 2002). Kestrels have adapted well to human presence and are often noted nesting in agricultural, suburban, and urban settings (Smallwood and Bird 2002).

Herron et al. (1985) reported that American Kestrels favor riparian and agricultural habitats in Nevada. Atlas records came from a broad variety of habitat types, and riparian habitat indeed ranked at the top, likely because it offers tree cavities and nearby open shrubland or agricultural areas for foraging.

CONSERVATION AND MANAGEMENT

Although the initial forest clearing and establishment of agriculture in North America probably increased American Kestrel numbers (Smallwood and Bird 2002), the species now seems to be declining in many parts of its range, including the West (Sauer et al. 2005). During the second half of the twentieth century, for instance, numbers declined in parts of northern California (McCaskie et al. 1988) and in the lower Colorado River Valley (Rosenberg et al. 1991). Contreras and Kindschy (1996) reported possible declines in Malheur County, Oregon, as well. Specific causes for the declines are not well understood, but it is likely that the loss of older, cavity-containing trees is a contributing factor (Herron et al. 1985). Also, modernized agricultural practices may eliminate potential nest sites, diminish the arthropod and mammalian prey base, and cause pesticide accumulation that can reduce reproductive success (Smallwood and Bird 2002).

The species remains abundant in Nevada, but Breeding Bird Survey data show a decline in recent decades (Sauer et al. 2005). The American Kestrel's ecological versatility may help to stabilize its numbers in the future, but additional research on threats to the species and its ecological requirements are clearly of interest in order to explain and address any further population decreases.

HABITAT USE

HABITAT	ATLAS BLOCKS	INCIDENTAL OBSERVATIONS
Agricultural	25 (6%)	3 (2%)
Alpine	2 (<1%)	0 (0%)
Barren	7 (2%)	1 (<1%)
Grassland	5 (1%)	0 (0%)
Mesquite	1 (<1%)	0 (0%)
Mojave	12 (3%)	10 (8%)
Montane Forest	16 (4%)	12 (9%)
Montane Parkland	3 (<1%)	0 (0%)
Montane Shrub	35 (9%)	3 (2%)
Open Water	0 (0%)	1 (<1%)
Pinyon-Juniper	61 (16%)	10 (8%)
Riparian	91 (24%)	44 (34%)
Salt Desert Scrub	14 (4%)	4 (3%)
Sagebrush Scrub	47 (12%)	12 (9%)
Sagebrush Steppe	34 (9%)	5 (4%)
Urban	32 (8%)	19 (15%)
Wetland	1 (<1%)	5 (4%)
TOTAL	386	129

Total numbers reported in the Habitat Use, Breeding Status, and Abundance tables may differ from each other (see pp. 22–23 for details). Percentage sums may differ slightly from 100% due to rounding.

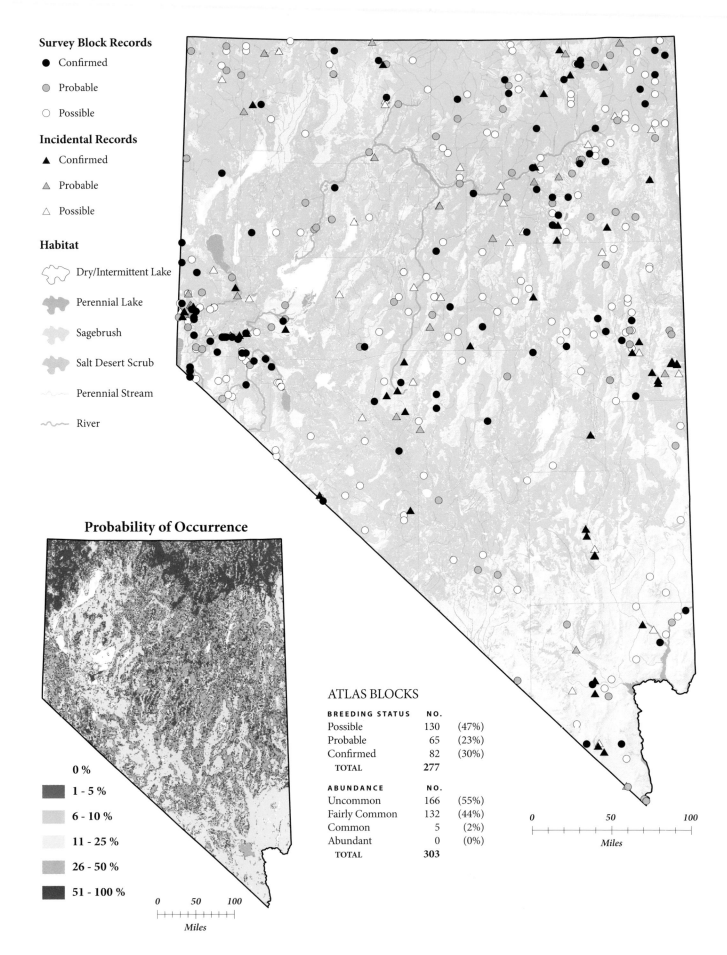

Survey Block Records

- ● Confirmed
- ● Probable
- ○ Possible

Incidental Records

- ▲ Confirmed
- ▲ Probable
- △ Possible

Habitat

- Dry/Intermittent Lake
- Perennial Lake
- Sagebrush
- Salt Desert Scrub
- Perennial Stream
- River

Probability of Occurrence

- 0 %
- 1 - 5 %
- 6 - 10 %
- 11 - 25 %
- 26 - 50 %
- 51 - 100 %

0   50   100
Miles

ATLAS BLOCKS

BREEDING STATUS	NO.	
Possible	130	(47%)
Probable	65	(23%)
Confirmed	82	(30%)
TOTAL	277	

ABUNDANCE	NO.	
Uncommon	166	(55%)
Fairly Common	132	(44%)
Common	5	(2%)
Abundant	0	(0%)
TOTAL	303	

0   50   100
Miles

PEREGRINE FALCON
Falco peregrinus

The Peregrine Falcon, a true world citizen and master aerialist, was nearly extirpated from large parts of its global and North American ranges by the use of DDT and other pesticides in the 1950s–1970s. Peregrine Falcons were probably never common in Nevada, and were apparently extirpated by the mid-twentieth century, even before the pesticide era. More recently, however, they have returned to southern and eastern Nevada, and a pair has even nested atop a high-rise casino on the Las Vegas Strip.

DISTRIBUTION

Peregrine Falcons were confirmed as breeders at eleven sites in southern and eastern Nevada. Eight of these records came from the Colorado River or its tributaries, especially the cliff walls surrounding Lake Mead. Elsewhere, nests were found in the Snake Range, the Spring Mountains, and Las Vegas. Nearly all of the confirmed breeding records were provided by the Nevada Department of Wildlife and Lake Mead National Recreation Area, based on their ongoing survey work. In addition to the confirmed records, possible breeders were noted in the Ruby Mountains, at several sites in southeastern Nevada, and near a historic breeding site at Walker Lake in Mineral County. Many of these records could have involved wandering individuals.

Although the historic distribution of the Peregrine Falcon in Nevada is largely unknown (Herron et al. 1985), the species may never have been common. Old breeding records exist from Pyramid and Walker lakes and from Elko County, but Peregrine Falcons were seemingly extirpated in Nevada by the early 1950s (Ryser 1985, Alcorn 1988). Ryser (1985) speculated about several possible explanations for this phenomenon, including persecution and climate change.

Throughout the West, the Peregrine's current breeding distribution is largely a product of natural recovery coupled with reintroductions. Together these have produced an interesting situation whereby Peregrine Falcons may now be more common in Nevada than at any time during the twentieth century. Recovery is also proceeding well in neighboring states, and Peregrine Falcons now breed in scattered locations in California, Oregon, and Idaho (Small 1994, Stephens and Sturts 1998, Adamus et al. 2001), and somewhat more widely in Utah and Arizona (White et al. 2002, Burger in Corman and Wise-Gervais 2005:156–157).

The handful of atlas records came from a variety of habitat types, no doubt reflecting the Peregrine Falcon's propensity for peregrinating. The predictive map therefore tends to indicate where birds are likely to be seen, rather than where they might actually nest. All known nest sites in Nevada have been on cliff ledges or high buildings, which the model cannot distinguish. Herron et al. (1985) reported that nesting in Nevada always occurs in the vicinity of a lake, wetland, or river system.

CONSERVATION AND MANAGEMENT

It is paradoxical that one of the most widely distributed land birds in the world was endangered until recently. Persecution and pesticides brought the Peregrine Falcon's numbers to dangerous lows, and it was listed as a federally Endangered species in 1970. The Peregrine's recent recovery is thought to be related to the DDT ban of the 1970s, supplemented by reintroductions in some areas (Herron et al. 1985, White et al. 2002). Populations recovered sufficiently to delist the species in 1999, and all indications are that the Peregrine Falcon continues to increase in numbers and is repopulating areas of its range where it was extirpated (White et al. 2002).

The Nevada Department of Wildlife began a reintroduction effort in 1984 (Herron et al. 1985), and intensive monitoring continues today (C. Tomlinson and R. Haley, pers. comm.). The prospects for the Peregrine Falcon in Nevada are probably good: it is willing to nest in diverse locations, and it is now the beneficiary of good publicity and careful monitoring.

HABITAT USE

HABITAT	ATLAS BLOCKS	INCIDENTAL OBSERVATIONS
Alpine	2 (25%)	0 (0%)
Barren	0 (0%)	1 (6%)
Mojave	0 (0%)	1 (6%)
Montane Forest	2 (25%)	2 (12%)
Montane Parkland	1 (13%)	0 (0%)
Montane Shrub	0 (0%)	1 (6%)
Open Water	1 (13%)	1 (6%)
Pinyon-Juniper	0 (0%)	3 (18%)
Riparian	2 (25%)	4 (24%)
Sagebrush Scrub	0 (0%)	1 (6%)
Sagebrush Steppe	0 (0%)	1 (6%)
Urban	0 (0%)	2 (12%)
TOTAL	8	17

Total numbers reported in the Habitat Use, Breeding Status, and Abundance tables may differ from each other (see pp. 22–23 for details). Percentage sums may differ slightly from 100% due to rounding.

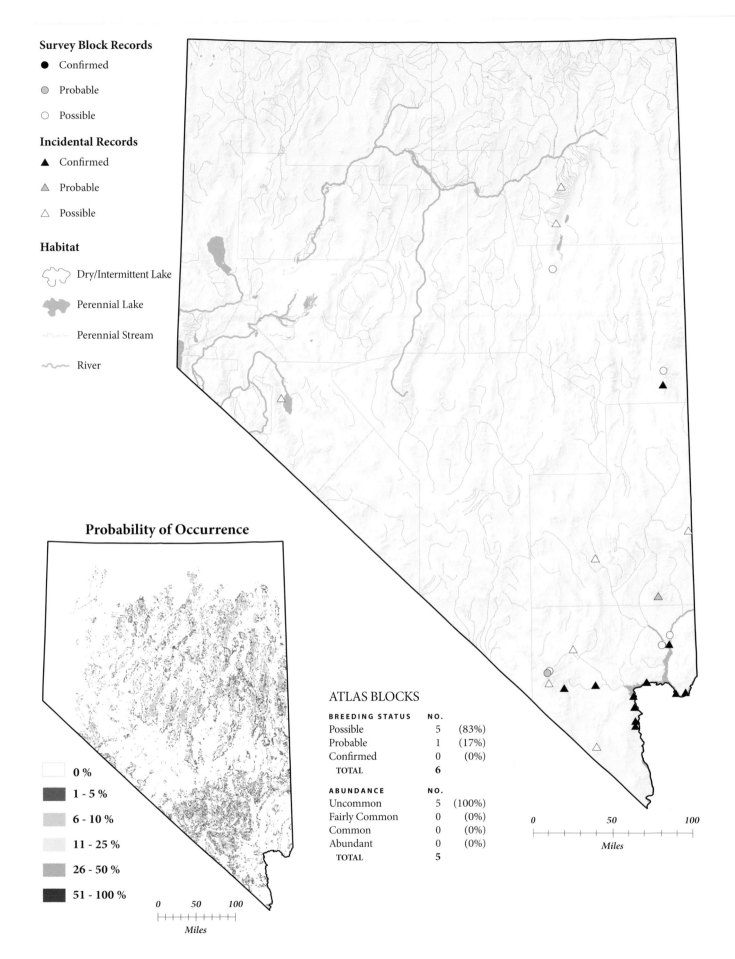

Survey Block Records

● Confirmed

◉ Probable

○ Possible

Incidental Records

▲ Confirmed

△ Probable

△ Possible

Habitat

Dry/Intermittent Lake

Perennial Lake

Perennial Stream

River

Probability of Occurrence

0 %

1 - 5 %

6 - 10 %

11 - 25 %

26 - 50 %

51 - 100 %

0   50   100

Miles

ATLAS BLOCKS

BREEDING STATUS	NO.	
Possible	5	(83%)
Probable	1	(17%)
Confirmed	0	(0%)
TOTAL	6	

ABUNDANCE	NO.	
Uncommon	5	(100%)
Fairly Common	0	(0%)
Common	0	(0%)
Abundant	0	(0%)
TOTAL	5	

0   50   100

Miles

PRAIRIE FALCON
Falco mexicanus

The Prairie Falcon's earth-toned plumage renders it inconspicuous, yet this bird seems appropriately dressed for the lonely rimrocks and jagged outcroppings of Nevada's backcountry. The Prairie Falcon is well adapted to life in arid environments and is often seen far from riparian or wetland habitats. Sagebrush or salt desert scrub suits it just fine, and it is one of the few Nevada species that atlas workers often classified as occurring in "barren" habitat.

DISTRIBUTION

Prairie Falcons were distributed fairly evenly across Nevada, as the map's prediction of a uniform chance of occurrence indicates. The breeding confirmation rate was just 16%, no doubt because of the inaccessibility of nest sites. This falcon's present-day distribution in Nevada has probably changed little from earlier times (Linsdale 1936, Ryser 1985, Alcorn 1988). A quantitative analysis by Herron et al. (1985) resulted in 425 known nesting territories and an estimated twelve hundred pairs statewide. Prairie Falcons appear to become slightly less common as one goes from northern to southern Nevada (Neel 1999a), although they were widely noted in the south by the late-nineteenth-century Death Valley Expedition (Fisher 1893).

Southwestern Idaho has the highest known breeding density of Prairie Falcons (Steenhof 1998, Paige and Ritter 1999). The species' range extends through eastern Oregon, Utah, Arizona, and in open habitats away from the coast in California (Behle et al. 1985, Small 1994, Adamus et al. 2001, Moors in Corman and Wise-Gervais 2005:158–159).

Prairie Falcons forage for small mammals and birds in a variety of open habitats, and locations where they were seen generally shared an austere element. Sagebrush scrub, barren lands, salt desert scrub, and sagebrush steppe were most commonly reported by atlas workers, and there is a moderate probability that this species breeds across nearly all of Nevada. Nests are almost always located on cliffs, canyons, or rocky ledges, and the availability of nest sites may impose local limits on nesting (Steenhof 1998). The predictive model could not specifically identify such sites.

CONSERVATION AND MANAGEMENT

As is the case with many raptors, conservation of Prairie Falcons seems largely a matter of protecting nest sites from disturbance or destruction, preventing illegal shooting, managing harvest of birds for falconry, and managing rangelands to ensure an adequate prey base. Agricultural development, heavy grazing, and altered fire regimes in particular have the potential to reduce prey populations (Steenhof 1998, Dobkin and Sauder 2004). Prairie Falcons are sensitive to bioaccumulation of pesticides, but their mixed menu of birds and mammals exposed them to substantially lower doses of DDT and other chemicals than was the case for the strictly bird-eating Peregrine Falcon. The negative effects on Prairie Falcon populations in the 1950s and 1960s were thus less severe and more local in nature (Steenhof 1998, Dobkin and Sauder 2004). An interesting note: it has been suggested that Prairie Falcons once appropriated nest sites vacated by declining Peregrine Falcon populations (Ryser 1985).

Overall, Prairie Falcon populations in the West appear to have been stable over recent years, though declines may have occurred in the 1960s and 1970s (Paige and Ritter 1999, Dobkin and Sauder 2004). The limited population data available suggest a recent slow decline in the Basin and Range region of the Intermountain West (Sauer et al. 2005). In contrast, Herron et al. (1985) reported Nevada's Prairie Falcon population to be stable. Nevada may carry a high degree of responsibility for the Prairie Falcon—as much as 20% of the world's population may breed here (Neel 1999a). It is thus not surprising that the Prairie Falcon is a Nevada Partners in Flight Priority species, with special focus on preserving its cliff habitats (Neel 1999a).

HABITAT USE

HABITAT	ATLAS BLOCKS	INCIDENTAL OBSERVATIONS
Agricultural	3 (3%)	1 (3%)
Alpine	3 (3%)	0 (0%)
Barren	14 (14%)	7 (20%)
Grassland	6 (6%)	4 (11%)
Mojave	6 (6%)	0 (0%)
Montane Forest	1 (1%)	0 (0%)
Montane Shrub	7 (7%)	1 (3%)
Open Water	1 (1%)	1 (3%)
Pinyon-Juniper	6 (6%)	0 (0%)
Riparian	5 (5%)	3 (9%)
Salt Desert Scrub	12 (12%)	2 (6%)
Sagebrush Scrub	22 (22%)	12 (34%)
Sagebrush Steppe	10 (10%)	2 (6%)
Urban	0 (0%)	1 (3%)
Wetland	3 (3%)	1 (3%)
TOTAL	99	35

Total numbers reported in the Habitat Use, Breeding Status, and Abundance tables may differ from each other (see pp. 22–23 for details). Percentage sums may differ slightly from 100% due to rounding.

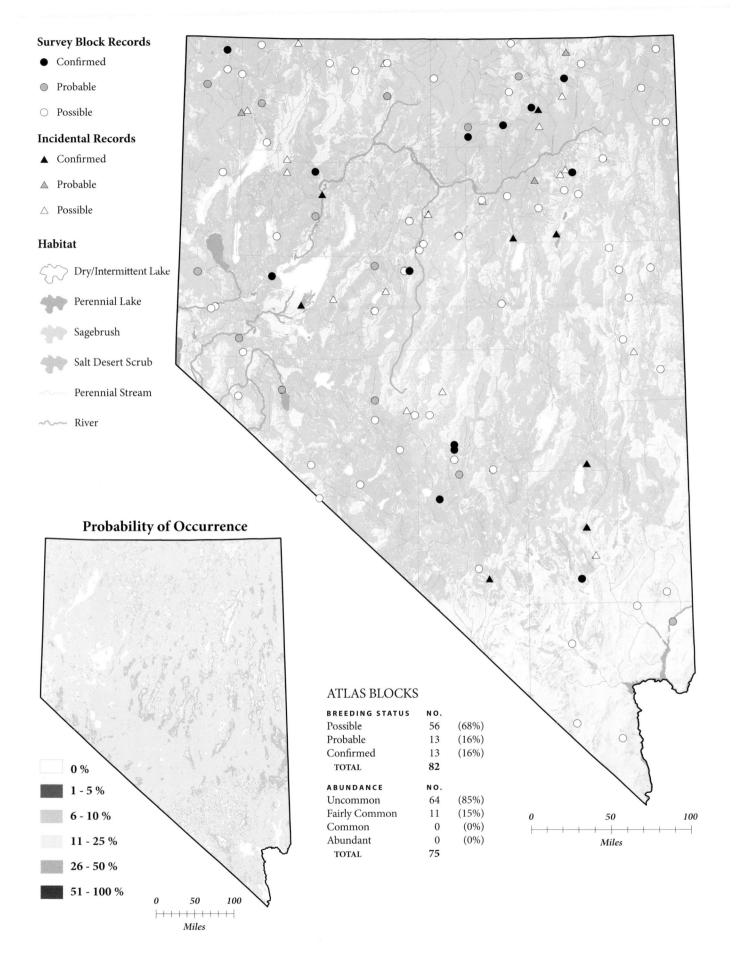

Survey Block Records

- ● Confirmed
- ◐ Probable
- ○ Possible

Incidental Records

- ▲ Confirmed
- ▲ Probable
- △ Possible

Habitat

- Dry/Intermittent Lake
- Perennial Lake
- Sagebrush
- Salt Desert Scrub
- Perennial Stream
- River

Probability of Occurrence

- ☐ 0 %
- ▨ 1 - 5 %
- ▨ 6 - 10 %
- ▨ 11 - 25 %
- ▨ 26 - 50 %
- ▨ 51 - 100 %

0   50   100
Miles

ATLAS BLOCKS

BREEDING STATUS	NO.	
Possible	56	(68%)
Probable	13	(16%)
Confirmed	13	(16%)
TOTAL	82	

ABUNDANCE	NO.	
Uncommon	64	(85%)
Fairly Common	11	(15%)
Common	0	(0%)
Abundant	0	(0%)
TOTAL	75	

0   50   100
Miles

VIRGINIA RAIL
Rallus limicola

One of the greatest inducements for a birder to visit a cattail marsh is the possibility of finding Virginia Rail chicks—dumpy little fluff balls, black as night. Most of the time, Virginia Rails are notoriously difficult to spot, even when one is calling from a tiny patch of marsh vegetation. Family groups, however, sometimes wander out onto the mud at the edge of a cattail bed, where they seem curiously oblivious to the presence of human observers. Because the birds are so secretive, though, information about breeding in Nevada is difficult to obtain.

DISTRIBUTION

Atlas workers found Virginia Rails at scattered sites throughout Nevada. Most records came from lowland marshes in southern and western Nevada, but this may reflect the high concentration of birders in these areas. Breeding was confirmed in almost one-third of the blocks in which the species was found. Observers estimated that multiple pairs were present in most blocks where the species occurred, but only one block was estimated to have more than ten pairs.

The widespread but scattered distribution of atlas records is largely consistent with earlier accounts. Linsdale (1936) cited observations from several areas, but breeding had not yet been confirmed at that time. Alcorn (1988) reported confirmed and suspected breeding records throughout Nevada, and Ryser (1985) included the entire Great Basin in the species' range. Despite the low numbers found in most atlas blocks, Virginia Rails can be locally common—for example, in the Lahontan Valley (Chisholm and Neel 2002) and at Ruby Lake, which evidently can have as many as six hundred individuals (Alcorn 1988).

The scattered distribution of Virginia Rail records is typical of the species throughout its breeding range, which covers roughly the northern half of the United States from coast to coast (Conway 1995). Virginia Rails breed widely in suitable marsh habitats within all of the states surrounding Nevada (Behle et al. 1985, Small 1994, Stephens and Sturts 1998, Adamus et al. 2001, Burger in Corman and Wise-Gervais 2005:164–165). Breeding in the

lower Colorado River Valley is apparently of fairly recent origin and was first confirmed in 1978 (Rosenberg et al. 1991).

The Virginia Rail's basic needs are marshes with shallow standing water, dense emergent vegetation such as cattails and bulrushes, and some exposed mud with adequate invertebrate prey (Conway 1995). Most atlas records came from wetlands, but the species also occurs in wet meadows and along streams in Nevada (Alcorn 1988). These birds are often plentiful in large marshes, but they also inhabit small wetlands such as cattail ponds within riparian woodlands and small marshes created by agricultural runoff (Garrett and Dunn 1981, Smith et al. 1997). Virginia Rails are prone to wander (Conway 1995) and could reasonably be expected in wetlands of any size anywhere in Nevada. This propensity likely accounts for the pattern shown by the predictive map, which indicates a small chance of occurrence in many lower-elevation areas of Nevada wherever a small marsh is embedded within the predominant habitat.

CONSERVATION AND MANAGEMENT

Because Virginia Rails are so secretive by nature, it is difficult to estimate their numbers or population trends. Standardized surveys that use tape recorders to elicit calls have become more commonplace recently (Conway and Gibbs 2005), and even loud clapping may elicit a response from an otherwise reclusive rail. No amount of secrecy, however, can protect the Virginia Rail from outright destruction of its habitat. Populations have declined because of wetland losses in California (Garrett and Dunn 1981) and elsewhere (Conway 1995). On the other hand, Virginia Rails respond quite well to habitat improvement initiatives. The species has benefited from the creation of stable marsh habitats in the lower Colorado River Valley (Rosenberg et al. 1991) and from irrigation projects and farm ponds in Colorado (Pantle in Kingery 1998:156–157).

HABITAT USE

HABITAT	ATLAS BLOCKS	INCIDENTAL OBSERVATIONS
Riparian	10 (36%)	3 (25%)
Urban	1 (4%)	0 (0%)
Wetland	17 (61%)	9 (75%)
TOTAL	28	12

Total numbers reported in the Habitat Use, Breeding Status, and Abundance tables may differ from each other (see pp. 22–23 for details). Percentage sums may differ slightly from 100% due to rounding.

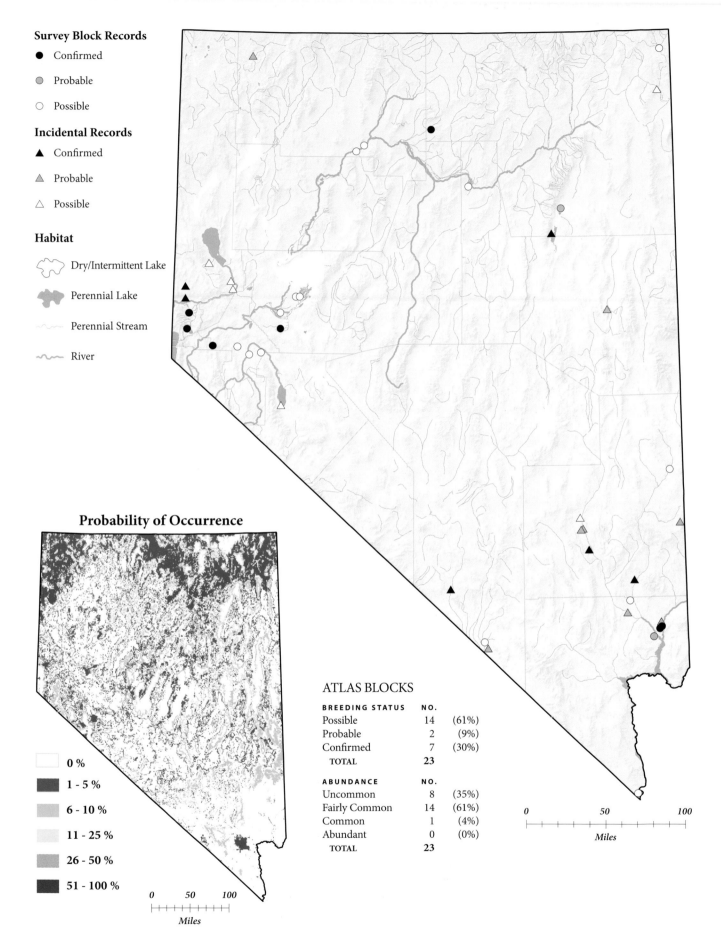

Survey Block Records

- ● Confirmed
- ● Probable
- ○ Possible

Incidental Records

- ▲ Confirmed
- ▲ Probable
- △ Possible

Habitat

Dry/Intermittent Lake

Perennial Lake

Perennial Stream

River

Probability of Occurrence

- 0 %
- 1 - 5 %
- 6 - 10 %
- 11 - 25 %
- 26 - 50 %
- 51 - 100 %

0   50   100

Miles

ATLAS BLOCKS

BREEDING STATUS	NO.	
Possible	14	(61%)
Probable	2	(9%)
Confirmed	7	(30%)
TOTAL	23	

ABUNDANCE	NO.	
Uncommon	8	(35%)
Fairly Common	14	(61%)
Common	1	(4%)
Abundant	0	(0%)
TOTAL	23	

0   50   100

Miles

SORA
Porzana carolina

Easy to hear but devilishly difficult to see, the Sora is a widespread and fairly common denizen of the wetlands of northern and eastern Nevada. Soras can be found in extensive marshes in places such as the Lahontan and Ruby valleys, but they are also found in smaller wetlands near cities and towns, and at the margins of agricultural areas. They occur along the Colorado River in southern Nevada, but it is unclear how commonly they breed in this region. Because Soras are difficult to find, it is possible that they are more widespread than they seem, with small numbers occurring in isolated marshes throughout the state.

DISTRIBUTION

Soras were found in twenty-six atlas blocks and at several other sites. They were widely recorded in the northern half of the state, and there were scattered records along the entire eastern border. The especially dry desert regions of south-central Nevada produced no records—a pattern common for many of our waterbirds—and the species was not confirmed as a breeder in the river drainages of southern Nevada. Multiple pairs were estimated for most of the blocks where Soras were found, and the majority of records came from wetland habitats, as expected. The atlas results largely match the account provided by Linsdale (1936), who described breeding as being restricted to the northern and eastern parts of Nevada. Alcorn (1988) hinted at possible breeding in the south and reported a high count of twelve hundred young Soras at Ruby Lake in 1983.

The Sora's large North American breeding range includes much of Canada and the northern United States (Melvin and Gibbs 1996). The species is widespread to the north and east of Nevada, and also breeds throughout the Central Valley of California and spottily in Arizona (Behle et al. 1985, Small 1994, Stephens and Sturts 1998, Adamus et al. 2001, Shrout in Corman and Wise-Gervais 2005:166–167).

The Sora's breeding habitat requirements are similar to those of the Virginia Rail: freshwater marshes with dense emergent vegetation (Melvin and Gibbs 1996). Finer-grained habitat differences between the Sora and Virginia Rail were summarized by Johnson and Dinsmore (1986). The predictive map shows a potential distribution quite similar to that of the Virginia Rail, with a relatively low probability of occurrence throughout the valleys of the Great Basin and a few isolated spots farther south. Soras are predicted to be slightly more common in the north, but even there the chance of finding them is quite low, reflecting the rarity of wetlands. The model also predicts a moderately high chance of occurrence in parts of southern Nevada, but this result is likely an artifact of the model's simplicity; there were no confirmed breeding records from anywhere in the south.

CONSERVATION AND MANAGEMENT

As with the Virginia Rail, one of the main challenges for Sora conservation is the acquisition of basic population data for this secretive and easily overlooked bird. The status of the species has been difficult to determine in Nevada (Alcorn 1988) and elsewhere (Garrett and Dunn 1981, Small 1994). For instance, there was no evidence of breeding in the lower Colorado River Valley prior to 1991 (Rosenberg et al. 1991), despite the apparent availability of suitable habitat. Even in places where Soras are common, such as southeastern Oregon (Littlefield 1990) and much of Utah (Behle et al. 1985), they are easy to miss.

Protection and enhancement of wetlands that have dense emergent vegetation are probably the most important elements in Sora conservation, which should also include basic monitoring of populations. The species also responds well to the creation of new wetlands (Melvin and Gibbs 1996).

HABITAT USE

HABITAT	ATLAS BLOCKS	INCIDENTAL OBSERVATIONS
Agricultural	2 (6%)	0 (0%)
Open Water	1 (3%)	0 (0%)
Riparian	6 (19%)	0 (0%)
Wetland	22 (71%)	6 (100%)
TOTAL	31	6

Total numbers reported in the Habitat Use, Breeding Status, and Abundance tables may differ from each other (see pp. 22–23 for details). Percentage sums may differ slightly from 100% due to rounding.

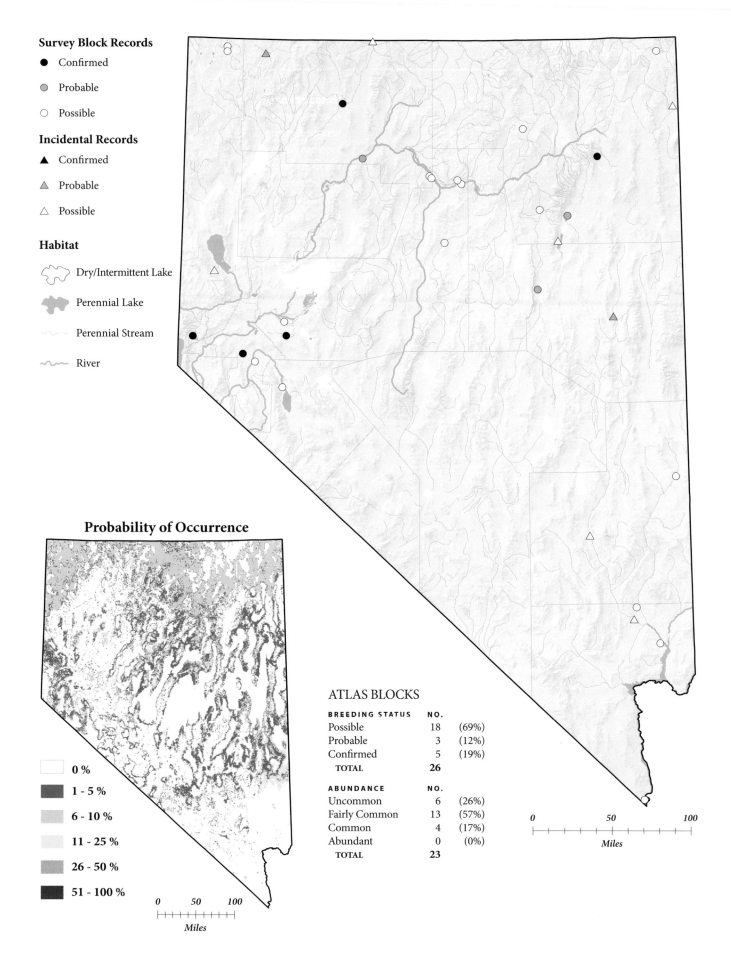

Survey Block Records

● Confirmed

● Probable

○ Possible

Incidental Records

▲ Confirmed

▲ Probable

△ Possible

Habitat

Dry/Intermittent Lake

Perennial Lake

Perennial Stream

River

Probability of Occurrence

☐ 0 %

■ 1 - 5 %

■ 6 - 10 %

☐ 11 - 25 %

■ 26 - 50 %

■ 51 - 100 %

0   50   100
Miles

ATLAS BLOCKS

BREEDING STATUS	NO.	
Possible	18	(69%)
Probable	3	(12%)
Confirmed	5	(19%)
TOTAL	26	

ABUNDANCE	NO.	
Uncommon	6	(26%)
Fairly Common	13	(57%)
Common	4	(17%)
Abundant	0	(0%)
TOTAL	23	

0   50   100
Miles

COMMON MOORHEN
Gallinula chloropus

The Common Moorhen is widely distributed in the eastern United States, Mexico, and over several continents, but the species breeds only sporadically in the West and barely makes it into Nevada. It has been documented at a few locations in northern Nevada and somewhat more broadly in the lowlands of southern Nevada. Birds of freshwater marshes (Bannor and Kiviat 2002), Common Moorhens are willing to use heavily altered habitats such as diversion ditches and reservoirs. In southern Nevada, they seem to be easiest to find at the intensively managed wetlands around the Las Vegas Wash.

DISTRIBUTION

Atlas records for Common Moorhens came from single locations in Lyon and Lincoln counties and from three scattered sites in Clark County. Breeding was confirmed only at the Henderson Bird Viewing Preserve in the Las Vegas Wash, and only one pair was estimated to be present in each of the two blocks for which abundance estimates were provided. Not surprisingly, all records were from wetland or riparian habitats.

The distribution revealed by the atlas data is more restricted than that previously documented for the Common Moorhen, although the difference may be a result of the difficulties involved in observing this species. In northern Nevada, Common Moorhens have been documented as breeders in the Lahontan and Ruby valleys (Ryser 1985, Alcorn 1988, Chisholm and Neel 2002); Tule Springs Park in southern Nevada has also been listed as a breeding site (Alcorn 1988). No birds were reported from any of these locations during the atlas fieldwork, however. Earlier accounts either characterize the Common Moorhen as resident throughout Nevada or indicate that it is at least locally common (Alcorn 1988, Titus 2003). In contrast, atlas records suggest a distribution more like that described by Linsdale (1936), who provided only one record for the state, in far southern Clark County,

and Ryser (1985:114), who characterized Common Moorhens as a "casual member of the Great Basin avifauna."

Common Moorhens are primarily eastern birds, and their distribution in the West has a southern orientation (Bannor and Kiviat 2002). They are fairly common breeders in the lower Colorado River Valley and occur locally throughout southern Arizona (Rosenberg et al. 1991, Wise-Gervais in Corman and Wise-Gervais 2005:168–169). Breeding also occurs in parts of southern and central California (Small 1994), and uncommonly in the marshes of northern Utah (Behle et al. 1985, UOSBRC 2004).

In the West, Common Moorhens seem to occur often in highly modified wetlands such as agricultural canals, wastewater ponds, and sewage ponds (Behle et al. 1985, Rosenberg et al. 1991, Burridge 1995:63, Chisholm and Neel 2002). The predictive map suggests a good chance of finding this species wherever wetland habitat exists, but the high probability shown for western Nevada's wetlands is almost certainly an overestimate, probably an artifact of the small number of wetland blocks on which the predictions were based. Most of the atlas field records came from southern Nevada.

CONSERVATION AND MANAGEMENT

The Common Moorhen's present distribution in Nevada is rather peripheral, and its long-term prospects in the state are uncertain. Population declines in southern California and Utah have been attributed to the destruction of freshwater marshes (Garrett and Dunn 1981, Behle et al. 1985, Gallagher 1997). On the other hand, Common Moorhens readily use alternate habitats when these become available (see Bannor and Kiviat 2002). In Pennsylvania, for example, Common Moorhens have increased greatly in numbers and geographic extent along with the proliferation of artificial wetlands throughout the state (Leberman in Brauning 1992:128–129). In Nevada, protection of both natural and created wetlands and continued efforts to better document this species' occurrence patterns seem warranted.

HABITAT USE

HABITAT	ATLAS BLOCKS	INCIDENTAL OBSERVATIONS
Riparian	1 (33%)	0 (0%)
Wetland	2 (67%)	2 (100%)
TOTAL	3	2

Total numbers reported in the Habitat Use, Breeding Status, and Abundance tables may differ from each other (see pp. 22–23 for details). Percentage sums may differ slightly from 100% due to rounding.

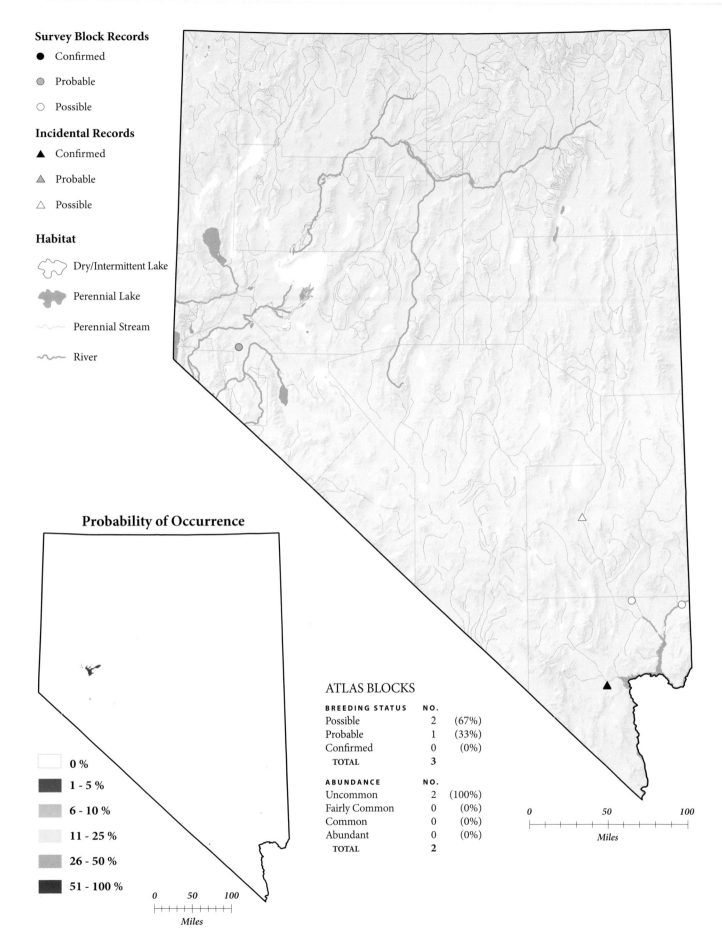

Survey Block Records

● Confirmed

◉ Probable

○ Possible

Incidental Records

▲ Confirmed

△ Probable

△ Possible

Habitat

Dry/Intermittent Lake

Perennial Lake

Perennial Stream

River

Probability of Occurrence

☐ 0 %

■ 1 - 5 %

■ 6 - 10 %

☐ 11 - 25 %

■ 26 - 50 %

■ 51 - 100 %

0   50   100
Miles

ATLAS BLOCKS

BREEDING STATUS	NO.	
Possible	2	(67%)
Probable	1	(33%)
Confirmed	0	(0%)
TOTAL	3	

ABUNDANCE	NO.	
Uncommon	2	(100%)
Fairly Common	0	(0%)
Common	0	(0%)
Abundant	0	(0%)
TOTAL	2	

0   50   100
Miles

AMERICAN COOT
Fulica americana

It is easy to forget that American Coots are in the same family as the rails. Unlike their secretive and reclusive relatives, coots are conspicuous, noisy, and familiar members of Nevada's breeding avifauna, and they are also abundant year-round residents. Large numbers can be seen at Virginia Lake in Reno and at Sunset Park in Las Vegas, vast concentrations occur in the wetlands of the Lahontan and Ruby valleys, and lesser numbers occur at many other locations across the state. Although the American Coot is a wetland-obligate species, its Nevada populations appear to be quite healthy. Indeed, American Coots are regarded by some as pests, and "mud hens" are the scourge of many a golf course owner.

DISTRIBUTION

The American Coot was widely recorded in northern and eastern Nevada, and much more sparsely in Nye and Esmeralda counties, where wetland birds were generally scarce. Breeding was confirmed in half of the blocks where coots were found, and they were often quite numerous, with more than a third of the occupied blocks estimated to have in excess of ten pairs, and a few blocks having more than one hundred.

Nearly all earlier writers noted the abundance and ubiquity of the American Coot in Nevada. Linsdale (1936:51) wrote that it is "present on nearly every pool of water, whether small or large," and Ryser (1985:114) said it can be "found at virtually every pond, lake, and marsh in the Great Basin." Alcorn (1988) described coots as common to abundant statewide, and Chisholm and Neel (2002) regarded them as extremely common in the Lahontan Valley. Just to our north, in Harney County, Oregon, the coot may be the most abundant bird at Malheur National Wildlife Refuge (Littlefield 1990).

The American Coot breeds in wetlands throughout the West

and in portions of the Great Plains and upper Midwest (Brisbin and Mowbray 2002). Coots prefer some emergent vegetation to be present, but beyond that just about any size or type of wetland will do (Brisbin and Mowbray 2002). Breeding is widespread in all of our neighboring states, where coots are also generally year-round residents (Behle et al. 1985, Small 1994, Stephens and Sturts 1998, Adamus et al. 2001, Brisbin and Mowbray 2002, Clark in Corman and Wise-Gervais 2005:170–171). Fall numbers can be especially high in Nevada and the surrounding region when resident birds are supplemented by migrants from farther north.

Almost all atlas records were associated with aquatic habitats, and the remainder almost certainly came from small wetlands embedded within the reported habitats. The large area over which a moderate chance of finding the species is predicted reflects the fact that coots will use even tiny wetlands in large expanses of terrestrial habitat. The higher probabilities predicted for foothills areas, however, probably result from the model's inability to distinguish between lowland and montane riparian habitats.

CONSERVATION AND MANAGEMENT

As common as coots are now, they were once much more common. Beginning in the late 1800s, loss of wetlands—and possibly hunting—caused substantial declines, and populations have stabilized only during the last several decades (Brisbin and Mowbray 2002). Although adaptable, omnivorous, and tolerant of human activity, the American Coot continues to experience modest declines in western North America (Sauer et al. 2005). Little long-term trend information is available for Nevada's breeding populations. Regardless of trends, the species clearly remains very numerous in certain areas. For instance, between seven hundred and one thousand pairs were counted annually at Ruby Lake National Wildlife Refuge during the atlas years (USFWS 2001), and counts of up to 200,000 birds have been recorded in the Lahontan Valley in the fall (Chisholm and Neel 2002).

HABITAT USE

HABITAT	ATLAS BLOCKS	INCIDENTAL OBSERVATIONS
Agricultural	2 (2%)	0 (0%)
Open Water	32 (30%)	24 (38%)
Riparian	25 (23%)	6 (10%)
Salt Desert Scrub	1 (1%)	0 (0%)
Urban	2 (2%)	0 (0%)
Wetland	45 (42%)	33 (52%)
TOTAL	107	63

Total numbers reported in the Habitat Use, Breeding Status, and Abundance tables may differ from each other (see pp. 22–23 for details). Percentage sums may differ slightly from 100% due to rounding.

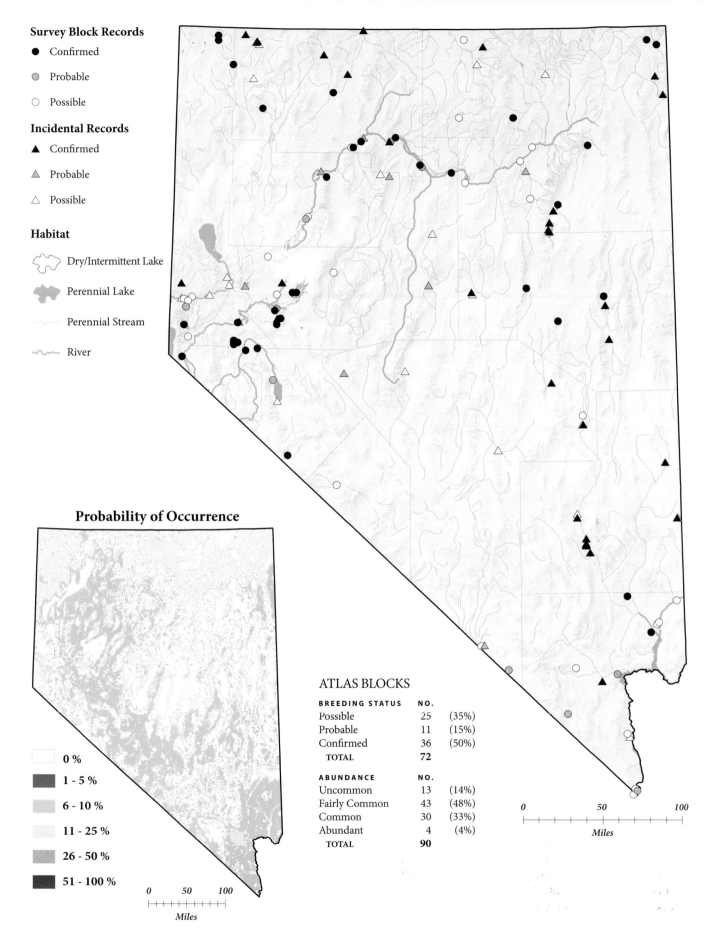

Survey Block Records

- **●** Confirmed
- **●** Probable (gray)
- **○** Possible

Incidental Records

- **▲** Confirmed
- **▲** Probable (gray)
- **△** Possible

Habitat

- Dry/Intermittent Lake
- Perennial Lake
- Perennial Stream
- River

Probability of Occurrence

- 0 %
- 1 - 5 %
- 6 - 10 %
- 11 - 25 %
- 26 - 50 %
- 51 - 100 %

0   50   100
Miles

ATLAS BLOCKS

BREEDING STATUS	NO.	
Possible	25	(35%)
Probable	11	(15%)
Confirmed	36	(50%)
TOTAL	**72**	

ABUNDANCE	NO.	
Uncommon	13	(14%)
Fairly Common	43	(48%)
Common	30	(33%)
Abundant	4	(4%)
TOTAL	**90**	

0   50   100
Miles

SANDHILL CRANE
Grus canadensis

RN'02

Few sights are more thrilling than the spectacle of cranes dancing—it is part bacchanalia, part ballet. And few sounds are more evocative than the stentorian bugling of Sandhill Cranes hidden by the morning mist. Indeed, every aspect of the Sandhill Crane's natural history seems to inspire admiration. While venerated by many, the Sandhill Crane has suffered from human encroachment on its breeding grounds in Nevada and elsewhere. It is an adaptable species, however, and has recently recolonized isolated wetlands and farmlands throughout its former Nevada range.

DISTRIBUTION

Sandhill Cranes were found in twenty-two blocks and at twenty-three other locations. Most records came from northeastern Nevada, but there were scattered sightings in the far northwestern corner of the state and a single confirmed breeding pair in the Carson Valley. Additional records from Lincoln and Clark counties were presumed to have involved nonbreeding birds and were not mapped. As expected for a large, sparsely distributed breeder, Sandhill Cranes were not numerous anywhere, and fewer than ten pairs were reported for all of the blocks where cranes occurred.

The Sandhill Crane was widespread in northern Nevada during the nineteenth century (Ryser 1985), but by the mid-1930s it bred in only a handful of locales in the northeastern corner (Linsdale 1936). Although the species was still thought to be restricted to that part of the state in the late twentieth century (Ryser 1985, Alcorn 1988, Neel 1999a), the atlas records from western Nevada suggest a recent expansion of the range. Given

the state's sparsely distributed population of birders, it is also possible that the species was simply overlooked in the northwest in the recent past.

Population trends in Nevada seem to mirror patterns elsewhere in the West. The Sandhill Crane was formerly much more common as a breeder in Utah than it is now (Behle 1985, Behle et al. 1985), and it has disappeared almost completely from the eastern slope of the Sierra Nevada (Gaines 1992). Elsewhere in northern California, Sandhill Cranes are local breeders (McCaskie et al. 1988, Small 1994). In Oregon, they nest locally east of the Cascades, especially in Malheur and Harney counties (Littlefield 1990, Contreras and Kindschy 1996, Adamus et al. 2001). Cranes are widespread in Idaho, but breeding has not been confirmed in the areas immediately adjacent to Nevada (Stephens and Sturts 1998).

Most atlas records came from aquatic or agricultural habitats, and the predictive map suggests a very good chance of finding Sandhill Cranes in these habitats in northern Nevada. At the microhabitat level, Sandhill Cranes are usually found in hay meadows within river valley floodplains (Neel 1999a).

CONSERVATION AND MANAGEMENT

The Sandhill Crane is a Nevada Partners in Flight Priority species (Neel 1999a). Primary concerns for the North American population are the Sandhill Crane's low reproductive rate and the loss of habitat at the major staging areas where birds concentrate outside the breeding season (Tacha et al. 1992). All of Nevada's breeding Sandhill Cranes appear to belong to the *tabida* subspecies, the type specimen for which was collected in Elko County (Alcorn 1988).

Sandhill Cranes nest primarily on private land in Nevada, making stewardship by landowners the primary venue for species conservation. The species has clear affinities for agricultural areas both during the breeding season and when staging during migration (Neel 1999a), and recent increases in Arizona have been attributed to changing agricultural practices (Rosenberg et al. 1991).

HABITAT USE

HABITAT	ATLAS BLOCKS	INCIDENTAL OBSERVATIONS
Agricultural	5 (17%)	10 (46%)
Grassland	4 (13%)	1 (5%)
Montane Shrub	1 (3%)	0 (0%)
Riparian	6 (20%)	3 (14%)
Sagebrush Scrub	0 (0%)	2 (9%)
Wetland	14 (47%)	6 (27%)
TOTAL	30	22

Total numbers reported in the Habitat Use, Breeding Status, and Abundance tables may differ from each other (see pp. 22–23 for details). Percentage sums may differ slightly from 100% due to rounding.

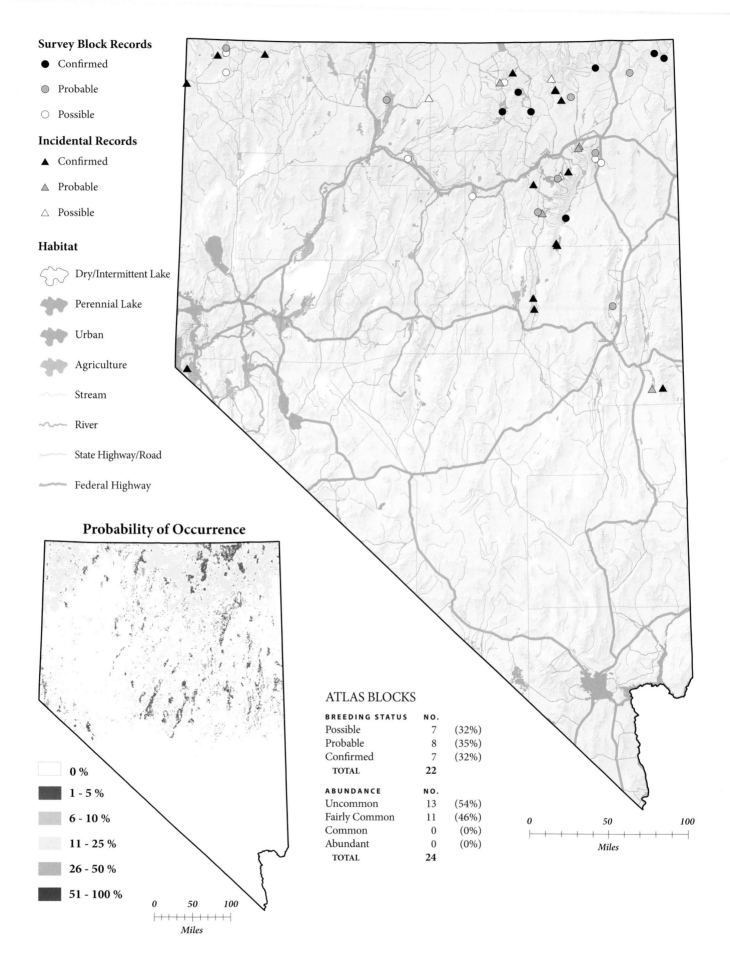

Survey Block Records

● Confirmed

● Probable

○ Possible

Incidental Records

▲ Confirmed

▲ Probable

△ Possible

Habitat

Dry/Intermittent Lake

Perennial Lake

Urban

Agriculture

Stream

River

State Highway/Road

Federal Highway

Probability of Occurrence

☐ 0 %

■ 1 - 5 %

■ 6 - 10 %

☐ 11 - 25 %

■ 26 - 50 %

■ 51 - 100 %

0   50   100
Miles

ATLAS BLOCKS

BREEDING STATUS	NO.	
Possible	7	(32%)
Probable	8	(35%)
Confirmed	7	(32%)
TOTAL	22	
ABUNDANCE	NO.	
Uncommon	13	(54%)
Fairly Common	11	(46%)
Common	0	(0%)
Abundant	0	(0%)
TOTAL	24	

0   50   100
Miles

SNOWY PLOVER
Charadrius alexandrinus

The white-hot playas of Nevada—sun-baked, foul smelling, and featureless—are so inhospitable to life that they seem almost post-apocalyptic. Yet, they are home to one of the most delicate and diminutive of Nevada's nesting shorebirds, the Snowy Plover. Even when there is sparse vegetation nearby to offer some shade, the Snowy Plover typically places its nest right out in the open—in a rudimentary scrape that is fully exposed to sun, wind, and predators.

DISTRIBUTION

The atlas fieldwork produced only four breeding records for the Snowy Plover, and all came from the Lahontan Valley. Breeding was confirmed at each site, and in the one block where abundance was estimated there were thought to be more than ten pairs. The older literature portrays Snowy Plovers as being more widespread than the atlas data suggest. Although he did not mention definite breeders, Linsdale (1936) reported sightings from the Humboldt Valley in Humboldt County, Pyramid Lake in Washoe County, and Walker Lake in Mineral County. Ryser (1985) considered the species to be locally common at lakes along the western edge of the Great Basin, and Alcorn (1988) referred to records from throughout the state. The most recent assessment of the region's shorebird populations estimates that a total of about three hundred pairs nest in the Lahontan Valley and Humboldt Sink (Oring et al. 2005).

The Snowy Plover is likewise very locally distributed in the states adjacent to Nevada. Much of the U.S. population occurs in the Intermountain West, however, and several important sites for this species lie close to Nevada's western border. For example, Mono and Owens lakes in California and Lake Abert and the Harney Basin in Oregon all hold large numbers (Page and Stenzel 1981, Oring et al. 2005). To the east, the Great Salt Lake in Utah supports several thousand Snowy Plovers and is probably the most important site for the species in North America (Paton et al. 1992, Page et al. 1995, Paton and Bachman 1997, Oring et al. 2005).

It is difficult to convey the harshness of the environment in which the Snowy Plover rears its young. In Nevada, the plovers re-quire hypersaline playas with minimum vegetation (Neel 1999a). For example, one of the easiest sites at which to see the species in Nevada, Soda Lake in Churchill County, is nothing more than bleached flotsam, spent firecracker shells, and alkaline-encrusted gleaming white ground.

CONSERVATION AND MANAGEMENT

Although the Snowy Plover has an extensive range worldwide, it is very patchily distributed in the United States. Numbers at Honey Lake in Lassen County, just across the border between Nevada and California, have declined considerably in recent years (Oring et al. 2005), and the population in Nevada is generally thought to be in decline. To the west, the Pacific Coast population is listed as Threatened under the U.S. Endangered Species Act. As a consequence of its precarious status throughout the region, the species has received Priority ranking from Nevada Partners in Flight (Neel 1999a) and is considered Highly Imperiled in the U.S. Shorebird Conservation Plan (Brown et al. 2001).

The main problems that nesting Snowy Plovers face on Pacific Coast beaches—human disturbance, nest loss due to trampling, and predators attracted by humans—are probably not so severe in the Great Basin. Predators undoubtedly take their toll, and where their densities are elevated above historic levels may pose a problem. Extremes in water availability also are likely to pose important limitations. High water can flood suitable breeding areas in the playas. On the other hand, Snowy Plovers feed largely on brine flies in the Great Basin (Neel 1999a), and extremely dry conditions may reduce their prey supply. The current status of the state's Snowy Plovers is not especially well known because many birds occur in remote, inaccessible areas. More frequent visits by birders to the state's playas would certainly help us to better monitor this imperiled shorebird.

HABITAT USE

HABITAT	ATLAS BLOCKS	INCIDENTAL OBSERVATIONS
Barren	0	1 (50%)
Wetland	1 (100%)	1 (50%)
TOTAL	1	2

Total numbers reported in the Habitat Use, Breeding Status, and Abundance tables may differ from each other (see pp. 22–23 for details). Percentage sums may differ slightly from 100% due to rounding.

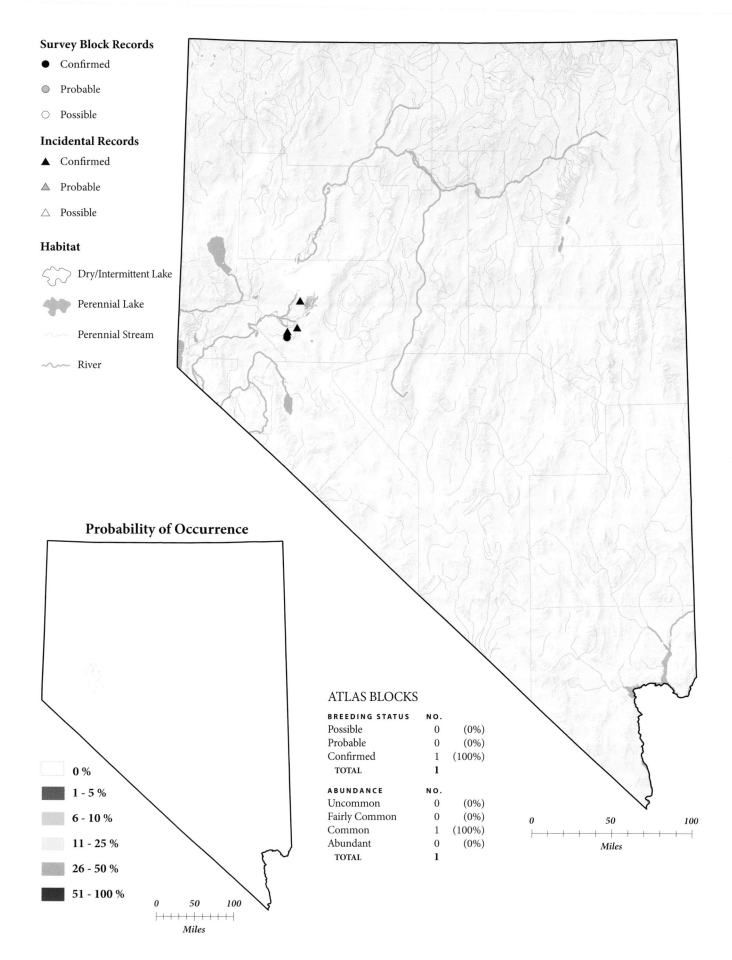

Survey Block Records

● Confirmed

◉ Probable

○ Possible

Incidental Records

▲ Confirmed

▲ Probable

△ Possible

Habitat

Dry/Intermittent Lake

Perennial Lake

Perennial Stream

River

Probability of Occurrence

☐ 0 %

■ 1 - 5 %

■ 6 - 10 %

☐ 11 - 25 %

■ 26 - 50 %

■ 51 - 100 %

0   50   100
Miles

ATLAS BLOCKS

BREEDING STATUS	NO.	
Possible	0	(0%)
Probable	0	(0%)
Confirmed	1	(100%)
TOTAL	1	

ABUNDANCE	NO.	
Uncommon	0	(0%)
Fairly Common	0	(0%)
Common	1	(100%)
Abundant	0	(0%)
TOTAL	1	

0   50   100
Miles

KILLDEER
Charadrius vociferus

Shrill sentry of farmyards, parking lots, and ball fields, the Killdeer is commonplace, conspicuous, and quite spectacular in its own way. Killdeer are found throughout the state and can be numerous in places. Their basic requirement for nesting is an open area near water of almost any sort—even an ornamental fountain or rainwater catchment will do. Killdeer can be found in open country far from human habitation, but they probably reach their highest densities in Nevada in altered habitats such as the outskirts of towns and irrigated farmland.

DISTRIBUTION

The atlas records came from throughout Nevada, although they were relatively scarce in the arid south-central portion of the state. Especially dense clusters of records came from the northeastern quadrant and from the drainages of west-central Nevada. Fieldworkers estimated multiple pairs to be present in the vast majority of occupied blocks, and in excess of ten pairs were thought to occur in more than forty blocks. Breeding was confirmed in almost half of the blocks in which the species was found.

The Killdeer has long been considered common and widespread in Nevada, as it is in all of the surrounding states (Behle et al. 1985, Small 1994, Stephens and Sturts 1998, Jackson and Jackson 2000, Adamus et al. 2001, Martin in Corman and Wise-Gervais 2005:174–175). Ryser (1985) described the Killdeer as the most widely distributed shorebird in the Great Basin, Alcorn (1988) labeled it a widespread resident, and Linsdale (1936) called it a numerous summer resident. Killdeer are very common in well-watered areas with a long history of human habitation, such as the Lahontan Valley (Chisholm and Neel 2002). Even in the Mojave Desert of southern Nevada, the Killdeer has long been considered the commonest shorebird (Fisher 1893), and atlas fieldwork produced numerous records along southern Nevada's river systems and around Las Vegas.

Killdeer are generally found near water, but they commonly use human-altered habitats and will nest far into the uplands.

Most atlas records came from wetlands or riparian habitat, but there were also many records from urban and agricultural areas. It is not at all unusual to find a clutch in the middle of a gravel road or parking lot (Rosenberg et al. 1991, Gallagher 1997). The predictive map reflects the general pattern of increasing abundance to the north, and also shows that Killdeer are most likely to be found in urban, agricultural, or aquatic habitats within the state.

CONSERVATION AND MANAGEMENT

The rapid expansion of humans into formerly "pristine" areas during the twentieth century probably brought a concomitant increase in Killdeer numbers, and evidently expanded the species' range as well (Jackson and Jackson 2000). During the last third of the twentieth century, however, the species declined substantially in many areas, including much of the West (Sanzenbacher and Haig 2001, Sauer et al. 2005). The reasons for the recent declines are not clear, but changes in agricultural practices and general degradation of wetland habitat may play an important role. In the past, Killdeer were taken in great numbers by market hunters, but the species is now legally protected from harvest (Jackson and Jackson 2000).

Killdeer numbers are generally high in human-altered habitats, and special management usually is not required in such environments (Gallagher 1997). Killdeer do, however, face challenges in the human-altered habitats with which they are so closely associated. They are susceptible to predation, which is often high in urban areas, and to pesticide poisoning; and with their propensity for nesting on gravel roads, they often get run over by cars (Jackson and Jackson 2000). Highly developed areas may thus actually function as population sinks (Jackson and Jackson 2000).

HABITAT USE

HABITAT	ATLAS BLOCKS	INCIDENTAL OBSERVATIONS
Agricultural	30 (10%)	2 (2%)
Barren	10 (3%)	2 (2%)
Grassland	4 (1%)	2 (2%)
Mojave	2 (<1%)	0 (0%)
Montane Forest	0 (0%)	1 (<1%)
Montane Shrub	2 (<1%)	2 (2%)
Open Water	11 (4%)	7 (6%)
Riparian	98 (34%)	22 (21%)
Salt Desert Scrub	11 (4%)	5 (5%)
Sagebrush Scrub	9 (3%)	5 (5%)
Sagebrush Steppe	16 (6%)	3 (3%)
Urban	34 (11%)	26 (25%)
Wetland	64 (22%)	28 (26%)
TOTAL	291	105

Total numbers reported in the Habitat Use, Breeding Status, and Abundance tables may differ from each other (see pp. 22–23 for details). Percentage sums may differ slightly from 100% due to rounding.

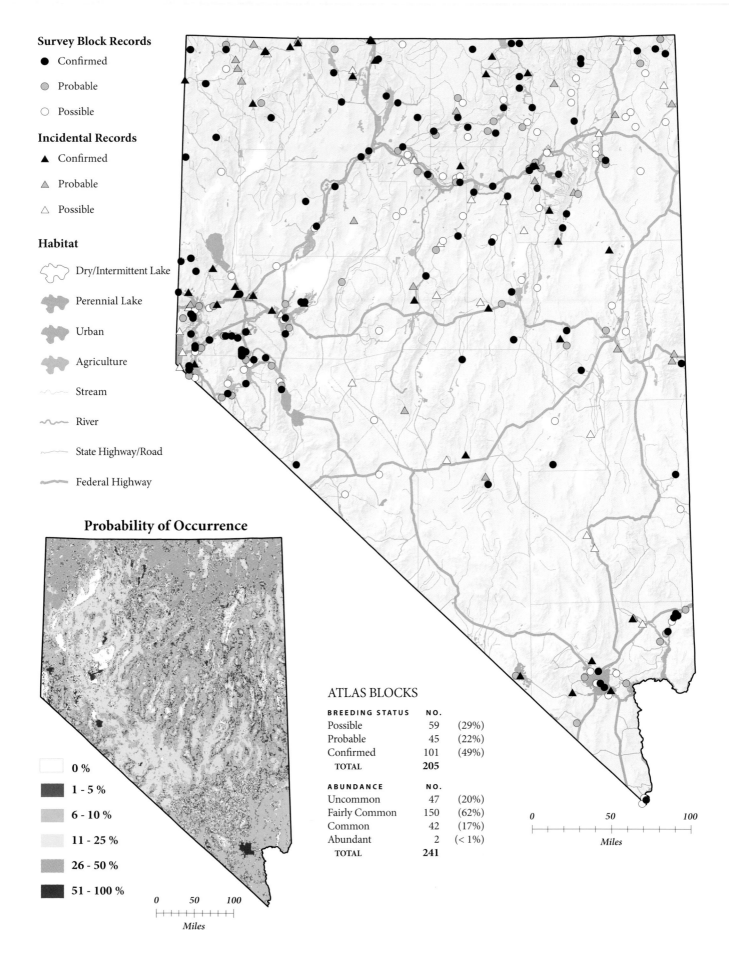

Survey Block Records

- ● Confirmed
- ● Probable
- ○ Possible

Incidental Records

- ▲ Confirmed
- ▲ Probable
- △ Possible

Habitat

- Dry/Intermittent Lake
- Perennial Lake
- Urban
- Agriculture
- Stream
- River
- State Highway/Road
- Federal Highway

Probability of Occurrence

- 0 %
- 1 - 5 %
- 6 - 10 %
- 11 - 25 %
- 26 - 50 %
- 51 - 100 %

0   50   100
Miles

ATLAS BLOCKS

BREEDING STATUS	NO.	
Possible	59	(29%)
Probable	45	(22%)
Confirmed	101	(49%)
TOTAL	205	

ABUNDANCE	NO.	
Uncommon	47	(20%)
Fairly Common	150	(62%)
Common	42	(17%)
Abundant	2	(< 1%)
TOTAL	241	

0   50   100
Miles

BLACK-NECKED STILT
Himantopus mexicanus

One of the state's most striking and conspicuous breeding shorebirds, the Black-necked Stilt can be found in pristine marshes and remnant wetlands far from human habitation; but it will also use sewage treatment facilities, golf course ponds, and irrigation ditches if conditions are suitable. Whatever the venue, stilts defend their nests with impressive aerial displays and piercing alarm notes.

DISTRIBUTION

The Black-necked Stilt occurs throughout the wetlands of western North America—where its range has been expanding to the north—and along the East and Gulf coasts (Robinson et al. 1999). The northern Great Basin is the species' stronghold, and more than half of the entire population is thought to occur there, with the bulk at the Great Salt Lake in Utah (Oring et al. 2005).

The Lahontan Valley wetlands are another important stronghold for the species (Chisholm and Neel 2002), and breeding also occurs at scattered sites in the valleys of northern Nevada, ranging from Sheldon National Wildlife Refuge in the northwest to sewage treatment ponds near Reno to the Ruby Valley marshes in the east. Scattered observations, and one confirmed breeding record, also came from southern Nevada. In adjacent states, stilts also nest in high numbers at various sites in northeastern California and southeastern Oregon (Oring et al. 2005).

Black-necked Stilts are often found in association with American Avocets, but stilts tend to occupy less saline wetlands and nest in areas with more emergent vegetation (Robinson et al. 1999). Linsdale (1936:57) described the Black-necked Stilt as "limited to borders of alkali pools," but this does not convey the versatility of a species that will nest in wet pastures, borrow ditches, and small ephemeral ponds as well as large, well-managed wetlands. Not surprisingly, almost all atlas observations came from aquatic habitats, and most were from wetlands. There were a few observations from agricultural habitat, which presumably referred either to birds nesting in a pasture or to birds at a small wetland surrounded by farmland.

The predictive map reflects an increasing chance of finding this species as one moves north in the state, as well as its absence at higher elevations. Although low, the estimated probabilities in the southern half of the state likely still overestimate the true occurrence as indicated by the distribution of field records. Nonetheless, it is probably accurate to say that Black-necked Stilts could turn up almost anywhere in the lowlands of Nevada where suitable habitat is available.

CONSERVATION AND MANAGEMENT

The Black-necked Stilt is considered a Priority species in the Intermountain West Regional Shorebird Conservation Plan because such a large proportion of the population nests in the region (Oring et al. 2005). Stilts are susceptible to selenium accumulation and salinization, both of which commonly occur in wetlands that lie near irrigated farmland in the arid West. Contaminated water can hamper the survival of stilt chicks (Rubega and Robinson 1996, Robinson et al. 1999).

Continent-wide, Black-necked Stilt populations seem to be relatively stable (Robinson et al. 1999, Sauer et al. 2005), but the only extensive data come from the Breeding Bird Survey, which is not well suited to monitor wetland species that often breed far from roads. In California, the species has increased in the San Francisco area but has decreased sharply in the Central Valley (Shuford 1993) and in southern coastal areas (Gallagher 1997). About a thousand pairs are currently estimated to breed in the Lahontan Valley wetlands, representing a substantial decline since earlier in the twentieth century (Neel and Henry 1997, Neel in Oring et al. 2005). Predation rates on eggs and chicks at this site were extremely high in the mid-1990s (M. Rubega, pers. comm.); however, it is unclear whether predation is an important factor determining overall population size.

HABITAT USE

HABITAT	ATLAS BLOCKS	INCIDENTAL OBSERVATIONS
Agricultural	2 (6%)	1 (5%)
Open Water	5 (15%)	2 (10%)
Riparian	2 (6%)	1 (5%)
Wetland	24 (73%)	16 (80%)
TOTAL	33	20

Total numbers reported in the Habitat Use, Breeding Status, and Abundance tables may differ from each other (see pp. 22–23 for details). Percentage sums may differ slightly from 100% due to rounding.

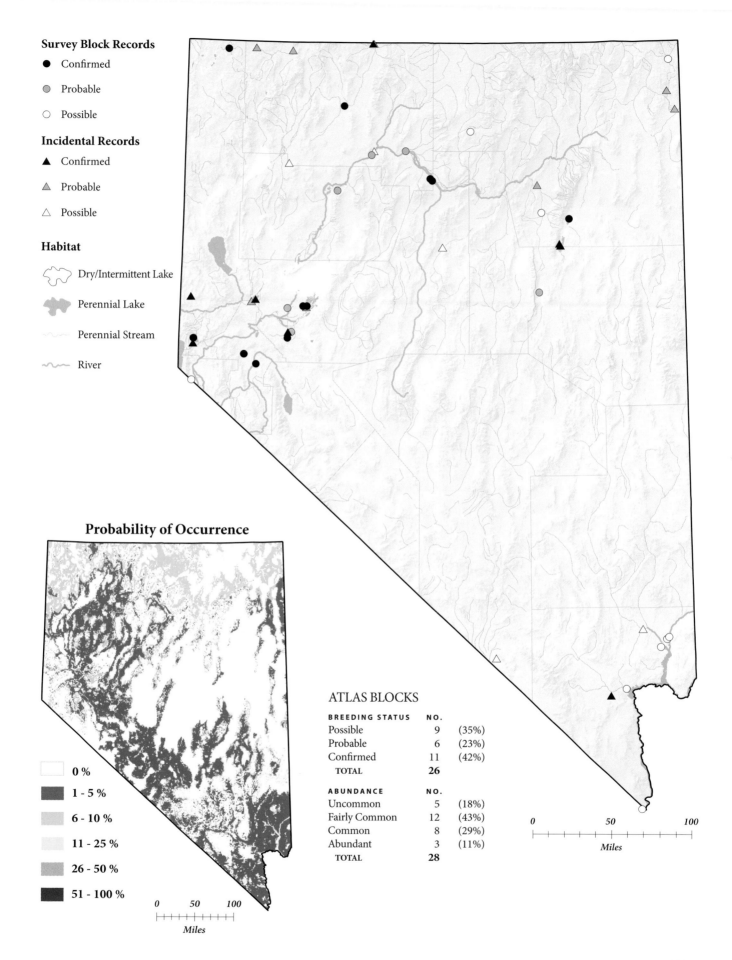

Survey Block Records

● Confirmed

● Probable

○ Possible

Incidental Records

▲ Confirmed

▲ Probable

△ Possible

Habitat

Dry/Intermittent Lake

Perennial Lake

Perennial Stream

River

Probability of Occurrence

☐ 0 %

■ 1 - 5 %

■ 6 - 10 %

☐ 11 - 25 %

■ 26 - 50 %

■ 51 - 100 %

0   50   100
Miles

ATLAS BLOCKS

BREEDING STATUS	NO.	
Possible	9	(35%)
Probable	6	(23%)
Confirmed	11	(42%)
TOTAL	26	

ABUNDANCE	NO.	
Uncommon	5	(18%)
Fairly Common	12	(43%)
Common	8	(29%)
Abundant	3	(11%)
TOTAL	28	

0   50   100
Miles

AMERICAN AVOCET
Recurvirostra americana

Ryser (1985:180) said of the American Avocet that it "enlivens the bleakest yet most characteristic of Great Basin summer landscapes—the shallow, foul bodies of alkaline or brackish water and their fringing flats of mud, alkali, or salt." With its black-and-white body, orange head and neck, blue legs, and oddly recurved bill, the avocet is one of our most stunning breeding shorebirds, and its delicate beauty seems strangely out of place in the sinks, sewage ponds, and stinking seeps that it calls home.

DISTRIBUTION

The American Avocet is fairly widespread in western North America, occurring in large numbers at interior wetlands throughout the northern Great Basin and Great Plains, as well as in coastal and valley marshes in California (Robinson et al. 1997). Alcorn (1988) called the avocet a common summer resident in Nevada, and the Great Basin is thought to contain a substantial proportion of the world's population (Ryser 1985, Oring et al. 2005).

Like Black-necked Stilts, avocets turned up in wetlands all across northern Nevada and at a few sites in the south, especially along the Colorado River. Although the two species often nest side-by-side, avocets are somewhat more broadly distributed in Nevada, reflecting their greater willingness to use more sparsely vegetated and saline wetlands. The distinction is evident on the predictive maps for the two species, with avocets having a slightly higher likelihood of occurrence in many areas. Even avocets, however, need fresh water to raise their young, which cannot survive in a strictly saltwater environment (Rubega and Robinson 1996; M. Rubega, unpubl. data). Consequently, breeding on salt playas generally occurs in the vicinity of freshwater seeps or near stream inflows.

Approximately five thousand pairs of avocets are estimated to nest in the Lahontan Valley wetlands, and this area supports many thousand birds more during migration (Chisholm and Neel 2002, Oring et al. 2005). Radiotelemetry studies have shown that Avocets regularly move between Great Basin wetlands after breeding but before the fall migration (Plissner et al. 2000). The purpose of these movements is unknown, but the birds may be searching for safe places with abundant food where they can stay during molting or gathering information about suitable breeding locations for future nesting attempts (Plissner et al. 2000).

CONSERVATION AND MANAGEMENT

Avocets are not especially site-faithful (Robinson and Oring 1997), a quality that, combined with their propensity to wander after breeding, makes them well suited for colonizing newly available or ephemeral habitats. Consequently, they are not unusual in human-created wetlands such as wastewater ponds (e.g., Burridge in Burridge 1995:69). Avocets also respond quickly to wetland creation, as has been demonstrated at the Jay Dow Sr. Wetlands just across the Nevada border near Honey Lake, California, where they have been extensively studied (e.g., Robinson and Oring 1997; Plissner et al. 1999, 2000; Johnson et al. 2003).

These same characteristics make American Avocet populations difficult to monitor, however, especially in the arid West, where locations of suitable habitat change from year to year depending on precipitation patterns (Reed et al. 1997). This is exacerbated by the difficulty of accessing some sites such as the Humboldt Sink, which can hold substantial numbers of avocets when conditions are suitable.

Avocet populations are not known to have declined in Nevada, or in the surrounding states, but water diversions and loss of wetlands have likely affected the species (Robinson et al. 1997). Because such a large proportion of the world's population occurs in the region, American Avocets are considered a Nevada Partners in Flight Priority species. Neel (1999a) and the Intermountain West Regional Shorebird Conservation Plan (Oring et al. 2005) recommend a variety of specific conservation actions.

HABITAT USE

HABITAT	ATLAS BLOCKS	INCIDENTAL OBSERVATIONS
Agricultural	1 (2%)	0 (0%)
Barren	1 (2%)	1 (4%)
Open Water	11 (21%)	5 (21%)
Riparian	3 (6%)	0 (0%)
Wetland	36 (69%)	18 (75%)
TOTAL	52	24

Total numbers reported in the Habitat Use, Breeding Status, and Abundance tables may differ from each other (see pp. 22–23 for details). Percentage sums may differ slightly from 100% due to rounding.

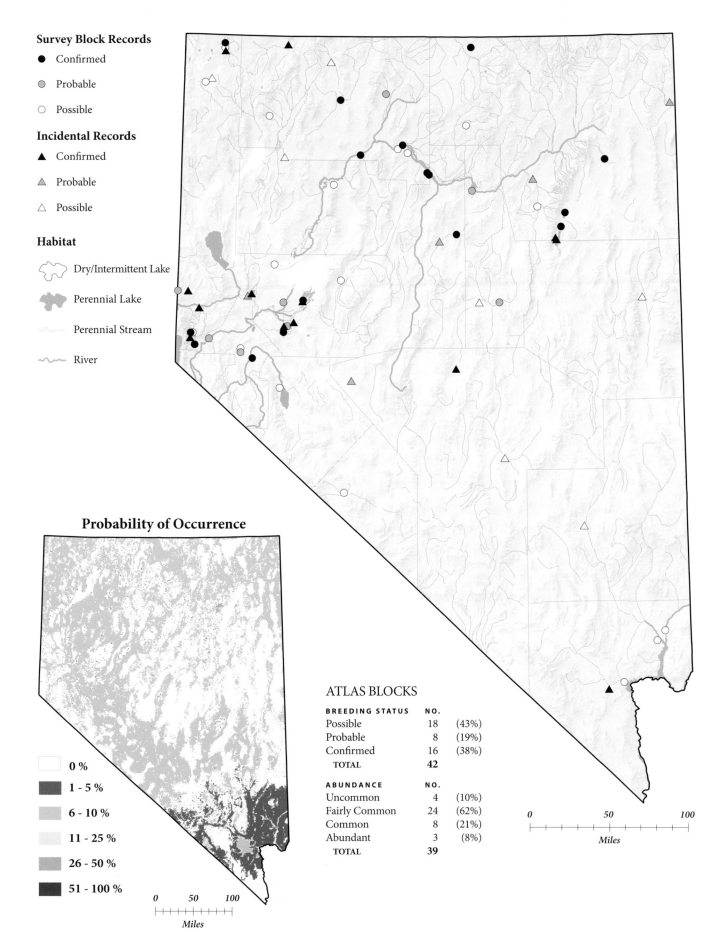

Survey Block Records

● Confirmed

● Probable

○ Possible

Incidental Records

▲ Confirmed

▲ Probable

△ Possible

Habitat

Dry/Intermittent Lake

Perennial Lake

Perennial Stream

River

Probability of Occurrence

0 %

1 - 5 %

6 - 10 %

11 - 25 %

26 - 50 %

51 - 100 %

0   50   100

Miles

ATLAS BLOCKS

BREEDING STATUS	NO.	
Possible	18	(43%)
Probable	8	(19%)
Confirmed	16	(38%)
TOTAL	**42**	

ABUNDANCE	NO.	
Uncommon	4	(10%)
Fairly Common	24	(62%)
Common	8	(21%)
Abundant	3	(8%)
TOTAL	**39**	

0   50   100

Miles

WILLET
Tringa semipalmata

Like most other shorebirds that breed in Nevada, Willets generally nest above latitude 37°N in the state. Within their Nevada range, most Willets occur in the valley wetlands near Reno and Fallon, or in the wetlands and grasslands of Elko County and adjoining counties. The Willet is a fairly versatile species, and atlas surveyors found this noisy bird in a variety of open-country habitats, as long as there was water nearby.

DISTRIBUTION

Willets in Nevada belong to the *inornatus* subspecies, which is widespread in central and western North America. Known as the "Western" Willet, it reaches its southern breeding limit in Nevada, and is distinct from the subspecies that occurs along the eastern U.S. coast (Lowther et al. 2001). Western Willets also breed in northeastern California, southeastern Oregon, southern Idaho, far northern Utah, and northeastward into the northern Great Plains (Behle et al. 1985, Small 1994, Stephens and Sturts 1998, Adamus et al. 2001, Lowther et al. 2001).

Willets tend to be concentrated in two general areas of northern Nevada, each associated with different river systems. Breeding was confirmed at ten locations in the north-central to northeastern portion of the state, where birds are often found in wetlands associated with the Humboldt River system. In contrast, breeding was not confirmed in the Truckee and Carson river systems, despite several probable breeding records in this area. Evidence from other sources, however, suggests that breeding does occur in this part of the state (e.g., Linsdale 1936, Alcorn 1988, Chisholm and Neel 2002).

Willets can be found in wetlands of varying salinity and vegetation composition, and in upland habitats ranging from agricultural fields to native prairies (Lowther et al. 2001). The predictive map for the Willet is similar to those for Wilson's Phalarope and Wilson's Snipe—two other shorebirds frequently found in wet meadow habitats. Willets are predicted to occur most commonly in wetlands, but the model also estimates a moderate chance of finding them in agricultural lands and sagebrush steppe across much of northern Nevada. In the western Great Basin, they will nest out in the sagebrush, commuting on a daily basis to nearby wetlands to feed (Haig et al. 2002). Willets also take their chicks to wetlands after hatching, and close inspection of the data map shows that almost all records lie near a water body of some sort.

CONSERVATION AND MANAGEMENT

The Breeding Bird Survey does not indicate a significant change in overall Willet numbers (Sauer et al. 2005); however, the species is probably not well sampled by these roadside surveys in much of its range. The eastern subspecies of the Willet has increased in range and in overall numbers in some areas (e.g., Peterson in Andrle and Carroll 1988:154–155) as it has recovered from former declines linked to market hunting. The situation for the western subspecies is not so well known. Declines have been detected in Canada in recent decades (Sauer et al. 2005), and the widespread loss of grasslands over the longer term may have had an impact throughout the West (Lowther et al. 2001).

Ryan and Renken (1987) provided a good summary of the conservation challenges for the *inornatus* subspecies in the northern Great Plains, and this work may have relevance in Nevada. Modest agricultural enterprises appear to be compatible with the bird's needs, but intensive agricultural operations are generally not. Overall the *inornatus* population is considered to be a species of Moderate Concern in the U.S. Shorebird Conservation Plan (Brown et al. 2001), but better monitoring is warranted. Recent research also suggests that Great Basin Willets might constitute a population segment that is distinct from others in the West (Haig et al. 2002), providing additional reason to learn more about the species' status in the state.

HABITAT USE

HABITAT	ATLAS BLOCKS	INCIDENTAL OBSERVATIONS
Agricultural	3 (7%)	1 (5%)
Barren	1 (2%)	0 (0%)
Grassland	2 (4%)	1 (5%)
Open Water	1 (2%)	2 (11%)
Riparian	9 (20%)	1 (5%)
Salt Desert Scrub	2 (4%)	1 (5%)
Sagebrush Scrub	0 (0%)	1 (5%)
Sagebrush Steppe	1 (2%)	0 (0%)
Wetland	26 (58%)	12 (63%)
TOTAL	45	19

Total numbers reported in the Habitat Use, Breeding Status, and Abundance tables may differ from each other (see pp. 22–23 for details). Percentage sums may differ slightly from 100% due to rounding.

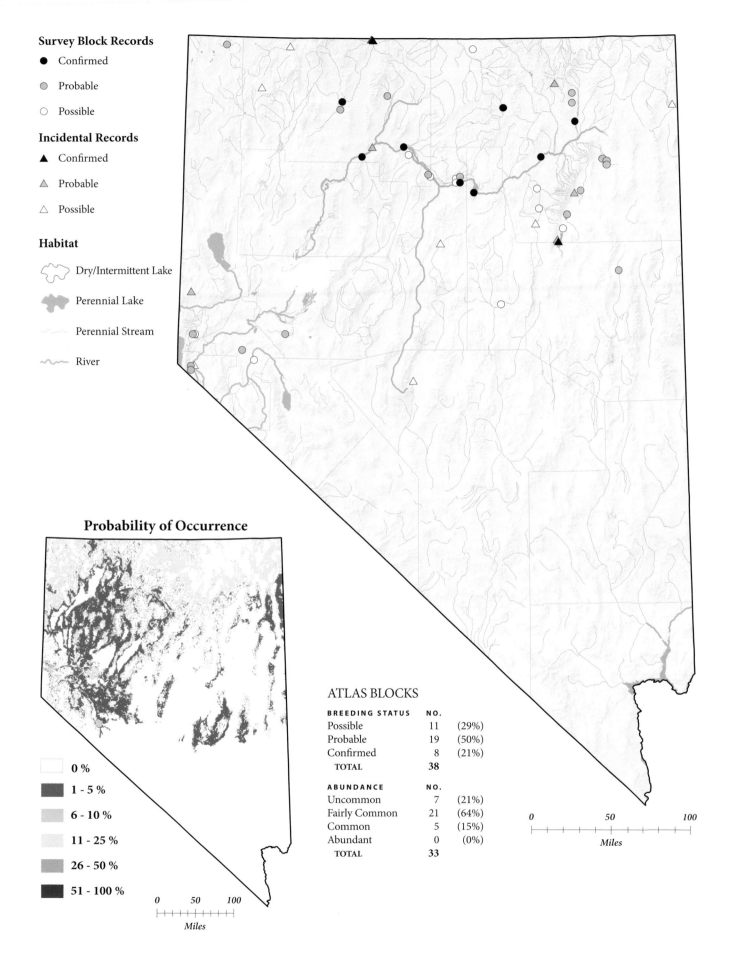

Survey Block Records

● Confirmed

● Probable

○ Possible

Incidental Records

▲ Confirmed

▲ Probable

△ Possible

Habitat

Dry/Intermittent Lake

Perennial Lake

Perennial Stream

River

Probability of Occurrence

☐ 0 %

■ 1 - 5 %

■ 6 - 10 %

■ 11 - 25 %

■ 26 - 50 %

■ 51 - 100 %

0   50   100

Miles

ATLAS BLOCKS

BREEDING STATUS	NO.	
Possible	11	(29%)
Probable	19	(50%)
Confirmed	8	(21%)
TOTAL	38	

ABUNDANCE	NO.	
Uncommon	7	(21%)
Fairly Common	21	(64%)
Common	5	(15%)
Abundant	0	(0%)
TOTAL	33	

0   50   100

Miles

SPOTTED SANDPIPER
Actitis macularia

At times it seems as though every stretch of shoreline in Nevada has a resident Spotted Sandpiper. These small shorebirds nest along the shores of our largest lakes, near the pebbly shoals of cold mountain streams, and even along the edges of high mountain tarns near the tree line. But they are not restricted to wilderness and can also be found in highly altered or artificial aquatic habitats such as sewage treatment plants, irrigation ditches, and farm ponds.

DISTRIBUTION

Spotted Sandpipers were widespread in the state, with confirmed breeding as far south as Las Vegas, but the bulk of the nesting records came from Elko County and the drainages of west-central Nevada. The species was virtually absent from the south-central part of the state, however, where aquatic habitats are few and scattered. Breeding confirmation rates were low, presumably reflecting the difficulty of finding nests, which are often placed in dense vegetation (Oring et al. 1997), or observing the precocial young. Multiple pairs were thought to be present in about 70% of the blocks where Spotted Sandpipers occurred, and a few blocks were estimated to have more than ten pairs.

The Spotted Sandpiper has long been regarded as widespread and common in Nevada. Among the shorebirds, Ryser (1985) reported it to be second only to the Killdeer in abundance and extent of distribution in the Great Basin, and the atlas data support this view. Linsdale (1936) considered it to be a common summer resident throughout the state, and Fisher (1893) noted that it was common even in the lowlands of the Mojave Desert.

The North American breeding range of the Spotted Sandpiper is extensive (Oring et al. 1997), and the species is found widely in the states to the north and east of Nevada (Behle et al. 1985, Stephens and Sturts 1998, Adamus et al. 2001). Spotted Sandpipers are widespread in California as well, but are fairly lo-

cal as breeders in the south and away from the mountains (Garrett and Dunn 1981, Small 1994, Oring et al. 1997). They are less common as breeders south of Nevada, with no direct evidence of nesting along the lower Colorado River (Rosenberg et al. 1991) and a patchy breeding range in Arizona (Wise-Gervais in Corman and Wise-Gervais 2005:180–181).

As expected, almost all Spotted Sandpiper reports came from aquatic habitats. The predictive map reflects a low to moderate probability of finding the species throughout most of the state, with the higher odds in the northernmost reaches and in riparian areas. Atlas records suggest that the actual occurrence is lower in the northwestern corner of the state and in the mountains of central Nevada than the predictions suggest, presumably because the model does not capture the fine details of the species' habitat needs.

CONSERVATION AND MANAGEMENT

Breeding Bird Survey data indicate that Spotted Sandpiper populations are stable in North America as a whole, although local declines have been reported in some areas (Sauer et al. 2005). This survey is not especially well suited to census most waterbirds, though, and few data are available from Nevada, so conclusions should be drawn with caution.

Numbers appear to have declined in the Lahontan Valley, perhaps because of reduced flows in the lower Carson River (Chisholm and Neel 2002), and there may be a need for management at a local level. Like other widespread but sparsely distributed shorebirds, the Spotted Sandpiper could easily decline without anyone noticing unless a concerted effort is made to survey the habitats that it uses. Riparian areas are particularly important for this species in the Intermountain West (Oring et al. 2005), and detailed monitoring of riparian habitats would go a long way toward addressing this concern.

HABITAT USE

HABITAT	ATLAS BLOCKS	INCIDENTAL OBSERVATIONS
Agricultural	1 (1%)	0 (0%)
Alpine	2 (2%)	1 (4%)
Barren	1 (1%)	0 (0%)
Montane Forest	0 (0%)	1 (4%)
Open Water	6 (7%)	2 (7%)
Riparian	48 (57%)	16 (57%)
Sagebrush Scrub	1 (1%)	0 (0%)
Urban	1 (1%)	1 (4%)
Wetland	24 (29%)	7 (25%)
TOTAL	84	28

Total numbers reported in the Habitat Use, Breeding Status, and Abundance tables may differ from each other (see pp. 22–23 for details). Percentage sums may differ slightly from 100% due to rounding.

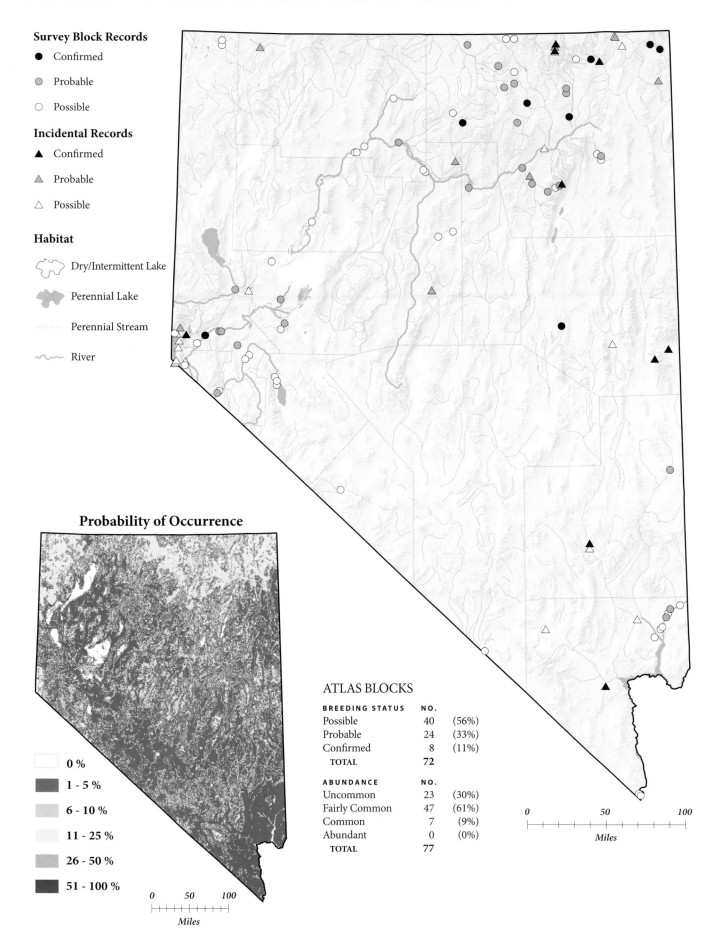

Survey Block Records

● Confirmed

● Probable

○ Possible

Incidental Records

▲ Confirmed

▲ Probable

△ Possible

Habitat

Dry/Intermittent Lake

Perennial Lake

Perennial Stream

River

Probability of Occurrence

	0 %
	1 - 5 %
	6 - 10 %
	11 - 25 %
	26 - 50 %
	51 - 100 %

0   50   100

Miles

ATLAS BLOCKS

BREEDING STATUS	NO.	
Possible	40	(56%)
Probable	24	(33%)
Confirmed	8	(11%)
TOTAL	72	

ABUNDANCE	NO.	
Uncommon	23	(30%)
Fairly Common	47	(61%)
Common	7	(9%)
Abundant	0	(0%)
TOTAL	77	

0   50   100

Miles

LONG-BILLED CURLEW

Numenius americanus

Rh '02

The striking courtship displays and wailing calls of the Long-billed Curlew evoke the desolate grasslands of northern Nevada where it breeds. Atlas workers found Long-billed Curlews to be somewhat more common—and certainly more widespread—than expected, but gratification at this result must be tempered by the species' long-term, rangewide population decline. Protection and management of this spectacular shorebird should therefore be a high conservation priority in the state.

DISTRIBUTION

Long-billed Curlews were primarily found north of about latitude 39°N, matching earlier descriptions of the species' range (Linsdale 1936). Reviews published prior to the atlas survey listed confirmed breeding at only four Nevada locations (Sheldon National Wildlife Refuge in Washoe County, Ruby Lake National Wildlife Refuge in Elko County, Lahontan Valley in Churchill County, and Fish Creek Ranch in Eureka County), with summer occurrence at other scattered sites in northern Nevada (Alcorn 1988, Neel 1999a). Atlas workers, however, found Long-billed Curlews in forty-three widely distributed blocks, and confirmed breeding at fourteen locations.

Most of the records came from the Humboldt River system, but White Pine and Churchill counties produced several as well. Not surprising for a sparsely distributed species with large breeding territories (Dugger and Dugger 2002), abundance was generally low: Only one pair was thought to be present in more than a quarter of the blocks for which abundance was estimated, and very few blocks were thought to contain more than ten pairs.

Curlews breed throughout the Intermountain West, and the region is considered critical for the species' survival (Oring et al. 2005). They are found across much of eastern Oregon, southern Idaho, and northern Utah (Behle et al. 1985, Stephens and Sturts 1998, Adamus et al. 2001). In California, nesting is largely restricted to the northeastern corner (Small 1994), and there is only very occasional breeding in Arizona (Corman in Corman and Wise-Gervais 2005:591).

The atlas results agree with Alcorn's (1988) description of the Long-billed Curlew as a summer resident in meadows and pastures. Most habitat records came from wetland, grassland, and agricultural areas, and the predictive map reflects the moderately high chance of finding Long-billed Curlews in these habitats.

CONSERVATION AND MANAGEMENT

Beginning in the late nineteenth century, the Long-billed Curlew experienced a long-term range retraction and population decline in North America, largely due to overhunting and conversion of land for agriculture (Ryser 1985, Dugger and Dugger 2002, Dobkin and Sauder 2004). Currently, the bird is considered Highly Imperiled in the U.S. Shorebird Conservation Plan (Brown et al. 2001). Although the northern Great Basin is not in the core of the species' range, it is thought to support an important portion of the breeding population (Neel 1999a; Young and Oring 2006), and Nevada Partners in Flight therefore recognizes the Long-billed Curlew as a Priority species (Neel 1999a).

Our understanding of the Long-billed Curlew's habitat needs remains incomplete. Although agriculture has been associated with the species' demise in many areas (Smith et al. 1997, Dugger and Dugger 2002, Dobkin and Sauder 2004), Long-billed Curlews are common in irrigated hayfields in Nevada. Recent studies in the Ruby Valley, for example, suggest that they may fare better in this habitat than in the more arid areas where most of the earlier studies were done (Hartman and Oring 2006). High levels of predation on eggs and young also may be an important management concern in some areas (Neel 1999a, Young and Oring 2006). Improved monitoring and studies of the effects of grazing, haying, and predation are high priorities for future work (Oring et al. 2005). Fortunately, populations in most of the West seem to have stabilized (Paige and Ritter 1999, Sauer et al. 2005), and some authors report that Long-billed Curlew numbers are on the rise (Dobkin and Sauder 2004).

HABITAT USE

HABITAT	ATLAS BLOCKS	INCIDENTAL OBSERVATIONS
Agricultural	8 (14%)	2 (17%)
Grassland	16 (29%)	3 (25%)
Open Water	1 (2%)	1 (8%)
Riparian	8 (14%)	0 (0%)
Salt Desert Scrub	2 (4%)	0 (0%)
Sagebrush Scrub	3 (6%)	1 (8%)
Urban	0 (0%)	1 (8%)
Wetland	17 (31%)	4 (33%)
TOTAL	55	12

Total numbers reported in the Habitat Use, Breeding Status, and Abundance tables may differ from each other (see pp. 22–23 for details). Percentage sums may differ slightly from 100% due to rounding.

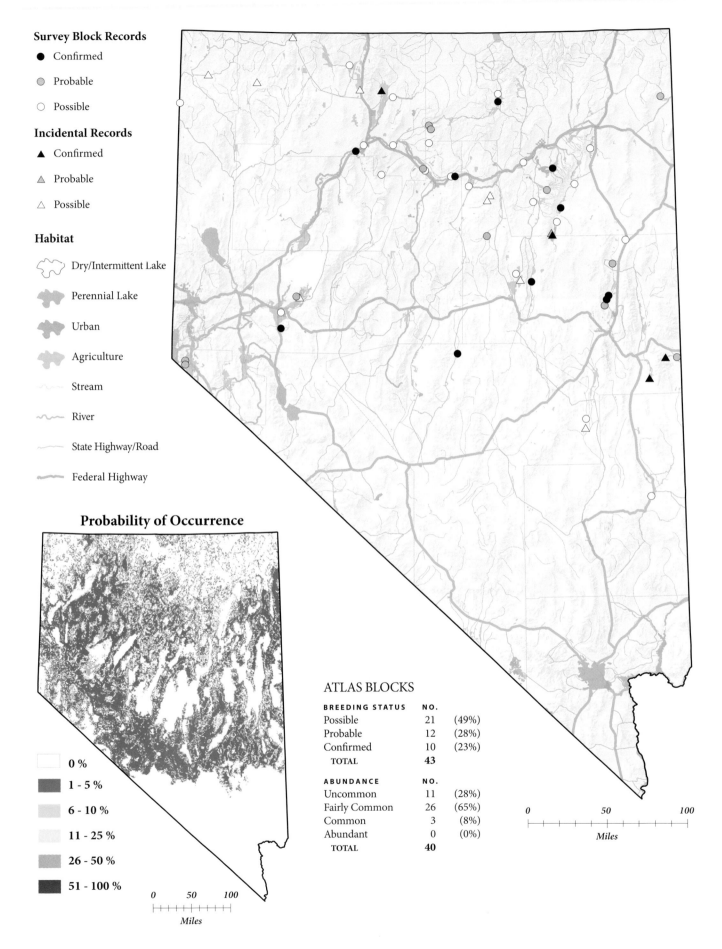

Survey Block Records

● Confirmed

● Probable

○ Possible

Incidental Records

▲ Confirmed

△ Probable

△ Possible

Habitat

Dry/Intermittent Lake

Perennial Lake

Urban

Agriculture

Stream

River

State Highway/Road

Federal Highway

Probability of Occurrence

☐ 0 %

▨ 1 - 5 %

▨ 6 - 10 %

▨ 11 - 25 %

▨ 26 - 50 %

▨ 51 - 100 %

0   50   100
Miles

ATLAS BLOCKS

BREEDING STATUS	NO.	
Possible	21	(49%)
Probable	12	(28%)
Confirmed	10	(23%)
TOTAL	43	

ABUNDANCE	NO.	
Uncommon	11	(28%)
Fairly Common	26	(65%)
Common	3	(8%)
Abundant	0	(0%)
TOTAL	40	

0   50   100
Miles

WILSON'S SNIPE
Gallinago delicata

RN'02

The winnowing of Wilson's Snipes imparts an air of enchantment to the marshy hinterlands of northern Nevada. Snipes can still be heard in increasingly urban areas, but their mournful call seems to carry a special urgency in the face of wetland conversion and habitat loss. Because Wilson's Snipes are secretive and often crepuscular, they may have been missed in some of the state's smaller wetlands.

DISTRIBUTION

Atlas workers found Wilson's Snipes at widely scattered locales throughout Nevada. The densest clusters of records came from west-central Nevada and from Elko County. Few reports came from the southern half of the state, and breeding was not confirmed south of Douglas County. There were no records at all from Nye County. Breeding was difficult to confirm, but multiple pairs were thought to occur in most of the blocks in which the species was found.

Atlas results were consistent with earlier studies in the state, which considered Wilson's Snipe to be a fairly common summer resident in northern Nevada (Ryser 1985, Alcorn 1988). Linsdale (1936) reported that the southernmost summer records came from latitude 39°N in Smoky Valley, Nye County.

The breeding range of this species lies largely to the north of Nevada and extends all the way to the Arctic Circle (Mueller 1999). Wilson's Snipe is sparsely distributed and localized as a breeder in most of California (Small 1994) and nests only rarely south of Nevada (Mueller 1999, Corman in Corman and Wise-Gervais 2005:182–183). It is a widespread breeder in Oregon

(Adamus et al. 2001) and a common resident in the southeastern counties near Nevada (Littlefield 1990, Contreras and Kindschy 1996). It is a common resident in Utah (Behle et al. 1985) as well, and a confirmed or suspected breeder throughout Idaho (Stephens and Sturts 1998).

Wilson's Snipe breeds mainly in marshes and wet meadows, and most atlas records came from wetland and riparian habitats. Many records also came from agricultural areas, and they presumably referred to birds observed in wet pastures or flooded fields (Chisholm and Neel 2002). The predictive model shows that Wilson's Snipes are most likely to be found in agricultural, wetland, and riparian habitats in more northern areas of the state. The map probably overemphasizes the importance of sagebrush habitats, where the birds usually occur only in small flooded areas embedded within the dominant habitat.

CONSERVATION AND MANAGEMENT

Wilson's Snipe is a widespread species and can be locally common. The North American population is thought to number about two million birds (Morrison et al. 2000) and appears to be relatively stable (Sauer et al. 2005). Breeding Wilson's Snipes can be difficult to enumerate (Mueller 1999), however, and populations in some areas may be underestimated as a result. Moreover, little monitoring is conducted in the boreal and tundra zones of Canada and Alaska, where a substantial portion of the population nests.

Local declines have occurred in some areas of the West. The species was eliminated as a breeder from parts of southern California during the twentieth century (Garrett and Dunn 1981), and declines have also been reported from Oregon and Washington (Smith et al. 1997, Sauer et al. 2005). Little monitoring information is available from Nevada, and it is unclear how the species is faring here. Maintaining areas of short, flooded vegetation is likely to be the most effective management strategy for Wilson's Snipe in the region, but much remains to be learned about the management of wetlands and agricultural areas for this and other shorebirds.

HABITAT USE

HABITAT	ATLAS BLOCKS	INCIDENTAL OBSERVATIONS
Agricultural	12 (17%)	1 (13%)
Grassland	1 (1%)	0 (0%)
Montane Parkland	1 (1%)	0 (0%)
Pinyon-Juniper	1 (1%)	0 (0%)
Riparian	24 (34%)	2 (25%)
Urban	2 (3%)	0 (0%)
Wetland	30 (42%)	5 (63%)
TOTAL	71	8

Total numbers reported in the Habitat Use, Breeding Status, and Abundance tables may differ from each other (see pp. 22–23 for details). Percentage sums may differ slightly from 100% due to rounding.

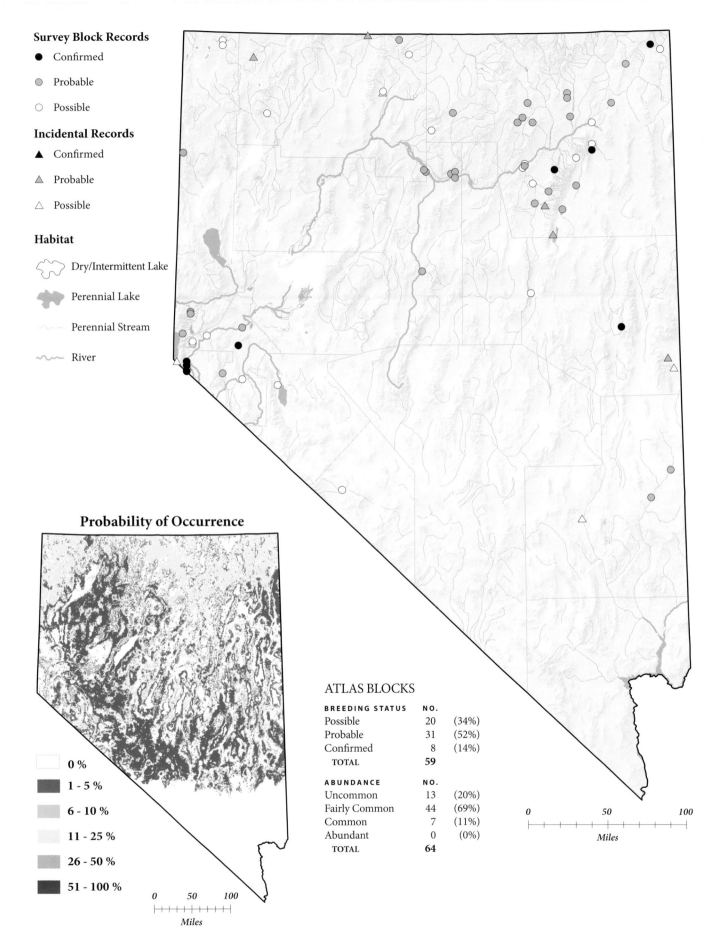

Survey Block Records

● Confirmed

● Probable

○ Possible

Incidental Records

▲ Confirmed

▲ Probable

△ Possible

Habitat

Dry/Intermittent Lake

Perennial Lake

Perennial Stream

River

Probability of Occurrence

☐ 0 %

■ 1 - 5 %

■ 6 - 10 %

☐ 11 - 25 %

☐ 26 - 50 %

■ 51 - 100 %

0   50   100
Miles

ATLAS BLOCKS

BREEDING STATUS	NO.	
Possible	20	(34%)
Probable	31	(52%)
Confirmed	8	(14%)
TOTAL	59	

ABUNDANCE	NO.	
Uncommon	13	(20%)
Fairly Common	44	(69%)
Common	7	(11%)
Abundant	0	(0%)
TOTAL	64	

0   50   100
Miles

WILSON'S PHALAROPE
Phalaropus tricolor

Wilson's Phalarope is a versatile and opportunistic breeder in the marshes and wet meadows of northern Nevada, although it breeds within a fairly narrow time window. Many birds are still migrating north around Memorial Day, and some are already moving south before the end of June. The nest is usually placed in dense—albeit fairly short—vegetation, and the precocial chicks usually stay concealed under cover.

DISTRIBUTION

Wilson's Phalaropes were found at widely scattered locations across the northern half of the state and were absent south of northern Nye and Mineral counties. Breeding was confirmed in almost 40% of the occupied blocks, and more than a quarter of those blocks were estimated to have more than ten pairs.

Older accounts of the state's avifauna present a picture similar to the atlas results (Linsdale 1936, Alcorn 1988), and Ryser (1985) considered Wilson's Phalarope to be a common, and often abundant, breeder in the Great Basin. Historical records suggest that thousands of young were produced in the Lahontan Valley in the 1970s, but numbers were much lower there during the atlas years (Chisholm and Neel 2002).

Nevada lies at the southwestern edge of the breeding range. Wilson's Phalarope breeds throughout eastern Oregon, much of Idaho, and northern Utah (Behle et al. 1985, Stephens and Sturts 1998, Adamus et al. 2001), but the bulk of its range lies in the northern Great Plains (Colwell and Jehl 1994). The species breeds in northeastern California all the way to the Nevada border (Small 1994), and very rarely in Arizona (Corman in Corman and Wise-Gervais 2005:602).

Wilson's Phalarope is a bird of wet, grassy places. Atlas records came from northern Nevada's major wetland complexes, as well as numerous sites along major and minor drainages. The few atlas records that did not come from wetland areas probably involved birds nesting in small wetlands located within other habitats. Because small marshes were sometimes found within blocks assigned to other habitats, the predictive model identified all areas where such wetlands could occur as potential habitat. Thus,

the predictive map reflects the probability with which areas are likely both to have suitable marshes and to have those marshes occupied by breeding phalaropes. The predicted distribution is restricted to the northern half of the state, with decreasing probabilities of occurrence in the high mountains and on the lower valley floors.

CONSERVATION AND MANAGEMENT

Although Wilson's Phalarope ranges widely as a breeder in western and central North America, the Great Basin stands out as a region of special significance. Breeding populations in the Great Basin are certainly important, but it is the region's hypersaline lakes—used by molting and staging birds—that are critical to populations migrating through from many other regions (Mahoney and Jehl 1985, Oring et al. 2005).

Population trends for this species are not well known, although declines have been reported at some migratory stopovers (Jehl 1994). The status in Nevada is not well known either, although numbers declined in the Lahontan Valley during the dry years of the late 1980s and early 1990s. As flooded habitats have recovered, phalaropes have returned to these wetlands, but not in their former breeding numbers (Chisholm and Neel 2002). Continuing efforts to restore wetlands and wet meadows throughout the major river systems of northern Nevada probably constitute the most important conservation strategy for this species.

HABITAT USE

HABITAT	ATLAS BLOCKS	INCIDENTAL OBSERVATIONS
Agricultural	2 (6%)	0 (0%)
Grassland	1 (3%)	0 (0%)
Open Water	4 (11%)	5 (31%)
Riparian	1 (3%)	1 (6%)
Sagebrush Scrub	1 (3%)	0 (0%)
Wetland	26 (74%)	10 (63%)
TOTAL	35	16

Total numbers reported in the Habitat Use, Breeding Status, and Abundance tables may differ from each other (see pp. 22–23 for details). Percentage sums may differ slightly from 100% due to rounding.

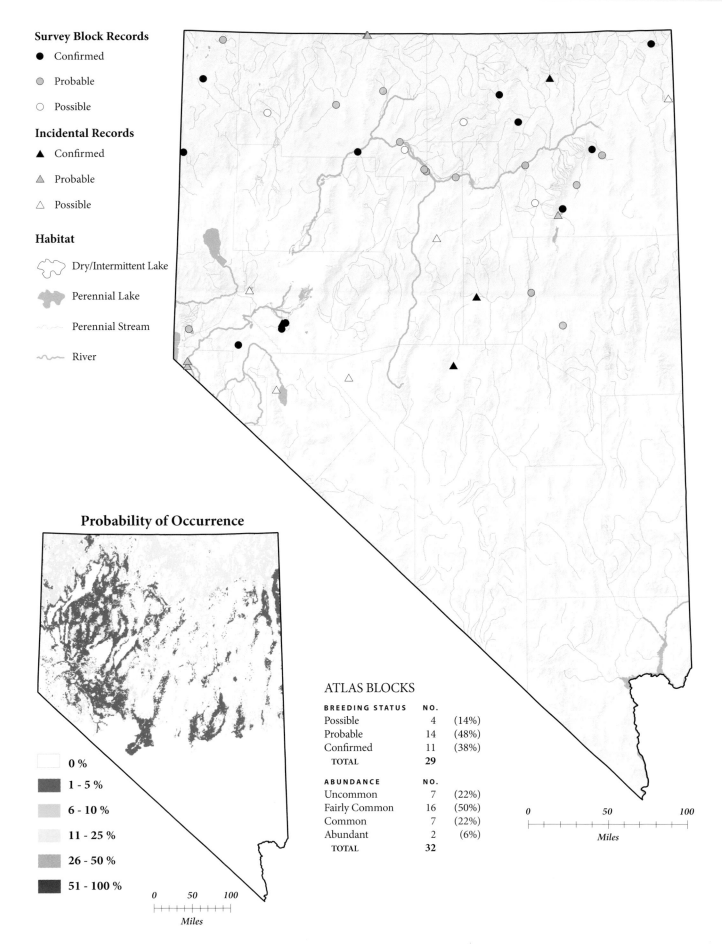

Survey Block Records

- ● Confirmed
- ● Probable
- ○ Possible

Incidental Records

- ▲ Confirmed
- ▲ Probable
- △ Possible

Habitat

Dry/Intermittent Lake

Perennial Lake

Perennial Stream

River

Probability of Occurrence

- 0 %
- 1 - 5 %
- 6 - 10 %
- 11 - 25 %
- 26 - 50 %
- 51 - 100 %

0   50   100
Miles

ATLAS BLOCKS

BREEDING STATUS	NO.	
Possible	4	(14%)
Probable	14	(48%)
Confirmed	11	(38%)
TOTAL	29	
ABUNDANCE	NO.	
Uncommon	7	(22%)
Fairly Common	16	(50%)
Common	7	(22%)
Abundant	2	(6%)
TOTAL	32	

0   50   100
Miles

FRANKLIN'S GULL

Larus pipixcan

A Franklin's Gull in breeding plumage is a striking bird: under-parts suffused in sunset pink, a jet black hood, and striking white eye crescents. During much of the past century these graceful gulls were no more than occasional wanderers into Nevada. They began breeding in the Lahontan Valley during recent decades, and they continue to nest there irregularly when water conditions are favorable.

DISTRIBUTION

Two widely spaced confirmed breeding records for Franklin's Gull were recorded during the atlas years. One block record came from Carson Lake, where a small colony was present in 1997 and possibly in subsequent years, and an incidental breeding record came from Franklin Lake in the Ruby Valley. In addition, uncon-firmed breeding was noted in a block along the Quinn River in northern Nevada.

Franklin's Gull expanded its range westward into the Great Basin during the twentieth century. Linsdale (1936) did not mention the species, which apparently first bred in Nevada in the 1970s (Ryser 1986, Alcorn 1988). The best-known nesting area is the Lahontan Valley, where Franklin's Gull is said to be "ir-regularly common" in summer (Chisholm and Neel 2002:102). Breeding in that area has varied with water levels and peaked at fifty nesting pairs (Alcorn 1988, Chisholm and Neel 2002). The Ruby Valley breeding colony, which was first recorded during at-las fieldwork, is probably more recent.

Nevada is peripheral to the core breeding range of Franklin's Gull, which lies largely in the northern Great Plains (Burger and

Gochfeld 1994). Breeding colonies are also found in southern Idaho, around the Great Salt Lake in northern Utah, and in southeastern Oregon (Behle et al. 1985, Burger and Gochfeld 1994, Stephens and Sturts 1998, Adamus et al. 2001). Breeding has also been confirmed in northern California (Small 1994), and recently fledged young have been observed at the Honey Lake Wildlife Area close to the northwest Nevada border (C. Elphick, pers. comm.). The predictive map's indication of a low chance of finding breeding Franklin's Gulls in many areas of north-central Nevada is probably an overestimate. Given the species' recent colonization of the state, however, the predictive map might help focus future surveys on areas with likely breeding sites.

The nesting habitat of Franklin's Gulls differs from that of Nevada's two other breeding gulls. While California and Ring-billed Gulls place their nests on dry ground—usually on is-lands—near water, Franklin's Gulls build a nest of floating vege-tation on the water (Burger and Gochfeld 1994). The locations of California and Ring-billed Gull colonies also tend to be more stable, while Franklin's Gulls may select different sites from year to year (Burger and Gochfeld 1994), presumably depending on water levels.

CONSERVATION AND MANAGEMENT

Numbers of Franklin's Gulls have probably declined substan-tially in parts of the Great Plains since historical times (Burger and Gochfeld 1994). More recently, there seems to be little evi-dence for widespread population changes (Burger and Gochfeld 1994, Sauer et al. 2005), although Franklin's Gull numbers do seem to increase locally when large wetlands are created within their range (Burger and Gochfeld 1994). Franklin's Gulls are sen-sitive to human disturbance during colony site selection and brooding, and breeding success is probably higher where there is little human access (Burger and Gochfeld 1994). Careful man-agement of water levels probably also helps this species. Because of its need for appropriate wetland conditions, Franklin's Gull is considered a species of Moderate Concern in the North Ameri-can Waterbird Conservation Plan (Kushlan et al. 2002) and is a species of High Concern in the Intermountain West (Ivey and Herziger 2005). Nonetheless, Franklin's Gulls are apparently on the increase in Nevada, and observers should keep looking for new colonies.

HABITAT USE

HABITAT	ATLAS BLOCKS	INCIDENTAL OBSERVATIONS
Wetland	3 (100%)	1 (100%)
TOTAL	3	1

Total numbers reported in the Habitat Use, Breeding Status, and Abun-dance tables may differ from each other (see pp. 22–23 for details). Percentage sums may differ slightly from 100% due to rounding.

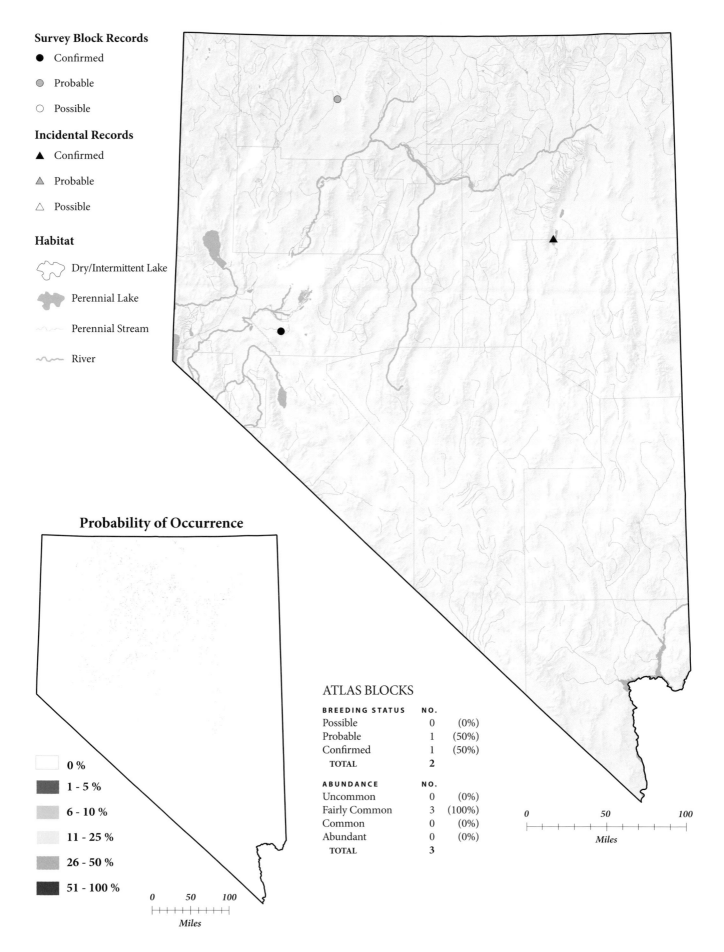

Survey Block Records

● Confirmed

◒ Probable

○ Possible

Incidental Records

▲ Confirmed

△ Probable

△ Possible

Habitat

Dry/Intermittent Lake

Perennial Lake

Perennial Stream

River

Probability of Occurrence

☐ 0 %

▨ 1 - 5 %

▨ 6 - 10 %

☐ 11 - 25 %

▨ 26 - 50 %

■ 51 - 100 %

0   50   100
Miles

ATLAS BLOCKS

BREEDING STATUS	NO.	
Possible	0	(0%)
Probable	1	(50%)
Confirmed	1	(50%)
TOTAL	2	

ABUNDANCE	NO.	
Uncommon	0	(0%)
Fairly Common	3	(100%)
Common	0	(0%)
Abundant	0	(0%)
TOTAL	3	

0   50   100
Miles

RING-BILLED GULL

Larus delawarensis

This bold visitor to fast food restaurant parking lots is a familiar sight in Nevada, yet its status as a breeder in the state is very limited. The bulk of Nevada's summer Ring-billed Gulls are non-breeding visitors that have forsaken the frenetic pace of the crowded breeding colonies farther north for easy living at strip malls and roadside rest stops. The Ring-billed Gull is a good example of a species with a summer distribution that is considerably broader than its breeding distribution.

DISTRIBUTION

The only confirmed breeding colony found by atlas workers was on Anaho Island, a protected site in Pyramid Lake. Breeding was suspected, but not actually confirmed, at Franklin Lake in the northern part of Ruby Valley. Possible breeding was noted at two other locations, but these records probably referred to summer visitors.

The Ring-billed Gull is another recent addition to Nevada's breeding avifauna. Linsdale (1936) listed no breeding records for the state, and Ryser (1985) suggested that the only Great Basin breeding colonies lie outside Nevada's boundaries. Alcorn (1988) described a small breeding colony on an island in Lahontan Reservoir, which has recently supported up to three hundred nests (Chisholm and Neel 2002). This colony was also active during the atlas years (L. Neel, pers. comm.). The small colony discovered by atlas workers at Pyramid Lake appears to be more recent, as does the species' presence in Ruby Valley.

The bulk of the Ring-billed Gull's population breeds north and northeast of Nevada, mostly in the southern half of Canada (Ryder 1993). The species is widely seen in California during the summer months (Small 1994), but breeding occurs south only into Lassen County (McCaskie et al. 1988). In Oregon, Ring-billed Gulls breed locally at a few lakes and along the Columbia River (Adamus et al. 2001). There are also scattered breeding colonies in Idaho (Stephens and Sturts 1998). Breeding had not been confirmed in Utah at the time Behle et al. (1985) published their checklist of Utah birds.

In the western United States, foraging Ring-billed Gulls are frequently associated with human-created habitats such as dumps, sewage ponds, farms, and urban centers (Ryder 1993, Small 1994, Chisholm and Neel 2002), but they also commonly hunt for fish. Ring-billed Gulls nest on flat ground near water, most often on islands (Ryder 1993). Occurrence data in Nevada are barely sufficient for generating a predictive map. Also, the main predictors of occurrence—large water bodies with islands and artificial food sources—are difficult to capture in a model and are therefore poorly represented on the predictive map.

CONSERVATION AND MANAGEMENT

Although many people do not realize it, Ring-billed Gull populations are actually in the process of recovering. By the early twentieth century, populations had been greatly depleted by the millinery and egg-collecting traditions (Ryder 1993). Present threats to the species are probably few, although some local populations have evidently declined with decreasing water quality; for example, at California's Salton Sea (Small 1994). Recent steady and substantial population increases in the West are often attributed to the growing bounty of urban landfills and other human-created food sources (Alcorn 1988, Ryder 1993).

Despite the Ring-billed Gull's conspicuousness, its status as a breeder in Nevada and other parts of the West is not well documented. Ring-billed Gulls tend to share nesting colonies with California Gulls and may sometimes go undetected in this setting (Ryser 1985, Alcorn 1988, Ryder 1993). Given recent population trends, the presence of additional colonies and an increase in population size are possible in Nevada, and birders and wildlife managers should be on the lookout for nesting Ring-billed Gulls throughout the northern part of the state.

HABITAT USE

HABITAT	ATLAS BLOCKS	INCIDENTAL OBSERVATIONS
Open Water	1 (50%)	2 (67%)
Wetland	1 (50%)	1 (33%)
TOTAL	2	3

Total numbers reported in the Habitat Use, Breeding Status, and Abundance tables may differ from each other (see pp. 22–23 for details). Percentage sums may differ slightly from 100% due to rounding.

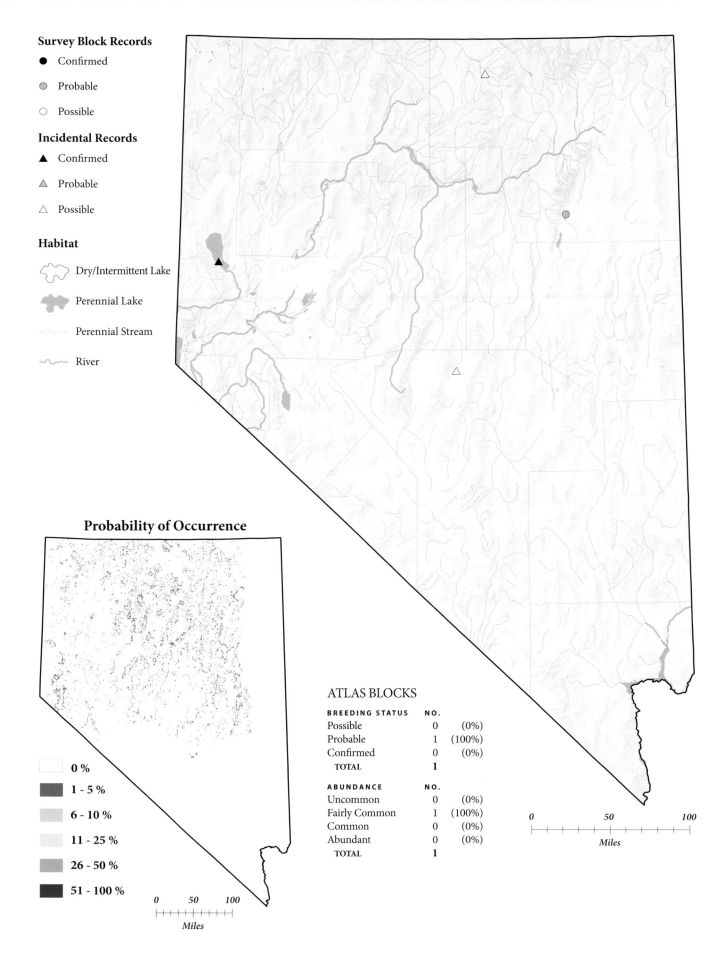

Survey Block Records

● Confirmed

◔ Probable

○ Possible

Incidental Records

▲ Confirmed

△ Probable

△ Possible

Habitat

Dry/Intermittent Lake

Perennial Lake

Perennial Stream

River

Probability of Occurrence

0 %

1 - 5 %

6 - 10 %

11 - 25 %

26 - 50 %

51 - 100 %

0   50   100
Miles

ATLAS BLOCKS

BREEDING STATUS	NO.	
Possible	0	(0%)
Probable	1	(100%)
Confirmed	0	(0%)
TOTAL	1	

ABUNDANCE	NO.	
Uncommon	0	(0%)
Fairly Common	1	(100%)
Common	0	(0%)
Abundant	0	(0%)
TOTAL	1	

0   50   100
Miles

CALIFORNIA GULL
Larus californicus

Virginia Lake in central Reno is a favorite haunt of joggers, power walkers, roller-bladers, picnickers, and a raucous colony of California Gulls. Most of these birds probably feed at nearby dumps, drive-through restaurants, and golf courses. Elsewhere in western Nevada, California Gull colonies have been recorded at Lahontan Reservoir, the Carson Sink, Pyramid Lake, and the Ruby Valley. What all of these sites—even Virginia Lake—provide for the gulls are protected islands in the middle of standing water.

DISTRIBUTION

Although atlas workers recorded California Gulls in widespread locations, most of the records referred to nonbreeding summer visitors or foraging wanderers. Only four actual colonies were documented during the atlas years: one on Anaho Island in Pyramid Lake, one in Virginia Lake in Reno, and two in the Ruby Valley. Each of these colonies contained fewer than one hundred pairs.

Although many California Gulls winter along the Pacific Coast, the species' breeding range lies mainly in the northern Great Plains of the United States and Canada and in the northern Intermountain West (Winkler 1996). There are numerous colonies in Idaho and in the Great Salt Lake area of Utah, and a few in eastern Oregon (Behle et al. 1985, Stephens and Sturts 1998, Adamus et al. 2001). In California, most breeding occurs in the northern interior part of the state, with the largest colony—numbering tens of thousands of birds—just across the Nevada border at Mono Lake (Winkler and Shuford 1988, Jehl 1994).

The California Gull is the most common breeding gull in the Great Basin, even though it occurs mainly around the region's margins (Ryser 1985). The Anaho Island colony has long been es-

tablished in Nevada and was the only one reported by Linsdale (1936). The colonies at Virginia Lake and Lahontan Reservoir, although more recent, seem well established (Alcorn 1988, Chisholm and Neel 2002). California Gulls are opportunistic nesters; for example, a colony of thirty-six hundred pairs suddenly appeared in the Carson Sink on dunes surrounded by floodwaters during the wet years of 1986–1987 (Chisholm and Neel 2002).

Although California Gulls are generalist foragers, colonies are usually found only on islands surrounded by water deep enough to deter predators (Winkler 1996). The predictive model used for the atlas project cannot capture such fine-scale associations; it simply identifies areas that contain habitat in which California Gulls were seen during summer. Even so, it does indicate the very low probability of finding breeding gulls almost anywhere in the state.

CONSERVATION AND MANAGEMENT

The global population of California Gulls probably increased during the twentieth century (Conover 1983), promoted by the availability of new islands in reservoirs and the proliferation of human-supplied foraging sites in agricultural areas, garbage dumps, and towns (Winkler 1996). Numbers of birds in individual colonies can fluctuate greatly, however, as a result of changing local conditions. Mono Lake is a case in point. When low lake levels linked a formerly isolated island to the mainland, coyotes were able to reach the nesting colony. High levels of predation ensued, and the gulls abandoned the colony in favor of small offshore islets that remained protected (Small 1994). In similar, though perhaps less dramatic, fashion, the locations and sizes of Nevada's California Gull colonies have also fluctuated greatly over the years (Ryser 1985). Because of this sensitivity to changing conditions, California Gulls merit a designation of Moderate Concern in both the North American and Intermountain West waterbird conservation plans (Kushlan et al. 2002, Ivey and Herziger 2005).

HABITAT USE

HABITAT	ATLAS BLOCKS	INCIDENTAL OBSERVATIONS
Open Water	1 (25%)	3 (25%)
Riparian	1 (25%)	1 (8%)
Urban	0 (0%)	3 (25%)
Wetland	2 (50%)	5 (42%)
TOTAL	4	12

Total numbers reported in the Habitat Use, Breeding Status, and Abundance tables may differ from each other (see pp. 22–23 for details). Percentage sums may differ slightly from 100% due to rounding.

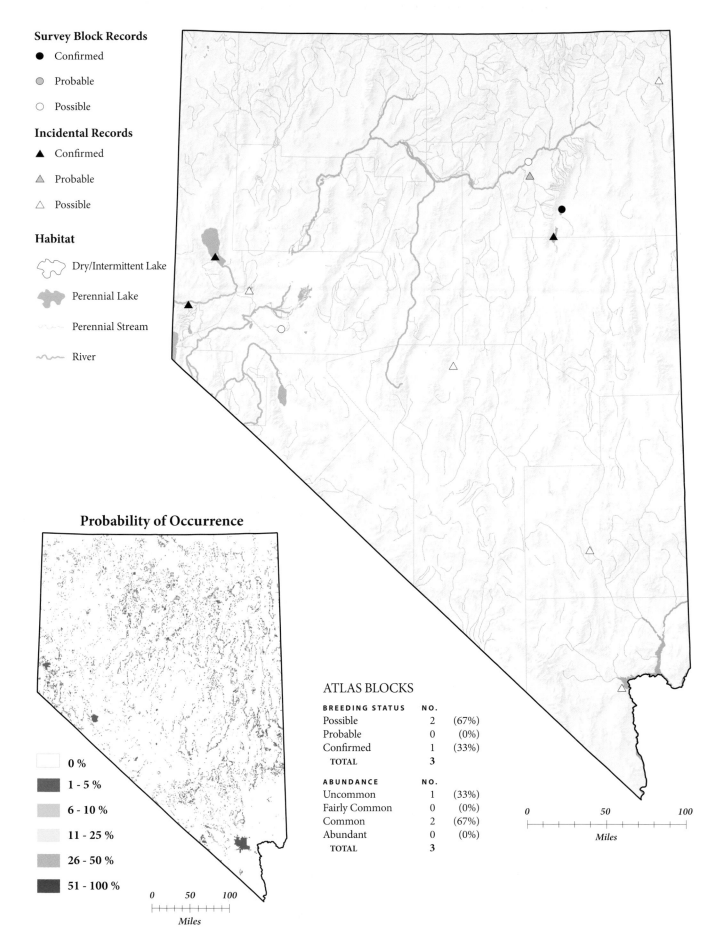

Survey Block Records

● Confirmed

◐ Probable

○ Possible

Incidental Records

▲ Confirmed

△ Probable

△ Possible

Habitat

⬡ Dry/Intermittent Lake

▨ Perennial Lake

〰 Perennial Stream

〰 River

Probability of Occurrence

☐ 0 %

■ 1 - 5 %

▨ 6 - 10 %

▨ 11 - 25 %

▨ 26 - 50 %

■ 51 - 100 %

0   50   100
Miles

ATLAS BLOCKS

BREEDING STATUS	NO.	
Possible	2	(67%)
Probable	0	(0%)
Confirmed	1	(33%)
TOTAL	3	

ABUNDANCE	NO.	
Uncommon	1	(33%)
Fairly Common	0	(0%)
Common	2	(67%)
Abundant	0	(0%)
TOTAL	3	

0   50   100
Miles

CASPIAN TERN
Hydroprogne caspia

RN'02

The mother of all terns—as big as a California Gull and with a massive red bill—the Caspian Tern can be found around larger lakes during the summer months. The few breeding colonies in Nevada are currently limited to the western part of the state. Caspian Tern colonies tend to be rather ephemeral, however, and the possibility of new nest sites in eastern or even southern Nevada should not be ruled out.

DISTRIBUTION

Atlas records came from several of the major river drainages of northern Nevada, from Sheldon National Wildlife Refuge and the Ruby Valley, and even from the Muddy River of southern Nevada. The only confirmed breeding records, however, came from Sheldon and Pyramid Lake. Both locations hosted modest colonies numbering well under one hundred pairs.

The Caspian Tern seems never to have been an especially prominent part of Nevada's breeding avifauna. Ryser (1985) mentioned colonies only at Lahontan Reservoir and Pyramid Lake, the latter of long standing (Linsdale 1936). Alcorn (1988) reported additional colonies in the Lahontan Valley, and several hundred pairs have nested there in some years (Chisholm and Neel 2002). Although atlas workers did not report these colonies, it is likely that breeding still occurs in the less accessible portions of the valley (L. Neel, pers. comm.). In general, the location and success rate of breeding colonies vary considerably in response to fluctuating water levels and other local site conditions; even the Pyramid Lake site fails to support breeding in some years (Alcorn 1988).

The worldwide range of the Caspian Tern is extensive but also highly disjunct, with breeding occurring at widely spaced locations that are suitable for colonies. In the states neighboring Nevada, colonies exist at the terminal lakes of northern Utah, along the Snake River and elsewhere in Idaho, and at a number of interior and coastal sites in California and Oregon (Behle et al. 1985, Small 1994, Stephens and Sturts 1998, Cuthbert and Wires 1999, Adamus et al. 2001).

In the Great Basin, Caspian Terns usually nest on islands in lakes, often interspersed among California Gulls (Ryser 1985). Rangewide, however, colonies also occur in other settings, including estuaries, salt marshes, barrier islands, rivers, and freshwater lakes (Cuthbert and Wires 1999). As is the case with the other colonial waterbirds, the predictive model does not capture the specific nesting requirements of this species; potential breeding sites are likely to be limited to the largest lakes and marshes, where predator-safe conditions occur.

CONSERVATION AND MANAGEMENT

North American populations of the Caspian Tern increased in number and geographic extent during the twentieth century, in large part because key nesting areas were protected and enhanced (Cuthbert and Wires 1999). In the West, range expansions and population increases have been documented in California (Gallagher 1997) and Oregon (Contreras and Kindschy 1996). There have been local declines, too, for reasons ranging from water pollution at the Salton Sea (Small 1994) to predation by California Gulls at Mono Lake (Gaines 1992).

Caspian Tern populations fluctuate in response to disturbance, predation levels, and availability of nesting habitat (Cuthbert and Wires 1999). A good case study in Nevada comes from the Carson Sink, where Caspian Tern fortunes seem to be tied directly to water runoff. As many as 475 nests have been found in high-water years (Chisholm and Neel 2002), but the species is completely absent at other times (L. Neel, pers. comm.). In the Intermountain West as a whole, the Caspian Tern is considered a species of Low Concern in the Intermountain West Waterbird Conservation Plan (Ivey and Herziger 2005).

HABITAT USE

HABITAT	ATLAS BLOCKS	INCIDENTAL OBSERVATIONS
Open Water	3 (43%)	1 (50%)
Wetland	4 (57%)	1 (50%)
TOTAL	7	2

Total numbers reported in the Habitat Use, Breeding Status, and Abundance tables may differ from each other (see pp. 22–23 for details). Percentage sums may differ slightly from 100% due to rounding.

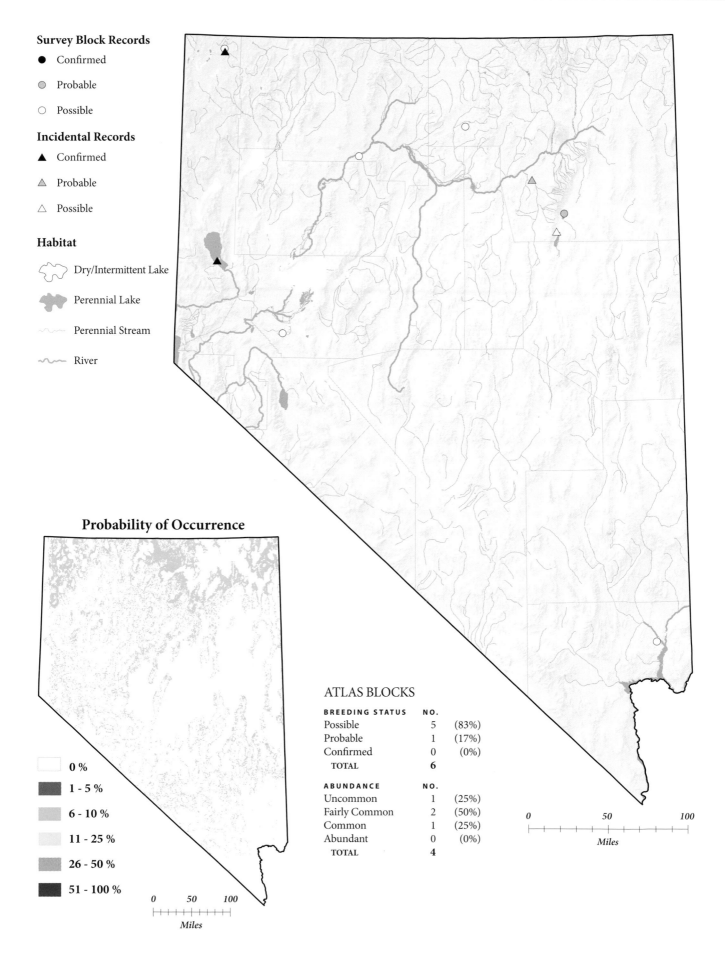

Survey Block Records

- **●** Confirmed
- **◐** Probable
- **○** Possible

Incidental Records

- **▲** Confirmed
- **△** Probable
- **△** Possible

Habitat

- Dry/Intermittent Lake
- Perennial Lake
- Perennial Stream
- River

Probability of Occurrence

- 0 %
- 1 - 5 %
- 6 - 10 %
- 11 - 25 %
- 26 - 50 %
- 51 - 100 %

0   50   100
Miles

ATLAS BLOCKS

BREEDING STATUS	NO.	
Possible	5	(83%)
Probable	1	(17%)
Confirmed	0	(0%)
TOTAL	6	

ABUNDANCE	NO.	
Uncommon	1	(25%)
Fairly Common	2	(50%)
Common	1	(25%)
Abundant	0	(0%)
TOTAL	4	

0   50   100
Miles

FORSTER'S TERN
Sterna forsteri

RN '02

In Nevada, Forster's Tern is most often seen as a shimmering, silvery presence hovering above an irrigation ditch, lake, or cattail marsh. This long-tailed, long-winged aerialist is especially adept at plunge-diving for minnows and other small fish, but it will hawk for flying insects, too. Forster's Tern is an opportunistic breeder, and its numbers vary from place to place and from year to year as a result of changing habitat conditions.

DISTRIBUTION

Forster's Terns were found in nine atlas blocks and at eight other sites. All confirmed breeding records were from the northern half of the state: at Sheldon National Wildlife Refuge and at Washoe Lake at opposite ends of Washoe County; at two sites in the Lahontan Valley; and at three sites in the Ruby Valley. Colonies may consist of just a few pairs (McNicholl et al. 2001), so unconfirmed records could represent either nonbreeding individuals or foraging individuals from small breeding colonies that were not detected by atlas workers.

Forster's Tern has usually been described as common and widespread in Nevada. It was reported breeding at or near all the confirmed atlas sites except Sheldon by Linsdale (1936), Alcorn (1988), and Chishom and Neel (2002). These authors also recorded possible breeding at a few additional sites, including Pyramid Lake and the historical marshes of the Humboldt River. Atlas data generally indicated lower abundances than those suggested by these earlier accounts.

Forster's Terns breed across the northern Great Plains and the northern Great Basin, with additional breeding sites along the coasts of North America and at scattered locations elsewhere (McNicholl et al. 2001). In neighboring states, breeding is regular in northern Utah, southern Idaho, and southeastern Oregon (Behle et al. 1985, Stephens and Sturts 1998, Adamus et al. 2001), and occurs at scattered locations along the coast and in the valley marshes of California (Small 1994).

During the summer, Forster's Terns can be found in fresh, brackish, and saltwater marshes. Nests tend to be located on marshy shorelines within clumps of vegetation, on small islands, or on mats of floating vegetation (McNicholl et al. 2001). As is true for Nevada's other gulls and terns, the reported habitat types capture these requirements only in the broadest sense. As a result, the predictive map likely overrepresents the distribution of suitable breeding sites, especially in the southern half of the state, where the model would have predicted no chance of finding the species were it not for the possible breeding records in southern Nye and Lincoln counties—both of which might represent nonbreeding migrants.

CONSERVATION AND MANAGEMENT

The overall population status of Forster's Tern is unclear (McNicholl et al. 2001). The species was once harvested in great numbers for millinery use of its feathers (Ryser 1985). Its reproductive success can vary tremendously from year to year, both locally and regionally, depending on wetland conditions. As such, Forster's Tern is a species of Moderate Concern in both the North American and the Intermountain West waterbird conservation plans (Kushlan et al. 2002, Ivey and Herziger 2005).

Nevada trends for Forster's Terns are difficult to assess. The historical literature suggests that a few historical breeding sites may have been lost, but most of the older accounts cannot be reliably compared with current information. In the Lahontan Valley, numbers have dropped by about 60% since the 1950s; only about 150–200 pairs were recorded during the atlas years (Chisholm and Neel 2002). The main culprit apparently has been habitat loss, in particular the elimination of winter flows from Lahontan Reservoir for hydropower.

HABITAT USE

HABITAT	ATLAS BLOCKS	INCIDENTAL OBSERVATIONS
Grassland	0 (0%)	1 (13%)
Open Water	3 (23%)	1 (13%)
Riparian	1 (8%)	0 (0%)
Wetland	9 (69%)	6 (75%)
TOTAL	13	8

Total numbers reported in the Habitat Use, Breeding Status, and Abundance tables may differ from each other (see pp. 22–23 for details). Percentage sums may differ slightly from 100% due to rounding.

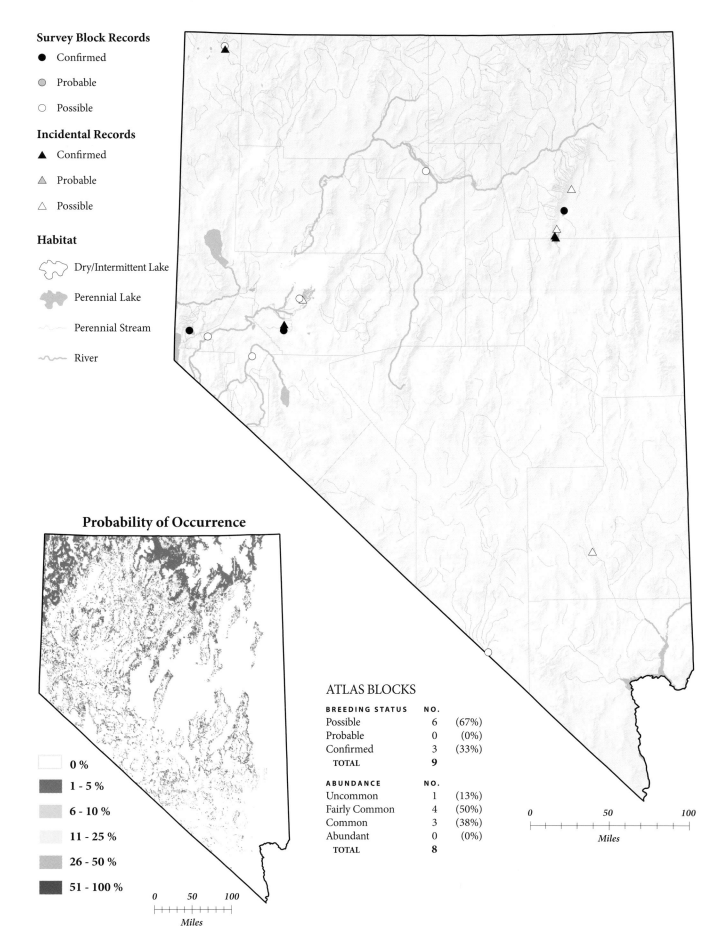

Survey Block Records
- ● Confirmed
- ◉ Probable
- ○ Possible

Incidental Records
- ▲ Confirmed
- ◮ Probable
- △ Possible

Habitat
- Dry/Intermittent Lake
- Perennial Lake
- Perennial Stream
- River

Probability of Occurrence

- ☐ 0 %
- 1 - 5 %
- 6 - 10 %
- 11 - 25 %
- 26 - 50 %
- 51 - 100 %

0   50   100
Miles

ATLAS BLOCKS

BREEDING STATUS	NO.	
Possible	6	(67%)
Probable	0	(0%)
Confirmed	3	(33%)
TOTAL	9	

ABUNDANCE	NO.	
Uncommon	1	(13%)
Fairly Common	4	(50%)
Common	3	(38%)
Abundant	0	(0%)
TOTAL	8	

0   50   100
Miles

BLACK TERN
Chlidonias niger

Dark and diminutive, the Black Tern presents an almost swallow-like gestalt as it forages for insects in the fly-infested airspace above Nevada's lowland marshes and lakes. The Black Tern has probably never been especially common in Nevada, and breeding sites are few and far between. Rangewide, the species is in decline in many areas where suitable wetland habitat has been lost.

DISTRIBUTION

Black Terns were reported in only four atlas blocks and at seven additional locations. Breeding was confirmed at three Ruby Valley locations and at single locations in the Lahontan and Pahranagat valleys. Breeding Black Terns had been noted in all of these locations previously, with presumed breeders also recorded at Washoe Lake and Sheldon National Wildlife Refuge in northwestern Nevada (Linsdale 1936, Alcorn 1988, Chisholm and Neel 2002). Neel (1999a) considered the Mason Valley Wildlife Management Area to be the southernmost Black Tern breeding location in the state, but confirmed breeding in the Pahranagat Valley during the atlas project extends the species' Nevada range to the south.

In North America, the Black Tern breeds across the northern tier of the United States and much of south-central Canada (Dunn and Agro 1995). The species breeds locally in southern and eastern Oregon, in northern Utah, and at several Idaho locations (Behle et al. 1985, Stephens and Sturts 1998, Adamus et al. 2001). In California, breeding occurs mostly in the northeastern corner of the state and in the Central Valley; many former breeding sites—especially in the south—have been lost (Shuford et al. 2001).

Black Terns and the other "marsh terns" of the genus *Chlidonias* usually nest semicolonially in shallow freshwater marshes with emergent vegetation (e.g., Shuford et al. 2001). Other types of wetlands or riparian areas, such as lake margins or river islands, may also provide nesting habitat (Dunn and Agro 1995). Wetlands suitable for nesting are fairly scarce in Nevada, and the predictive map undoubtedly overestimates the extent of the species' range. On the other hand, the map does indicate the potential—albeit with low probability—for Black Terns to occur wherever suitable aquatic habitat is found within the valleys of the Great Basin.

CONSERVATION AND MANAGEMENT

Black Tern populations exhibit significant fluctuations in response to changing wetland conditions. Individual nests are easily destroyed by changes in water level, for instance (Dunn and Agro 1995). Beyond these year-to-year fluctuations, long-term declines appear to have occurred in many areas (Dunn and Agro 1995), probably due to habitat loss. Marsh drainage and expansion of agriculture were the main sources of habitat loss in the Central Valley of California (Shuford et al. 2001). Black Terns readily accept artificial habitats, however, and can be found in settings such as rice fields, sewage ponds, and restored wetlands (Dunn and Agro 1995).

In the Lahontan Valley, the species has declined in the latter half of the twentieth century due to reduction of water releases from Lahontan Reservoir for hydropower (Chisholm and Neel 2002). Numbers may have also declined elsewhere in the state because of habitat loss, diminished water quality, and pesticides (Neel 1999a). For all these reasons, Black Terns are ranked as waterbirds of Moderate Concern in the North American Waterbird Conservation Plan (Kushlan et al. 2002) and of High Concern in the Intermountain West Waterbird Conservation Plan (Ivey and Herziger 2005), and are a Partners in Flight Priority species in Nevada (Neel 1999a).

HABITAT USE

HABITAT	ATLAS BLOCKS	INCIDENTAL OBSERVATIONS
Agricultural	0 (0%)	1 (13%)
Open Water	3 (50%)	2 (25%)
Riparian	0 (0%)	1 (13%)
Wetland	3 (50%)	4 (50%)
TOTAL	6	8

Total numbers reported in the Habitat Use, Breeding Status, and Abundance tables may differ from each other (see pp. 22–23 for details). Percentage sums may differ slightly from 100% due to rounding.

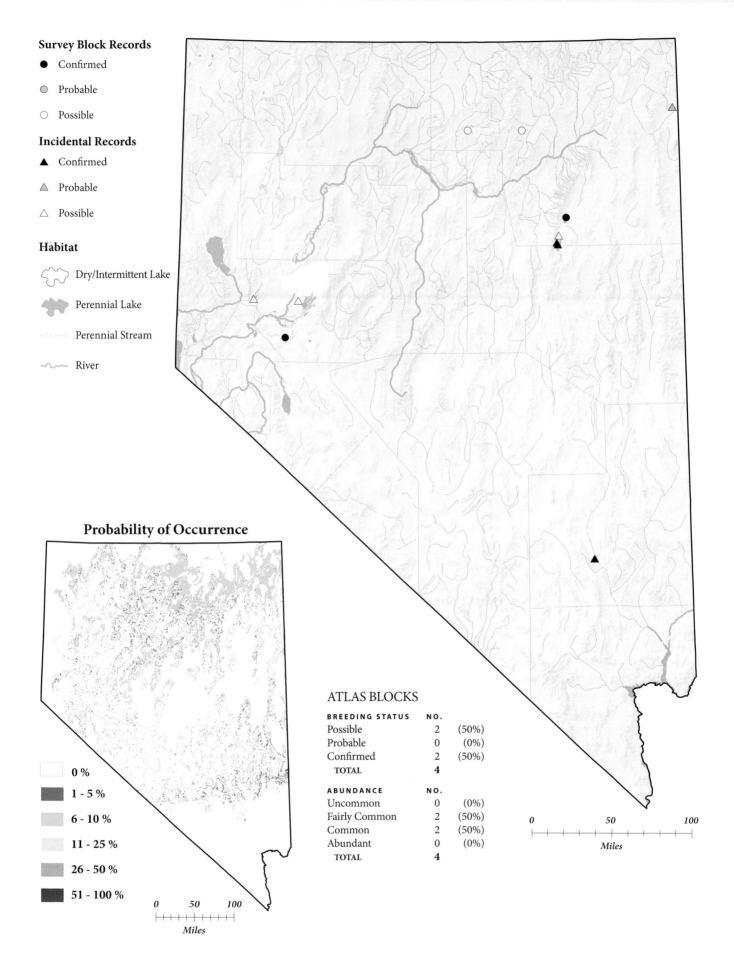

Survey Block Records

● Confirmed

◐ Probable

○ Possible

Incidental Records

▲ Confirmed

▲ Probable

△ Possible

Habitat

Dry/Intermittent Lake

Perennial Lake

Perennial Stream

River

Probability of Occurrence

☐ 0 %

■ 1 - 5 %

■ 6 - 10 %

☐ 11 - 25 %

■ 26 - 50 %

■ 51 - 100 %

0   50   100
Miles

ATLAS BLOCKS

BREEDING STATUS	NO.	
Possible	2	(50%)
Probable	0	(0%)
Confirmed	2	(50%)
TOTAL	4	
ABUNDANCE	NO.	
Uncommon	0	(0%)
Fairly Common	2	(50%)
Common	2	(50%)
Abundant	0	(0%)
TOTAL	4	

0   50   100
Miles

ROCK PIGEON

Columba livia

The North American range of the introduced Rock Pigeon—formerly called the Rock Dove—is usually depicted as a gigantic blob stretching from sea to shining sea. In reality, however, this species is far from ubiquitous in Nevada. Almost all confirmed atlas records came from the Reno–Carson City and Las Vegas metropolitan areas or along the Interstate 80 corridor. Rock Pigeons no doubt breed elsewhere in the state, but it is clear that urban centers and major roadways provide the greater part of their habitat. Despite its rather disreputable reputation in some circles, the Rock Pigeon has played a crucial role in our understanding of many aspects of bird biology, from endocrinology to navigation, and has even played a significant role in military history.

DISTRIBUTION

First brought to North America by European settlers in the early seventeenth century, the Rock Pigeon now occurs in all corners of the United States and Mexico (Johnston 1992). The most interesting thing about the Rock Pigeon's distribution in Nevada may be the places where it does not occur. There were no records at all in a large swath across the central part of the state that encompassed all of White Pine, Lincoln, Eureka, Mineral, and Esmeralda counties and most of Nye County.

Presumably because it is not a native species, there is little historical information on the Rock Pigeon in Nevada. Earlier writers, such as Linsdale (1936), did not mention pigeons, and it is unclear when this Eurasian species first arrived in the state (Ryser 1985). Today, the species is closely associated with developed areas, and most atlas records were from urban habitats. The remaining records were primarily from agricultural and riparian areas. Rock Pigeons that were recorded outside the major urban centers were found in proximity to smaller towns, truck stops, or outlying ranches.

The model's predictions suggest a very low chance of the Rock Pigeon being found in most areas of Nevada, and the predictive map nicely highlights the state's urban areas as hotspots. Pigeons do occur away from human habitation in some parts of their range, however. Indeed, their native habitat, prior to domestication, included rugged cliffs in coastal and upland areas (Johnston 1992). In California, they can be found around coastal sea cliffs and in interior mountain ranges (Small 1994). In southern Nevada, Rock Pigeons breed in Red Rock Canyon and on the high cliffs overlooking Lake Mojave and the Colorado River (Alcorn 1988), and farther north they nest on some cliffs that surround the Lahontan Valley (Chisholm and Neel 2002). Cliffs are available in great abundance in Nevada, and it is interesting that Rock Pigeons seem to use only a very small subset of them.

In the states surrounding Nevada, the Rock Pigeon is widely but unevenly distributed. In Utah, it is common in cities but uncommon elsewhere (Behle et al. 1985). There are confirmed breeding records from scattered locations in southern Idaho (Stephens and Sturts 1998), and in Oregon the species is locally abundant (Contreras 1999, Adamus et al. 2001). In California (Small 1994) and Arizona (Wise-Gervais in Corman and Wise-Gervais 2005:184–185), Rock Pigeons are locally common to abundant around areas of human habitation.

CONSERVATION AND MANAGEMENT

It is ironic that the Rock Pigeon is one of the most intensively studied of all bird species, and yet many details about its breeding range and movements remain unknown to birders. The species is apparently scarcer overall in Nevada than anywhere else in the forty-eight contiguous states (Robbins et al. 1986, Sauer et al. 2005), but it can be very common where it does occur. More than half of the atlas blocks were estimated to contain more than ten nesting pairs, and several were thought to have in excess of a hundred. Continued monitoring in Nevada should shed light on whether Rock Pigeon populations are increasing in the state or if there is a declining trend as is true in a few other regions (Sauer et al. 2005).

HABITAT USE

HABITAT	ATLAS BLOCKS	INCIDENTAL OBSERVATIONS
Agricultural	16 (18%)	0 (0%)
Barren	3 (3%)	0 (0%)
Mojave	0 (0%)	1 (3%)
Riparian	12 (14%)	9 (26%)
Salt Desert Scrub	1 (1%)	0 (0%)
Sagebrush Scrub	2 (2%)	1 (3%)
Sagebrush Steppe	2 (2%)	1 (3%)
Urban	52 (59%)	23 (66%)
TOTAL	88	35

Total numbers reported in the Habitat Use, Breeding Status, and Abundance tables may differ from each other (see pp. 22–23 for details). Percentage sums may differ slightly from 100% due to rounding.

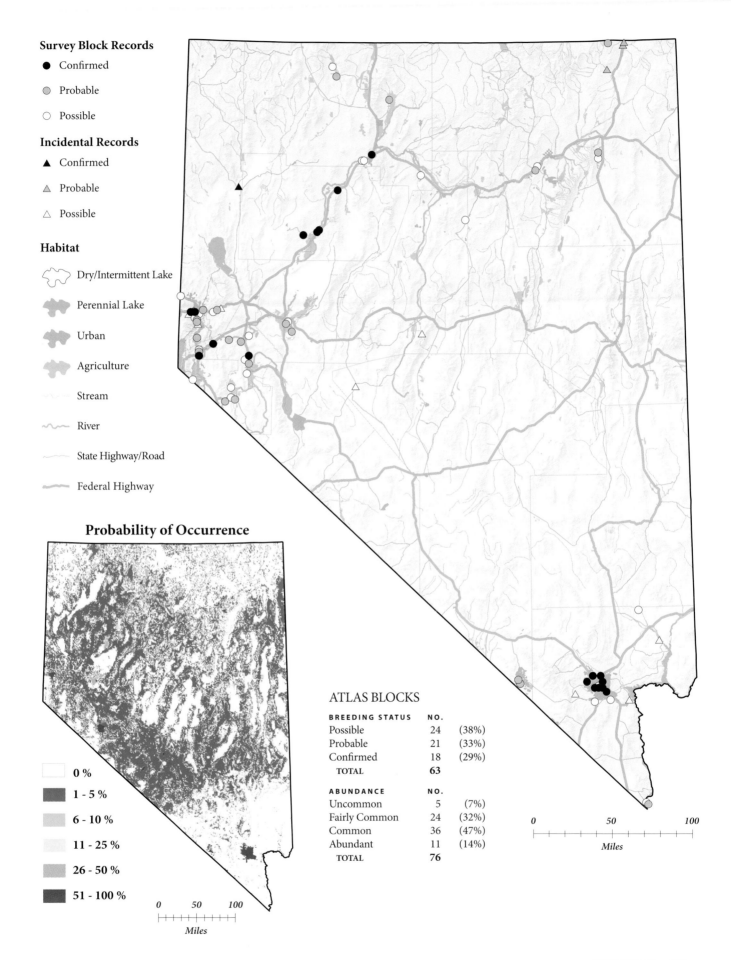

Survey Block Records
● Confirmed
● Probable
○ Possible

Incidental Records
▲ Confirmed
△ Probable
△ Possible

Habitat
◊ Dry/Intermittent Lake

Perennial Lake

Urban

Agriculture

Stream

River

State Highway/Road

Federal Highway

Probability of Occurrence

0 %
1 - 5 %
6 - 10 %
11 - 25 %
26 - 50 %
51 - 100 %

0   50   100
Miles

ATLAS BLOCKS

BREEDING STATUS	NO.	
Possible	24	(38%)
Probable	21	(33%)
Confirmed	18	(29%)
TOTAL	63	

ABUNDANCE	NO.	
Uncommon	5	(7%)
Fairly Common	24	(32%)
Common	36	(47%)
Abundant	11	(14%)
TOTAL	76	

0   50   100
Miles

BAND-TAILED PIGEON
Patagioenas fasciata

Until recently, the Band-tailed Pigeon and Rock Pigeon were considered members of the same genus, and superficially they look quite similar. But the two species conjure up wholly different feelings in the minds of birders. The Rock Pigeon is an emblem of ordinariness, a common sight in business districts and back alleys. The Band-tailed Pigeon, in contrast, is a bird of the wilderness and an inhabitant of tall trees on the steep slopes of some of Nevada's most rugged mountains. Unlike Rock Pigeons, which are widespread and faring well in North America, Band-tailed Pigeons have a restricted distribution in the state and are generally declining.

DISTRIBUTION

Most atlas records of Band-tailed Pigeons came from either the Carson Range of western Nevada or the vicinity of Mount Charleston in Clark County. Interestingly, these two groupings may represent separate subspecies, with the southern population most closely related to the birds of the southern Rocky Mountains, and the Carson Range birds more closely aligned with Pacific Coast populations. Fieldworkers estimated more than one pair to be present in more than two-thirds of the occupied atlas blocks, and in several cases estimates exceeded ten pairs. Nesting was confirmed at only three sites, all in the Carson Range, but was strongly suspected in the Spring Mountains in southern Nevada. An unresolved and interesting question is whether the single observation of probable breeding in northern Lincoln County and other isolated sightings from around the state represent nonbreeding nomads or additional small breeding populations.

Most records came from montane forest habitat, as expected. The predictive map indicates a potential for breeding at higher elevations across the center of the state, but our field observations suggest that most of these ranges may be too isolated to support breeding populations. Atlas data also suggest that Band-tailed Pigeons are more widespread in Nevada than the earlier literature indicates. Linsdale (1936) gave no records anywhere in the state, and Ryser (1985) considered the species to be absent as a breeder from the entire Great Basin. Alcorn (1988) called the Band-tailed Pigeon an uncommon resident in limited areas of southern and western Nevada but did not provide any definite breeding records.

Band-tailed Pigeons are common breeders in western Oregon (Adamus et al. 2001) and are found throughout the mountains of California (Small 1994, Keppie and Braun 2000). In contrast, they are scarce or absent as breeders in eastern Oregon (Contreras 1999, Adamus et al. 2001) and apparently do not breed at all in Idaho (Stephens and Sturts 1998). To the east and south of Nevada, they are uncommon summer residents at middle elevations in Utah (Behle et al. 1985), and they occur locally in montane areas of Arizona (Keppie and Braun 2000, Martin in Corman and Wise-Gervais 2005:186–187).

CONSERVATION AND MANAGEMENT

Band-tailed Pigeon numbers in North America appear to have declined substantially during historic times, probably due to overharvesting before protective measures were implemented in the early 1900s (Keppie and Braun 2000). Declines are still continuing throughout much of the range, although their causes are not clear. Hunting continues in several states, but bag limits keep it at a much lower level than in the past (Keppie and Braun 2000).

The status of the Band-tailed Pigeon in Nevada is not well known, and monitoring is limited in scope as the species occurs only peripherally in the state. The Band-tailed Pigeon is a Partners in Flight Watch List species for the Sierra Nevada and southwestern regions, with the conservation goal of doubling these populations (Rich et al. 2004). From the perspective of basic research, a key question that remains unanswered about the Nevada birds is the subspecies identity of the Band-tailed Pigeons that nest in Clark County (D. M. Keppie, pers. comm.).

HABITAT USE

HABITAT	ATLAS BLOCKS	INCIDENTAL OBSERVATIONS
Montane Forest	21 (70%)	7 (100%)
Montane Parkland	2 (7%)	0 (0%)
Pinyon-Juniper	2 (7%)	0 (0%)
Riparian	1 (3%)	0 (0%)
Urban	4 (13%)	0 (0%)
TOTAL	30	7

Total numbers reported in the Habitat Use, Breeding Status, and Abundance tables may differ from each other (see pp. 22–23 for details). Percentage sums may differ slightly from 100% due to rounding.

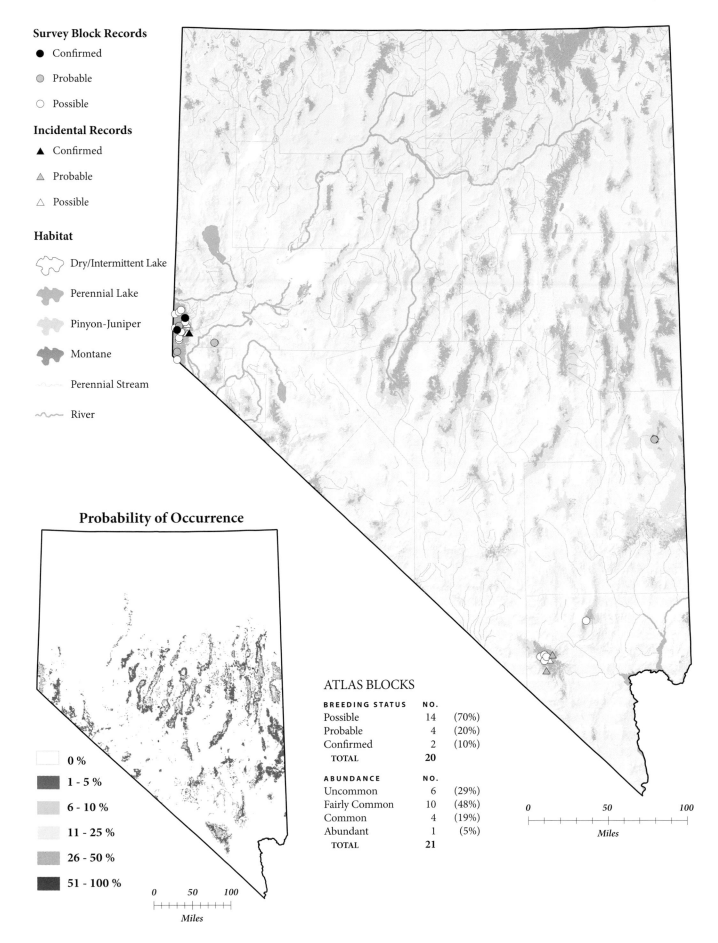

Survey Block Records

● Confirmed

● Probable

○ Possible

Incidental Records

▲ Confirmed

△ Probable

△ Possible

Habitat

Dry/Intermittent Lake

Perennial Lake

Pinyon-Juniper

Montane

Perennial Stream

River

Probability of Occurrence

☐	0 %
▨	1 - 5 %
▨	6 - 10 %
☐	11 - 25 %
▨	26 - 50 %
■	51 - 100 %

0   50   100
Miles

ATLAS BLOCKS

BREEDING STATUS	NO.	
Possible	14	(70%)
Probable	4	(20%)
Confirmed	2	(10%)
TOTAL	**20**	

ABUNDANCE	NO.	
Uncommon	6	(29%)
Fairly Common	10	(48%)
Common	4	(19%)
Abundant	1	(5%)
TOTAL	**21**	

0   50   100
Miles

WHITE-WINGED DOVE
Zenaida asiatica

Even among Nevada's desert-dwelling bird species, the White-winged Dove stands out as especially well suited for extreme heat. It is a bird of the south and is quite common around the low-elevation town of Laughlin—consistently the hottest place in Nevada. The species also occurs north of Las Vegas to just below the Lincoln County line. The White-winged Dove may be undergoing a population increase and range expansion in Nevada, however, and the possibility of future breeders well to the north of its present Nevada range should not be discounted.

DISTRIBUTION

White-winged Doves were recorded in nine atlas blocks and at eight other locations, all of which fell in Clark County or extreme southern Nye County. Breeding was confirmed only in the far southern tip of the state, in and around Laughlin. Reported numbers varied, but more than half of the occupied blocks were estimated to have ten or more pairs.

The White-winged Dove appears to have been in southern Nevada for many years, with reports dating back at least to the 1930s (Linsdale 1936). Alcorn (1988) cited breeding from as far north as Las Vegas, but he emphasized that these doves are really common only in the hot bottomlands farther south.

The vast bulk of the White-winged Dove's range lies south of Nevada, and the species is a locally abundant breeder in the lower Colorado River Valley (Rosenberg et al. 1991). In California it breeds in the hot lower Sonoran and Colorado deserts (Small 1994), and in Utah it is a rare summer resident found only in the southwestern corner of the state (Behle et al. 1985). North of Nevada, the White-winged Dove is strictly a vagrant (Stephens and Sturts 1998, Contreras 1999).

Atlas records came largely from mesquite, riparian, and salt desert scrub habitats, with relatively few reports from Mojave habitat. Nonetheless, the predictive map suggests that the species could potentially occur throughout much of the Mojave Desert, presumably because small patches of these other habitats are often embedded within Mojave habitat. White-winged Doves also do well in various human-created habitats. For instance, in southern California they can be found in citrus orchards and eucalyptus groves (Garrett and Dunn 1981), and in Arizona they are among the few species that breed in tamarisk monocultures (Rosenberg et al. 1991). Southern Nevada presents a wide array of urbanized and exotic floristic offerings, and the species may increasingly be found in such settings in the future.

CONSERVATION AND MANAGEMENT

The White-winged Dove expanded its range in the latter half of the twentieth century, but the overall population trends are not clear. Breeding Bird Survey data suggest that range expansion has been coupled with population increases in some areas (Sauer et al. 2005). White-winged Doves often nest colonially, though, which may lessen the ability of extensive surveys such as the Breeding Bird Survey to accurately measure populations and their trends.

Like other doves, White-wingeds seem capable of coexisting with people, and they can be expected at least to hold their own in the face of urbanization in the West. In Nevada, the White-winged Dove may be poised for a range expansion. Conditions are apparently suitable throughout southern Nevada, and sightings well to the north of Clark County seem to have increased (Floyd and Stackhouse 2001). On the other hand, the overall status of the White-winged Dove in the state appears to have remained stable for much of the past century despite considerable range expansion elsewhere along the northern edge of its range.

HABITAT USE

HABITAT	ATLAS BLOCKS	INCIDENTAL OBSERVATIONS
Mesquite	7 (30%)	0 (0%)
Mojave	2 (9%)	2 (29%)
Riparian	8 (35%)	3 (43%)
Salt Desert Scrub	4 (17%)	0 (0%)
Urban	2 (9%)	2 (29%)
TOTAL	23	7

Total numbers reported in the Habitat Use, Breeding Status, and Abundance tables may differ from each other (see pp. 22–23 for details). Percentage sums may differ slightly from 100% due to rounding.

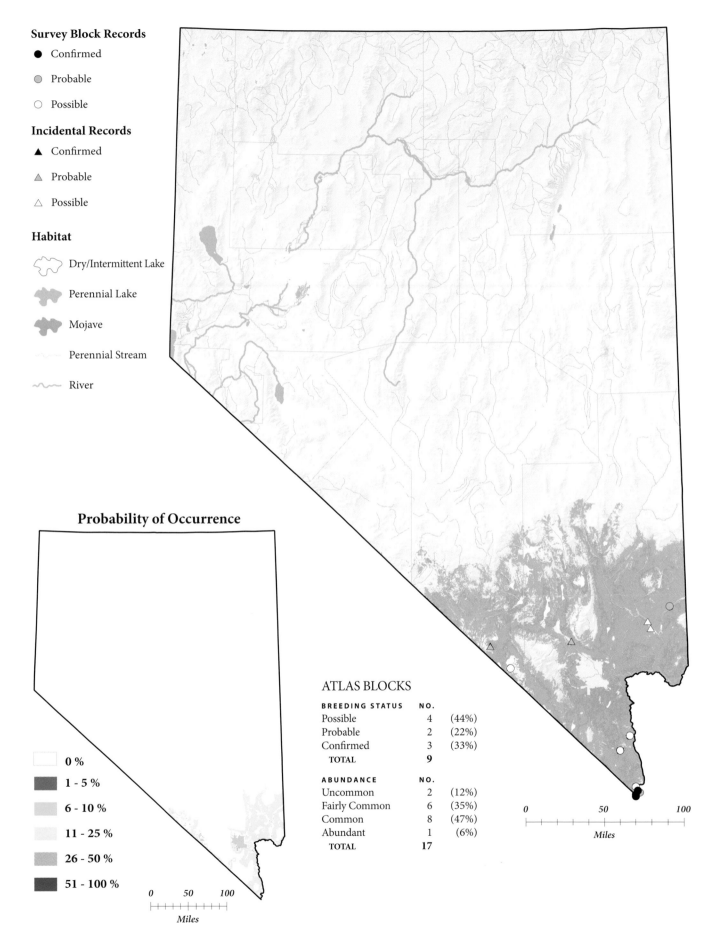

Survey Block Records

● Confirmed

◉ Probable

○ Possible

Incidental Records

▲ Confirmed

△ Probable

△ Possible

Habitat

Dry/Intermittent Lake

Perennial Lake

Mojave

Perennial Stream

River

Probability of Occurrence

☐ 0 %

■ 1 - 5 %

■ 6 - 10 %

☐ 11 - 25 %

■ 26 - 50 %

■ 51 - 100 %

0   50   100

Miles

ATLAS BLOCKS

BREEDING STATUS	NO.	
Possible	4	(44%)
Probable	2	(22%)
Confirmed	3	(33%)
TOTAL	**9**	

ABUNDANCE	NO.	
Uncommon	2	(12%)
Fairly Common	6	(35%)
Common	8	(47%)
Abundant	1	(6%)
TOTAL	**17**	

0   50   100

Miles

MOURNING DOVE
Zenaida macroura

The Mourning Dove is one of the most widely distributed birds in Nevada. It breeds throughout the length and breadth of the state, from the low deserts to the high mountains. Its telltale cooing is a familiar sound in the remote desert scrublands and lonely pinyon-juniper woodlands of the Nevada hinterlands, but it can also be heard above the din of traffic and human chatter in residential and urban areas. Just about the only places where the Mourning Dove cannot be found are desert playas, the thickly wooded spruce-fir slopes of Nevada's higher mountains, and their alpine summits.

DISTRIBUTION

The Mourning Dove has long been characterized as a widespread and common breeder in Nevada (Linsdale 1936, Alcorn 1988), and Ryser (1985) noted that its status in that regard has changed little since the nineteenth century. Atlas workers found Mourning Doves all over the state in a wide variety of habitats. The breeding confirmation rate was relatively low, just 18%. Abundance, however, was fairly high; more than 40% of the occupied blocks were estimated to have at least ten pairs. The species' ubiquity and habitat versatility are accurately reflected on the predictive map, and Mourning Doves are among the few species in Nevada that have a high probability of breeding nearly everywhere.

Mourning Doves breed across the entire continental United States (Mirarchi and Baskett 1994) and are common and widespread in California, Oregon, Idaho, Utah, and Arizona (Behle et al. 1985, Rosenberg et al. 1991, Small 1994, Stephens and Sturts 1998, Adamus et al. 2001, Clark in Corman and Wise-Gervais 2005:192–193). In the West, Mourning Doves generally avoid thickly forested areas but can otherwise be found in a wide variety of habitats (Mirarchi and Baskett 1994). Atlas fieldworkers most often found Mourning Doves in riparian zones, but observations were made in an extremely broad variety of habitat types, and breeding was suspected in most of these.

CONSERVATION AND MANAGEMENT

One of the most abundant birds in North America and a direct beneficiary of human changes to the landscape, the Mourning Dove has expanded its range and increased in numbers since Europeans arrived in North America (Mirarchi and Baskett 1994). Continent-wide, the population has been relatively stable in recent decades, but the pattern is more complex locally. Populations have generally declined in the West, for instance, but in Nevada there has been an overall increase in numbers (Sauer et al. 2005).

Despite its ubiquity, the Mourning Dove is a species of management interest because of its popularity as a game bird. Hunters take more than seventy million Mourning Doves each year in North America, yet the species seems resilient to hunting pressure, and hunting probably does not limit populations (Mirarchi and Baskett 1994). Presumably, the key to the Mourning Dove's success is its long breeding season, which allows multiple broods per year.

In Nevada, the Mourning Dove is also of interest because it has undergone a transformation from summer visitor to permanent resident. Linsdale (1936) described the species as restricted to extreme southern Nevada during winter, and Fisher (1893) considered it to be migratory even in the warm Mojave Desert. Now, however, Mourning Doves are seen during winter throughout Nevada, and can be locally common in places.

HABITAT USE

HABITAT	ATLAS BLOCKS	INCIDENTAL OBSERVATIONS
Agricultural	31 (5%)	4 (2%)
Ash	7 (1%)	0 (0%)
Barren	3 (<1%)	0 (0%)
Grassland	17 (3%)	3 (2%)
Mesquite	17 (3%)	2 (<1%)
Mojave	35 (5%)	22 (13%)
Montane Forest	24 (4%)	13 (7%)
Montane Parkland	7 (1%)	1 (<1%)
Montane Shrub	31 (5%)	9 (5%)
Open Water	0 (0%)	1 (<1%)
Pinyon-Juniper	99 (15%)	11 (6%)
Riparian	141 (21%)	56 (32%)
Salt Desert Scrub	47 (7%)	2 (<1%)
Sagebrush Scrub	61 (9%)	6 (3%)
Sagebrush Steppe	77 (12%)	11 (6%)
Urban	62 (9%)	32 (18%)
Wetland	5 (<1%)	3 (2%)
TOTAL	664	176

Total numbers reported in the Habitat Use, Breeding Status, and Abundance tables may differ from each other (see pp. 22–23 for details). Percentage sums may differ slightly from 100% due to rounding.

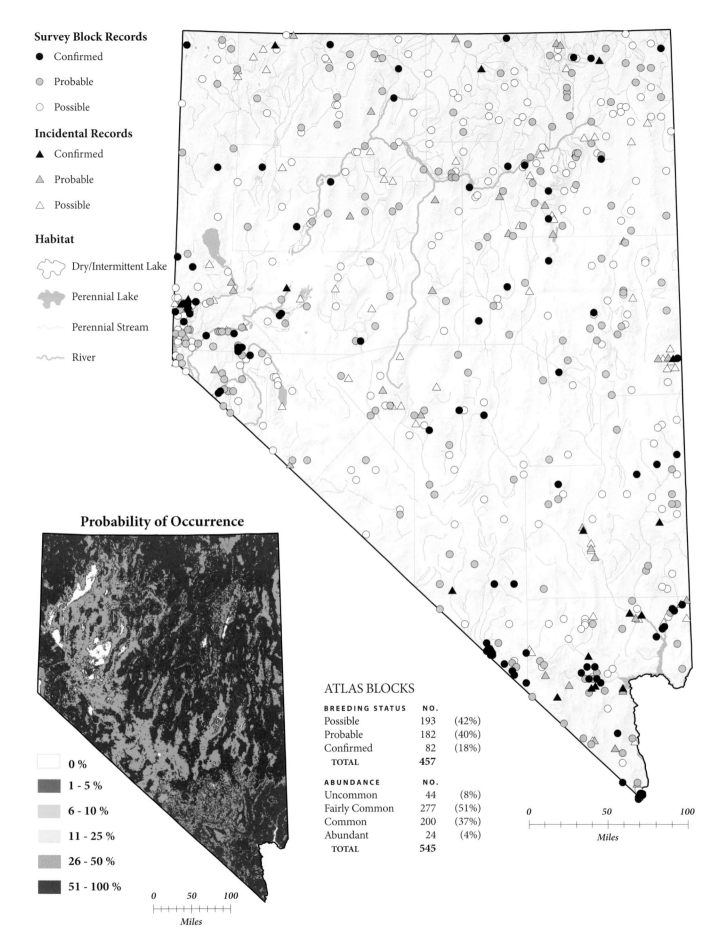

Probability of Occurrence

Survey Block Records

● Confirmed
● Probable
○ Possible

Incidental Records

▲ Confirmed
▲ Probable
△ Possible

Habitat

Dry/Intermittent Lake

Perennial Lake

Perennial Stream

River

0 %

1 - 5 %

6 - 10 %

11 - 25 %

26 - 50 %

51 - 100 %

0   50   100
Miles

ATLAS BLOCKS

BREEDING STATUS	NO.	
Possible	193	(42%)
Probable	182	(40%)
Confirmed	82	(18%)
TOTAL	457	

ABUNDANCE	NO.	
Uncommon	44	(8%)
Fairly Common	277	(51%)
Common	200	(37%)
Abundant	24	(4%)
TOTAL	545	

0   50   100
Miles

INCA DOVE
Columbina inca

On warm afternoons in early spring, the singsong cooing of the Inca Dove drifts down from shade trees in some of southern Nevada's residential areas. Like White-winged Doves, Inca Doves are restricted to the southernmost portion of the state, but unlike White-winged Doves they are almost exclusively reported from urban areas. The abundance of the Inca Dove is easy to underestimate because the bird is so small and inconspicuous. It clearly favors human company, however, and is one of the most successfully urbanized of southern Nevada's bird species. The Inca Dove appears to be colonizing new breeding areas, too, and it will be interesting to see how far north the species eventually expands its range.

DISTRIBUTION

The Inca Dove is a relatively recent arrival in Nevada. Linsdale (1936) did not mention the species, which is said to have first occurred in the state in 1952 (Alcorn 1988). All atlas records for the Inca Dove came from Clark County, primarily from the Las Vegas Valley, and both of the confirmed breeding records were from the city of Las Vegas.

In Nevada, the Inca Dove is frequently found in areas inhabited by humans, and all of the block records and most of the incidental observations came from urban habitats. The Inca Dove's concentration in urban areas is also apparent from the predictive map, which essentially highlights the greater Las Vegas area. This pattern also matches the habitats the species uses elsewhere in its range (Mueller 1992). For instance, Inca Doves are restricted to residential areas in the lower Colorado River Valley (Rosenberg et al. 1991) and in the hot southern deserts of California (Small 1994). Where Inca Doves were recorded, there were generally multiple pairs.

Inca Doves reach the northern limit of their present range in Nevada. This little dove is now widespread in southern Arizona (especially the Phoenix area) by virtue of a natural invasion from the south beginning in the late nineteenth century (Corman in Corman and Wise-Gervais 2005:194–195). The species first appeared in California in 1948 (Small 1994). Behle et al. (1985) considered the Inca Dove to be casual in Utah, and it is a recent local breeder in the southwestern part of the state (Floyd 2001). Inca Doves have never been reported in Oregon (Contreras 1999, Adamus et al. 2001) or Idaho (Stephens and Sturts 1998).

CONSERVATION AND MANAGEMENT

Like most dove species in North America, Inca Doves have fared well in recent decades. Their range has expanded both north into the United States and south within Central America, and their numbers have also increased in many regions within their historic range (Mueller 1992, Sauer et al. 2005). Inca Doves' association with human dwellings has greatly facilitated their success, and their spread in the American Southwest has been linked directly to modern human incursions into the region. The distribution of Inca Doves is still patchy, but local densities can be quite high (Garrett and Dunn 1981).

The Inca Dove is not considered to be a conservation priority, and it will probably continue to prosper in southern Nevada as long as the region is inhabited by humans. Key questions include whether numbers will continue to increase in currently occupied areas and whether northward range expansion will continue. Local population increases, especially around Las Vegas, seem likely. Northward expansion may reach its limits near the northern extent of the Mojave Desert because Inca Doves are considered relatively intolerant of cold (Mueller 1992).

HABITAT USE

HABITAT	ATLAS BLOCKS	INCIDENTAL OBSERVATIONS
Mojave	0 (0%)	1 (11%)
Riparian	0 (0%)	2 (22%)
Urban	7 (100%)	6 (67%)
TOTAL	7	9

Total numbers reported in the Habitat Use, Breeding Status, and Abundance tables may differ from each other (see pp. 22–23 for details). Percentage sums may differ slightly from 100% due to rounding.

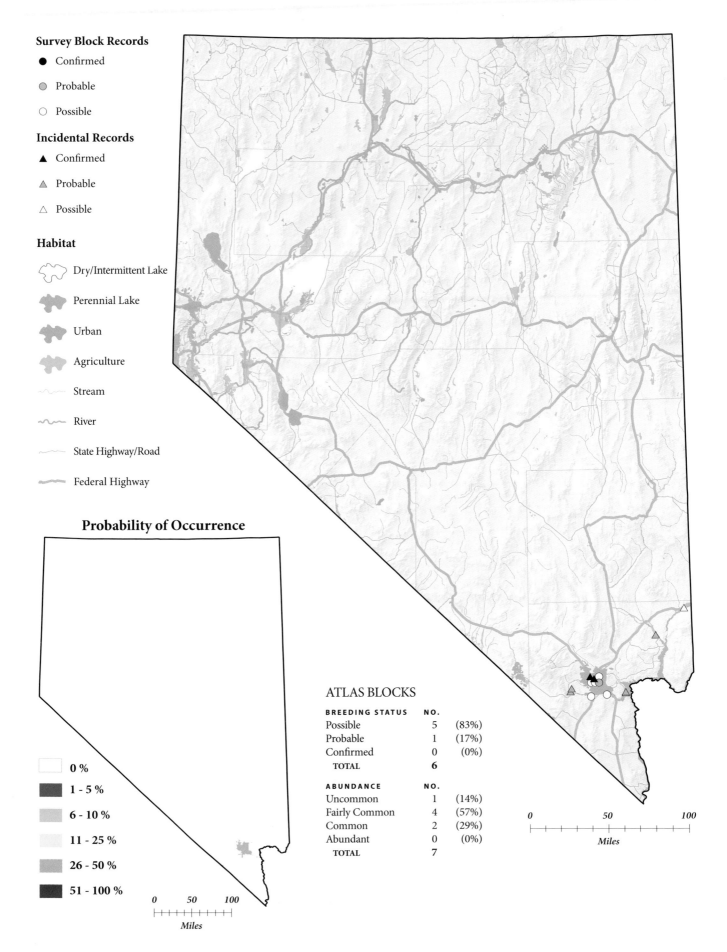

Survey Block Records

● Confirmed

◉ Probable

○ Possible

Incidental Records

▲ Confirmed

▲ Probable

△ Possible

Habitat

Dry/Intermittent Lake

Perennial Lake

Urban

Agriculture

Stream

River

State Highway/Road

Federal Highway

Probability of Occurrence

0 %

1 - 5 %

6 - 10 %

11 - 25 %

26 - 50 %

51 - 100 %

0   50   100
Miles

ATLAS BLOCKS

BREEDING STATUS	NO.	
Possible	5	(83%)
Probable	1	(17%)
Confirmed	0	(0%)
TOTAL	6	

ABUNDANCE	NO.	
Uncommon	1	(14%)
Fairly Common	4	(57%)
Common	2	(29%)
Abundant	0	(0%)
TOTAL	7	

0   50   100
Miles

YELLOW-BILLED CUCKOO
Coccyzus americanus

When the atlas project began, birders expected to find Yellow-billed Cuckoos breeding along the Carson River in western Nevada. As it turned out, there were only two cuckoo records in western Nevada, and neither involved confirmed breeding. Breeding was, however, documented along the upper Muddy River—a wildly productive drainage that birders did not visit much until the fieldwork was concluding. Here and in a few other places in southern Nevada, the mocking call of the cuckoo can be heard right through the hottest days of June and July.

DISTRIBUTION

Atlas workers found Yellow-billed Cuckoos in four blocks and at seven other sites, with most records coming from southern Nevada. Breeding was confirmed only on a private ranch along the upper Muddy River where birds were seen carrying food to nestlings, and where several nests were discovered after the atlas project concluded. Nesting was also suspected to occur at Ash Meadows and Pahranagat national wildlife refuges and along the Virgin River, but this could not be confirmed during the atlas years. The northern Nevada atlas records likely referred to non-breeding birds, although the intriguing sighting of a probable breeder in Elko County deserves attention in future surveys.

The atlas findings revised previous views of the Yellow-billed Cuckoo's status in Nevada. Alcorn (1988) gave records from western and southern Nevada only, and Neel (1999a) stated that the species last nested in southern Nevada in 1979. Cuckoos were observed nearly annually along the lower Carson River of western Nevada during the decade prior to the atlas (Chisholm and Neel 2002), but were seen rarely after that. Tomlinson (2001) summarized recent additional records for the Oasis Valley, Overton Wildlife Management Area, the Pahranagat Valley, the Virgin River, and Warm Springs Ranch.

The Yellow-billed Cuckoo is still widespread in the eastern United States and is common in the Southeast, but it is now uncommon or rare throughout the West (Hughes 1999, Dobkin and Sauder 2004). It occurs very locally in California (Small 1994), and numbers have declined in the lower Colorado River Valley (Rosenberg et al. 1991). The species is also declining and

rare in Utah (Behle et al. 1985) and Arizona (Corman in Corman and Wise-Gervais 2005:202–203), does not breed in Oregon (Adamus et al. 2001), and is extremely rare in southern Idaho (Dobkin 1994, Stephens and Sturts 1998).

In the arid West, Yellow-billed Cuckoos are riparian specialists (Hughes 1999). The location of breeding territories may be influenced by habitat patch size, plant species composition, vegetation density, canopy cover, and distance to water (Laymon and Halterman 1989, Halterman 1991, 2002, SWCA 2005). Cuckoos are typically found in large stands of mature, dense willows, but they also use smaller patches of mesquite, tamarisk, hackberry, and other woody vegetation (Laymon and Halterman 1989, Corman and Magill 2000, Halterman 2003). Given the small number of atlas records, the predictive map for this species should be viewed with caution, but it does suggest other potential locations where cuckoos could occur in the state.

CONSERVATION AND MANAGEMENT

Yellow-billed Cuckoo populations in western North America have declined dramatically from their former numbers, primarily because of the loss or degradation of high-quality riparian habitats (Laymon 1998, Hughes 1999, Dobkin and Sauder 2004). The cuckoo is a Partners in Flight Priority species in Nevada (Neel 1999a) and a Covered species of both the Clark County Multiple Species Habitat Conservation Plan (Clark County 2000) and the Lower Colorado River Multi-Species Conservation Program (BOR-LCR 2004). In 2001, the western subspecies *occidentalis* became a candidate for federal listing under the Endangered Species Act (*Federal Register* 2001). Management recommendations for the Yellow-billed Cuckoo in Nevada focus on riparian habitat protection and restoration (Neel 1999a). Careful monitoring is also recommended, and requires specific methods because this species is a late-season, secretive, and disturbance-sensitive breeder (Hughes 1999).

HABITAT USE

HABITAT	ATLAS BLOCKS	INCIDENTAL OBSERVATIONS
Agricultural	0 (0%)	1 (13%)
Mesquite	0 (0%)	1 (13%)
Riparian	4 (80%)	6 (75%)
Wetland	1 (20%)	0 (0%)
TOTAL	5	8

Total numbers reported in the Habitat Use, Breeding Status, and Abundance tables may differ from each other (see pp. 22–23 for details). Percentage sums may differ slightly from 100% due to rounding.

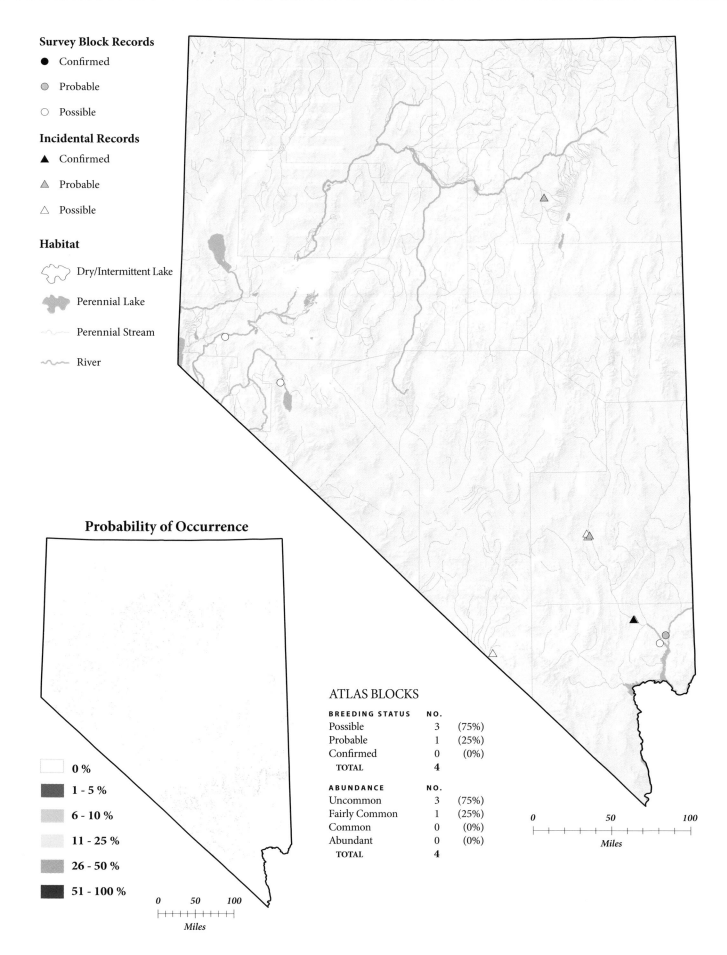

Survey Block Records

● Confirmed

◉ Probable

○ Possible

Incidental Records

▲ Confirmed

△ Probable

△ Possible

Habitat

Dry/Intermittent Lake

Perennial Lake

Perennial Stream

River

Probability of Occurrence

☐ 0 %

▨ 1 - 5 %

▨ 6 - 10 %

▨ 11 - 25 %

▨ 26 - 50 %

▨ 51 - 100 %

0   50   100
Miles

ATLAS BLOCKS

BREEDING STATUS	NO.	
Possible	3	(75%)
Probable	1	(25%)
Confirmed	0	(0%)
TOTAL	**4**	

ABUNDANCE	NO.	
Uncommon	3	(75%)
Fairly Common	1	(25%)
Common	0	(0%)
Abundant	0	(0%)
TOTAL	**4**	

0   50   100
Miles

GREATER ROADRUNNER

Geococcyx californianus

The Greater Roadrunner is perhaps the most beloved bird in the American Southwest. This large, comical bird has an undeniable mystique, and it comes as little surprise that the *Birds of North America* account (Hughes 1996) devotes an entire section to its folklore. Roadrunners are sufficiently widespread to be familiar to a great many people, but they are nowhere so common as to be ordinary. In the scrublands of southern Nevada, the sighting of a roadrunner always elicits a smile.

DISTRIBUTION

All atlas records for the Greater Roadrunner came from Clark, Lincoln, and Nye counties, and all lay within the boundaries of the Mojave Desert. Observers estimated more than one pair to be present in about three-quarters of the blocks for which abundance estimates were supplied, but the species was not especially common anywhere. Roadrunners were confirmed as breeders in only about a quarter of the blocks in which they were found, but they are sedentary birds and probably nested at most sites where they were reported.

The present status of the roadrunner in Nevada has probably changed little since earlier times. Fisher (1893) considered it very common around Ash Meadows, the Las Vegas Valley, the Colorado River lowlands, and north to Oasis Valley in southern Nye County. Linsdale (1936) gave records as far north as Esmeralda County, but no roadrunners were found there during the atlas years.

The roadrunner is at the northern edge of its range in southern Nevada. There are no records for the species from Oregon (Contreras 1999, Adamus et al. 2001) or Idaho (Stephens and Sturts 1998), and occurrences anywhere in the Great Basin are exceptional (Ryser 1985). The species is resident in southern California and occurs north into the Central Valley (Small 1994, Hughes 1996), but it is restricted to the southwestern part of Utah (Behle et al. 1985). Roadrunners breed throughout Arizona but are most widespread in the southwestern and central portions (Averill-Murray in Corman and Wise-Gervais 2005:204–205).

Roadrunners are birds of desert scrublands, but in Nevada they were often noted in desert riparian corridors and even in isolated remnant patches of suitable habitat. There were atlas records, for example, from Sunset Park and the Reserve Casino in urban Las Vegas. The predictive map even suggests that roadrunners are more likely to be found in urban areas than in surrounding desert, though not as likely as in riparian habitat. This result appears at odds with the view that roadrunners have declined wherever urbanization has been especially heavy (Hughes 1996), but it is possible that frequent occurrences in xeriscaped suburban settings (which are classified as urban habitat in this study) account for the difference. Additionally, many urban blocks might still have contained small patches of suitable habitat in which roadrunners remained during the atlas years. Whether such fragmented populations can persist over the long term, even if urbanization does not convert the remaining patches, is uncertain.

CONSERVATION AND MANAGEMENT

The roadrunner is often portrayed as an adaptable and expanding species, with recent incursions into areas north of its current range cited as an example of its versatility (Hughes 1996). In California, however, the species has disappeared from the northern Central Valley (McCaskie et al. 1988) and northern coastal regions (Shuford 1993) and is declining statewide (Sauer et al. 2005). These losses are probably connected with agricultural and urban development of native habitats (Small 1994). Throughout the rest of its U.S. range the roadrunner appears to have a stable population (Sauer et al. 2005), and it is not considered a conservation priority (Hughes 1996). Further research clarifying the species' response to urbanization would be valuable.

HABITAT USE

HABITAT	ATLAS BLOCKS	INCIDENTAL OBSERVATIONS
Grassland	1 (3%)	0 (0%)
Mesquite	4 (10%)	2 (6%)
Mojave	13 (33%)	4 (13%)
Montane Parkland	0 (0%)	1 (3%)
Open Water	0 (0%)	1 (3%)
Pinyon-Juniper	1 (3%)	1 (3%)
Riparian	12 (30%)	13 (42%)
Salt Desert Scrub	3 (8%)	2 (6%)
Urban	6 (15%)	5 (16%)
Wetland	0 (0%)	2 (6%)
TOTAL	40	31

Total numbers reported in the Habitat Use, Breeding Status, and Abundance tables may differ from each other (see pp. 22–23 for details). Percentage sums may differ slightly from 100% due to rounding.

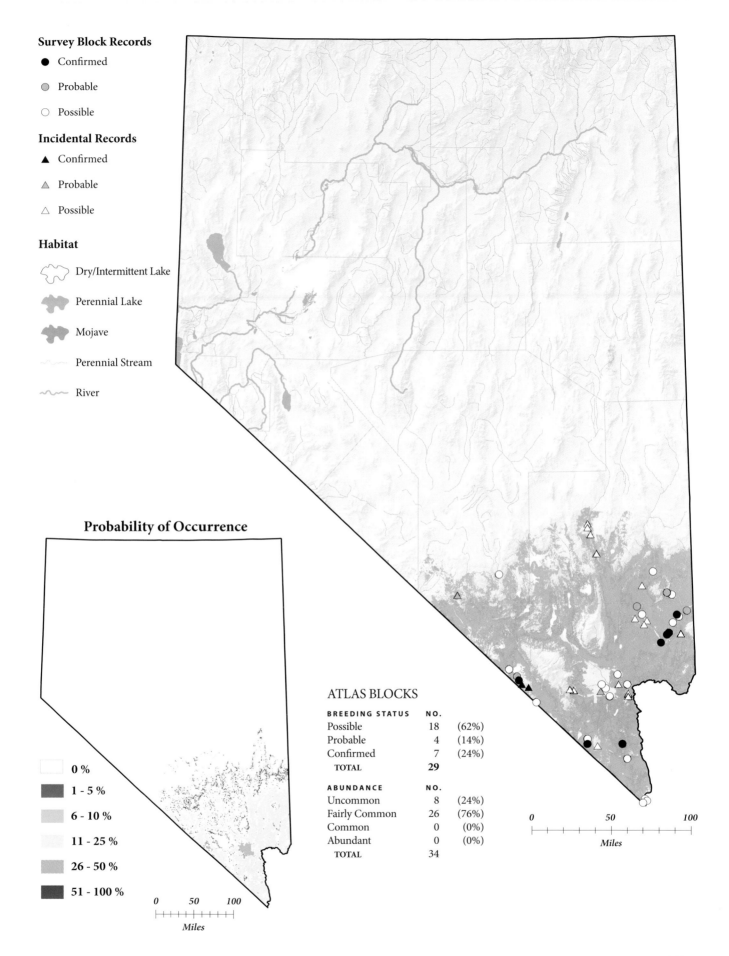

Survey Block Records

● Confirmed

◔ Probable

○ Possible

Incidental Records

▲ Confirmed

△ Probable

△ Possible

Habitat

Dry/Intermittent Lake

Perennial Lake

Mojave

Perennial Stream

River

Probability of Occurrence

☐ 0 %

1 - 5 %

6 - 10 %

11 - 25 %

26 - 50 %

51 - 100 %

0   50   100
Miles

ATLAS BLOCKS

BREEDING STATUS	NO.	
Possible	18	(62%)
Probable	4	(14%)
Confirmed	7	(24%)
TOTAL	29	

ABUNDANCE	NO.	
Uncommon	8	(24%)
Fairly Common	26	(76%)
Common	0	(0%)
Abundant	0	(0%)
TOTAL	34	

0   50   100
Miles

BARN OWL
Tyto alba

The Barn Owl is the consummate avian spook. This strictly nocturnal denizen of church steeples, cemetery outbuildings, and long-dead cottonwoods is most often seen as a flash of white in the night sky as it hunts down the mice that are its usual prey. Barn Owls are among the best-studied nocturnal birds, not only by ecologists, but also by neurobiologists, who have focused on this owl's excellent low-light vision and acute hearing (Marti 1992).

DISTRIBUTION

Barn Owls were found primarily along the river systems of northwestern and southern Nevada. Most confirmed breeding occurred in the northwest, a pattern that conforms closely to Ryser's (1985) impression that the species is more common in the western than the eastern Great Basin.

The Barn Owl is one of the most widely distributed terrestrial birds in the world (Marti 1992), and it is widespread throughout the West, including all the states surrounding Nevada (Behle et al. 1985, Rosenberg et al. 1991, Small 1994, Adamus et al. 2001, Wise-Gervais in Corman and Wise-Gervais 2005:206–207) except Idaho, where it is limited to the southern and western regions (Stephens and Sturts 1998). The Barn Owl's distribution in all of these states seems to mirror the situation in Nevada: fairly common in certain areas but relatively uncommon overall.

For nesting and roosting, Barn Owls require a secluded cavity in the vicinity of open country that is suitable for hunting (Marti 1992). Dead trees far out in the remotest desert scrublands are acceptable, and so are nooks and crannies in large warehouses near urban centers. In the Lahontan Valley, haystacks and barns are also used (Chisholm and Neel 2002). Atlas fieldworkers most

often reported riparian, urban, and agricultural habitats—which all present a good mix of nesting and foraging opportunities—as Barn Owl haunts. This species is generally not found in heavily forested and mountainous areas, as is indicated on the predictive map, which also shows that the chance of finding breeding Barn Owls is typically low throughout the state. The highest probabilities are predicted to be in scattered agricultural areas, in the foothills of the Spring Mountains, and across a sizable swath of the northern Mojave Desert and southeastern Nevada. The latter area does not appear to have much in the way of resources attractive to Barn Owls, however, and actual occurrence data from the atlas and other sources suggest that predictions in this part of the state are overestimates.

CONSERVATION AND MANAGEMENT

Barn Owls were persecuted in Nevada in the past, but populations are thought to be stable today (Herron et al. 1985). The species has a patchy distribution in the state, though, and is difficult to survey accurately because of its nocturnal habits. Declines have been reported in some parts of the U.S. range, especially the Midwest and Northeast, where Barn Owls are considered endangered in several states, but other regions report stable or even increasing populations (Marti 1992). Barn Owl numbers may have increased in recent years in the northern Great Basin (Littlefield 1990), but local population declines have been reported in California (Shuford 1993). Woodland clearing, agricultural irrigation, and availability of artificial nesting structures have probably had positive effects on populations. Barn Owls have also responded favorably to nest box supplementation in some areas (Marti 1992). Elsewhere, some agricultural practices, including pesticide use, may have contributed to declines (Marti 1992). A clearer understanding of the species' distribution, trends, and ecological requirements would be helpful in formulating conservation recommendations for Nevada.

HABITAT USE

HABITAT	ATLAS BLOCKS	INCIDENTAL OBSERVATIONS
Agricultural	2 (10%)	2 (14%)
Grassland	1 (5%)	0 (0%)
Mojave	1 (5%)	0 (0%)
Open Water	0 (0%)	2 (14%)
Pinyon-Juniper	1 (5%)	1 (7%)
Riparian	10 (50%)	3 (21%)
Sagebrush Scrub	1 (5%)	1 (7%)
Urban	3 (15%)	4 (29%)
Wetland	1 (5%)	1 (7%)
TOTAL	20	14

Total numbers reported in the Habitat Use, Breeding Status, and Abundance tables may differ from each other (see pp. 22–23 for details). Percentage sums may differ slightly from 100% due to rounding.

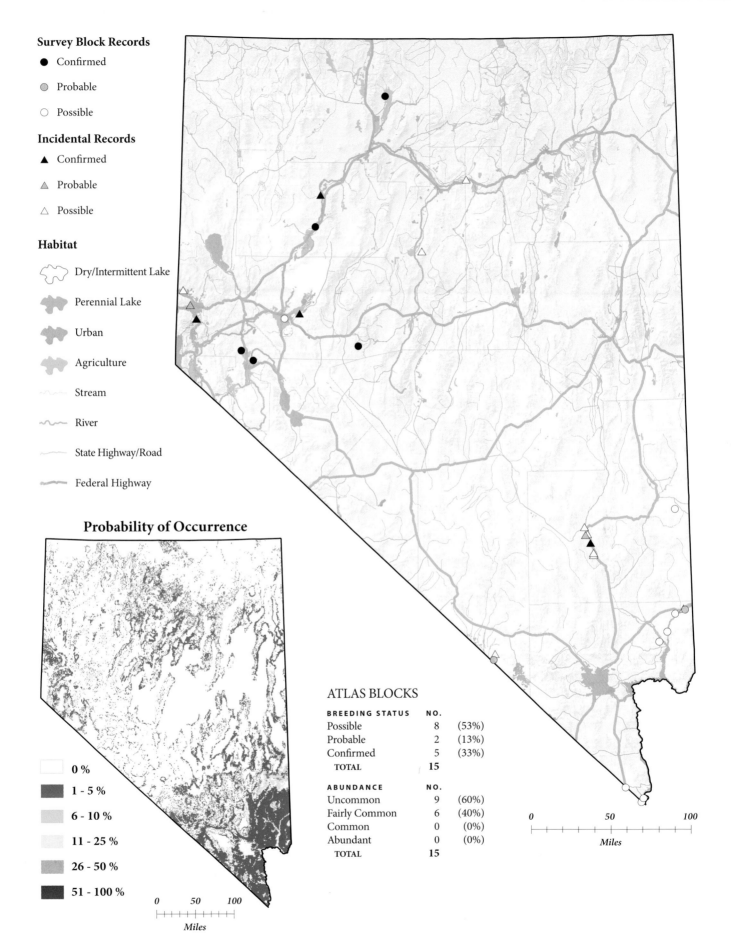

Survey Block Records

● Confirmed

◐ Probable

○ Possible

Incidental Records

▲ Confirmed

△ Probable

△ Possible

Habitat

Dry/Intermittent Lake

Perennial Lake

Urban

Agriculture

Stream

River

State Highway/Road

Federal Highway

Probability of Occurrence

0 %

1 - 5 %

6 - 10 %

11 - 25 %

26 - 50 %

51 - 100 %

0   50   100
Miles

ATLAS BLOCKS

BREEDING STATUS	NO.	
Possible	8	(53%)
Probable	2	(13%)
Confirmed	5	(33%)
TOTAL	15	
ABUNDANCE	NO.	
Uncommon	9	(60%)
Fairly Common	6	(40%)
Common	0	(0%)
Abundant	0	(0%)
TOTAL	15	

0   50   100
Miles

FLAMMULATED OWL
Otus flammeolus

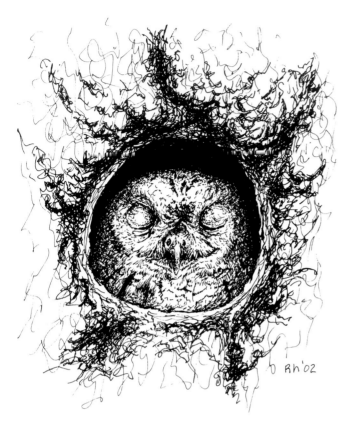

Lucky late-night visitors to the conifer forests of Nevada's mountains may hear the monosyllabic utterances of the Flammulated Owl. The bird's call—a single *hoo* repeated over and over—is deep voiced and carries far, yet the Flammulated Owl is no larger than a small thrush. Many Nevada birders have never even seen this tiny insectivore, and much is unknown about its basic biology and distribution in the state.

DISTRIBUTION

The breeding range of the Flammulated Owl was poorly known until the 1950s, and many specifics remain to be documented (McCallum 1994). The species is locally common in California, Arizona, and Oregon, and also occurs in parts of Utah and Idaho (Behle et al. 1985, Small 1994, Stephens and Sturts 1998, Adamus et al. 2001, Wise-Gervais in Corman and Wise-Gervais 2005:208–209). The Nevada distribution includes the Santa Rosa, Jarbidge, Spring, Schell Creek, Quinn Canyon, White Pine, Sheep, Clover, Snake, Highland, and Carson ranges (Dunham et al. 1996). Atlas data came from several of these ranges, but our records clearly provide only a partial description of where this species occurs. Birds were reported from four atlas blocks and from eight other sites, but breeding was confirmed only in the Sheep Range.

The Flammulated Owl is usually a bird of open coniferous forests (McCallum 1994), and in Nevada it has been found in areas dominated by ponderosa pine, limber pine, Jeffrey pine, white fir, and subalpine fir (Dunham et al. 1996). Most atlas records came from montane forest. Dead trees containing woodpecker holes and an abundant supply of invertebrate prey, including beetles and moths, are important for this cavity-nesting species (Neel 1999a, Arsenault et al. 2003).

Given the difficulty of surveying Flammulated Owls, it is likely that additional nesting areas remain to be found in Nevada. The predictive map identifies locations where there seems to be a good chance of finding them, and indeed, earlier studies found Flammulated Owls in many of these areas (Dunham et al. 1996). The map might overpredict the probability of occurrence in the center of the state, where the species has not previously been reported and where the most appropriate types of montane forest do not occur (Dunham et al. 1996). Additional areas that most warrant surveying include those identified both by the predictive map and by Dunham et al. (1996).

CONSERVATION AND MANAGEMENT

Rangewide, Flammulated Owl population trends are largely unknown because reliable monitoring data are lacking. Potential threats to the species include selective logging of standing deadwood on the breeding grounds and deforestation on the wintering grounds (McCallum 1994), both of which are compounded by the species' low reproductive rate and the perils associated with long-distance migration.

The Flammulated Owl is a Partners in Flight Watch List species (Rich et al. 2004) and a Nevada Priority species (Neel 1999a). Timber harvest has been identified as a potential threat in Nevada, and forest management that preserves the old-growth character of known or suspected breeding grounds is recommended (Neel 1999a). A statewide monitoring effort for nocturnal birds would greatly enhance our understanding of this species' responses to habitat change.

HABITAT USE

HABITAT	ATLAS BLOCKS	INCIDENTAL OBSERVATIONS
Montane Forest	4 (57%)	6 (100%)
Montane Parkland	1 (14%)	0 (0%)
Pinyon-Juniper	1 (14%)	0 (0%)
Sagebrush Scrub	1 (14%)	0 (0%)
TOTAL	7	6

Total numbers reported in the Habitat Use, Breeding Status, and Abundance tables may differ from each other (see pp. 22–23 for details). Percentage sums may differ slightly from 100% due to rounding.

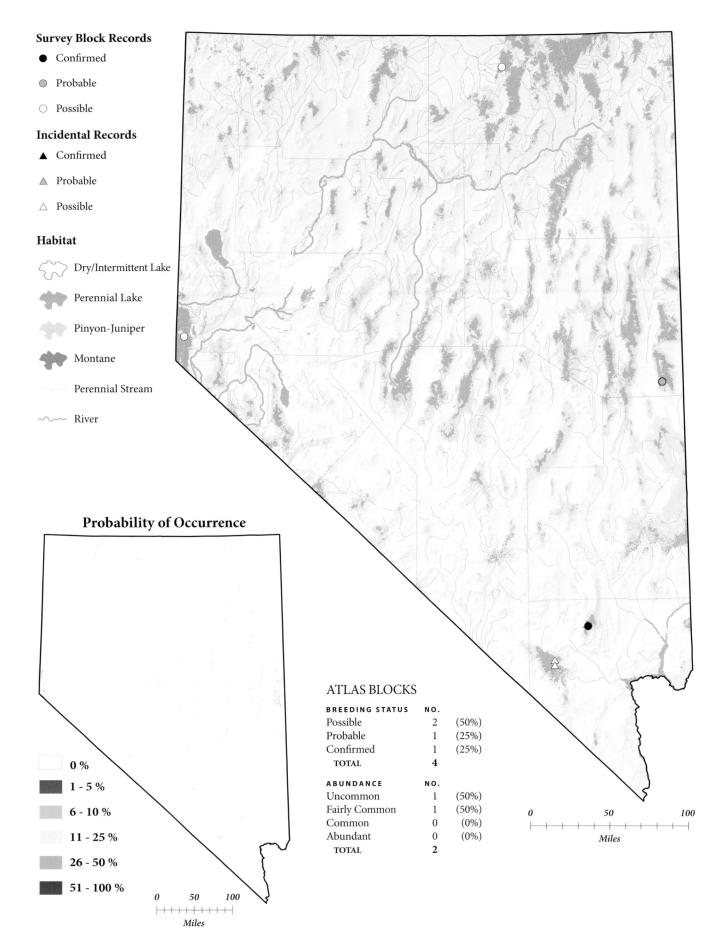

Survey Block Records

● Confirmed

◐ Probable

○ Possible

Incidental Records

▲ Confirmed

△ Probable

△ Possible

Habitat

Dry/Intermittent Lake

Perennial Lake

Pinyon-Juniper

Montane

Perennial Stream

River

Probability of Occurrence

0 %

1 - 5 %

6 - 10 %

11 - 25 %

26 - 50 %

51 - 100 %

0   50   100

Miles

ATLAS BLOCKS

BREEDING STATUS	NO.	
Possible	2	(50%)
Probable	1	(25%)
Confirmed	1	(25%)
TOTAL	4	

ABUNDANCE	NO.	
Uncommon	1	(50%)
Fairly Common	1	(50%)
Common	0	(0%)
Abundant	0	(0%)
TOTAL	2	

0   50   100

Miles

WESTERN SCREECH-OWL
Megascops kennicottii

Even though Western Screech-Owls are considered common throughout much of their range, they are surprisingly hard to find in Nevada. More often than not, dedicated owl searchers simply could not find them in what seemed like perfectly suitable habitat in the lowlands and foothills. The few places in Nevada that did harbor screech-owls tended to be characterized by open or broken woodlands along riparian corridors or in pinyon-juniper-covered foothills. Areas with old cottonwoods seem to be especially favored by these little gray owls with their telltale "bouncing-ball" songs.

DISTRIBUTION

Atlas workers came across Western Screech-Owls at seventeen locations in Nevada, with breeding confirmed at three. Single pairs were most commonly encountered. The records occurred throughout the state, indicating that the species is widespread, albeit sporadically distributed and uncommon. Earlier writers also attested to the statewide scarcity of screech-owls: Linsdale (1936) provided sparse records for southern Nevada only, and Herron et al. (1985) and Alcorn (1988) concurred that the Western Screech-Owl is uncommon in Nevada. On the other hand, screech-owls are easily overlooked, as indicated by the lack of confirmed nesting records from the Lahontan Valley during the atlas project despite the species' known use of nest boxes in that area (Chisholm and Neel 2002).

For the most part, the Western Screech-Owl is more common in surrounding states and throughout the West than it appears

to be in Nevada. It is fairly widespread and common in Arizona (Rosenberg et al. 1991, Wise-Gervais in Corman and Wise-Gervais 2005:210–211), Utah (Behle et al. 1985), Oregon (Adamus et al. 2001), and southern and western Idaho (Stephens and Sturts 1998). The California distribution is complex and patchy, but these owls are moderately common in suitable habitat (Small 1994).

The Western Screech-Owl's breeding habitat is varied, but lower-altitude riparian areas and other deciduous woodlands are most commonly used, and nesting cavities are required (Cannings and Angell 2001). In Nevada, atlas workers found Western Screech-Owls most often in riparian habitat and in pinyon-juniper woodlands. Herron et al. (1985) emphasized their rather catholic habitat associations, so even in a dry state like Nevada it is likely an oversimplification to regard Western Screech-Owls as a strictly riparian species. The predictive map shows the probability of encountering screech-owls to be highest in mountain forests and in lowland areas—such as riparian and agricultural lands—where scattered trees are most likely to be found. Not surprisingly, given its reliance on trees with cavities, the species is predicted to be very unlikely in most of the state and absent from the lower-elevation salt desert and Mojave habitats.

CONSERVATION AND MANAGEMENT

Although the Western Screech-Owl is not usually thought of as a conservation concern, there may be various threats to local populations. Numbers remain high in certain areas of the lower Colorado River Valley, for example, but habitat loss is reported to have caused net population reductions (Rosenberg et al. 1991). In southern California, populations have been extirpated locally in agricultural valleys (Garrett and Dunn 1981). Wholesale loss of riparian habitat is very likely to harm this species. Other habitat changes may be more complex, though: Shuford (1993) discussed how logging and urbanization may have a mix of negative and positive effects on Western Screech-Owl populations.

Herron et al. (1985) predicted stable populations for the Western Screech-Owl in Nevada because suitable habitat is available. Screech-owls readily accept artificial nest sites (Cannings and Angell 2001), which could mitigate the loss of natural cavities in some areas. Management of riparian areas to retain or improve their natural vegetation characteristics, however, is probably the best conservation strategy.

HABITAT USE

HABITAT	ATLAS BLOCKS	INCIDENTAL OBSERVATIONS
Mesquite	1 (7%)	0 (0%)
Montane Forest	1 (7%)	0 (0%)
Montane Shrub	0 (0%)	1 (25%)
Pinyon-Juniper	4 (28%)	2 (50%)
Riparian	7 (50%)	1 (25%)
Urban	1 (7%)	0 (0%)
TOTAL	14	4

Total numbers reported in the Habitat Use, Breeding Status, and Abundance tables may differ from each other (see pp. 22–23 for details). Percentage sums may differ slightly from 100% due to rounding.

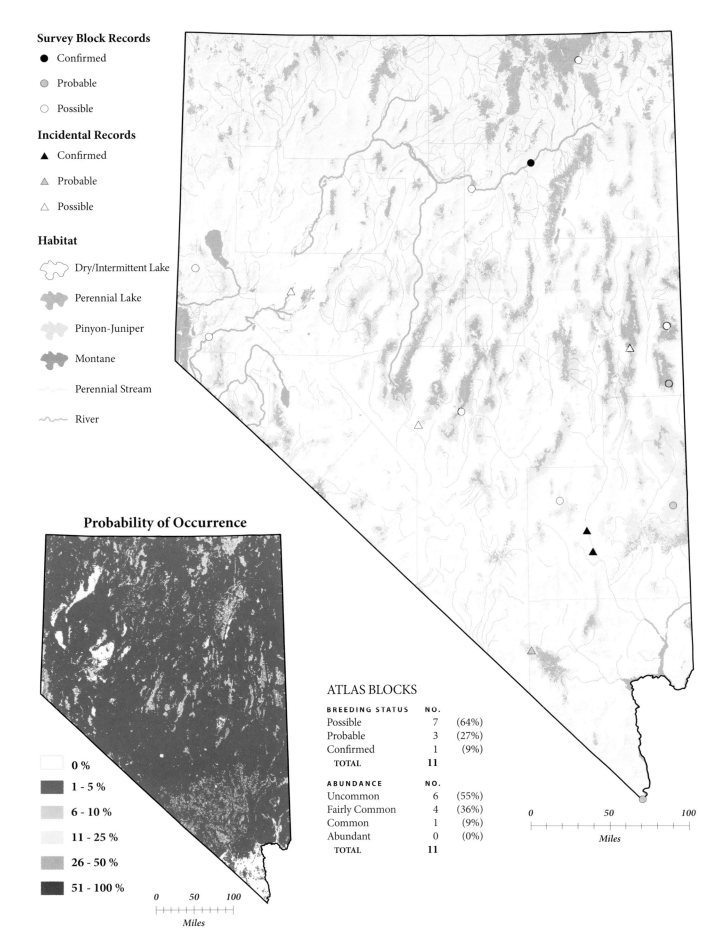

Survey Block Records

- ● Confirmed
- ◐ Probable
- ○ Possible

Incidental Records

- ▲ Confirmed
- △ Probable
- △ Possible

Habitat

- Dry/Intermittent Lake
- Perennial Lake
- Pinyon-Juniper
- Montane
- Perennial Stream
- River

Probability of Occurrence

- ☐ 0 %
- 1 - 5 %
- 6 - 10 %
- 11 - 25 %
- 26 - 50 %
- 51 - 100 %

0   50   100

Miles

ATLAS BLOCKS

BREEDING STATUS	NO.	
Possible	7	(64%)
Probable	3	(27%)
Confirmed	1	(9%)
TOTAL	11	

ABUNDANCE	NO.	
Uncommon	6	(55%)
Fairly Common	4	(36%)
Common	1	(9%)
Abundant	0	(0%)
TOTAL	11	

0   50   100

Miles

GREAT HORNED OWL

Bubo virginianus

Even before the winter solstice, the mellow hoots of courting Great Horned Owls echo through the Nevada night. There are few habitat types that do not support this very large and most versatile of owls, from the mountains to the valley floors. And Great Horned Owls are not at all averse to human company, as demonstrated by many successful nesting attempts in Reno and other urban centers.

DISTRIBUTION

The Great Horned Owl has an extensive distribution in the New World. It occurs from Alaska to Tierra del Fuego, from the Atlantic to the Pacific (Houston et al. 1998), and throughout all of the states surrounding Nevada (Behle et al. 1985, Small 1994, Stephens and Sturts 1998, Adamus et al. 2001, Wise-Gervais in Corman and Wise-Gervais 2005:214–215). Great Horned Owls were reported throughout Nevada as well, and in more atlas blocks than any other owl species. Nesting was confirmed at many locations. Other treatments of the species describe a similarly extensive range in the state (Linsdale 1936, Herron et al. 1985, Ryser 1985, Alcorn 1988). As a top predator, though, this owl was nowhere reported to be especially numerous, and 71% of the abundance estimates had just a single pair in a block.

Great Horned Owls are habitat generalists (Houston et al. 1998). Ryser (1985:263) noted that they occur in "every conceivable type of desert and montane habitat," and it is easier to list the few habitats from which they are absent than the many in which they are found (Gaines 1992). They nest in trees, cliffs, buildings, and on platforms (Houston et al. 1998), which may explain the paucity of records from treeless areas of the desert. They also nest in or near towns, where they may expand their prey base to include city fare such as domestic cats. Especially high Great Horned Owl densities may occur near agricultural lands and in riparian zones (Herron et al. 1985). The atlas data supported a strong association with riparian areas, which accounted for about 40% of the records. The predictive map also indicates a moderate, though somewhat lower, chance of finding Great Horned Owls in agricultural and urban habitats. The map also suggests that these owls are more likely to be found in northern parts of the state and are least likely to occur in lowland salt desert and Mojave habitats.

CONSERVATION AND MANAGEMENT

The Great Horned Owl is thinly distributed across a very wide geographic expanse (Houston et al. 1998). As is true of owls in general, its nocturnal tendencies make reliable population trend estimates difficult to obtain. Monitoring is further complicated by the tendency of Great Horned Owl populations to fluctuate with prey abundance (Houston et al. 1998). The available evidence, however, suggests that populations are relatively stable (Sauer et al. 2005).

Great Horned Owls seem remarkably adaptable in the face of habitat alteration as long as nest sites remain available, and they are not at all skittish about the approach of humans (Houston et al. 1998). The species has benefited from plantings of large trees (Contreras and Kindschy 1996) and from forest thinning (Houston et al. 1998). Few threats to the Great Horned Owl have been identified, but they include vehicle collisions, electrocution, persecution, and accidental pesticide poisoning. None of these factors is thought to limit populations, however, and the species is generally faring well across its range, with no active management required (Houston et al. 1998). There are no indications of problems in Nevada, either, and as Herron et al. (1985) concluded, Great Horned Owls likely have a bright future in the state.

HABITAT USE

HABITAT	ATLAS BLOCKS	INCIDENTAL OBSERVATIONS
Agricultural	4 (4%)	1 (2%)
Alpine	1 (1%)	0 (0%)
Barren	1 (1%)	1 (2%)
Grassland	2 (2%)	0 (0%)
Mesquite	1 (1%)	0 (0%)
Mojave	2 (2%)	1 (2%)
Montane Forest	10 (10%)	3 (6%)
Montane Parkland	1 (1%)	1 (2%)
Montane Shrub	2 (2%)	2 (4%)
Open Water	0 (0%)	2 (4%)
Pinyon-Juniper	19 (19%)	1 (2%)
Riparian	37 (37%)	23 (44%)
Salt Desert Scrub	1 (1%)	1 (2%)
Sagebrush Scrub	4 (4%)	4 (8%)
Sagebrush Steppe	2 (2%)	0 (0%)
Urban	11 (11%)	11 (21%)
Wetland	1 (1%)	1 (2%)
TOTAL	99	52

Total numbers reported in the Habitat Use, Breeding Status, and Abundance tables may differ from each other (see pp. 22–23 for details). Percentage sums may differ slightly from 100% due to rounding.

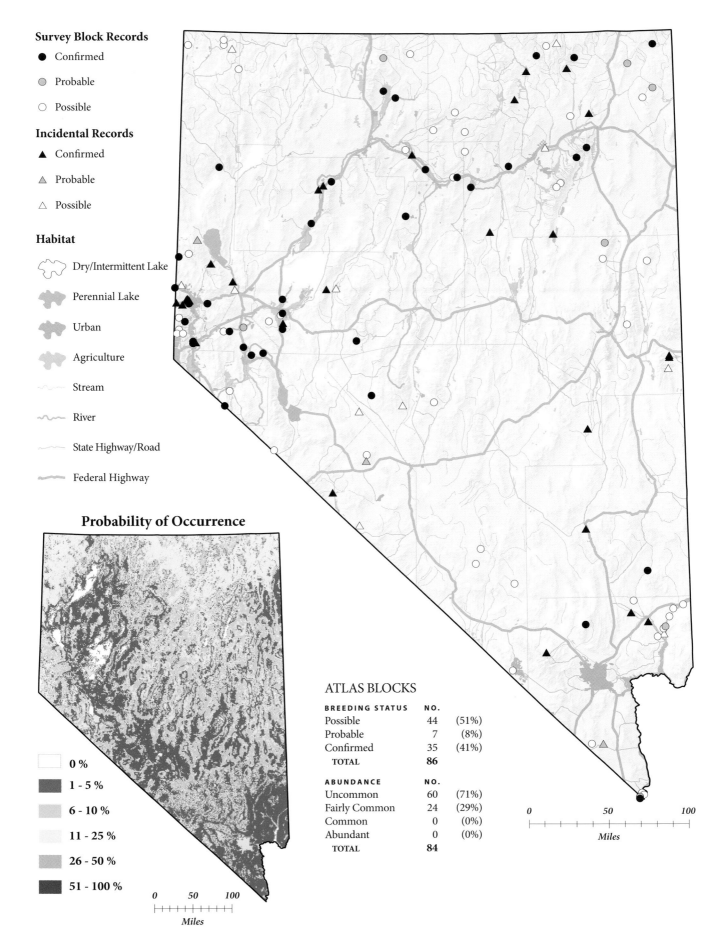

Survey Block Records

● Confirmed
◐ Probable
○ Possible

Incidental Records

▲ Confirmed
△ Probable
△ Possible

Habitat

Dry/Intermittent Lake

Perennial Lake

Urban

Agriculture

Stream

River

State Highway/Road

Federal Highway

Probability of Occurrence

0 %
1 - 5 %
6 - 10 %
11 - 25 %
26 - 50 %
51 - 100 %

0   50   100
Miles

ATLAS BLOCKS

BREEDING STATUS	NO.	
Possible	44	(51%)
Probable	7	(8%)
Confirmed	35	(41%)
TOTAL	86	

ABUNDANCE	NO.	
Uncommon	60	(71%)
Fairly Common	24	(29%)
Common	0	(0%)
Abundant	0	(0%)
TOTAL	84	

0   50   100
Miles

NORTHERN PYGMY-OWL

Glaucidium gnoma

RN '02

The pugnacious little Northern Pygmy-Owl lives in the montane forests of Nevada but is not often encountered. This diurnally active owl can be lured in by imitating its monosyllabic song— a procedure also likely to arouse the ire of seething swarms of creepers, chickadees, and nuthatches—but more often, bumping into a pygmy-owl is entirely serendipitous: a sudden face-to-face encounter with a slightly nonplussed miniature owl in broad daylight.

DISTRIBUTION

Northern Pygmy-Owls were most frequently reported from the Carson Range of far western Nevada, and this was the only area where breeding was confirmed. Other authors have also identified the Carson Range, and to a lesser extent east-central Nevada, as important areas for pygmy-owls in Nevada (Herron et al. 1985, Ryser 1985, Alcorn 1988). A handful of additional records were scattered throughout the state, even in the interior, where some older accounts suggest the species does not occur (Holt and Petersen 2000). Given the overall paucity of records, it is interesting that two or more pairs were estimated to be present in more than three-fourths of the blocks where the species occurred.

Throughout the West, the pygmy-owl is generally considered uncommon. This seems especially true in Nevada, and pygmy-owls appear to be more abundant and widespread in our neighboring states than they are here (see Holt and Petersen 2000). In fact, the areas where the species is least abundant are generally those closest to Nevada, including southeastern Oregon, south-

western Idaho, west-central Utah, and the Mojave regions of California and Arizona (Small 1994, Stephens and Sturts 1998, Holt and Petersen 2000, Adamus et al. 2001, Corman in Corman and Wise-Gervais 2005:216–217).

Pygmy-owls are forest birds, and they usually occur in the mountains. They tolerate a wide variety of cover types and conditions, but do require tree cavities for nesting (Holt and Petersen 2000). In the southern Sierra Nevada, they can be found among oaks and conifers, and even in riparian hardwoods and abandoned orchards (Gaines 1992). In Utah, pygmy-owls have been found in forested areas that run the gamut from lowland cottonwood groves to montane coniferous woods (Behle et al. 1985). In Nevada, they tend to be seen in "edge" habitats where the forest abuts a clearing or meadow (Herron et al. 1985). Three-quarters of all atlas records were associated with pinyon-juniper or montane forests. Pygmy-owls are predicted to occur in suitable habitat throughout the state, but the chances of encountering them anywhere are low.

CONSERVATION AND MANAGEMENT

Little is known about the status and distribution of the Northern Pygmy-Owl in Nevada, hampering attempts to address its conservation needs. The effects of logging, for example, are difficult to gauge: it may open up habitat for foraging, but dead tree removal and large-scale reduction of woodlands are presumably harmful (Shuford 1993, Holt and Petersen 2000). As is the case for most owls, the substantial challenge of even finding these birds makes it difficult to draw conclusions about population trends (e.g., Sauer et al. 2005).

The pygmy-owl's fairly sedentary behavior and the naturally fragmented nature of its habitats in Nevada could make the species susceptible to local extirpation. Herron et al. (1985), though, noted that there have been few surveys in the state and indicated that the Northern Pygmy-Owl may be more common than is generally assumed. Only dedicated studies focused on pygmy-owls will allow us to reach a comprehensive understanding of the species' conservation status in Nevada and determine its ecological requirements.

HABITAT USE

HABITAT	ATLAS BLOCKS	INCIDENTAL OBSERVATIONS
Montane Forest	3 (25%)	1 (100%)
Montane Parkland	1 (8%)	0 (0%)
Pinyon-Juniper	6 (50%)	0 (0%)
Riparian	1 (8%)	0 (0%)
Sagebrush Steppe	1 (8%)	0 (0%)
TOTAL	12	1

Total numbers reported in the Habitat Use, Breeding Status, and Abundance tables may differ from each other (see pp. 22–23 for details). Percentage sums may differ slightly from 100% due to rounding.

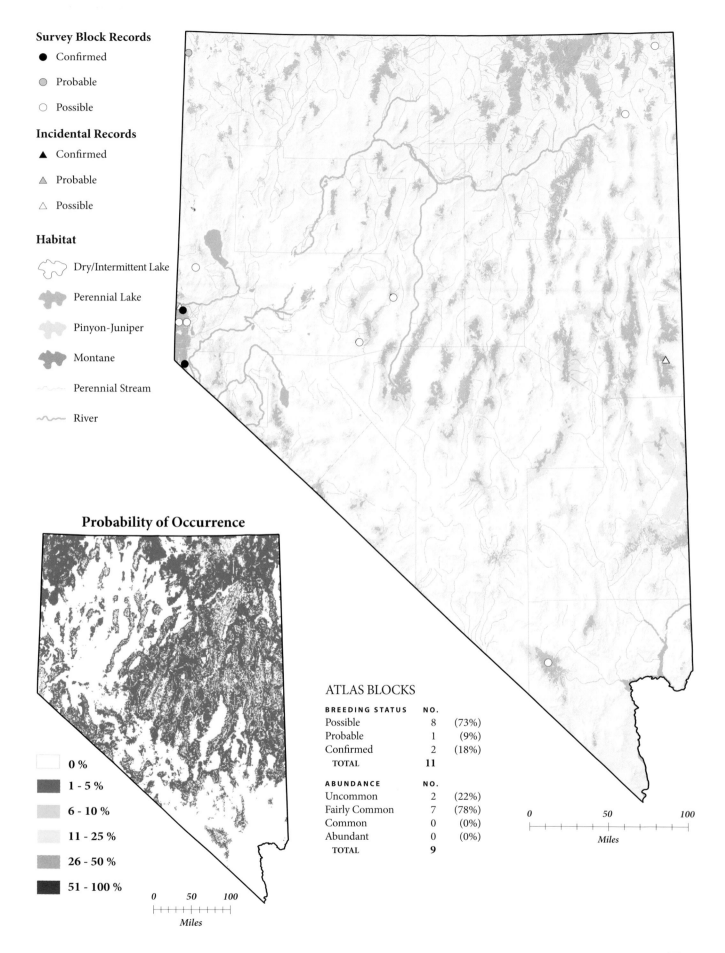

Survey Block Records

● Confirmed

◐ Probable

○ Possible

Incidental Records

▲ Confirmed

▲ Probable

△ Possible

Habitat

Dry/Intermittent Lake

Perennial Lake

Pinyon-Juniper

Montane

Perennial Stream

River

Probability of Occurrence

☐ 0 %

■ 1 - 5 %

■ 6 - 10 %

☐ 11 - 25 %

■ 26 - 50 %

■ 51 - 100 %

0   50   100

Miles

ATLAS BLOCKS

BREEDING STATUS	NO.	
Possible	8	(73%)
Probable	1	(9%)
Confirmed	2	(18%)
TOTAL	11	

ABUNDANCE	NO.	
Uncommon	2	(22%)
Fairly Common	7	(78%)
Common	0	(0%)
Abundant	0	(0%)
TOTAL	9	

0   50   100

Miles

BURROWING OWL

Athene cunicularia

Most people think of owls as solitary, arboreal, and nocturnal—as ghostly voices in the night woods. It is thus an understatement to say that the Burrowing Owl is an unusual owl. It is a long-legged, ground-dwelling, often diurnal, semicolonial resident of treeless deserts and grasslands, not to mention golf courses and airports. There is also an especially fierce quality about Burrowing Owls, with their glowering golden eyes. Burrowing Owls often occur in close proximity to people, but numbers have declined sharply in areas where their habitat has been destroyed.

DISTRIBUTION

Burrowing Owls were widely but infrequently recorded throughout Nevada during the atlas survey. In most locations, only a single pair was detected. Recent research indicates that surveys conducted during the day systematically underestimate abundance (D. Crowe, pers. comm.), however, and Burrowing Owls may be more common in the Great Basin than is usually assumed (Herron et al. 1985, Ryser 1985). Outside Nevada, Burrowing Owls breed in southeastern Oregon, southern Idaho, parts of California, and throughout Utah and Arizona (Behle et al. 1985, Small 1994, Stephens and Sturts 1998, Adamus et al. 2001, Martin in Corman and Wise-Gervais 2005:222–223).

Burrowing Owls prefer open, arid, treeless landscapes with low vegetation (Haug et al. 1993, Dobkin and Sauder 2004), habitats in plentiful supply in Nevada. The predictive map confirms a widespread distribution, although the probability of occurrence in any particular spot is predicted to be low. The Burrowing Owl is most common where suitable burrows, typically dug by mammals, are available for nesting (Haug et al. 1993). Be-

yond this habitat requirement, atlas data suggest that the owls use a variety of lower-elevation habitat types in roughly equal proportions. Burrowing Owls are fairly tolerant of humans, and often breed in or adjacent to agricultural lands, on golf courses, and along roadsides (Neel 1999a, Dobkin and Sauder 2004), but the future of this ground dweller is questionable in highly disturbed areas.

CONSERVATION AND MANAGEMENT

The Burrowing Owl is considered a species of conservation concern throughout much of the West (Paige and Ritter 1999), and historical population declines have been recorded in many areas, including Utah (Behle et al. 1985), Oregon (Littlefield 1990), and California (Small 1994). Conversion of suitable habitat for human use and reduction of burrowing mammal populations have been identified as causes of declines (Paige and Ritter 1999). More recent data suggest that populations have stabilized and may even be increasing in parts of the West (Dobkin and Sauder 2004, Sauer et al. 2005), though the species' overall range may still be contracting.

Burrowing Owl population trends in Nevada are unclear. Herron et al. (1985) characterized populations as relatively stable, but Alcorn (1988) provided anecdotal accounts of declines in some areas. The Burrowing Owl is a Nevada Partners in Flight Priority species (Neel 1999a) and an Evaluation species for the Clark County Multiple Species Habitat Conservation Plan (Clark County 2000) because of concerns about loss and degradation of habitats due to urbanization. Aside from maintaining suitable breeding habitat, additional monitoring and assessment of potential threats such as motorized off-road recreation have been recommended (Neel 1999a). A detailed conservation plan has been developed for Burrowing Owls (Klute et al. 2003). Fortunately, Burrowing Owls often respond positively to habitat enhancement such as the installation of artificial nest structures (Neel 1999a), and may benefit from local interest in maintaining Burrowing Owl colonies (Haug et al. 1993).

HABITAT USE

HABITAT	ATLAS BLOCKS	INCIDENTAL OBSERVATIONS
Agricultural	3 (10%)	3 (9%)
Grassland	8 (26%)	6 (19%)
Mojave	1 (3%)	0 (0%)
Salt Desert Scrub	6 (19%)	7 (22%)
Sagebrush Scrub	9 (29%)	7 (22%)
Sagebrush Steppe	3 (10%)	9 (28%)
Wetland	1 (3%)	0 (0%)
TOTAL	31	32

Total numbers reported in the Habitat Use, Breeding Status, and Abundance tables may differ from each other (see pp. 22–23 for details). Percentage sums may differ slightly from 100% due to rounding.

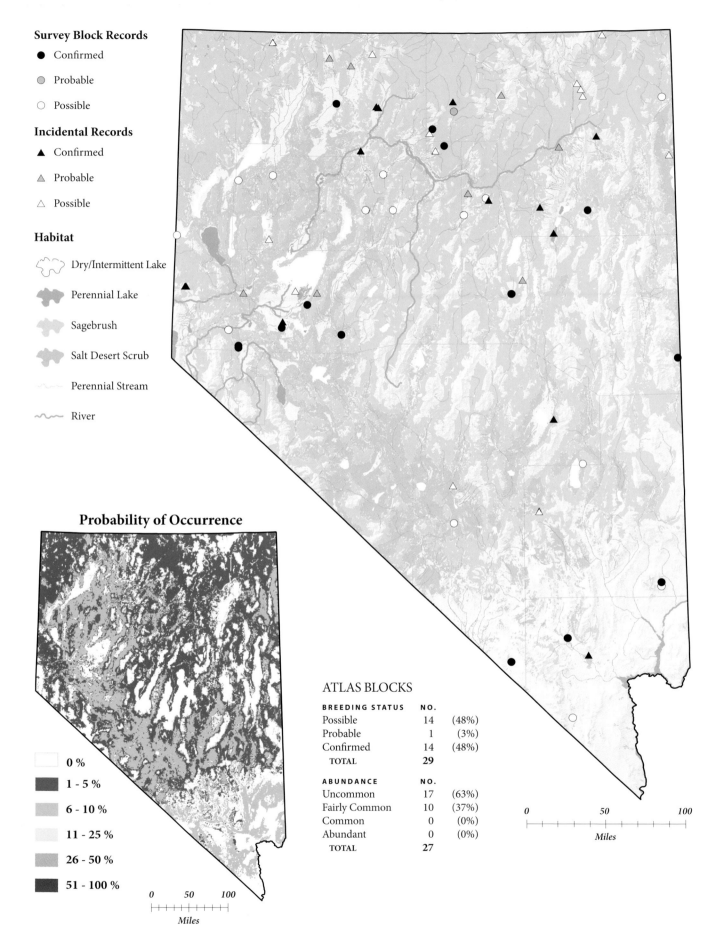

Survey Block Records

- ● Confirmed
- ◐ Probable
- ○ Possible

Incidental Records

- ▲ Confirmed
- △ Probable
- △ Possible

Habitat

- Dry/Intermittent Lake
- Perennial Lake
- Sagebrush
- Salt Desert Scrub
- Perennial Stream
- River

Probability of Occurrence

- 0 %
- 1 - 5 %
- 6 - 10 %
- 11 - 25 %
- 26 - 50 %
- 51 - 100 %

0   50   100
Miles

ATLAS BLOCKS

BREEDING STATUS	NO.	
Possible	14	(48%)
Probable	1	(3%)
Confirmed	14	(48%)
TOTAL	29	

ABUNDANCE	NO.	
Uncommon	17	(63%)
Fairly Common	10	(37%)
Common	0	(0%)
Abundant	0	(0%)
TOTAL	27	

0   50   100
Miles

LONG-EARED OWL
Asio otus

The Long-eared Owl is widespread, fairly common, easily approached, and yet, paradoxically, quite hard to find. It roosts by day in dense vegetation, especially in riparian areas, and hunts by night over grasslands and shrublands. The best time to see a Long-eared Owl is at dusk, when individuals or small groups emerge from streamside thickets to hunt the rolling hills for small mammals. Still evenings in early spring are especially propitious, because the owls are very vocal at this time. No utterance is too weird for a Long-eared Owl on its breeding grounds: hooting, hissing, whistling, and wild wailing sounds are all a part of its repertoire.

DISTRIBUTION

Atlas workers recorded Long-eared Owls at forty-one locations across the state, and a handful of owl enthusiasts supplied many of the records. The species was found widely throughout Nevada, but there were no areas of especially high concentrations. Single pairs were observed in most of the occupied blocks, and breeding was confirmed for about half of the records. Both Linsdale (1936) and Ryser (1985) described the Long-eared as one of Nevada's most common owls.

The Long-eared Owl occurs throughout much of the Northern Hemisphere (Marks et al. 1994) and is common but patchily distributed in much of the West. In adjacent states, it is perhaps most common in Utah (Behle et al. 1985) and least common in California, where its range is restricted mainly to the Sierra Nevada and parts of the coast (Small 1994). In Oregon, Long-eared Owls nest mainly east of the Cascades (Adamus et al. 2001), but in Idaho and Arizona, their breeding range is nearly statewide (Stephens and Sturts 1998, Friederici in Corman and Wise-Gervais 2005:226–227).

Long-eared Owls require a combination of dense, woody vegetation for roosting and open country for hunting (Marks et al. 1994), habitat features that are often juxtaposed at the riparian-upland interface in Nevada. Pinyon-juniper woodlands are also regularly used. Native vegetation is probably not a requirement, though; in southern California, for example, Long-eared Owls can be found in plantings around ranches (Garrett and Dunn 1981), and in Oregon they sometimes nest in Russian olives (Contreras and Kindschy 1996). This array of breeding habitats produces a predictive map quite similar to that of the Great Horned Owl (which often co-occurs with the Long-eared Owl in Nevada), except that the Long-eared Owl is predicted to be completely absent from salt desert scrub.

CONSERVATION AND MANAGEMENT

The status of the Long-eared Owl in North America is not well known. Numbers fluctuate from year to year, and population trends are difficult to determine for this elusive and somewhat nomadic species (Marks et al. 1994). Numbers have apparently declined in California (Shuford 1993, Bloom 1994), especially in the north (McCaskie et al. 1988). Loss of riparian vegetation is detrimental (Marks et al. 1994) because it provides the bulk of the suitable nesting habitat in the arid West. Conversely, plantings in desert environments and around ranches have proven to be acceptable nesting habitat (Garrett and Dunn 1981).

Nevada populations of the Long-eared Owl are considered to be stable based on past assessments (Herron et al. 1985), and the status of this species is not thought to have changed much as the region has been increasingly settled and disturbed (Ryser 1985). Given the paucity of historical baseline data on Long-eared Owls and their riparian woodland habitats, however, it seems prudent to consider this species a worthwhile focus for future work. A high-quality monitoring program for all of Nevada's nocturnal birds would be especially valuable.

HABITAT USE

HABITAT	ATLAS BLOCKS	INCIDENTAL OBSERVATIONS
Agricultural	1 (3%)	0 (0%)
Grassland	1 (3%)	0 (0%)
Mesquite	1 (3%)	0 (0%)
Montane Forest	1 (3%)	0 (0%)
Montane Shrub	2 (5%)	0 (0%)
Pinyon-Juniper	11 (28%)	1 (25%)
Riparian	19 (48%)	3 (75%)
Salt Desert Scrub	1 (3%)	0 (0%)
Sagebrush Scrub	3 (8%)	0 (0%)
TOTAL	40	4

Total numbers reported in the Habitat Use, Breeding Status, and Abundance tables may differ from each other (see pp. 22–23 for details). Percentage sums may differ slightly from 100% due to rounding.

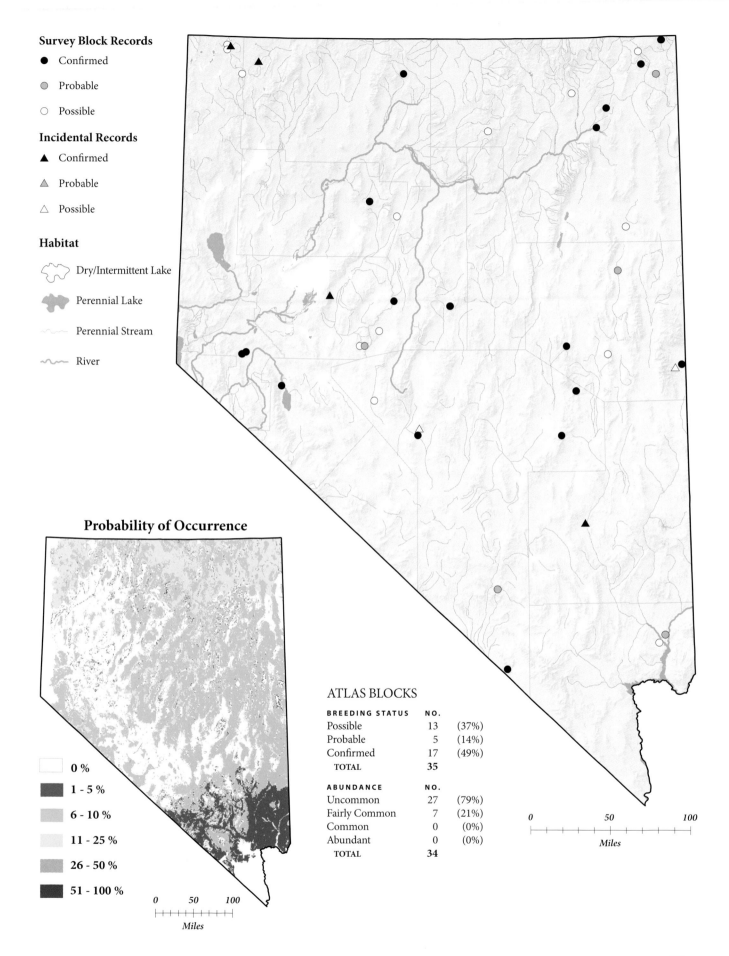

Survey Block Records

● Confirmed

● Probable

○ Possible

Incidental Records

▲ Confirmed

▲ Probable

△ Possible

Habitat

Dry/Intermittent Lake

Perennial Lake

Perennial Stream

River

Probability of Occurrence

☐ 0 %

1 - 5 %

6 - 10 %

11 - 25 %

26 - 50 %

51 - 100 %

0   50   100
Miles

ATLAS BLOCKS

BREEDING STATUS	NO.	
Possible	13	(37%)
Probable	5	(14%)
Confirmed	17	(49%)
TOTAL	35	
ABUNDANCE	NO.	
Uncommon	27	(79%)
Fairly Common	7	(21%)
Common	0	(0%)
Abundant	0	(0%)
TOTAL	34	

0   50   100
Miles

SHORT-EARED OWL
Asio flammeus

For most of the year, this crepuscular counterpart of the Northern Harrier is a no-nonsense bird. It searches diligently for small mammals in shrublands and grasslands, pausing only occasionally to rest on a tussock or haystack. During its spring mating displays, however, the Short-eared Owl is given to flights of fancy that are rivaled by few other birds in its open-country haunts. Territorial males fly to great heights and then plummet to the earth in a wild display of wing-clapping and full-body gyrations.

DISTRIBUTION

Atlas data indicate that Short-eared Owls have a northerly distribution in the state, although breeding occurred as far south as the Pahranagat Valley in Lincoln County. The southern boundary of the species' breeding range passes through Nevada, California (Small 1994), and northern Utah (Holt and Leasure 1993), and atlas data suggest that this limit is a bit farther south in Nevada than is typically portrayed (Holt and Leasure 1993, Sibley 2003). To our north, the Short-eared Owl breeds mainly in the eastern part of Oregon (Adamus et al. 2001) and throughout most of Idaho (Stephens and Sturts 1998). Earlier reports about the Short-eared Owl's distribution in Nevada are generally consistent with the atlas data (Herron et al. 1985, Alcorn 1988).

Short-eared Owls were found in a variety of mostly treeless habitats, from shrublands and grasslands to wetlands, as would be expected of a species that uses open country both for hunting and for nesting (Holt and Leasure 1993). Short-eared Owls are often associated with habitats that support healthy populations of small mammals (Paige and Ritter 1999), particularly voles of

the genus *Microtus,* which are favored prey when available (Holt and Leasure 1993). The predictive map reflects the Short-eared Owl's northern distribution in the state, where there is a moderate chance of occurrence throughout the sagebrush steppe. Areas with low to moderate probability of occurrence predominate in the lower elevations in central Nevada, but in the higher montane woodlands, this owl is predicted to be rare or absent.

CONSERVATION AND MANAGEMENT

Short-eared Owl numbers have declined throughout the species' North American range (Holt and Leasure 1993)—most severely in the Northeast but also in the West (Paige and Ritter 1999, Sauer et al. 2005). Declines have been documented, or suggested, for Utah (Behle et al. 1985) and for various areas of California (e.g., Gaines 1992, Shuford 1993). Habitat loss resulting from heavy livestock grazing, conversion to agricultural areas, recreational development, and wetland destruction has been implicated as the probable cause of the declines (Gaines 1992, Holt and Leasure 1993, Paige and Ritter 1999). The Short-eared Owl sometimes benefits from habitat alterations, though, and seems to respond favorably to waterfowl management programs that encourage the dense shoreline vegetation that supports healthy vole populations (Paige and Ritter 1999).

Whether Short-eared Owls have actually declined in Nevada is not known. Complicating the assessment is the fact that population size can vary considerably from year to year in response to prey populations (Holt and Leasure 1993, Chisholm and Neel 2002). Herron et al. (1985) pointed out that the species has been only cursorily surveyed in Nevada, and that loss of wetlands has probably been detrimental. Ongoing large-scale declines, uncertainty about the Short-eared Owl's status in Nevada, and threats to its breeding habitats have resulted in a Partners in Flight Priority ranking for the species in Nevada (Neel 1999a) and a Watch List designation rangewide (Rich et al. 2004).

HABITAT USE

HABITAT	ATLAS BLOCKS	INCIDENTAL OBSERVATIONS
Agricultural	2 (4%)	1 (5%)
Grassland	11 (23%)	7 (33%)
Montane Shrub	1 (2%)	0 (0%)
Pinyon-Juniper	1 (2%)	0 (0%)
Riparian	6 (13%)	3 (14%)
Salt Desert Scrub	3 (6%)	0 (0%)
Sagebrush Scrub	9 (19%)	5 (24%)
Sagebrush Steppe	6 (13%)	4 (19%)
Urban	1 (2%)	0 (0%)
Wetland	8 (17%)	1 (5%)
TOTAL	48	21

Total numbers reported in the Habitat Use, Breeding Status, and Abundance tables may differ from each other (see pp. 22–23 for details). Percentage sums may differ slightly from 100% due to rounding.

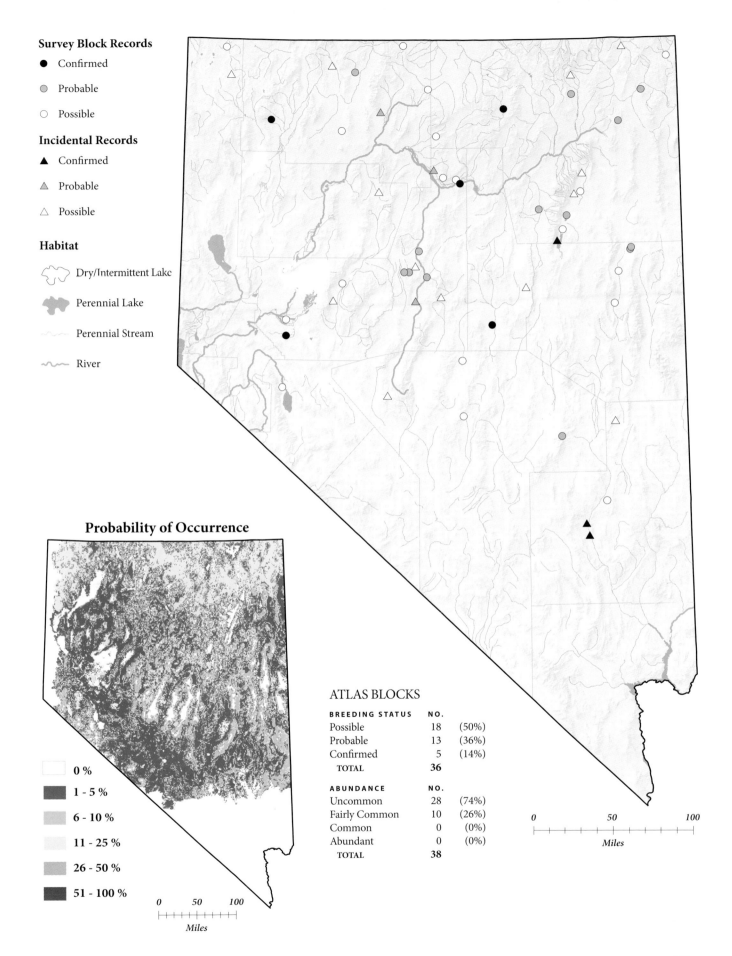

Survey Block Records

● Confirmed

● Probable

○ Possible

Incidental Records

▲ Confirmed

▲ Probable

△ Possible

Habitat

⬡ Dry/Intermittent Lake

⬢ Perennial Lake

〜 Perennial Stream

〜 River

Probability of Occurrence

☐ 0 %

■ 1 - 5 %

■ 6 - 10 %

☐ 11 - 25 %

■ 26 - 50 %

■ 51 - 100 %

0   50   100
Miles

ATLAS BLOCKS

BREEDING STATUS	NO.	
Possible	18	(50%)
Probable	13	(36%)
Confirmed	5	(14%)
TOTAL	36	

ABUNDANCE	NO.	
Uncommon	28	(74%)
Fairly Common	10	(26%)
Common	0	(0%)
Abundant	0	(0%)
TOTAL	38	

0   50   100
Miles

NORTHERN SAW-WHET OWL
Aegolius acadicus

Rh'02

Some birds are named for their appearance, others for their habits, and still others to memorialize naturalists. Many are named for their sounds, but only the Northern Saw-whet Owl is named after the sound of a tool being sharpened! Saw-whet owls are widespread in much of their northerly range (Cannings 1993), but probably not in Nevada. These small, mouse-eating raptors are either absent or undetected in many areas of the state, and they can be hard to find even in ideal habitat—mountain woodlands with plenty of dead trees to provide nesting cavities. Not surprisingly, much of the species' natural history remains little studied.

DISTRIBUTION

Atlas surveys produced only a few records for the Northern Saw-whet Owl: four in atlas blocks and three incidental observations. The records came from several distantly spaced mountain ranges, and breeding was confirmed only at a site in the Jarbidge area. Alcorn (1988) listed observations throughout much of the state, but it is unclear how many of these records represented potential breeders. Immediately prior to the atlas project, in 1995 and 1996, saw-whets were found breeding at several sites in the Toiyabe Range, and earlier in the 1990s they were also found in the White Pine Range and in Great Basin National Park during the nesting season (S. Dunham, pers. comm.).

Both Herron et al. (1985) and Ryser (1985) characterized the Northern Saw-whet Owl as an uncommon and sporadically distributed summer resident in Nevada, but Alcorn (1988) speculated that the species is probably more common than the records imply. This owl breeds more extensively in Oregon, California, Idaho, Utah, and Arizona than in Nevada, but it is still considered relatively uncommon in those states (Behle et al. 1985, Cannings 1993, Small 1994, Stephens and Sturts 1998, Adamus et al. 2001, Wise-Gervais in Corman and Wise-Gervais 2005:228–229).

Saw-whet owls breed in diverse types of forests and woodlands. Coniferous forests appear to be the favored habitat in Nevada, but riparian woodlands are also used, especially in more arid landscapes, where they represent the main source of nesting trees (Herron et al. 1985). In such settings the owls often forage in adjacent open country (Cannings 1993). The limited atlas data support this notion. The predictive map reflects the saw-whet owl's relative scarcity and restriction to higher elevations, but also suggests that there are many potentially suitable areas where breeding has yet to be confirmed.

CONSERVATION AND MANAGEMENT

Much remains to be discovered about the basic requirements of the Northern Saw-whet Owl. Even where it is comparatively well known, such as in California, its status is uncertain (Small 1994). Some speculate that the species is slowly declining because of habitat loss (Cannings 1993), but population trend data are unavailable to substantiate that view. Saw-whet owls are cavity nesters; thus, extensive removal of dead and dying trees is likely to harm the species (Cannings 1993). For example, apparent declines in California have been tentatively linked to logging activity (Shuford 1993).

Because of their small size and nocturnal habits, saw-whet owls have probably gone undetected in many parts of Nevada. Much of what we know about their distribution, habitat requirements, and movements in the state is thus conjectural and incomplete (Herron et al. 1985). Management actions that are probably beneficial include protection of big trees and standing deadwood, where cavities most often occur. Installation of nest boxes has been suggested as a management strategy for some areas (Cannings 1993). This might be worth considering in Nevada, not only to bolster populations, but also to facilitate monitoring.

HABITAT USE

HABITAT	ATLAS BLOCKS	INCIDENTAL OBSERVATIONS
Montane Forest	2 (50%)	2 (40%)
Montane Parkland	0 (0%)	1 (20%)
Pinyon-Juniper	1 (25%)	0 (0%)
Riparian	1 (25%)	2 (40%)
TOTAL	4	5

Total numbers reported in the Habitat Use, Breeding Status, and Abundance tables may differ from each other (see pp. 22–23 for details). Percentage sums may differ slightly from 100% due to rounding.

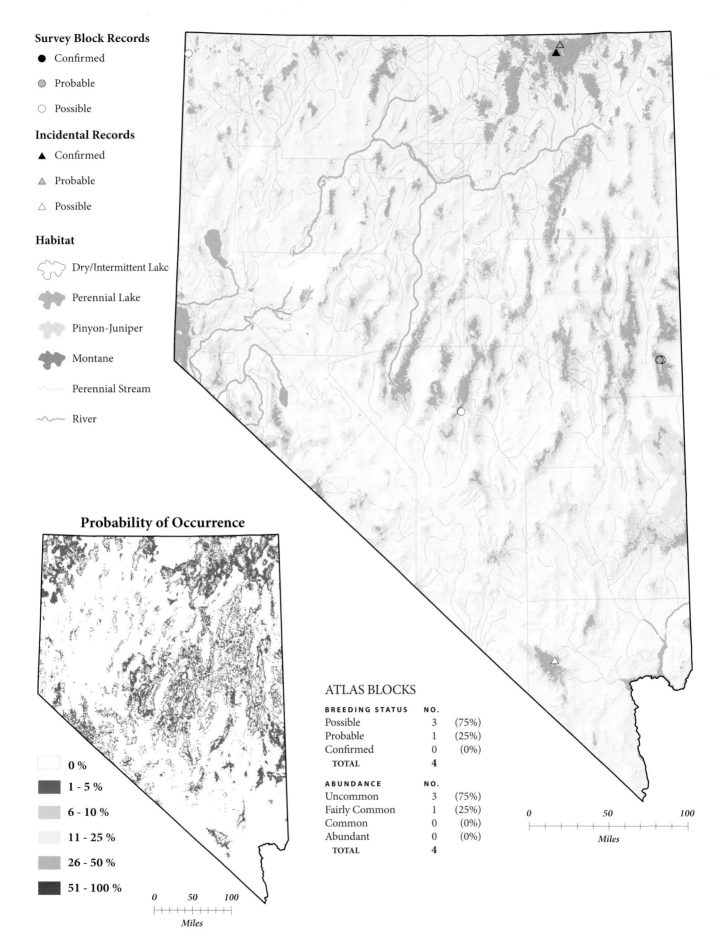

Survey Block Records

● Confirmed

◉ Probable

○ Possible

Incidental Records

▲ Confirmed

△ Probable

△ Possible

Habitat

⬡ Dry/Intermittent Lake

⬡ Perennial Lake

⬡ Pinyon-Juniper

⬡ Montane

〜 Perennial Stream

〜 River

Probability of Occurrence

☐ 0 %

■ 1 - 5 %

■ 6 - 10 %

☐ 11 - 25 %

■ 26 - 50 %

■ 51 - 100 %

0   50   100
Miles

ATLAS BLOCKS

BREEDING STATUS	NO.	
Possible	3	(75%)
Probable	1	(25%)
Confirmed	0	(0%)
TOTAL	4	
ABUNDANCE	NO.	
Uncommon	3	(75%)
Fairly Common	1	(25%)
Common	0	(0%)
Abundant	0	(0%)
TOTAL	4	

0   50   100
Miles

LESSER NIGHTHAWK
Chordeiles acutipennis

The aftermath of an evening thunderstorm in the Mojave Desert is one of Nevada's most entrancing, but fleeting, natural spectacles. Sweet-smelling creosotebush perfumes the air, tarantulas and scorpions emerge from their flooded burrows, kangaroo rats scurry about the desert gravel, and one can almost sense long-dormant seeds springing to life. The real show-stopper, though, is the Lesser Nighthawk, whose maniacal cackling rises from the desert and imparts a rollicking, bawdy quality to the evening.

DISTRIBUTION
Most of the atlas records for the Lesser Nighthawk came from Clark County. There were outlying clusters of records from Pahranagat and Ash Meadows national wildlife refuges in southern Lincoln and Nye counties, and two isolated occurrences of possible breeders farther north in Nye County. Confirmation of breeding was limited to one atlas block and two incidental observations in three widely spaced locations. That the Lesser Nighthawk's breeding distribution in Nevada is limited to the south is well established (Linsdale 1936). Ryser (1985) and Alcorn (1988) also reported occasional sightings farther north, however, in Elko and Churchill counties. Because the species is so similar in appearance to the Common Nighthawk, which inhabits central and northern Nevada, the range limits of (or potential overlap between) the two species are not precisely known. There is some evidence of elevational separation in areas of overlap, with the Lesser Nighthawk occurring at lower elevations (Latta and Baltz 1997).

The Lesser Nighthawk breeds in the arid desert lowlands of southern California, Nevada, Arizona, New Mexico, and Texas, and southward into Central and South America (Rosenberg et al. 1991, Small 1994, Latta and Baltz 1997, Wise-Gervais in Corman and Wise-Gervais 2005:230–231). Northward range extensions have been reported in California's Central Valley and in far southwestern Utah along the Virgin River (Behle et al. 1985, Small 1994), but there is no clear evidence that a similar phenomenon is occurring in Nevada.

About two-thirds of the habitat records from atlas blocks came from distinctly southern habitats, including Mojave, mesquite, and ash. The Lesser Nighthawk is especially partial to creosotebush-dominated lowlands, and its distribution largely matches that of creosotebush (Latta and Baltz 1997). Although Lesser Nighthawks roost and nest on the ground in arid deserts, they tend to concentrate near riparian areas, where there are abundant flying insects to eat. It is thus not surprising that most of the remaining atlas records came from riparian habitat. Far more of the incidental records came from riparian habitat than was true for block records, illustrating the biases that can appear when sites are sampled based on birders' preferences rather than a careful study design. The predictive map shows a moderate probability of finding this species throughout the Mojave Desert, with a higher likelihood in areas of riparian or mesquite habitat, and a diminishing chance toward the boundary with the Great Basin.

CONSERVATION AND MANAGEMENT
The main research need for this species is better information on its natural history and ecology, which are currently poorly understood. Lesser Nighthawks seem to be locally common in most of their range, and there appear to have been no widespread changes in populations (Latta and Baltz 1997, Sauer et al. 2005). Local declines have been reported from urbanized portions of California (Gallagher 1997). Few specific threats have been identified apart from the paving over of the desert (Latta and Baltz 1997). There is currently no conservation concern for the Lesser Nighthawk (Latta and Baltz 1997), and it is probably secure in Nevada at the present time. Parts of its southern Nevada haunts are undergoing substantial changes, however, and periodic monitoring is merited.

HABITAT USE

HABITAT	ATLAS BLOCKS	INCIDENTAL OBSERVATIONS
Agricultural	0 (0%)	2 (10%)
Ash	3 (10%)	0 (0%)
Mesquite	8 (28%)	1 (5%)
Mojave	8 (28%)	3 (15%)
Pinyon-Juniper	1 (3%)	0 (0%)
Riparian	8 (28%)	12 (60%)
Salt Desert Scrub	1 (3%)	1 (5%)
Urban	0 (0%)	1 (5%)
TOTAL	29	20

Total numbers reported in the Habitat Use, Breeding Status, and Abundance tables may differ from each other (see pp. 22–23 for details). Percentage sums may differ slightly from 100% due to rounding.

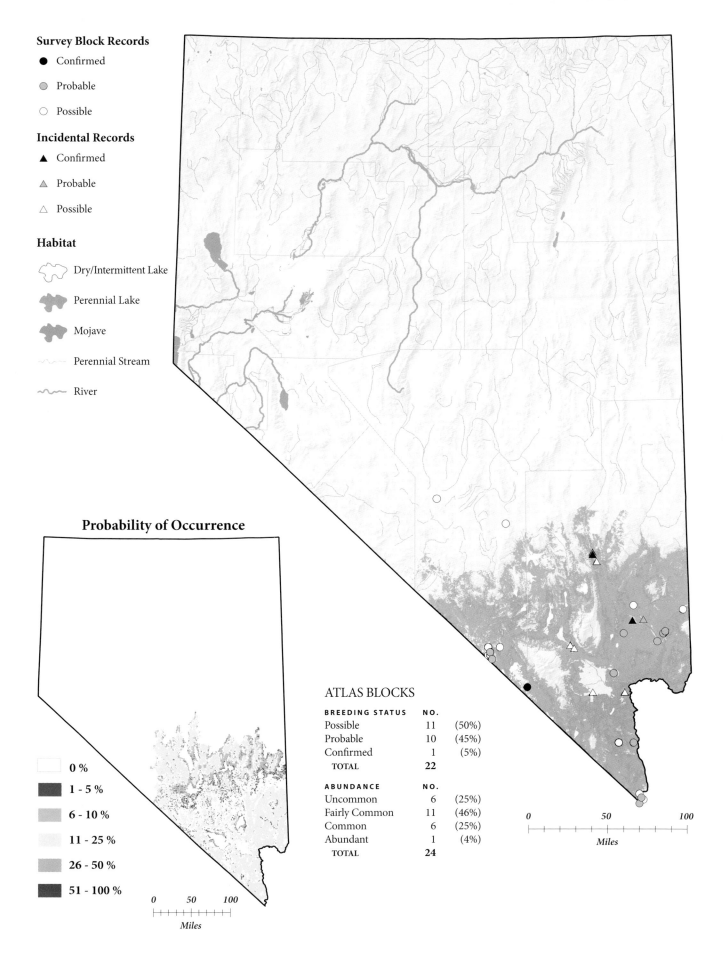

Survey Block Records

● Confirmed

● Probable

○ Possible

Incidental Records

▲ Confirmed

▲ Probable

△ Possible

Habitat

Dry/Intermittent Lake

Perennial Lake

Mojave

Perennial Stream

River

Probability of Occurrence

☐	0 %
■	1 - 5 %
■	6 - 10 %
☐	11 - 25 %
■	26 - 50 %
■	51 - 100 %

0   50   100

Miles

ATLAS BLOCKS

BREEDING STATUS	NO.	
Possible	11	(50%)
Probable	10	(45%)
Confirmed	1	(5%)
TOTAL	22	

ABUNDANCE	NO.	
Uncommon	6	(25%)
Fairly Common	11	(46%)
Common	6	(25%)
Abundant	1	(4%)
TOTAL	24	

0   50   100

Miles

COMMON NIGHTHAWK
Chordeiles minor

RN'02

In Nevada, the Common Nighthawk is truly a bird of summer. This least nocturnal of the state's goatsuckers is not really common in Nevada until Memorial Day, and most individuals depart for their South American winter range well before Labor Day. But during their brief summer stint they are familiar sights throughout the northern two-thirds of the state. They are equally common in lonely sagebrush and juniper landscapes and around the bright lights of ball parks and strip malls. Whatever the venue, their darting pursuit of flying insects and their spectacular dive-and-boom courtship displays are show-stopping performances.

DISTRIBUTION
Atlas workers found Common Nighthawks throughout the portion of the state north of the thirty-eighth parallel. The three southern observations noted on the map probably represent migrants still moving northward. This observed distribution does not differ appreciably from that noted in earlier accounts (Linsdale 1936, Ryser 1985, Alcorn 1988). There is apparently little overlap between the breeding ranges of Common and Lesser Nighthawks in Nevada, but this pattern needs to be confirmed through additional midsummer fieldwork. As was also the case with other goatsuckers, breeding was difficult to confirm during the atlas project.

Common Nighthawks breed throughout the coterminous United States and well into Canada, with the notable exceptions of southern Nevada, coastal and southern California, and southwestern Arizona (Rosenberg et al. 1991, Small 1994, Poulin et al. 1996, Wise-Gervais in Corman and Wise-Gervais 2005:232–233). North and east of Nevada, the species is a common summer resident in Utah, Idaho, and Oregon (Behle et al. 1985, Stephens and Sturts 1998, Adamus et al. 2001).

Common Nighthawks forage in flight over a wide variety of habitats, including open woodlands, barren areas, grasslands, shrublands, farms, and cities. For breeding they require level, cobble-strewn ground on which to lay their eggs; Common Nighthawks sometimes even nest on flat, pebbled rooftops (Poulin et al. 1996). Atlas habitat records were diverse, and the species is predicted to have a moderately high chance of occurring in most of central and northern Nevada. Probabilities are generally higher toward the north, matching the northerly geographic range of the species, and in mid-elevation habitats. The map probably predicts occurrence of this species too far south in Nevada, however, because of the two records from southern Nye County, which may have involved nonbreeders.

CONSERVATION AND MANAGEMENT
Although they are not easy to monitor, Common Nighthawks appear to be suffering sustained declines throughout North America (Poulin et al. 1996). Utah, Idaho, and Arizona seem to be part of this declining trend, but the situation in Nevada is unclear. Increased predation by corvids, increased use of insecticides, habitat loss, collisions with automobiles, and even the decreasing use of pebbled rooftops in favor of smoother coverings have been named as potential factors in the declines (Poulin et al. 1996, Versaw in Kingery 1998:232–233). No single conclusive explanation has emerged, however, and it is possible that the decreasing numbers reflect the cumulative effects of a variety of smaller factors. Possible impacts suffered during migration or wintering have not been investigated in depth. Thus, even though Common Nighthawks remain widespread and fairly common in northern Nevada, we should continue to monitor them carefully.

HABITAT USE

HABITAT	ATLAS BLOCKS	INCIDENTAL OBSERVATIONS
Agricultural	10 (4%)	0 (0%)
Alpine	1 (<1%)	0 (0%)
Barren	3 (1%)	0 (0%)
Grassland	10 (4%)	4 (8%)
Mesquite	1 (<1%)	0 (0%)
Montane Forest	9 (3%)	2 (4%)
Montane Parkland	1 (<1%)	0 (0%)
Montane Shrub	11 (4%)	0 (0%)
Open Water	5 (2%)	2 (4%)
Pinyon-Juniper	49 (18%)	4 (8%)
Riparian	25 (9%)	13 (27%)
Salt Desert Scrub	23 (8%)	0 (0%)
Sagebrush Scrub	48 (17%)	8 (16%)
Sagebrush Steppe	60 (22%)	5 (10%)
Urban	15 (5%)	7 (14%)
Wetland	8 (3%)	4 (8%)
TOTAL	279	49

Total numbers reported in the Habitat Use, Breeding Status, and Abundance tables may differ from each other (see pp. 22–23 for details). Percentage sums may differ slightly from 100% due to rounding.

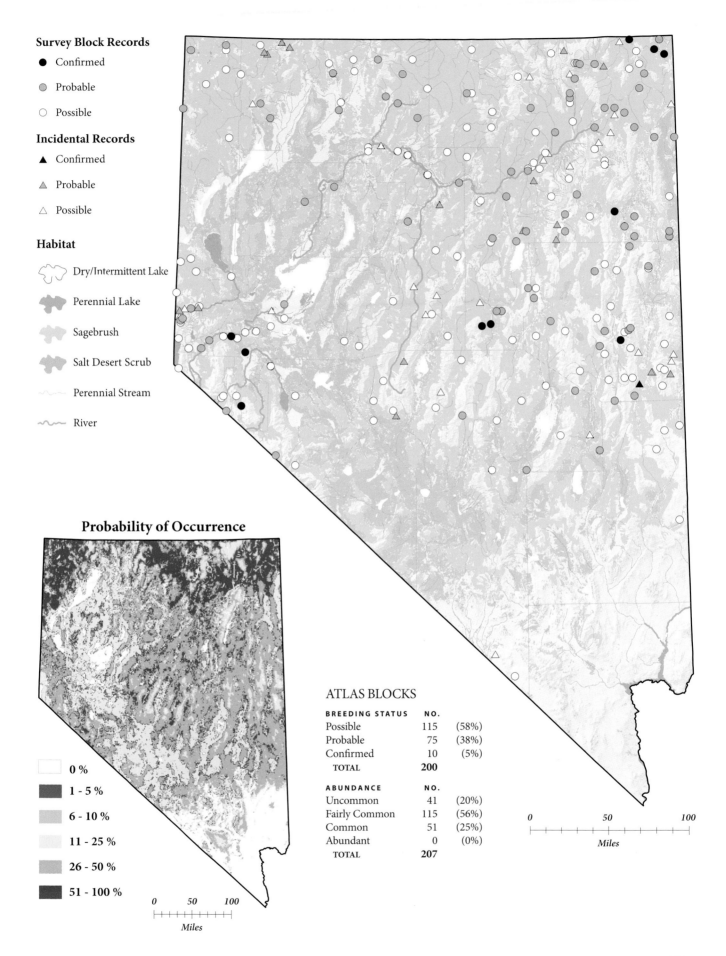

Survey Block Records

● Confirmed

● Probable

○ Possible

Incidental Records

▲ Confirmed

▲ Probable

△ Possible

Habitat

Dry/Intermittent Lake

Perennial Lake

Sagebrush

Salt Desert Scrub

Perennial Stream

River

Probability of Occurrence

☐	0 %
■	1 - 5 %
■	6 - 10 %
■	11 - 25 %
■	26 - 50 %
■	51 - 100 %

0   50   100

Miles

ATLAS BLOCKS

BREEDING STATUS	NO.	
Possible	115	(58%)
Probable	75	(38%)
Confirmed	10	(5%)
TOTAL	**200**	

ABUNDANCE	NO.	
Uncommon	41	(20%)
Fairly Common	115	(56%)
Common	51	(25%)
Abundant	0	(0%)
TOTAL	**207**	

0   50   100

Miles

COMMON POORWILL
Phalaenoptilus nuttallii

The dirt roads of Nevada's backcountry at nightfall offer several predictable sights: cattle guards, bullet-riddled signs, hopping kangaroo rats, and the bright-orange eye shine of Common Poorwills sitting right in the middle of the road. For most birders, the poorwill is only a fleeting glimpse on a dark road or a mournful nighttime voice drifting down from the arid hillsides. Adding to their mystique, Common Poorwills are highly unusual among birds in their ability to lower their body temperature, seemingly at will, and enter a state of torpor that can last for days. But poorwills are not just well adapted to the cold of desert nights; they are also physiologically able to tolerate the blazing heat of the day.

DISTRIBUTION

Atlas workers found Common Poorwills in scattered locations throughout the state, with most of the probable and confirmed breeding records coming from the northeastern quadrant. The species was found to be largely absent from the salt desert flats of west-central Nevada, although it is possible that the combination of inaccessible military land and inhospitable surroundings played a role by reducing the amount of night surveying in these areas. The atlas-generated distribution broadly matches those given in earlier accounts (Linsdale 1936, Alcorn 1988). Breeding was exceptionally difficult to confirm in this species.

The breeding range of the poorwill stretches through most of the western United States, adjacent portions of Mexico, and just into southern Canada. Poorwill occurrences tend to be relatively infrequent and localized in the north and east of its range, where the species is at least partly migratory, and more contiguous in the south, where it is a year-round resident (Woods et al. 2005). In the states surrounding Nevada, poorwills are found throughout much of California but are largely absent from the Central Valley and northern coastal region (Small 1994). They

also breed in eastern Oregon (Adamus et al. 2001), southern Idaho (Stephens and Sturts 1998), and throughout Utah and Arizona (Behle et al. 1985, Averill-Murray in Corman and Wise-Gervais 2005:234–235).

Poorwills nest on bare ground in shrublands, grasslands, and other open areas at low to intermediate elevations. Although they avoid dense forest, they do inhabit open woodlands and forest edges. The breeding habitat is similar to that of the Common Nighthawk, although poorwills tend to nest in areas with denser ground cover (see Woods et al. 2005). Poorwills are predicted to occur throughout most of the state except in the salt desert lowlands of west-central Nevada, and the chance of finding them appears to be highest in the shrub steppe region of northern Nevada.

CONSERVATION AND MANAGEMENT

Within its U.S. range, the Common Poorwill is nowhere considered to be at risk (Woods et al. 2005). Breeding Bird Survey data are of little help in determining the status and population trends of this nocturnal bird (Sauer et al. 2005). The effects on the species of landscape changes due to logging, grazing, and altered fire regimes are poorly understood (see Woods et al. 2005). In Nevada, an opportunity currently exists to examine the possible impacts of these factors on the Common Poorwill and other birds. The extensive pinyon-juniper woodlands of Lincoln County are considered by several land managers to be unnaturally overgrown, and prescribed burns, selective logging, and catastrophic fires are already occurring in these areas (L. Neel, pers. comm.). Poorwill monitoring is difficult, but it would be worth the effort to better understand the effects of land management practices on this low-profile species.

HABITAT USE

HABITAT	ATLAS BLOCKS	INCIDENTAL OBSERVATIONS
Barren	4 (3%)	2 (4%)
Grassland	2 (2%)	0 (0%)
Mojave	7 (5%)	4 (7%)
Montane Forest	12 (9%)	6 (11%)
Montane Parkland	4 (3%)	1 (2%)
Montane Shrub	14 (11%)	7 (12%)
Pinyon-Juniper	25 (20%)	13 (23%)
Riparian	6 (5%)	11 (19%)
Salt Desert Scrub	5 (4%)	0 (0%)
Sagebrush Scrub	15 (12%)	5 (9%)
Sagebrush Steppe	32 (25%)	6 (11%)
Urban	1 (<1%)	0 (0%)
Wetland	1 (<1%)	2 (4%)
TOTAL	128	57

Total numbers reported in the Habitat Use, Breeding Status, and Abundance tables may differ from each other (see pp. 22–23 for details). Percentage sums may differ slightly from 100% due to rounding.

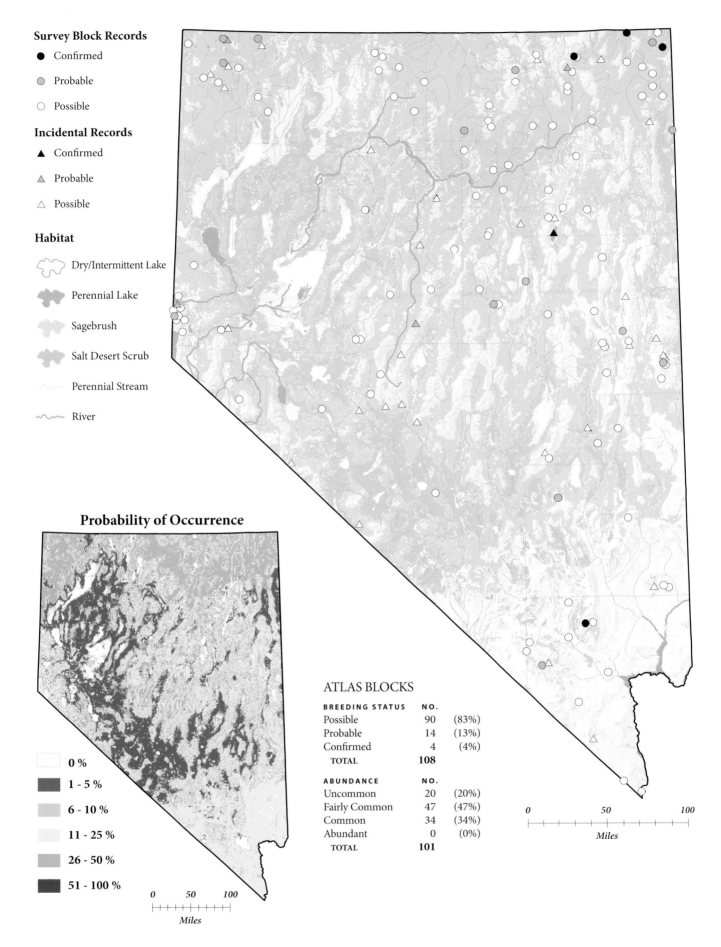

Survey Block Records

● Confirmed

◐ Probable

○ Possible

Incidental Records

▲ Confirmed

△ Probable

△ Possible

Habitat

Dry/Intermittent Lake

Perennial Lake

Sagebrush

Salt Desert Scrub

Perennial Stream

River

Probability of Occurrence

☐ 0 %

■ 1 - 5 %

■ 6 - 10 %

☐ 11 - 25 %

■ 26 - 50 %

■ 51 - 100 %

0   50   100
Miles

ATLAS BLOCKS

BREEDING STATUS	NO.	
Possible	90	(83%)
Probable	14	(13%)
Confirmed	4	(4%)
TOTAL	**108**	

ABUNDANCE	NO.	
Uncommon	20	(20%)
Fairly Common	47	(47%)
Common	34	(34%)
Abundant	0	(0%)
TOTAL	**101**	

0   50   100
Miles

WHIP-POOR-WILL
Caprimulgus vociferus

One of the many allures of birding in the montane "islands" of southern Nevada is the possibility of finding a bird that occurs nowhere else in the state. The Whip-poor-will, with its small breeding population in the Sheep Range, is a perfect example. Multiple birds are also seen from time to time in the Spring Mountains, and it seems possible that they breed there as well. In any event, the species clearly is at the limit of its distribution in southern Nevada, and its continued presence here is by no means guaranteed.

DISTRIBUTION

The Whip-poor-will is primarily an eastern species. The western populations, which breed from Mexico into the Southwest, are disjunct from the more extensive eastern populations and represent a different suite of subspecies (Cink 2002). The only atlas records came from the Sheep Range of northern Clark County. Multiple pairs were found in one atlas block over two years, and there was one incidental observation from another site within this mountain range. In the West, the Whip-poor-will appears to have expanded its range northward and westward during the last decades of the twentieth century, and the Nevada population may be a result of this range extension. Linsdale's (1936) account of Nevada birds does not mention the species. There were numerous records by the time Alcorn (1988) published *The Birds of Nevada,* however, starting with the discovery of Whip-poor-wills in the Sheep Range in the 1960s (Johnson 1965).

Nevada's Whip-poor-wills belong to the distinctive *arizonae* subspecies, which some authors think merits full species status (Cink 2002). This subspecies also breeds in southeastern Arizona, southwestern New Mexico, and western Texas (AOU 1998, Cink 2002, Corman in Corman and Wise-Gervais 2005:238–239). Whip-poor-wills were first recorded in southern California in the 1960s, but it has been hard to confirm breeding in that state despite records in several mountain ranges (Small 1994, Cink 2002).

Western Whip-poor-wills are crepuscular inhabitants of dense montane woodlands, where they nest on plant litter on the ground. Cink (2002) argued that a relatively open understory is more important to the birds than specific forest types. In Arizona, however, Whip-poor-wills reportedly prefer to nest beneath dense understory shrubs (Corman in Corman and Wise-Gervais 2005:238–239). The species' two known Nevada haunts in the Sheep Range and Spring Mountains are characterized by a fairly mesic and diverse forest community. According to the predictive model, these barely discernable spots are the only areas in the state where Whip-poor-wills are likely to occur.

CONSERVATION AND MANAGEMENT

Whip-poor-wills are poorly studied and difficult to survey by virtue of their nocturnal habits. Eastern populations appear to have been steadily decreasing in recent decades (Sauer et al. 2005). Many explanations have been invoked—including forest maturation following earlier clearing, expansion of agriculture, sheep grazing, urbanization, and changes in the insect prey base—but actual demonstrations of these effects are known only from small regions (Cink 2002).

The *arizonae* subspecies appears to be expanding northward and westward (Small 1994, Cink 2002). This expansion is recent, and the Whip-poor-will's numbers are low in recently colonized areas (e.g., Garrett and Dunn 1981). The species appears to be barely established in Nevada, and its numbers probably fluctuate from year to year. Local expansions have also been noted in the Whip-poor-will's historical Arizona range (Corman in Corman and Wise-Gervais 2005:238–239).

It is interesting to contemplate the future of this species in Nevada. Is the Whip-poor-will's foray into our state temporary, or is it the beginning of a well-established breeding presence? Will the resounding call of the Whip-poor-will continue to grace the evenings in our southern mountains?

HABITAT USE

HABITAT	ATLAS BLOCKS	INCIDENTAL OBSERVATIONS
Montane Forest	1 (25%)	1 (100%)
Montane Parkland	1 (25%)	0 (0%)
Pinyon-Juniper	1 (25%)	0 (0%)
Sagebrush Scrub	1 (25%)	0 (0%)
TOTAL	4	1

Total numbers reported in the Habitat Use, Breeding Status, and Abundance tables may differ from each other (see pp. 22–23 for details). Percentage sums may differ slightly from 100% due to rounding.

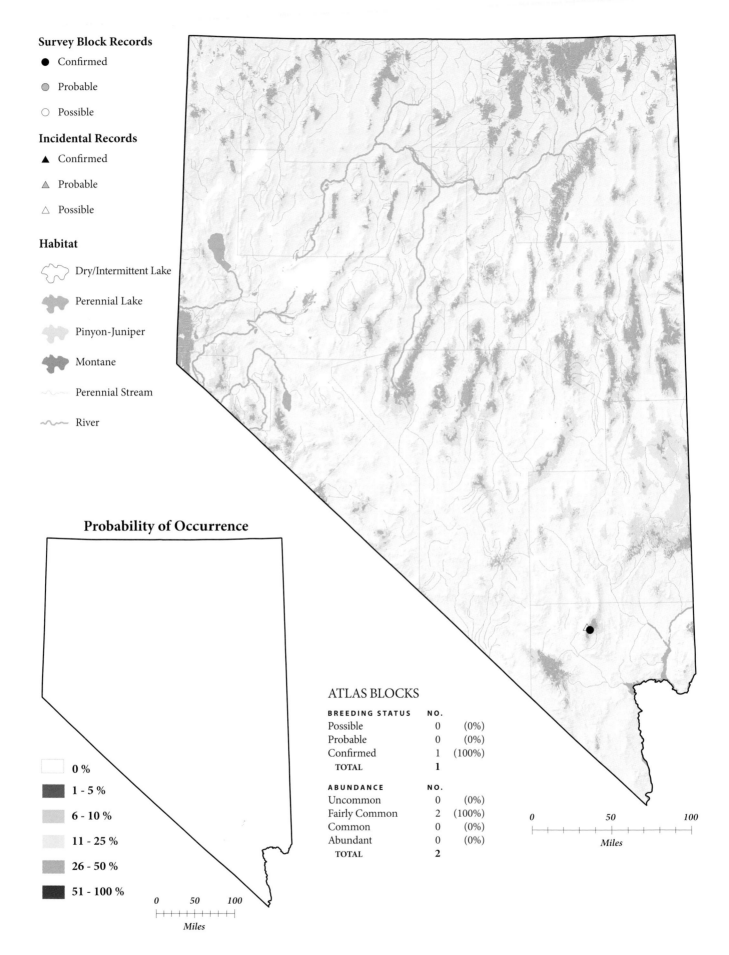

Survey Block Records

● Confirmed

● Probable

○ Possible

Incidental Records

▲ Confirmed

▲ Probable

△ Possible

Habitat

Dry/Intermittent Lake

Perennial Lake

Pinyon-Juniper

Montane

Perennial Stream

River

Probability of Occurrence

	0 %
	1 - 5 %
	6 - 10 %
	11 - 25 %
	26 - 50 %
	51 - 100 %

0   50   100
Miles

ATLAS BLOCKS

BREEDING STATUS	NO.	
Possible	0	(0%)
Probable	0	(0%)
Confirmed	1	(100%)
TOTAL	1	

ABUNDANCE	NO.	
Uncommon	0	(0%)
Fairly Common	2	(100%)
Common	0	(0%)
Abundant	0	(0%)
TOTAL	2	

0   50   100
Miles

WHITE-THROATED SWIFT
Aeronautes saxatalis

A bird of steep cliffs and lonely rimrocks, the fast-flying White-throated Swift is frequently encountered but poorly known. Foraging individuals are observed often, but only a handful of birders have ever seen these birds at the nest. White-throated Swifts typically nest in rock crevices, but in some areas they have also begun to use gaps in buildings and in other artificial structures. It will be interesting to see if White-throated Swifts make inroads into Nevada's urban areas in the years to come.

DISTRIBUTION

Although White-throated Swifts were reported throughout the state, there was a decided concentration of records in the southern and eastern counties, especially Clark County, where the swifts reside year-round. Abundance estimates were relatively high, but breeding was rarely confirmed—not surprising, given the wide-ranging foraging flights and cryptic nesting habits of this species. White-throated Swifts were common in the desert valleys and mountains of southern Nevada at the turn of the century (Fisher 1893), and Linsdale (1936) and Alcorn (1988) added numerous records from other parts of the state in subsequent years.

White-throated Swifts breed widely in the western United States and Mexico. In California, there is a summer population in the Sierra Nevada, and resident populations occur along the coast and in the south (Small 1994, Ryan and Collins 2000), although only in the winter along the lower Colorado River (Rosenberg et al. 1991). White-throated Swifts breed locally in eastern Oregon (Adamus et al. 2001) and widely throughout Idaho, Utah, and Arizona (Behle et al. 1985, Stephens and Sturts 1998, Ryan and Collins 2000, Wise-Gervais in Corman and Wise-Gervais 2005:240–241).

During the breeding season, White-throated Swifts can be found virtually anywhere within foraging range of appropriate nesting areas, which include cliffs, canyons, rock outcrops, and talus fields from low deserts to the alpine zone. Suitable sites are often used for decades or longer (Ryan and Collins 2000). Given the abundance of bare-rock outcroppings in the Mojave Desert, it is not surprising that the species was reported from Mojave habitat more than any other habitat type. In general, the chance of finding White-throated Swifts is predicted to be higher in the Mojave Desert than in the Great Basin, but the greatest chance is expected on the higher-elevation slopes of the state's mountain ranges. Presumably, this pattern indicates the distribution of suitable rocky habitat for nesting.

CONSERVATION AND MANAGEMENT

White-throated Swift populations appear to be stable overall, but local decreases and increases have occurred (Shuford 1993, Ryan and Collins 2000, Sauer et al. 2005). It is difficult to quantify population trends because of the tremendous distances foraging birds cover and because breeding colonies are patchily distributed. Potential threats, ranging from the effects of pesticides on the swifts' insect prey to rock climbing, are mostly speculative, and much remains unknown about this bird's biology (Ryan and Collins 2000). Because the species has a small estimated population size globally, and because of its highly specific nesting habitat requirements, Partners in Flight has designated it a Watch List species across its range (Rich et al. 2004).

In Nevada, populations appear to be stable, although data are very limited. Observers should be on the lookout for increased nesting in urban areas, as has occurred in California. In particular, the city of Las Vegas seems a propitious site for the White-throated Swift. The species is already widespread throughout the surrounding deserts and hills of Clark County, and given the pace of urban development in the Las Vegas Valley, swifts may soon explore alternative breeding sites in the cities.

HABITAT USE

HABITAT	ATLAS BLOCKS	INCIDENTAL OBSERVATIONS
Alpine	6 (10%)	0 (0%)
Ash	1 (2%)	0 (0%)
Barren	6 (10%)	4 (13%)
Mesquite	1 (2%)	0 (0%)
Mojave	13 (21%)	8 (26%)
Montane Forest	3 (5%)	2 (6%)
Montane Parkland	3 (5%)	0 (0%)
Montane Shrub	3 (5%)	2 (6%)
Pinyon-Juniper	6 (10%)	1 (3%)
Riparian	8 (13%)	9 (29%)
Sagebrush Scrub	8 (13%)	1 (3%)
Sagebrush Steppe	5 (8%)	0 (0%)
Urban	0 (0%)	2 (6%)
Wetland	0 (0%)	2 (6%)
TOTAL	63	31

Total numbers reported in the Habitat Use, Breeding Status, and Abundance tables may differ from each other (see pp. 22–23 for details). Percentage sums may differ slightly from 100% due to rounding.

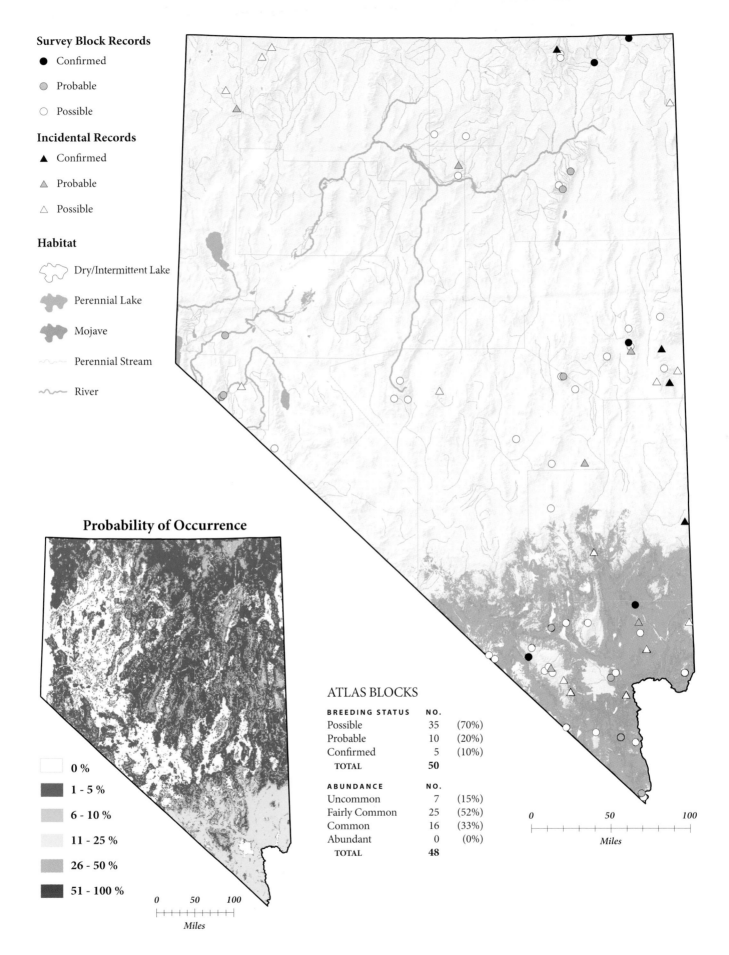

Survey Block Records

● Confirmed

◐ Probable

○ Possible

Incidental Records

▲ Confirmed

△ Probable

△ Possible

Habitat

Dry/Intermittent Lake

Perennial Lake

Mojave

Perennial Stream

River

Probability of Occurrence

0 %

1 - 5 %

6 - 10 %

11 - 25 %

26 - 50 %

51 - 100 %

0   50   100
Miles

ATLAS BLOCKS

BREEDING STATUS	NO.	
Possible	35	(70%)
Probable	10	(20%)
Confirmed	5	(10%)
TOTAL	**50**	

ABUNDANCE	NO.	
Uncommon	7	(15%)
Fairly Common	25	(52%)
Common	16	(33%)
Abundant	0	(0%)
TOTAL	**48**	

0   50   100
Miles

BLACK-CHINNED HUMMINGBIRD

Archilochus alexandri

The Black-chinned Hummingbird is the most common breeding hummingbird in most Nevada towns and cities. But it is not just a backyard bird; it is noted for its widespread abundance and generalized habitat use. In the Great Basin, Black-chinned Hummingbirds are usually found at lower elevations than are Broad-tailed and Calliope Hummingbirds. And in the lowland deserts of the south, the Black-chinned is a less common breeder than Costa's Hummingbird, and maybe even Anna's Hummingbird, except in the vicinity of Las Vegas.

DISTRIBUTION

Atlas workers found the Black-chinned Hummingbird to be widely but unevenly distributed throughout the state. Breeding confirmations were relatively few—as was true for hummingbirds generally—particularly in northern Nevada. Despite the lack of confirmed records, breeding was considered very likely to have occurred in some of these areas, especially around Reno. Earlier accounts were of little help in pinpointing breeding locations but did confirm the scattered summer presence of the species across the state (Linsdale 1936, Ryser 1985, Alcorn 1988).

The breeding range of the Black-chinned Hummingbird stretches from the interior Pacific Northwest through the Intermountain West to northeastern Mexico (Baltosser and Russell 2000). Breeding is scattered in eastern Oregon and more regular throughout Idaho and Utah (Behle et al. 1985, Stephens and Sturts 1998, Baltosser and Russell 2000, Adamus et al. 2001). Black-chinned Hummingbirds are especially widespread and abundant in Arizona, including the lower Colorado River region (Rosenberg et al. 1991, Wise-Gervais in Corman and Wise-Gervais 2005:256–257). In California, Black-chinned Hummingbirds breed along the state's eastern border, but breeding is more frequent in the Central Valley and along the southern coast (Small 1994).

Away from urban areas, Black-chinned Hummingbirds use many different types of woodlands, particularly mesic ones (Bal-

tosser and Russell 2000). In Nevada, riparian and urban records clearly predominated among atlas records. Many riparian records were from smaller streams, and the species was notably rare in the Humboldt River system. In the Lahontan Valley, Black-chinned Hummingbirds seemed to avoid the exotic vegetation that has invaded many riparian areas (Chisholm and Neel 2002), but conversely, breeding was common even in tamarisk stands along the lower Colorado River (Rosenberg et al. 1991). The predictive map shows the species' tendency to occur in urban areas and in the foothills of mountain ranges, with much lower probabilities both on the valley floors and at higher elevations.

CONSERVATION AND MANAGEMENT

Black-chinned Hummingbirds are thought to have expanded their range and increased in number in recent years. Evidence for these trends consists primarily of comparisons between historical and modern observations rather than standard monitoring schemes such as the Breeding Bird Survey (Baltosser and Russell 2000). Such comparative data, however, may be skewed by an increasing number of knowledgeable birders who can reliably identify hummingbirds by species. Certainly Black-chinneds have capitalized on the proliferation of residential landscaping and artificial feeders, and these factors are likely responsible in part for reported increases in range and abundance. Factors affecting populations in native habitats are poorly understood, as are the effects of the increasing urbanization of hummingbird populations on interspecific competition.

HABITAT USE

HABITAT	ATLAS BLOCKS	INCIDENTAL OBSERVATIONS
Agricultural	1 (2%)	2 (5%)
Grassland	1 (2%)	0 (0%)
Mesquite	2 (4%)	1 (3%)
Mojave	1 (2%)	5 (13%)
Montane Forest	7 (14%)	2 (5%)
Montane Parkland	1 (2%)	0 (0%)
Montane Shrub	3 (6%)	1 (3%)
Pinyon-Juniper	7 (14%)	2 (5%)
Riparian	12 (24%)	16 (41%)
Salt Desert Scrub	1 (2%)	1 (3%)
Sagebrush Scrub	3 (6%)	1 (3%)
Sagebrush Steppe	1 (2%)	0 (0%)
Urban	10 (20%)	7 (18%)
Wetland	0 (0%)	1 (3%)
TOTAL	50	39

Total numbers reported in the Habitat Use, Breeding Status, and Abundance tables may differ from each other (see pp. 22–23 for details). Percentage sums may differ slightly from 100% due to rounding.

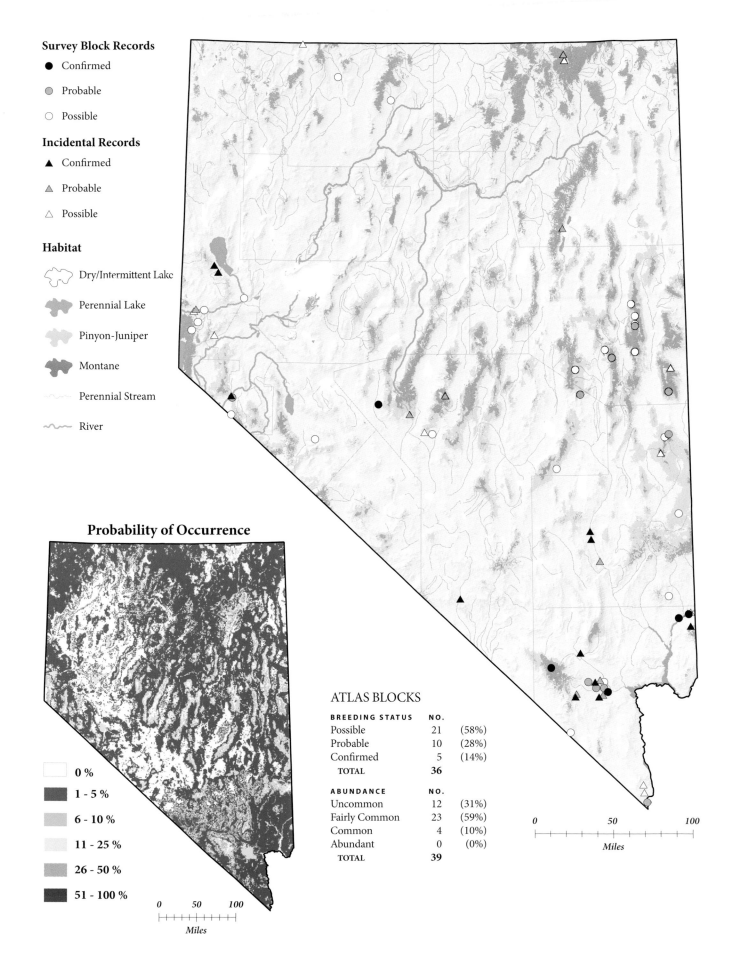

Survey Block Records
● Confirmed
◉ Probable
○ Possible

Incidental Records
▲ Confirmed
▲ Probable
△ Possible

Habitat
Dry/Intermittent Lake
Perennial Lake
Pinyon-Juniper
Montane
Perennial Stream
River

Probability of Occurrence

0 %
1 - 5 %
6 - 10 %
11 - 25 %
26 - 50 %
51 - 100 %

0   50   100
Miles

ATLAS BLOCKS

BREEDING STATUS	NO.	
Possible	21	(58%)
Probable	10	(28%)
Confirmed	5	(14%)
TOTAL	36	

ABUNDANCE	NO.	
Uncommon	12	(31%)
Fairly Common	23	(59%)
Common	4	(10%)
Abundant	0	(0%)
TOTAL	39	

0   50   100
Miles

ANNA'S HUMMINGBIRD
Calypte anna

Once rare in the state, this bejeweled pixie is now a fairly common inhabitant of southern Nevada. There are also occasional reports of individuals farther north, especially in the Reno area, where breeding has not yet been confirmed. Anna's Hummingbird often flourishes in human-altered landscapes, especially in residential districts that have exotic flowering plants and hummingbird feeders.

DISTRIBUTION

Most atlas records for Anna's Hummingbird came from Clark County, and the bird's predicted breeding occurrence is limited to that region. Breeding was confirmed in habitats as disparate as the shady pine forests of the Spring Mountains and the trailer parks of Laughlin. Anna's Hummingbirds were also observed in Reno's Oxbow Park during the atlas years. These birds were assumed at the time to be nonbreeders and thus are not mapped here. Although territorial behavior was noted, it was thought to involve food defense. Intriguingly, however, just after the atlas project was completed, a juvenile Anna's Hummingbird was observed at a feeder in Reno (R. Rovansek, pers. comm.).

The residency of Anna's Hummingbirds in Nevada is clearly recent. Linsdale (1936) did not include the species among the state's residents. Even Russell (1996) did not include it as a breeder in the state, although Alcorn (1988) had previously reported multiple observations from Clark County and the Reno area. Anna's Hummingbird's historical breeding range was probably limited to the western half of southern California and northern Baja California (Russell 1996). By the 1960s, the species had expanded north along the coast to southern British Columbia (Small 1994, Russell 1996, Adamus et al. 2001). Breeding was also noted in Arizona by the 1960s (Corman in Corman and Wise-Gervais 2005:258–259). Breeders first appeared in Nevada in the 1980s, but the species has only recently been characterized as an established breeder in the state (see Sibley 2003, Titus 2003). Anna's Hummingbirds have an unusual nonbreeding range that extends southward, northward, and inland from the nesting grounds, perhaps setting the stage for further breeding range extensions. Floyd (2001), for instance, reported that Anna's Hummingbirds are beginning to appear in Utah.

Historically, native chaparral was the nesting habitat of Anna's Hummingbird in California. The species' range expansions, however, are most likely driven by the increasing availability of exotic flowering plants and hummingbird feeders (Russell 1996), which the birds now frequent throughout their range (Garrett and Dunn 1981). Once a breeding population is established in a new area, Anna's Hummingbirds readily make use of adjacent native habitats as well as residential areas. Interestingly, the atlas records from southern Nevada were primarily from native habitats, with only two urban records. The predictive map shows the highest likelihood of occurrence in the more mesic portions of southern Nevada. Curiously, the species is predicted to be absent from the Las Vegas area, in contrast to its reported use of urban areas elsewhere.

CONSERVATION AND MANAGEMENT

Anna's Hummingbirds will probably continue to expand their range and prosper, even in heavily developed areas of California (Shuford 1993, Gallagher 1997) and rapidly urbanizing areas of Arizona (Rosenberg et al. 1991). Despite the difficulty of counting hummingbirds accurately, Breeding Bird Survey data show clear evidence of continuing population increases (Sauer et al. 2005).

Anna's Hummingbird deserves additional study in Nevada to determine whether breeding does actually occur in the Reno–Lake Tahoe area. Ryser (1985) attributed Reno sightings to post-breeding movements of birds from central California, but this is unconfirmed. Another interesting question is whether Anna's breed in southern Nevada in winter, as they do in southern California (Gallagher 1997). Future monitoring efforts should watch for further expansion of populations as southern desert areas continue to be developed.

HABITAT USE

HABITAT	ATLAS BLOCKS	INCIDENTAL OBSERVATIONS
Mesquite	2 (18%)	0 (0%)
Mojave	4 (36%)	0 (0%)
Montane Forest	1 (9%)	0 (0%)
Pinyon-Juniper	0 (0%)	2 (67%)
Riparian	2 (18%)	0 (0%)
Salt Desert Scrub	1 (9%)	0 (0%)
Urban	1 (9%)	1 (33%)
TOTAL	11	3

Total numbers reported in the Habitat Use, Breeding Status, and Abundance tables may differ from each other (see pp. 22–23 for details). Percentage sums may differ slightly from 100% due to rounding.

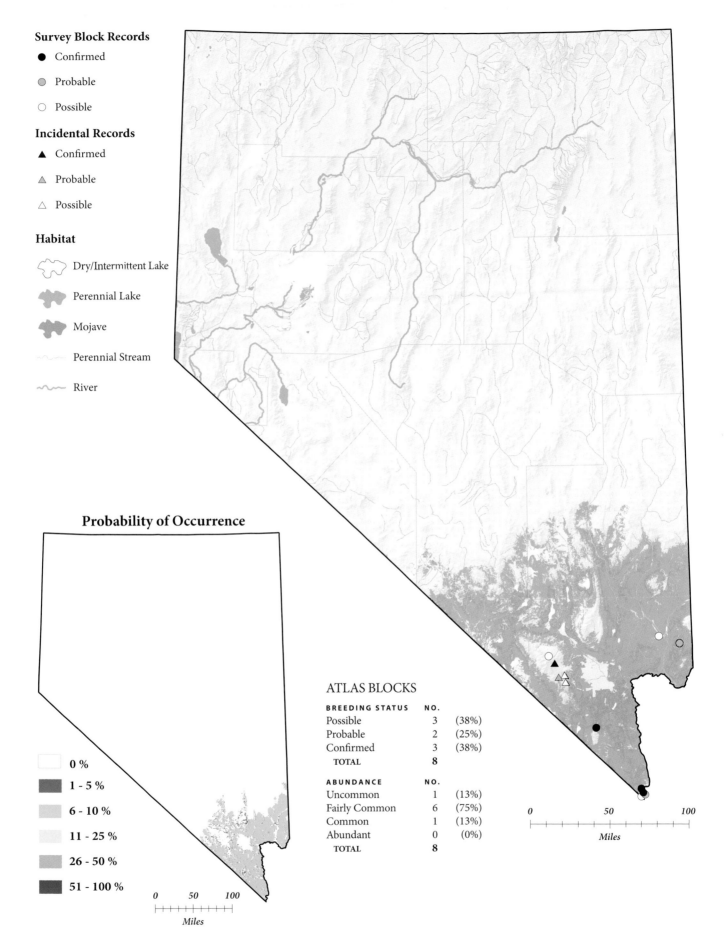

Survey Block Records

● Confirmed

◐ Probable

○ Possible

Incidental Records

▲ Confirmed

△ Probable

△ Possible

Habitat

Dry/Intermittent Lake

Perennial Lake

Mojave

Perennial Stream

River

Probability of Occurrence

☐ 0 %

■ 1 - 5 %

■ 6 - 10 %

☐ 11 - 25 %

■ 26 - 50 %

■ 51 - 100 %

0   50   100
Miles

ATLAS BLOCKS

BREEDING STATUS	NO.	
Possible	3	(38%)
Probable	2	(25%)
Confirmed	3	(38%)
TOTAL	**8**	

ABUNDANCE	NO.	
Uncommon	1	(13%)
Fairly Common	6	(75%)
Common	1	(13%)
Abundant	0	(0%)
TOTAL	**8**	

0   50   100
Miles

COSTA'S HUMMINGBIRD
Calypte costae

In the desert washes of southern Nevada, Costa's Hummingbirds flit among flowering plants such as creosotebush and desert willow. If the light is just right, the male's purple gorget glistens like amethyst. Costa's Hummingbird is truly a bird of the desert, with far less of a presence in residential neighborhoods and montane habitats than most other hummingbirds. Though its range may be expanding northward in Nevada, the species is declining in its core range in the Mojave and Sonoran deserts.

DISTRIBUTION

Atlas workers found Costa's Hummingbirds throughout southern Nevada, with scattered records along the western edge of the state all the way to Pyramid Lake. The species' early breeding season and tendency to nest in extensive arid scrublands probably explain why breeding was confirmed in only six locations. Atlas results suggest that Costa's Hummingbirds are more widespread and somewhat more common in Nevada than was previously reported. Linsdale (1936) and Alcorn (1988) regarded this hummingbird as uncommon in the south and provided few records from farther north. In recent years, however, Costa's Hummingbirds have been reported with greater frequency in the northern part of the state, even becoming somewhat regular near Pyramid Lake (Chisholm and Neel 2002).

The breeding range extends from southern Nevada and adjacent portions of California and Utah, down the lower Colorado River Valley into Baja California, and into southwestern Arizona and northwestern Mexico (Behle et al. 1985, Rosenberg et al. 1991, Small 1994, Baltosser and Scott 1996, Corman in Corman and Wise-Gervais 2005:260–261). Costa's Hummingbird is noted for vagrant movements northward, and may be expanding its range north in California (Small 1994). It is also reported from the Pacific Northwest, and has attempted nesting in Oregon (Contreras 1999).

Costa's Hummingbirds are most often found in arid environments that receive winter rain and produce early-season flowers (Baltosser and Scott 1996). They occur in desert washes and around Joshua trees, but also use springs and riparian vegetation (Rosenberg et al. 1991). The predictive map shows a good chance of finding them in areas with Mojave habitat, where Costa's is the most likely breeding hummingbird. Farther north the probability is much lower, and—based on actual records—the model may overestimate the area over which Costa's Hummingbirds might occur in east-central Nevada.

CONSERVATION AND MANAGEMENT

Significant concern exists for Costa's Hummingbird populations in their core range in the Sonoran and Mojave deserts. Monitoring data are limited, but they suggest declining populations (Sauer et al. 2005). Historically, the loss of coastal scrub habitat has probably also caused declines. The main current threat to the species is the degradation and loss of native desert shrub communities and native flowering plants (Baltosser and Scott 1996). In contrast to some other hummingbirds, Costa's has not expanded appreciably into residential areas that feature exotic flowering plants (Baltosser and Scott 1996), as indicated both by the low number of atlas block records from urban habitats and by the model's prediction that Costa's Hummingbirds are very unlikely to nest in urban Las Vegas. Because of its limited geographic distribution and population declines, Costa's Hummingbird is a Partners in Flight Watch List species (Rich et al. 2004).

Costa's Hummingbird deserves further study to determine current trends in Nevada. Interactions and possible hybridization with the recently arrived Anna's Hummingbird offer other interesting lines of research (Baltosser and Scott 1996, Gallagher 1997), as do an apparent range expansion into northern Nevada and possible occurrences in central Nevada.

HABITAT USE

HABITAT	ATLAS BLOCKS	INCIDENTAL OBSERVATIONS
Ash	3 (8%)	0 (0%)
Grassland	1 (3%)	0 (0%)
Mesquite	5 (14%)	0 (0%)
Mojave	16 (44%)	6 (43%)
Open Water	0 (0%)	1 (7%)
Pinyon-Juniper	2 (6%)	0 (0%)
Riparian	4 (11%)	2 (14%)
Salt Desert Scrub	2 (6%)	0 (0%)
Sagebrush Scrub	3 (8%)	0 (0%)
Urban	0 (0%)	5 (36%)
TOTAL	36	14

Total numbers reported in the Habitat Use, Breeding Status, and Abundance tables may differ from each other (see pp. 22–23 for details). Percentage sums may differ slightly from 100% due to rounding.

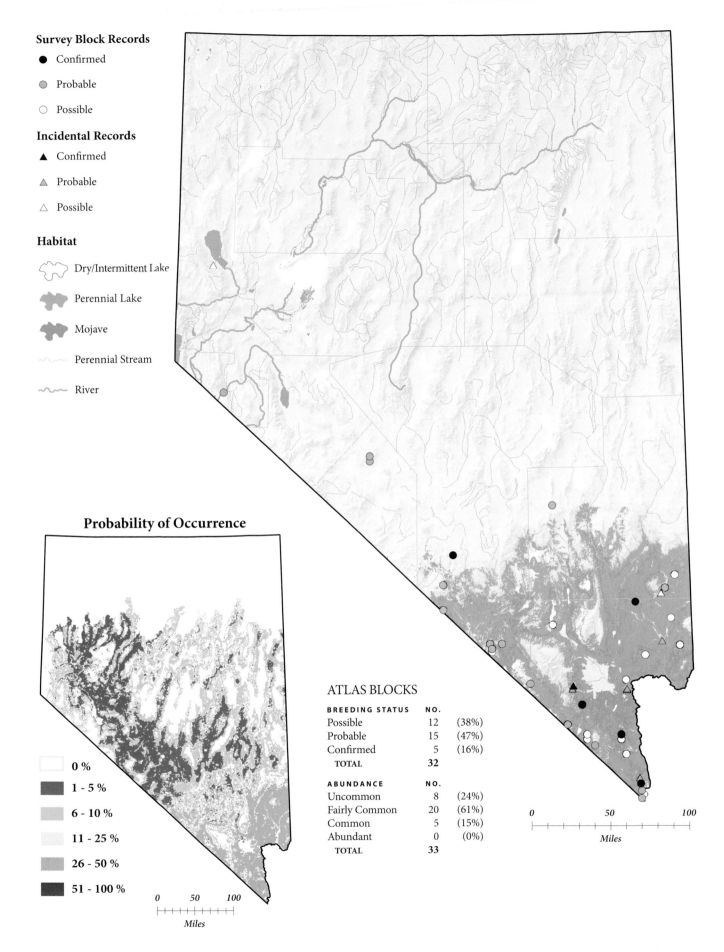

Survey Block Records

- ● Confirmed
- ◐ Probable
- ○ Possible

Incidental Records

- ▲ Confirmed
- ◭ Probable
- △ Possible

Habitat

- Dry/Intermittent Lake
- Perennial Lake
- Mojave
- Perennial Stream
- River

Probability of Occurrence

- ☐ 0 %
- 1 - 5 %
- 6 - 10 %
- 11 - 25 %
- 26 - 50 %
- 51 - 100 %

0   50   100
Miles

ATLAS BLOCKS

BREEDING STATUS	NO.	
Possible	12	(38%)
Probable	15	(47%)
Confirmed	5	(16%)
TOTAL	**32**	

ABUNDANCE	NO.	
Uncommon	8	(24%)
Fairly Common	20	(61%)
Common	5	(15%)
Abundant	0	(0%)
TOTAL	**33**	

0   50   100
Miles

CALLIOPE HUMMINGBIRD
Stellula calliope

The Calliope Hummingbird is both the tiniest bird in North America and the world's smallest long-distance migrant bird; some individuals travel thousands of miles each year (Calder and Calder 1994). It is also something of a physiological wonder, given its chilly montane breeding grounds. Although Calliopes occur in the mountains across most of northern Nevada, they are easiest to find in flower-filled streamside clearings or open pine woodlands of the Carson Range.

DISTRIBUTION

Atlas records for the Calliope Hummingbird were mostly concentrated in far western Nevada and in the northeastern mountain ranges, with breeding confirmed in both areas. Most of the remaining records came from other northern and eastern ranges, and Clark County produced a single record of a possible breeder. Additional confirmed breeding records came from the Pyramid Lake area and the southern Toiyabe Range. Earlier accounts describe Calliope Hummingbirds as widespread but scattered in Nevada, with most records coming from the north (Linsdale 1936, Alcorn 1988). Neel (1999a), however, reported breeding as far south as the Spring Mountains.

The breeding range extends through the higher-elevation portions of the northwestern United States and southwestern Canada (Calder and Calder 1994). In nearby states, breeding is concentrated in the Sierra Nevada, northern California, central and northeastern Oregon, northern Idaho, and the mountainous areas of Utah (Behle et al. 1985, Calder and Calder 1994, Small 1994, Stephens and Sturts 1998, Adamus et al. 2001).

The Calliope Hummingbird breeds primarily in montane forests (Calder and Calder 1994). Upper-elevation riparian woodlands are especially favored in Nevada and in other southern parts of the breeding range (Ryser 1985). The species has been documented as a breeder at elevations as low as the Lahontan Valley (Chisholm and Neel 2002), where it even forages in sagebrush habitat (Neel 1999a). Calliopes are also seen in urban areas, but not as commonly as many other hummingbirds. The predictive map suggests a distribution superficially similar to that for the Black-chinned Hummingbird. Both species are likely to be found in the montane forests of the Great Basin, but Calliopes are most likely to be found at higher elevations. The single possible breeding record from Clark County extends the predicted range farther to the south than is perhaps accurate, especially as the observation could simply have involved a migrating bird.

CONSERVATION AND MANAGEMENT

Calliope Hummingbird populations appear stable, both in numbers and distribution (Calder and Calder 1994, Sauer et al. 2005). There are recent records of birds overwintering in the southern United States, including Arizona, far from their traditional winter range in southern Mexico (Calder and Calder 1994). Very little has been reported about the effects of forest management on Calliope Hummingbirds, but selective logging could be beneficial by generating forest clearings enriched with flowering plants (Calder and Calder 1994).

Although the Calliope Hummingbird is not considered threatened by most measures, Partners in Flight recognizes it as a Stewardship species and as a Nevada Priority species (Neel 1999a, Rich et al. 2004). Part of the rationale for these designations is the lack of reliable trend data, especially in Nevada, and a primary recommendation is improved monitoring. Management strategies have not been determined and may not be warranted, but as is true for all hummingbirds, much remains to be learned about this species.

HABITAT USE

HABITAT	ATLAS BLOCKS	INCIDENTAL OBSERVATIONS
Alpine	1 (3%)	0 (0%)
Barren	1 (3%)	0 (0%)
Montane Forest	6 (19%)	2 (14%)
Montane Parkland	2 (6%)	0 (0%)
Montane Shrub	5 (16%)	1 (7%)
Pinyon-Juniper	0 (0%)	1 (7%)
Riparian	15 (47%)	10 (71%)
Urban	1 (3%)	0 (0%)
Wetland	1 (3%)	0 (0%)
TOTAL	32	14

Total numbers reported in the Habitat Use, Breeding Status, and Abundance tables may differ from each other (see pp. 22–23 for details). Percentage sums may differ slightly from 100% due to rounding.

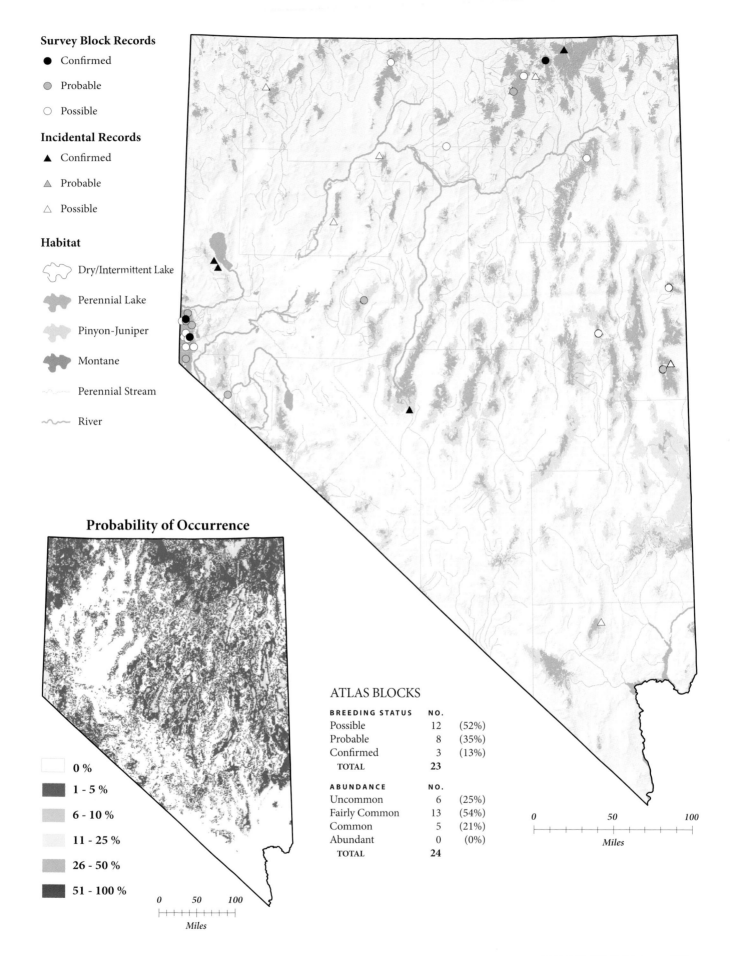

Survey Block Records

● Confirmed

◉ Probable

○ Possible

Incidental Records

▲ Confirmed

▲ Probable

△ Possible

Habitat

Dry/Intermittent Lake

Perennial Lake

Pinyon-Juniper

Montane

Perennial Stream

River

Probability of Occurrence

0 %

1 - 5 %

6 - 10 %

11 - 25 %

26 - 50 %

51 - 100 %

0   50   100
Miles

ATLAS BLOCKS

BREEDING STATUS	NO.	
Possible	12	(52%)
Probable	8	(35%)
Confirmed	3	(13%)
TOTAL	23	

ABUNDANCE	NO.	
Uncommon	6	(25%)
Fairly Common	13	(54%)
Common	5	(21%)
Abundant	0	(0%)
TOTAL	24	

0   50   100
Miles

BROAD-TAILED HUMMINGBIRD

Selasphorus platycercus

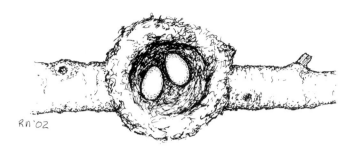

RN '02

The characteristic metallic wing-trill of the Broad-tailed Hummingbird is a common summer sound in the high meadows and forests of Nevada. This species generally occurs at higher elevations and is probably the most common breeding hummingbird in the state. Despite this, it is tricky to get a satisfactory glimpse of one of these little insectivores, which are grass green above with a deep rose gorget. Finding a nest—usually placed on top of a branch near a stream—is even more of a challenge.

DISTRIBUTION

Atlas workers found Broad-tailed Hummingbirds in most of Nevada's higher mountains, but records were few and breeding confirmations were almost nonexistent in the western portion of the state. The confirmation rate was fairly low, as it was for all Nevada hummingbirds. Older accounts describe a similar distribution, with this species regarded as common in the eastern and central mountains and scarce or absent to the west (Linsdale 1936, Ryser 1985, Alcorn 1988).

Nevada is near the western limit of the Broad-tailed Hummingbird's range. Its occurrence as a breeder in California is limited (Small 1994), although Fisher (1893) reported the species to be common on the western slope of the Sierra Nevada in the late 1800s. The breeding range extends throughout the Rocky Mountain and Intermountain regions and north to southern Montana. Breeding probably occurs in Oregon, but it is unconfirmed, and sightings are sparsely distributed (Adamus et al. 2001). Breeding is well established in limited portions of central Idaho and widespread in Utah and Arizona (Behle et al. 1985, Calder and Calder 1992, Stephens and Sturts 1998, Grossi and Corman in Corman and Wise-Gervais 2005:262–263). Lowland breeding records are unusual throughout the species' range.

The Broad-tailed Hummingbird is most often found in montane meadows and forest openings, but it also ranges down into more xeric woodlands at intermediate elevations. Insects are an important component of the species' diet during breeding (Calder and Calder 1992), as is true for all hummingbirds. In addition to the usual flower nectar, Broad-tailed Hummingbirds also sometimes feed at the sap wells dug by Red-naped Sapsuckers (Ehrlich and Daily 1988). In Nevada, the species is commonly found along mountain streams (Linsdale 1936), and a third of the reported habitats in atlas blocks were riparian. Most other atlas records came from montane forest, montane shrublands, or pinyon-juniper woodland. The bird's montane breeding orientation is particularly apparent on the predictive map, which shows the odds of finding it to be highest at middle elevations. The likelihood of occurrence in the northern Mojave Desert lowlands is an overprediction because the model took into account a few unconfirmed low-elevation breeding records in the region.

CONSERVATION AND MANAGEMENT

Few conservation challenges seem to exist for the Broad-tailed Hummingbird. Populations do fluctuate a great deal from year to year (Boyle in Kingery 1998:244–245), but no long-term declines or increases have been demonstrated (Sauer et al. 2005). Little is known about how land management practices affect Broad-tailed Hummingbirds.

In Nevada, no specific management strategies for the Broad-tailed Hummingbird have been developed, and none are thought to be required. Past monitoring of this species—and many other montane birds—has not been very effective, and current assessments are hampered by the lack of survey data. Questions about the basic status of the species also remain. For instance, it is still not known whether populations in far western Nevada are breeding regularly.

HABITAT USE

HABITAT	ATLAS BLOCKS	INCIDENTAL OBSERVATIONS
Agricultural	1 (<1%)	0 (0%)
Alpine	5 (3%)	0 (0%)
Montane Forest	40 (21%)	26 (28%)
Montane Parkland	5 (3%)	2 (2%)
Montane Shrub	30 (16%)	6 (7%)
Pinyon-Juniper	32 (17%)	11 (12%)
Riparian	63 (34%)	41 (45%)
Sagebrush Scrub	8 (4%)	2 (2%)
Sagebrush Steppe	2 (1%)	1 (1%)
Urban	1 (<1%)	2 (2%)
Wetland	1 (<1%)	1 (1%)
TOTAL	188	92

Total numbers reported in the Habitat Use, Breeding Status, and Abundance tables may differ from each other (see pp. 22–23 for details). Percentage sums may differ slightly from 100% due to rounding.

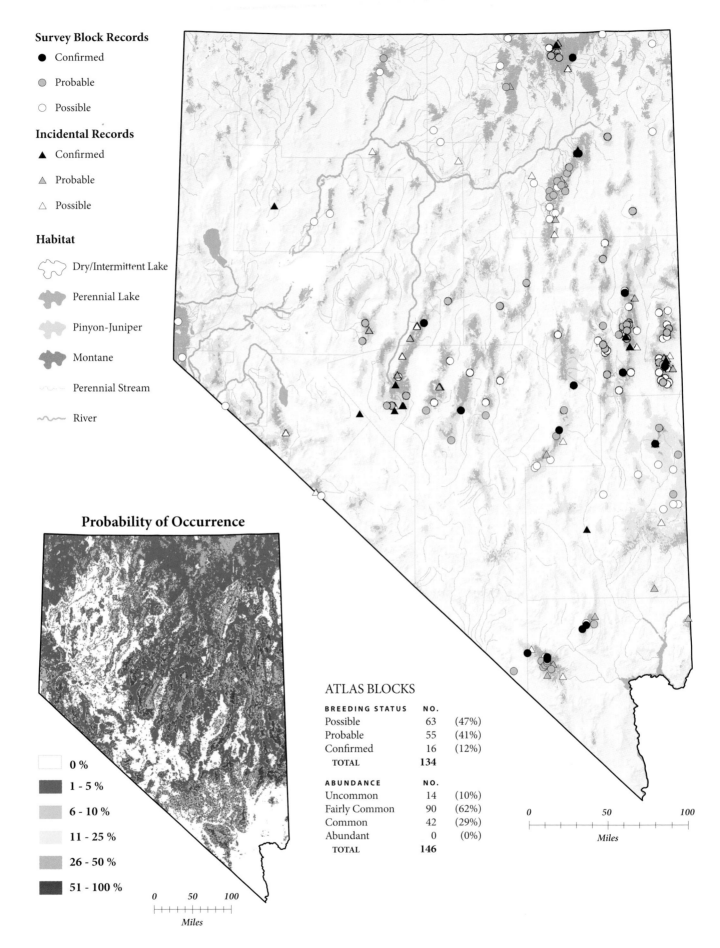

Survey Block Records

● Confirmed

◉ Probable

○ Possible

Incidental Records

▲ Confirmed

△ Probable

△ Possible

Habitat

Dry/Intermittent Lake

Perennial Lake

Pinyon-Juniper

Montane

Perennial Stream

River

Probability of Occurrence

◻ 0 %

1 - 5 %

6 - 10 %

11 - 25 %

26 - 50 %

51 - 100 %

0   50   100
Miles

ATLAS BLOCKS

BREEDING STATUS	NO.	
Possible	63	(47%)
Probable	55	(41%)
Confirmed	16	(12%)
TOTAL	**134**	

ABUNDANCE	NO.	
Uncommon	14	(10%)
Fairly Common	90	(62%)
Common	42	(29%)
Abundant	0	(0%)
TOTAL	**146**	

0   50   100
Miles

BELTED KINGFISHER
Ceryle alcyon

There is a certain tranquility associated with slow-moving rivers and quiet farm ponds. Muskrats glide through the water, herons stalk about the reedy edges, and leaves whisper faintly in the summer breeze. A bull-headed Belted Kingfisher sits on a broken snag and seems at one with the quietude of the surroundings—until it erupts from its perch with a piercing rattle, courses wildly about the tules and cattails, and then splashes headfirst into the fish-filled water.

DISTRIBUTION
Belted Kingfishers occur in nearly all of North America, but the Nevada atlas data show a decided concentration of records in the west-central and northeastern portions of the state—and these were also the only areas in which breeding was confirmed. Most of the records came from lowland riparian sites. Scattered records also occurred down Nevada's eastern boundary all the way to Clark County. Older accounts are not very specific about the Belted Kingfisher's strongholds in the state, but they concur that the species is not very common (Linsdale 1936, Ryser 1985, Alcorn 1988). The predictive map is consistent with this assessment in that the chances of finding the species are generally estimated to be low. Not surprisingly, the likelihood of finding Belted Kingfishers is predicted to be highest where riparian habitats are present. Urban habitats are also highlighted, probably because many towns are situated along rivers. The high probabilities predicted for Las Vegas and other urban areas that are not associated with major rivers, especially in the south, are thus probably incorrect.

In Nevada's neighboring states, Belted Kingfishers are most common and widely distributed to the north, in Oregon and Idaho (Stephens and Sturts 1998, Adamus et al. 2001). Breeding is less common in the northern parts of California and Utah (Behle et al. 1985, Small 1994), and is very limited in the southern deserts of those states and in Arizona (Corman in Corman and Wise-Gervais 2005:266–267).

As a fish-eating bird, the Belted Kingfisher is closely tied to water bodies. In Nevada, kingfishers tend to be found near larger streams and rivers (Linsdale 1936), but atlas records also came from farm ponds, city lakes, and wetlands. Kingfishers require clear water with an unobstructed view of their aquatic prey. They tend to avoid muddy or turbid waters and areas with dense emergent vegetation (Hamas 1994). Kingfishers nest in burrows they dig into steep earthen banks, as close to the water as possible, and they require suitable hunting perches overlooking the water (Hamas 1994, Dobkin and Sauder 2004).

CONSERVATION AND MANAGEMENT
Human influences on the Belted Kingfisher are multifaceted. Creation of reservoirs has probably been beneficial, while degradation of riparian habitats has probably been harmful (Shuford 1993). In marked contrast to many other fish eaters, Belted Kingfishers have been spared serious effects from environmental contaminants, possibly because they eat relatively small fish (Hamas 1994). Nonetheless, monitoring data suggest that the species is slowly declining rangewide, though this pattern is less pronounced in the West than elsewhere (Sauer et al. 2005). Dobkin and Sauder (2004) suggested that Belted Kingfishers are also declining slowly in the West, possibly due to vegetation changes in riparian areas and stream bank erosion. Nest site availability appears to be a limiting factor in the southern Sierra Nevada (Gaines 1992) and possibly elsewhere.

The Belted Kingfisher is probably secure in much of Nevada, at least for the short term. Pairs nesting near urban areas should be monitored, however, to detect effects of the continued expansion of Nevada's cities and likely impacts of that expansion on wetlands. Breeding birds are also sensitive to human disturbance (Hamas 1994), a threat that is likely to increase as urban areas spread.

HABITAT USE

HABITAT	ATLAS BLOCKS	INCIDENTAL OBSERVATIONS
Agricultural	2 (5%)	0 (0%)
Open Water	1 (3%)	1 (7%)
Riparian	28 (70%)	11 (79%)
Urban	2 (5%)	1 (7%)
Wetland	7 (18%)	1 (7%)
TOTAL	40	14

Total numbers reported in the Habitat Use, Breeding Status, and Abundance tables may differ from each other (see pp. 22–23 for details). Percentage sums may differ slightly from 100% due to rounding.

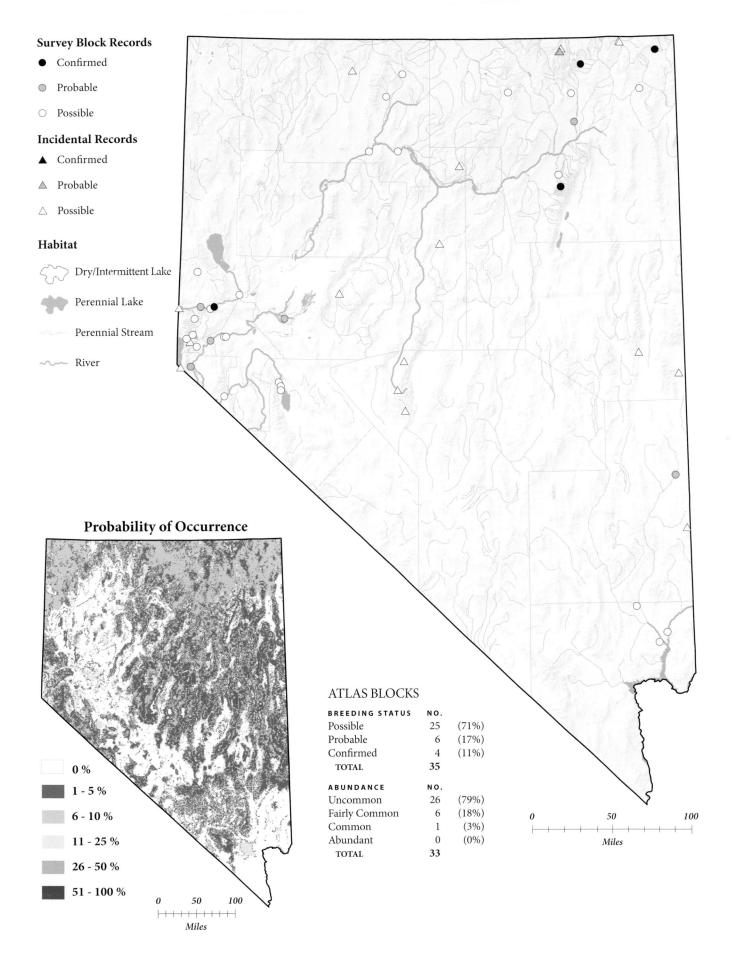

Survey Block Records
● Confirmed
◔ Probable
○ Possible

Incidental Records
▲ Confirmed
◮ Probable
△ Possible

Habitat
⬚ Dry/Intermittent Lake
▨ Perennial Lake
～ Perennial Stream
〜 River

Probability of Occurrence

☐ 0 %
■ 1 - 5 %
▨ 6 - 10 %
▨ 11 - 25 %
▨ 26 - 50 %
■ 51 - 100 %

0   50   100
|—+—+—+—+—|
Miles

ATLAS BLOCKS

BREEDING STATUS	NO.	
Possible	25	(71%)
Probable	6	(17%)
Confirmed	4	(11%)
TOTAL	**35**	

ABUNDANCE	NO.	
Uncommon	26	(79%)
Fairly Common	6	(18%)
Common	1	(3%)
Abundant	0	(0%)
TOTAL	**33**	

0   50   100
|—+—+—+—+—|
Miles

LEWIS'S WOODPECKER
Melanerpes lewis

Lewis's Woodpecker is a conspicuous and visually striking inhabitant of open woodlands in northern Nevada. The sight of one of these iridescent woodpeckers chasing flies in an otherwise quiet forest is especially memorable. Recently burned coniferous woodlands are favored habitats in much of the range. In Nevada, atlas workers found Lewis's Woodpeckers most often in riparian woodlands. A healthy supply of flying insects is a basic requirement, as is sufficient open space for their aerobatic foraging. They nest in dead trees.

DISTRIBUTION

Lewis's Woodpecker is a migratory species with a patchy breeding distribution in northern Nevada. It was found in thirty-five blocks and twenty-five other locations, with breeding confirmed at twenty-three sites. Records were concentrated at higher elevations in Elko County—especially in the Ruby Mountains and the mountains around Jarbidge. Elsewhere there was a smaller cluster of records in the Carson Range of far western Nevada and isolated records in the center of the state. The predictive map suggests the potential for this species to occur in other interior ranges, but not with an especially high probability.

Although apparently disjunct, the Nevada range of Lewis's Woodpeckers fits neatly into the species' range in adjacent states. It is found throughout southwestern Idaho immediately north of Elko County, and throughout the Sierra Nevada in California. In western Utah and southeastern Oregon, the species is not as common (Tobalske 1997, Adamus et al. 2001). Overall, the pattern in the West appears to combine regular occurrence throughout the more extensive mountain ranges and distinct rarity in the more isolated ranges of the central Great Basin.

Multiple pairs of Lewis's Woodpeckers were estimated to be present in most of the blocks where they were found. About half of the records were from riparian habitat, and most of the remainder came from a variety of montane habitats.

CONSERVATION AND MANAGEMENT

The conservation status of Lewis's Woodpecker is difficult to assess because populations tend to be patchy and to undergo large fluctuations in local abundance (Bock 1970). Although there have been increases in some areas, there has been an overall decline in the population since the 1960s (Tobalske 1997, Sauer et al. 2005). Lewis's Woodpecker is a Partners in Flight Priority species in Nevada (Neel 1999a) and a Watch List species for the Intermountain West (Rich et al. 2004) because its habitats are considered threatened and the population size is generally low. Declines throughout the range have been attributed to several factors, although the loss of open forests—especially ponderosa pine—and long-term fire suppression seem to be central problems (Tobalske 1997). Burned coniferous woodlands and open riparian woodlands constitute primary habitats for this species, and a relatively intact grass or shrub understory is important for providing abundant insects (Neel 1999a).

Saab and Vierling (2001) found significantly lower reproductive success—due to higher predation rates—in riparian forest in Colorado than in burned ponderosa pine forest in Idaho, and suggested that agricultural and residential development near the riparian sites had increased predator densities. Human encroachment on montane riparian areas is still less severe in northern Nevada than in Colorado, however, and riparian areas might provide better habitat in Nevada than the Colorado data suggest. Nonetheless, the Colorado-Idaho study indicates that better documentation of Lewis's Woodpecker breeding success in Nevada is warranted, especially in the context of continuing development in the mountains of Elko and Washoe counties.

HABITAT USE

HABITAT	ATLAS BLOCKS	INCIDENTAL OBSERVATIONS
Agricultural	1 (2%)	2 (9%)
Alpine	1 (2%)	0 (0%)
Montane Forest	7 (16%)	6 (25%)
Montane Parkland	4 (9%)	0 (0%)
Montane Shrub	5 (12%)	1 (4%)
Pinyon-Juniper	1 (2%)	1 (4%)
Riparian	21 (49%)	13 (54%)
Sagebrush Scrub	2 (5%)	1 (4%)
Urban	1 (2%)	0 (0%)
TOTAL	43	24

Total numbers reported in the Habitat Use, Breeding Status, and Abundance tables may differ from each other (see pp. 22–23 for details). Percentage sums may differ slightly from 100% due to rounding.

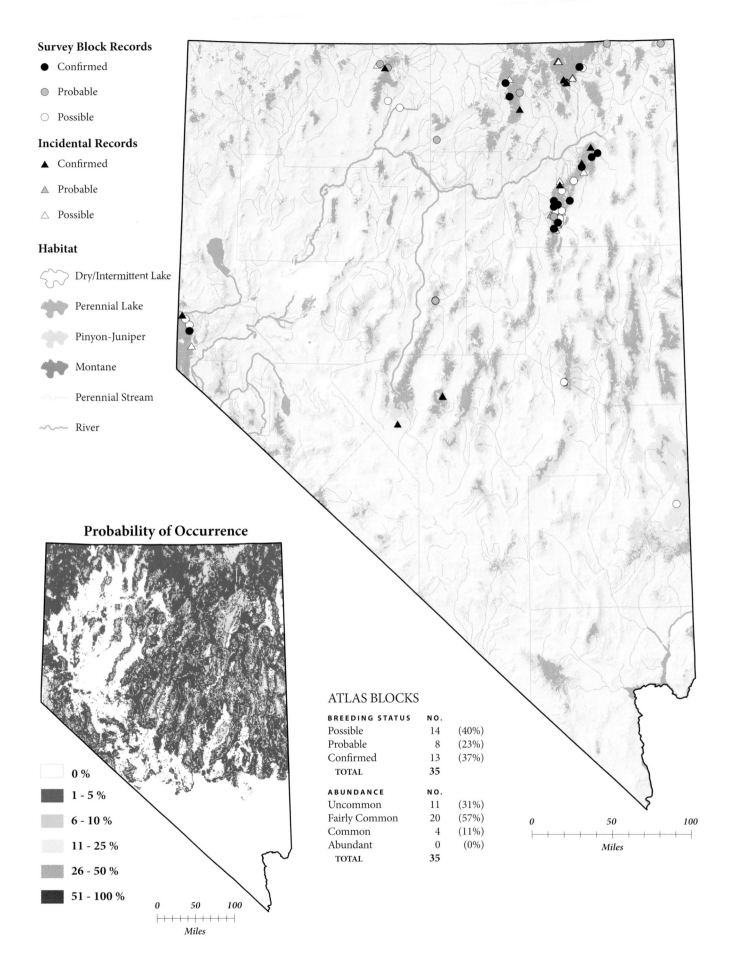

Survey Block Records
- ● Confirmed
- ● Probable
- ○ Possible

Incidental Records
- ▲ Confirmed
- ▲ Probable
- △ Possible

Habitat
- Dry/Intermittent Lake
- Perennial Lake
- Pinyon-Juniper
- Montane
- Perennial Stream
- River

Probability of Occurrence

- ☐ 0 %
- 1 - 5 %
- 6 - 10 %
- 11 - 25 %
- 26 - 50 %
- 51 - 100 %

0   50   100
Miles

ATLAS BLOCKS

BREEDING STATUS	NO.	
Possible	14	(40%)
Probable	8	(23%)
Confirmed	13	(37%)
TOTAL	35	

ABUNDANCE	NO.	
Uncommon	11	(31%)
Fairly Common	20	(57%)
Common	4	(11%)
Abundant	0	(0%)
TOTAL	35	

0   50   100
Miles

WILLIAMSON'S SAPSUCKER
Sphyrapicus thyroideus

Williamson's Sapsucker is a fairly uncommon and poorly known species throughout its western North American range. Indeed, for twenty years after it was discovered the species was so poorly known that the males and females were classified as separate species. The confusion arose from the unusually extreme differences in the plumage patterns of the two sexes. Williamson's Sapsucker is a bird of high-elevation forests, and in Nevada it is restricted to a few of the higher mountain ranges around the periphery of the state.

DISTRIBUTION

Atlas workers found Williamson's Sapsuckers in sixteen atlas blocks and eleven additional locations; the species was confirmed as a breeder at nine sites. The records were from three widely separated areas: the Carson Range in the far west, the high country of White Pine County in the east, and the Spring Mountains in the south. Almost all reports came from montane forest habitat, although there were also records from riparian habitat. This mixture of habitats accords with other studies, which suggest that Williamson's Sapsuckers use pines for foraging but nest primarily in aspens (e.g., Crockett and Hadow 1975, Dobbs et al. 1997).

Older accounts describe Williamson's Sapsucker as uncommon in the western and northeastern portions of the state and absent from the areas in between (Johnson 1975, Ryser 1985). Atlas results are largely consistent with this pattern but indicate an even greater tendency toward a "feast or famine" distribution. The species' occurrence in the Carson Range is not surprising, because Williamson's Sapsuckers are found at higher elevations throughout the Sierra Nevada (Small 1994) and the Cascade Range (Adamus et al. 2001). The eastern and southern locations are fairly isolated, as the species is not very widespread as a breeder in southern Idaho and Utah (Behle et al. 1985, Stephens and Sturts 1998), and is only spottily distributed in Arizona (Latta and Corman in Corman and Wise-Gervais 2005:276–277). Sapsuckers are more migratory than most North American woodpeckers, and this propensity for long-distance movements may help these isolated populations to persist.

Our model indicates some chance of finding Williamson's Sapsuckers in the higher parts of central Nevada, suggesting that these mountains contain suitable habitat. The relatively small area of habitat in these ranges, combined with their isolation, reduces the likelihood that populations will persist, though, and probably accounts for the lack of records from these areas. The model does not predict Williamson's Sapsuckers to occur in the northernmost parts of Nevada, because there were no atlas records from the northern edge of the state. There are historical records from the Jarbidge Mountains and elsewhere in Elko County (Alcorn 1988), however, suggesting that future surveys in this area would be worthwhile.

CONSERVATION AND MANAGEMENT

Williamson's Sapsucker is an uncommon and often sparsely distributed species whose population status and conservation needs have received relatively little attention (Dobbs et al. 1997). Although earlier studies have suggested population declines, Breeding Bird Survey data suggest long-term stability of populations (Sauer et al. 2005). The species' remote breeding habitats make it difficult to know whether survey data are representative of the overall population (Dobbs et al. 1997).

Although typically found in coniferous forests, Williamson's Sapsuckers often seek out aspens for nesting and frequently use dead trees (Dobbs et al. 1997). Mixed conifer and deciduous forest, with diverse and complex structure and standing deadwood, is therefore likely to be the key habitat requirement.

HABITAT USE

HABITAT	ATLAS BLOCKS	INCIDENTAL OBSERVATIONS
Montane Forest	17 (90%)	9 (90%)
Riparian	2 (10%)	1 (10%)
TOTAL	19	10

Total numbers reported in the Habitat Use, Breeding Status, and Abundance tables may differ from each other (see pp. 22–23 for details). Percentage sums may differ slightly from 100% due to rounding.

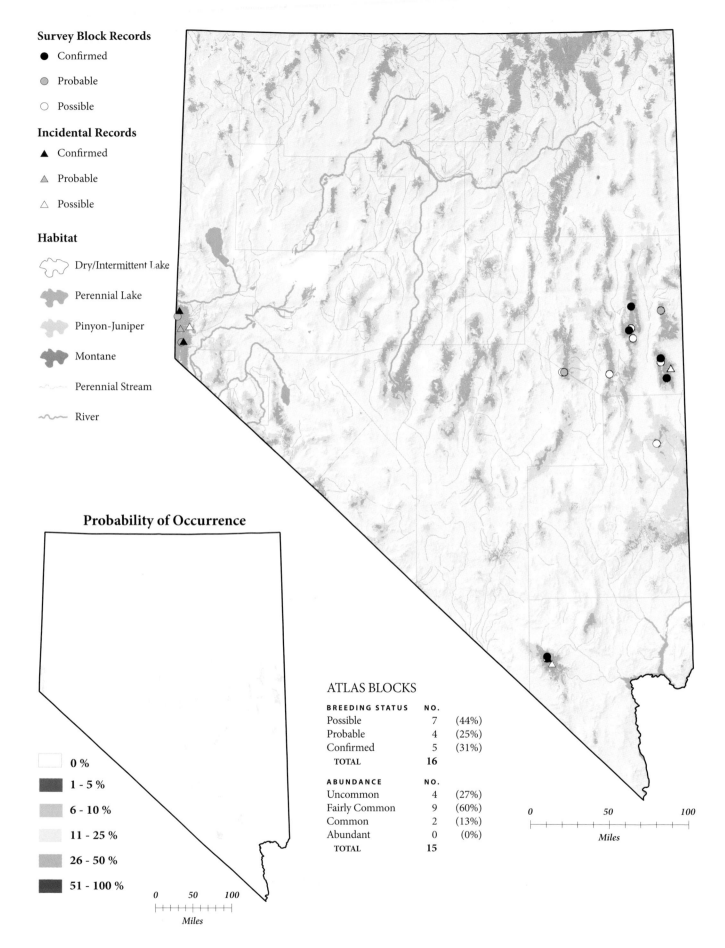

Survey Block Records

● Confirmed

● Probable

○ Possible

Incidental Records

▲ Confirmed

△ Probable

△ Possible

Habitat

Dry/Intermittent Lake

Perennial Lake

Pinyon-Juniper

Montane

Perennial Stream

River

Probability of Occurrence

◻ 0 %

■ 1 - 5 %

■ 6 - 10 %

■ 11 - 25 %

■ 26 - 50 %

■ 51 - 100 %

0   50   100
Miles

ATLAS BLOCKS

BREEDING STATUS	NO.	
Possible	7	(44%)
Probable	4	(25%)
Confirmed	5	(31%)
TOTAL	16	

ABUNDANCE	NO.	
Uncommon	4	(27%)
Fairly Common	9	(60%)
Common	2	(13%)
Abundant	0	(0%)
TOTAL	15	

0   50   100
Miles

RED-NAPED SAPSUCKER

Sphyrapicus nuchalis

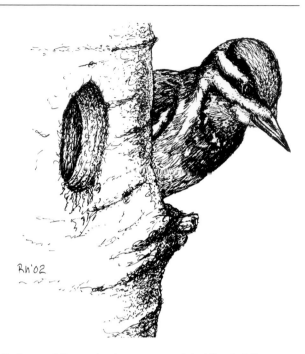

Rh'02

The Red-naped Sapsucker is a common inhabitant of the aspen forests of northern Nevada. It is especially widespread in the northeastern part of the state, where its distinctive *rallentando* drumming is a common sound in the mountain forests. The Red-naped Sapsucker also appears to be a "keystone" species that provides valuable resources for a variety of other animals (Fleury 1998). Species as diverse as Mountain Bluebirds and Northern Saw-whet Owls use abandoned sapsucker nest holes, and many insects and other birds feed at the sap wells that sapsuckers create in the bark of aspens and willows.

DISTRIBUTION

Red-naped Sapsuckers were found in fifty-eight atlas blocks and at twenty-six other sites. Records were concentrated in Elko and White Pine counties, and a few confirmed breeders were found as far west as the northern California border and as far south as Clark County. Unlike most woodpeckers, Red-naped Sapsuckers are also found throughout the center of the state. Only Northern Flickers and Hairy Woodpeckers have more extensive ranges in Nevada, and most other woodpeckers apparently lack the dispersal ability to colonize and populate the state's more isolated ranges. With their conspicuous nests and unretiring behavior, Red-naped Sapsuckers were relatively easy to confirm as breeders in almost half of the blocks in which they occurred. Most of these blocks were also estimated to contain multiple pairs, some even with more than ten.

In neighboring states, the Red-naped Sapsucker breeds throughout most of Idaho (Stephens and Sturts 1998), in central and northeastern Oregon (Adamus et al. 2001), in scattered locations in eastern California (Small 1994), in eastern Arizona

(Sitko and Corman in Corman and Wise-Gervais 2005:278–279), and in Utah (Behle et al. 1985). Red-naped Sapsuckers regularly interbreed with Red-breasted Sapsuckers (which occur mostly to the west), and for a long time the two taxa, along with the Yellow-bellied Sapsucker of eastern and northern North America, were considered a single species (Walters et al. 2002).

Most atlas records of Red-naped Sapsuckers came from montane forest or riparian habitats. The predictive map highlights the high chance of finding this species in the drainages of major mountain ranges and the very low chance of finding it elsewhere. Although Red-naped Sapsuckers can be found in various forest types (Walters et al. 2002), an aspen or aspen-willow component must be present (Crockett and Hadow 1975), and quaking aspen is the predominant nest tree throughout much of the species' range (Walters et al. 2002).

CONSERVATION AND MANAGEMENT

Red-naped Sapsuckers seem to be holding their own in Nevada, although solid population monitoring data are lacking. According to Breeding Bird Survey data, populations are stable in the surrounding region (Sauer et al. 2005), and the atlas confirmed that this species is widespread and relatively common in appropriate habitat in Nevada. Because of its globally restricted distribution, Partners in Flight has designated the Red-naped Sapsucker an Intermountain West Stewardship species (Rich et al. 2004).

Although originally found across central Nevada (Linsdale 1936), the Red-naped Sapsucker was absent from parts of its current range—notably the Toiyabe Range—during the mid-twentieth century, when much of the quaking aspen was logged (Dobkin and Wilcox 1986). This vulnerability persists today (Walters et al. 2002), but the return of sapsuckers to the Toiyabes (Fleury 2000) suggests that the species can recover as long as its habitat is available.

The conservation of Red-naped Sapsuckers is important to several other bird species because of sapsuckers' apparent keystone role in western montane ecosystems (Dobkin and Wilcox 1986, Ehrlich and Daily 1988). Sapsucker wells are thought to be especially important for foraging hummingbirds, and their nest holes are reused by a range of secondary cavity-nesting species (Ryser 1985, Barrett in Kingery 1998:256–257, Fleury 1998, Walters et al. 2002).

HABITAT USE

HABITAT	ATLAS BLOCKS	INCIDENTAL OBSERVATIONS
Montane Forest	40 (47%)	16 (50%)
Montane Parkland	5 (6%)	0 (0%)
Montane Shrub	5 (6%)	1 (3%)
Pinyon-Juniper	3 (3%)	2 (6%)
Riparian	29 (34%)	13 (41%)
Sagebrush Scrub	1 (1%)	0 (0%)
Sagebrush Steppe	1 (1%)	0 (0%)
Urban	2 (2%)	0 (0%)
TOTAL	86	32

Total numbers reported in the Habitat Use, Breeding Status, and Abundance tables may differ from each other (see pp. 22–23 for details). Percentage sums may differ slightly from 100% due to rounding.

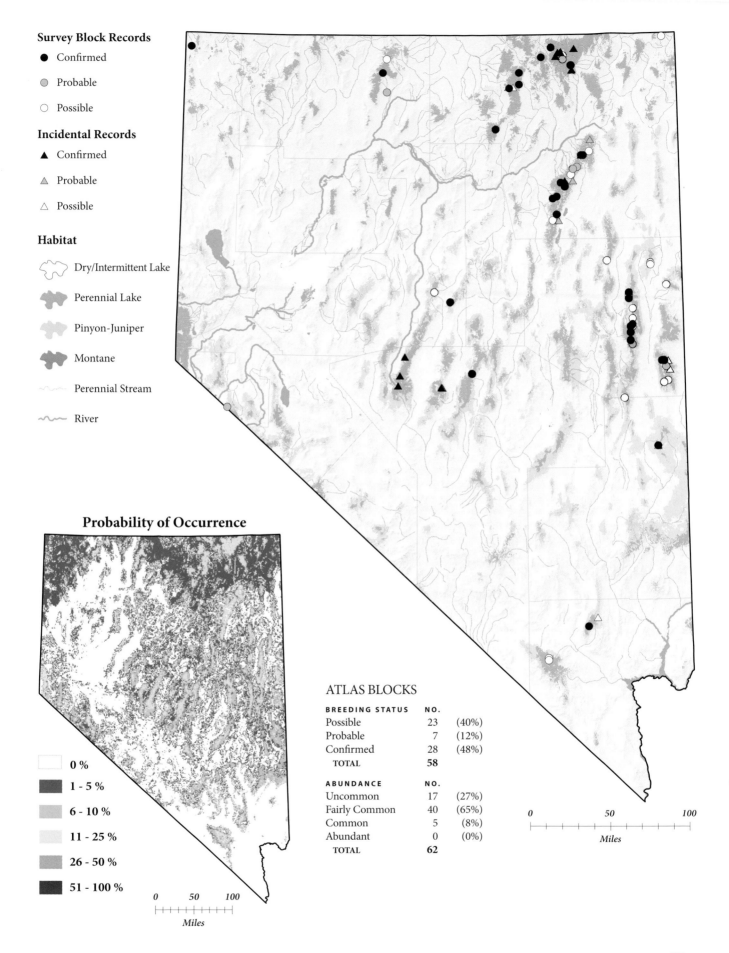

Survey Block Records

● Confirmed

● Probable

○ Possible

Incidental Records

▲ Confirmed

▲ Probable

△ Possible

Habitat

Dry/Intermittent Lake

Perennial Lake

Pinyon-Juniper

Montane

Perennial Stream

River

Probability of Occurrence

	0 %
	1 - 5 %
	6 - 10 %
	11 - 25 %
	26 - 50 %
	51 - 100 %

0   50   100

Miles

ATLAS BLOCKS

BREEDING STATUS	NO.	
Possible	23	(40%)
Probable	7	(12%)
Confirmed	28	(48%)
TOTAL	**58**	

ABUNDANCE	NO.	
Uncommon	17	(27%)
Fairly Common	40	(65%)
Common	5	(8%)
Abundant	0	(0%)
TOTAL	**62**	

0   50   100

Miles

RED-BREASTED SAPSUCKER

Sphyrapicus ruber

The breeding range of this beautiful woodpecker barely enters Nevada. Red-breasted Sapsuckers are not hard to find, however, along the California border from southern Washoe County to southern Lyon County. Nevada breeding records all seem to be of the *daggetti* subspecies. Where their ranges overlap, Red-breasted Sapsuckers frequently hybridize with Red-naped Sapsuckers, and the two forms were long considered to be a single species that also included the Yellow-bellied Sapsucker. Because of this taxonomic history, many historical records refer to the old collective name of "Yellow-bellied Sapsuckers" and do not differentiate among the three forms.

DISTRIBUTION

Red-breasted Sapsuckers were found in twenty-two blocks, the bulk of which were located in the Carson Range of far western Nevada, where many species typical of the Sierra Nevada are found. Less expected were records from two Clark County blocks and isolated records from Humboldt, Elko, Lander, and Nye counties. There was also a well-documented case of a hybrid pair from Eureka County, more than 125 miles (200 kilometers) east of the Carson Range (not mapped here): the female was a Red-breasted Sapsucker, the male a Red-naped Sapsucker, and the pair fledged at least two young. The predictive map suggests a moderate chance of finding Red-breasted Sapsuckers in the larger mountain ranges across the state. In reality, however, this likelihood probably diminishes substantially as one moves away from the California border, while the probability of finding hybrids probably increases going east.

In the states surrounding Nevada, the species breeds throughout much of northern California and in montane areas farther south (Small 1994); in western Oregon (Adamus et al. 2001) and northward through the Cascade and coastal ranges well into Canada (Walters et al. 2002); but appears to be absent from Idaho, Utah, Arizona, and the lower Colorado River Valley (Burleigh 1972, Behle et al. 1985, Rosenberg et al. 1991, Stephens and Sturts 1998, Corman and Wise-Gervais 2005). Not surprisingly, most atlas records of Red-breasted Sapsuckers came from montane forest or riparian habitats, and where they occurred, there were frequently multiple pairs per block. Overall, despite the differences in geographical distribution, the Red-breasted Sapsucker's habitat use and abundance patterns were remarkably similar to those of the Red-naped Sapsucker.

Understanding the historical distributions of Red-breasted and Red-naped Sapsuckers in Nevada is difficult because much of the older literature treats the two taxa as a single species (e.g., Ryser 1985, Alcorn 1988). Nonetheless, Linsdale (1936) did discuss the two forms separately and provided a picture that is not too different from what the atlas data show, with Red-breasted Sapsuckers found primarily in the vicinity of Lake Tahoe and Red-napeds elsewhere in the state. Although Alcorn (1988) referred primarily to the combined "Yellow-bellied Sapsucker," he did cite specific records of Red-breasted Sapsuckers from the mountains of Elko and Clark counties, and there are also early records of hybrids from southern Nevada (Johnson 1965). These reports suggest that the occurrence of Red-breasted Sapsuckers well to the east and south of their core Nevada range, while fairly unusual, is not a new phenomenon.

CONSERVATION AND MANAGEMENT

Most large-scale population trend data do not differentiate among the three forms in the "Yellow-bellied Sapsucker" group (Sauer et al. 2005). Data from California, though, undoubtedly dominated by the *daggetti* subspecies of Red-breasted Sapsuckers, suggest a population decline during the latter third of the twentieth century (Walters et al. 2002, Sauer et al. 2005).

The Red-breasted Sapsucker is at the periphery of its range in Nevada and is not especially common in the state. Its status is likely tied to forest management policies and practices throughout the multistate montane ecosystems of the Sierra Nevada, Cascade, and coastal ranges.

HABITAT USE

HABITAT	ATLAS BLOCKS	INCIDENTAL OBSERVATIONS
Montane Forest	12 (46%)	2 (33%)
Montane Parkland	4 (15%)	0 (0%)
Montane Shrub	0 (0%)	1 (17%)
Riparian	9 (35%)	3 (50%)
Urban	1 (4%)	0 (0%)
TOTAL	26	6

Total numbers reported in the Habitat Use, Breeding Status, and Abundance tables may differ from each other (see pp. 22–23 for details). Percentage sums may differ slightly from 100% due to rounding.

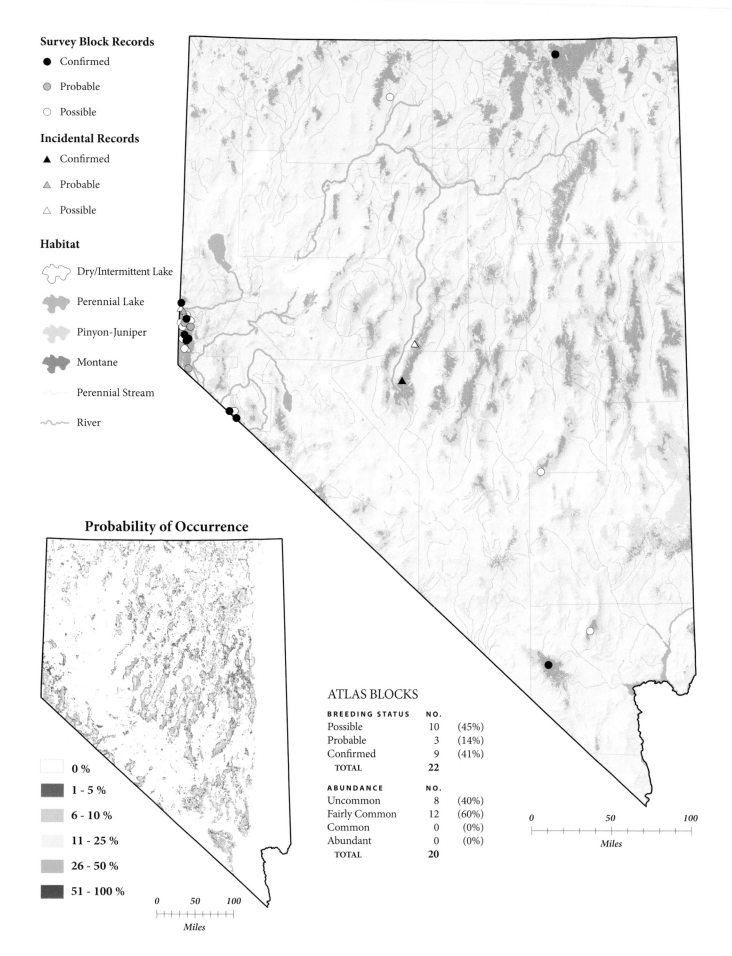

Survey Block Records

- ● Confirmed
- ◐ Probable
- ○ Possible

Incidental Records

- ▲ Confirmed
- △ Probable
- △ Possible

Habitat

- Dry/Intermittent Lake
- Perennial Lake
- Pinyon-Juniper
- Montane
- Perennial Stream
- River

Probability of Occurrence

- ☐ 0 %
- 1 - 5 %
- 6 - 10 %
- 11 - 25 %
- 26 - 50 %
- 51 - 100 %

0   50   100
Miles

ATLAS BLOCKS

BREEDING STATUS	NO.	
Possible	10	(45%)
Probable	3	(14%)
Confirmed	9	(41%)
TOTAL	22	

ABUNDANCE	NO.	
Uncommon	8	(40%)
Fairly Common	12	(60%)
Common	0	(0%)
Abundant	0	(0%)
TOTAL	20	

0   50   100
Miles

LADDER-BACKED WOODPECKER
Picoides scalaris

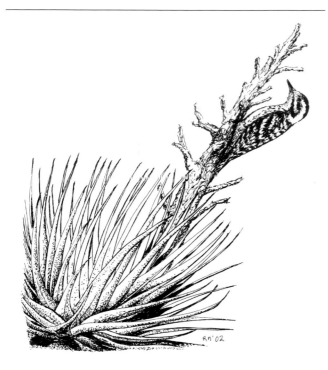

The Ladder-backed Woodpecker provided one of the most exciting discoveries of the atlas project. Before the atlas project was initiated, the species was assumed to reach the northern limit of its range at the rather obvious break point between the creosotebush-dominated Mojave Desert and the sagebrush-and-greasewood-dominated Great Basin. But atlas workers found previously undocumented populations of Ladder-backed Woodpeckers, along with other Mojave Desert breeders, well to the north near the towns of Tonopah and Goldfield. Except for an old record from Elko County (Alcorn 1988), there are no earlier sightings from this far north in the state.

DISTRIBUTION

The southwestern U.S. breeding range of the Ladder-backed Woodpecker extends across southern California, southern Nevada, southwestern Utah, southern Arizona, and thence eastward to Texas and south into Mexico (Behle et al. 1985, Small 1994, Lowther 2001, McCarthey in Corman and Wise-Gervais 2005: 280–281). Atlas workers found them in twenty-five blocks, and there were thirty-five incidental observations. The birds near Tonopah accounted for four records, and the bulk of the remaining records came from Clark County or adjacent areas of Nye and Lincoln counties.

The occurrence of birds near Tonopah was a great surprise because previous Nevada breeding records were confined to the Mojave Desert (Linsdale 1936, Alcorn 1988). In hindsight, however, the discovery was less surprising because most of the birds were found on Nellis Air Force Base, in an area where there is very limited access due to military operations. Moreover, these

birds were found in an extensive Joshua tree forest that covers more than 38 square miles (100 square kilometers). Ladder-backed Woodpeckers favor Joshua trees throughout the core of their Nevada range and in nearby parts of southern California (Lowther 2001). Elsewhere in southern Nevada, these wood-peckers can be found in old-growth mesquite stands and in desert washes.

Whether Ladder-backed Woodpeckers have been present near Tonopah for a long time or this peripheral population instead represents a recent range expansion is not certain. The inaccessibility of the area, the abundance there of the bird's preferred habitat, and the co-occurrence of another warm-desert species—the Cactus Wren—all suggest that this outlying community of Mojave Desert breeders has been present but undetected for quite some time.

The predictive map suggests a moderate chance of finding Ladder-backed Woodpeckers throughout the Mojave Desert. The map highlights riparian habitats, but only riparian patches in lower-elevation areas are truly likely to harbor Ladder-backed Woodpeckers. The distribution of actual records suggests that the model overestimates the probability of occurrence in the drainages that flow through the foothills of major mountain ranges.

CONSERVATION AND MANAGEMENT

The Ladder-backed Woodpecker reaches the northern limit of its range in southern Nevada. Populations appear to be declining rangewide, although the decline is more pronounced in the eastern portion of the species' range, especially in Texas, than farther west (Sauer et al. 2005).

The species is not well studied, and much of its biology remains poorly known or is largely inferred from other closely related species (Lowther 2001). Consequently, the reasons for declines and the best methods for managing populations are unclear. Like other woodpeckers, the Ladder-backed requires suitable sites for building nest holes, which may make the protection of old trees, Joshua trees, and large cactuses throughout the Mojave and Sonoran deserts a key factor in determining its long-term prospects. Destruction of well-established patches of riparian, mesquite, and Joshua tree habitat is likely to be especially harmful to the species in Nevada.

HABITAT USE

HABITAT	ATLAS BLOCKS	INCIDENTAL OBSERVATIONS
Mesquite	2 (6%)	1 (3%)
Mojave	13 (36%)	17 (52%)
Montane Parkland	0 (0%)	1 (3%)
Montane Shrub	0 (0%)	1 (3%)
Pinyon-Juniper	0 (0%)	1 (3%)
Riparian	19 (53%)	8 (24%)
Salt Desert Scrub	0 (0%)	1 (3%)
Sagebrush Scrub	2 (6%)	2 (6%)
Urban	0 (0%)	1 (3%)
TOTAL	36	33

Total numbers reported in the Habitat Use, Breeding Status, and Abundance tables may differ from each other (see pp. 22–23 for details). Percentage sums may differ slightly from 100% due to rounding.

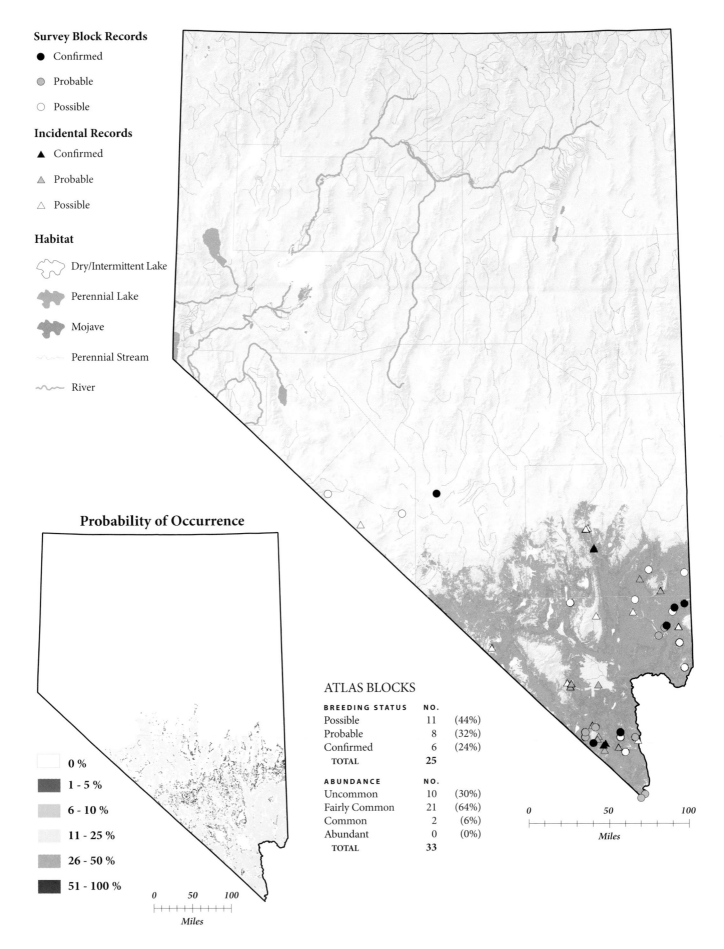

Survey Block Records

● Confirmed

◐ Probable

○ Possible

Incidental Records

▲ Confirmed

△ Probable

△ Possible

Habitat

⬡ Dry/Intermittent Lake

⬢ Perennial Lake

⬢ Mojave

〜 Perennial Stream

〜 River

Probability of Occurrence

☐ 0 %

■ 1 - 5 %

▨ 6 - 10 %

▨ 11 - 25 %

▨ 26 - 50 %

■ 51 - 100 %

0   50   100
Miles

ATLAS BLOCKS

BREEDING STATUS	NO.	
Possible	11	(44%)
Probable	8	(32%)
Confirmed	6	(24%)
TOTAL	25	

ABUNDANCE	NO.	
Uncommon	10	(30%)
Fairly Common	21	(64%)
Common	2	(6%)
Abundant	0	(0%)
TOTAL	33	

0   50   100
Miles

DOWNY WOODPECKER
Picoides pubescens

Although it is frequently portrayed as widespread across the West, the inconspicuous Downy Woodpecker has a decidedly patchy distribution in Nevada. Breeding records were concentrated in western Nevada, in the high mountain ranges of White Pine County, and in the national forests along the Idaho border. A bird of deciduous or mixed deciduous-coniferous forests, the Downy Woodpecker is typically found near rivers and other waterways, but it also occurs in other habitats with large shade trees. Suburban parks, golf courses, and backyards are among the spots often used. The Downy Woodpecker is resourceful and forages widely—on a variety of trees and even on goldenrod galls and corn stalks. Nevertheless, its breeding range is fundamentally restricted by the availability of old dead trees or tree limbs for nest sites.

DISTRIBUTION

Downy Woodpeckers were found in twenty-eight blocks. More than one pair was reported in more than half of the blocks where these woodpeckers occurred, but there were only two blocks where observers estimated more than ten pairs. Records were split fairly evenly among three habitat types. About a third of the reports came from riparian areas, closely followed by montane forest and urban habitat. Many of the riparian and urban blocks were in the Truckee and Carson river drainages in western Nevada, which harbor significant stands of large trees. Moreover, many of the montane forest blocks contained aspen and other broadleaved trees, suggesting that the Downy Woodpecker generally eschews the pure conifer stands typically used by its close cousin, the Hairy Woodpecker. The same habitat use pattern has been described for the species in Washington (Smith et al. 1997).

In adjoining states, the Downy Woodpecker is widespread as a breeder in Idaho and Utah, and in nondesert areas of Oregon and California (Behle et al. 1985, Small 1994, Stephens and Sturts 1998, Adamus et al. 2001). It is fairly scarce as a breeder in Arizona and is absent from the lower Colorado River Valley (Rosenberg et al. 1991, Corman in Corman and Wise-Gervais 2005:282–283).

Although the central and southern parts of Nevada produced widely scattered records, there was little evidence of breeding away from the species' strongholds in northeastern and western Nevada. The paucity of records from the center of the state is perhaps surprising, and one might expect to find this species along the Humboldt River tributaries or in some of the larger mountain ranges. For instance, research in the 1980s documented Downy Woodpeckers in some of the larger canyons of the Toiyabe Range (Dobkin and Wilcox 1986). Nonetheless, our predictive model largely matches the observed distribution, with a relatively small portion of the state identified as suitable for this species. Perhaps the habitat patches present in other areas are too small and isolated to support this largely sedentary woodpecker.

CONSERVATION AND MANAGEMENT

Rangewide, Downy Woodpecker populations have been stable since the 1960s (Sauer et al. 2005) and the species is generally not a conservation concern. In Nevada, however, the species is locally distributed, probably because broadleaved forest habitat is naturally fragmented in the state. Additionally, these highly sedentary woodpeckers could be vulnerable to local extirpation. The current extent of the species' distribution, however, does not seem to differ substantially from that of the early twentieth century (Linsdale 1936). Habitat restoration in riparian areas, for instance along the Truckee River, is expected to benefit this species and may even allow it to solidify and extend its range in the state.

HABITAT USE

HABITAT	ATLAS BLOCKS	INCIDENTAL OBSERVATIONS
Agricultural	1 (3%)	0 (0%)
Montane Forest	10 (25%)	4 (19%)
Montane Parkland	1 (3%)	0 (0%)
Pinyon-Juniper	1 (3%)	1 (5%)
Riparian	15 (38%)	14 (67%)
Sagebrush Scrub	1 (3%)	0 (0%)
Urban	10 (25%)	2 (9%)
TOTAL	39	21

Total numbers reported in the Habitat Use, Breeding Status, and Abundance tables may differ from each other (see pp. 22–23 for details). Percentage sums may differ slightly from 100% due to rounding.

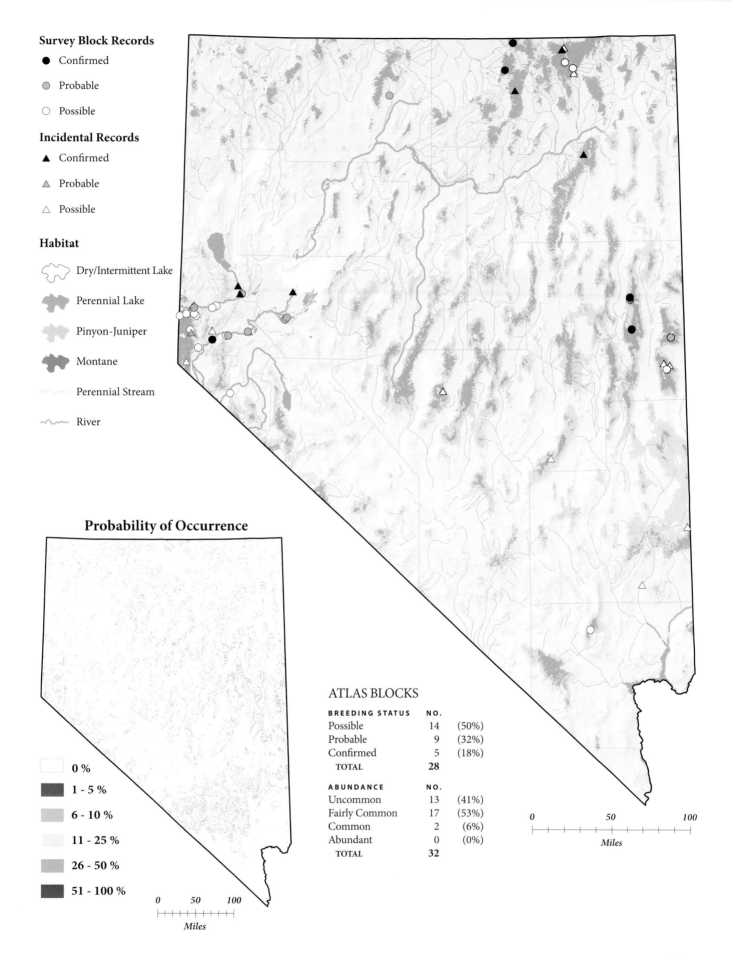

Survey Block Records

- ● Confirmed
- ● Probable
- ○ Possible

Incidental Records

- ▲ Confirmed
- ▲ Probable
- △ Possible

Habitat

Dry/Intermittent Lake

Perennial Lake

Pinyon-Juniper

Montane

Perennial Stream

River

Probability of Occurrence

- 0 %
- 1 - 5 %
- 6 - 10 %
- 11 - 25 %
- 26 - 50 %
- 51 - 100 %

0   50   100
Miles

ATLAS BLOCKS

BREEDING STATUS	NO.	
Possible	14	(50%)
Probable	9	(32%)
Confirmed	5	(18%)
TOTAL	28	

ABUNDANCE	NO.	
Uncommon	13	(41%)
Fairly Common	17	(53%)
Common	2	(6%)
Abundant	0	(0%)
TOTAL	32	

0   50   100
Miles

HAIRY WOODPECKER

Picoides villosus

Wherever there are extensive coniferous forests in Nevada, there are likely to be at least a few pairs of Hairy Woodpeckers. With the exception of the Northern Flicker, this species is probably the most widespread woodpecker in the state. Hairy Woodpeckers are quite vocal, and their telltale calls carry far through Nevada's woodlands. Despite its ubiquity, much remains unknown about the species in Nevada. In particular, three different subspecies have been recognized in the state (Linsdale 1936), yet their range boundaries and the extent of their intergradation are not well understood.

DISTRIBUTION

Hairy Woodpeckers were confirmed as breeders throughout the center of the state in the Carson Range of western Nevada; in some northern ranges, including the Jarbidge and Ruby Mountains; and in the southern Sheep Range and Spring Mountains. Many unconfirmed records also came from these areas, and there were additional possible and probable records from Humboldt and Lincoln counties.

Before the atlas survey was initiated, the status of the Hairy Woodpecker in Nevada had been described only in general terms. Alcorn (1988) considered the species to be resident but not abundant in mountain ranges statewide, and Ryser (1985) described it as a common permanent resident throughout the Great Basin.

Accounts from the surrounding states generally portray the Hairy Woodpecker as fairly common to common (e.g., Monson and Phillips 1981, Small 1994), although its range is more restricted to mountainous habitats in the south than in the north.

Hairy Woodpeckers are common across much of Oregon, although they are uncommon in the counties closest to Nevada (Littlefield 1990, Contreras and Kindschy 1996, Adamus et al. 2001). The species is a common permanent resident in Utah (Behle et al. 1985) and in much of Arizona except the southwestern part of the state (Grossi and Corman in Corman and Wise-Gervais 2005:284–285), and it is confirmed or suspected to breed in most of Idaho (Stephens and Sturts 1998).

Habitat use of the western populations is surprisingly complex. In parts of Arizona (Monson and Phillips 1981) and California (Gallagher 1997), Hairy Woodpeckers tend to favor coniferous forests. In Utah, however, the species is reportedly common in cottonwood forests at middle elevations (Behle et al. 1985). More than half of the Nevada atlas block records came from montane forest, and Hairys were the only woodpeckers other than flickers to occur frequently in pinyon-juniper habitat. Most remaining atlas records came from riparian areas. Interestingly, data from incidental observations suggest that Hairy Woodpeckers are more likely to be found in riparian habitats than in pinyon-juniper, while data from the randomly selected blocks suggest much more equal use of the two habitats. This difference likely arose from biases in the way incidental observations were collected, once again illustrating the value of habitat-stratified random sampling.

CONSERVATION AND MANAGEMENT

Continent-wide, Hairy Woodpecker populations are increasing (Sauer et al. 2005) and the species is not currently a conservation priority, even though there are reports of local declines (Jackson et al. 2002). In general, the Hairy Woodpecker's main requirement appears to be stands of mature forest with dead or dying trees for nesting (Small 1994). In some regions, Hairy Woodpeckers also seem to require relatively large forest tracts and may suffer when forests are fragmented by logging or development (Jackson et al. 2002). On the other hand, their abundance across the central mountain ranges of Nevada suggests that they are better able to cope with the naturally fragmented forests of Nevada than are most other woodpecker species in the state.

HABITAT USE

HABITAT	ATLAS BLOCKS	INCIDENTAL OBSERVATIONS
Alpine	1 (1%)	0 (0%)
Barren	0 (0%)	1 (3%)
Montane Forest	75 (52%)	16 (46%)
Montane Parkland	8 (6%)	0 (0%)
Montane Shrub	6 (4%)	4 (11%)
Pinyon-Juniper	25 (18%)	1 (3%)
Riparian	24 (16%)	12 (34%)
Sagebrush Scrub	1 (1%)	0 (0%)
Sagebrush Steppe	0 (0%)	1 (3%)
Urban	3 (2%)	0 (0%)
TOTAL	143	35

Total numbers reported in the Habitat Use, Breeding Status, and Abundance tables may differ from each other (see pp. 22–23 for details). Percentage sums may differ slightly from 100% due to rounding.

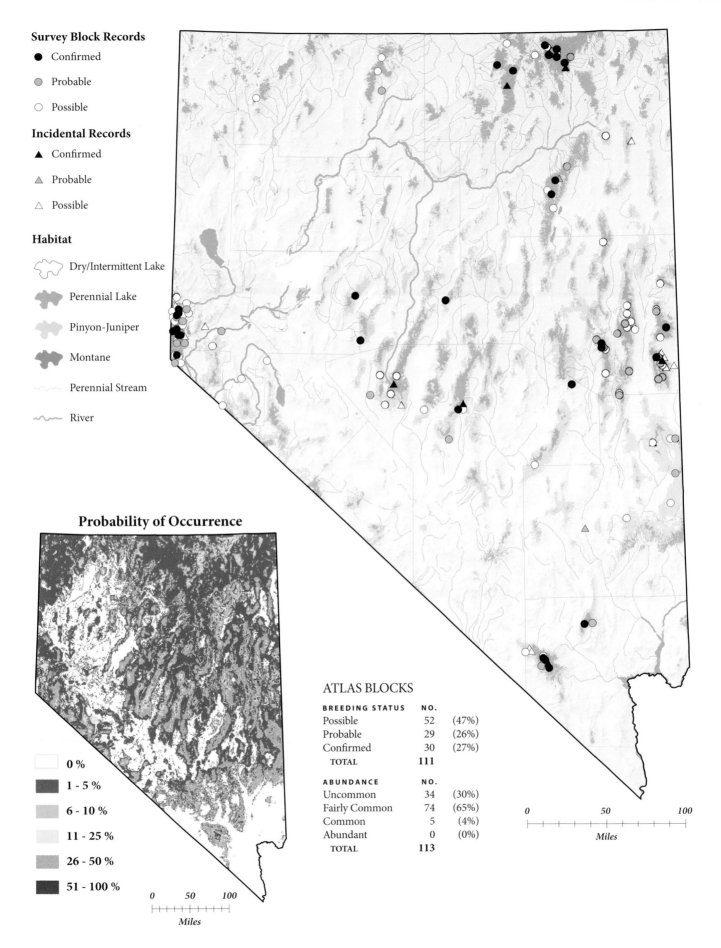

Survey Block Records

- ● Confirmed
- ● Probable
- ○ Possible

Incidental Records

- ▲ Confirmed
- ▲ Probable
- △ Possible

Habitat

- Dry/Intermittent Lake
- Perennial Lake
- Pinyon-Juniper
- Montane
- Perennial Stream
- River

Probability of Occurrence

- 0 %
- 1 - 5 %
- 6 - 10 %
- 11 - 25 %
- 26 - 50 %
- 51 - 100 %

0   50   100

Miles

ATLAS BLOCKS

BREEDING STATUS	NO.	
Possible	52	(47%)
Probable	29	(26%)
Confirmed	30	(27%)
TOTAL	111	

ABUNDANCE	NO.	
Uncommon	34	(30%)
Fairly Common	74	(65%)
Common	5	(4%)
Abundant	0	(0%)
TOTAL	113	

0   50   100

Miles

WHITE-HEADED WOODPECKER
Picoides albolarvatus

The White-headed Woodpecker is found in the thin sliver of Sierra Nevada coniferous forest that barely grazes Nevada's border around Lake Tahoe, and nowhere else in the state. Closely associated with ponderosa and Jeffrey pines, the species can be found throughout the wilderness areas of the Sierra Nevada and is at least sometimes tolerant of humans. For instance, the town of Incline Village in Washoe County was home to a pair that successfully raised young in a cavity in a wooden pylon. The nest, placed right in the middle of a busy shopping center parking lot at roughly chest height, was surely one of the most-watched White-headed Woodpecker nests in history.

DISTRIBUTION

The atlas results are consistent with older accounts of the White-headed Woodpecker's distribution in Nevada (Ryser 1985, Garrett et al. 1996). The species is restricted to the pine forests of the far western United States, where it is found from southern California north to southern British Columbia. In Nevada, all known records are from the Carson Range in Washoe, Carson, and Douglas counties.

The species was found in ten blocks but was confirmed as a breeder in only two. Most records came from montane forest or montane parkland habitats, with the remaining two records coming from urban sites. In the blocks where they were found, White-headed Woodpeckers were usually estimated to be present in multiple pairs.

Although the Carson Range is its Nevada stronghold, the White-headed Woodpecker was not detected in almost two-thirds of the blocks sampled in that area. The region's steep to-pography, which made many blocks difficult to access and survey thoroughly, may have contributed to this result. Fieldworkers also felt that most of the White-headed Woodpeckers they encountered were fairly shy and retiring, which probably also reduced detection rates. Nonetheless, the predictive map indicates a good chance of finding the species throughout the Carson Range.

CONSERVATION AND MANAGEMENT

White-headed Woodpeckers are found primarily in mixed-conifer forests, which are common throughout the Carson Range. The species is relatively poorly studied compared with other woodpeckers, and its high-elevation habitats make effective monitoring difficult (Garrett et al. 1996). The monitoring data that do exist suggest that populations are generally increasing (Sauer et al. 2005).

White-headed Woodpeckers depend on conifers both for nesting and for food; much of their diet consists of conifer seeds (Ligon 1973, Garrett et al. 1996). Although they are often associated with ponderosa pine, White-headed Woodpeckers tend to be most common in areas with multiple pine species, presumably because these conditions increase the chance that at least one tree species will have a good seed crop in any given year. The relationship between White-headed Woodpeckers and conifers is further complicated by differences in conifer use between the sexes and at different times of the year (Morrison and With 1987).

Nests are usually placed in large-diameter, usually dead, standing trees, although downed trees are sometimes used (Garrett et al. 1996). The frequent use of older and dead trees suggests that in addition to a mixture of conifer species, White-headed Woodpeckers require mixed-age forest stands. Due to its limited range and small global population size, the White-headed Woodpecker is a Partners in Flight Watch List species and a Priority species in Nevada (Neel 1999a, Rich et al. 2004). Forest management that would remove dead trees or reduce the forest's structural or species diversity is likely to be harmful for White-headed Woodpeckers.

HABITAT USE

HABITAT	ATLAS BLOCKS	INCIDENTAL OBSERVATIONS
Montane Forest	11 (74%)	0 (0%)
Montane Parkland	2 (13%)	0 (0%)
Urban	2 (13%)	0 (0%)
TOTAL	15	0

Total numbers reported in the Habitat Use, Breeding Status, and Abundance tables may differ from each other (see pp. 22–23 for details). Percentage sums may differ slightly from 100% due to rounding.

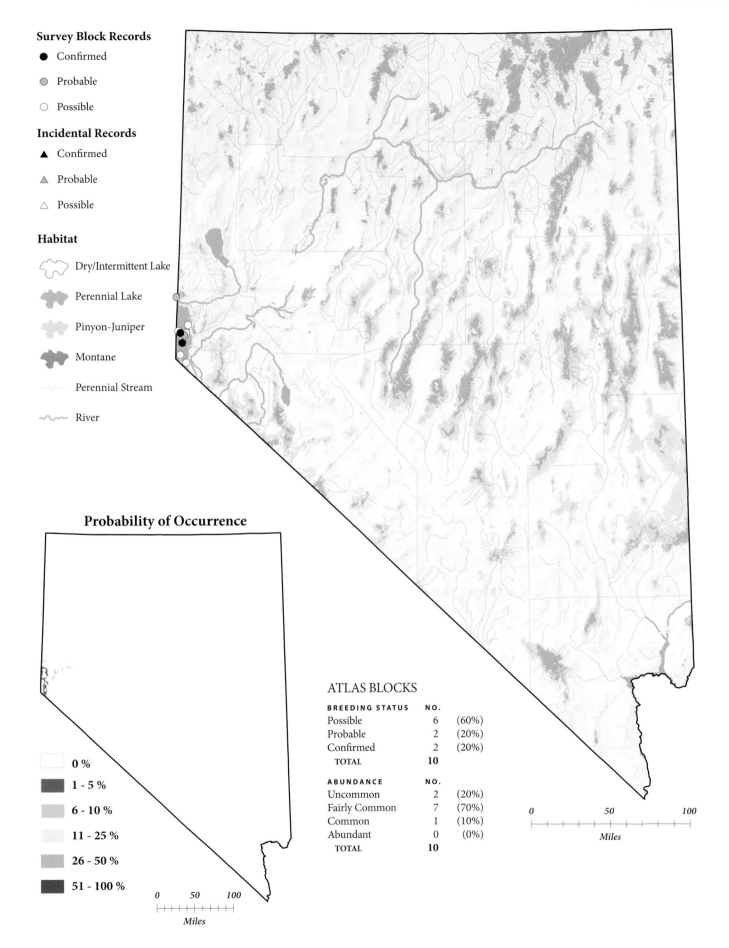

Survey Block Records

● Confirmed

◉ Probable

○ Possible

Incidental Records

▲ Confirmed

△ Probable

△ Possible

Habitat

Dry/Intermittent Lake

Perennial Lake

Pinyon-Juniper

Montane

Perennial Stream

River

Probability of Occurrence

0 %

1 - 5 %

6 - 10 %

11 - 25 %

26 - 50 %

51 - 100 %

0   50   100
Miles

ATLAS BLOCKS

BREEDING STATUS	NO.	
Possible	6	(60%)
Probable	2	(20%)
Confirmed	2	(20%)
TOTAL	10	

ABUNDANCE	NO.	
Uncommon	2	(20%)
Fairly Common	7	(70%)
Common	1	(10%)
Abundant	0	(0%)
TOTAL	10	

0   50   100
Miles

AMERICAN THREE-TOED WOODPECKER

Picoides dorsalis

The American Three-toed Woodpecker is primarily a bird of the vast boreal forests of Canada and Alaska, but its range also extends down the Rocky Mountain chain as far south as New Mexico. Although the European and American forms were once considered to be the same species, the two were recently split on the basis of genetic and vocalization differences (AOU 2003). In Nevada, the American Three-toed Woodpecker is found only in the high-elevation spruce-fir forests of the Snake Range in southeastern White Pine County, where it is not common but has been established for a long time. The Nevada population is small and very isolated, and the species' long-term prospects in the state are thus uncertain.

DISTRIBUTION

American Three-toed Woodpeckers were found in three blocks in the Snake Range. Two of these blocks had birds in at least two of the survey years, and breeding was confirmed. Possible breeding was reported from a third block, and there was an incidental record of confirmed breeding nearby. Although the woodpeckers were not common anywhere, observers estimated more than a single pair in both blocks where breeding was confirmed. All records came from montane forest habitat.

Records of the American Three-toed Woodpecker's presence in Nevada date back to at least the 1930s (Linsdale 1936), and it is interesting to speculate why it does not occur elsewhere in the state. The answer may be that the spruce-fir habitat it typically uses (see Bock and Bock 1974, Hill 2002) is restricted in Nevada to just a few ranges in the vicinity of Great Basin National Park (Charlet 1996). Neel (1999a) raised the possibility that the

species may occur in other nearby ranges with spruce-fir assemblages, such as the Schell Creek and Cherry Creek ranges. Currently, there is no evidence for the presence of American Three-toed Woodpeckers in these areas, but the species occupies remote high-elevation sites that are infrequently visited by birders and it might easily have been missed. Although these woodpeckers are rare in the state as a whole, the map predicts a good chance of finding them within the restricted area where they occur. These predictions, however, are based on very little data.

CONSERVATION AND MANAGEMENT

The American Three-toed Woodpecker is not encountered frequently anywhere within its extensive range (Hill 2002), probably due to a combination of its preference for inaccessible habitats and its intrinsically low population densities. It is relatively sedentary and not as prone to wandering or periodic irruptions as its close cousin, the Black-backed Woodpecker (Yunick 1985). Given its sedentary nature and the naturally fragmented distribution of suitable high-elevation nesting sites in the Great Basin, the species is probably not a likely candidate for range expansion in Nevada and is susceptible to local extirpation.

While its distribution in Nevada may be constrained primarily by natural biogeographic processes (Johnson 1975), the American Three-toed Woodpecker does face anthropogenic threats as well. Researchers working at the southern limit of the species' continental range have demonstrated that these woodpeckers are attracted to forests with a high proportion of dead and dying trees, including those affected by fire, windstorms, or insect outbreaks (Ryser 1985, Versaw in Kingery 1998:264–265)—conditions discouraged by modern forest management practices. Fortunately for the Nevada population, much of its habitat here (including that within Great Basin National Park and Mount Moriah Wilderness Area) is not subjected to harvest-oriented forest management.

The American Three-toed Woodpecker is a Priority species of Nevada Partners in Flight (Neel 1999a). The Nevada population is at the periphery of the species' range, though, and in a geographically broader view of conservation priorities its protection in Nevada ranks lower than that of many other species in the state.

HABITAT USE

HABITAT	ATLAS BLOCKS	INCIDENTAL OBSERVATIONS
Montane Forest	5 (100%)	1 (100%)
TOTAL	5	1

Total numbers reported in the Habitat Use, Breeding Status, and Abundance tables may differ from each other (see pp. 22–23 for details). Percentage sums may differ slightly from 100% due to rounding.

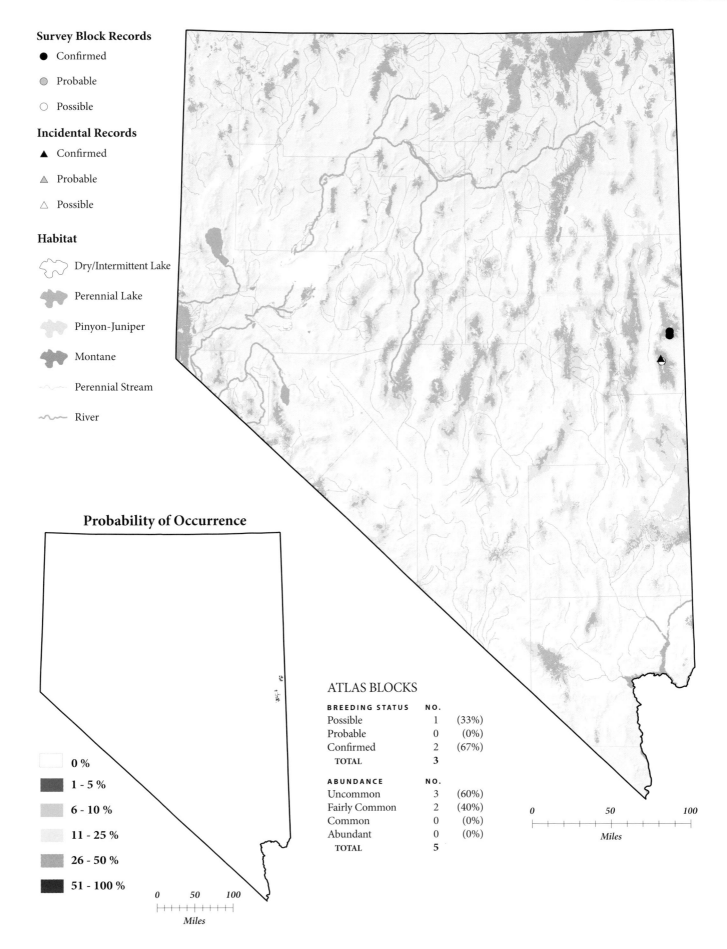

Survey Block Records

● Confirmed

◉ Probable

○ Possible

Incidental Records

▲ Confirmed

△ Probable

△ Possible

Habitat

Dry/Intermittent Lake

Perennial Lake

Pinyon-Juniper

Montane

Perennial Stream

River

Probability of Occurrence

0 %

1 - 5 %

6 - 10 %

11 - 25 %

26 - 50 %

51 - 100 %

0   50   100
Miles

ATLAS BLOCKS

BREEDING STATUS	NO.	
Possible	1	(33%)
Probable	0	(0%)
Confirmed	2	(67%)
TOTAL	3	

ABUNDANCE	NO.	
Uncommon	3	(60%)
Fairly Common	2	(40%)
Common	0	(0%)
Abundant	0	(0%)
TOTAL	5	

0   50   100
Miles

BLACK-BACKED WOODPECKER
Picoides arcticus

The Carson Range of far western Nevada hosts an assemblage of breeding bird species that are rarely found elsewhere in the state. Some, like Cassin's Vireo and the Nashville Warbler, are quite easy to observe; others, like the White-headed Woodpecker and Pine Grosbeak, are more difficult; among the rarest of all is the Black-backed Woodpecker. This woodpecker is not particularly averse to humans, however, and can be viewed at close range along trails through high-elevation lodgepole pine forests. Spooner Lake and Galena Creek parks have been semi-reliable spots for finding this species in recent years; however, there are still very few breeding records in the state.

DISTRIBUTION

There were five sightings of the Black-backed Woodpecker during the atlas period—three were from blocks, and all five were in the Carson Range. The only confirmed breeding record involved a sighting of a fledged young in 1998 at Galena Creek Park in southern Washoe County, a fairly low-elevation site (ca. 6,500 feet, or 2,000 meters) close to the desert-forest ecotone. In 2002, after the atlas fieldwork was completed, nests were found in the eastern Lake Tahoe Basin at Marlette Lake in Washoe County and along Logan House Creek in Douglas County (Richardson 2003). Nesting has continued at both sites (W. Richardson, pers. comm.). A family group with dependent fledglings was also

observed during a point count in recently burned Ash Canyon in 2005 (D. Booth, pers. comm.).

Nevada records of the Black-backed Woodpecker are rare in the earlier literature. Linsdale (1936) cited only three from the nineteenth century, and Alcorn (1988) provided only a few additional sightings. There are no verified records from sites away from the Carson Range (Ryser 1985). Most Nevada sightings occur in the fall and winter (Richardson 2003).

The Black-backed Woodpecker is a rare and local resident in the mountains of northern California (Small 1994). In Oregon, it is uncommon and also somewhat local in the central and northeastern mountain ranges (Contreras 1999, Adamus et al. 2001). In Idaho, breeding is concentrated in the northwest, far from the Nevada border (Stephens and Sturts 1998). There are no records of the Black-backed Woodpecker from Utah (Behle et al. 1985) or Arizona (Monson and Phillips 1981).

All atlas records came from montane forest habitats, in keeping with the species' use of coniferous forests throughout its range (Dixon and Saab 2000). In the Sierra Nevada, Black-backed Woodpeckers seem to be especially partial to lodgepole pines (Gaines 1992, Small 1994) around clearings and mountain meadows (J. Nachlinger, pers. comm.). Like the woodpecker, lodgepole pines are restricted to the western edge of Nevada (Charlet 1996).

CONSERVATION AND MANAGEMENT

The Black-backed Woodpecker is known as a postfire specialist (Dixon and Saab 2000), and the species probably also benefits from other perturbations—such as insect outbreaks or floods—that kill conifers. Black-backed Woodpecker numbers often increase dramatically following fires, and they may spread into previously unoccupied areas following a forest burn. They have been seen after recent burns in the Carson Range, and at least one family with dependent young was observed in a burned area. How strongly the species responds to recent forest fires in Nevada, however, is still unclear. Nests found at two sites in the Carson Range in 2002 were not in recently burned areas, but they were close to large stands of trees killed by beetle infestations (Richardson 2003). Research done elsewhere in the species' range (Dixon and Saab 2000) indicates that reducing fire suppression activities and restricting postfire timber salvage benefits the species.

HABITAT USE

HABITAT	ATLAS BLOCKS	INCIDENTAL OBSERVATIONS
Montane Forest	3 (100%)	2 (100%)
TOTAL	3	2

Total numbers reported in the Habitat Use, Breeding Status, and Abundance tables may differ from each other (see pp. 22–23 for details). Percentage sums may differ slightly from 100% due to rounding.

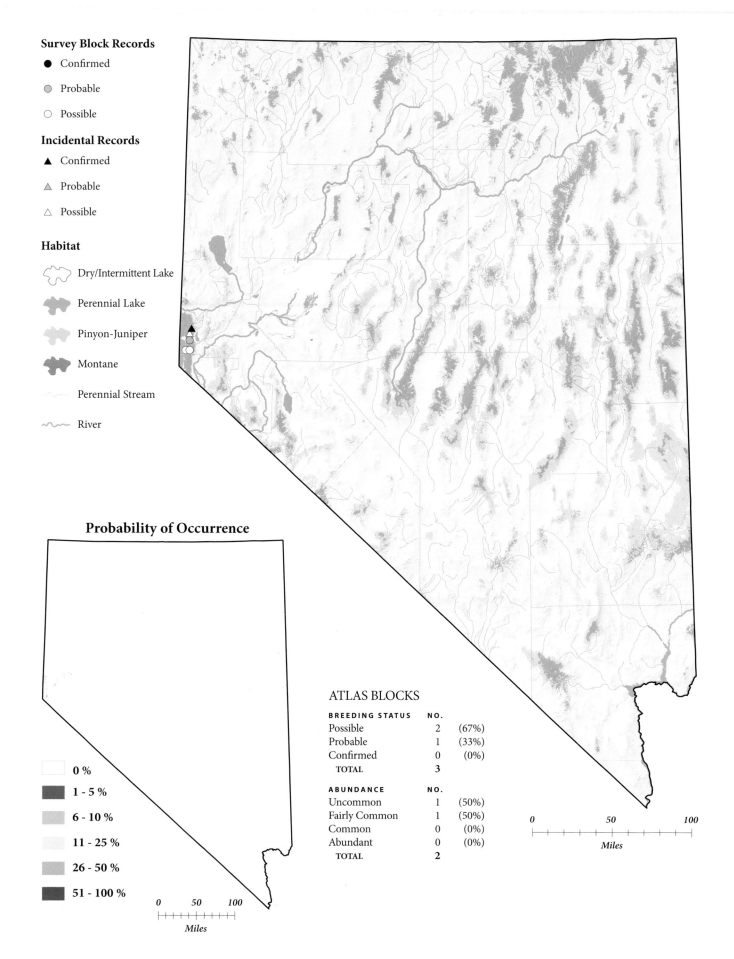

Survey Block Records

● Confirmed

◉ Probable

○ Possible

Incidental Records

▲ Confirmed

△ Probable

△ Possible

Habitat

⬡ Dry/Intermittent Lake

⬢ Perennial Lake

⬢ Pinyon-Juniper

⬢ Montane

∿ Perennial Stream

∿ River

Probability of Occurrence

☐ 0 %

■ 1 - 5 %

■ 6 - 10 %

☐ 11 - 25 %

■ 26 - 50 %

■ 51 - 100 %

0   50   100
Miles

ATLAS BLOCKS

BREEDING STATUS	NO.	
Possible	2	(67%)
Probable	1	(33%)
Confirmed	0	(0%)
TOTAL	3	

ABUNDANCE	NO.	
Uncommon	1	(50%)
Fairly Common	1	(50%)
Common	0	(0%)
Abundant	0	(0%)
TOTAL	2	

0   50   100
Miles

NORTHERN FLICKER
Colaptes auratus

Nevada has a diverse woodpecker fauna, but most species are restricted to small areas near the state's periphery. With its strident calls, bizarre courtship dances, and clownish plumage, the Northern Flicker is an exception to this generalization: it is by far our most widespread and numerous woodpecker. During the breeding season, Northern Flickers can be found almost anyplace that has a dead tree or two, from montane forests to isolated woodlands in the desert, city parks, and older residential neighborhoods. Nonetheless, the species is declining throughout North America, and the maintenance of dead trees and tree limbs seems key to its long-term persistence.

DISTRIBUTION

Atlas workers found Northern Flickers throughout Nevada, although most confirmed breeding records came from the northern two-thirds of the state. Overall, breeding was confirmed in more than a third of the 340 blocks in which Northern Flickers were found, and multiple pairs were thought to be present in most blocks for which abundance estimates were supplied.

The Northern Flicker has long been considered common in Nevada (Linsdale 1936, Alcorn 1988). In his survey of the Great Basin, Ryser (1985) called this species the most abundant and widespread woodpecker in the region and remarked that its status had changed little since the nineteenth century. Members of the Death Valley Expedition of the late nineteenth century saw the Northern Flicker in many places (Fisher 1893). The Northern Flicker is reported to be a common breeder in the states surrounding Nevada as well, except in the Sonoran Desert lowlands (Behle et al. 1985, Rosenberg et al. 1991, Small 1994, Stephens and Sturts 1998, Adamus et al. 2001, Spence in Corman and Wise-Gervais 2005:290–291).

Woodlands of all sorts seem to harbor breeding Northern Flickers (Moore 1995), and the species occurred in a wide variety of habitats during the atlas period. Most records, however, came from montane forest, pinyon-juniper, and riparian habitats. Consequently, Northern Flickers are predicted to occur with high probability throughout the higher-elevation portions of the state, but to be less likely to occur in lowland areas such as the salt deserts of southern and western Nevada.

CONSERVATION AND MANAGEMENT

Although the Northern Flicker is a widespread and adaptable species, its numbers have declined rangewide in recent decades (Sauer et al. 2005). Clarifying which factors are causing Northern Flickers to decline is particularly important in the Great Basin, because flickers play a keystone role as excavators of nest holes that are subsequently used by a range of secondary cavity nesters, including Tree Swallows and bluebirds (e.g., Dobkin et al. 1995).

Competition with European Starlings is often cited as a cause of the species' decline (e.g., Gallagher 1997), but hard evidence for this is weak (Moore 1995). Starlings are certainly capable of displacing Northern Flickers from nest holes (Weitzel 1988), but it is unclear how often this occurs. Starlings breed early in the season, and it is possible that the evicted woodpeckers could re-nest before the breeding season ended (Moore 1995). Another possible explanation for the widespread decline is the loss of suitable nest sites, especially the removal of dead branches and trees (Moore 1995). In the Midwest, Northern Flickers have evidently benefited from the planting of hedgerows and shelterbelts (Moore 1995), and small woodlots around ranches and irrigation projects may have had similar effects in Nevada. On the other hand, relatively few atlas records came from agricultural or other low-elevation settings, despite the Northern Flicker's flexibility in habitat use. In Nevada, therefore, protection of standing dead-wood in forested mountain habitats and riparian areas may be the best strategy for long-term protection of the species until better information becomes available.

HABITAT USE

HABITAT	ATLAS BLOCKS	INCIDENTAL OBSERVATIONS
Agricultural	8 (2%)	1 (1%)
Alpine	5 (1%)	0 (0%)
Grassland	0 (0%)	1 (1%)
Mesquite	1 (<1%)	1 (1%)
Mojave	3 (1%)	3 (2%)
Montane Forest	134 (26%)	36 (26%)
Montane Parkland	14 (3%)	4 (3%)
Montane Shrub	39 (8%)	10 (7%)
Pinyon-Juniper	128 (25%)	14 (10%)
Riparian	134 (26%)	57 (41%)
Salt Desert Scrub	2 (<1%)	0 (0%)
Sagebrush Scrub	10 (2%)	3 (2%)
Sagebrush Steppe	14 (3%)	3 (2%)
Urban	18 (4%)	6 (4%)
Wetland	2 (<1%)	0 (0%)
TOTAL	512	139

Total numbers reported in the Habitat Use, Breeding Status, and Abundance tables may differ from each other (see pp. 22–23 for details). Percentage sums may differ slightly from 100% due to rounding.

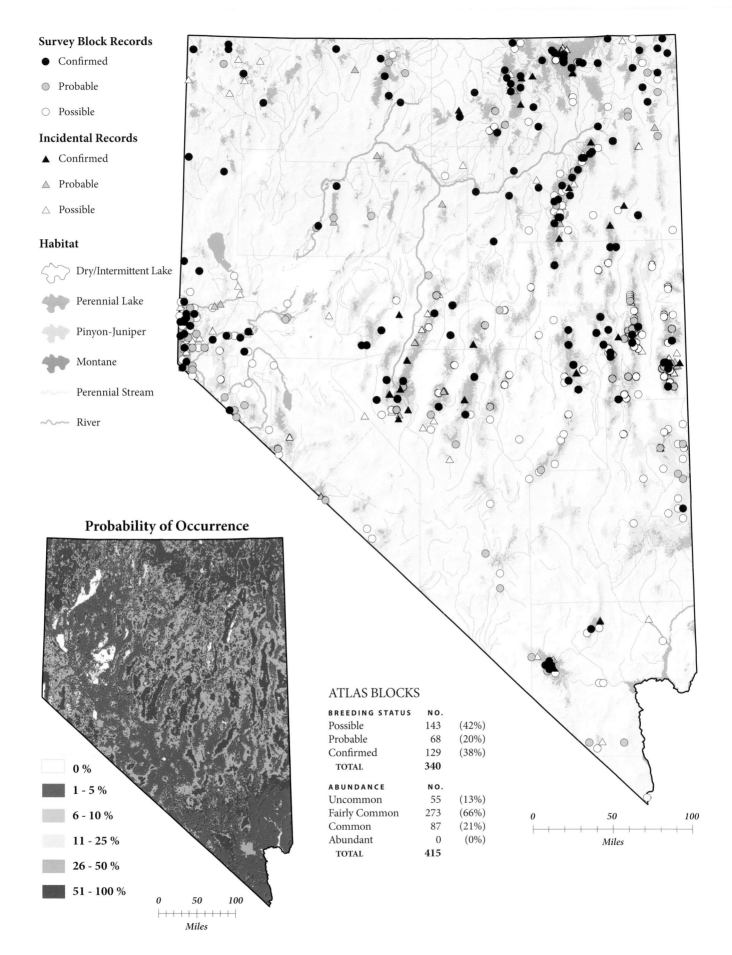

Survey Block Records

● Confirmed

◐ Probable

○ Possible

Incidental Records

▲ Confirmed

△ Probable

△ Possible

Habitat

Dry/Intermittent Lake

Perennial Lake

Pinyon-Juniper

Montane

Perennial Stream

River

Probability of Occurrence

0 %

1 - 5 %

6 - 10 %

11 - 25 %

26 - 50 %

51 - 100 %

0   50   100
Miles

ATLAS BLOCKS

BREEDING STATUS	NO.	
Possible	143	(42%)
Probable	68	(20%)
Confirmed	129	(38%)
TOTAL	340	

ABUNDANCE	NO.	
Uncommon	55	(13%)
Fairly Common	273	(66%)
Common	87	(21%)
Abundant	0	(0%)
TOTAL	415	

0   50   100
Miles

GILDED FLICKER
Colaptes chrysoides

The Gilded Flicker was one of the more unexpected finds during the atlas project. Before that, the nearest confirmed breeding grounds were in west-central Arizona and California's San Bernardino County. The Nevada observations, taken from a Joshua tree and yucca forest between the southern Nevada town of Searchlight and the California border, provided Nevada's first breeding confirmation for the species.

DISTRIBUTION

For the most part, Gilded Flickers are restricted to the biologically rich Sonoran Desert communities of southern Arizona and northwestern Mexico (Corman in Corman and Wise-Gervais 2005:292–293). The species does occur northward along the Colorado River and into the dry Mojave Desert of southern California, but the northern limit of its distribution is not well known. This uncertainty probably arises because Gilded Flickers were long classified as a subspecies of Northern Flicker and thus did not arouse much interest among birders. Another obstacle to better understanding the limits of the Gilded Flicker's distribution is the hot, dry, and generally inhospitable terrain in which it occurs.

Atlas workers first found Gilded Flickers in 1997, and by 2000 there were records from at least seven locations within a small area of southern Clark County. Breeding was confirmed at only one site, although probable breeding was recorded at five of the other sites. Multiple pairs were thought to be present in some places; for instance, up to twenty Gilded Flickers were present in the foothills of the southern El Dorado Mountains (R. Saval, pers. comm.). All reports came from Mojave habitat.

CONSERVATION AND MANAGEMENT

Until recently, Gilded Flickers were considered a subspecies of the much more widespread Northern Flicker, with which they hybridize. The two species were split, however, because hybridization is relatively rare and the two forms use distinct habitats (AOU 1995). The population status and trends of the Gilded Flicker are not well known because most analyses combine the data for all flickers. Earlier studies suggested that the U.S. population is faring well, although concerns have been raised about the situation in Mexico (Moore 1995). More recent analysis of Breeding Bird Survey data, however, suggests that flickers in southern Arizona (which are presumably mostly Gilded Flickers) are declining just like flickers are elsewhere in North America (Sauer et al. 2005).

The greatest threat to Gilded Flickers is probably deterioration and loss of its favored saguaro cactus habitat. This is not a problem in Nevada, where saguaro does not occur, although availability of suitable nesting cavities is still likely to be important. The Gilded Flicker is a Covered species under the Lower Colorado River Multi-species Conservation Program (BOR-LCR 2004). Given how little we know about this species in Nevada, it is difficult to devise conservation strategies. On the one hand, the Nevada portion of the Gilded Flicker's range may be of only marginal importance for the global population of the species. On the other hand, the species has a small overall range, which may mean that it should receive attention wherever it occurs. Moreover, because of its reliance in Nevada on mature Joshua trees and yuccas for nesting cavities, the Gilded Flicker may serve as a good indicator species of intact, healthy Mojave habitat.

It is unclear whether Gilded Flickers are a recent addition to the Nevada avifauna or have simply gone undetected in the past. In the short term, identifying the limits of the species' range and determining whether there is range overlap and interbreeding with Northern Flickers (see Grudzien et al. 1987) are probably the next steps toward understanding the status of the species in the state.

HABITAT USE

HABITAT	ATLAS BLOCKS	INCIDENTAL OBSERVATIONS
Mojave	1 (100%)	6 (100%)
TOTAL	1	6

Total numbers reported in the Habitat Use, Breeding Status, and Abundance tables may differ from each other (see pp. 22–23 for details). Percentage sums may differ slightly from 100% due to rounding.

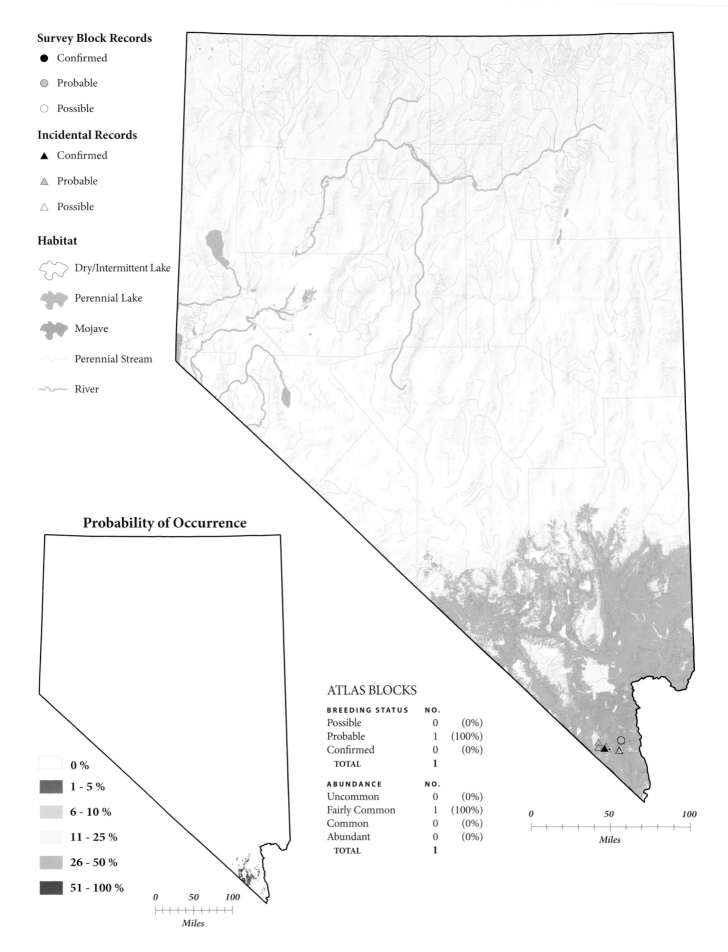

Survey Block Records

● Confirmed

● Probable

○ Possible

Incidental Records

▲ Confirmed

△ Probable

△ Possible

Habitat

Dry/Intermittent Lake

Perennial Lake

Mojave

Perennial Stream

River

Probability of Occurrence

☐ 0 %

■ 1 - 5 %

■ 6 - 10 %

11 - 25 %

26 - 50 %

■ 51 - 100 %

0   50   100

Miles

ATLAS BLOCKS

BREEDING STATUS	NO.	
Possible	0	(0%)
Probable	1	(100%)
Confirmed	0	(0%)
TOTAL	**1**	

ABUNDANCE	NO.	
Uncommon	0	(0%)
Fairly Common	1	(100%)
Common	0	(0%)
Abundant	0	(0%)
TOTAL	**1**	

0   50   100

Miles

PILEATED WOODPECKER
Dryocopus pileatus

The Pileated Woodpecker frustrated birders and atlas organizers alike because it was not confirmed as a breeder during the atlas project despite strong suspicions that it was nesting in the Carson Range of far western Nevada. Birders have continued to observe this large woodpecker at sites close to the California border since the fieldwork ended, and it was finally confirmed in 2006.

DISTRIBUTION

Like the White-headed Woodpecker, Red-breasted Sapsucker, and Black-backed Woodpecker, the Pileated Woodpecker is found throughout the Sierra Nevada but barely enters Nevada. All three atlas records came from mixed conifer forests immediately to the east of Lake Tahoe in the Carson Range.

The most compelling evidence of breeding that was observed during the atlas period was a partially excavated nest hole with the characteristic shape of a Pileated Woodpecker's nest found at Little Valley in Washoe County in 1997 (J. Eidel, pers. comm.) and a pair accompanied by a fledgling seen in the vicinity of the Chimney Beach Trail at Lake Tahoe in August 1999 (G. Scyphers, pers. comm.). There have been numerous other sightings in the same general area in subsequent years, including one of a family group near Logan House Creek in 2005 (W. Richardson, pers. comm.), culminating in 2006 in the discovery of an occupied nest near the state line in the Carso Range (D. Catalano, pers. comm.). Whether these sightings represent recent movements of birds into the state or just more people searching for them is unclear.

There were no reports of the species from elsewhere in the state, and none appear in accounts published before the atlas project (Ryser 1985, Alcorn 1988). With the exception of the mountains of northern California, Pileated Woodpeckers do not occur in nearby portions of neighboring states (Bull and Jackson 1995), but they do occupy western and northeastern Oregon and northern Idaho (Stephens and Sturts 1998, Adamus et al. 2001).

CONSERVATION AND MANAGEMENT

The Pileated Woodpecker is one of the happier stories in the annals of American conservation history. This large and striking bird is an emblem of wilderness—and it is currently increasing in abundance and expanding its already extensive North American range (Sauer et al. 2005). Two characteristics account for most of the Pileated Woodpecker's success: it readily accepts second-growth forest habitats, and it is fairly tolerant of humans. In Washington State, for example, Pileated Woodpeckers can even be found in forested sections of city parks (Smith et al. 1997).

At present, the Pileated Woodpecker's situation in Nevada appears tenuous, with sightings limited to one area right on the border with California. Conditions in the Carson Range do seem appropriate for continued breeding, however, and, barring any major natural or anthropogenic changes of the Sierra Nevada's forests, we can be cautiously optimistic that additional surveys will confirm the Pileated Woodpecker is a regular, albeit rare, breeder in Nevada, perhaps even throughout the Carson Range.

HABITAT USE

HABITAT	ATLAS BLOCKS	INCIDENTAL OBSERVATIONS
Montane Forest	2 (100%)	0 (0%)
TOTAL	2	0

Total numbers reported in the Habitat Use, Breeding Status, and Abundance tables may differ from each other (see pp. 22–23 for details). Percentage sums may differ slightly from 100% due to rounding.

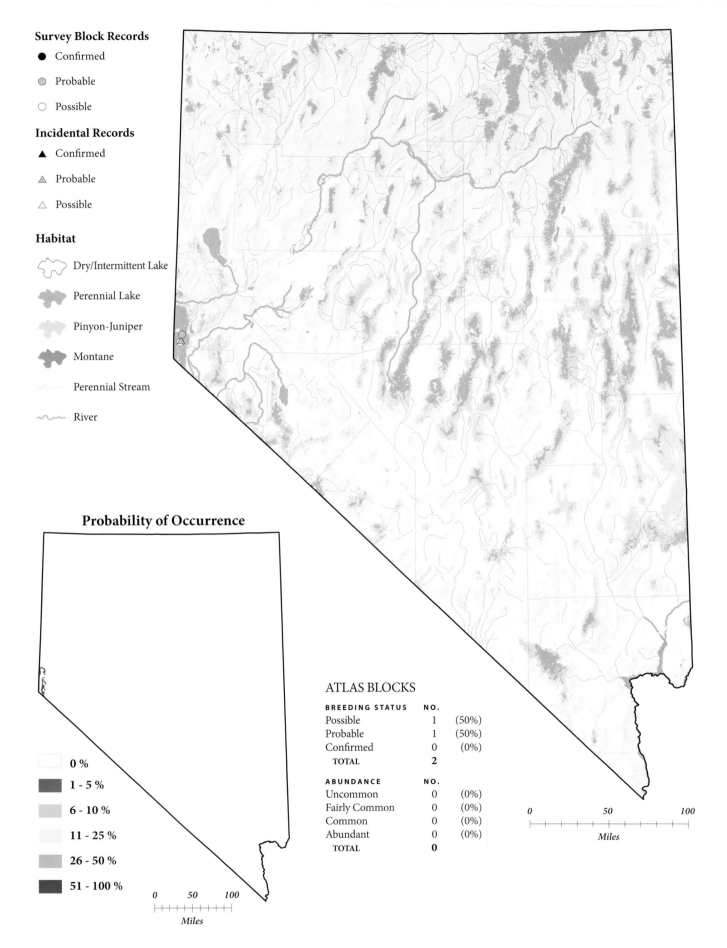

Survey Block Records

● Confirmed

◐ Probable

○ Possible

Incidental Records

▲ Confirmed

△ Probable

△ Possible

Habitat

⬡ Dry/Intermittent Lake

⬢ Perennial Lake

⬢ Pinyon-Juniper

⬢ Montane

〜 Perennial Stream

〜 River

Probability of Occurrence

▢ 0 %

■ 1 - 5 %

▨ 6 - 10 %

▢ 11 - 25 %

▨ 26 - 50 %

■ 51 - 100 %

0   50   100
Miles

ATLAS BLOCKS

BREEDING STATUS	NO.	
Possible	1	(50%)
Probable	1	(50%)
Confirmed	0	(0%)
TOTAL	2	

ABUNDANCE	NO.	
Uncommon	0	(0%)
Fairly Common	0	(0%)
Common	0	(0%)
Abundant	0	(0%)
TOTAL	0	

0   50   100
Miles

OLIVE-SIDED FLYCATCHER

Contopus cooperi

The Olive-sided Flycatcher is not much to look at: bob-tailed and bull-headed, medium-sized, grayish, and rather devoid of field marks. But it is distinctive in other ways. The Olive-sided Fly-catcher's odd song (*quick, three beers!*) is diagnostic and easily learned, and its stereotypical behavior of perching at the very top of tall snags is remarkably consistent. Olive-sided Flycatchers are found in coniferous forests, particularly near openings, at middle to high elevations in Nevada, and there is concern about recent declines in their numbers.

DISTRIBUTION

Atlas workers found Olive-sided Flycatchers at a variety of locations across Nevada, but breeding was confirmed in only three distantly separated mountain ranges: the Carson Range, the Jarbidge Mountains, and the Schell Creek Range. More than half of the records were from the Carson Range. Olive-sided Flycatcher breeding was confirmed in only 10% of the blocks, and 40% of the blocks where the species occurred were thought to contain only a single pair. This result agrees with earlier reports indicating that the Olive-sided Flycatcher is uncommon in Nevada (Linsdale 1936, Ryser 1985, Alcorn 1988), and has been since historical times (Fisher 1893). The species seems to be more common and widespread to the north in Idaho and Oregon (Stephens and Sturts 1998, Marshall et al. 2003, Adamus et al. 2001), but the situation in California, Utah, and Arizona more closely approxi-

mates that in Nevada (Behle et al. 1985, Small 1994, Wise-Gervais in Corman and Wise-Gervais 2005:296–297).

With such low densities, it is possible that evidence of breeding was simply overlooked in some locations. Reinforcing this possibility, Neel's (1999a) overview for Nevada gives breeding records along the western edge of the state from the Carson Range to the Spring Mountains, and along the eastern edge of the state from the Jarbidge Mountains to the Snake Range. The predictive map indicates other places that might harbor breeders. Atlas records from lower foothills and valley floors, which probably reflect migrants rather than breeders, caused the model to predict a small chance of breeding across much of the mid-elevation extent of the state.

The Olive-sided Flycatcher is typically portrayed as a boreal forest specialist. In the Great Basin, it usually breeds in landscapes where firs, spruces, or both are interspersed with forest openings (Ryser 1985, Altman and Sallabanks 2000). Atlas habitat data confirm this pattern. In Arizona, Olive-sided Flycatchers are more common in extensive pine forests than in peripheral oak habitats (Monson and Phillips 1981, Wise-Gervais in Corman and Wise-Gervais 2005:296–297), and in California, they range from montane forests all the way down to sea level in eucalyptus groves (Small 1994). Contrary to earlier suggestions, recent evidence indicates that Olive-sided Flycatchers do not prefer recently burned areas (Meehan and George 2003).

CONSERVATION AND MANAGEMENT

Significant population declines and range contractions in recent decades (Altman and Sallabanks 2000) are reasons for concern. In our region, the Olive-sided Flycatcher has declined in Oregon (Marshall et al. 2003) and in the central Sierra Nevada (Small 1994). Habitat loss on the breeding grounds and especially on the wintering grounds is thought to be the main culprit (Altman and Sallabanks 2000).

Partners in Flight has designated the Olive-sided Flycatcher a Priority species in Nevada (Neel 1999a) and a Watch List species rangewide (Rich et al. 2004), and wise management of Nevada's forested habitats is likely to be a key element in conserving the species here. Possible strategies in Nevada, where the basic status is still not well known (Neel 1999a), include management for open-canopy, old-growth forests (Neel 1999a), preservation of dead standing trees (Gaines 1992), and additional monitoring.

HABITAT USE

HABITAT	ATLAS BLOCKS	INCIDENTAL OBSERVATIONS
Montane Forest	17 (71%)	1 (20%)
Montane Parkland	1 (4%)	0 (0%)
Montane Shrub	3 (13%)	0 (0%)
Pinyon-Juniper	2 (8%)	0 (0%)
Riparian	1 (4%)	3 (60%)
Wetland	0 (0%)	1 (20%)
TOTAL	24	5

Total numbers reported in the Habitat Use, Breeding Status, and Abundance tables may differ from each other (see pp. 22–23 for details). Percentage sums may differ slightly from 100% due to rounding.

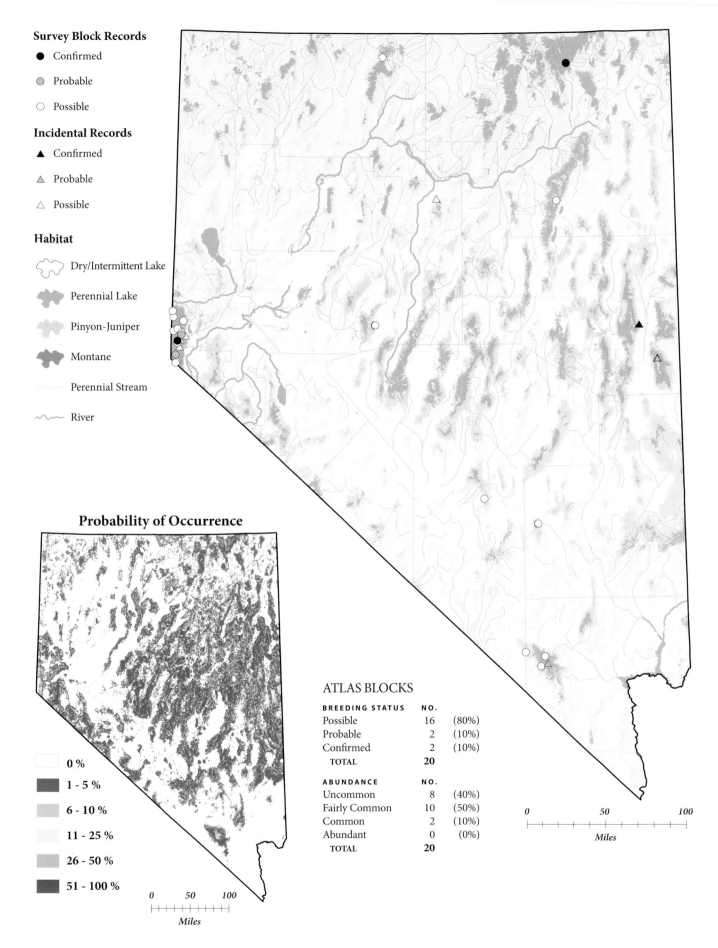

Survey Block Records

- ● Confirmed
- ◐ Probable
- ○ Possible

Incidental Records

- ▲ Confirmed
- △ Probable
- △ Possible

Habitat

- Dry/Intermittent Lake
- Perennial Lake
- Pinyon-Juniper
- Montane
- Perennial Stream
- River

Probability of Occurrence

- 0 %
- 1 - 5 %
- 6 - 10 %
- 11 - 25 %
- 26 - 50 %
- 51 - 100 %

0   50   100
Miles

ATLAS BLOCKS

BREEDING STATUS	NO.	
Possible	16	(80%)
Probable	2	(10%)
Confirmed	2	(10%)
TOTAL	**20**	

ABUNDANCE	NO.	
Uncommon	8	(40%)
Fairly Common	10	(50%)
Common	2	(10%)
Abundant	0	(0%)
TOTAL	**20**	

0   50   100
Miles

WESTERN WOOD-PEWEE
Contopus sordidulus

There is perhaps no better place to start a summer morning than at a campground in the Nevada mountains. Even before it is light the birds are stirring: Pine Siskins chattering in the tall pines, Warbling Vireos calling from the broadleaved trees, and right in the campground itself the haunting dawn song of the Western Wood-Pewee. The Wood-Pewee's performance may last for just a few minutes, but that does not mean he is silent for the rest of the morning. Even during the heat of the afternoon, when most other birds are disinclined to sing, the lazy *peeeeeerrr . . .* of the wood-pewee enlivens the montane forests.

DISTRIBUTION

Atlas workers found Western Wood-Pewees at many locations throughout Nevada, but breeding was confirmed in only 13% of the blocks where they occurred. Widespread breeding was found only in the Carson Range of far western Nevada; elsewhere, there were scattered breeding records as far north as the Jarbidge Mountains and as far south as the Spring Mountains. Western Wood-Pewees were not especially common anywhere, and more than three-quarters of the blocks in which they were found were estimated to have fewer than ten pairs.

Earlier writers considered the Western Wood-Pewee to be widely distributed in Nevada but provided few details on its ecological and elevational limits (Linsdale 1936, Ryser 1985, Alcorn 1988). Linsdale (1936) stated that the species can be found wherever there are trees, although Bemis and Rising (1999) reported it to be absent from dense forests. Wood-pewees are widespread as breeders in the forests of northern California, Oregon, Idaho, Utah, and all but southwestern Arizona (Behle et al. 1985, Small 1994, Stephens and Sturts 1998, Adamus et al. 2001, Wise-Gervais in Corman and Wise-Gervais 2005:300–301). In the southern Sierra Nevada, it is the most abundant, ubiquitous, and conspicuous flycatcher present (Gaines 1992).

Most atlas records came from montane forests or riparian habitat. While few atlas records were from lowland riparian woodlands, the species was reported to be abundant along the lower Truckee River in 1868 and has apparently declined in numbers since then (Ammon 2002). This historical observation and reports from other parts of the range (Bemis and Rising 1999) suggest that Wood-Pewees may well inhabit lowland riparian areas in Nevada where habitat conditions remain suitable. The predictive map indicates a greater chance of finding breeding Western Wood-Pewees with increasing elevation, but it also predicts at least some probability of finding them throughout most of the state except in the far south and in the lowland salt deserts of western Nevada.

CONSERVATION AND MANAGEMENT

Because of its generalist habits and apparently healthy populations, the Western Wood-Pewee is not usually thought of as a conservation concern, although the national Breeding Bird Survey indicates recent population declines in the Sierra Nevada region (Sauer et al. 2005). The Western Wood-Pewee is a champion long-distance migrant, though, and thus faces all the perils associated with sustained migrations and potential habitat impacts on both wintering and breeding grounds.

In Nevada, we still need basic information on the status of the Western Wood-Pewee. In particular, we know relatively little about its occurrence in lowland riparian habitat. Given possible declines in this habitat in Nevada (Ammon 2002) and the occurrence of wood-pewees along low-elevation streams in Utah (Behle et al. 1985), an investigation of the potential of riparian habitat restoration to benefit this species would be worthwhile.

HABITAT USE

HABITAT	ATLAS BLOCKS	INCIDENTAL OBSERVATIONS
Agricultural	3 (2%)	0 (0%)
Ash	2 (1%)	0 (0%)
Montane Forest	60 (44%)	14 (25%)
Montane Parkland	4 (3%)	4 (7%)
Montane Shrub	3 (2%)	3 (5%)
Pinyon-Juniper	16 (11%)	5 (9%)
Riparian	35 (26%)	27 (47%)
Sagebrush Scrub	3 (2%)	0 (0%)
Sagebrush Steppe	3 (2%)	0 (0%)
Urban	7 (5%)	3 (5%)
Wetland	0 (0%)	1 (2%)
TOTAL	136	57

Total numbers reported in the Habitat Use, Breeding Status, and Abundance tables may differ from each other (see pp. 22–23 for details). Percentage sums may differ slightly from 100% due to rounding.

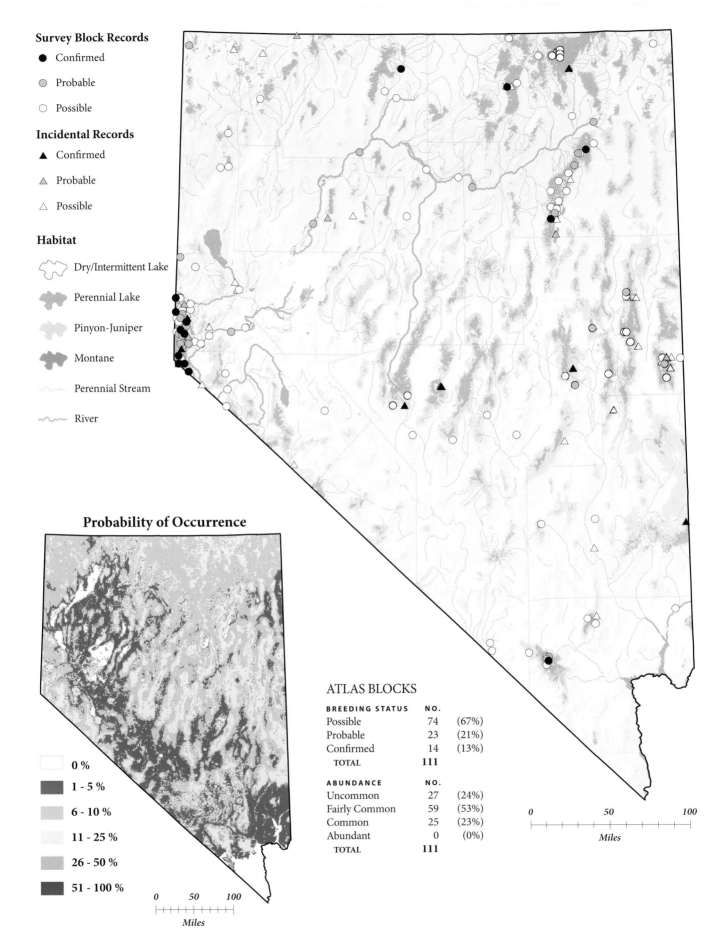

Survey Block Records

- ● Confirmed
- ◐ Probable
- ○ Possible

Incidental Records

- ▲ Confirmed
- ▲ Probable
- △ Possible

Habitat

- Dry/Intermittent Lake
- Perennial Lake
- Pinyon-Juniper
- Montane
- Perennial Stream
- River

Probability of Occurrence

- 0 %
- 1 - 5 %
- 6 - 10 %
- 11 - 25 %
- 26 - 50 %
- 51 - 100 %

0   50   100

Miles

ATLAS BLOCKS

BREEDING STATUS	NO.	
Possible	74	(67%)
Probable	23	(21%)
Confirmed	14	(13%)
TOTAL	**111**	

ABUNDANCE	NO.	
Uncommon	27	(24%)
Fairly Common	59	(53%)
Common	25	(23%)
Abundant	0	(0%)
TOTAL	**111**	

0   50   100

Miles

WILLOW FLYCATCHER
Empidonax trailli

Although it was once probably widespread and fairly common in Nevada's riparian areas, today the Willow Flycatcher breeds only in scattered locations throughout the state. Two of the three recognized subspecies that occur in Nevada (*adastus* and *brewsteri*) are at risk, and the third, the Southwestern Willow Flycatcher (*extimus*), was federally listed as Endangered in 1995 (USFWS 1995). Willow Flycatcher conservation is pursued particularly vigorously in southern Nevada, but given the many pressures on riparian areas everywhere, prospects for this species in our desert state remain uncertain.

DISTRIBUTION

The Willow Flycatcher has the broadest North American distribution of its genus, extending north into Canada and from coast to coast (Sedgwick 2000). In Nevada's neighboring states, the species is fairly common to the north (Stephens and Sturts 1998, Adamus et al. 2001) and east (Behle et al.1985), and is uncommon in the Sierra Nevada to the west (Small 1994). The imperiled Southwestern Willow Flycatcher occurs in southern California (Small 1994), southern Utah (Sedgwick 2000), Arizona (McCarthey in Corman and Wise-Gervais 2005:302–303), New Mexico (Sedgwick 2000), and southern Nevada (McKernan and Braden 2002, McLeod et al. 2005).

For nesting, Willow Flycatchers favor moist or inundated shrubby areas and riparian forests, especially those with willows (Sedgwick 2000). Appropriate breeding sites include wet meadows, pond margins, and occasionally uplands in mesic regions

(Bergstrom 1998, Sedgwick 2000, Dobkin and Sauder 2004). In the Southwest, *extimus* is restricted to riparian habitats (Voget 1998) and has a particular affinity for low-elevation rivers and streams with a large floodplain that is subject to inundation.

Most atlas observations came from the south and involved the subspecies *extimus,* with scattered observations in the north that were probably *brewsteri* to the far west and *adastus* elsewhere. The predictive map suggests a possibility of encountering the species throughout Nevada, but that possibility is very low overall, and especially in central Nevada. This low probability reflects both the scarcity of suitable riparian habitat in Nevada's landscapes and the rarity of breeding Willow Flycatchers in our region.

CONSERVATION AND MANAGEMENT

Willow Flycatcher populations have declined significantly in the West (Dobkin and Sauder 2004, Sauer et al. 2005). The main cause cited in the declines is habitat degradation, often due to livestock grazing, water developments, channelization, and flood control; cowbird parasitism and invasions by exotic plants have also been implicated (see Sedgwick 2000 for details). Historically common along the lower Colorado River, *extimus* was believed to be extirpated there by the early 1990s (Rosenberg et al. 1991). More recently, nesting has been documented in the lower reaches of the Muddy and Virgin rivers, the Pahranagat Valley, and other parts of the lower Colorado River region (Swett 1999, McCarthey in Corman and Wise-Gervais 2005:302–303, McLeod et al. 2005). Interestingly, a substantial portion of the nests in these areas have been found in stands of introduced tamarisk.

Although not as imperiled rangewide as *extimus,* the other two subspecies may also be susceptible to impacts from habitat changes (Small 1994, Bergstrom 1998). Evidence from the lower Truckee River, for instance, suggests that northern Nevada populations have declined greatly as the human presence has increased over the past century (Ridgway 1877, Ammon 2002). The Willow Flycatcher is a Partners in Flight Watch List species throughout its range (Rich et al. 2004) and a Priority species in Nevada (Neel 1999a). In addition to its endangered status, *extimus* is a Covered species of both the Clark County Multiple Species Habitat Conservation Plan (Clark County 2000) and the Lower Colorado River Multi-Species Conservation Program (BOR-LCR 2004). Maintenance and active restoration of suitable nesting habitat are likely to be the most effective conservation strategies for this species (Sedgwick 2000).

HABITAT USE

HABITAT	ATLAS BLOCKS	INCIDENTAL OBSERVATIONS
Agricultural	0 (0%)	2 (9%)
Mesquite	0 (0%)	2 (9%)
Mojave	0 (0%)	1 (4%)
Open Water	0 (0%)	1 (4%)
Riparian	13 (100%)	15 (65%)
Wetland	0 (0%)	2 (9%)
TOTAL	13	23

Total numbers reported in the Habitat Use, Breeding Status, and Abundance tables may differ from each other (see pp. 22–23 for details). Percentage sums may differ slightly from 100% due to rounding.

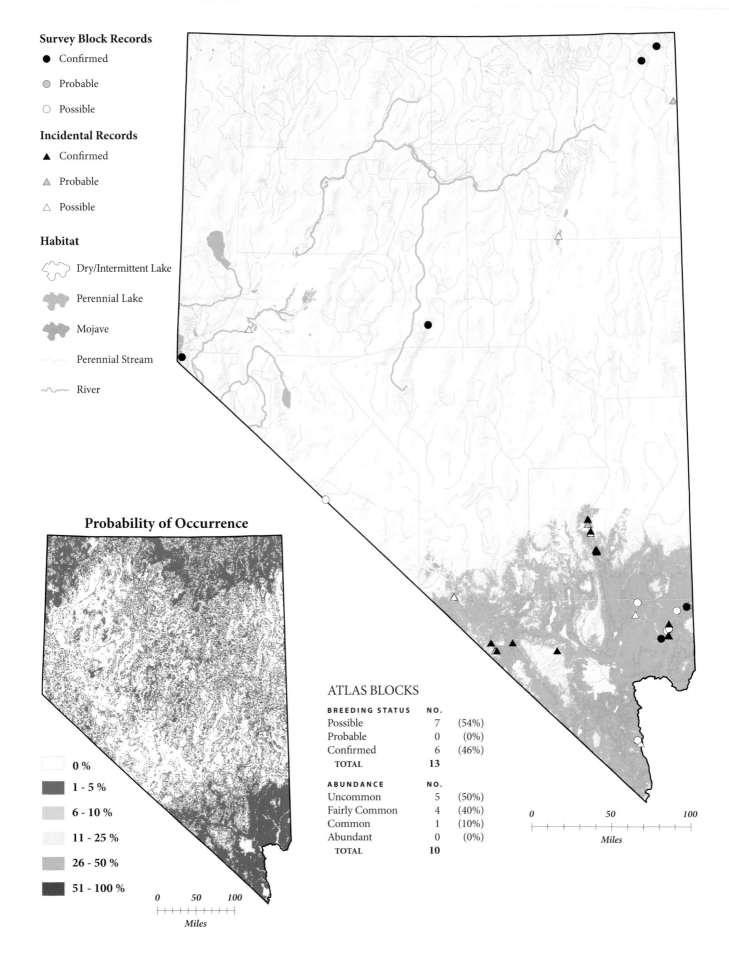

Survey Block Records
● Confirmed
◑ Probable
○ Possible

Incidental Records
▲ Confirmed
△ Probable
△ Possible

Habitat

⬡ Dry/Intermittent Lake

⬤ Perennial Lake

⬤ Mojave

Perennial Stream

River

Probability of Occurrence

☐	0 %
■	1 - 5 %
▨	6 - 10 %
☐	11 - 25 %
▨	26 - 50 %
■	51 - 100 %

0   50   100
Miles

ATLAS BLOCKS

BREEDING STATUS	NO.	
Possible	7	(54%)
Probable	0	(0%)
Confirmed	6	(46%)
TOTAL	13	

ABUNDANCE	NO.	
Uncommon	5	(50%)
Fairly Common	4	(40%)
Common	1	(10%)
Abundant	0	(0%)
TOTAL	10	

0   50   100
Miles

HAMMOND'S FLYCATCHER
Empidonax hammondii

Compared with the similar Dusky Flycatcher, Hammond's Flycatcher is a bird that likes heights. It breeds at high elevations in coniferous forests in Nevada, and its nests are typically placed high in the dense boughs of a tall spruce or fir. In Nevada's montane forests, Hammond's Flycatchers are characteristic, although not especially common, breeders.

DISTRIBUTION

Hammond's Flycatchers were reported from thirty-two atlas blocks. Nests are difficult to detect in the treetops, and as a consequence, breeding was confirmed in only six blocks (19%) and at one additional location. There was a major cluster of records in Nevada's eastern mountains, and smaller clusters were reported in the Jarbidge Mountains in northern Elko County and the mountainous areas west of Reno and west of Las Vegas. The remaining atlas records were from scattered locations throughout the state.

Linsdale (1936) characterized this flycatcher as a possible summer resident in the high mountains of the Carson Range in western Nevada but did not mention breeding in the eastern part of the state. Ryser (1985) and Alcorn (1988) both recognized breeding on Nevada's western margins and in scattered interior mountain ranges, which corresponds more closely to the atlas findings. Sedgwick (1994) reported breeding in eastern Nevada, but none of the earlier authors noted breeding in the south, where atlas fieldwork produced one confirmed and one probable breeding record.

Hammond's Flycatchers are found mainly in upper-elevation forests in Nevada, but farther north in Washington they are birds of mid-elevation forests (Smith et al. 1997). In neighboring Oregon and Idaho, Hammond's Flycatcher is a common breeder (Stephens and Sturts 1998, Adamus et al. 2001), and it can be fairly common in northern California, too (Small 1944). The species' breeding presence in Utah and Arizona is far more limited (Behle et al. 1985, Corman in Corman and Wise-Gervais 2005:592). The predictive map reinforces Hammond's Flycatcher's strong association with higher-elevation forests in Nevada by suggesting virtually no chance of finding this species away from the tallest mountains. Even in the mountains, the probability of occurrence is usually low, but it is greatest in the largest and highest ranges.

More than four in five abundance estimates were for fewer than ten pairs in a block, indicating that Hammond's Flycatcher is a fairly uncommon bird in the state; certainly it is the least common of the *Empidonax* species that breed here. It is also one of the more difficult of Nevada's species to identify, and its quiet demeanor may cause observers to overlook or ignore it.

CONSERVATION AND MANAGEMENT

Hammond's Flycatcher gravitates toward the same sorts of forests that attract the logging industry: large, intact tracts of old growth with a sparse understory (Sakai and Noon 1991, Sedgwick 1994, Sedgwick in Kingery 1998:274–275). In contrast to the situation in other parts of the species' range, this does not create a significant land management conflict in Nevada, since such forests are mostly remote and logging is limited in them. Should logging become more prevalent, however, it could have particularly negative effects on our Hammond's Flycatchers. The species is naturally at the edge of its geographical and ecological ranges here, and its numbers already tend to be lower than they are farther north. Accurate monitoring of this species is complicated by the potential for confusion with the Dusky Flycatcher, and even if the proper identification is made, studying this bird's nests and estimating its abundance remain difficult.

HABITAT USE

HABITAT	ATLAS BLOCKS	INCIDENTAL OBSERVATIONS
Montane Forest	27 (77%)	2 (40%)
Montane Parkland	1 (3%)	0 (0%)
Montane Shrub	3 (9%)	0 (0%)
Pinyon-Juniper	1 (3%)	0 (0%)
Riparian	3 (9%)	3 (60%)
TOTAL	35	5

Total numbers reported in the Habitat Use, Breeding Status, and Abundance tables may differ from each other (see pp. 22–23 for details). Percentage sums may differ slightly from 100% due to rounding.

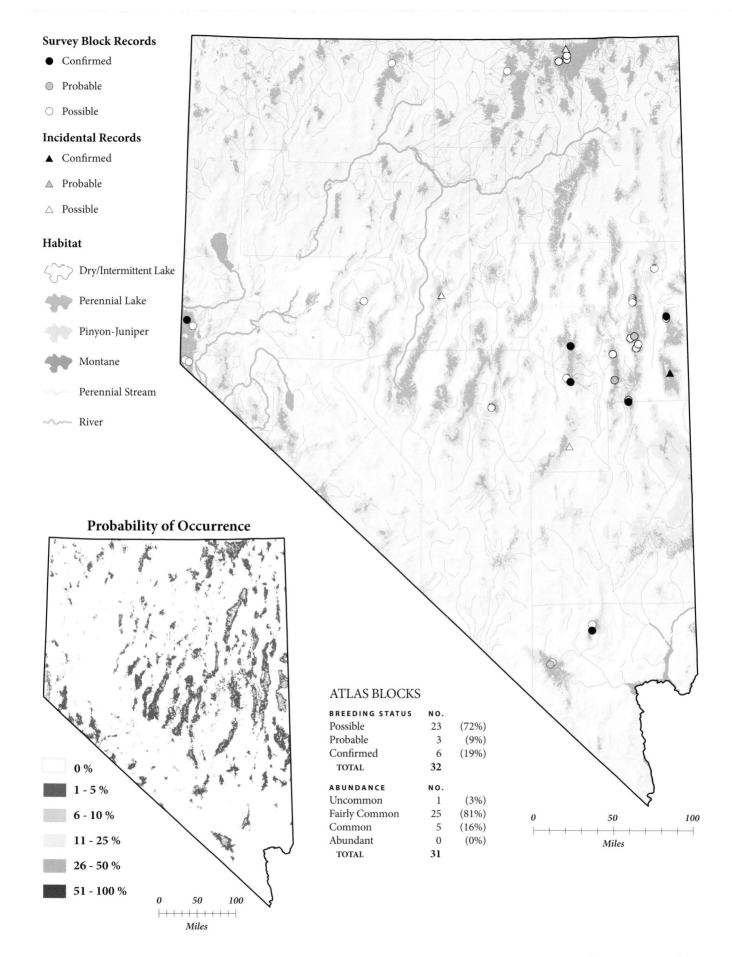

Survey Block Records

● Confirmed

◉ Probable

○ Possible

Incidental Records

▲ Confirmed

△ Probable

△ Possible

Habitat

Dry/Intermittent Lake

Perennial Lake

Pinyon-Juniper

Montane

Perennial Stream

River

Probability of Occurrence

☐	0 %
▨	1 - 5 %
▨	6 - 10 %
▨	11 - 25 %
▨	26 - 50 %
▨	51 - 100 %

0   50   100
Miles

ATLAS BLOCKS

BREEDING STATUS	NO.	
Possible	23	(72%)
Probable	3	(9%)
Confirmed	6	(19%)
TOTAL	32	

ABUNDANCE	NO.	
Uncommon	1	(3%)
Fairly Common	25	(81%)
Common	5	(16%)
Abundant	0	(0%)
TOTAL	31	

0   50   100
Miles

GRAY FLYCATCHER
Empidonax wrightii

Most American birders think of *Empidonax* flycatchers as birds of boreal forests, riparian corridors, or second-growth woodlots. But across a large swath of the Great Basin, in a land of sagebrush and Sage Thrashers, live flourishing populations of the bleached-out, tail-wagging Gray Flycatcher, by far Nevada's most numerous *Empidonax* flycatcher and the one most likely to be encountered throughout the state.

DISTRIBUTION

Atlas workers found Gray Flycatchers breeding at widely scattered localities throughout Nevada. Overall, breeding was confirmed in about a quarter of the blocks where Gray Flycatchers occurred, and the species was usually reported to be numerous: nearly half of the atlas blocks in which it was found had more than ten pairs.

Other authors (Alcorn 1988, Neel 1999a, Chisholm and Neel 2002) have been consistent in characterizing the Gray Flycatcher as a widespread and common species in Nevada. In continental terms, its breeding range is restricted to the Intermountain West, centered in the Great Basin. Gray Flycatchers barely range into California, but where they do, they are fairly common (Small 1994). They are common in southeastern Oregon and are common summer residents in the northern part of Arizona (Adamus et al. 2001, Wise-Gervais in Corman and Wise-Gervais 2005: 304–305). Gray Flycatchers are common summer residents statewide in Utah but barely make it into southern Idaho (Behle et al. 1985, Stephens and Sturts 1998).

Habitat use in Nevada is a complex matter (Dobkin and Sauder 2004). In northern Nevada, the Gray Flycatcher occurs in old-growth stands of "tall sage"; in central Nevada, it uses sage-

brush interspersed with mountain mahogany, junipers, and pinyon pines; and in southern Nevada, Joshua trees are added to the mix (Neel 1999a). Similar latitudinal gradients in habitat use occur in California (Garrett and Dunn 1981, McCaskie et al. 1988), while in Utah and Arizona the Gray Flycatcher is regarded as a pinyon-juniper species (Pavlackey and Anderson 2001, Wise-Gervais in Corman and Wise-Gervais 2005:304–305). Although Gray Flycatchers are generally associated with dry landscapes, atlas workers also found them in riparian and higher-elevation habitats. These patterns are reflected in the predictive map, which shows Gray Flycatchers absent only from the most barren and arid parts of the state. Still, Gray Flycatchers are most likely to occur through the northern tier of Nevada, where habitat comprised of large, mature sagebrush with interspersed pinyon-juniper trees is most widespread.

CONSERVATION AND MANAGEMENT

Until recently, the Gray Flycatcher was not particularly well known. Sterling (1999) described the taxonomic mayhem and basic biological befuddlement that have characterized study of this species. Currently, Gray Flycatchers seem to be relatively common and increasing in western North America (Sterling 1999, Dobkin and Sauder 2004). Partners in Flight nevertheless ranks the Gray Flycatcher as a Priority species for Nevada and a Stewardship species for the Intermountain West, because Nevada and adjoining states harbor the majority of the world's population (Neel 1999a, Rich et al. 2004). Protection of, and management for, old-growth sagebrush is important for the Gray Flycatcher (Sterling 1999, Dobkin and Sauder 2004), and Partners in Flight has recommended this as a management strategy for the species in Nevada (Neel 1999a). Cowbird parasitism and clear-cutting of pinyon-juniper may also be detrimental to Gray Flycatchers (Neel 1999a, Sterling 1999, Dobkin and Sauder 2004).

HABITAT USE

HABITAT	ATLAS BLOCKS	INCIDENTAL OBSERVATIONS
Barren	0 (0%)	1 (3%)
Mojave	0 (0%)	1 (3%)
Montane Forest	5 (2%)	2 (5%)
Montane Parkland	1 (<1%)	0 (0%)
Montane Shrub	18 (9%)	8 (21%)
Pinyon-Juniper	82 (39%)	11 (29%)
Riparian	18 (9%)	7 (18%)
Salt Desert Scrub	3 (1%)	0 (0%)
Sagebrush Scrub	20 (10%)	4 (11%)
Sagebrush Steppe	62 (30%)	3 (8%)
Urban	0 (0%)	1 (3%)
TOTAL	209	38

Total numbers reported in the Habitat Use, Breeding Status, and Abundance tables may differ from each other (see pp. 22–23 for details). Percentage sums may differ slightly from 100% due to rounding.

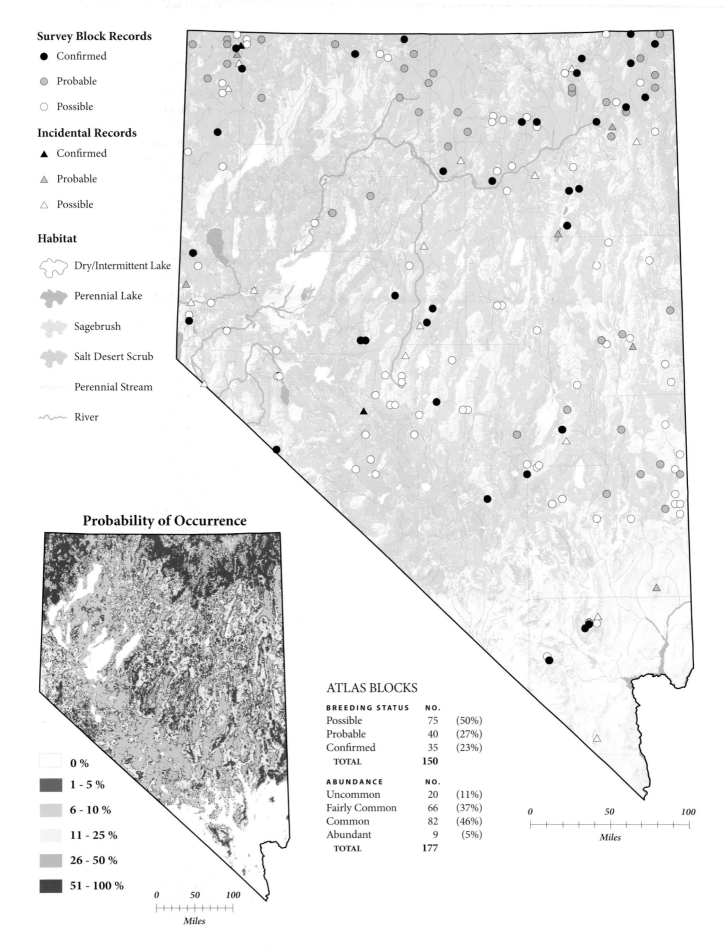

Survey Block Records

● Confirmed

● Probable

○ Possible

Incidental Records

▲ Confirmed

▲ Probable

△ Possible

Habitat

Dry/Intermittent Lake

Perennial Lake

Sagebrush

Salt Desert Scrub

Perennial Stream

River

Probability of Occurrence

0 %

1 - 5 %

6 - 10 %

11 - 25 %

26 - 50 %

51 - 100 %

0   50   100

Miles

ATLAS BLOCKS

BREEDING STATUS	NO.	
Possible	75	(50%)
Probable	40	(27%)
Confirmed	35	(23%)
TOTAL	**150**	
ABUNDANCE	NO.	
Uncommon	20	(11%)
Fairly Common	66	(37%)
Common	82	(46%)
Abundant	9	(5%)
TOTAL	**177**	

0   50   100

Miles

DUSKY FLYCATCHER
Empidonax oberholseri

The geographic ranges of the Dusky, Hammond's, and Gray Flycatchers broadly overlap, but within this general region, their breeding habitat requirements are distinctive. Gray Flycatchers occur in pinyon-juniper and sagebrush-dominated habitats, and Hammond's Flycatchers tend to be found in higher-elevation montane forests. Dusky Flycatchers, in contrast, occupy various Nevada woodlands, including broadleaved shrublands, open coniferous forests with a shrubby undergrowth, and riparian areas. They occur across a broad elevation gradient as well, ranging from timberline nearly down to the valley floors. Unlike the very similar Hammond's Flycatcher, Dusky Flycatchers usually nest in shrubby forest undergrowth or in riparian willow thickets (Sedgwick 1993, Liebezeit and George 2002).

DISTRIBUTION

Dusky Flycatchers were found in seventy-seven atlas blocks, with more than half of the records coming from Elko and White Pine counties. Another stronghold for this species was the Carson Range in far western Nevada, with records from twelve blocks. The remaining records came from montane and riparian sites scattered throughout northern and central Nevada, along with confirmed breeding records in the Spring Mountains and Sheep Range of southern Nevada. Atlas results, in conjunction with earlier Nevada accounts (Linsdale 1936, Ryser 1985, Alcorn 1988), suggest that several published range maps (in Sedgwick 1993, for example) should be amended to include parts of central and southern Nevada.

Breeding was confirmed in 45% of the occupied atlas blocks, and nesting populations sometimes reached fairly high densities.

For instance, observers estimated more than ten pairs in about a third of the blocks for which abundance was recorded. Most breeding records came from montane forest, montane shrubland, and higher-elevation riparian habitats. The predictive map reflects this association by showing a moderately high chance of finding the species in forested montane areas. The model probably overrepresents the breeding occurrence of Dusky Flycatchers in lower-elevation sagebrush, even though it predicts only a very low chance of finding them in these areas. At least three factors might account for this overprediction. First, small patches of suitable nesting habitat may have been embedded within surrounding sagebrush habitat, resulting in misclassifications in the original data set. Second, Dusky Flycatchers migrate late in the spring (Smith et al. 1997), and some late migrants seen in nonbreeding habitat could have been mistaken for potential breeders. Third, Gray Flycatchers might occasionally have been misidentified as Dusky Flycatchers, despite the fact that *Empidonax* records received particularly careful scrutiny during the review of atlas data. Dusky Flycatchers tend to be fairly common to the west, north, and east of Nevada (Behle et al. 1985, Small 1994, Stephens and Sturts 1998, Adamus et al. 2001), although their habitat use may differ somewhat in surrounding states (Sedgwick 1993).

CONSERVATION AND MANAGEMENT

The outlook for the Dusky Flycatcher is comparatively bright in most of its range. It is tolerant of forest clearing and, unlike some of its close relatives, uses a variety of deciduous and coniferous woodlands. Moreover, although the species is declining in the northern part of its range in Canada, U.S. populations have shown steady increases over the last few decades (Sauer et al. 2005). Nonetheless, Partners in Flight designated the Dusky Flycatcher a Stewardship species in the Intermountain West because of its high degree of dependence on montane forests, which within our region occur in relatively limited and isolated patches (Rich et al. 2004).

HABITAT USE

HABITAT	ATLAS BLOCKS	INCIDENTAL OBSERVATIONS
Alpine	3 (3%)	0 (0%)
Montane Forest	34 (34%)	18 (56%)
Montane Parkland	5 (5%)	1 (3%)
Montane Shrub	20 (20%)	7 (22%)
Pinyon-Juniper	9 (9%)	1 (3%)
Riparian	25 (25%)	5 (16%)
Sagebrush Scrub	3 (3%)	0 (0%)
Sagebrush Steppe	1 (1%)	0 (0%)
TOTAL	100	32

Total numbers reported in the Habitat Use, Breeding Status, and Abundance tables may differ from each other (see pp. 22–23 for details). Percentage sums may differ slightly from 100% due to rounding.

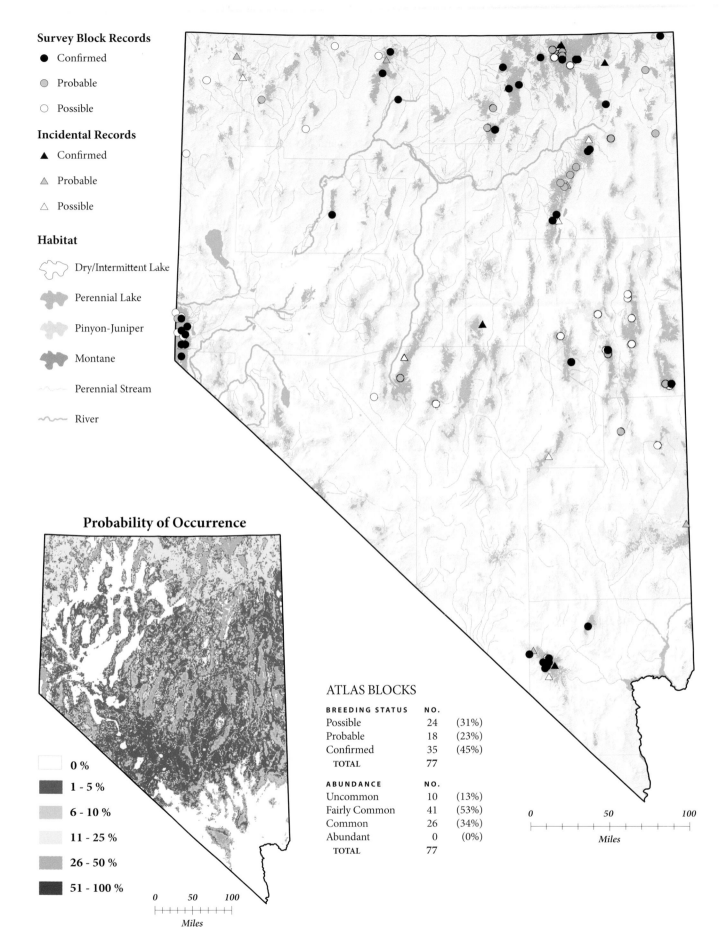

Survey Block Records

● Confirmed

○ Probable

○ Possible

Incidental Records

▲ Confirmed

△ Probable

△ Possible

Habitat

Dry/Intermittent Lake

Perennial Lake

Pinyon-Juniper

Montane

Perennial Stream

River

Probability of Occurrence

☐ 0 %

1 - 5 %

6 - 10 %

11 - 25 %

26 - 50 %

51 - 100 %

0   50   100

Miles

ATLAS BLOCKS

BREEDING STATUS	NO.	
Possible	24	(31%)
Probable	18	(23%)
Confirmed	35	(45%)
TOTAL	77	

ABUNDANCE	NO.	
Uncommon	10	(13%)
Fairly Common	41	(53%)
Common	26	(34%)
Abundant	0	(0%)
TOTAL	77	

0   50   100

Miles

CORDILLERAN FLYCATCHER

Empidonax occidentalis

The Cordilleran Flycatcher is a common breeder in northeastern Nevada and in isolated mountain ranges elsewhere in the state. We still have much to learn about this species, which was only recently split from the Pacific-slope Flycatcher of far western North America. Field identification of the two species that make up the Flycatcher "Western" complex is extremely difficult, and the situation was further complicated in Nevada by the discovery of breeding-season birds far from the core breeding ranges of either the Cordilleran or the Pacific-slope Flycatcher. Our present understanding is that the Cordilleran is the only breeding Flycatcher "Western" in Nevada, but the possibility that the Pacific-slope also breeds here should not be dismissed (see p. 542).

DISTRIBUTION

Most atlas records for the Cordilleran Flycatcher came from the mountains of Elko and White Pine counties in northeastern Nevada, with scattered records from other mountains throughout the state. The confirmation rate for breeding was quite low, and all confirmed records were from the northeast. Atlas workers found the Cordilleran Flycatcher to be fairly common in the blocks where it occurred, with at least ten pairs in one-fifth of the occupied blocks and multiple pairs in most others.

Observations in the Spring Mountains and Sheep Range of Clark County and in the extreme northwest were especially intriguing because of their disjunct nature. Although breeding could not be confirmed in these areas, birds have been regularly sighted in the southern mountains for several years (P. Lehman, pers. comm.). An isolated population of (presumably Cordilleran) Flycatchers "Western" was earlier reported from the Virgin Mountains (Johnson 1973) in Clark County, but the atlas survey did not produce additional records in this area. There were also reports of possible Pacific-slope Flycatchers in the Carson Range of far western Nevada and the wooded drainages of extreme northwestern Nevada during the atlas project. Moreover, banding data from the river valleys of western Nevada indicate that many of our migrant Flycatchers "Western" are in fact Pacific-slopes (E. Ammon, pers. observ.). Given the current uncertainties about the range limits of these species, it is possible that both breed in Nevada. In neighboring Oregon, for example, the relative ranges of Pacific-slope and Cordilleran Flycatchers also are still being worked out (Contreras and Kindschy 1996, Contreras 1999, Adamus et al. 2001). Earlier observations in Nevada (Ryser 1985, Alcorn 1988) were made prior to the taxonomic split and are thus of limited utility in resolving these questions.

In Nevada, Cordilleran Flycatchers are found primarily in heavily wooded, mesic forests in mountains and riparian zones. The predictive map shows the association with montane habitats, but even in these areas the likelihood of occurrence is not high. The larger swath of low-elevation habitat where the predictions suggest there is a very low chance of finding these flycatchers is probably a by-product of migrant birds being misclassified as potential breeders (or perhaps rare misidentifications of Gray Flycatchers).

CONSERVATION AND MANAGEMENT

The Cordilleran Flycatcher is rarely cited as a conservation concern, and population trend data show a mixed pattern for the Flycatcher "Western" complex (Sauer et al. 2005). Sedgwick (in Kingery 1998:280–281) suggested, however, that the species is potentially vulnerable to logging despite its versatile use of microhabitats that include disturbed areas. In Nevada, the Cordilleran Flycatcher also deserves special attention directed toward clarifying its distribution and habitat needs relative to those of the Pacific-slope Flycatcher.

HABITAT USE

HABITAT	ATLAS BLOCKS	INCIDENTAL OBSERVATIONS
Alpine	2 (3%)	0 (0%)
Montane Forest	35 (44%)	9 (35%)
Montane Parkland	2 (3%)	1 (4%)
Montane Shrub	11 (14%)	0 (0%)
Pinyon-Juniper	5 (6%)	1 (4%)
Riparian	24 (30%)	13 (50%)
Salt Desert Scrub	1 (1%)	0 (0%)
Sagebrush Steppe	0 (0%)	1 (4%)
Urban	0 (0%)	1 (4%)
TOTAL	80	26

Total numbers reported in the Habitat Use, Breeding Status, and Abundance tables may differ from each other (see pp. 22–23 for details). Percentage sums may differ slightly from 100% due to rounding.

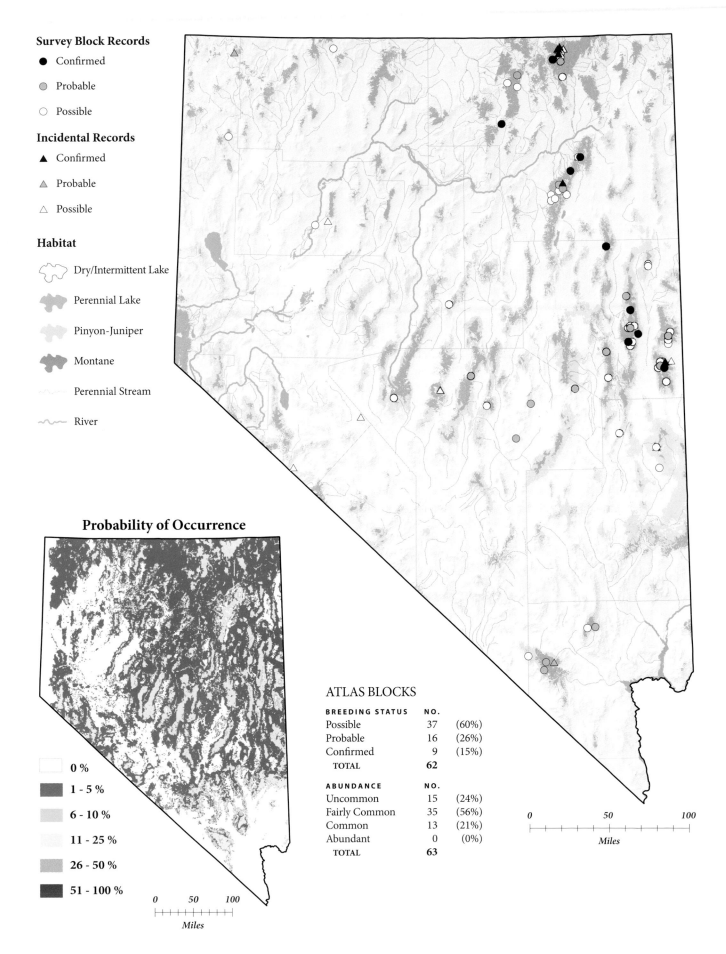

Survey Block Records

- ● Confirmed
- ◐ Probable
- ○ Possible

Incidental Records

- ▲ Confirmed
- △ Probable
- △ Possible

Habitat

- Dry/Intermittent Lake
- Perennial Lake
- Pinyon-Juniper
- Montane
- Perennial Stream
- River

Probability of Occurrence

- 0 %
- 1 - 5 %
- 6 - 10 %
- 11 - 25 %
- 26 - 50 %
- 51 - 100 %

0   50   100
Miles

ATLAS BLOCKS

BREEDING STATUS	NO.	
Possible	37	(60%)
Probable	16	(26%)
Confirmed	9	(15%)
TOTAL	62	

ABUNDANCE	NO.	
Uncommon	15	(24%)
Fairly Common	35	(56%)
Common	13	(21%)
Abundant	0	(0%)
TOTAL	63	

0   50   100
Miles

BLACK PHOEBE
Sayornis nigricans

An opportunistic breeder with a penchant for showing up in unexpected places, the Black Phoebe confounded atlas workers throughout the southern two-thirds of Nevada by being absent from several seemingly suitable locations. There were multiple confirmations of breeding in western Nevada, however, well beyond the species' core range. The Black Phoebe's life history is intimately connected with water, and it can be found in the vicinity of streams, ponds, rivers, and lakes. It was even found breeding in various settings at Hoover Dam.

DISTRIBUTION

Black Phoebes were found in eighteen atlas blocks and at a number of incidental locations. They were fairly widespread but uncommon in southern Nevada, mainly along the larger rivers and at Ash Meadows National Wildlife Refuge. Interestingly, a second cluster of records, including four breeding confirmations, came from the Walker and Carson river drainages of Douglas and Lyon counties in western Nevada—far north of the Black Phoebe's core breeding range. There were no areas where the Black Phoebe was especially numerous: more than 95% of the blocks for which abundance was reported were estimated to have fewer than ten pairs. Breeding was confirmed in slightly more than half of the blocks where the species was seen.

The Black Phoebe's range is mostly confined to the warm desert regions along the border with Mexico, but it also breeds throughout much of lowland California and even into southwestern Oregon (Small 1994, Wolf 1997, Wise-Gervais in Corman and Wise-Gervais 2005:312–313). Breeders also extend across southern Nevada into southwestern Utah (Behle et al. 1985). Earlier studies in Nevada portray the species as an uncom-

mon resident in the south and an occasional wanderer farther north (Linsdale 1936, Ryser 1985, Alcorn 1988), which corresponds well to atlas findings. Black Phoebes are also well known for their extensive postbreeding wanderings. Chisholm and Neel's (2002:126) report that "they occur in every season and with little apparent pattern" in the Lahontan Valley is an apt description. Black Phoebes also appear to be expanding their breeding range northward, as has been recently reported in Colorado (Faulkner et al. 2005).

Black Phoebes are almost always found associated with water. The predictive model, as expected, shows that Black Phoebes are most likely to occur along waterways in the southern two-thirds of the state. The low probability of their occurrence across the dry deserts of far southern Nevada reflects the region's general dearth of smaller riparian areas. Despite the wandering tendencies of Black Phoebes, predicted occurrences north of the Mojave Desert and east of the Carson-Walker drainages may be unrealistic given the isolated and fragmented distribution of suitable habitat in this region.

CONSERVATION AND MANAGEMENT

The status of the Black Phoebe in Nevada is enigmatic: on the one hand, this species seems to breed opportunistically far from its core range; on the other hand, it is somewhat scarce and localized even in southern Nevada, where its likelihood of occurrence is greatest. No significant trends have been detected in population numbers (Sauer et al. 2005). The proliferation of human structures near water has benefited these phoebes by providing abundant sheltered vertical surfaces on which females can build their mud nests. Drainage of wetlands, water diversions, and intensive degradation of riparian vegetation, however, can be detrimental (Wolf 1997). Because of its irregularity in Nevada, its strict dependence on aquatic habitats, and its intriguing lack of respect for its "established" breeding range, the Black Phoebe bears watching in Nevada.

HABITAT USE

HABITAT	ATLAS BLOCKS	INCIDENTAL OBSERVATIONS
Agricultural	2 (8%)	1 (4%)
Ash	1 (4%)	0 (0%)
Grassland	1 (4%)	0 (0%)
Mojave	0 (0%)	1 (4%)
Open Water	0 (0%)	1 (4%)
Riparian	18 (75%)	16 (70%)
Urban	1 (4%)	3 (13%)
Wetland	1 (4%)	1 (4%)
TOTAL	24	23

Total numbers reported in the Habitat Use, Breeding Status, and Abundance tables may differ from each other (see pp. 22–23 for details). Percentage sums may differ slightly from 100% due to rounding.

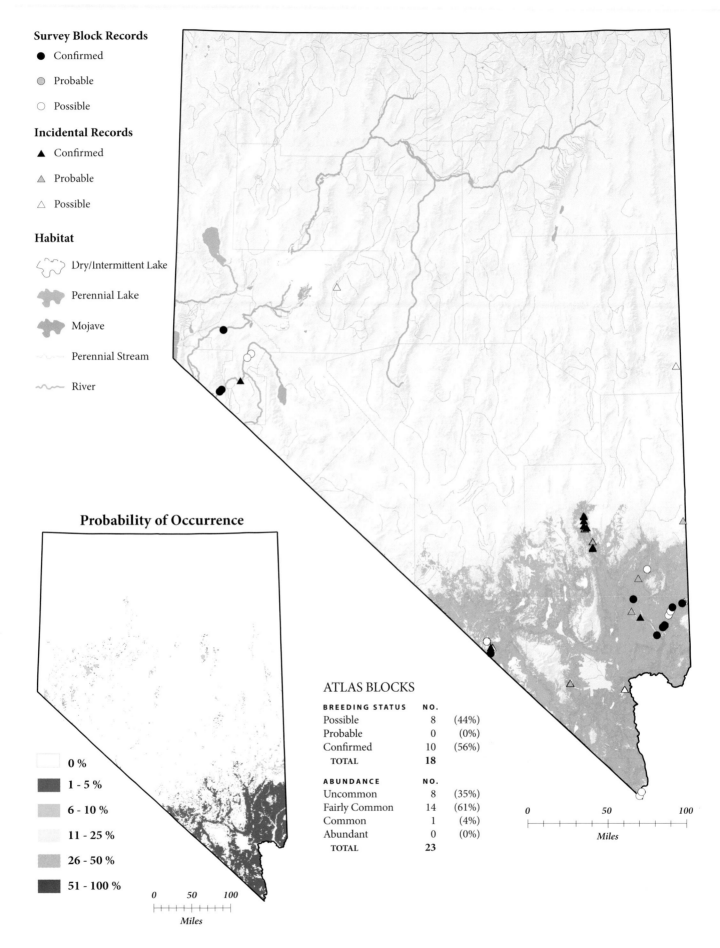

Survey Block Records

● Confirmed

● Probable

○ Possible

Incidental Records

▲ Confirmed

▲ Probable

△ Possible

Habitat

Dry/Intermittent Lake

Perennial Lake

Mojave

Perennial Stream

River

Probability of Occurrence

☐	0 %
■	1 - 5 %
▦	6 - 10 %
▨	11 - 25 %
▩	26 - 50 %
■	51 - 100 %

0   50   100
Miles

ATLAS BLOCKS

BREEDING STATUS	NO.	
Possible	8	(44%)
Probable	0	(0%)
Confirmed	10	(56%)
TOTAL	18	

ABUNDANCE	NO.	
Uncommon	8	(35%)
Fairly Common	14	(61%)
Common	1	(4%)
Abundant	0	(0%)
TOTAL	23	

0   50   100
Miles

SAY'S PHOEBE
Sayornis saya

Much of Nevada's mystique centers on its loneliness and desolation: distant mountain peaks, vast arid scrublands, and the occasional weather-beaten old shanty or farm building. Say's Phoebe—a frequent inhabitant of long-abandoned outbuildings—epitomizes this lonely landscape. This phoebe is attired in muted earth tones, and its simple song carries an element of longing and melancholy. It can be found throughout much of Nevada but seems to be especially prevalent in the south. Say's Phoebe is also the only tyrannid flycatcher regularly found in the western Great Basin in the depths of winter.

DISTRIBUTION

Say's Phoebes were found at a variety of locations in Nevada. The greatest concentration of sightings was in and around Clark County, and there were relatively few records from the east-central part of the state. Breeding was confirmed in more than a quarter of the atlas blocks in which the species was found, and more than half of these blocks were estimated to have multiple pairs.

Earlier writers variously described Say's Phoebe as uncommon to common in the Great Basin (Linsdale 1936, Ryser 1985, Alcorn 1988, Chisholm and Neel 2002), and as common throughout the warm desert lowlands of the south (Fisher 1893). In surrounding states, it is fairly common in much of California (Small 1994) and throughout Arizona (McCarthey in Corman and Wise-Gervais 2005:314–315), common in the drier eastern half of Oregon (Adamus et al. 2001) and throughout Utah (Behle et al. 1985), and breeds widely but rather sporadically in Idaho (Stephens and Sturts 1998).

Suitable nesting habitat for Say's Phoebe abounds in the arid West (Schukman and Wolf 1998). Linsdale (1936) reported Say's Phoebes to occur in most of Nevada's valleys and lower mountain slopes, and Ryser (1985) noted that they prefer localities with few trees and sparse, low shrub cover. Atlas workers found them almost everywhere in Nevada except the higher-elevation forested habitats. The predictive map likewise shows moderate to high probabilities of their occurrence throughout the state. The chance of occurrence is generally predicted to be greater in the Mojave Desert and in lowlands than elsewhere, although the Las Vegas area is an exception. While they are often found in abandoned structures, Say's Phoebes are not real backyard birds.

CONSERVATION AND MANAGEMENT

Say's Phoebes have long been loosely associated with humans—for example, they used to be found around adobe ruins in the lower Colorado River Valley (Rosenberg et al. 1991)—and seem to have adapted well to European settlement of the desert West (Littlefield 1990). Rangewide, their populations have recently increased (Sauer et al. 2005), and the species is probably in good shape in Nevada. Although they willingly use various human-built structures as nest sites, the more general effects of urbanization are unclear (Shukman and Wolf 1998). Our predictive map suggests that the species is less common in urban areas than in native Mojave habitats, and further investigation of this prediction in southern Nevada would add valuable knowledge about the conservation status of this species.

HABITAT USE

HABITAT	ATLAS BLOCKS	INCIDENTAL OBSERVATIONS
Agricultural	9 (5%)	2 (4%)
Ash	1 (<1%)	0 (0%)
Barren	8 (4%)	2 (4%)
Mesquite	3 (2%)	0 (0%)
Mojave	36 (19%)	11 (20%)
Montane Parkland	0 (0%)	1 (2%)
Montane Shrub	2 (1%)	3 (5%)
Open Water	1 (<1%)	1 (2%)
Pinyon-Juniper	11 (6%)	1 (2%)
Riparian	28 (15%)	15 (27%)
Salt Desert Scrub	15 (8%)	1 (2%)
Sagebrush Scrub	35 (19%)	4 (7%)
Sagebrush Steppe	9 (5%)	1 (2%)
Urban	27 (14%)	9 (16%)
Wetland	2 (1%)	4 (7%)
TOTAL	187	55

Total numbers reported in the Habitat Use, Breeding Status, and Abundance tables may differ from each other (see pp. 22–23 for details). Percentage sums may differ slightly from 100% due to rounding.

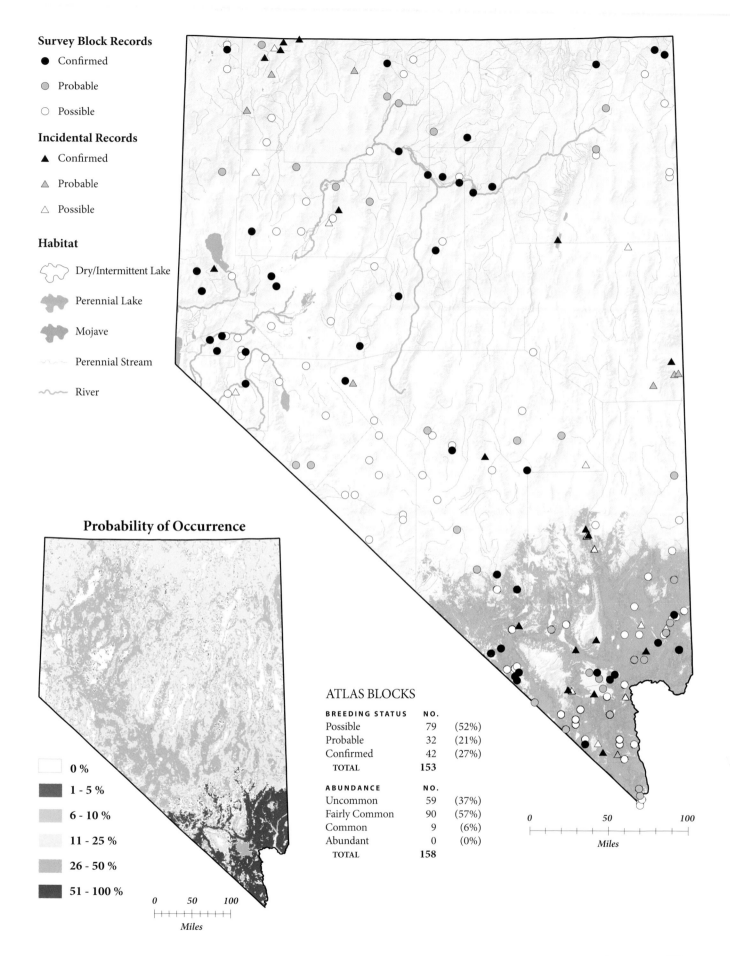

Survey Block Records

● Confirmed

● Probable

○ Possible

Incidental Records

▲ Confirmed

▲ Probable

△ Possible

Habitat

Dry/Intermittent Lake

Perennial Lake

Mojave

Perennial Stream

River

Probability of Occurrence

☐ 0 %

■ 1 - 5 %

■ 6 - 10 %

☐ 11 - 25 %

■ 26 - 50 %

■ 51 - 100 %

0   50   100
Miles

ATLAS BLOCKS

BREEDING STATUS	NO.	
Possible	79	(52%)
Probable	32	(21%)
Confirmed	42	(27%)
TOTAL	153	

ABUNDANCE	NO.	
Uncommon	59	(37%)
Fairly Common	90	(57%)
Common	9	(6%)
Abundant	0	(0%)
TOTAL	158	

0   50   100
Miles

VERMILION FLYCATCHER
Pyrocephalus rubinus

Although atlas workers at the Mormon Ranch in Clark County had no trouble finding Vermilion Flycatchers to count, the same was not true for fieldworkers elsewhere in the state. Sullivan and Titus's (2001) estimate of twenty-five or more breeding pairs in the area around the Muddy River's headwaters that includes Mormon Ranch raised the prospect that the species may breed commonly elsewhere in southern Nevada, but Vermilion Flycatchers were recorded from only a few other locales during the atlas project. It remains possible, of course, that they have yet to be discovered in other small, hard-to-access riparian woodlands of the Mojave Desert. These brilliantly colored birds are not shy, and their spectacular territorial activities provide a special treat for those lucky enough to find themselves in the Vermilion Flycatchers' breeding grounds.

DISTRIBUTION

Vermilion Flycatchers were detected in only two atlas blocks—one in southwestern Nye County and one in northeastern Clark County—but were not numerous in either. Breeding was confirmed with three incidental records from the Virgin and Muddy rivers and from the eastern foothills of the Spring Mountains. Vermilion Flycatchers reach the northern limit of both their breeding and year-round ranges in southern Nevada (Wolf and Jones 2000, Sullivan and Titus 2001). Linsdale (1936) reported two Vermilion Flycatcher records from Nye County, and in the mid-1960s the species was noted as far north as the Pahranagat Valley (Alcorn 1988). Vermilion Flycatchers are occasionally seen as far north into the Great Basin as the towns of Reno and Fallon (Ryser 1985, Alcorn 1988).

Vermilion Flycatchers breed commonly in the southern and central regions of Arizona, but they occur only locally in western Arizona, southern California, and throughout the Colorado River Valley (Rosenberg et al. 1991, Small 1994, Averill-Murray and Corman in Corman and Wise-Gervais 2005:316–317). They rarely breed in Utah (Behle et al. 1985).

Vermilion Flycatchers are usually found close to water in woodlands dominated by cottonwoods, willows, mesquite, or ash (Rosenberg et al. 1991). They also sometimes use agricultural areas, golf courses, and parks, if water is nearby (Small 1994). The Vermilion Flycatcher habitats reported during the atlas project reflect this mixture. The predictive map supports the notion that the chance of seeing Vermilion Flycatchers is very low in most places, but there are a few small areas where there are good odds of finding them. These areas occur where potentially suitable habitat, such as mesquite, is located close to water, and they are generally so small that they are barely discernible on the predictive map.

CONSERVATION AND MANAGEMENT

Data on Vermilion Flycatcher populations, regionally and nationally, are too scanty to allow clear conclusions (Sauer et al. 2005). In the lower Colorado River Valley, however, Vermilion Flycatcher populations reportedly declined in the mid-1900s (Rosenberg et al. 1991). Although they will sometimes use human-created habitats, habitat loss is probably the primary threat, and conservation of riparian areas is a top management priority (Wolf and Jones 2000). For these reasons, they are a Covered species in the Clark County Multiple Species Habitat Conservation Plan (Clark County 2000) and the Lower Colorado River Multi-Species Conservation Program (BOR-LCR 2004).

In Nevada, it is important to know whether Vermilion Flycatchers breed in previously overlooked pockets of riparian habitat, and the predictive model can help to identify where those additional patches might be found. This effort may be especially useful for refining our understanding of Vermilion Flycatcher habitat requirements and for prioritizing areas for monitoring and conservation. Meanwhile, protection and restoration of riparian woodlands in southern Nevada are the most promising conservation strategies for this species.

HABITAT USE

HABITAT	ATLAS BLOCKS	INCIDENTAL OBSERVATIONS
Agricultural	0 (0%)	1 (13%)
Ash	1 (33%)	0 (0%)
Mesquite	1 (33%)	0 (0%)
Riparian	0 (0%)	5 (63%)
Urban	1 (33%)	2 (25%)
TOTAL	3	8

Total numbers reported in the Habitat Use, Breeding Status, and Abundance tables may differ from each other (see pp. 22–23 for details). Percentage sums may differ slightly from 100% due to rounding.

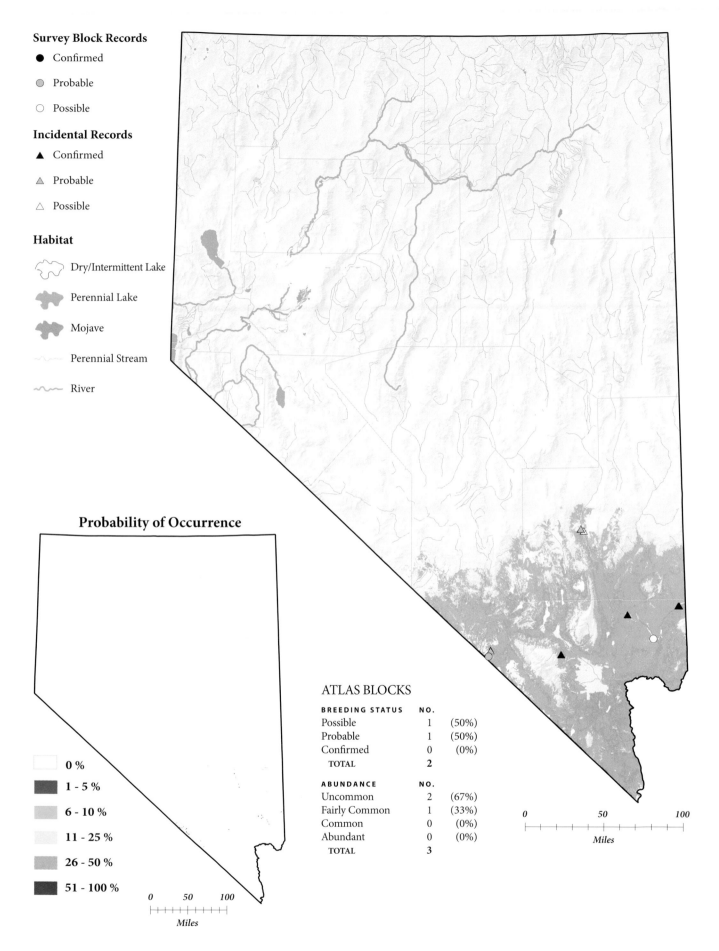

Survey Block Records

● Confirmed

◐ Probable

○ Possible

Incidental Records

▲ Confirmed

△ Probable

△ Possible

Habitat

Dry/Intermittent Lake

Perennial Lake

Mojave

Perennial Stream

River

Probability of Occurrence

◻ 0 %

■ 1 - 5 %

▨ 6 - 10 %

▨ 11 - 25 %

▨ 26 - 50 %

■ 51 - 100 %

0   50   100
Miles

ATLAS BLOCKS

BREEDING STATUS	NO.	
Possible	1	(50%)
Probable	1	(50%)
Confirmed	0	(0%)
TOTAL	2	

ABUNDANCE	NO.	
Uncommon	2	(67%)
Fairly Common	1	(33%)
Common	0	(0%)
Abundant	0	(0%)
TOTAL	3	

0   50   100
Miles

ASH-THROATED FLYCATCHER
Myiarchus cinerascens

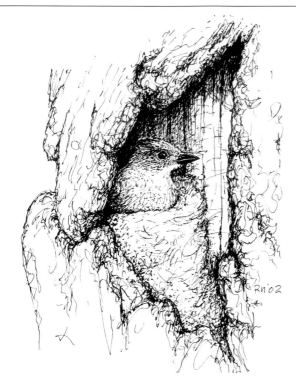

The spunky Ash-throated Flycatcher is one of the most spirited inhabitants of Nevada's woodlands. Right through the heat of the day it can be seen hawking for insects and chasing away intruders. On the rare occasions when an Ash-throated Flycatcher sits still long enough for careful study, it can be seen to be a very handsome bird—with a sulfurous suffusion across the breast, bright rufous accents in the wings and tail, and crown feathers that are often raised into a distinctive crest.

DISTRIBUTION

Ash-throated Flycatchers were found throughout Nevada, but there was an obvious concentration of sightings in the southern part of the state—a pattern that is made especially clear by the predictive map. The species was confirmed as a breeder in thirty-five blocks, twenty-four of which were in Clark, Lincoln, and southern Nye counties. Where Ash-throated Flycatchers were present, they were numerous, with multiple pairs estimated in more than 80% of the blocks. The Ash-throated Flycatcher's present distribution in the state is consistent with that reported in the earlier literature. Linsdale (1936) described the species as common in southern Nevada but less so toward the north. Fisher (1893) also considered these flycatchers common in southern Nevada, while Ryser (1985) considered them somewhat less common in the Great Basin.

The latitudinal gradient in abundance that the Ash-throated Flycatcher exhibits in Nevada is also evident in the surrounding states. The species is common in Arizona (Wise-Gervais in Corman and Wise-Gervais 2005:320–321), including the lower Col-

orado River Valley (Rosenberg et al. 1991), southern California (Garrett and Dunn 1981), and in southern and eastern Utah (Behle et al. 1985). In contrast, it is considered only fairly common or locally common in central and northern California (Small 1994), northern and western Utah (Behle et al. 1985), eastern and central Oregon (Contreras 1999, Adamus et al. 2001), and southern Idaho (Stephens and Sturts 1998).

Ash-throated Flycatchers can be found in diverse types of open woodlands, where they nest in trees and shrubs large enough to support the woodpecker-excavated cavities that the species often uses for nesting. They typically inhabit pinyon-juniper or riparian woodlands in the Great Basin, while in the Mojave Desert, Joshua tree, mesquite, ash, and catclaw woodlands are all used. In the Great Basin, Ash-throated Flycatchers are most likely to be found in riparian areas and at elevations that support the preferred open woodlands, as the predictive map shows. The highest probability of finding the species in Nevada is in the Mojave region, closer to the core of its range. The Ash-throated Flycatcher appears to avoid cities like Las Vegas and Pahrump, and generally does not accept exotic landscaping as suitable habitat (Cardiff and Dittmann 2002).

CONSERVATION AND MANAGEMENT

Obligate cavity nesters like the Ash-throated Flycatcher present conservation challenges on several fronts. Nest sites can be limited, causing competition with other cavity nesters. But Ash-throated Flycatchers are also willing to accept a variety of artificial nest sites, including old drainpipes and bird boxes (Ryser 1985).

Nevada Partners in Flight has designated the Ash-throated Flycatcher a Priority species (Neel 1999a), primarily because it serves as a good indicator of intact Mojave and lowland riparian habitats. Its overall numbers appear to be increasing, both in Nevada and rangewide (Sauer et al. 2005), but local declines may be occurring (Chisholm and Neel 2002). Habitat management that protects and promotes mature, native riparian trees could have the greatest positive impact for this species (Neel 1999a).

HABITAT USE

HABITAT	ATLAS BLOCKS	INCIDENTAL OBSERVATIONS
Agricultural	0 (0%)	1 (1%)
Ash	11 (7%)	0 (0%)
Barren	1 (1%)	0 (0%)
Mesquite	11 (7%)	2 (2%)
Mojave	36 (19%)	36 (44%)
Montane Forest	1 (1%)	1 (1%)
Montane Shrub	4 (2%)	3 (4%)
Pinyon-Juniper	51 (32%)	9 (11%)
Riparian	25 (16%)	19 (24%)
Salt Desert Scrub	4 (3%)	1 (1%)
Sagebrush Scrub	13 (8%)	6 (7%)
Sagebrush Steppe	0 (0%)	1 (1%)
Urban	2 (1%)	0 (0%)
Wetland	0 (0%)	3 (4%)
TOTAL	159	82

Total numbers reported in the Habitat Use, Breeding Status, and Abundance tables may differ from each other (see pp. 22–23 for details). Percentage sums may differ slightly from 100% due to rounding.

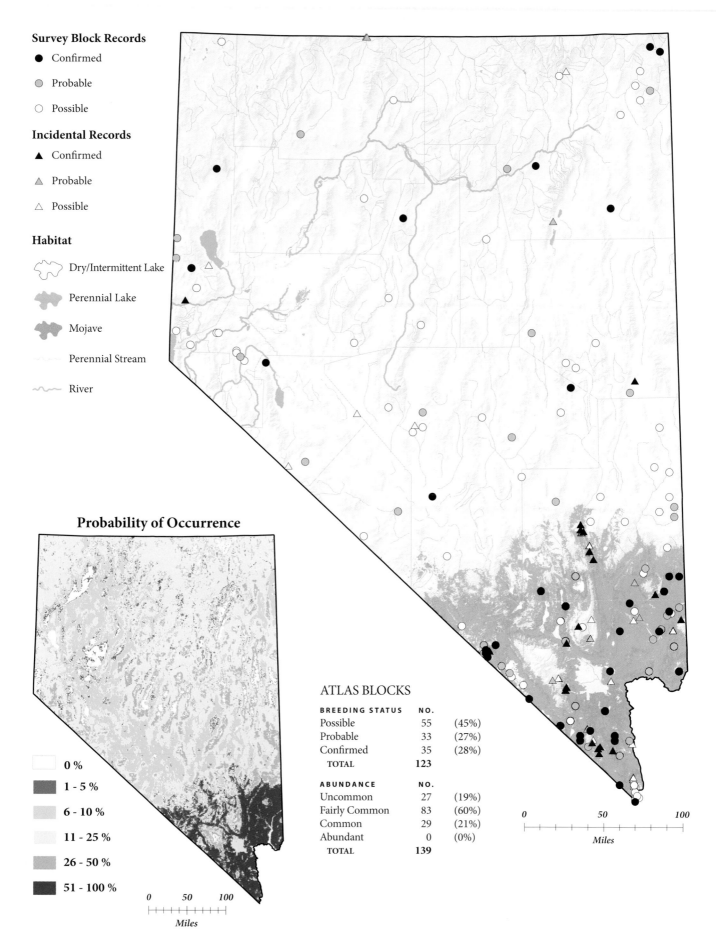

Survey Block Records

- ● Confirmed
- ◐ Probable
- ○ Possible

Incidental Records

- ▲ Confirmed
- △ Probable
- △ Possible

Habitat

- Dry/Intermittent Lake
- Perennial Lake
- Mojave
- Perennial Stream
- River

Probability of Occurrence

- ☐ 0 %
- 1 - 5 %
- 6 - 10 %
- 11 - 25 %
- 26 - 50 %
- 51 - 100 %

0   50   100
Miles

ATLAS BLOCKS

BREEDING STATUS	NO.	
Possible	55	(45%)
Probable	33	(27%)
Confirmed	35	(28%)
TOTAL	123	

ABUNDANCE	NO.	
Uncommon	27	(19%)
Fairly Common	83	(60%)
Common	29	(21%)
Abundant	0	(0%)
TOTAL	139	

0   50   100
Miles

BROWN-CRESTED FLYCATCHER

Myiarchus tyrannulus

Until recently the Brown-crested Flycatcher was barely known in Nevada, and only a few published records existed for the state. Atlas workers were thus somewhat surprised to find probable or confirmed breeders at several riparian sites in southern Nevada. It is unclear if Brown-crested Flycatchers have recently spread into Nevada or were previously overlooked because of their similarity to the more abundant Ash-throated Flycatcher.

DISTRIBUTION

Brown-crested Flycatchers were found in one atlas block in southwestern Nye County and in three blocks along the Virgin River in Clark County. Incidental records were also obtained from the Pahranagat Valley and from the foothills of the Spring Mountains, with breeding confirmed in both locations. Brown-crested Flycatchers were not numerous anywhere; only a single pair or individual was seen in three of the four blocks where they occurred, and fewer than ten pairs were estimated to be in the other block. The predictive map reflects their general scarcity by suggesting a low probability of detecting Brown-crested Flycatchers in most of the Mojave Desert, although there is a higher chance of finding them along the region's larger watercourses.

The Brown-crested Flycatcher is not mentioned at all in Linsdale's (1936, 1951) accounts of Nevada birds. Alcorn (1988) con-

sidered it an uncommon summer resident in southern Nevada. This flycatcher reaches the northern edge of its breeding range in the Mojave Desert (Cardiff and Dittmann 2000), and it rarely occurs in extreme southwestern Utah (Behle et al. 1985). It is a rare and local summer visitor in southern California (Small 1994), and a common but local breeder in the lower Colorado River Valley (Rosenberg et al. 1991) and in the southern and central parts of Arizona (Wise-Gervais in Corman and Wise-Gervais 2005:322–323).

The Brown-crested Flycatcher's main habitat requirement is large, natural nest cavities; usually these have been excavated by woodpeckers (Cardiff and Dittmann 2000). In the southwestern deserts, the species is most likely to find suitable nest cavities in giant cactus or in mature riparian forests. Cottonwood, willow, and mesquite woodlands are the traditional riparian vegetation for nest sites (Cardiff and Dittmann 2000), but Brown-crested Flycatchers may also be found in nonnative vegetation (Rosenberg et al. 1991). They tend to favor habitats that are more heavily vegetated than those used by Ash-throated Flycatchers (Wise-Gervais in Corman and Wise-Gervais 2005:320–323). The two species overlap broadly, though, and it is not unusual to see both in the same grove of trees.

CONSERVATION AND MANAGEMENT

The conservation status of the Brown-crested Flycatcher in Nevada has not been established. No population trend data exist for the state, but elsewhere in the U.S. range its numbers have been increasing (Sauer et al. 2005). In some parts of the Southwest, destruction of riparian habitat has been associated with local population declines, although the Brown-crested Flycatcher's willingness to use residential areas and urban parks may offset some of those effects locally (Rosenberg et al. 1991).

Along the northern edge of its range the Brown-crested Flycatcher appears to be a recent arrival (Cardiff and Dittmann 2000), and it may still be on the increase in Nevada. This possibility warrants further study, especially in areas such as the Joshua tree woodlands around Searchlight, where Gilded Flickers—a good indicator of breeding populations of Brown-crested Flycatchers elsewhere (Cardiff and Dittmann 2000)—are found. Meanwhile, because it is primarily associated with groves of old trees along waterways, this species may be locally vulnerable to riparian degradation and to the replacement of native forests with cavity-deficient, invasive woodlands.

HABITAT USE

HABITAT	ATLAS BLOCKS	INCIDENTAL OBSERVATIONS
Agricultural	0 (0%)	1 (13%)
Ash	1 (25%)	0 (0%)
Mesquite	0 (0%)	1 (13%)
Mojave	0 (0%)	1 (13%)
Riparian	3 (75%)	5 (63%)
TOTAL	4	8

Total numbers reported in the Habitat Use, Breeding Status, and Abundance tables may differ from each other (see pp. 22–23 for details). Percentage sums may differ slightly from 100% due to rounding.

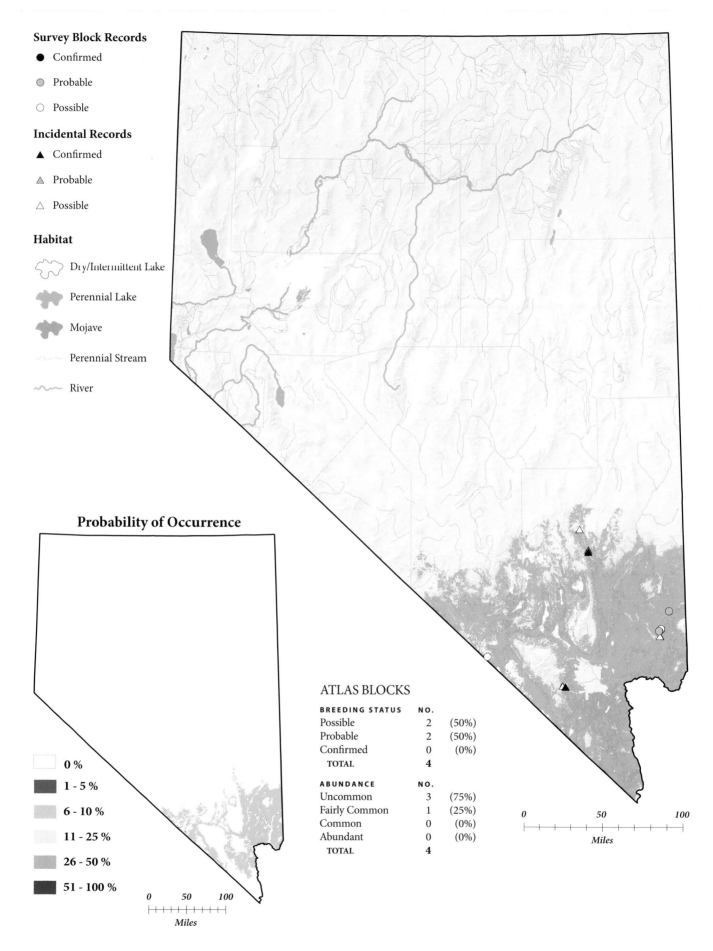

Survey Block Records

● Confirmed

◐ Probable

○ Possible

Incidental Records

▲ Confirmed

△ Probable

△ Possible

Habitat

Dry/Intermittent Lake

Perennial Lake

Mojave

Perennial Stream

River

Probability of Occurrence

☐ 0 %

■ 1 - 5 %

■ 6 - 10 %

□ 11 - 25 %

■ 26 - 50 %

■ 51 - 100 %

0   50   100
Miles

ATLAS BLOCKS

BREEDING STATUS	NO.	
Possible	2	(50%)
Probable	2	(50%)
Confirmed	0	(0%)
TOTAL	4	

ABUNDANCE	NO.	
Uncommon	3	(75%)
Fairly Common	1	(25%)
Common	0	(0%)
Abundant	0	(0%)
TOTAL	4	

0   50   100
Miles

CASSIN'S KINGBIRD
Tyrannus vociferans

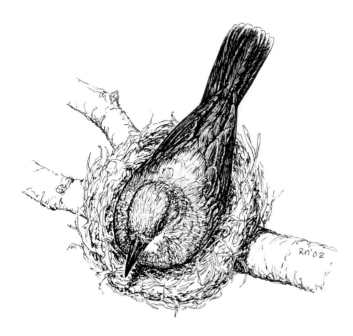

Kingbirds with bright yellow bellies are a dime a dozen in southern Nevada. They are so numerous, in fact, that it is easy to take them for granted and assume that they are all the same species. And, indeed, most of them are Western Kingbirds. But in luxurious stands of mature trees at middle elevations in the Mojave Desert, a yellow-bellied kingbird might just be the much less common Cassin's Kingbird. Cassin's Kingbirds are darker overall than Westerns, with a contrasting white throat and a highly distinctive *ch'BOO!* call.

DISTRIBUTION

Cassin's Kingbirds were reported from only two atlas blocks, both in southern Nevada. Additional incidental records, three of confirmed breeders, occurred throughout Clark County and southern Lincoln County. Few Nevada records for Cassin's Kingbird existed prior to the atlas fieldwork. The first state record was reported in 1938 (Linsdale 1951, Alcorn 1988). Although Cassin's Kingbird is primarily a bird of southern Nevada, it is occasionally seen north into the Great Basin (Ryser 1985).

Most of the breeding range lies to the south and east of Nevada. Cassin's Kingbird is accidental in Idaho and Oregon (Stephens and Sturts 1998, Adamus et al. 2001), but it occurs regularly in a disjunct population that covers parts of Wyoming, Montana, and South Dakota (Tweit and Tweit 2000). Unlike many warm-desert species, it is curiously rare in the lower Colorado River Valley, southeastern California, and southwestern Arizona. It is more common in the rest of Arizona and in southern Utah and southwestern and lowland California (Behle et al. 1985, Rosenberg et al. 1991, Tweit and Tweit 2000, Wise-Gervais in Corman and Wise-Gervais 2005:328–329).

Cassin's Kingbirds require large nesting trees near open foraging areas. An Arizona study found that they are more likely to be found in riparian woodlands than are Western Kingbirds, although in open habitats the two species largely overlap (Blancher and Robertson 1984). Most Nevada atlas records came from riparian areas; interestingly, none came from pinyon-juniper woodlands, which this species commonly uses elsewhere (Tweit and Tweit 2000). The predictive map should be viewed with caution as it is based on very few records, but it does illustrate Cassin's Kingbirds' avoidance of the lower Colorado River system and an overall low probability of finding them throughout the Mojave Desert portion of Nevada. Future surveys will be needed to reveal whether this species is indeed more likely to occur in human-modified areas than elsewhere in Nevada, as indicated by the slightly higher probability of finding them near Las Vegas and Pahrump.

CONSERVATION AND MANAGEMENT

Cassin's Kingbird populations appear stable, both nationally and regionally (Sauer et al. 2005). Cassin's Kingbirds generally are pickier about their nest sites than other kingbirds (Tweit and Tweit 2000), but they are sufficiently versatile to use ornamental trees around human habitations (Wise-Gervais in Corman and Wise-Gervais 2005:328–329). Determining the basic habitat needs and nesting requirements of this kingbird would be a valuable addition to our knowledge of the species in this region. The conventional wisdom is that Cassin's Kingbird is more of a habitat specialist than the Western Kingbird in Nevada; however, this observation remains to be quantitatively established.

Because Cassin's Kingbird populations are fairly secure rangewide, despite being rather scarce in Nevada, no general conservation actions seem warranted at the present time. In the southern woodlands where Cassin's Kingbirds do occur, however, they offer birders an enticing rarity and may thus be of some management interest.

HABITAT USE

HABITAT	ATLAS BLOCKS	INCIDENTAL OBSERVATIONS
Agricultural	0 (0%)	1 (8%)
Mojave	1 (50%)	2 (17%)
Open Water	0 (0%)	1 (8%)
Riparian	0 (0%)	6 (50%)
Urban	1 (50%)	1 (8%)
Wetland	0 (0%)	1 (8%)
TOTAL	2	12

Total numbers reported in the Habitat Use, Breeding Status, and Abundance tables may differ from each other (see pp. 22–23 for details). Percentage sums may differ slightly from 100% due to rounding.

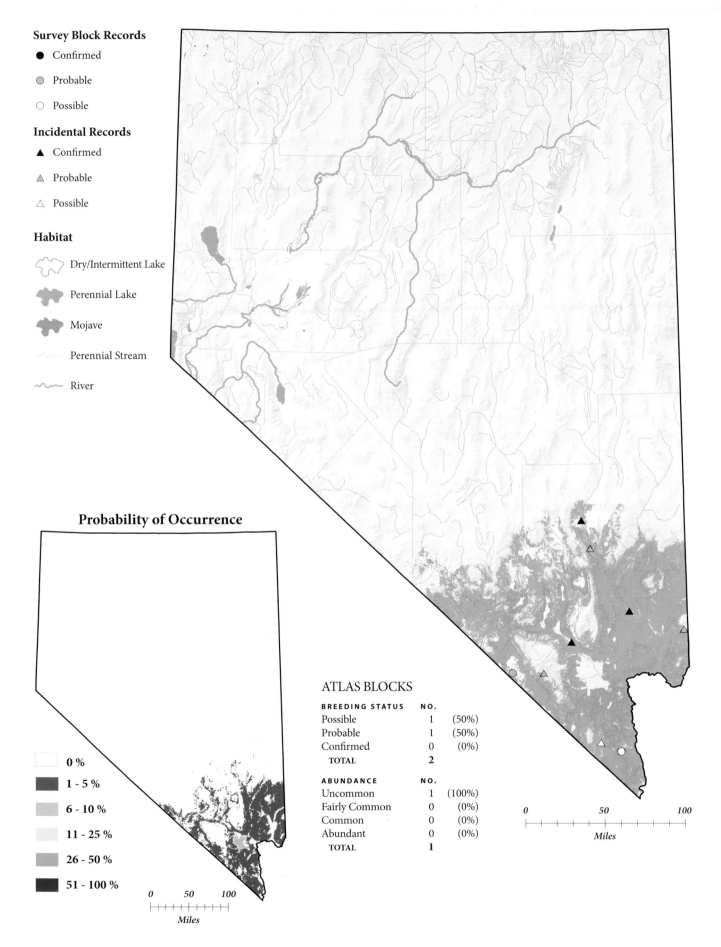

Survey Block Records

● Confirmed

◉ Probable

○ Possible

Incidental Records

▲ Confirmed

△ Probable

△ Possible

Habitat

Dry/Intermittent Lake

Perennial Lake

Mojave

Perennial Stream

River

Probability of Occurrence

☐ 0 %

■ 1 - 5 %

■ 6 - 10 %

☐ 11 - 25 %

■ 26 - 50 %

■ 51 - 100 %

0   50   100
Miles

ATLAS BLOCKS

BREEDING STATUS	NO.	
Possible	1	(50%)
Probable	1	(50%)
Confirmed	0	(0%)
TOTAL	2	

ABUNDANCE	NO.	
Uncommon	1	(100%)
Fairly Common	0	(0%)
Common	0	(0%)
Abundant	0	(0%)
TOTAL	1	

0   50   100
Miles

WESTERN KINGBIRD
Tyrannus verticalis

Feisty and fretful, the Western Kingbird is a familiar sight on telephone wires and road signs in Nevada. It breeds wherever there is open country with scattered trees; open juniper woodlands in sagebrush country, agricultural settings, and plantings around ranch buildings are all acceptable. Western Kingbirds are common around our larger towns and cities, too. Most of them do not return to breed until late April, and fall migration is often well under way by July.

DISTRIBUTION

Atlas workers found Western Kingbirds all over Nevada and confirmed breeding at many locations throughout the state. Most of the confirmed breeding records came from valley locations such as Ash Meadows and Pahranagat national wildlife refuges, the Virgin and Humboldt river drainages, and the Mason and Lahontan valleys. The species was rated as fairly common in more than half of the instances in which abundance codes were supplied, and at least ten pairs were estimated to occur in about a fifth of these blocks.

Earlier workers, beginning with those of the Death Valley Expedition, universally characterized the Western Kingbird as widespread and common in Nevada, especially at low and middle elevations (Fisher 1893, Linsdale 1936, Ryser 1985, Alcorn 1988, Chisholm and Neel 2002). Western Kingbirds are also common in California (Small 1994); Arizona (Wise-Gervais in Corman and Wise-Gervais 2005:332–333), including the hot lowlands of the lower Colorado River Valley (Rosenberg et al. 1991); Utah, with the greatest abundance in the southern part of the state (Behle et al. 1985); Oregon (Adamus et al. 2001); and Idaho (Stephens and Sturts 1998).

In general, Western Kingbirds favor open habitats with lookouts of any conceivable sort—telephone poles, windmills, and even tufa towers will do (Gaines 1992)—and they will nest in trees, shrubs, or on artificial structures (Gamble and Bergin 1996). In Nevada, they are most often found nesting in riparian habitats, on ranches and farms, and along the edges of towns and cities (Ryser 1985). In southern California, they can be found in virtually any habitat with scattered tall trees, and in the lower Colorado River Valley, they favor tall trees and human residences in agricultural areas (Rosenberg et al. 1991). This relationship is reflected on the predictive map, which highlights riparian, agricultural, and urban areas—all habitats where tall deciduous trees abound. The map further reflects the Western Kingbird's ubiquity and habitat flexibility by showing a moderate probability of finding the species across most of the state, and lower probabilities of finding it in the mountains.

CONSERVATION AND MANAGEMENT

Unlike many other long-distance passerine migrants, the Western Kingbird has increased in numbers, expanded its range, and generally benefited from human activity (Gamble and Bergin 1996). Local increases may be attributable to exotic plantings or artificial structures (Shuford 1993).

In Nevada, the status of the Western Kingbird has probably not changed since the nineteenth century, and may even have improved. Ryser (1985) considered its numbers to have changed little since Ridgway's surveys in the 1860s, but Littlefield (1990) argued that the species has benefited from settlement of the Great Basin. The Western Kingbird certainly is tolerant of urban areas (Gamble and Bergin 1996), and it is possible that continued urbanization in Nevada will have a positive impact on its numbers in the state.

HABITAT USE

HABITAT	ATLAS BLOCKS	INCIDENTAL OBSERVATIONS
Agricultural	30 (13%)	11 (9%)
Ash	8 (4%)	0 (0%)
Barren	1 (<1%)	1 (<1%)
Grassland	5 (2%)	2 (2%)
Mesquite	3 (1%)	3 (3%)
Mojave	12 (5%)	8 (7%)
Montane Forest	2 (<1%)	0 (0%)
Montane Parkland	1 (<1%)	1 (<1%)
Montane Shrub	2 (<1%)	1 (<1%)
Open Water	0 (0%)	2 (2%)
Pinyon-Juniper	17 (8%)	0 (0%)
Riparian	53 (23%)	36 (31%)
Salt Desert Scrub	11 (5%)	1 (<1%)
Sagebrush Scrub	28 (12%)	6 (5%)
Sagebrush Steppe	12 (5%)	4 (3%)
Urban	39 (17%)	34 (29%)
Wetland	4 (2%)	7 (6%)
TOTAL	228	117

Total numbers reported in the Habitat Use, Breeding Status, and Abundance tables may differ from each other (see pp. 22–23 for details). Percentage sums may differ slightly from 100% due to rounding.

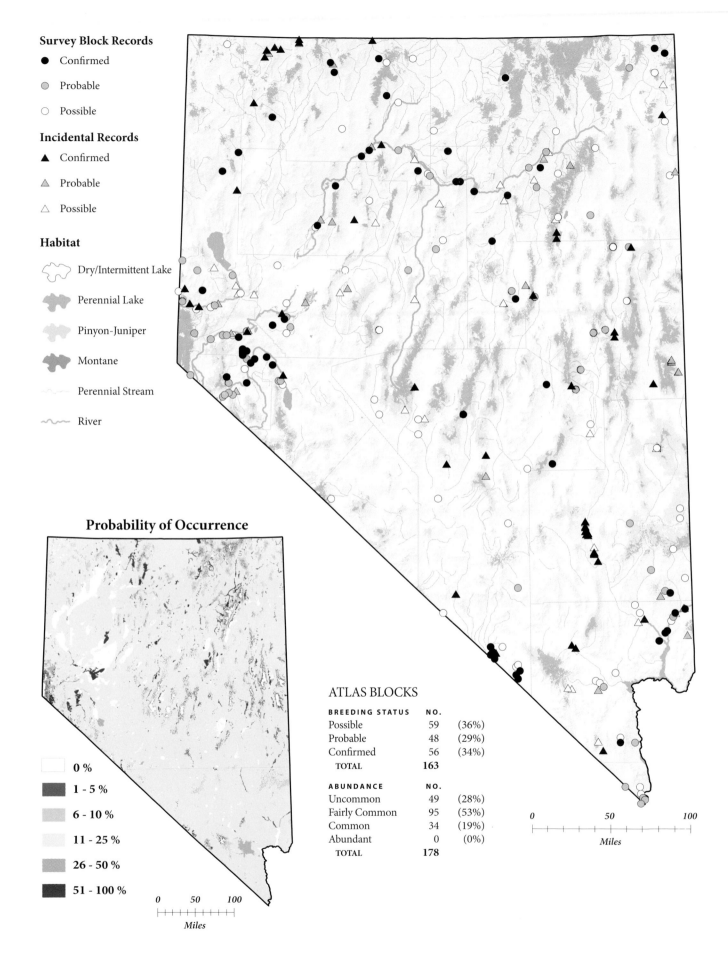

Survey Block Records

● Confirmed

● Probable

○ Possible

Incidental Records

▲ Confirmed

▲ Probable

△ Possible

Habitat

Dry/Intermittent Lake

Perennial Lake

Pinyon-Juniper

Montane

Perennial Stream

River

Probability of Occurrence

☐	0 %
■	1 - 5 %
■	6 - 10 %
☐	11 - 25 %
■	26 - 50 %
■	51 - 100 %

0   50   100
Miles

ATLAS BLOCKS

BREEDING STATUS	NO.	
Possible	59	(36%)
Probable	48	(29%)
Confirmed	56	(34%)
TOTAL	163	

ABUNDANCE	NO.	
Uncommon	49	(28%)
Fairly Common	95	(53%)
Common	34	(19%)
Abundant	0	(0%)
TOTAL	178	

0   50   100
Miles

EASTERN KINGBIRD
Tyrannus tyrannus

The backcountry road through Lamoille near Elko offers the jaded birder a welcome detour from the sagebrush sea of northern Nevada. Here, one can find lush hayfields, lovely old homesteads, and dense tangles of broadleaved vegetation at every stream crossing. Savannah Sparrows sing from the fence posts, Bobolinks babble in the tall grasses, and Sandhill Cranes prance about the freshly mown pastures. It is even possible to find Eastern Kingbirds here, at the extreme limit of their distribution. They are few in number but hard to miss because they perch out in the open and frequently produce wild bursts of sputtering alarm notes.

DISTRIBUTION

Atlas workers found Eastern Kingbirds at eight sites in the north and northeast, the historic breeding range of the species in Nevada. Only one site had confirmed breeding: an incidental record in the Lamoille Valley, in a grove of trees along Soldier Creek. Linsdale (1936) and Alcorn (1988) reported that Eastern Kingbirds occur throughout the northern part of Nevada, although the records they listed were few and scattered. Ryser (1985) considered the Eastern Kingbird to be largely absent from Nevada, despite the presence of local populations just to our north and east.

The summer distribution of the Eastern Kingbird in the states surrounding Nevada is complex. It is rare in California but is suspected to breed in some of the northern counties that adjoin Nevada (Grinnell and Miller 1944, Garrett and Dunn 1981, Small 1994). Breeding has been reported in the northeastern corner of Arizona (Monson and Phillips 1981), far from the Nevada border, but the *Arizona Breeding Bird Atlas* project did not report breeding in the state. The Eastern Kingbird is locally common in eastern Oregon, but not in the portions of the state closest to Nevada (Adamus et al. 2001). The species breeds in much of Idaho, including the south (Stephens and Sturts 1998). In Utah, Eastern Kingbirds are uncommon in the north and even less common toward the Nevada border (Behle et al. 1985).

Eastern Kingbirds occupy a variety of disturbed, edge, and second-growth habitats. The availability of water may also influence their occurrence (Murphy 1996). It is easy to see why the Lamoille area is suitable for breeding Eastern Kingbirds: it is fairly well watered and is abundantly supplied with woodland corridors and riparian groves. The predictive map reflects the scattered occurrence of similar suitable habitats in Nevada within or near the Eastern Kingbird's known geographical range limits.

CONSERVATION AND MANAGEMENT

Eastern Kingbird numbers are declining throughout much of North America, especially in the eastern heart of the range (Murphy 1996, Sauer et al. 2005). The species seems to be less common now in Malheur County, Oregon, than it was at the beginning of the twentieth century (Contreras and Kindschy 1996), and it probably has declined in Utah (Behle et al. 1985). Loss of both riparian habitat and traditionally farmed agricultural lands are reported to be prime factors in the declines (Murphy 1996).

Based on the anecdotal information presented by Linsdale (1936) and Alcorn (1988), it would appear that the Eastern Kingbird's range in Nevada has contracted somewhat. Whether the factors driving declines elsewhere are relevant in Nevada is unclear. Elko County still supports a reasonably extensive network of farming communities, and this may explain why the Eastern Kingbird has managed to hold on there. Nevada Partners in Flight has emphasized the benefit of traditional farming practices for a number of Nevada's breeding bird species (Neel 1999a), and the Eastern Kingbird may well be one of these.

HABITAT USE

HABITAT	ATLAS BLOCKS	INCIDENTAL OBSERVATIONS
Agricultural	2 (22%)	2 (40%)
Grassland	1 (11%)	0 (0%)
Riparian	3 (33%)	3 (60%)
Sagebrush Scrub	1 (11%)	0 (0%)
Urban	2 (22%)	0 (0%)
TOTAL	9	5

Total numbers reported in the Habitat Use, Breeding Status, and Abundance tables may differ from each other (see pp. 22–23 for details). Percentage sums may differ slightly from 100% due to rounding.

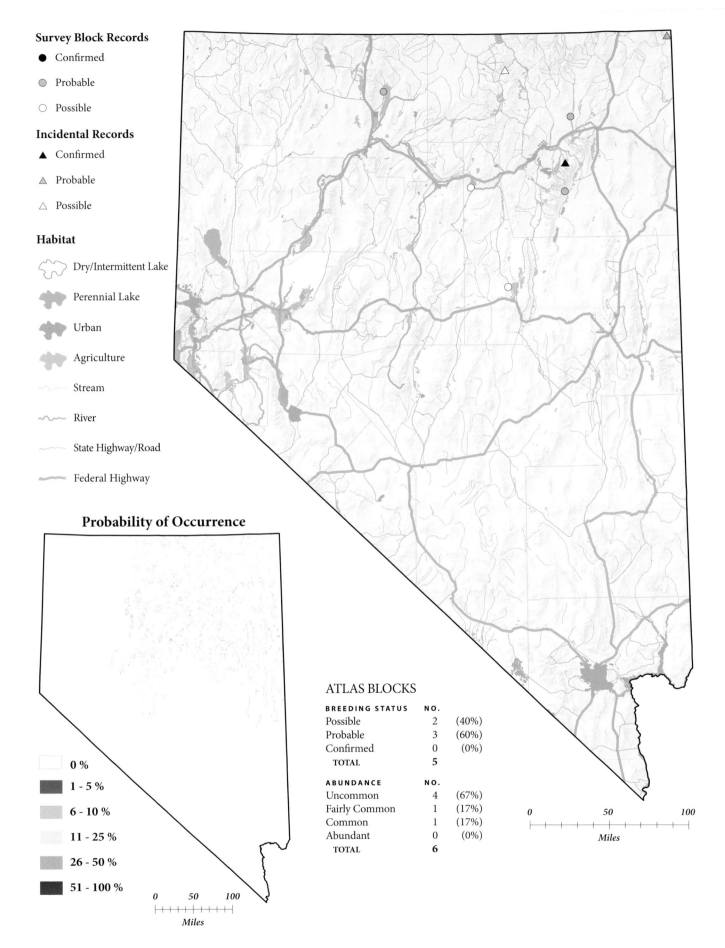

Survey Block Records

● Confirmed

◉ Probable

○ Possible

Incidental Records

▲ Confirmed

△ Probable

△ Possible

Habitat

Dry/Intermittent Lake

Perennial Lake

Urban

Agriculture

Stream

River

State Highway/Road

Federal Highway

Probability of Occurrence

☐ 0 %

■ 1 - 5 %

▨ 6 - 10 %

☐ 11 - 25 %

▨ 26 - 50 %

■ 51 - 100 %

0   50   100

Miles

ATLAS BLOCKS

BREEDING STATUS	NO.	
Possible	2	(40%)
Probable	3	(60%)
Confirmed	0	(0%)
TOTAL	**5**	

ABUNDANCE	NO.	
Uncommon	4	(67%)
Fairly Common	1	(17%)
Common	1	(17%)
Abundant	0	(0%)
TOTAL	**6**	

0   50   100

Miles

LOGGERHEAD SHRIKE
Lanius ludovicianus

Everything about the Loggerhead Shrike seems menacing: its striking pattern, its sharply hooked beak, and its apparent indifference to the close approach of curious naturalists. But it is most renowned for its gruesome habit of impaling prey items (usually large arthropods but occasionally small vertebrates) on thorns or barbed wire. The Loggerhead Shrike is a familiar, but always fascinating, sight in most of Nevada's open shrublands.

DISTRIBUTION

Atlas workers found Loggerhead Shrikes in almost a third of all atlas blocks, and widely distributed around the state. Shrikes were confirmed as breeders in 35% of the blocks in which they were found, and atlas workers reported multiple pairs in 67% of the blocks for which abundance estimates were supplied.

Loggerhead Shrikes breed across most of the United States except for the upper Midwest, the Northeast, and the far Northwest (Yosef 1996). Earlier writers portrayed the Loggerhead Shrike as widespread and common in Nevada (Fisher 1893, Linsdale 1936, Ryser 1985, Alcorn 1988), across the Great Basin into Utah (Behle et al. 1985), in California (Small 1994), in Arizona (Wise-Gervais in Corman and Wise-Gervais 2005:336–337), and in the lowlands of the Colorado River Valley (Rosenberg et al. 1991). The species is limited to eastern Oregon (Adamus et al. 2001) and is more common in Idaho's southern half than farther north in the state (Stephens and Sturts 1998).

For breeding habitat, western populations of the Loggerhead Shrike tend to favor arid open country with just a few perches or lookouts (Yosef 1996). Ryser (1985) listed desert shrublands, juniper and pinyon-juniper woodlands, mountain mahogany stands, and the outskirts of ranches and towns as suitable haunts. Scattered small trees and taller shrubs appear to be important for breeding birds (Neel 1999a). The predictive map shows a high probability of occurrence across most of the state, with lower probabilities only in forests, higher mountains, barren zones, and urban areas.

CONSERVATION AND MANAGEMENT

Loggerhead Shrike populations have declined throughout their range over the last half-century (Sauer et al. 2005), and various government agencies and research institutions have listed the species as sensitive (Yosef 1996, Gallagher 1997). Declines have been especially pronounced in the East, where the species had earlier expanded in the wake of European settlement (Dobkin and Sauder 2004). The major factors in the decline are thought to include pesticides, habitat loss, and winter mortality (Paige and Ritter 1999).

Loggerhead Shrike populations also show significant declines in the Intermountain West (Sauer et al. 2005). Dobkin and Sauder (2004) noted that declines in shrub steppe regions were particularly severe, and suggested conversion of shrub steppe for grazing and agriculture, altered fire regimes, and invasion of annual grasses as contributing factors. Paige and Ritter (1999) concurred, but added that the causes of these declines are hypothetical and have not been well demonstrated.

Survey coverage in Nevada has been too limited to determine trends within the state. Certain areas of Nevada support high densities of shrikes, but the reasons are not clear. Rangeland spraying of pesticides is a current concern, but direct human disturbance is presently not (Neel 1999a). The consensus in Nevada seems to be that further study and additional monitoring are warranted. In the meantime, given ongoing declines in the Intermountain West, the Loggerhead Shrike has been designated a Nevada Partners in Flight Priority species (Neel 1999a).

HABITAT USE

HABITAT	ATLAS BLOCKS	INCIDENTAL OBSERVATIONS
Agricultural	9 (3%)	4 (6%)
Ash	1 (<1%)	0 (0%)
Barren	1 (<1%)	0 (0%)
Grassland	4 (1%)	1 (2%)
Mesquite	7 (2%)	1 (2%)
Mojave	41 (14%)	18 (29%)
Montane Forest	0 (0%)	1 (2%)
Montane Shrub	3 (1%)	1 (2%)
Pinyon-Juniper	36 (12%)	2 (3%)
Riparian	7 (2%)	10 (16%)
Salt Desert Scrub	46 (16%)	5 (8%)
Sagebrush Scrub	77 (26%)	12 (19%)
Sagebrush Steppe	55 (19%)	4 (6%)
Urban	3 (1%)	3 (5%)
Wetland	5 (2%)	1 (2%)
TOTAL	295	63

Total numbers reported in the Habitat Use, Breeding Status, and Abundance tables may differ from each other (see pp. 22–23 for details). Percentage sums may differ slightly from 100% due to rounding.

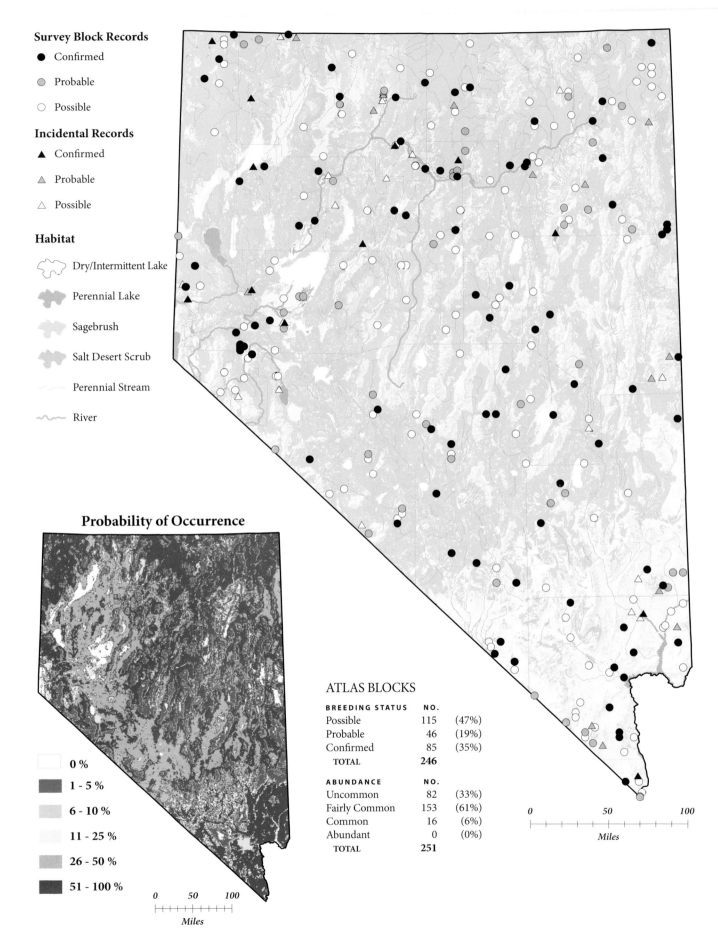

Survey Block Records

- ● Confirmed
- ● Probable
- ○ Possible

Incidental Records

- ▲ Confirmed
- ▲ Probable
- △ Possible

Habitat

Dry/Intermittent Lake

Perennial Lake

Sagebrush

Salt Desert Scrub

Perennial Stream

River

Probability of Occurrence

- 0 %
- 1 - 5 %
- 6 - 10 %
- 11 - 25 %
- 26 - 50 %
- 51 - 100 %

0   50   100

Miles

ATLAS BLOCKS

BREEDING STATUS	NO.	
Possible	115	(47%)
Probable	46	(19%)
Confirmed	85	(35%)
TOTAL	**246**	

ABUNDANCE	NO.	
Uncommon	82	(33%)
Fairly Common	153	(61%)
Common	16	(6%)
Abundant	0	(0%)
TOTAL	**251**	

0   50   100

Miles

BELL'S VIREO
Vireo bellii

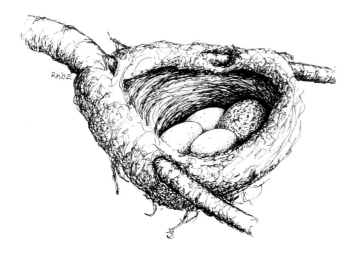

Right through the midday heat, the peculiar and distinctive chatter of Bell's Vireo can be heard coming from dense willow and tamarisk stands around washes and ponds in southern Nevada. Unfortunately, as is the case elsewhere in the species' range, degradation and loss of habitat have caused population declines (Brown 1993). Thus, conservation of the species—specifically the Arizona race of Bell's Vireo (subspecies *arizonae*) that breeds in southern Nevada—requires our attention, especially along the Colorado River and its tributaries.

DISTRIBUTION
Bell's Vireos were found in seventeen blocks in southern Nevada and were fairly numerous; more than half of the blocks were estimated to have in excess of ten pairs. Breeding was confirmed in 35% of these blocks, all of which were located along rivers. Bell's Vireo has been difficult to confirm as a breeder in other states, primarily because nests tend to be in dense, often impenetrable vegetation (Cecil in Jackson et al. 1996:302–303, Jacobs and Wilson 1997, Chace in Kingery 1998:302–303).

Linsdale (1936) reported Bell's Vireo to be limited to the southern tip of Clark County, but Alcorn (1988) also cited records from northern Clark County, including the town of Mesquite. In areas adjoining Nevada, Bell's Vireo breeds in the Colorado River system of southwestern Utah (Behle et al. 1985)—although breeding populations of the *arizonae* subspecies were reportedly decimated in the lower Colorado River Valley in the mid-1900s (Rosenberg et al. 1991)—and in much of Arizona (Averill-Murray and Corman in Corman and Wise-Gervais 2005:338–339). The *pusillus* subspecies, the Least Bell's Vireo, nests in southern California and northern Baja California, Mexico (Brown 1993). Interestingly, a single specimen of that now-endangered subspecies was collected in the Ash Meadows area during the Death Valley Expedition in the late 1800s (Fisher 1893, Linsdale 1936).

Dense understory vegetation is critical nesting habitat for Bell's Vireos (Brown 1993). They use mesquite and willow stands (Rosenberg et al. 1991), as well as tamarisk in some areas (Brown and Trosset 1989). Atlas workers found Bell's Vireos primarily in riparian habitat, with most other records coming from mesquite habitat. This specialized habitat use is also illustrated in the predictive map, which suggests a moderate to high chance of finding Bell's Vireos along major watercourses in southern Nevada and a low probability of encountering them in the rest of the Mojave Desert, where riparian habitat is scarce. Although it is not fully evident on the predictive map, most breeding populations in Nevada are associated with the Colorado River and its tributaries. Like many southern species, the chance of occurrence in the northernmost Mojave is probably overestimated due to occasional records of the species in poorly sampled habitats.

CONSERVATION AND MANAGEMENT
Population declines have been reported rangewide, particularly in the Southwest (Sauer et al. 2005). The *pusillus* subspecies, which occurs in California and Mexico, was federally listed as Endangered in 1986 as a result of recent declines, habitat threats, and cowbird parasitism (USFWS 1986, Brown 1993); these are also potential concerns in Nevada. There has been very little monitoring of Bell's Vireo populations in the state, but the species' specialized habitat use is cause for concern. Bell's Vireo is a Covered species under the Clark County Multiple Species Habitat Conservation Plan (Clark County 2000) and the Lower Colorado River Multi-species Conservation Program (BOR-LCR 2004). It is also a Partners in Flight Watch List species that warrants immediate conservation action (Rich et al. 2004).

Evidence about the impacts of cowbirds and water management regimes on breeding Bell's Vireos is mixed (Brown et al. 1983, Brown 1993, Robinson et al. 1995, Chace and Cruz 1996, Jacobs and Wilson 1997, Chace in Kingery 1998:302–303). The most promising conservation strategies involve recovery and active restoration of large patches of suitable riparian habitat (see also Kus 1998, Budnik et al. 2002).

HABITAT USE

HABITAT	ATLAS BLOCKS	INCIDENTAL OBSERVATIONS
Mesquite	6 (19%)	2 (12%)
Mojave	0 (0%)	2 (12%)
Montane Forest	1 (3%)	0 (0%)
Riparian	24 (75%)	12 (71%)
Wetland	1 (3%)	1 (6%)
TOTAL	32	17

Total numbers reported in the Habitat Use, Breeding Status, and Abundance tables may differ from each other (see pp. 22–23 for details). Percentage sums may differ slightly from 100% due to rounding.

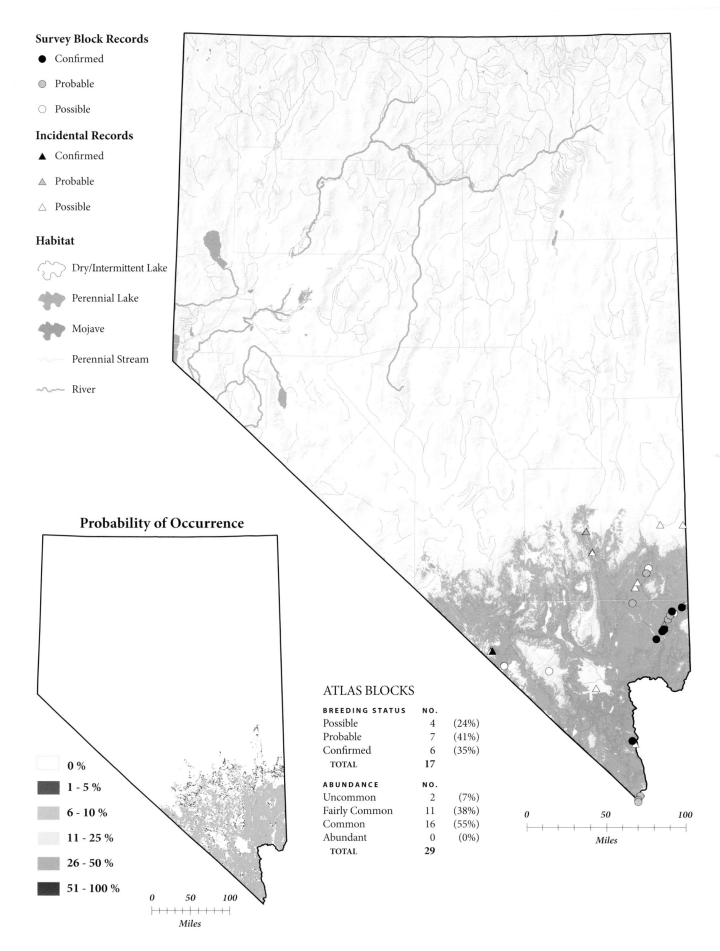

Survey Block Records

● Confirmed

◉ Probable

○ Possible

Incidental Records

▲ Confirmed

▲ Probable

△ Possible

Habitat

Dry/Intermittent Lake

Perennial Lake

Mojave

Perennial Stream

River

Probability of Occurrence

0 %

1 - 5 %

6 - 10 %

11 - 25 %

26 - 50 %

51 - 100 %

0   50   100

Miles

ATLAS BLOCKS

BREEDING STATUS	NO.	
Possible	4	(24%)
Probable	7	(41%)
Confirmed	6	(35%)
TOTAL	17	

ABUNDANCE	NO.	
Uncommon	2	(7%)
Fairly Common	11	(38%)
Common	16	(55%)
Abundant	0	(0%)
TOTAL	29	

0   50   100

Miles

GRAY VIREO
Vireo vicinior

An easily overlooked species throughout its range in the southwestern deserts, the Gray Vireo turned out to be more common than had been anticipated in Nevada—but not where many observers were expecting to find it. Instead of being restricted to far southern Nevada, the Gray Vireo was also found farther north in Lincoln and northeastern Nye counties. Breeding Gray Vireos seem to favor arid situations, especially narrow belts of mid-elevation pinyon-juniper woodlands. Such habitats tend to be in remote areas, and it is thus no surprise that many birders know little about this species. Although the Gray Vireo is visually nondescript, its loud song and peculiar call note are readily detected, and the bird was found to be relatively common within its limited Nevada range.

DISTRIBUTION

Gray Vireos were found in twenty-three atlas blocks, eighteen of which formed a distinct cluster in Lincoln and Nye counties. The remaining records were scattered in areas to the north and south, including the Spring Mountains and Sheep Range of Clark County, where area birders have long been familiar with the species. The atlas results and predictive model portray the overall Nevada range of the Gray Vireo as being slightly more northerly than previously described (Johnson 1972, Barlow et al. 1999, Neel 1999a), and it now seems safe to say that Gray Vireos are widespread at least as far north as the southern border of White Pine County. The possibility that the range may extend even farther northward, as suggested by several unconfirmed records and the predictive map, deserves investigation. Ryser (1985), Alcorn (1988), and the more recent Nevada Bird Count program have all reported Gray Vireos from the Toiyabe Mountains of central Nevada, and the atlas effort produced possible breeding records from the Ruby Mountains and the Cherry Creek Range.

The Gray Vireo's continental breeding range is mostly encompassed by the Four Corners states; Arizona and southern Utah form the core areas for the species. Nevada populations are somewhat peripheral, and additional isolated breeding populations are present in southern California (Behle et al. 1985, Small 1994, Barlow et al. 1999, Corman in Corman and Wise-Gervais 2005:340–341).

Gray Vireos were found primarily in pinyon-juniper woodlands and in habitats that commonly border or intersperse with pinyon-juniper. Gray Vireo breeding was confirmed in only 13% of the atlas blocks in which the species was found. Although confirming breeding was relatively difficult, Gray Vireos were not uncommon locally; more than half of the atlas blocks where the species was present were estimated to have more than ten pairs.

CONSERVATION AND MANAGEMENT

A striking feature of the Gray Vireo is its compact geographical range and specialized habitat requirements. It requires open, mature pinyon-juniper woodlands that are especially warm and dry and have at least some shrub or scrub understory (Barlow et al. 1999). Gray Vireos occur spottily throughout their range (Ligon 1961), and the total breeding population is relatively small (Rich et al. 2004). In Nevada, the species often has doughnut-shaped local distributions that correspond to the narrow band of mid-elevation pinyon-juniper woodlands that encircle mountains; some of these are visible on the predictive map. The Gray Vireo's winter range is restricted almost entirely to northwestern Mexico (Barlow 1980), where it occurs primarily in association with elephant trees (Bates 1992).

Available trend data suggest that Gray Vireo populations are stable, although the limited number of surveys in appropriate habitat leaves room for doubt (Dobkin and Sauder 2004, Sauer et al. 2005), especially in Nevada. Because of its restricted geographical range and low population size, the Gray Vireo is a Partners in Flight Priority Species for Nevada and a Stewardship species for the Southwest and Intermountain West (Neel 1999a, Rich et al. 2004). The species has been reported to face potential threats from cowbird parasitism (DeSante and George 1994) and heavy grazing (Wauer 1977).

HABITAT USE

HABITAT	ATLAS BLOCKS	INCIDENTAL OBSERVATIONS
Mesquite	0 (0%)	1 (5%)
Mojave	0 (0%)	1 (5%)
Montane Forest	2 (7%)	1 (5%)
Montane Parkland	0 (0%)	1 (5%)
Montane Shrub	2 (7%)	5 (25%)
Pinyon-Juniper	21 (75%)	9 (45%)
Riparian	2 (7%)	0 (0%)
Sagebrush Scrub	1 (4%)	2 (10%)
TOTAL	28	20

Total numbers reported in the Habitat Use, Breeding Status, and Abundance tables may differ from each other (see pp. 22–23 for details). Percentage sums may differ slightly from 100% due to rounding.

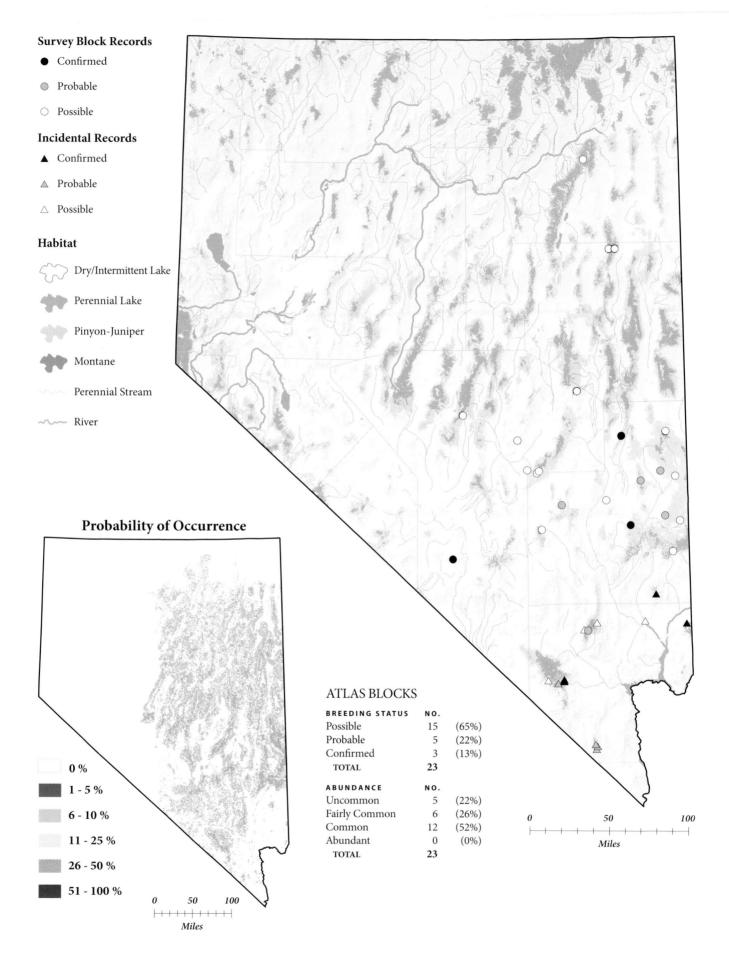

Survey Block Records

● Confirmed

● Probable

○ Possible

Incidental Records

▲ Confirmed

▲ Probable

△ Possible

Habitat

Dry/Intermittent Lake

Perennial Lake

Pinyon-Juniper

Montane

Perennial Stream

River

Probability of Occurrence

☐ 0 %

■ 1 - 5 %

■ 6 - 10 %

11 - 25 %

■ 26 - 50 %

■ 51 - 100 %

0   50   100
Miles

ATLAS BLOCKS

BREEDING STATUS	NO.	
Possible	15	(65%)
Probable	5	(22%)
Confirmed	3	(13%)
TOTAL	23	

ABUNDANCE	NO.	
Uncommon	5	(22%)
Fairly Common	6	(26%)
Common	12	(52%)
Abundant	0	(0%)
TOTAL	23	

0   50   100
Miles

PLUMBEOUS VIREO
Vireo plumbeus

When the atlas project began, Plumbeous and Cassin's Vireos and the Blue-headed Vireo of the East were collectively known as a single species. Until the "Solitary Vireo" was split up, partway through the project, the atlas reports referred only to that single species. About twenty records have thus been omitted from the maps and analyses because of uncertainty about the identity of the vireo involved. The pattern that emerged from the remaining records indicates that the Plumbeous Vireo is a widespread and at times common bird in arid mountains throughout the state.

DISTRIBUTION
Plumbeous Vireos were found in forty blocks, primarily in eastern and central Nevada, although several records also came from the west. Other than in the Sheep Range and Spring Mountains, the species was not especially common, and abundance estimates averaged lower than those for the Gray Vireo, which often occurred in the same areas. The western and central records and those from northwestern Nye County occur in possible areas of sympatry between Plumbeous and Cassin's Vireos. Most "Solitary Vireo" records, which are not shown here, matched the distribution of the Plumbeous Vireo. Distributional information in Curson and Goguen 1998 supports the possibility of Plumbeous Vireos in the western and central regions. However, the presence of Cassin's Vireos in the Carson Range (Chisholm and Neel 2002) and the difficulty in distinguishing between the two newly defined species make the need for further data collection and evaluation clear.

Although much of the historical literature does not distinguish between Plumbeous and Cassin's Vireos, Linsdale (1936) did provide separate information for the two taxa. Surprisingly, his account implies that the Plumbeous Vireo has a smaller Nevada range and is confined to the northeastern portion of the state. The current understanding of the division between the two vireos, however, accords closely with the distributions of other closely related taxonomic pairs—Nashville and Virginia's Warblers, Red-breasted and Red-naped Sapsuckers, and the "Thick-billed" and "Slate-colored" Fox Sparrows—that meet in the Great Basin. In surrounding states, the Plumbeous Vireo is fairly widespread in Utah and Arizona (Curson and Goguen 1998, Wise-Gervais in Corman and Wise-Gervais 2005:342–243) and more sporadic in California (Small 1994) and Oregon (Adamus et al. 2001).

Plumbeous Vireo records came primarily from pinyon-juniper habitat, followed in frequency by montane and riparian woodlands. The habitats used by Plumbeous Vireos in Nevada thus seem to be intermediate between the generally more mesic habitats of Warbling Vireos and the usually more arid habitats of Gray Vireos. Our predictive map suggests that the species is most likely to be found in the pinyon-juniper woodlands that encircle the state's larger mountain ranges. Low chances of Plumbeous Vireo occurrence are also predicted farther to the north and west than the species is often assumed to range (Curson and Goguen 1998). Given the identification challenges mentioned above, it is also possible that some of the records toward the periphery of the range involve misidentified birds, although all were carefully screened. Such records could have generated overestimates of predicted occurrences at the limits of the distribution.

CONSERVATION AND MANAGEMENT
Atlas data confirm that the Plumbeous Vireo is at least as widespread as Alcorn (1988) described it to be, and also confirm reports that the species is holding its own or even expanding its breeding range (Curson and Goguen 1998, Sauer et al. 2005); for instance, it seems to be a recent colonizer of southern California (Garrett and Dunn 1981). An immediate research priority is to fine-tune our knowledge of the distribution in Nevada and elsewhere. In areas of potential overlap with Cassin's Vireo, such as western Nevada and southern Oregon (Contreras and Kindschy 1996), the ranges of both species need clarification.

HABITAT USE

HABITAT	ATLAS BLOCKS	INCIDENTAL OBSERVATIONS
Montane Forest	13 (25%)	10 (43%)
Montane Parkland	1 (2%)	0 (0%)
Montane Shrub	1 (2%)	2 (9%)
Pinyon Juniper	25 (48%)	6 (26%)
Riparian	10 (19%)	5 (22%)
Sagebrush Scrub	2 (4%)	0 (0%)
TOTAL	52	23

Total numbers reported in the Habitat Use, Breeding Status, and Abundance tables may differ from each other (see pp. 22–23 for details). Percentage sums may differ slightly from 100% due to rounding.

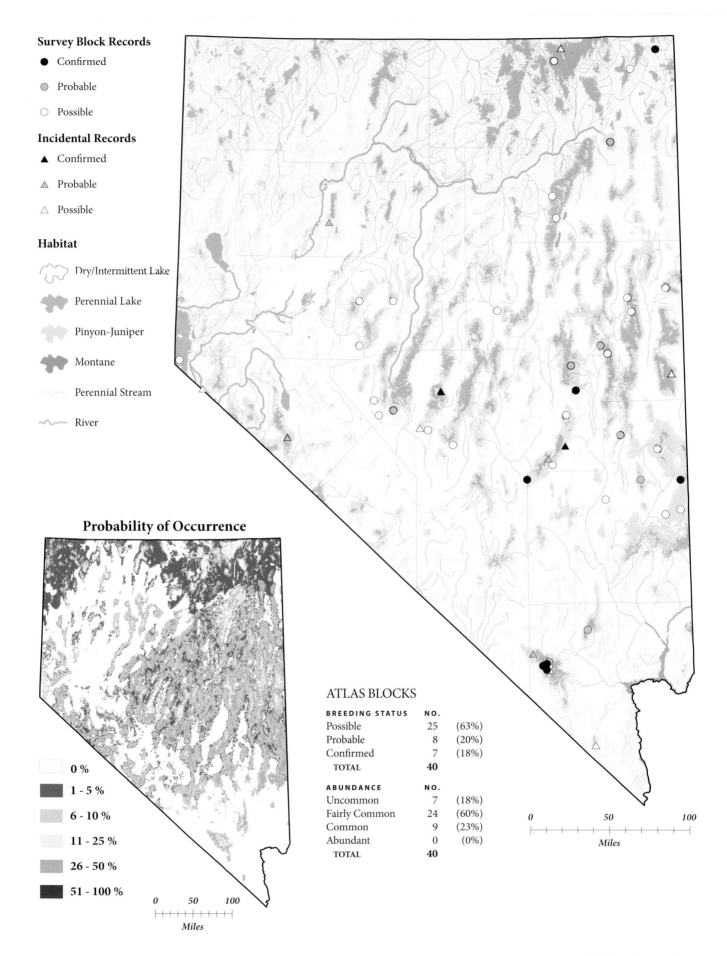

Survey Block Records

● Confirmed

◐ Probable

○ Possible

Incidental Records

▲ Confirmed

△ Probable

△ Possible

Habitat

Dry/Intermittent Lake

Perennial Lake

Pinyon-Juniper

Montane

Perennial Stream

River

Probability of Occurrence

☐ 0 %

■ 1 - 5 %

■ 6 - 10 %

■ 11 - 25 %

■ 26 - 50 %

■ 51 - 100 %

0   50   100

Miles

ATLAS BLOCKS

BREEDING STATUS	NO.	
Possible	25	(63%)
Probable	8	(20%)
Confirmed	7	(18%)
TOTAL	40	

ABUNDANCE	NO.	
Uncommon	7	(18%)
Fairly Common	24	(60%)
Common	9	(23%)
Abundant	0	(0%)
TOTAL	40	

0   50   100

Miles

CASSIN'S VIREO
Vireo cassinii

Cassin's Vireo was literally unknown to many birders at the beginning of the atlas project. At the time, Cassin's and Plumbeous Vireos were lumped into a single species known as the "Solitary Vireo." Understandably, especially given the challenges of distinguishing the two new species in the field, many atlas workers simply wrote "Solitary Vireo" on their field cards. Records that could not be assigned to either of the new species were omitted from our analyses. Although it was thus impossible to determine the precise distributions of Cassin's and Plumbeous Vireos in Nevada, it is clear that Cassin's Vireo is a widespread breeder in the Carson Range of far western Nevada and is part of the distinctive Sierra Nevada avifauna that barely enters the state from California.

DISTRIBUTION

Cassin's Vireos breed mainly in the Pacific Northwest, northern Rockies, and through California (Goguen and Curson 2002). In Nevada, they were recorded in twelve blocks, nine of which were in the Carson Range. The remaining three block records and an incidental observation were from Nye and Esmeralda counties, and could have involved late migrants or misidentified birds. If these records do represent breeding Cassin's Vireos, however, then south-central Nevada, in addition to the Carson Range, should be investigated for possible sympatry between Cassin's and Plumbeous Vireos.

The distribution of confirmed records is consistent with those given in Linsdale 1936 and Chisholm and Neel 2002, which characterize the bird as a summer resident in the mountains of western Nevada. Interestingly, Linsdale (1936) also cited records of Cassin's Vireo from across much of the state. Most of these sightings were made during migration periods, but they nonetheless suggest that breeding outside the Carson Range may be possible. Alcorn (1988) also noted a presumed breeder from Potosi Mountain in Clark County. Despite intensive atlas work in this region, however, the only "Solitary Vireos" found in this range, or elsewhere in southern Nevada, were Plumbeous Vireos. Revisiting old museum specimens to verify their identity might help to clarify the true range limits of Cassin's and Plumbeous Vireos. Our predictive map suggests a moderately good chance of finding Cassin's Vireos well into central Nevada. This result, however, clearly depends on the reliability of the four possible breeding records from Nye and Esmeralda counties. Potentially, the true breeding range is much narrower—restricted to the far west—and the true chance of occurrence within this area is much higher than indicated.

Cassin's Vireos in Nevada were found mostly in montane forest, but they also turned up in pinyon-juniper and riparian forests. Elsewhere, for example in southern California, this species also occurs in oak and mixed woodlands (Garrett and Dunn 1981, Goguen and Curson 2002). Observers estimated just one pair per block in more than half of the blocks where Cassin's Vireos occurred, and nesting densities were quite low.

CONSERVATION AND MANAGEMENT

The conservation status of Cassin's Vireo will likely require careful reexamination as observers learn to distinguish it from the Plumbeous Vireo and discover more about the basic biology of both species in this region. Rangewide, Cassin's Vireo populations appear to be increasing, especially to the north (Sauer et al. 2005). Threats to breeding habitat in Nevada are probably minimal at present because the species seems to tolerate a variety of woodland types, as it does elsewhere in its range. Cassin's Vireos appear to have declined in southern California, however, and possibly elsewhere (Gallagher 1997), and population trends for western Nevada remain to be determined through continuous monitoring.

HABITAT USE

HABITAT	ATLAS BLOCKS	INCIDENTAL OBSERVATIONS
Montane Forest	7 (50%)	0 (0%)
Montane Parkland	2 (14%)	0 (0%)
Pinyon-Juniper	3 (21%)	1 (50%)
Riparian	2 (14%)	1 (50%)
TOTAL	14	2

Total numbers reported in the Habitat Use, Breeding Status, and Abundance tables may differ from each other (see pp. 22–23 for details). Percentage sums may differ slightly from 100% due to rounding.

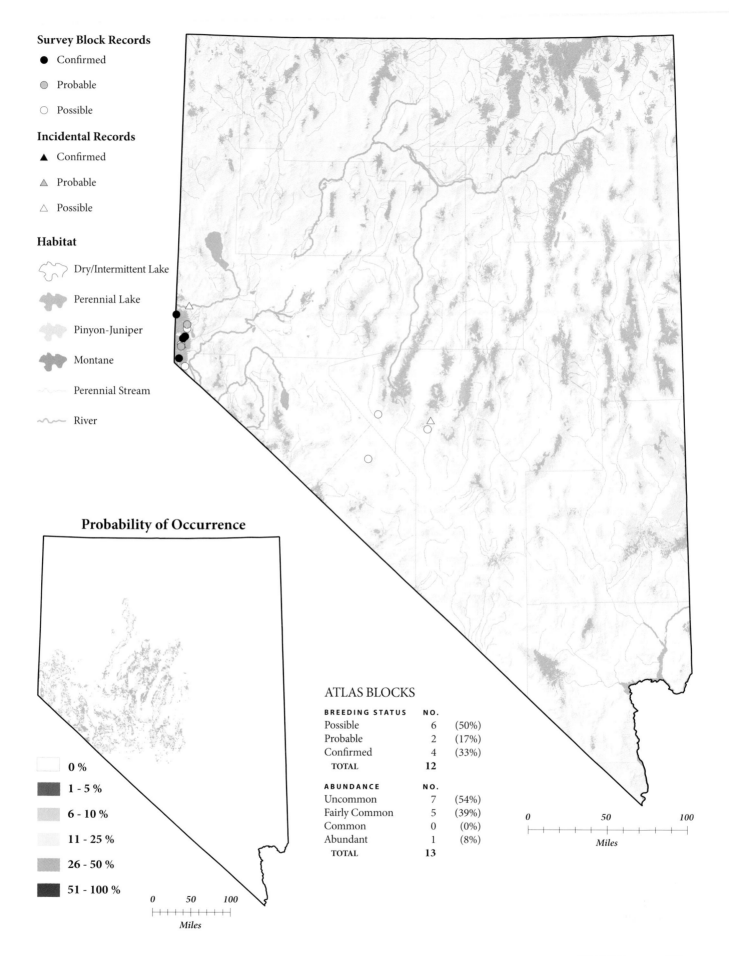

Survey Block Records

● Confirmed

◐ Probable

○ Possible

Incidental Records

▲ Confirmed

▲ Probable

△ Possible

Habitat

⬡ Dry/Intermittent Lake

Perennial Lake

Pinyon-Juniper

Montane

Perennial Stream

River

Probability of Occurrence

☐ 0 %

■ 1 - 5 %

■ 6 - 10 %

11 - 25 %

26 - 50 %

■ 51 - 100 %

0   50   100
Miles

ATLAS BLOCKS

BREEDING STATUS	NO.	
Possible	6	(50%)
Probable	2	(17%)
Confirmed	4	(33%)
TOTAL	12	

ABUNDANCE	NO.	
Uncommon	7	(54%)
Fairly Common	5	(39%)
Common	0	(0%)
Abundant	1	(8%)
TOTAL	13	

0   50   100
Miles

WARBLING VIREO
Vireo gilvus

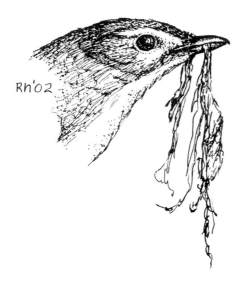

The lazy, rambling song of the Warbling Vireo is a distinctive sound of summer along Nevada's mountain streams. The species is especially numerous around creeks and wooded ponds, but it also can be found in conifer forests, aspen woodlands, foothills, and out into the valleys along the major rivers. Despite its widespread distribution in Nevada, there is some conservation concern about the Warbling Vireo.

DISTRIBUTION

Atlas workers found Warbling Vireos throughout the state, with most confirmed breeding records coming from the mountains. Most sites were estimated to have multiple pairs, and just over a third of the blocks had more than ten pairs. Earlier writers considered the Warbling Vireo a common bird in Nevada (Linsdale 1936, Ryser 1985).

The Warbling Vireo is also common in several surrounding states, including Oregon (Adamus et al. 2001), Utah (Behle et al. 1985), and Idaho (Stephens and Sturts 1998). It is a fairly common breeder in California (Small 1994) and is one of the most ubiquitous birds of the southern Sierra Nevada (Gaines 1992). In Arizona, it is moderately widespread but is absent from the lower Colorado River Valley (Rosenberg et al. 1991, Averill-Murray and Corman in Corman and Wise-Gervais 2005:346–347).

The habitat associations of the Warbling Vireo are not straightforward, although it is clear that the species is most likely to be found in montane forests and riparian areas. The extent to which Warbling Vireos use coniferous or lowland broadleaved forests is variable, though (Alcorn 1988, Gardali and Ballard 2000). In Utah, they make extensive use of aspen woodlands and can be found breeding right down into the cottonwood groves of valley bottoms (Behle et al. 1985). In the Sierra Nevada, they favor riparian woodlands but may also breed in pure coniferous forests (Gaines 1992, Small 1994).

The predictive map primarily reflects the Warbling Vireo's affinity for montane forests, including those associated with mountain streams, and suggests a high probability of finding the species in such areas. The low probability of finding Warbling Vireos in most other habitats reflects the fact that small patches of riparian habitat are scattered throughout areas such as the sagebrush steppe across the northern tier of the state. The high probability predicted for the lowlands of the northern Mojave Desert is clearly an error given the species' habitat use and the distribution of real observations.

CONSERVATION AND MANAGEMENT

Warbling Vireos are common and widespread in North America, and populations overall are on the increase (Sauer et al. 2005). Locally, there have been declines, though; for example, the species no longer breeds in California's Central Valley (McCaskie et al. 1988) and is decreasing in some parts of southern California (Garrett and Dunn 1981, Gallagher 1997). Increased nest predation and habitat loss seem to be the primary causes of these declines, although cowbird parasitism may also play a role (Gardali and Ballard 2000).

Population trends in Nevada are not well known, but the relative scarcity of the species in valleys is interesting. Evidence suggests that the Warbling Vireo was historically abundant in lowland riparian areas such as the lower Truckee River (Ridgway 1877), where it has declined since settlement (Klebenow and Oakleaf 1984, Ammon 2002). Just to our north, in Malheur County, Oregon, Warbling Vireos are inexplicably absent or rare in areas with seemingly appropriate habitat (Contreras and Kindschy 1996). This species' association with riparian gallery forests and its apparent sensitivity to riparian habitat change suggest that it may be a useful species for monitoring riparian restoration projects or for evaluating riparian habitat conditions.

HABITAT USE

HABITAT	ATLAS BLOCKS	INCIDENTAL OBSERVATIONS
Alpine	3 (1%)	0 (0%)
Mesquite	0 (0%)	2 (2%)
Montane Forest	74 (33%)	29 (30%)
Montane Parkland	3 (1%)	0 (0%)
Montane Shrub	14 (6%)	2 (2%)
Pinyon-Juniper	12 (5%)	7 (7%)
Riparian	116 (51%)	56 (57%)
Sagebrush Scrub	1 (<1%)	0 (0%)
Sagebrush Steppe	1 (<1%)	2 (2%)
Urban	2 (<1%)	0 (0%)
Wetland	1 (<1%)	0 (0%)
TOTAL	227	98

Total numbers reported in the Habitat Use, Breeding Status, and Abundance tables may differ from each other (see pp. 22–23 for details). Percentage sums may differ slightly from 100% due to rounding.

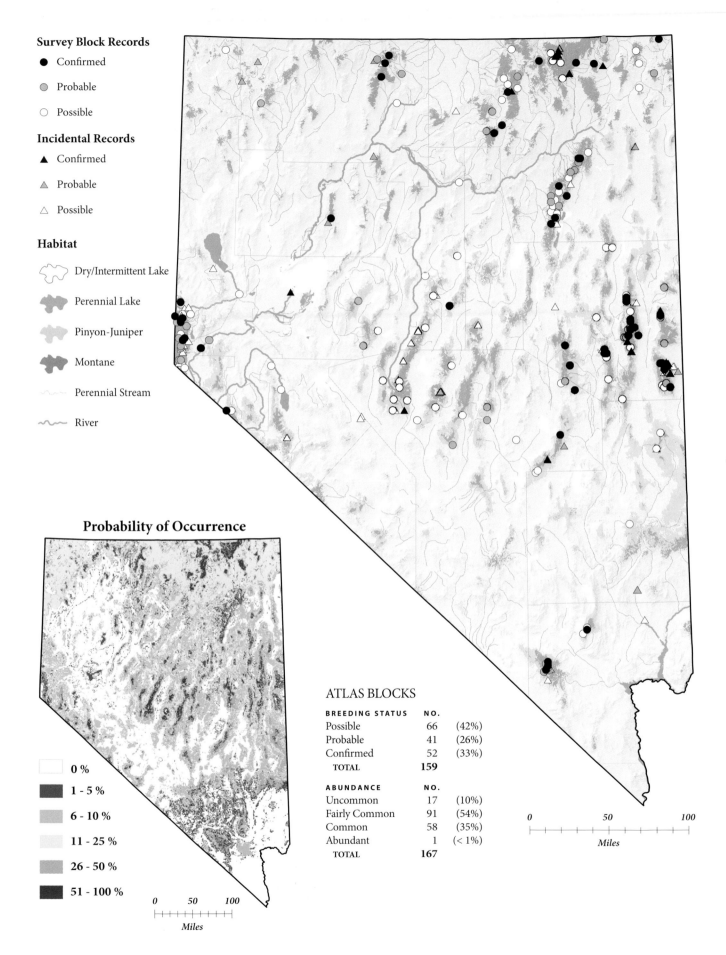

Survey Block Records

● Confirmed

● Probable

○ Possible

Incidental Records

▲ Confirmed

▲ Probable

△ Possible

Habitat

Dry/Intermittent Lake

Perennial Lake

Pinyon-Juniper

Montane

Perennial Stream

River

Probability of Occurrence

☐ 0 %

■ 1 - 5 %

■ 6 - 10 %

☐ 11 - 25 %

■ 26 - 50 %

■ 51 - 100 %

0   50   100

Miles

ATLAS BLOCKS

BREEDING STATUS	NO.	
Possible	66	(42%)
Probable	41	(26%)
Confirmed	52	(33%)
TOTAL	**159**	

ABUNDANCE	NO.	
Uncommon	17	(10%)
Fairly Common	91	(54%)
Common	58	(35%)
Abundant	1	(< 1%)
TOTAL	**167**	

0   50   100

Miles

STELLER'S JAY
Cyanocitta stelleri

Wherever Steller's Jays occur, there can be no doubt of their presence. Bold and sassy, they are given to catcalls, screeching, and a variety of other wild outbursts. And they are, frankly, a cinch to identify. Steller's Jays can be found in habitats as diverse as busy urban parks in Reno and remote mountain forests in Great Basin National Park. And yet they are curiously absent from similar habitats in much of the state—including the mountain ranges of northern and central Nevada.

DISTRIBUTION

Atlas workers found Steller's Jays in three discrete clumps: in the mountains and adjacent river valleys of far western Nevada, in the extensive montane pine forests of east-central Nevada, and in the Spring Mountains of southern Nevada. This apparently disjunct distribution mirrors the broader range pattern of the species. Two separate populations extend south from the boreal forests of Canada—one through the mountain ranges of the West Coast, the other through the Rocky Mountains. Birds in western Nevada belong to the *frontalis* subspecies group, while those in eastern—and most likely those in southern—Nevada are from the interior *macrolopha* group. Two outlying records of possible breeding in northern Nevada (see map) probably involved nonbreeding wanderers. Breeding was confirmed in about one-third of the occupied blocks, and multiple pairs were generally thought to be present.

The atlas results are generally consistent with earlier Nevada studies (e.g., Linsdale 1936, Ryser 1985). One exception is the apparent present-day absence of Steller's Jays from the Sheep Range—an area in which they formerly were documented as breeders (Linsdale 1936). Steller's Jays are common in appropriate habitat in most of the states surrounding Nevada, including the uplands of California (Small 1994) and Arizona (Sitko in

Corman and Wise-Gervais 2005:350–351). In contrast to its patchy distribution in Nevada, the species is common in virtually all of the higher mountains in Utah (Behle et al. 1985). In both Oregon and Idaho it is common everywhere except in the regions closest to Nevada (Littlefield 1990, Contreras and Kindschy 1996, Stephens and Sturts 1998, Adamus et al. 2001).

Not surprisingly, nearly two-thirds of atlas records were from montane forest. Steller's Jays are most typical of pine forests in mountainous areas (Greene et al. 1998), and the predictive map shows a high chance of finding them in this habitat. This prediction, however, does not reflect the scarcity of the species in apparently suitable habitat in the center of the state, which presumably results from the isolation of the interior ranges and the lack of immigration from the larger populations that fringe the state. Steller's Jays do range out into dry pinyon woodlands on occasion, though (Gaines 1992), and even occur in low-elevation towns and cities with ornamental landscaping (Small 1994).

CONSERVATION AND MANAGEMENT

Steller's Jay populations are currently stable (Sauer et al. 2005), and the species seems to tolerate, and indeed may benefit from, certain human-related changes to the landscape. These jays appear to be expanding into urban areas in California (McCaskie et al. 1988, Small 1994), they are eager visitors to bird feeders, and they may have benefited from edge habitats created by logging (Shuford 1993).

The working assumption in Nevada, and throughout the species' range (Greene et al. 1998), has been that Steller's Jay populations are stable and sedentary. Recent incursions into the Reno area, however, and reported range expansions in California suggest that the hypothesis of population stability in Nevada may require reevaluation. Also worthy of investigation is the status of the different subspecies of Steller's Jay that breed in Nevada. The taxonomy of this species is complex (Behle 1985), and the situation in Nevada has not been revised since the time of Linsdale (1936).

HABITAT USE

HABITAT	ATLAS BLOCKS	INCIDENTAL OBSERVATIONS
Alpine	1 (<1%)	0 (0%)
Montane Forest	87 (65%)	14 (64%)
Montane Parkland	7 (5%)	1 (5%)
Montane Shrub	10 (8%)	2 (9%)
Pinyon-Juniper	15 (12%)	1 (5%)
Riparian	10 (8%)	4 (18%)
Urban	3 (2%)	0 (0%)
TOTAL	133	22

Total numbers reported in the Habitat Use, Breeding Status, and Abundance tables may differ from each other (see pp. 22–23 for details). Percentage sums may differ slightly from 100% due to rounding.

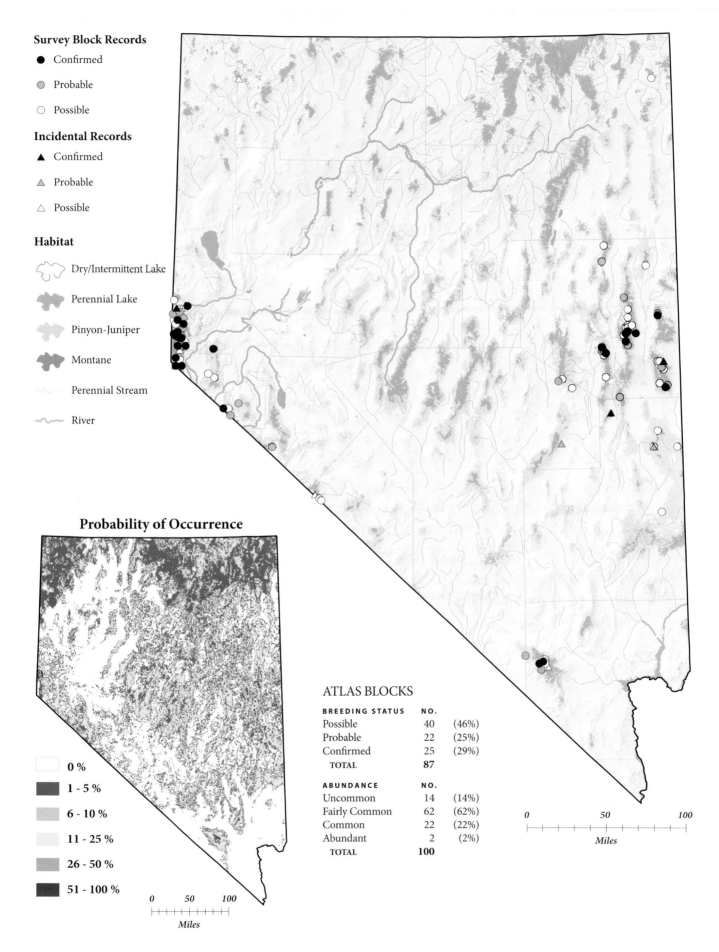

Survey Block Records

● Confirmed

● Probable

○ Possible

Incidental Records

▲ Confirmed

▲ Probable

△ Possible

Habitat

Dry/Intermittent Lake

Perennial Lake

Pinyon-Juniper

Montane

Perennial Stream

River

Probability of Occurrence

	0 %
	1 - 5 %
	6 - 10 %
	11 - 25 %
	26 - 50 %
	51 - 100 %

0   50   100

Miles

ATLAS BLOCKS

BREEDING STATUS	NO.	
Possible	40	(46%)
Probable	22	(25%)
Confirmed	25	(29%)
TOTAL	87	

ABUNDANCE	NO.	
Uncommon	14	(14%)
Fairly Common	62	(62%)
Common	22	(22%)
Abundant	2	(2%)
TOTAL	100	

0   50   100

Miles

WESTERN SCRUB-JAY
Aphelocoma californica

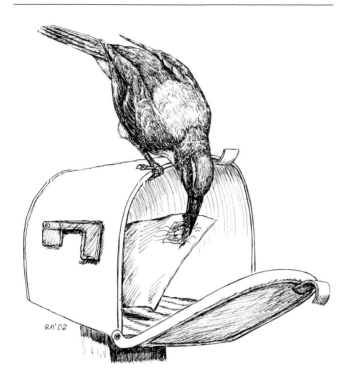

Western Scrub-Jays come in two "personality types" in Nevada. In the remote pinyon-juniper woodlands of much of the state they appear reticent and aloof. Eschewing human company, they simply go about the business of finding food and raising their young. But around Reno and Carson City, Western Scrub-Jays comport themselves in an entirely different manner. They are flashy and conspicuous inhabitants of residential districts, especially where there are planted shrubs, bird feeders, and trash cans. These behavioral distinctions are accompanied by appreciable differences in plumage and vocalizations.

DISTRIBUTION

Breeding Western Scrub-Jays were found south to the McCullough Range and north into the far western and eastern corners of the state, but not in the north-central region. Their frequency of occurrence peaked in the mountain ranges across the center of the state. Breeding was confirmed in more than a third of the blocks in which the birds were found. Estimated abundances were most often intermediate, although they exceeded ten pairs in about a third of the blocks for which abundance was reported.

Scrub-jays have long been known to be widespread in Nevada, but earlier authors gave mixed reports on their abundance (Linsdale 1936, 1951; Ryser 1985; Alcorn 1988). In surrounding states, the Western Scrub-Jay is a common or very common breeder in much of California (Garrett and Dunn 1981, Small 1994), Arizona (Wise-Gervais in Corman and Wise-Gervais 2005:352–353), and Utah (Behle et al. 1985), but is absent from the lower Colorado River Valley (Rosenberg et al. 1991). Al-

though frequently found in western Oregon, the species does not breed in the eastern half of the state, including the areas closest to Nevada (Littlefield 1990, Contreras and Kindschy 1996, Adamus et al. 2001). Western Scrub-Jays also are largely absent from all of Idaho except the extreme south (Stephens and Sturts 1998).

As its common name implies, the Western Scrub-Jay generally favors scrubby habitats. In the Great Basin, pinyon-juniper is preferred, but there is considerable variation in habitat use throughout the range (Curry et al. 2002). In addition to their widely known association with residential areas, Western Scrub-Jays use montane forests and shrublands, riparian zones, and sagebrush. The predictive map, as expected, shows a good chance of finding this species in areas that have pinyon-juniper and a slightly smaller chance in montane habitats. Areas in north-central Nevada where this species seems to be absent have at least some probability of scrub-jay presence based on the predictive model. Future surveys in this area are needed to tell whether these predictions hold true.

CONSERVATION AND MANAGEMENT

In general, Western Scrub-Jay populations appear to be in good shape, although long-term declines have occurred in the pinyon-juniper woodland regions of the West (Curry et al. 2002, Sauer et al. 2005). The California race is probably increasing and expanding (Shuford 1993), and range expansions have also been reported in several parts of Oregon (Marshall et al. 2003). Scrub-jays' frequent association with exotic plantings and residential development may explain local increases (e.g., Gallagher 1997). Meanwhile, the historical threat of persecution is by and large a thing of the past (Ryser 1985).

The overall status of the Western Scrub-Jay in Nevada appears to have changed relatively little during the past one hundred or more years (Ryser 1985). Scrub-jays that use native habitats, though, may face threats from landscape-level changes to pinyon-juniper communities (see Neel 1999a), while urban populations may expand along with expanding cities.

HABITAT USE

HABITAT	ATLAS BLOCKS	INCIDENTAL OBSERVATIONS
Agricultural	3 (1%)	0 (0%)
Mojave	0 (0%)	2 (3%)
Montane Forest	17 (7%)	6 (10%)
Montane Parkland	2 (1%)	1 (2%)
Montane Shrub	23 (10%)	10 (16%)
Pinyon-Juniper	125 (51%)	27 (43%)
Riparian	28 (12%)	10 (16%)
Sagebrush Scrub	14 (6%)	3 (5%)
Sagebrush Steppe	9 (4%)	1 (2%)
Urban	19 (8%)	3 (5%)
Wetland	1 (<1%)	0 (0%)
TOTAL	241	63

Total numbers reported in the Habitat Use, Breeding Status, and Abundance tables may differ from each other (see pp. 22–23 for details). Percentage sums may differ slightly from 100% due to rounding.

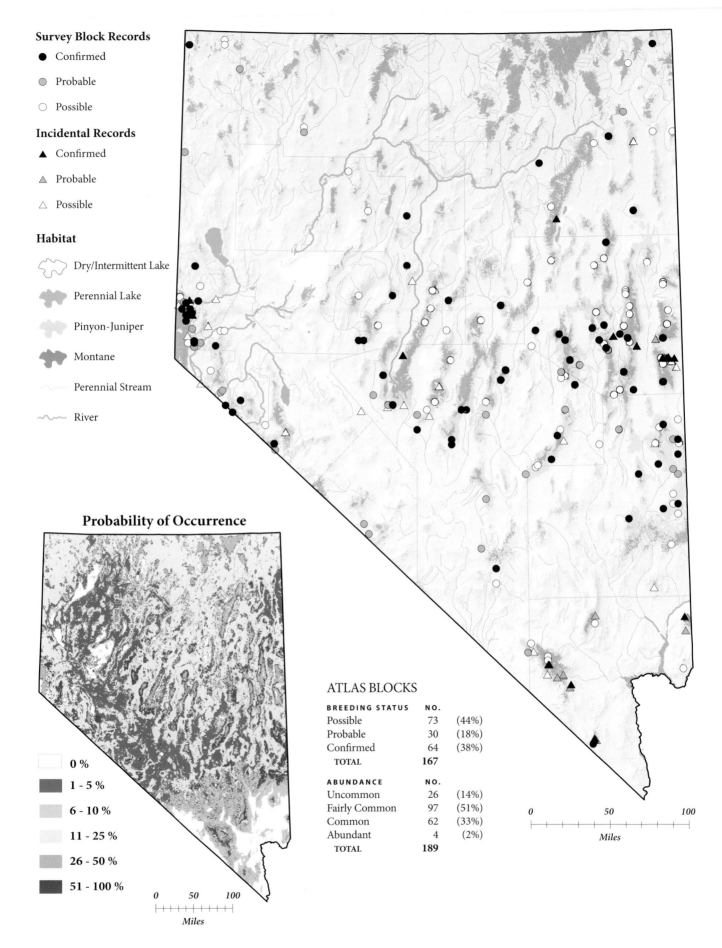

Survey Block Records

● Confirmed

● Probable

○ Possible

Incidental Records

▲ Confirmed

▲ Probable

△ Possible

Habitat

Dry/Intermittent Lake

Perennial Lake

Pinyon-Juniper

Montane

Perennial Stream

River

Probability of Occurrence

0 %

1 - 5 %

6 - 10 %

11 - 25 %

26 - 50 %

51 - 100 %

0   50   100

Miles

ATLAS BLOCKS

BREEDING STATUS	NO.	
Possible	73	(44%)
Probable	30	(18%)
Confirmed	64	(38%)
TOTAL	**167**	

ABUNDANCE	NO.	
Uncommon	26	(14%)
Fairly Common	97	(51%)
Common	62	(33%)
Abundant	4	(2%)
TOTAL	**189**	

0   50   100

Miles

PINYON JAY
Gymnorhinus cyanocephalus

The pinyon pine woodlands of Nevada are places of subtlety and solitude. One can amble all afternoon among the trees and hear little more than the twangy call notes of towhees and the distant drumming of a flicker. Then—suddenly and without warning—the landscape erupts with the shrieking of hundreds of cobalt-colored Pinyon Jays. They pass through quickly, like the summer thunderstorms typical of Nevada, and then they are gone. The towhees and flickers resume their quiet deliberations, and all returns to normal, except for the memory of an encounter with one of the most interesting of Nevada's bird species.

DISTRIBUTION

Pinyon Jays occur throughout Nevada except for the northwestern tier of counties. Breeding was not confirmed in northern Elko County, and the species was not reported north of a line running from Pyramid Lake to the northeastern corner of the state. Interestingly, this distribution corresponds closely to the distribution of pinyon pine in Nevada (cf. Charlet 1996). Pinyon Jays, as would be expected, tended to be numerous in the blocks where they occurred, with more than half of the abundance estimates exceeding ten pairs per block. Making accurate abundance estimates for this nomadic and highly gregarious species is especially difficult, though. A high proportion of records (60%) noted only possible breeding, which is not surprising given that Pinyon Jays breed in very early spring.

The Pinyon Jay has long been recognized as a common resident in much of Nevada (Fisher 1893, Linsdale 1936, Ryser 1985, Alcorn 1988), but its occurrence is more variable in the surrounding states. In California, Arizona, and Utah, it is variously described as uncommon to common but often irregular (Behle et al. 1985, McCaskie et al. 1988, Small 1994, Martin in Corman and Wise-Gervais 2005:356–357). It is locally common in central Oregon (Adamus et al. 2001) but occurs only rarely in the region closest to Nevada (Littlefield 1990, Contreras and Kindschy 1996). In Idaho, it breeds only in the southeast (Stephens and Sturts 1998).

As expected, pinyon-juniper was the most frequently reported habitat, and the predictive map shows the greatest chance of finding Pinyon Jays to be in the foothills of mountain ranges, where this habitat predominates. The species also forages in other habitats (Balda 2002), such as sagebrush shrublands (Garrett and Dunn 1981), and there is a smaller predicted chance of finding Pinyon Jays in these habitats. In areas where pinyon pines are in short supply, the birds may also use juniper or other woodland types for nesting and foraging (Ryser 1985, Contreras 1999, Balda 2002). The complete absence of Pinyon Jays from northwestern Nevada, despite their predicted occurrence there (albeit with a generally low probability), reflects the species' close association with pinyon pines and the fact that our habitat classifications did not distinguish between pinyon pine and juniper forests.

CONSERVATION AND MANAGEMENT

Pinyon Jay populations have declined significantly rangewide in recent decades (Sauer et al. 2005). Regional data are not sufficient to define trends in Nevada, and accurate monitoring of populations is exceedingly difficult due to the tendency of this species to occur in flocks that range widely within large home ranges.

Because of its tight affinity with pinyon pines (Balda 2002), the Pinyon Jay is a Partners in Flight Priority species in Nevada and a Watch List species in the Intermountain West, with the conservation goal of doubling the population size in the future (Rich et al. 2004). Although Pinyon Jays are presently common in much of the state, Nevada has a high regional responsibility for protecting the species. Cutting of mature pinyon pines, changes in forest management and fire regimes, and increasing encroachment on Pinyon Jay habitat may affect the long-term sustainability of this species. Landscape-level population analyses of this far-ranging and nomadic species are therefore warranted (Neel 1999a).

HABITAT USE

HABITAT	ATLAS BLOCKS	INCIDENTAL OBSERVATIONS
Barren	0 (0%)	1 (3%)
Mojave	0 (0%)	1 (3%)
Montane Forest	4 (3%)	2 (5%)
Montane Shrub	6 (4%)	4 (10%)
Pinyon-Juniper	125 (81%)	20 (52%)
Riparian	2 (1%)	4 (10%)
Salt Desert Scrub	1 (1%)	0 (0%)
Sagebrush Scrub	11 (7%)	2 (5%)
Sagebrush Steppe	4 (3%)	3 (8%)
Urban	0 (0%)	2 (5%)
TOTAL	153	39

Total numbers reported in the Habitat Use, Breeding Status, and Abundance tables may differ from each other (see pp. 22–23 for details). Percentage sums may differ slightly from 100% due to rounding.

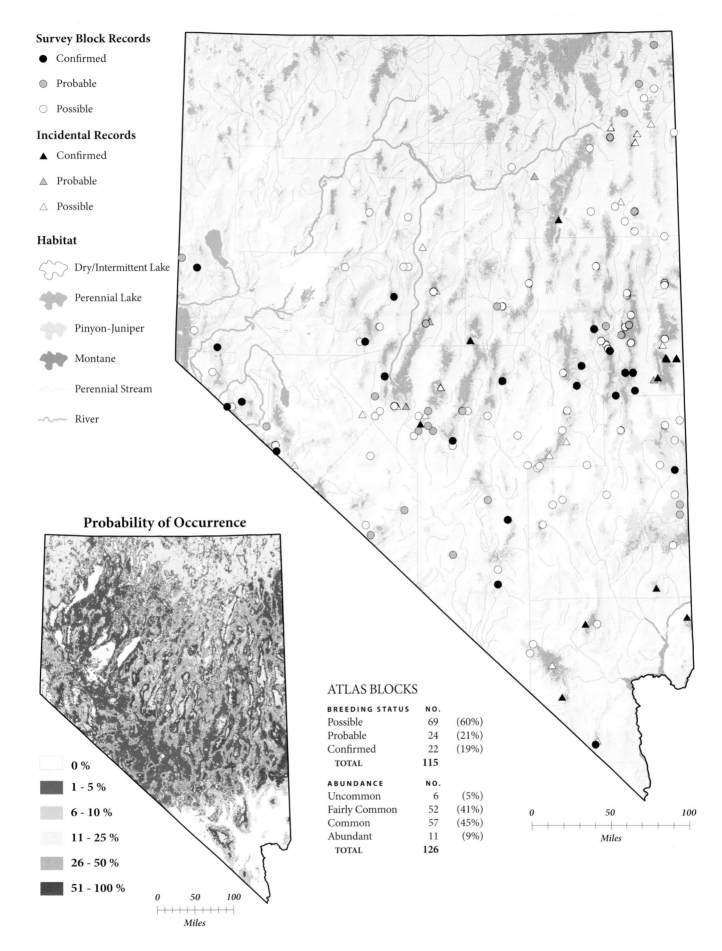

Survey Block Records

● Confirmed

● Probable

○ Possible

Incidental Records

▲ Confirmed

▲ Probable

△ Possible

Habitat

Dry/Intermittent Lake

Perennial Lake

Pinyon-Juniper

Montane

Perennial Stream

River

Probability of Occurrence

0 %

1 - 5 %

6 - 10 %

11 - 25 %

26 - 50 %

51 - 100 %

0   50   100
Miles

ATLAS BLOCKS

BREEDING STATUS	NO.	
Possible	69	(60%)
Probable	24	(21%)
Confirmed	22	(19%)
TOTAL	**115**	

ABUNDANCE	NO.	
Uncommon	6	(5%)
Fairly Common	52	(41%)
Common	57	(45%)
Abundant	11	(9%)
TOTAL	**126**	

0   50   100
Miles

CLARK'S NUTCRACKER

Nucifraga columbiana

It can be hard to avoid encountering the dapper and agreeable Clark's Nutcracker in the tall conifer forests of Nevada; this is a mountain bird if ever there was one. Family groups can also be found in hummingbird-and-flower-filled clearings, lonely pinyon-juniper woodlands, and in picnic areas and trailhead parking lots that are bustling with human activity. And while nutcrackers readily coexist with humans, they are also found in the remotest wilderness areas. On cold mornings high in the mountains—still deep with snow and months away from summer—the grating call notes of a distant Clark's Nutcracker may be the only sure sign of vertebrate life.

DISTRIBUTION

Atlas workers found Clark's Nutcrackers in most of the major mountain ranges of Nevada—as far north as the Jarbidge Range and as far south as Mount Charleston in the Spring Mountains, and east to west all the way from Lake Tahoe to the Utah border. Breeding was confirmed in about one-third of the blocks in which the species was found, and abundance estimates exceeded ten pairs in nearly one-fourth of them.

The atlas results are consistent with earlier occurrence data reported by both Linsdale (1936) and Alcorn (1988). Ryser (1985) considered Clark's Nutcracker to be common in the Great Basin, as did Fisher (1893) in the high mountains that the Death Valley Expedition explored. Clark's Nutcracker is distributed throughout the western mountains, and it is readily found in the higher-elevation portions of Arizona (Wise-Gervais in Corman and Wise-Gervais 2005:358–359), Oregon (Adamus et al. 2001),

Utah (Behle et al. 1985), California (Small 1994), and Idaho (Stephens and Sturts 1998).

Although the nutcracker ventures well above timberline into alpine scree and snowfields, it is more commonly seen in coniferous forests, and it ranges downward in elevation to pinyon-juniper woodlands (Tomback 1998). Well over half of the Clark's Nutcracker habitat records were from montane forest, with an additional 5% from alpine habitats. The predictive map reflects these associations clearly, showing the greatest chance of finding the species in montane forests throughout the state, and predicting it to be nearly absent from the lowlands and to occur with low to moderate likelihood in the foothills.

Breeding takes place very early in the spring (Gaines 1992, Tomback 1998), usually in high-elevation conifer forests (Garrett and Dunn 1981). Clark's Nutcrackers often begin an extensive downslope movement after breeding (Linsdale 1936, Alcorn 1988), and many midsummer sightings in mid- or low-elevation conifer forests may involve postbreeding wanderers.

CONSERVATION AND MANAGEMENT

Overall, Clark's Nutcracker populations are increasing (Sauer et al. 2005). Regional and annual population fluctuations can be extreme, however, obscuring underlying long-term trends. In any given area, the status of the nutcracker is largely determined by the availability of conifer seeds (Tomback 1998), and just as nutcrackers depend on conifers for food, conifers depend on nutcrackers to disperse their seeds (Vander Wall and Balda 1977).

The status of Clark's Nutcracker has changed little in Nevada since the nineteenth century (Ryser 1985), and there is no indication of major threats to the species at present, particularly on the breeding grounds. Possible threats in the lower-elevation sites to which nutcrackers move after breeding may warrant additional research, though. Individuals or small flocks can show up far from mountains (e.g., the lower Colorado River Valley; see Rosenberg et al. 1991) at any time of the year (Littlefield 1990), and this aspect of the species' biology may be important for its management in Nevada.

HABITAT USE

HABITAT	ATLAS BLOCKS	INCIDENTAL OBSERVATIONS
Alpine	11 (5%)	2 (4%)
Barren	2 (<1%)	0 (0%)
Montane Forest	139 (61%)	21 (45%)
Montane Parkland	12 (5%)	5 (11%)
Montane Shrub	24 (11%)	3 (6%)
Open Water	0 (0%)	1 (2%)
Pinyon-Juniper	28 (12%)	7 (15%)
Riparian	8 (4%)	4 (9%)
Sagebrush Scrub	3 (1%)	2 (4%)
Sagebrush Steppe	1 (<1%)	2 (4%)
TOTAL	228	47

Total numbers reported in the Habitat Use, Breeding Status, and Abundance tables may differ from each other (see pp. 22–23 for details). Percentage sums may differ slightly from 100% due to rounding.

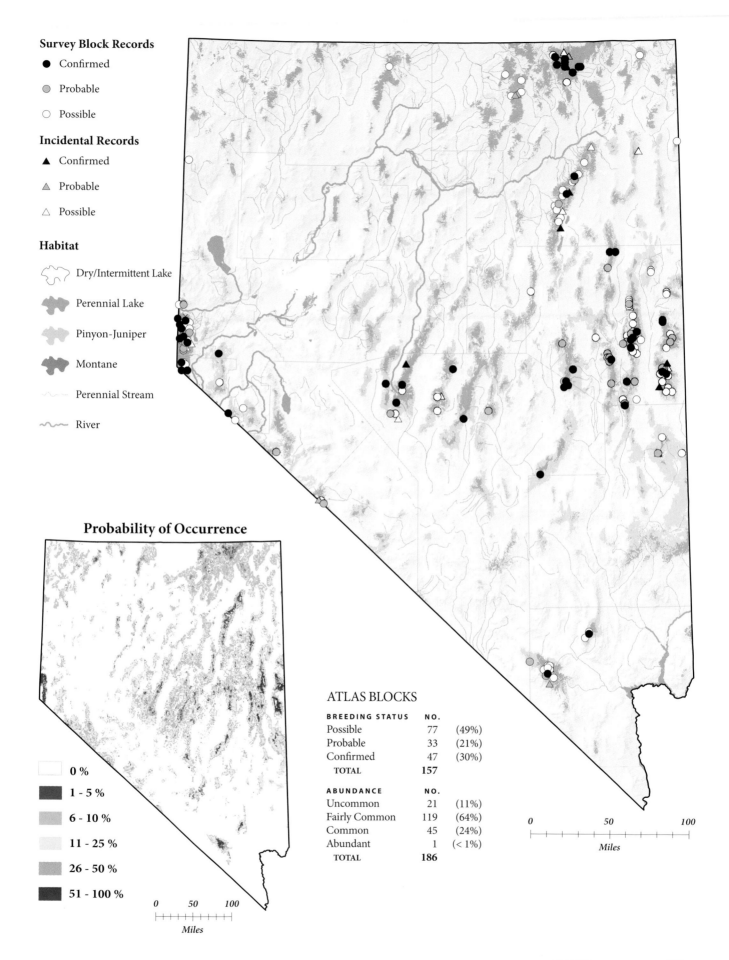

Survey Block Records
- ● Confirmed
- ⬤ Probable
- ○ Possible

Incidental Records
- ▲ Confirmed
- △ Probable
- △ Possible

Habitat
- Dry/Intermittent Lake
- Perennial Lake
- Pinyon-Juniper
- Montane
- Perennial Stream
- River

Probability of Occurrence

- 0 %
- 1 - 5 %
- 6 - 10 %
- 11 - 25 %
- 26 - 50 %
- 51 - 100 %

0   50   100
Miles

ATLAS BLOCKS

BREEDING STATUS	NO.	
Possible	77	(49%)
Probable	33	(21%)
Confirmed	47	(30%)
TOTAL	**157**	

ABUNDANCE	NO.	
Uncommon	21	(11%)
Fairly Common	119	(64%)
Common	45	(24%)
Abundant	1	(< 1%)
TOTAL	**186**	

0   50   100
Miles

BLACK-BILLED MAGPIE
Pica hudsonia

RH'02

First-time visitors to the West find the Black-billed Magpie to be one of the more enjoyable members of the bird fauna: strikingly plumaged and improbably proportioned, talkative and highly sociable, it is in every sense a spectacle. Magpies bring an element of panache to otherwise commonplace ranches and country roads. But the magpie is so common and conspicuous that longtime birders sometimes fail to pay it close attention. Magpies have a decided affinity for human company and are among the most characteristic elements of human-altered landscapes in northern Nevada.

DISTRIBUTION

Black-billed Magpies are widespread throughout northern Nevada but are completely absent south of about latitude 38°30′N, a boundary clearly reflected on the predictive map. Abundance estimates in occupied blocks suggest that most areas have multiple pairs, but very high densities are rare. Breeding was confirmed in about two-thirds of the blocks in which it was suspected, probably reflecting the ease of finding and identifying the magpie's large stick nests.

The atlas data uphold Ryser's (1985) assessment of the magpie as common and widespread in the Great Basin's valleys and foothills. Linsdale (1936) placed the southern limit of the species' Nevada range at latitude 37°N, almost one hundred miles (160 kilometers) south of the present limit. Alcorn (1988), too, implied a somewhat more southerly distribution than the atlas data indicate.

The Black-billed Magpie thrives in open country such as the shrub steppe regions of North America, and Nevada lies at the southwestern limit of its range. Magpies are common and widespread to our north and east (Behle et al. 1985, Trost 1999,

Adamus et al. 2001), but they are limited to the northeastern portions of both California (Small 1994) and Arizona (Corman in Corman and Wise-Gervais 2005:360–361).

Magpies were reported from a wide variety of habitats in Nevada, with riparian zones leading the way. In the Great Basin, the species is generally associated with wooded bottomlands and open terrain with at least some dense, shrubby vegetation (Behle et al. 1985). Magpies make effective use of anthropogenic food sources and can reach high densities around ranches (Ryser 1985). As the predictive map suggests, there is at least a moderate chance of finding them in most of northern Nevada.

CONSERVATION AND MANAGEMENT

Trends vary across the Black-billed Magpie's range, but populations appear stable overall (Trost 1999, Sauer et al. 2005). In the southwestern portion of the species' range, there may have been increases in California's Owens Valley (Small 1994), declines in parts of southeastern Oregon (Littlefield 1990), and historic declines followed by recent increases in northeastern Arizona (Corman in Corman and Wise-Gervais 2005:360–361). The "greening" of the desert, a result of plantings around houses and ranches (Small 1994), may be responsible for some population increases, while declines could be related to mowing, cutting, heavy grazing, and spraying in agricultural areas (Littlefield 1990). Also, the magpie is one of a handful of species that farmers sometimes persecute because of their depredations of game bird nests and fruit crops (Trost 1999).

In Nevada, the status of the Black-billed Magpie is unclear (Sauer et al. 2005). Numbers are thought to have increased (Ryser 1985), but they fluctuate so much that it is difficult to discern a clear pattern (Sauer et al. 2005). Curiously, the current southern range limit appears to lie farther north than older sources suggest (see Linsdale 1936). Climatic factors are thought to limit the Black-billed Magpie's range to northerly regions (Trost 1999); thus, future range expansions into southern Nevada seem unlikely, despite numerous magpie-friendly sites such as golf courses, housing developments, and highway rest stops.

HABITAT USE

HABITAT	ATLAS BLOCKS	INCIDENTAL OBSERVATIONS
Agricultural	25 (10%)	1 (2%)
Barren	3 (1%)	1 (2%)
Grassland	0 (0%)	1 (2%)
Montane Forest	5 (2%)	2 (3%)
Montane Shrub	20 (8%)	4 (6%)
Open Water	1 (<1%)	0 (0%)
Pinyon-Juniper	17 (7%)	1 (2%)
Riparian	92 (38%)	32 (48%)
Salt Desert Scrub	7 (3%)	1 (2%)
Sagebrush Scrub	23 (9%)	8 (12%)
Sagebrush Steppe	21 (9%)	3 (5%)
Urban	22 (9%)	7 (11%)
Wetland	9 (4%)	5 (8%)
TOTAL	245	66

Total numbers reported in the Habitat Use, Breeding Status, and Abundance tables may differ from each other (see pp. 22–23 for details). Percentage sums may differ slightly from 100% due to rounding.

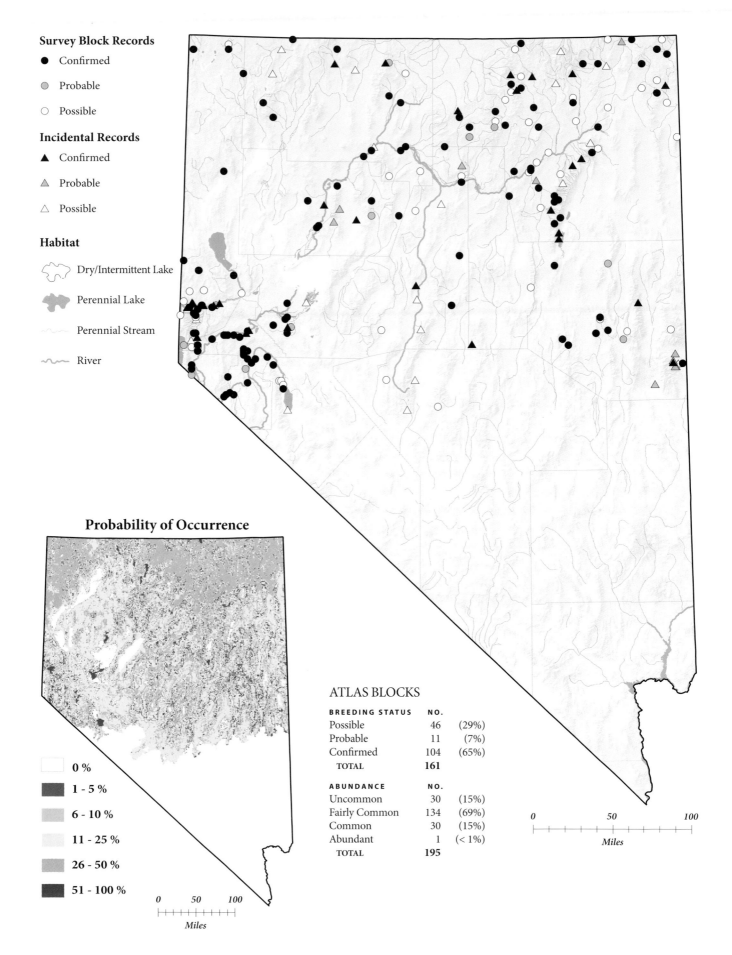

Survey Block Records

● Confirmed

● Probable

○ Possible

Incidental Records

▲ Confirmed

▲ Probable

△ Possible

Habitat

Dry/Intermittent Lake

Perennial Lake

Perennial Stream

River

Probability of Occurrence

☐ 0 %

■ 1 - 5 %

■ 6 - 10 %

☐ 11 - 25 %

■ 26 - 50 %

■ 51 - 100 %

0   50   100
Miles

ATLAS BLOCKS

BREEDING STATUS	NO.	
Possible	46	(29%)
Probable	11	(7%)
Confirmed	104	(65%)
TOTAL	161	

ABUNDANCE	NO.	
Uncommon	30	(15%)
Fairly Common	134	(69%)
Common	30	(15%)
Abundant	1	(< 1%)
TOTAL	195	

0   50   100
Miles

AMERICAN CROW
Corvus brachyrhynchos

Disliked by farmers and barely acknowledged by birders, the wily American Crow has quietly managed to become a common sight around towns, ranches, and farms all over Nevada. For nesting, it usually requires dense vegetation cover in tall trees— native or exotic. For food, it will accept virtually anything, although it looks on the offerings at municipal dumps and fast food restaurants with special favor. The secret to the American Crow's success may be that it has adapted splendidly to the presence of humans.

DISTRIBUTION

Atlas workers found American Crows across much of northern Nevada and at scattered locations in the south. Most sightings occurred in valley or foothills regions, and all of the southerly records came from riparian habitats such as lakeshores or river edges. Crows were most often present in intermediate abundances, and breeding was confirmed in 40% of the blocks occupied by the species.

As is the case elsewhere in its range (Verbeek and Caffrey 2002), the American Crow appears to have benefited from the increasing human presence in Nevada (Ryser 1985). Earlier writers portrayed American Crows as scarce and local in most of the state (Linsdale 1936, Ryser 1985). Alcorn (1988) considered the American Crow to be primarily a northern Nevada species but also noted its increasing numbers. Crows are generally common to the west and north of Nevada (Small 1994, Stephens and Sturts 1998, Adamus et al. 2001). In Arizona they occur primarily in the north (Wise-Gervais in Corman and Wise-Gervais 2005: 362–363), and in Utah they are most numerous in the east (Behle et al. 1985).

Crows use a variety of habitats, but they appear to be most successful in areas touched by human activity (Verbeek and Caffrey 2002). They are frequently found in agricultural areas in California (Small 1994), and cities and towns are also increasingly occupied (Gallagher 1997). In Nevada, American Crows may number into the hundreds on farms and ranches, and they have adopted city life, too, benefiting from the varied food supply and availability of water (Ryser 1985). Because American Crows were frequently noted in urban areas, the predictive map shows a high probability of their occurrence in and around Las Vegas, where none were actually observed during atlas surveys. This inconsistency may reflect the simple nature of our predictive model, which does not distinguish among areas within each broad habitat class; but it also might indicate an area where American Crows could appear in the future. The map also predicts a high chance of finding American Crows near Walker Lake, but that is because the habitat map used for predictions classifies the Hawthorne Army Depot as "urban"; in fact, this treeless area is probably poorly suited for the species.

CONSERVATION AND MANAGEMENT

Even though American Crows have faced—and still face—persecution, their populations have increased rangewide (Sauer et al. 2005). Although little quantitative information exists for Nevada, populations are probably increasing here as well (Alcorn 1988). Crow populations in the Lahontan Valley initially increased after the area was developed for farming, but have since declined with the closure of trash dumps and charnel pits (Chisholm and Neel 2002). Until recently, the only conservation concerns involving this species focused on its role as a predator on the eggs and young of other birds. The emergence of West Nile virus in North America in 1999, however, and American Crows' extreme vulnerability to the disease (Hochachka et al. 2004) have changed the way the species is viewed. Atlas data (and the Nevada Bird Count program implemented since the atlas work was completed) probably provide the best baseline information on the distribution of American Crows in the state and may prove helpful in evaluating the future impacts of the virus.

HABITAT USE

HABITAT	ATLAS BLOCKS	INCIDENTAL OBSERVATIONS
Agricultural	23 (18%)	2 (7%)
Barren	1 (<1%)	0 (0%)
Grassland	3 (2%)	0 (0%)
Mesquite	0 (0%)	1 (4%)
Montane Forest	3 (2%)	1 (4%)
Montane Parkland	2 (2%)	0 (0%)
Montane Shrub	3 (2%)	0 (0%)
Pinyon-Juniper	12 (9%)	0 (0%)
Riparian	51 (39%)	15 (54%)
Sagebrush Scrub	4 (3%)	0 (0%)
Sagebrush Steppe	5 (4%)	0 (0%)
Urban	22 (17%)	6 (21%)
Wetland	2 (2%)	3 (11%)
TOTAL	131	28

Total numbers reported in the Habitat Use, Breeding Status, and Abundance tables may differ from each other (see pp. 22–23 for details). Percentage sums may differ slightly from 100% due to rounding.

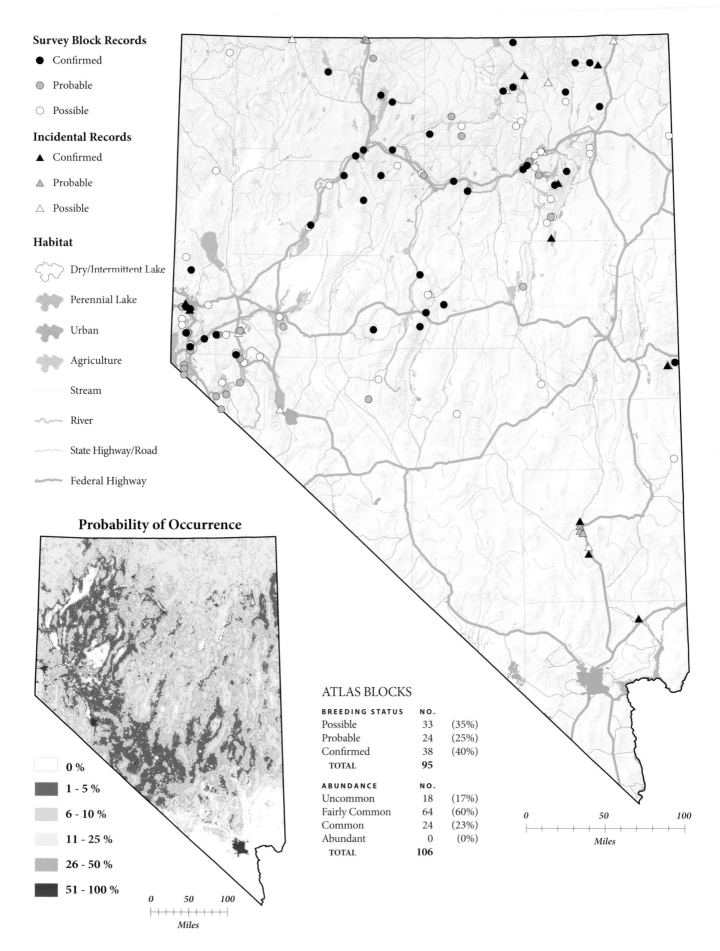

Survey Block Records

● Confirmed

● Probable

○ Possible

Incidental Records

▲ Confirmed

▲ Probable

△ Possible

Habitat

Dry/Intermittent Lake

Perennial Lake

Urban

Agriculture

Stream

River

State Highway/Road

Federal Highway

Probability of Occurrence

☐	0 %
■	1 - 5 %
■	6 - 10 %
■	11 - 25 %
■	26 - 50 %
■	51 - 100 %

0   50   100
Miles

ATLAS BLOCKS

BREEDING STATUS	NO.	
Possible	33	(35%)
Probable	24	(25%)
Confirmed	38	(40%)
TOTAL	**95**	

ABUNDANCE	NO.	
Uncommon	18	(17%)
Fairly Common	64	(60%)
Common	24	(23%)
Abundant	0	(0%)
TOTAL	**106**	

0   50   100
Miles

COMMON RAVEN
Corvus corax

In many cultures the Common Raven is considered a trickster, and now there is scientific evidence to prove it! Among other things, Common Ravens have been observed peeling identification labels off toxic waste drums, pecking holes in airplane wings, and stealing golf balls (Boarman and Heinrich 1999). For these and other reasons, some people consider ravens to be pests. Despite their "crimes," though, ravens reflect in many ways the human spirit in Nevada: bold, quirky, resourceful, desert loving, and rugged.

DISTRIBUTION

Common Ravens were found throughout the state in most habitat types. Breeding was confirmed in about a third of the blocks where nesting was suspected. Ravens appear to be locally less abundant than one might expect based on the ease with which they can be seen, and most blocks were estimated to contain only a few pairs. Still, ravens have long been described as widespread in Nevada (Fisher 1893, Linsdale 1936, Ryser 1985, Alcorn 1988), and they remain so; in fact, they were reported in more atlas blocks than any species except Brewer's Sparrow. They are also widespread in the states surrounding Nevada, although sometimes spottily distributed (Behle et al. 1985, Rosenberg et al. 1991, Small 1994, Stephens and Sturts 1998, Adamus et al. 2001, Wise-Gervais in Corman and Wise-Gervais 2005:366–367).

Common Ravens are extremely versatile in their use of different habitats (Boarman and Heinrich 1999). Unlike their cousins the crows, Common Ravens thrive in the hottest, remotest, and driest spots of the Mojave Desert, often in places where canopy cover is nowhere within sight. While they fare well in wilderness settings, Common Ravens also readily use resources that are by-products of human activity, such as power lines, garbage dumps,

campsites, and road kill (Gaines 1992). They have even begun to colonize some urban centers such as Los Angeles and Chicago (Boarman and Heinrich 1999). The Common Raven's ubiquity and flexibility are also illustrated by the predictive map, which indicates a high chance of finding this species almost everywhere in the state.

CONSERVATION AND MANAGEMENT

Management of the Common Raven is a complicated matter: in parts of the United States these birds are treated as pests, while elsewhere reintroduction efforts have been implemented to offset declining populations (Boarman and Heinrich 1999). Rangewide, and throughout most of the West, populations have increased in recent decades (Sauer et al. 2005). Population trends in Nevada are less clear, but based on Breeding Bird Survey data, the state's population is holding fairly steady (Sauer et al. 2005).

Ravens can come into conflict with species other than humans as well—sometimes species of concern to conservationists. In southern Nevada, for example, Common Ravens prey on juvenile desert tortoises (Boarman 2002), a species listed as Threatened under the Endangered Species Act. Raven predation is identified in the Desert Tortoise Recovery Plan as one of many threats to the desert tortoise, and measures to minimize mortality by Common Raven predation have been recommended, particularly for tortoise populations that occur in the vicinity of raven attractants such as garbage dumps and roads.

In addition to better understanding the effects of Nevada's ravens on other species, interested researchers could tackle the bird's taxonomic status. Recent evidence suggests that ravens in the far western United States are genetically very distinct from populations elsewhere in North America (Omland et al. 2000), raising interesting questions about how species boundaries are drawn.

HABITAT USE

HABITAT	ATLAS BLOCKS	INCIDENTAL OBSERVATIONS
Agricultural	28 (4%)	4 (2%)
Alpine	15 (2%)	2 (1%)
Ash	1 (<1%)	0 (0%)
Barren	29 (4%)	3 (2%)
Grassland	20 (3%)	3 (2%)
Mesquite	8 (1%)	1 (<1%)
Mojave	54 (7%)	22 (13%)
Montane Forest	40 (6%)	12 (7%)
Montane Parkland	6 (<1%)	2 (1%)
Montane Shrub	30 (4%)	9 (5%)
Open Water	1 (<1%)	2 (1%)
Pinyon Juniper	99 (14%)	14 (9%)
Riparian	50 (7%)	32 (20%)
Salt Desert Scrub	100 (14%)	10 (6%)
Sagebrush Scrub	114 (16%)	22 (13%)
Sagebrush Steppe	97 (13%)	9 (5%)
Urban	24 (3%)	13 (8%)
Wetland	7 (<1%)	4 (2%)
TOTAL	723	164

Total numbers reported in the Habitat Use, Breeding Status, and Abundance tables may differ from each other (see pp. 22–23 for details). Percentage sums may differ slightly from 100% due to rounding.

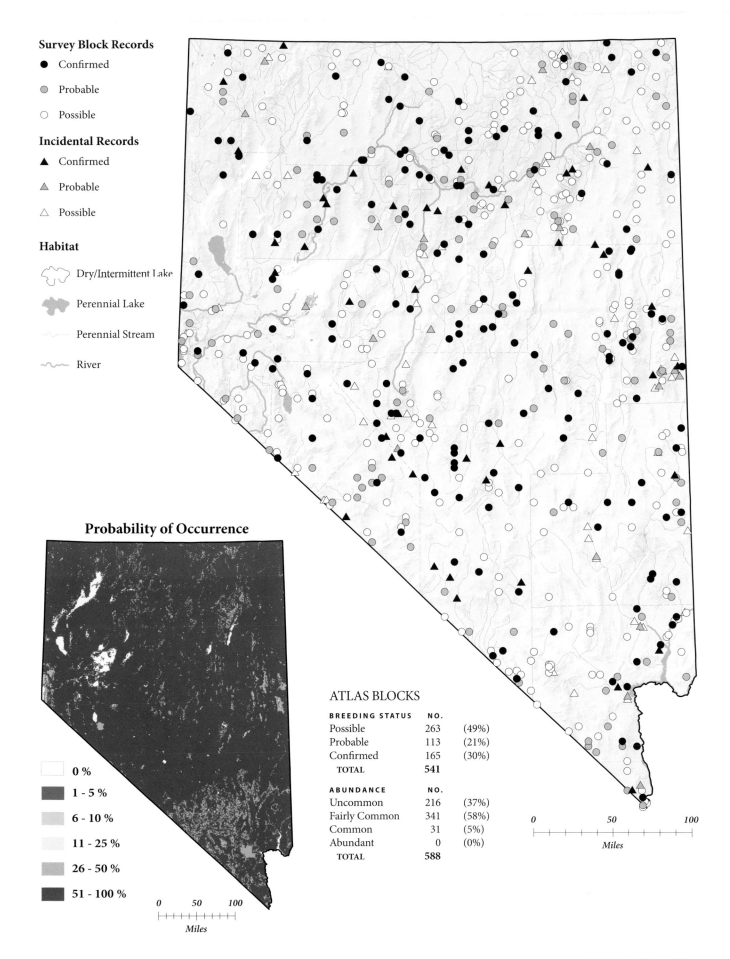

Survey Block Records

● Confirmed
● Probable
○ Possible

Incidental Records

▲ Confirmed
▲ Probable
△ Possible

Habitat

Dry/Intermittent Lake

Perennial Lake

Perennial Stream

River

Probability of Occurrence

☐ 0 %
■ 1 - 5 %
■ 6 - 10 %
☐ 11 - 25 %
■ 26 - 50 %
■ 51 - 100 %

0   50   100
Miles

ATLAS BLOCKS

BREEDING STATUS	NO.	
Possible	263	(49%)
Probable	113	(21%)
Confirmed	165	(30%)
TOTAL	**541**	

ABUNDANCE	NO.	
Uncommon	216	(37%)
Fairly Common	341	(58%)
Common	31	(5%)
Abundant	0	(0%)
TOTAL	**588**	

0   50   100
Miles

HORNED LARK

Eremophila alpestris

Eremophila means "desert lover," and that is a suitable epithet for the Horned Lark. It is one of the few species that is easy to find amid the white-hot creosotebush flats of far southern Nevada, in the bleak greasewood scrub of the central part of the state, and throughout the vast sagebrush steppe of northern Nevada. Horned Larks are not limited to the valley floors, however; they breed in sparsely vegetated landscapes from the lowlands of the Colorado River Valley to scree fields above the timberline.

DISTRIBUTION

Atlas workers found Horned Larks just about anywhere without trees. The one area of the state that supported rather few Horned Larks was Lincoln County, much of which is covered by pinyon-juniper woodlands. Breeding was confirmed in almost half of the blocks in which the species was found, and Horned Larks were typically quite numerous: abundance was estimated at greater than ten pairs (sometimes much greater) in nearly three-quarters of the occupied blocks.

The Horned Lark has long been recognized as widespread and common in Nevada (Linsdale 1936, Ryser 1985, Alcorn 1988, Chisholm and Neel 2002). The species is similarly numerous in surrounding states wherever suitable open habitat is present (Behle et al. 1985, Small 1994, Stephens and Sturts 1998, Adamus et al. 2001, Wise-Gervais in Corman and Wise-Gervais 2005:368–369). Horned Larks are also present throughout much of the rest of North America, as well as large portions of Eurasia (Beason 1995).

Beason's (1995:6) description of the breeding habitat of the Horned Lark as "open, generally barren country" is somewhat of an understatement given the utterly desolate nature of many of the Horned Lark's Nevada haunts. Likewise, Gaines (1992) indicated that its typically bleak habitat includes elements such as

windswept treeless plains, pumice flats, heavily grazed meadows, fell fields, and dirt roads. At a smaller habitat scale, Horned Larks favor low or widely scattered vegetation with interstices of bare ground (Holmes and Barton 2003, Dobkin and Sauder 2004).

The predictive map shows that Horned Larks are likely to breed nearly everywhere in Nevada, excepting only forests, cities, and alkali flats totally devoid of vegetation. The map suggests that they are slightly less likely to be found in the Mojave Desert—although they are still common there—than in more northern scrublands.

CONSERVATION AND MANAGEMENT

The Horned Lark is a very widely distributed species, and it is generally common in North America. Over the past two centuries its fortunes have waxed and waned in different regions—mainly in connection with the spread of agriculture and subsequent reversion of cultivated lands to second-growth habitats (Beason 1995, Dobkin and Sauder 2004). Substantial declines have occurred since the 1960s, both rangewide and in the West—particularly the Intermountain West (Sauer et al. 2005). The causes of this decline are not well understood, although it has been suggested that exotic annual grasses such as cheatgrass may be unsuitable for Horned Larks (Holmes and Barton 2003, Dobkin and Sauder 2004).

The Horned Lark is still common and abundant in Nevada. Ryser (1985) considered it to be the most abundant bird in the arid valleys of the Great Basin, and several atlas workers expressed a similar sentiment. Research into the causes of apparent declines in the West and in the state would be valuable, however. In some areas, such as the lower Colorado River Valley, the Horned Lark may come into (perceived) conflict with agricultural interests (Rosenberg et al. 1991), but such conflicts have not been reported in Nevada.

HABITAT USE

HABITAT	ATLAS BLOCKS	INCIDENTAL OBSERVATIONS
Agricultural	13 (3%)	0 (0%)
Alpine	7 (1%)	1 (3%)
Ash	3 (<1%)	0 (0%)
Barren	10 (2%)	0 (0%)
Grassland	57 (12%)	5 (13%)
Mojave	18 (4%)	7 (18%)
Montane Parkland	1 (<1%)	0 (0%)
Montane Shrub	5 (1%)	0 (0%)
Open Water	0 (0%)	1 (3%)
Pinyon-Juniper	1 (<1%)	0 (0%)
Riparian	6 (1%)	4 (10%)
Salt Desert Scrub	128 (27%)	2 (5%)
Sagebrush Scrub	110 (23%)	10 (26%)
Sagebrush Steppe	107 (23%)	7 (18%)
Urban	3 (<1%)	0 (0%)
Wetland	5 (1%)	2 (5%)
TOTAL	474	39

Total numbers reported in the Habitat Use, Breeding Status, and Abundance tables may differ from each other (see pp. 22–23 for details). Percentage sums may differ slightly from 100% due to rounding.

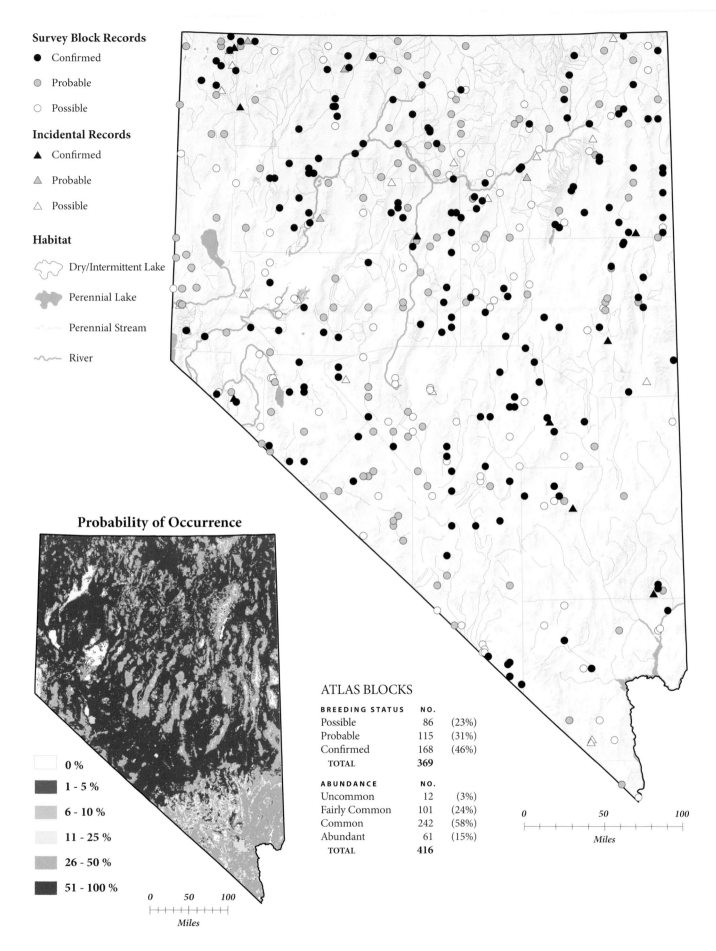

Survey Block Records

● Confirmed
● Probable
○ Possible

Incidental Records

▲ Confirmed
▲ Probable
△ Possible

Habitat

Dry/Intermittent Lake

Perennial Lake

Perennial Stream

River

Probability of Occurrence

0 %

1 - 5 %

6 - 10 %

11 - 25 %

26 - 50 %

51 - 100 %

0   50   100
Miles

ATLAS BLOCKS

BREEDING STATUS	NO.	
Possible	86	(23%)
Probable	115	(31%)
Confirmed	168	(46%)
TOTAL	369	

ABUNDANCE	NO.	
Uncommon	12	(3%)
Fairly Common	101	(24%)
Common	242	(58%)
Abundant	61	(15%)
TOTAL	416	

0   50   100
Miles

TREE SWALLOW
Tachycineta bicolor

The cheerful twittering of Tree Swallows is a sure sign of spring along the snow-fed streams of northern Nevada. Many of these birds are on their way to breeding grounds much farther north, but some stick around to breed. Individual pairs or small colonies are patchily distributed, and they favor riparian areas with standing deadwood and unused woodpecker cavities or nest boxes. Easily studied and fun to watch, Tree Swallows are widely admired for their insect-eating proclivities.

DISTRIBUTION

Although atlas workers found Tree Swallows throughout much of the state, some records may have involved migrants, particularly the possible breeders reported from southern Nevada, where breeding had not previously been noted. Nesting was confirmed at scattered locales in the state's more northerly regions, with the valleys around the Jackpot-Jarbidge area and Snake Mountains having local concentrations. Tree Swallows were most often estimated to occur in intermediate abundances of two to ten pairs per block.

The atlas results are generally consistent with the earlier literature from Nevada. Linsdale (1936) regarded the species as a summer resident in the state, and Alcorn (1988) noted breeding only in northern Nevada. Chisholm and Neel (2002) described the Tree Swallow as a common migrant but a rare breeder in the Lahontan Valley. Tree Swallows are spottily distributed as breeders in Arizona (Corman in Corman and Wise-Gervais 2005:372–373), are absent from the lower Colorado River Valley (Rosenberg et al. 1991), but are common summer residents across much of Utah (Behle et al. 1985), Oregon (Adamus et al. 2001), and Idaho (Stephens and Sturts 1998). In California, they summer in most regions except the deserts (Grinnell and Miller 1944).

On their breeding grounds, Tree Swallows require close proximity to insect-producing aquatic habitats for foraging and standing timber with woodpecker holes (Ryser, 1985, Alcorn 1988), snags with existing cavities, or nest boxes (Robertson et al. 1992) for nesting. The species can be found about towns and farmland (Contreras and Kindschy 1996) and in montane coniferous forests—anywhere their basic habitat needs are met (Behle et al. 1985). The predictive map shows a low chance of finding Tree Swallows breeding throughout most of the state. In the south—where none of the atlas records involved confirmed breeding—the true probabilities are probably even lower than the map suggests. The highest chance of finding the species is at middle elevations in the mountains, but even there they are much less likely to breed than are the closely related Violet-green Swallows.

CONSERVATION AND MANAGEMENT

Tree Swallow populations tend to be patchily distributed (Shuford 1993), which makes it difficult to obtain density estimates (Robertson et al. 1992). Local declines have been documented in some areas (e.g., Garrett and Dunn 1981), but overall, populations seem to be increasing (Sauer et al. 2005). Little if any trend information is available for Nevada, however. Availability of suitable nest cavities is likely an important element of the Tree Swallow's biology—and by extension its management needs.

Even though Nevada's nesting Tree Swallows are a miniscule fraction of the North American population (see Robertson et al. 1992), the species does present interesting local conservation opportunities. For example, nest boxes can both benefit local Tree Swallow populations and provide opportunities for public education and community involvement (Gallagher 1997).

HABITAT USE

HABITAT	ATLAS BLOCKS	INCIDENTAL OBSERVATIONS
Agricultural	3 (4%)	2 (6%)
Grassland	2 (3%)	0 (0%)
Montane Forest	15 (21%)	4 (12%)
Montane Shrub	4 (6%)	1 (3%)
Open Water	1 (1%)	3 (9%)
Pinyon-Juniper	2 (3%)	1 (3%)
Riparian	30 (42%)	17 (52%)
Salt Desert Scrub	1 (1%)	0 (0%)
Sagebrush Scrub	3 (4%)	1 (3%)
Sagebrush Steppe	2 (3%)	1 (3%)
Urban	7 (10%)	0 (0%)
Wetland	2 (3%)	3 (9%)
TOTAL	72	33

Total numbers reported in the Habitat Use, Breeding Status, and Abundance tables may differ from each other (see pp. 22–23 for details). Percentage sums may differ slightly from 100% due to rounding.

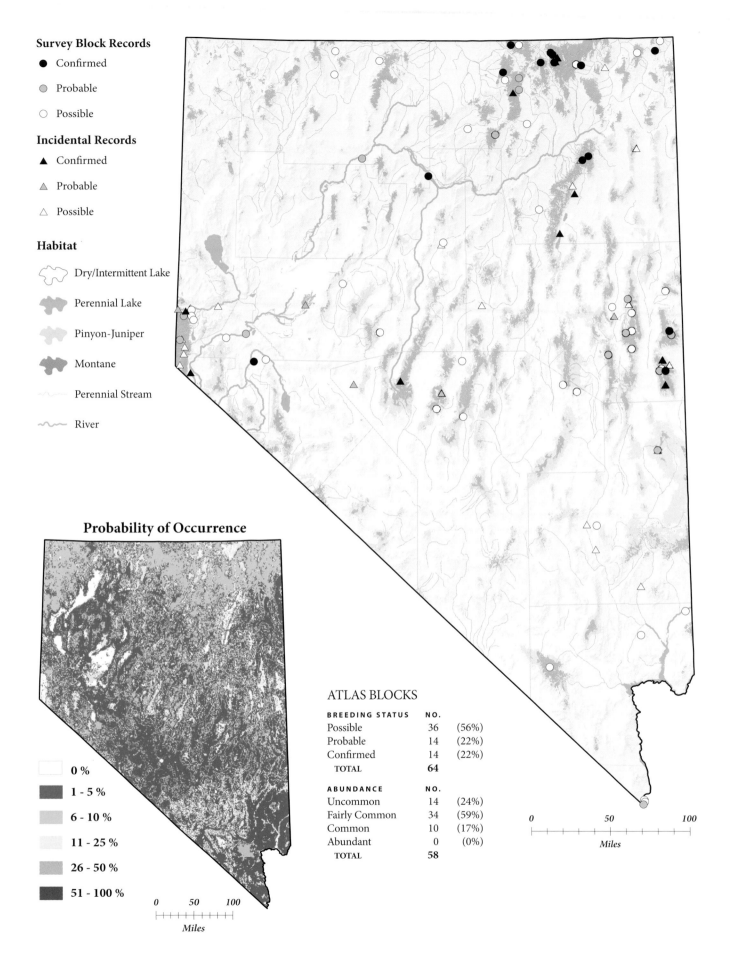

Survey Block Records

● Confirmed

● Probable

○ Possible

Incidental Records

▲ Confirmed

▲ Probable

△ Possible

Habitat

Dry/Intermittent Lake

Perennial Lake

Pinyon-Juniper

Montane

Perennial Stream

River

Probability of Occurrence

☐ 0 %

1 - 5 %

6 - 10 %

11 - 25 %

26 - 50 %

51 - 100 %

0   50   100

Miles

ATLAS BLOCKS

BREEDING STATUS	NO.	
Possible	36	(56%)
Probable	14	(22%)
Confirmed	14	(22%)
TOTAL	**64**	

ABUNDANCE	NO.	
Uncommon	14	(24%)
Fairly Common	34	(59%)
Common	10	(17%)
Abundant	0	(0%)
TOTAL	**58**	

0   50   100

Miles

VIOLET-GREEN SWALLOW
Tachycineta thalassina

This snow white and velvety green swallow is a common sight—and sound—in the foothills and highlands of Nevada. Violet-green Swallows nest wherever there are snags, cliffs, and other structures containing small cavities—along mountain streams, in dry pine woods, and in cliff sides and canyon walls. They also use artificial nest boxes, although perhaps not as readily as their closest relative, the Tree Swallow (Brown et al. 1992). Differences between these two species have not been well studied, and the two sometimes nest side by side (Pinkowski 1981). Violet-green Swallows often range far from their nest sites during foraging, making an exact description of their breeding distribution difficult.

DISTRIBUTION

Violet-green Swallows were found throughout the state, with most of the confirmed breeders reported from the mountains and canyons. Breeding was confirmed in 28% of the blocks in which the species was found. Many records of possible breeding may have represented individuals feeding far from their nests. In most occupied blocks, observers estimated that there were multiple pairs nesting, but generally fewer than ten.

Older treatments of the Violet-green Swallow in Nevada uniformly characterize the species as common. Some report that it is resident in the valleys (Ryser 1985), while others indicate that it breeds primarily in the mountains (Linsdale 1936). The Violet-green Swallow is absent as a breeder from the Lahontan Valley (Chisholm and Neel 2002), but atlas workers confirmed nesting at elevations as low as about 3,400 feet (1,000 meters). Violet-green Swallows are also common in the states surrounding Nevada (Behle et al. 1985, Small 1994, Stephens and Sturts 1998, Adamus et al. 2001, Corman in Corman and Wise-Gervais 2005:374–375).

The species was reported from a wide variety of atlas habitat types, several of which are presumably used only for foraging. The basic requirement for breeding is a cavity of virtually any sort—a woodpecker hole, a stone crevice, a gap in a building, and so on. Breeders are most often found in upland forest or woodland settings (Garrett and Dunn 1981), but they may nest all the way down to sea level (Small 1994) and in the lowest river valleys (Rosenberg et al. 1991).

The frequency of foraging habitat observations complicates predictions of breeding occurrence, and it is probably appropriate to regard the predictive map as an indication of the probability of encountering the species rather than the probability of local nesting. The actual breeding range is likely smaller than shown, occurring mostly in the subset of habitats where suitable cavities are available. Nonetheless, the prediction of an increasing chance of finding these swallows at higher elevations matches what is known about the species. Comparison of this predictive map with that for the Tree Swallow also shows that the Violet-green Swallow is far more likely to be found as a breeder in most of Nevada.

CONSERVATION AND MANAGEMENT

The Violet-green Swallow is rarely thought of in connection with bird conservation, and the species has received relatively little attention from biologists (Brown et al. 1992). Rangewide, populations have remained stable, but significant declines have been detected in California (Sauer et al. 2005). Little information is available about trends in the Great Basin, and additional monitoring is clearly a priority. As secondary cavity nesters, Violet-green Swallows depend on dead trees for nest sites. Their nesting flexibility and willingness to use nest boxes, however, may make them less susceptible to loss of mature and dead trees than are some other cavity nesters.

HABITAT USE

HABITAT	ATLAS BLOCKS	INCIDENTAL OBSERVATIONS
Agricultural	1 (<1%)	0 (0%)
Alpine	9 (4%)	0 (0%)
Barren	20 (9%)	5 (6%)
Grassland	1 (<1%)	0 (0%)
Mojave	4 (2%)	4 (5%)
Montane Forest	37 (17%)	19 (24%)
Montane Parkland	6 (3%)	1 (1%)
Montane Shrub	21 (10%)	5 (6%)
Open Water	1 (<1%)	2 (3%)
Pinyon-Juniper	25 (12%)	3 (4%)
Riparian	54 (26%)	31 (40%)
Salt Desert Scrub	1 (<1%)	0 (0%)
Sagebrush Scrub	13 (6%)	1 (1%)
Sagebrush Steppe	10 (5%)	1 (1%)
Urban	3 (1%)	0 (0%)
Wetland	6 (3%)	6 (8%)
TOTAL	212	78

Total numbers reported in the Habitat Use, Breeding Status, and Abundance tables may differ from each other (see pp. 22–23 for details). Percentage sums may differ slightly from 100% due to rounding.

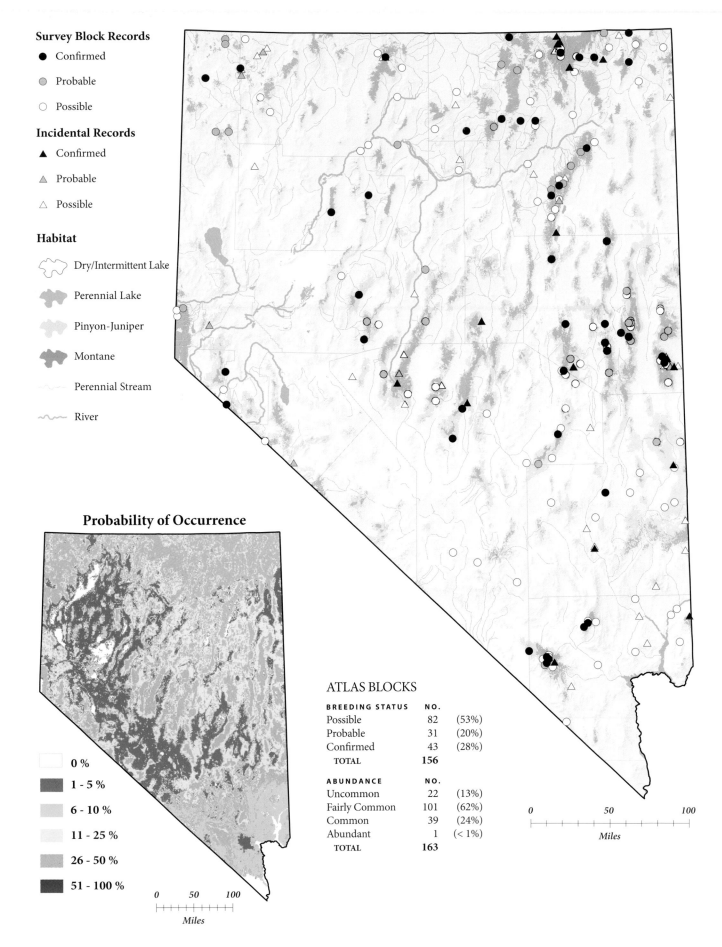

Survey Block Records

- ● Confirmed
- ● Probable (gray)
- ○ Possible

Incidental Records

- ▲ Confirmed
- ▲ Probable (gray)
- △ Possible

Habitat

- Dry/Intermittent Lake
- Perennial Lake
- Pinyon-Juniper
- Montane
- Perennial Stream
- River

Probability of Occurrence

- ☐ 0 %
- 1 - 5 %
- 6 - 10 %
- 11 - 25 %
- 26 - 50 %
- 51 - 100 %

0   50   100
Miles

ATLAS BLOCKS

BREEDING STATUS	NO.	
Possible	82	(53%)
Probable	31	(20%)
Confirmed	43	(28%)
TOTAL	**156**	

ABUNDANCE	NO.	
Uncommon	22	(13%)
Fairly Common	101	(62%)
Common	39	(24%)
Abundant	1	(< 1%)
TOTAL	**163**	

0   50   100
Miles

NORTHERN ROUGH-WINGED SWALLOW

Stelgidopteryx serripennis

Of all the birds that migrate to nest in Nevada, the Northern Rough-winged Swallow is among the earliest to return each year. Individuals or small flocks can be found along the Colorado River drainage as early as January, and nesting may be under way there by February (Rosenberg et al. 1991). Though it begins early, spring migration is a protracted affair for this bird, and the earliest arrivals in the northern part of Nevada are often not reported until April. By early May, breeders can be found in a variety of open habitats throughout the northern part of the state.

DISTRIBUTION

Atlas workers reported Northern Rough-winged Swallows across much of northern Nevada. The species was virtually absent from the dry deserts of south-central Nevada, including Nye and Esmeralda counties and adjacent areas. There were many records from the far south, in Clark and Lincoln counties, though the breeding confirmation rate was lower there than it was farther north.

The Northern Rough-winged Swallow's breeding distribution fills the continental United States and spills out just a little beyond, especially into Mexico (DeJong 1996). It is a common, though rather local, breeder in all of the states surrounding Nevada (Behle et al. 1985, Small 1994, Stephens and Sturts 1998, Adamus et al. 2001, Wise-Gervais in Corman and Wise-Gervais 2005:376–377). Its status in Nevada was somewhat unclear prior to the atlas survey. Linsdale (1936) called it a summer resident at a few localities, but Alcorn (1988) upgraded it to a common summer resident throughout the state. Ryser (1985) considered the

rough-winged swallow to be a common breeder in the Great Basin, while Chisholm and Neel (2002) noted that it breeds only in low numbers in the Lahontan Valley.

Just over half of the habitat records were riparian, with the remainder coming from a wide variety of upland and wetland habitats. This result reflects the fact that Northern Rough-winged Swallows nest in burrows in crumbling earthen walls, typically along stream banks, but forage widely in adjacent open areas (DeJong 1996). Although the predictive map suggests at least a small chance of finding this species in most of the state, the probabilities are greatest across the northern tier and in riparian areas near the base of mountain ranges. In addition, most of the southern Nevada records involved only possible breeders; if these were actually observations of late migrants, then the map would overestimate the true probability of encountering breeding swallows there.

CONSERVATION AND MANAGEMENT

Northern Rough-winged Swallows are widespread in North America, and they are generally thought to have benefited from the proliferation of artificial nesting sites provided by humans, such as road cuts, gravel pits, and buildings (DeJong 1996). The species is less colonial than the superficially similar Bank Swallow, and it does not necessarily require close proximity to water for nesting (DeJong 1996). Even in areas where rough-winged swallows seem scarce, they may in fact be common; these birds often go undetected because of their nondescript plumage and tendency to join larger flocks of Bank Swallows (DeJong 1996, Gallagher 1997).

Overall, Northern Rough-winged Swallow populations appear stable (Sauer et al. 2005), and their prospects in Nevada and elsewhere are probably good. Because they reuse nesting sites year after year (Chisholm and Neel 2002), protecting the known nest sites may be important for maintaining their populations.

HABITAT USE

HABITAT	ATLAS BLOCKS	INCIDENTAL OBSERVATIONS
Agricultural	16 (9%)	1 (3%)
Alpine	0 (0%)	1 (3%)
Barren	3 (2%)	2 (5%)
Grassland	4 (2%)	0 (0%)
Mesquite	4 (2%)	0 (0%)
Mojave	8 (4%)	4 (11%)
Montane Forest	2 (1%)	0 (0%)
Montane Shrub	2 (1%)	0 (0%)
Open Water	10 (5%)	2 (5%)
Pinyon-Juniper	2 (1%)	0 (0%)
Riparian	94 (51%)	25 (66%)
Salt Desert Scrub	1 (<1%)	0 (0%)
Sagebrush Scrub	11 (6%)	0 (0%)
Sagebrush Steppe	8 (4%)	1 (3%)
Urban	8 (4%)	1 (3%)
Wetland	13 (7%)	1 (3%)
TOTAL	186	38

Total numbers reported in the Habitat Use, Breeding Status, and Abundance tables may differ from each other (see pp. 22–23 for details). Percentage sums may differ slightly from 100% due to rounding.

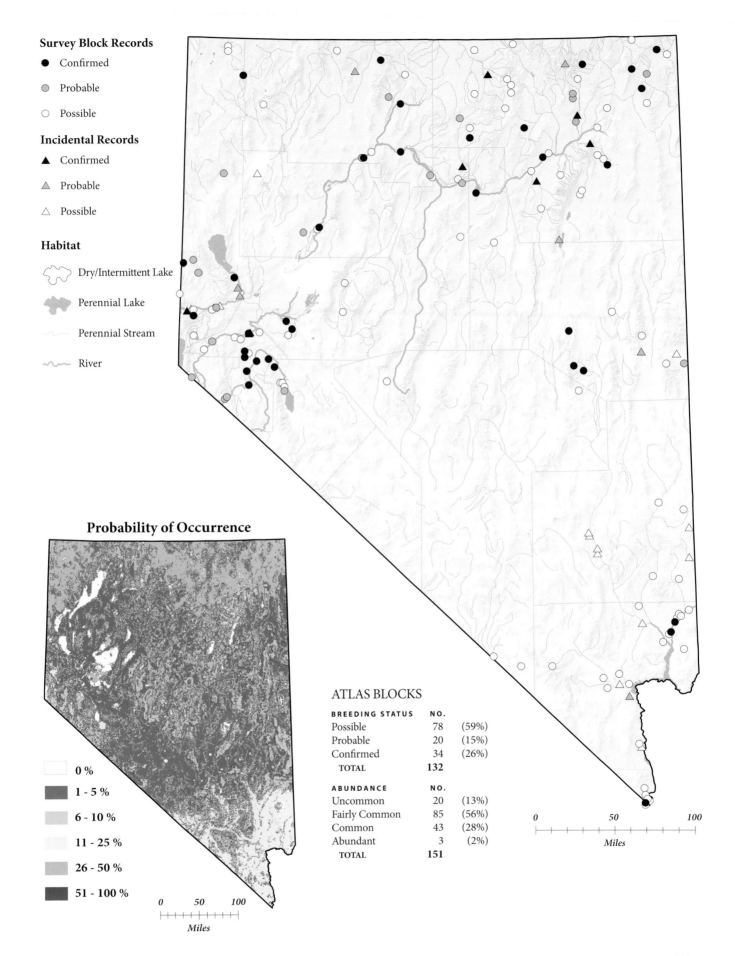

Survey Block Records

● Confirmed

● Probable

○ Possible

Incidental Records

▲ Confirmed

▲ Probable

△ Possible

Habitat

⬡ Dry/Intermittent Lake

⬢ Perennial Lake

〜 Perennial Stream

⁓ River

Probability of Occurrence

☐ 0 %

▨ 1 - 5 %

▨ 6 - 10 %

☐ 11 - 25 %

▨ 26 - 50 %

■ 51 - 100 %

0   50   100

Miles

ATLAS BLOCKS

BREEDING STATUS	NO.	
Possible	78	(59%)
Probable	20	(15%)
Confirmed	34	(26%)
TOTAL	132	

ABUNDANCE	NO.	
Uncommon	20	(13%)
Fairly Common	85	(56%)
Common	43	(28%)
Abundant	3	(2%)
TOTAL	151	

0   50   100

Miles

BANK SWALLOW
Riparia riparia

Although the Bank Swallow is one of the most widely distributed swallows in the world (Garrison 1999), it is probably the least widely distributed one in Nevada. Bank Swallows are not uncommon breeders here, but most of them are restricted to the major river systems of northern Nevada. Elsewhere in the state—away from rivers and streams—the species is largely absent except as a migrant. Their highly colonial nature and specialized nest site requirements make breeding Bank Swallows difficult to census and monitor, and they are at constant risk of local extirpation due to collapse of river banks or other events that render previously used colony sites unsuitable.

DISTRIBUTION

Although widely distributed on the continental scale, the Bank Swallow is a highly localized breeder throughout much of its range (Buckelew and Hall 1994, Garrison 1999). In northern and western Nevada, atlas workers found Bank Swallows primarily in the major drainages of the Humboldt, Truckee, Carson, and Walker rivers, and they were largely absent elsewhere. Although a few records from southern Nevada were classified as possible breeders, these probably involved migrants or wanderers.

Abundance was estimated at more than ten pairs in nearly half of the occupied blocks. It is interesting, and perhaps surprising, that only three blocks harbored large colonies of more than one hundred pairs. Large colonies are by nature very localized, however, and it is quite likely that others occurred in areas that atlas workers did not visit.

Bank Swallows reach the southern limit of their western summer distribution in northern Nevada. To the north and east, they are widespread though localized (Behle et al. 1985, Garrison 1999, Adamus et al. 2001). In California, the Bank Swallow is listed as a Threatened species because of its limited distribution in the northern part of the state (Garrison 1999).

Bank Swallows generally avoid forested headwaters at higher elevations and occur instead in open riparian zones at lower elevations, where moderately to severely eroded stream banks are available. These conditions are not especially common in Nevada, and it is therefore not surprising that there is a low chance of finding Bank Swallows in most of the state. The extension of the species' predicted occurrence into southern Nevada is based on three records, each of which involved only possible breeding; actual nesting may be even less likely in the south than the predictive map suggests.

CONSERVATION AND MANAGEMENT

Bank Swallows have been both beneficiaries and victims of human activity. They are willing to accept disturbed sites such as incised stream banks, and they can be found in entirely artificial settings such as sandy areas of gravel quarries (Palmer-Ball 1996, Garrison 1999). Conversely, they have suffered from pesticide application and streamside grazing (Hemesath and Wilson in Jackson et al. 1996:248–249).

Overall, Bank Swallow populations are stable in North America (Sauer et al. 2005), although local fluctuations do occur. In Nevada, Partners in Flight recognizes the Bank Swallow as a Priority species, with its sensitivity to disturbance cited as a particular concern (Neel 1999a). Specifically, because the species is colonial, a single disturbance can affect a large number of birds (Hoogland and Sherman 1976, Freer 1979). Given that the total number of Bank Swallow colonies in Nevada is probably small, it seems prudent to monitor this species carefully.

HABITAT USE

HABITAT	ATLAS BLOCKS	INCIDENTAL OBSERVATIONS
Agricultural	3 (9%)	0 (0%)
Barren	2 (6%)	1 (20%)
Grassland	1 (3%)	0 (0%)
Riparian	16 (48%)	4 (80%)
Sagebrush Scrub	1 (3%)	0 (0%)
Sagebrush Steppe	2 (6%)	0 (0%)
Urban	2 (6%)	0 (0%)
Wetland	6 (18%)	0 (0%)
TOTAL	33	5

Total numbers reported in the Habitat Use, Breeding Status, and Abundance tables may differ from each other (see pp. 22–23 for details). Percentage sums may differ slightly from 100% due to rounding.

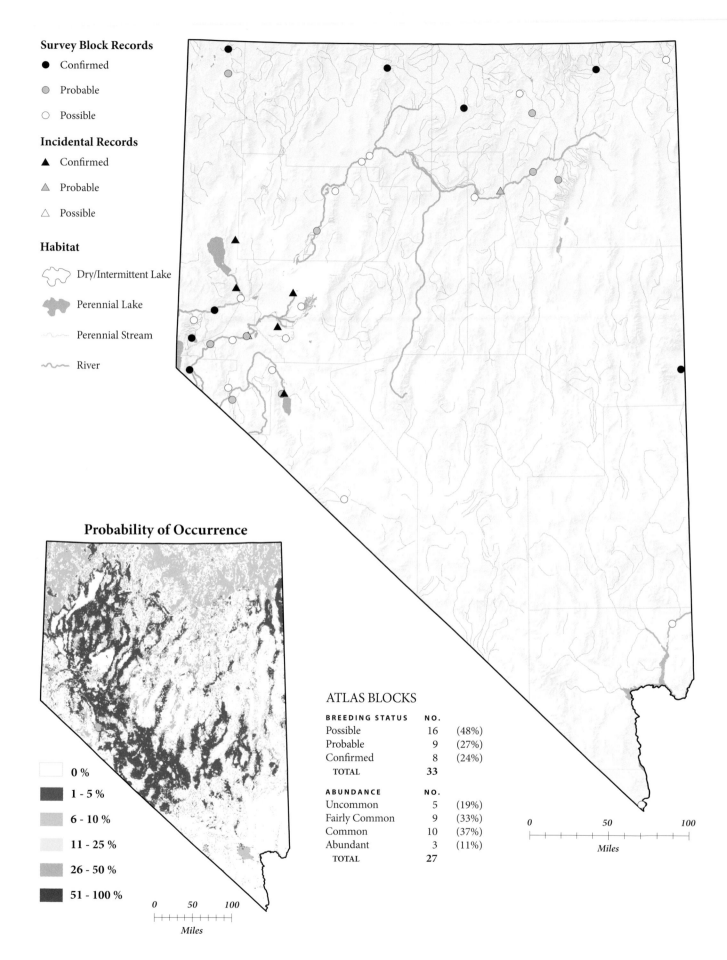

Survey Block Records

● Confirmed

● Probable

○ Possible

Incidental Records

▲ Confirmed

▲ Probable

△ Possible

Habitat

Dry/Intermittent Lake

Perennial Lake

Perennial Stream

River

Probability of Occurrence

	0 %
	1 - 5 %
	6 - 10 %
	11 - 25 %
	26 - 50 %
	51 - 100 %

0   50   100
Miles

ATLAS BLOCKS

BREEDING STATUS	NO.	
Possible	16	(48%)
Probable	9	(27%)
Confirmed	8	(24%)
TOTAL	33	

ABUNDANCE	NO.	
Uncommon	5	(19%)
Fairly Common	9	(33%)
Common	10	(37%)
Abundant	3	(11%)
TOTAL	27	

0   50   100
Miles

CLIFF SWALLOW
Petrochelidon pyrrhonota

The colorful Cliff Swallow is a social bird that has specialized needs for colony placement and maintenance—yet it is also one of our widest-ranging summer birds. Small foraging flocks or individuals can be found in diverse habitats: high above mountain forests, in the middle of the desert, and in towns and cities. Colonies tend to be located wherever there is placid water or moist meadows for foraging, wet mud for nest building, and human structures or cliffs for nest placement. Highway underpasses and their environs are often ideal in this regard, and on some stretches of Nevada's interstate highway system every culvert seems to harbor a Cliff Swallow colony.

DISTRIBUTION

Cliff Swallows were confirmed as breeders into the farthest corners of the state, although there were large regions with few or no observations, such as southern Nye County, central Washoe County, and western Pershing County. The greatest concentration of nesting colonies was along the Humboldt River and its tributaries, where bridges along Interstate 80 provide many nesting opportunities. Breeding was confirmed in about half of the blocks in which the species was recorded, and, as expected for this highly colonial species, relatively large concentrations of individuals were common—61% of occupied blocks were estimated to have more than ten pairs.

Cliff Swallows have long been considered the most common swallows in Nevada (Fisher 1893, Linsdale 1936, Ryser 1985). The atlas data, however, suggest that Violet-green Swallows may challenge Cliff Swallows for this title, and even Barn and Northern Rough-winged Swallows were reported from more atlas blocks than Cliff Swallows were. Because of their coloniality and use of human structures for nesting, Cliff Swallows apparently are more obvious to observers than are the other swallows. The Cliff Swallow is as common in the states surrounding Nevada as it is here, except in the hottest, driest deserts of southern California and southern Arizona (Behle et al. 1985, Small 1994, Stephens and Sturts 1998, Adamus et al. 2001, Wise-Gervais in Corman and Wise-Gervais 2005:378–379).

Historically, Cliff Swallows nested in canyons, cliffs, and escarpments (Brown and Brown 1995), but they have been won over by human structures in recent decades. The predictive map accurately illustrates that the likelihood of encountering the species increases as one moves north in the state, but our simplistic model was unable to capture the tight relationship between Cliff Swallows and areas that combine water and suitable nesting structures. A model designed specifically to consider the nesting habitat requirements of this species could doubtless produce even better predictions.

CONSERVATION AND MANAGEMENT

Overall, the breeding range of the Cliff Swallow expanded greatly during the twentieth century (Brown and Brown 1995), probably due to the increased availability of artificial structures. Currently, there seem to be no major threats to the species. Cliff Swallow colonies are vulnerable to human harassment (Shuford 1993), though, and competition with House Sparrows has been noted (Brown and Brown 1995).

In Nevada as well there are probably few threats to Cliff Swallows at the present time. In fact, the state's rapidly increasing human population may present the Cliff Swallow with new opportunities for expansion. Breeding Bird Survey data for Nevada show growth rates well in excess of 10% per year in recent decades, though this is based on just a handful of survey routes (Sauer et al. 2005). Local protection may be warranted for specific colonies, however, since Cliff Swallows are extremely loyal to their breeding sites (Brown and Brown 1995).

HABITAT USE

HABITAT	ATLAS BLOCKS	INCIDENTAL OBSERVATIONS
Agricultural	12 (7%)	1 (1%)
Barren	11 (6%)	3 (4%)
Grassland	1 (<1%)	2 (3%)
Mojave	5 (3%)	5 (6%)
Montane Forest	1 (<1%)	1 (1%)
Montane Shrub	4 (2%)	0 (0%)
Open Water	3 (2%)	3 (4%)
Pinyon-Juniper	4 (2%)	0 (0%)
Riparian	76 (43%)	31 (39%)
Salt Desert Scrub	6 (3%)	1 (1%)
Sagebrush Scrub	9 (5%)	7 (9%)
Sagebrush Steppe	13 (7%)	3 (4%)
Urban	21 (12%)	13 (16%)
Wetland	10 (6%)	10 (13%)
TOTAL	176	80

Total numbers reported in the Habitat Use, Breeding Status, and Abundance tables may differ from each other (see pp. 22–23 for details). Percentage sums may differ slightly from 100% due to rounding.

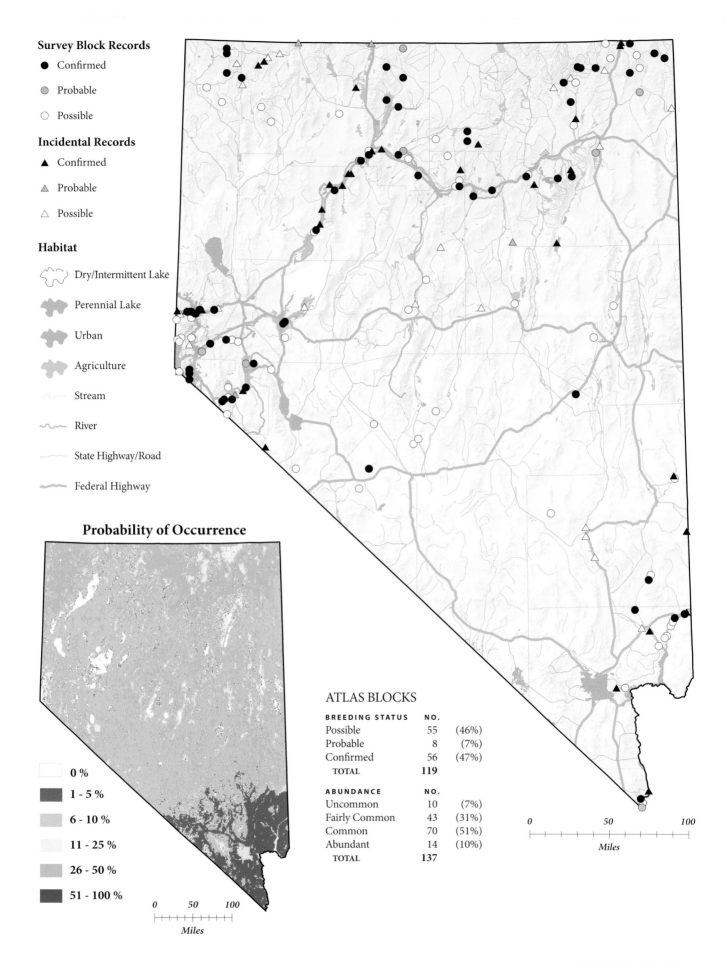

Survey Block Records

● Confirmed

● Probable

○ Possible

Incidental Records

▲ Confirmed

△ Probable

△ Possible

Habitat

Dry/Intermittent Lake

Perennial Lake

Urban

Agriculture

Stream

River

State Highway/Road

Federal Highway

Probability of Occurrence

	0 %
	1 - 5 %
	6 - 10 %
	11 - 25 %
	26 - 50 %
	51 - 100 %

0   50   100

Miles

ATLAS BLOCKS

BREEDING STATUS	NO.	
Possible	55	(46%)
Probable	8	(7%)
Confirmed	56	(47%)
TOTAL	119	

ABUNDANCE	NO.	
Uncommon	10	(7%)
Fairly Common	43	(31%)
Common	70	(51%)
Abundant	14	(10%)
TOTAL	137	

0   50   100

Miles

BARN SWALLOW
Hirundo rustica

The Barn Swallow seems to have been destined for human companionship. This globally distributed bird figures prominently in myths and legends, and it has adapted splendidly to human-altered landscapes. In Nevada, Barn Swallows can be found in agricultural valleys, in big cities, and at bridges and culverts in the middle of nowhere.

DISTRIBUTION

Migrant Barn Swallows can be seen throughout Nevada, but they are restricted as confirmed breeders to latitudes north of approximately 38°30′N. Although they tend to congregate where suitable structures occur, Barn Swallows are not as colonial as Cliff and Bank Swallows, and nearly three-quarters of the abundance estimates from occupied blocks suggested fewer than ten pairs per block.

The atlas results were not entirely consistent with the earlier literature from Nevada. Linsdale (1936) implied that the Barn Swallow is a summer resident throughout the state but provided no definitive breeding records from the south. Alcorn (1988), however, listed the Barn Swallow as a summer resident only in the northern half of Nevada, where it is considered a common breeder (Ryser 1985, Chisholm and Neel 2002).

Although the Barn Swallow is widespread in the temperate zone of the Northern Hemisphere, it is absent as a breeder from much of the desert Southwest. In Arizona, it breeds primarily in the southeast (Corman in Corman and Wise-Gervais 2005:380–381), and it is absent from much of southern California (Garrett and Dunn 1981, Small 1994) and the lower Colorado River Valley (Rosenberg et al. 1991). To our north and east, however, the Barn Swallow is widespread and abundant (Behle et al. 1985, Stephens and Sturts 1998, Adamus et al. 2001).

Historically, Barn Swallows nested in caves and on cliffs, but nowadays they primarily use human-built structures (Brown and Brown 1999). Most atlas records came from urban and riparian habitats—both of which are relatively limited in Nevada. Close inspection of the data also shows a string of records where Interstate 80 runs alongside the Humboldt River. This pattern may reflect the corridor's accessibility to birders, but it may also be that this region provides both riparian zones for foraging and highway overpasses and bridges that offer nest sites. The predictive map generally reflects the northerly, low-elevation distribution of the species in Nevada, with an especially high chance of occurrence in riparian, wetland, and urban habitats. The prediction that the species is highly likely to breed in the Las Vegas area and in other smaller patches in the south is clearly wrong; it is the result of a few records of possible breeding in far southern Nevada.

CONSERVATION AND MANAGEMENT

Over the long term, human activities have clearly benefited Barn Swallows in Nevada (Ryser 1985) and elsewhere, and the North American population may actually be several orders of magnitude larger than it was before European settlement (Brown and Brown 1999). Breeding Bird Survey data show a steady increase in numbers rangewide during the 1960s and 1970s, followed by a sustained decline (Sauer et al. 2005). The reasons for this recent decline are not clear. Barn Swallow populations do not seem to be regulated by nest site availability (although they may have been in the past), but they may be limited by severe weather (Brown and Brown 1999) and food shortages (Gallagher 1997). In some regions, competition with House Sparrows may be a problem (Brown and Brown 1999).

HABITAT USE

HABITAT	ATLAS BLOCKS	INCIDENTAL OBSERVATIONS
Agricultural	22 (12%)	1 (2%)
Grassland	2 (1%)	0 (0%)
Mesquite	1 (<1%)	0 (0%)
Mojave	2 (1%)	1 (2%)
Montane Forest	1 (<1%)	0 (0%)
Montane Shrub	1 (<1%)	1 (2%)
Open Water	6 (3%)	2 (3%)
Pinyon-Juniper	4 (2%)	0 (0%)
Riparian	62 (33%)	21 (32%)
Salt Desert Scrub	4 (2%)	1 (2%)
Sagebrush Scrub	11 (6%)	3 (5%)
Sagebrush Steppe	9 (5%)	2 (3%)
Urban	46 (24%)	22 (34%)
Wetland	17 (9%)	11 (17%)
TOTAL	188	65

Total numbers reported in the Habitat Use, Breeding Status, and Abundance tables may differ from each other (see pp. 22–23 for details). Percentage sums may differ slightly from 100% due to rounding.

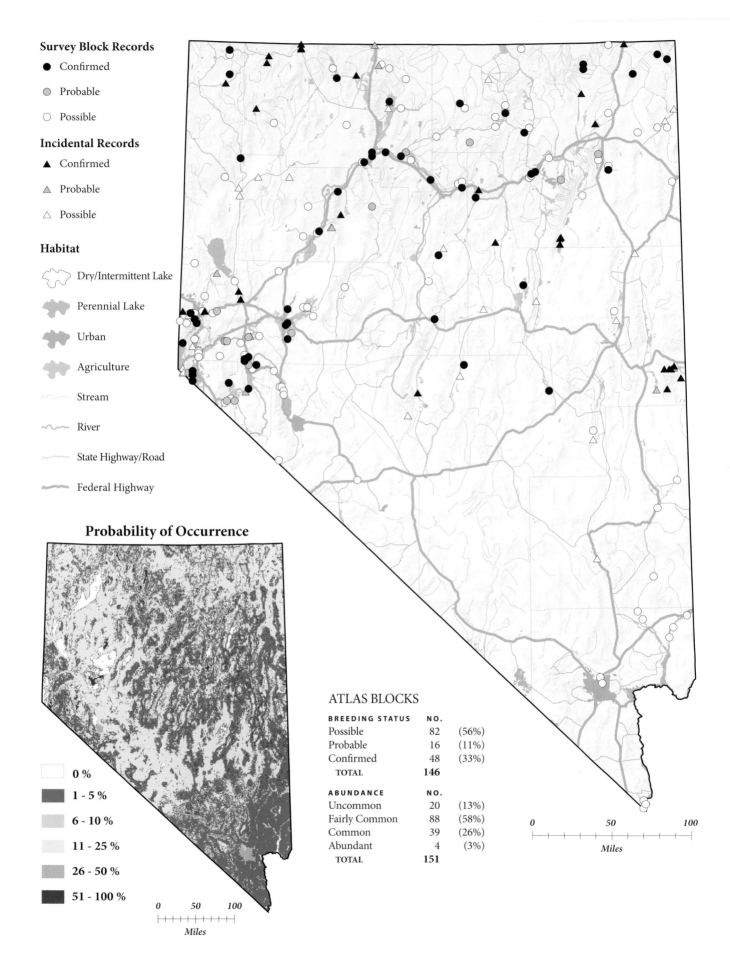

Survey Block Records

● Confirmed

● Probable

○ Possible

Incidental Records

▲ Confirmed

▲ Probable

△ Possible

Habitat

Dry/Intermittent Lake

Perennial Lake

Urban

Agriculture

Stream

River

State Highway/Road

Federal Highway

Probability of Occurrence

	0 %
	1 - 5 %
	6 - 10 %
	11 - 25 %
	26 - 50 %
	51 - 100 %

0   50   100
Miles

ATLAS BLOCKS

BREEDING STATUS	NO.	
Possible	82	(56%)
Probable	16	(11%)
Confirmed	48	(33%)
TOTAL	146	

ABUNDANCE	NO.	
Uncommon	20	(13%)
Fairly Common	88	(58%)
Common	39	(26%)
Abundant	4	(3%)
TOTAL	151	

0   50   100
Miles

BLACK-CAPPED CHICKADEE
Poecile atricapillus

Rh'02

In the riparian groves and woodlands of far northeastern Nevada, some of the chickadees don't sound quite right. Their call notes are less grating and husky than those of the widespread Mountain Chickadee, and their two-syllable songs are quite distinctive. On closer inspection, they are longer tailed, with frostier wings and a different head pattern. This is because they are Black-capped Chickadees at the southwestern limit of their range. They are not always easy to separate from Mountain Chickadees, though, and their status in Nevada is still not perfectly known.

DISTRIBUTION
All Black-capped Chickadee records from atlas blocks came from the lowlands and foothills of northeastern Elko County. Three incidental records came from east-central Nevada: one from the foothills of the Snake Range, where the species has been reported in the past (Ryser 1985), and the others from the White Pine Range. These records match assessments in the older literature, which refers to the Black-capped Chickadee primarily as a species of far northeastern Nevada (Linsdale 1936, Ryser 1985, Alcorn 1988).

The range of the Black-capped Chickadee lies primarily to our north and east. The species barely enters northern California and Arizona (Small 1994, Corman in Corman and Wise-Gervais 2005:599). Black-capped Chickadees are, however, common in Utah, Idaho, and Oregon (Behle et al. 1985, Stephens and Sturts 1998, Adamus et al. 2001).

Throughout North America, Black-capped Chickadees can be found in various woodland settings and in residential yards and parks (Smith 1993). Here in Nevada, atlas workers found Black-capped Chickadees mostly in riparian habitats, in keeping with the pattern that Ryser (1985) suggested. The predictive map indicates that the species has the potential to show up along much of the northeastern boundary of the state, although the probability of actually finding it in many areas is relatively low.

The accuracy of the model rests largely on whether patches of riparian habitat, most of them far too small to visualize on the map, are actually present within a given area.

CONSERVATION AND MANAGEMENT
The impacts of forestry on western populations of Black-capped Chickadees are mixed. In Colorado, snag removal and logging of mature trees have reduced the number of nest sites for this obligate cavity nester, and bird boxes appear insufficient to counter the deficit (Kingery in Kingery 1998:348–349). But logging has probably been beneficial in Washington, where clear-cutting has increased the amount of suitable foraging habitat (Smith et al. 1997).

The Black-capped Chickadee's limited distribution in Nevada does not indicate that it is at risk here, but simply that it is at the edge of its range. An interesting question concerns the stability of the range boundary, especially as populations in various parts of the West have expanded or contracted (Smith et al. 1997, Kingery in Kingery 1998:348–349, Jones and Rosenberg 2001). This pattern suggests that records south of Elko County should not be immediately dismissed, and that observers should be on the lookout for Black-capped Chickadees in previously unoccupied regions of the state.

HABITAT USE

HABITAT	ATLAS BLOCKS	INCIDENTAL OBSERVATIONS
Montane Forest	0 (0%)	2 (22%)
Montane Shrub	1 (20%)	0 (0%)
Pinyon Juniper	0 (0%)	1 (11%)
Riparian	4 (80%)	6 (67%)
TOTAL	5	9

Total numbers reported in the Habitat Use, Breeding Status, and Abundance tables may differ from each other (see pp. 22–23 for details). Percentage sums may differ slightly from 100% due to rounding.

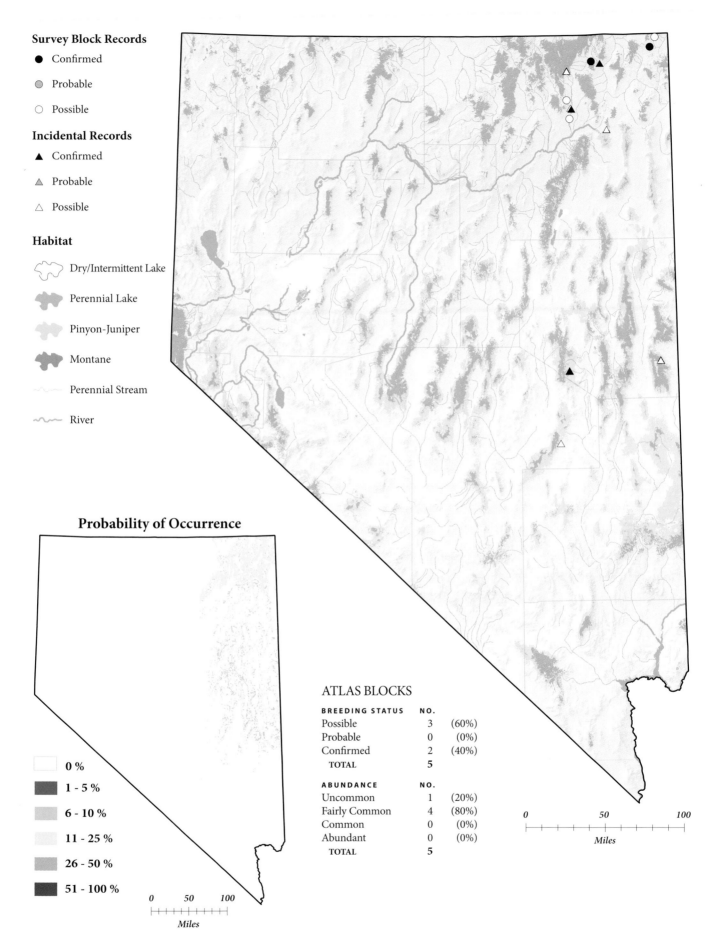

Survey Block Records

● Confirmed

● Probable

○ Possible

Incidental Records

▲ Confirmed

▲ Probable

△ Possible

Habitat

Dry/Intermittent Lake

Perennial Lake

Pinyon-Juniper

Montane

Perennial Stream

River

Probability of Occurrence

☐	0 %
■	1 - 5 %
■	6 - 10 %
☐	11 - 25 %
■	26 - 50 %
■	51 - 100 %

0   50   100

Miles

ATLAS BLOCKS

BREEDING STATUS	NO.	
Possible	3	(60%)
Probable	0	(0%)
Confirmed	2	(40%)
TOTAL	5	

ABUNDANCE	NO.	
Uncommon	1	(20%)
Fairly Common	4	(80%)
Common	0	(0%)
Abundant	0	(0%)
TOTAL	5	

0   50   100

Miles

MOUNTAIN CHICKADEE
Poecile gambeli

It is impossible for a visitor to Nevada's coniferous forests not to become acquainted with the Mountain Chickadee. Despite their diminutive size, Mountain Chickadees are active, conspicuous, and easily identified. Their telltale song—a tinny *cheeseburger!*—is unmistakable. Family groups can be found in the treetops as well as in understory vegetation, dense thickets, and clearings. And while they occur in remote wilderness areas, they also tolerate human company: ski resorts, campsites, and residential neighborhoods are often home to Mountain Chickadees.

DISTRIBUTION

Atlas workers found Mountain Chickadees in most of Nevada's major mountain ranges. Breeding was confirmed in about half of the blocks in which the species was recorded, but it is very likely that breeding occurred in most places where this largely sedentary species was found. Mountain Chickadees tended to be common; more than half of the abundance estimates exceeded ten pairs per block, and estimated densities of more than one hundred pairs were not unusual.

Mountain Chickadees have always been regarded as common and widely distributed in Nevada's mountains (Fisher 1983, Ryser 1985, Alcorn 1988). Linsdale (1936) reported two subspecies: *abbreviatus* in the northwestern mountains and *inyoensis* elsewhere. As a characteristic species of western montane forests, it is not surprising that the Mountain Chickadee is also widespread in the wooded uplands and mountains of California (Small 1994), Oregon (Contreras 1999, Adamus et al. 2001), Utah (Behle et al.

1985), Arizona (Sitko in Corman and Wise-Gervais 2005:382–383), and Idaho (Stephens and Sturts 1998).

This cavity-nesting species can be found in various forest habitats (Garrett and Dunn 1981) but generally uses those dominated by conifers (McCallum et al. 1999). In mixed coniferous-deciduous forests and riparian zones, Mountain Chickadees use conifers more than chance would predict (McCallum et al. 1999). Almost half of the atlas records came from montane forests, and about half of the remainder came from pinyon-juniper woodlands. Mountain Chickadees also range down into wooded valleys (Linsdale 1936), and breeding birds were frequently found in riparian zones. These habitat associations are reflected on the predictive map, which shows a high chance of finding the species throughout the state's mountain ranges and a smaller chance of finding it in the lower elevations, where it is generally restricted to riparian and other woodland habitats. Interestingly, the predictive map suggests a moderate to high likelihood of encountering breeding Mountain Chickadees in northwestern Nevada, despite the small number of records (none confirmed) in this area. In fact, northwestern Nevada and the adjoining portions of Oregon and Idaho form a distinct "hole" in the species' breeding range as presented by McCallum et al. (1999). Further surveys in the area, especially since probable breeding was noted during atlas fieldwork, would present an interesting test for the predictive model.

CONSERVATION AND MANAGEMENT

The Mountain Chickadee is an adaptable species that is rarely thought of as a conservation concern. Nonetheless, recent declines have been reported in the United States, particularly in the western regions (Sauer et al. 2005). In the Nevada landscape, the Mountain Chickadee's range is naturally fragmented. Local extirpations are therefore possible, especially given the sedentary nature of the species. The Mountain Chickadee's ecology in areas where it co-occurs with Black-capped Chickadees is worth further study, as is the species' apparent spread into lower-elevation urban sites. Reno birders have been seeing Mountain Chickadees in recent years, and nesting has been noted at several locations in this rapidly growing city. Whether or not this is a recent or long-established phenomenon is unclear.

HABITAT USE

HABITAT	ATLAS BLOCKS	INCIDENTAL OBSERVATIONS
Alpine	9 (3%)	0 (0%)
Barren	2 (<1%)	0 (0%)
Montane Forest	151 (44%)	36 (46%)
Montane Parkland	11 (3%)	1 (1%)
Montane Shrub	32 (9%)	3 (4%)
Pinyon-Juniper	103 (30%)	17 (22%)
Riparian	33 (10%)	20 (25%)
Sagebrush Scrub	0 (0%)	1 (1%)
Sagebrush Steppe	1 (<1%)	1 (1%)
Urban	2 (<1%)	0 (0%)
TOTAL	344	79

Total numbers reported in the Habitat Use, Breeding Status, and Abundance tables may differ from each other (see pp. 22–23 for details). Percentage sums may differ slightly from 100% due to rounding.

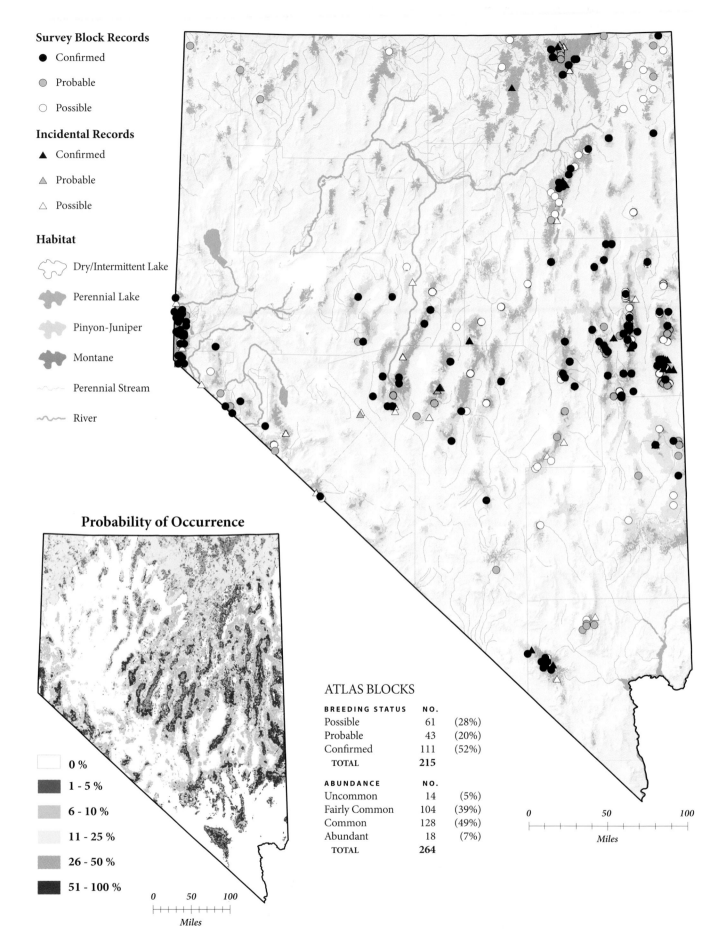

Survey Block Records

● Confirmed

● Probable

○ Possible

Incidental Records

▲ Confirmed

△ Probable

△ Possible

Habitat

Dry/Intermittent Lake

Perennial Lake

Pinyon-Juniper

Montane

Perennial Stream

River

Probability of Occurrence

☐ 0 %

■ 1 - 5 %

■ 6 - 10 %

☐ 11 - 25 %

■ 26 - 50 %

■ 51 - 100 %

0   50   100

Miles

ATLAS BLOCKS

BREEDING STATUS	NO.	
Possible	61	(28%)
Probable	43	(20%)
Confirmed	111	(52%)
TOTAL	215	

ABUNDANCE	NO.	
Uncommon	14	(5%)
Fairly Common	104	(39%)
Common	128	(49%)
Abundant	18	(7%)
TOTAL	264	

0   50   100

Miles

JUNIPER TITMOUSE

Baeolophus ridgwayi

The bird formerly known as the "Plain Titmouse" was well named: it is drab to the point of being featureless. The new name for the form that occurs in Nevada is similarly apt because Juniper Titmice are almost always found in arid pinyon-juniper woodlands. The Juniper Titmouse is at or near the western limits of its poorly defined range in Nevada, and there are unresolved questions about its vocalizations, behaviors, and natural history—especially for populations in the western and southern parts of the state.

DISTRIBUTION

Juniper Titmice were widespread but spottily distributed everywhere in Nevada except north of a line running from Pyramid Lake to the northeastern corner of the state, where there were only two observations. Incidentally, the Pinyon Jay—another arid woodland specialist in Nevada—showed a very similar distribution pattern. Juniper Titmice were found in about one in ten atlas blocks and were confirmed as breeders in 38% of them. This is a sedentary species, however, and it is likely that titmice were breeding in most of the areas where they were observed, especially given that most occupied sites (88%) were thought to contain multiple pairs.

Earlier authors described the Nevada distribution of Juniper Titmice in rather general terms (Fisher 1893, Linsdale 1936, Ryser 1985, Alcorn 1988). Neel's (1999a) conclusion that the species is found primarily south of Interstate 80 is supported by the atlas data. Within this region, the Juniper Titmouse is known for its

patchy distribution (Cicero 1996, Neel 1999a), which matches the distribution of its favored habitat.

Because the Juniper Titmouse was only recently recognized as distinct from the Oak Titmouse (Cicero 1996, AOU 1998, Cicero 2000), the western limits of its range are not perfectly known. It occurs west into the desert mountains of eastern California (Garrett and Dunn 1981, Cicero 2000), and it is the common titmouse on the eastern slope of the southern Sierra Nevada (Gaines 1992, Cicero 2000). The Juniper Titmouse is locally distributed and uncommon in southern Oregon (Adamus et al. 2001), and it ranges only into extreme southern Idaho (Stephens and Sturts 1998). In Arizona, it is fairly common in appropriate habitat (LaRue in Corman and Wise-Gervais 2005:388–389), and it is a common permanent resident in Utah (Behle et al. 1985).

It comes as little surprise that almost 90% of the atlas block records for the Juniper Titmouse came from pinyon-juniper habitats. One potential exception to this pattern came from the oak groves in the eastern foothills of the Spring Mountains. Here the titmice sound like Oak Titmice, look like Oak Titmice, and can be seen feeding among the oaks (T. Floyd, pers. observ.)—yet the location is well outside the described range of that species. Determining whether these birds are Juniper Titmice or a disjunct population of Oak Titmice will require further study. The predictive map suggests a high chance of finding Juniper Titmice in mid-elevation locations where pinyon-juniper habitat dominates, with decreasing probabilities as one moves either to lower or higher elevations.

CONSERVATION AND MANAGEMENT

Rangewide, Juniper Titmouse populations appear to be stable, although there have been some declines in the eastern part of the range (Sauer et al. 2005). Juniper Titmice tend to spend their time in dense cover (Ryser 1985), and thus favor areas of pinyon-juniper with closed canopies (Neel 1999a, Cicero 2000, Pavlacky and Anderson 2001). Since they are cavity nesters, older trees with snags and heart rot are also important (Neel 1999a, Pavlacky and Anderson 2001). Because of its patchy distribution and strong preference for old-growth woodlands, Nevada Partners in Flight has designated the Juniper Titmouse a Priority species (Neel 1999a).

HABITAT USE

HABITAT	ATLAS BLOCKS	INCIDENTAL OBSERVATIONS
Montane Forest	2 (2%)	1 (3%)
Montane Parkland	0 (0%)	2 (6%)
Montane Shrub	3 (3%)	5 (15%)
Pinyon Juniper	85 (88%)	17 (52%)
Riparian	0 (0%)	6 (18%)
Salt Desert Scrub	1 (1%)	0 (0%)
Sagebrush Scrub	3 (3%)	0 (0%)
Sagebrush Steppe	1 (1%)	1 (3%)
Urban	2 (2%)	1 (3%)
TOTAL	97	33

Total numbers reported in the Habitat Use, Breeding Status, and Abundance tables may differ from each other (see pp. 22–23 for details). Percentage sums may differ slightly from 100% due to rounding.

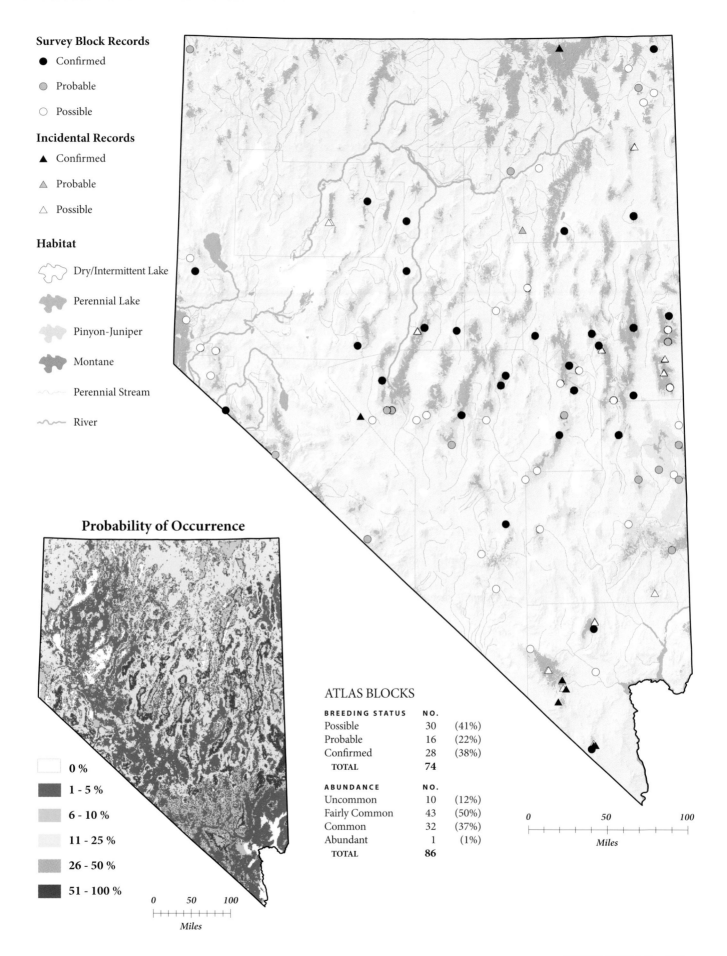

Survey Block Records

- ● Confirmed
- ● Probable
- ○ Possible

Incidental Records

- ▲ Confirmed
- ▲ Probable
- △ Possible

Habitat

- Dry/Intermittent Lake
- Perennial Lake
- Pinyon-Juniper
- Montane
- Perennial Stream
- River

Probability of Occurrence

- 0 %
- 1 - 5 %
- 6 - 10 %
- 11 - 25 %
- 26 - 50 %
- 51 - 100 %

0   50   100
Miles

ATLAS BLOCKS

BREEDING STATUS	NO.	
Possible	30	(41%)
Probable	16	(22%)
Confirmed	28	(38%)
TOTAL	74	

ABUNDANCE	NO.	
Uncommon	10	(12%)
Fairly Common	43	(50%)
Common	32	(37%)
Abundant	1	(1%)
TOTAL	86	

0   50   100
Miles

VERDIN
Auriparus flaviceps

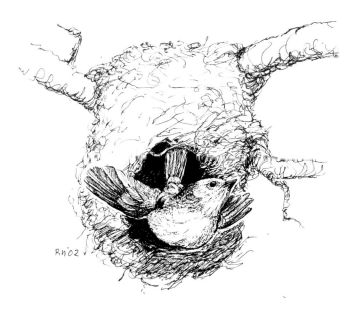

Small it may be, but the yellow-faced Verdin makes a big impression. It is a bundle of energy flitting through the thorny desert shrublands and riparian corridors of southern Nevada. Here, Verdins are conspicuous and easy to confirm as breeders; they call incessantly and build highly visible globular nests, both for roosting and for breeding, that seem too large for such a tiny bird. Verdins, along with Phainopeplas, Gambel's Quail, and Cactus Wrens, are among the characteristic birds of Nevada's Mojave Desert region, and many birders from more northerly climes head south just to add these species to their lists.

DISTRIBUTION
Verdins were widely distributed and often numerous in the lowlands of southern Nevada, where they are year-round residents. Southern Nevada lies at the northern limit of the species' range, which extends south into California and Arizona, and east through the southern U.S. border states and into Mexico (Behle et al. 1985, Small 1994, Webster 1999, Wise-Gervais in Corman and Wise-Gervais 2005:390–391). Historically, Verdins were thought to be restricted to the Virgin and Colorado river valleys in Nevada (Linsdale 1936, 1951), but Guillon et al. (1959) recognized that they are more widespread.

Verdins favor brushy, thorny vegetation, and often use tamarisk, creosotebush, scrub oaks, and other similar scrubby plants (Fisher 1893, Gullion et al. 1959, Garrett and Dunn 1981, Webster 1999). It is not surprising, then, that Verdin records came primarily from the lowland habitat types of the Mojave Desert: Mojave, riparian, and mesquite. They used mesquite and ash habitats disproportionately often, but—as the predictive map indicates—there is a very good chance of finding Verdins throughout the Mojave Desert lowlands. They also occur in ur-

ban centers, but as the predictive map indicates, the chance of finding them in this habitat is distinctly lower than it is in areas that have native habitats. Occurrence declines as the desert becomes more barren, and the species is also essentially absent from thick forests and high altitudes (Webster 1999).

CONSERVATION AND MANAGEMENT
The Verdin is often cited as a bird that is well suited for life in the desert (e.g., Austin 1976), where it tailors its microhabitat use to weather the extreme temperatures (Wolf and Walsberg 1996). Its habitat flexibility shines forth, however, in an anecdote provided by one atlas worker, who observed a Verdin in downtown Las Vegas drinking from a box of fruit juice and eating a discarded piece of chicken. Despite this apparent adaptability, trend data suggest that Verdin populations are declining at a substantial rate rangewide, and especially in the heart of their U.S. range in Arizona's Sonoran Desert (Sauer et al. 2005), although Wise-Gervais (in Corman and Wise-Gervais 2005:390–391) stated that these populations are stable.

The causes of the reported declines are not well understood (see Webster 1999). The lower chance of finding this species in urban habitat, however, points to urbanization of the desert as at least a factor that warrants future investigation. Partners in Flight recognized this by designating the Verdin a Stewardship species for the southwestern region (Rich et al. 2004). Future research may clarify the Verdin's status and population trends, and its ecological requirements in Nevada.

HABITAT USE

HABITAT	ATLAS BLOCKS	INCIDENTAL OBSERVATIONS
Agricultural	0 (0%)	1 (2%)
Ash	5 (6%)	0 (0%)
Mesquite	22 (26%)	5 (8%)
Mojave	33 (39%)	15 (23%)
Montane Parkland	0 (0%)	1 (2%)
Riparian	21 (25%)	23 (36%)
Salt Desert Scrub	0 (0%)	2 (3%)
Urban	2 (2%)	16 (25%)
Wetland	1 (1%)	1 (2%)
TOTAL	84	64

Total numbers reported in the Habitat Use, Breeding Status, and Abundance tables may differ from each other (see pp. 22–23 for details). Percentage sums may differ slightly from 100% due to rounding.

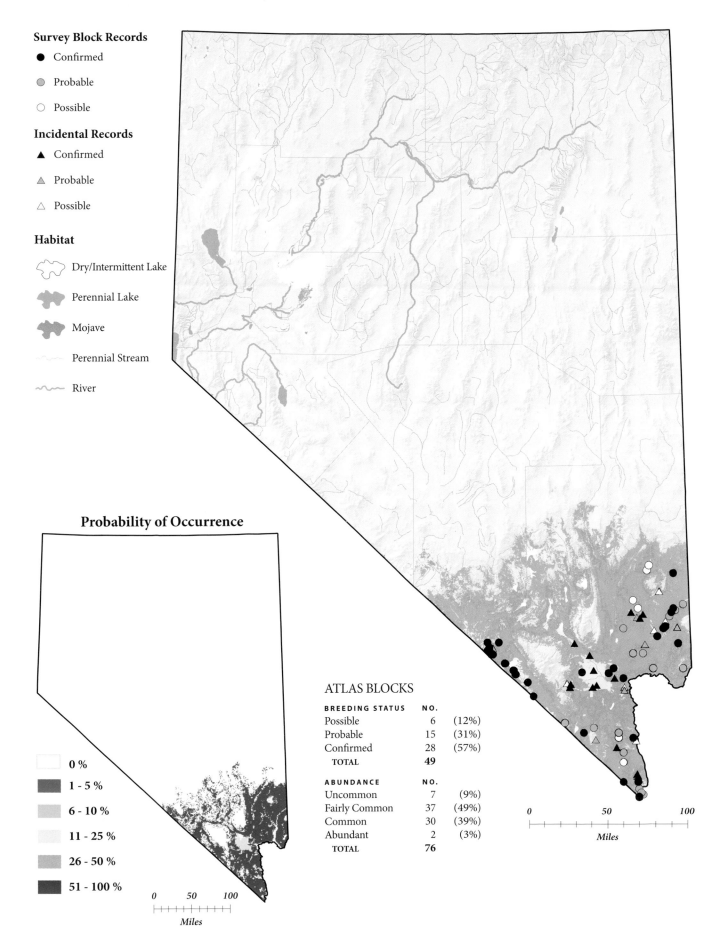

Survey Block Records

● Confirmed

◐ Probable

○ Possible

Incidental Records

▲ Confirmed

△ Probable

△ Possible

Habitat

⬡ Dry/Intermittent Lake

⬢ Perennial Lake

⬢ Mojave

〜 Perennial Stream

〜 River

Probability of Occurrence

☐ 0 %

▨ 1 - 5 %

▨ 6 - 10 %

▨ 11 - 25 %

▨ 26 - 50 %

▨ 51 - 100 %

0   50   100
Miles

ATLAS BLOCKS

BREEDING STATUS	NO.	
Possible	6	(12%)
Probable	15	(31%)
Confirmed	28	(57%)
TOTAL	49	

ABUNDANCE	NO.	
Uncommon	7	(9%)
Fairly Common	37	(49%)
Common	30	(39%)
Abundant	2	(3%)
TOTAL	76	

0   50   100
Miles

BUSHTIT
Psaltriparus minimus

Bushtits are best known during most of the year as roiling flocks of twittering little fluff-balls in dense streamside or foothills vegetation. During summer, they take on a rather different demeanor, as the roving bands break into pairs and tend discreetly to the matters of courtship and nest maintenance. And what a nest! As if to make up for their paltry size and drab plumage, Bushtits build huge hanging nests that are elaborately woven together with spider webs, mosses, and various plant parts.

DISTRIBUTION

Atlas workers found Bushtits throughout the state, although less frequently in the northwest. Breeding was confirmed in a relatively high (48%) proportion of blocks, probably owing to the ease of identifying active nests. Bushtits were also numerous: nearly half the occupied blocks were estimated to have more than ten pairs.

Earlier writers concurred that the Bushtit's Nevada range is extensive, and that it is associated in the summer with streamside thickets and dense upland vegetation in the lower elevations of our mountain ranges (Linsdale 1936, Ryser 1985, Alcorn 1988). The Bushtit tends to be absent where these habitats are absent or limited—for example, the species is not currently known to breed in the well-studied Lahontan Valley (Chisholm and Neel 2002).

In the surrounding states, the Bushtit's status is variable. It is a common resident in much of California (Grinnell and Miller 1944, Small 1994), but most of that state's Bushtits belong to the distinctive coastal race; interior Bushtits of the sort found in Nevada occur along the eastern slope of the Sierra Nevada (Gaines 1992). Bushtits are locally distributed in eastern Oregon and are restricted to the southernmost portions of Idaho (Stephens and Sturts 1998, Contreras 1999, Adamus et al. 2001). They are rather more common in both Utah and Arizona (Behle et al. 1985, Wise-Gervais in Corman and Wise-Gervais 2005:392–393).

Regardless of the season, Bushtits are closely tied to dense shrubbery (Sloane 2001). Over half of the atlas block records came from pinyon-juniper woodlands, with riparian and montane shrub habitats also being important. Degree of aridity does not seem to matter to Bushtits (Gaines 1992), which can be found in both lush riparian settings and dry scrublands in the hills. Although they are typically found in mountain foothills, they occasionally nest in the desert valleys (e.g., Monson and Phillips 1981) if suitable vegetative cover is available. The predictive map clearly reflects the species' general orientation toward foothills habitats, but the potential for finding this species in lower-elevation shrubland habitats is also evident. Although Bushtits have been characterized as readily adapting to suburban edge environments (Sloane 2001), atlas workers did not observe much evidence for the species' use of urban habitat (which would include suburbia).

CONSERVATION AND MANAGEMENT

Data from the Breeding Bird Survey show no clear rangewide trends for Bushtit populations (Sauer et al. 2005), which are difficult to assess because of the species' habit of wandering extensively in flocks (Sloane 2001). The situation in Nevada is poorly characterized for the same reason, and also because of insufficient survey coverage in the past. It is thought that Bushtits may once have bred in the Lahontan Valley, though they no longer do. Loss of riparian vegetation and buffaloberry is the probable cause of their absence in this region (Chisholm and Neel 2002). Bushtits still remain relatively common and widespread in Nevada, but additional monitoring is clearly warranted to quantify the population's status and trends.

HABITAT USE

HABITAT	ATLAS BLOCKS	INCIDENTAL OBSERVATIONS
Agricultural	0 (0%)	1 (1%)
Alpine	1 (<1%)	0 (0%)
Barren	1 (<1%)	0 (0%)
Mojave	2 (<1%)	1 (1%)
Montane Forest	17 (7%)	10 (14%)
Montane Parkland	6 (3%)	0 (0%)
Montane Shrub	24 (10%)	7 (10%)
Pinyon-Juniper	121 (53%)	23 (33%)
Riparian	43 (19%)	22 (31%)
Sagebrush Scrub	8 (3%)	3 (4%)
Sagebrush Steppe	7 (3%)	2 (3%)
Urban	0 (0%)	1 (1%)
TOTAL	230	70

Total numbers reported in the Habitat Use, Breeding Status, and Abundance tables may differ from each other (see pp. 22–23 for details). Percentage sums may differ slightly from 100% due to rounding.

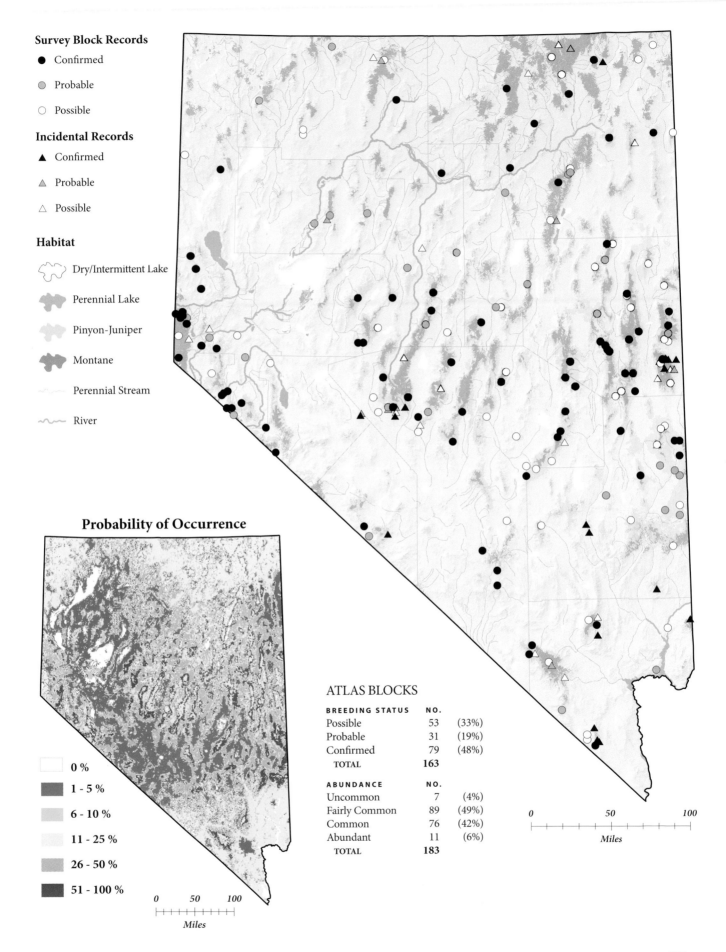

Survey Block Records

● Confirmed

● Probable

○ Possible

Incidental Records

▲ Confirmed

△ Probable

△ Possible

Habitat

Dry/Intermittent Lake

Perennial Lake

Pinyon-Juniper

Montane

Perennial Stream

River

Probability of Occurrence

☐	0 %
■	1 - 5 %
■	6 - 10 %
■	11 - 25 %
■	26 - 50 %
■	51 - 100 %

0   50   100

Miles

ATLAS BLOCKS

BREEDING STATUS	NO.	
Possible	53	(33%)
Probable	31	(19%)
Confirmed	79	(48%)
TOTAL	**163**	

ABUNDANCE	NO.	
Uncommon	7	(4%)
Fairly Common	89	(49%)
Common	76	(42%)
Abundant	11	(6%)
TOTAL	**183**	

0   50   100

Miles

RED-BREASTED NUTHATCH

Sitta canadensis

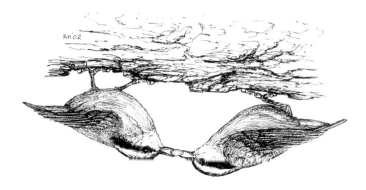

It is hard not to assign human emotions to the vocalizations of the Red-breasted Nuthatch, as its characteristic nasal call notes have an undeniably petulant quality. Red-breasted Nuthatches are the consummate whiners of the dense pine and fir forests of Nevada's taller mountain ranges. Sometimes they gripe with good cause, as when a hiker wanders too close or when a foraging individual accidentally stumbles upon a roosting owl. At other times, they grumble for no apparent reason at all. It is not uncommon for the stillness of a summer afternoon to be broken by the random outburst of a Red-breasted Nuthatch grousing about some imagined insult.

DISTRIBUTION

Atlas workers confirmed breeding by Red-breasted Nuthatches in four widely spaced groups of mountain ranges: the Carson Range in far western Nevada; the Egan, Schell Creek, and Snake ranges in the east; the Jarbidge Range in the north; and the Spring Mountains in the south. Records from other mountain ranges, such as the White Pine and Santa Rosa ranges, probably indicate breeding populations, although nesting was not actually confirmed there. Curiously, the species was largely absent from the central Nevada mountain ranges. Multiple pairs were estimated to occur in most occupied blocks, but abundance was rarely thought to exceed ten pairs in a block.

The atlas results for this species provide a useful addition to earlier reports of its breeding range in Nevada (Fisher 1893, Linsdale 1936, Alcorn 1988). Linsdale (1936) reported breeding only in northern Nevada, and Ghalambor and Martin (1999) considered the possibility of breeding in central Nevada to be uncertain.

While the Red-breasted Nuthatch is somewhat scarce and local in Nevada, it appears to be fairly widespread and numerous in the surrounding states, including California (Small 1994), Utah (Behle et al. 1985), Idaho (Stephens and Sturts 1998), and in Arizona's mountain ranges (Spence in Corman and Wise-Gervais 2005:394–395). In Oregon, the species is common in the north and west, but not in the southeastern shrub steppe bordering Nevada (Adamus et al. 2001).

In Nevada and elsewhere, the Red-breasted Nuthatch is most frequently encountered in boreal coniferous forests, and the vast majority of atlas records came from the montane forest habitat type. The predictive map illustrates this pattern, predicting the greatest chance of finding the species in the high-elevation conifer zone and a decreasing chance of finding it as one moves downslope. The prediction that this nuthatch has a moderate chance of occurring in mountain ranges in central Nevada seems not to be borne out by the field data. Given the long-distance irruptive movements that Red-breasted Nuthatches make between mountain ranges during the nonbreeding seasons, this lack of records is rather surprising.

CONSERVATION AND MANAGEMENT

Red-breasted Nuthatch populations can vary greatly from year to year and may be cyclic or irruptive in some areas (Shuford 1993, Ghalambor and Martin 1999). Rangewide, numbers are increasing, although this increase is most evident through southern Canada and in the Rocky Mountain region (Sauer et al. 2005). Red-breasted Nuthatches are obligate cavity nesters that require dead trees in advanced stages of decay (Ghalambor and Martin 1999). An additional factor for population health is the summer food supply available in the spruce, fir, and pine forests on which they depend (Littlefield 1990). Consequently, the protection of mature forests, managed so that deadwood remains standing, is likely to be important for this species.

HABITAT USE

HABITAT	ATLAS BLOCKS	INCIDENTAL OBSERVATIONS
Alpine	2 (2%)	0 (0%)
Montane Forest	69 (80%)	6 (55%)
Montane Parkland	2 (2%)	0 (0%)
Montane Shrub	2 (2%)	1 (9%)
Pinyon-Juniper	4 (4%)	1 (9%)
Riparian	7 (8%)	3 (27%)
TOTAL	86	11

Total numbers reported in the Habitat Use, Breeding Status, and Abundance tables may differ from each other (see pp. 22–23 for details). Percentage sums may differ slightly from 100% due to rounding.

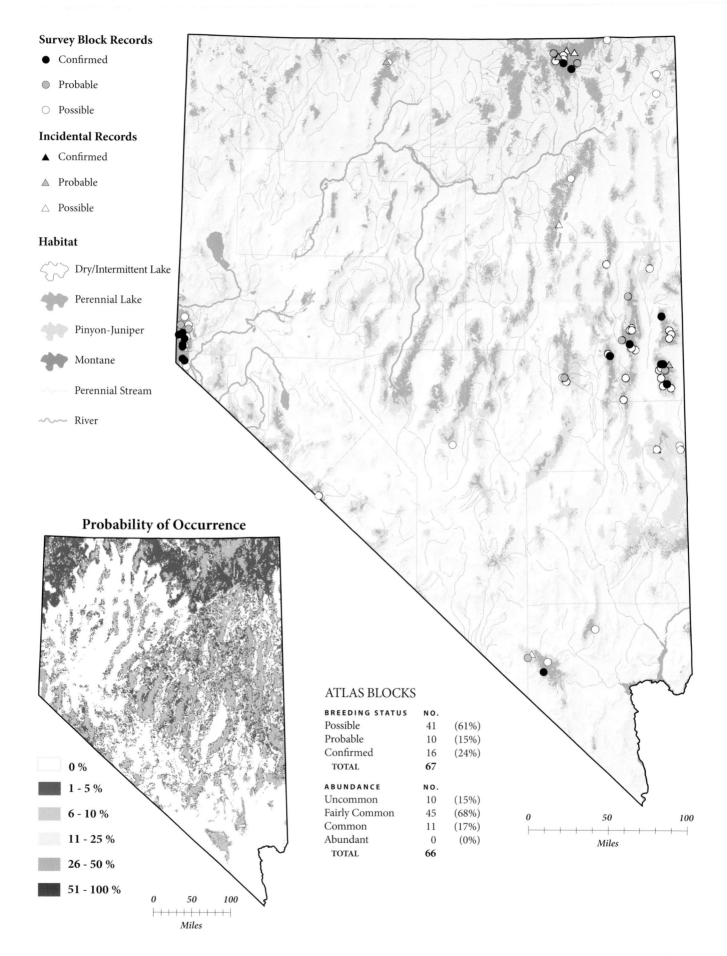

Survey Block Records

● Confirmed

● Probable

○ Possible

Incidental Records

▲ Confirmed

▲ Probable

△ Possible

Habitat

Dry/Intermittent Lake

Perennial Lake

Pinyon-Juniper

Montane

Perennial Stream

River

Probability of Occurrence

0 %

1 - 5 %

6 - 10 %

11 - 25 %

26 - 50 %

51 - 100 %

0   50   100

Miles

ATLAS BLOCKS

BREEDING STATUS	NO.	
Possible	41	(61%)
Probable	10	(15%)
Confirmed	16	(24%)
TOTAL	67	
ABUNDANCE	NO.	
Uncommon	10	(15%)
Fairly Common	45	(68%)
Common	11	(17%)
Abundant	0	(0%)
TOTAL	66	

0   50   100

Miles

WHITE-BREASTED NUTHATCH

Sitta carolinensis

There is no such thing as an idle White-breasted Nuthatch: individuals, pairs, or family groups are forever exploring the nooks and crannies of some aging tree, chattering amiably all the while. These likable and energetic birds are fairly common inhabitants of montane and pinyon-juniper forests throughout much of Nevada. But even though White-breasted Nuthatches are familiar and easily observed, questions remain about the full extent of their Nevada range, especially in the northern part of the state.

DISTRIBUTION

White-breasted Nuthatch records showed a distribution similar to those of the Red-breasted Nuthatch, with the majority falling in either the mountains of far western and far eastern Nevada or the Spring Mountains and Sheep Range of southern Nevada. The records were not as concentrated as those of the Red-breasted Nuthatch, however, and there were additional records of confirmed or suspected breeding in several central Nevada mountain ranges. There was no confirmed breeding north of latitude 40°N, although eastern Elko County produced one record of probable breeding. Over three-fourths of the occupied blocks were estimated to have multiple pairs, and 18% had more than ten pairs.

Linsdale (1936) listed two subspecies for the state: *tenuissima* in the western and central mountains and *nelsoni* in the eastern part of the state, but did not mention which subspecies occurs in Clark County. Alcorn (1988) described the species as uncommon and implied that although it was found throughout the state, it did not occur in all mountain ranges. Similarly, Ryser

(1985) reported that breeding was not known for several ranges in the northern part of the state. In California (Small 1994) and Arizona (Spence in Corman and Wise-Gervais 2005:396–397), the White-breasted Nuthatch is fairly common and broadly distributed in appropriate habitats. It has a patchier distribution in Utah (Behle et al. 1985), Idaho (Stephens and Sturts 1998), and Oregon (Adamus et al. 2001), and in all three states tends to be absent or rare in the regions that directly adjoin Nevada.

During the breeding season, White-breasted Nuthatches can be found in Nevada's montane forests, pinyon-juniper woodlands, and riparian groves (Ryser 1985). For the most part, they are birds of open woods or forest edges, and they avoid the deep forests in which Red-breasted Nuthatches can be found (Pravosudov and Grubb 1993). Nonetheless, they do not generally breed in the valleys of Nevada, even when wooded areas are present (e.g., see Chisholm and Neel 2002). The predictive map reflects this strong orientation toward montane habitats, and it accords well with the observed distribution of records. Determining whether the species actually occupies all interior ranges, where a high chance of occurrence is predicted, would be a useful topic for future investigation.

CONSERVATION AND MANAGEMENT

The White-breasted Nuthatch, like the Red-breasted, has shown sustained population increases throughout most of its range during the past few decades (Sauer et al. 2005). Little information exists for the Great Basin specifically, however, and regional declines have been noted in some parts of the West, notably in Washington State (Smith et al. 1997).

White-breasted Nuthatches are obligate cavity nesters, and they require large trees with rugose bark for foraging. Consequently, forest management practices that maintain mature forests with standing deadwood are important for this species. A topic of future research may also include the differences in voice, morphology, and habits that exist among the White-breasted Nuthatches from the Pacific, interior, and eastern subspecies groups (AOU 1998). The taxonomy and exact range limits of these groups, two of which occur in Nevada, are not well known, and they may deserve to be treated as distinct taxa.

HABITAT USE

HABITAT	ATLAS BLOCKS	INCIDENTAL OBSERVATIONS
Montane Forest	57 (56%)	19 (70%)
Montane Parkland	9 (9%)	0 (0%)
Montane Shrub	2 (2%)	1 (4%)
Pinyon-Juniper	24 (24%)	4 (15%)
Riparian	5 (5%)	3 (11%)
Sagebrush Scrub	1 (1%)	0 (0%)
Sagebrush Steppe	1 (1%)	0 (0%)
Urban	2 (2%)	0 (0%)
TOTAL	101	27

Total numbers reported in the Habitat Use, Breeding Status, and Abundance tables may differ from each other (see pp. 22–23 for details). Percentage sums may differ slightly from 100% due to rounding.

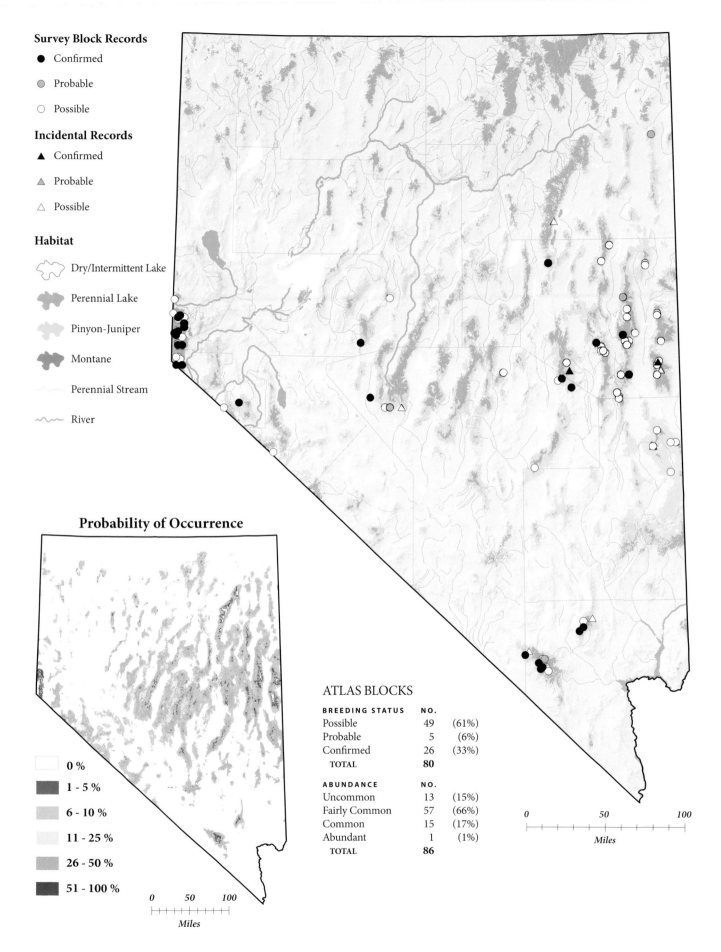

Survey Block Records

● Confirmed

● Probable

○ Possible

Incidental Records

▲ Confirmed

▲ Probable

△ Possible

Habitat

Dry/Intermittent Lake

Perennial Lake

Pinyon-Juniper

Montane

Perennial Stream

River

Probability of Occurrence

	0 %
	1 - 5 %
	6 - 10 %
	11 - 25 %
	26 - 50 %
	51 - 100 %

0   50   100

Miles

ATLAS BLOCKS

BREEDING STATUS	NO.	
Possible	49	(61%)
Probable	5	(6%)
Confirmed	26	(33%)
TOTAL	**80**	

ABUNDANCE	NO.	
Uncommon	13	(15%)
Fairly Common	57	(66%)
Common	15	(17%)
Abundant	1	(1%)
TOTAL	**86**	

0   50   100

Miles

PYGMY NUTHATCH

Sitta pygmaea

The Pygmy Nuthatch is a good example of a habitat specialist; its confirmed breeding range in Nevada is restricted to three mountain ranges that contain extensive, tall ponderosa pine forests: the Carson Range of the far west, and the Sheep Range and Spring Mountains of the south. The species breeds cooperatively, with young males frequently helping relatives to raise their young (Kingery and Ghalambor 2001). The Pygmy Nuthatch was fairly common in areas where appropriate habitat was available, and its noisy and sociable nature made it an easy species to detect.

DISTRIBUTION

The Pygmy Nuthatch has a feast-or-famine distribution in Nevada. It was found in only a small number of atlas blocks, but in relatively large numbers. Abundance estimates exceeded ten pairs in nearly one-third of the occupied blocks, and single pairs were relatively unusual. Additionally, the species had an exceptionally high breeding confirmation rate of 74%. Previous accounts of the Pygmy Nuthatch in Nevada suggested the presence of additional breeding populations in central Nevada (Alcorn 1988) and in the Snake Range of eastern Nevada (Linsdale 1936, Ryser 1985), but atlas workers found no evidence of breeding in either area.

Distribution patterns in Nevada largely match patterns known from neighboring states (Kingery and Ghalambor 2001). Specifically, in Arizona, Idaho, Utah, and Oregon, Pygmy Nuthatches occur in most places where substantial ponderosa pine forests exist (Behle et al. 1985, Stephens and Sturts 1998, Adamus et al. 2001, Spence in Corman and Wise-Gervais 2005:398–399). In

California, the species is mostly restricted to coniferous mountain areas and the central coastal region (Kingery and Ghalambor 2001).

The Pygmy Nuthatch occurs primarily in forests dominated by long-needled conifers (Kingery and Ghalambor 2001). In Nevada, most atlas records came from montane forest or montane parkland habitats that fell within the range of ponderosa pine (cf. Charlet 1996). The species is also found occasionally in pinyon pine stands, but mostly in southern Nevada, where the *canescens* subspecies occurs (Gullion et al. 1959). Because the Pygmy Nuthatch is a habitat specialist, predictions based on broad habitat classifications can be misleading in two ways. First, the model probably overpredicts the likelihood of occurrence in montane forests throughout the interior of the state. Less obviously, the likelihood of occurrence may be underestimated in areas where the preferred forest types are concentrated. In this case, a species-specific model based on the Nevada distribution of long-needled pines would probably provide more accurate predictions.

CONSERVATION AND MANAGEMENT

Where they occur, Pygmy Nuthatches are usually common and conspicuous, and therefore arouse little conservation concern. As habitat specialists, though, Pygmy Nuthatches are potentially vulnerable to outside forces detrimental to the forest types that they use. This risk is amplified since they require mature forest with dead trees for nesting and large cone-producing trees for foraging (Kingery and Ghalambor 2001)—the very sorts of trees that usually become rare in heavily logged regions (Ewell and Cruz 1998). There is no evidence that this threat is currently a problem in Nevada, however.

These nuthatches have interesting life history traits that make them excellent study subjects. For example, they engage in communal roosting and in cooperative breeding (Knorr 1957, Kingery and Ghalambor 2001), and they accept nest boxes (Bock and Fleck 1995). Birders here should be on the lookout for additional populations of Pygmy Nuthatches wherever there are significant stands of long-needled pine trees.

HABITAT USE

HABITAT	ATLAS BLOCKS	INCIDENTAL OBSERVATIONS
Montane Forest	23 (77%)	20 (91%)
Montane Parkland	5 (17%)	0 (0%)
Pinyon-Juniper	2 (7%)	1 (5%)
Riparian	0 (0%)	1 (5%)
TOTAL	30	22

Total numbers reported in the Habitat Use, Breeding Status, and Abundance tables may differ from each other (see pp. 22–23 for details). Percentage sums may differ slightly from 100% due to rounding.

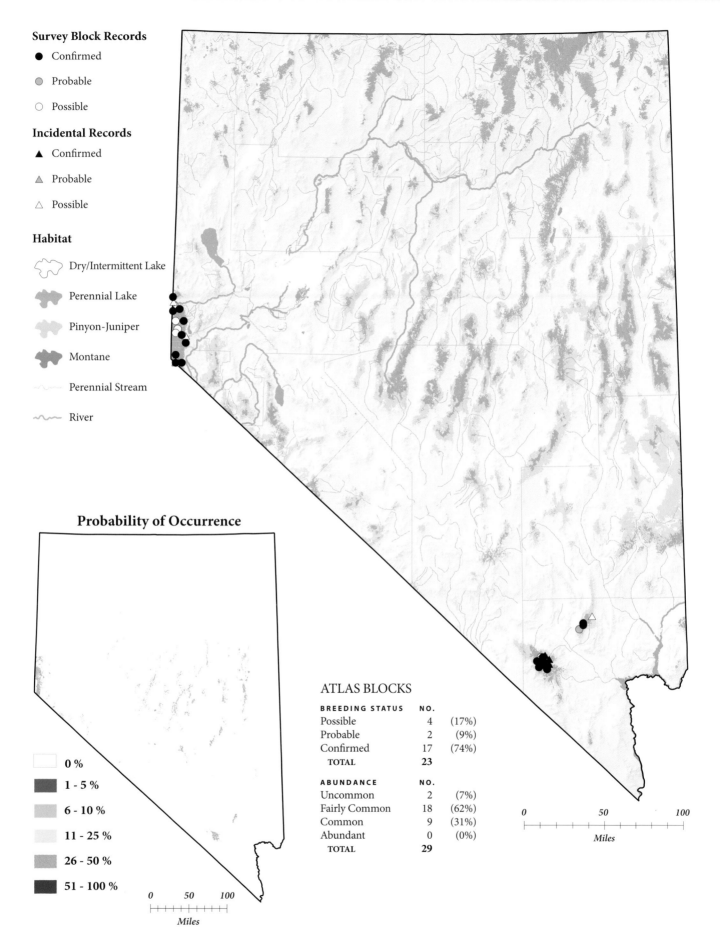

Survey Block Records

● Confirmed

◐ Probable

○ Possible

Incidental Records

▲ Confirmed

△ Probable

△ Possible

Habitat

Dry/Intermittent Lake

Perennial Lake

Pinyon-Juniper

Montane

Perennial Stream

River

Probability of Occurrence

☐	**0 %**
■	**1 - 5 %**
■	**6 - 10 %**
☐	**11 - 25 %**
■	**26 - 50 %**
■	**51 - 100 %**

0   50   100

Miles

ATLAS BLOCKS

BREEDING STATUS	NO.	
Possible	4	(17%)
Probable	2	(9%)
Confirmed	17	(74%)
TOTAL	**23**	

ABUNDANCE	NO.	
Uncommon	2	(7%)
Fairly Common	18	(62%)
Common	9	(31%)
Abundant	0	(0%)
TOTAL	**29**	

0   50   100

Miles

BROWN CREEPER
Certhia americana

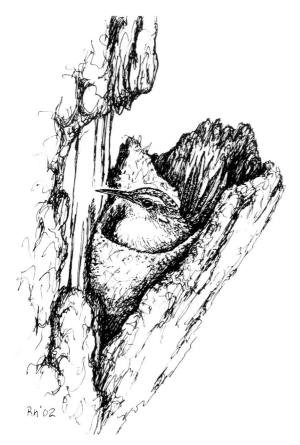

Rh'02

The Brown Creeper is part bird, part apparition. Its high-pitched calls are barely audible and easily missed. It is a master of camouflage and is hardly ever seen on a substrate other than tree bark—where it is almost invisible. Its delicate nest of mosses and spider webs is placed inconspicuously behind a flake of bark. Brown Creepers are fairly common in shady montane forests, however, and it is not unusual to catch a fleeting glimpse of one creeping upward around the main trunk and then fluttering weakly away into the silent pines.

DISTRIBUTION

Atlas records for the Brown Creeper were distributed unevenly throughout most, but not all, of Nevada's major mountain ranges. The densest clusters occurred in the Carson Range, the Spring Mountains, and the mountains of eastern White Pine County. Elsewhere, Brown Creepers were confirmed or suspected breeders in several ranges in central and northeastern Nevada. Breeding was confirmed in almost half of the blocks in which the species was found, and observers usually estimated multiple pairs to be present in occupied blocks.

The atlas findings are largely consistent with other accounts of the Brown Creeper's distribution in Nevada (Linsdale 1936, Ryser 1985, Alcorn 1988, Hejl et al. 2002b). Ryser (1985) reported that the species had not yet been confirmed as a breeder

in the mountains of central Nevada, but the atlas results agree with Alcorn (1988) and Hejl et al. (2002) in suggesting that Brown Creepers do breed in at least some of our interior ranges.

In all of the surrounding states, the Brown Creeper is typically a fairly common bird with a local distribution that closely matches that of its mature, well-shaded, closed-canopy forest habitat (Behle et al. 1985, Small 1994, Stephens and Sturts 1998, Adamus et al. 2001, Friederici in Corman and Wise-Gervais 2005:400–401). In terms of habitat use, tree structure is probably more important than species composition, and trees with large trunks and limbs appear to be favored (Shuford 1993). In the West, creepers are most often associated with coniferous forests, and in Nevada they are largely limited to high-elevation coniferous forests. The predictive map mirrors the observed distribution fairly closely, but it also points to mountain ranges in central and northern Nevada where ambitious birders might profitably seek to extend the species' known breeding distribution. In these areas, however, the model implies only a moderately good chance of finding Brown Creepers.

CONSERVATION AND MANAGEMENT

Overall, Brown Creeper numbers are thought to be stable in North America. Little information is available from Nevada, however, and there is evidence of recent declines in California (Sauer et al. 2005). Brown Creepers are typically found in mature forest stands, and they require large trees that provide good foraging habitat and suitable crevices behind the bark for their nests (Hejl et al. 2002b). Consequently, forestry practices that eliminate large trees and snags are likely to be harmful.

Atlas data indicate that Brown Creepers may be absent from several mountain ranges in central and northern Nevada. In some cases, this may simply be because the birds were difficult to detect (Gaines 1992, Hejl et al. 2002b). On the other hand, patches of suitable habitat in these ranges may not be large enough to support viable populations. Future research that better documents the Brown Creeper's distribution in these areas of Nevada and determines its minimum habitat requirements would be useful.

HABITAT USE

HABITAT	ATLAS BLOCKS	INCIDENTAL OBSERVATIONS
Alpine	1 (1%)	0 (0%)
Montane Forest	59 (80%)	17 (89%)
Montane Parkland	5 (7%)	1 (5%)
Pinyon-Juniper	2 (3%)	0 (0%)
Riparian	5 (7%)	1 (5%)
Urban	2 (3%)	0 (0%)
TOTAL	74	19

Total numbers reported in the Habitat Use, Breeding Status, and Abundance tables may differ from each other (see pp. 22–23 for details). Percentage sums may differ slightly from 100% due to rounding.

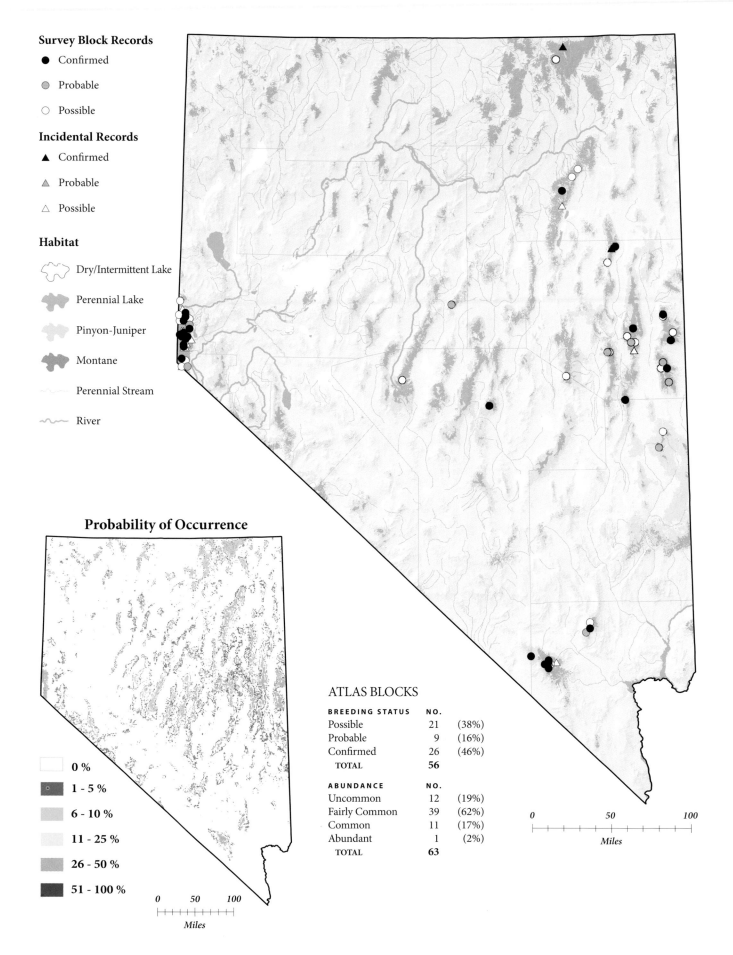

Survey Block Records

● Confirmed

● Probable

○ Possible

Incidental Records

▲ Confirmed

▲ Probable

△ Possible

Habitat

Dry/Intermittent Lake

Perennial Lake

Pinyon-Juniper

Montane

Perennial Stream

River

Probability of Occurrence

☐ 0 %

▨ 1 - 5 %

▨ 6 - 10 %

▨ 11 - 25 %

▨ 26 - 50 %

▨ 51 - 100 %

0   50   100
Miles

ATLAS BLOCKS

BREEDING STATUS	NO.	
Possible	21	(38%)
Probable	9	(16%)
Confirmed	26	(46%)
TOTAL	56	

ABUNDANCE	NO.	
Uncommon	12	(19%)
Fairly Common	39	(62%)
Common	11	(17%)
Abundant	1	(2%)
TOTAL	63	

0   50   100
Miles

CACTUS WREN
Campylorhynchus brunneicapillus

The Cactus Wren is a characteristic inhabitant of the warm deserts of southern Nevada. Its odd song can be heard at almost any time of the year, and at most times of the day and night. Even in the dead of summer Cactus Wrens are surprisingly active, and they were one of the strongest incentives for atlas workers to visit otherwise "lethargic" Mojave Desert blocks in Clark County. Given the Cactus Wren's conspicuousness, it was a surprising reward for atlas workers to come upon a previously undetected population of this species more than 60 miles (100 kilometers) north of its previously recognized range limit.

DISTRIBUTION

Cactus Wrens were present in thirty-five blocks located throughout southern Nevada. They were typically fairly numerous, with at least ten pairs estimated in more than a third of the occupied blocks. Breeding was confirmed in more than half of these blocks. The atlas results are largely consistent with earlier descriptions of the Cactus Wren's distribution in Nevada. By the 1930s, the species had been recorded in many locations in Clark County (Linsdale 1936, Van Rossem 1936), and Linsdale (1936) and Alcorn (1988) characterized it as a permanent resident of southern Nevada. One of the most interesting atlas records for this species came from the northern part of the Nellis Air Force Base near Tonopah, considerably north of the species' expected range. This area was especially notable because other Mojave Desert specialists—including the Ladder-backed Woodpecker—were also found there.

Like many other southwestern birds, Cactus Wrens reach the northern extent of their range in Nevada (Proudfoot et al. 2000). This range extends through extreme southwestern Utah, southeastern California, southern and western Arizona, and beyond into southern New Mexico, western Texas, and Mexico (Behle et al. 1985, Rosenberg et al. 1991, Small 1994, Proudfoot et al. 2000, Wise-Gervais in Corman and Wise-Gervais 2005:402–403).

Cactus Wrens can survive without access to water, and thus can inhabit some of our most arid scrub communities (Proudfoot et al. 2000). They often build their nests among thorns—for example, in cholla cactus, yucca, or desert riparian shrubs (Austin and Bradley 1971, Small 1994, Proudfoot et al. 2000). The atlas data match these patterns of habitat use, with the majority of Cactus Wrens reported from the expansive Mojave scrublands of southern Nevada. The high predicted probability of finding Cactus Wrens in most of the Mojave region reflects their ubiquity and their affinity for southern Nevada's most extensive habitat type. Yucca woodlands also extend somewhat into the southwestern portion of the Great Basin, and the map predicts a low chance of finding the wrens there. Notably, their occurrence is predicted to be much less likely in urban than in nonurban areas of southern Nevada.

CONSERVATION AND MANAGEMENT

Cactus Wren populations have been declining in the Mojave Desert and throughout the species' U.S. range (Sauer et al. 2005). Habitat loss to urbanization, agriculture, and fire are cited as likely causes for some declines (Proudfoot et al. 2000), but how these factors apply to Nevada populations is still unclear. Because it has a limited range, the Cactus Wren is a Partners in Flight Stewardship species (Rich et al. 2004). It is also an Evaluation species of the Clark County Multiple Species Habitat Conservation Plan (Clark County 2000), with the goal of gaining a better understanding of the threats and impacts associated with rapidly expanding urban areas. Additional research on the Cactus Wren's habitat requirements and the effects of urbanization, habitat fragmentation, and increased recreational activities may be of particular importance in southern Nevada.

HABITAT USE

HABITAT	ATLAS BLOCKS	INCIDENTAL OBSERVATIONS
Grassland	4 (10%)	0 (0%)
Mesquite	0 (0%)	1 (3%)
Mojave	34 (81%)	27 (73%)
Montane Parkland	0 (0%)	1 (3%)
Montane Shrub	1 (2%)	2 (5%)
Pinyon-Juniper	0 (0%)	1 (3%)
Riparian	0 (0%)	2 (5%)
Salt Desert Scrub	1 (2%)	1 (3%)
Sagebrush Scrub	1 (2%)	2 (5%)
Urban	1 (2%)	0 (0%)
TOTAL	42	37

Total numbers reported in the Habitat Use, Breeding Status, and Abundance tables may differ from each other (see pp. 22–23 for details). Percentage sums may differ slightly from 100% due to rounding.

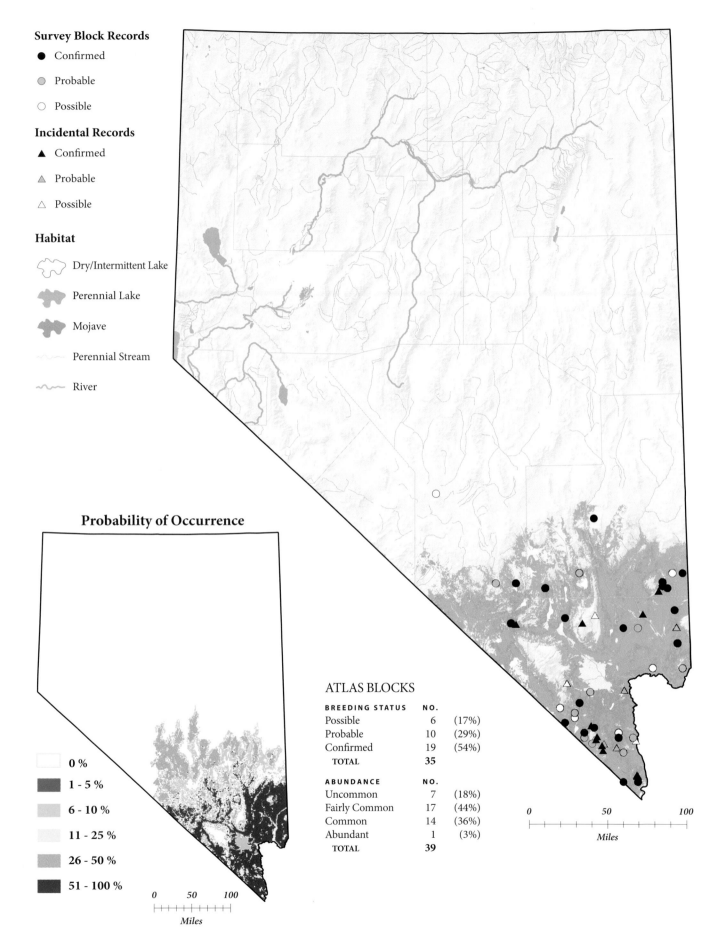

Survey Block Records

● Confirmed

◐ Probable

○ Possible

Incidental Records

▲ Confirmed

△ Probable

△ Possible

Habitat

Dry/Intermittent Lake

Perennial Lake

Mojave

Perennial Stream

River

Probability of Occurrence

0 %

1 - 5 %

6 - 10 %

11 - 25 %

26 - 50 %

51 - 100 %

0   50   100
Miles

ATLAS BLOCKS

BREEDING STATUS	NO.	
Possible	6	(17%)
Probable	10	(29%)
Confirmed	19	(54%)
TOTAL	**35**	

ABUNDANCE	NO.	
Uncommon	7	(18%)
Fairly Common	17	(44%)
Common	14	(36%)
Abundant	1	(3%)
TOTAL	**39**	

0   50   100
Miles

ROCK WREN
Salpinctes obsoletus

There seems to be no place in Nevada too arid or too barren for Rock Wrens. Among the most widespread birds in the state, Rock Wrens can be found year-round in rocky areas across the lowlands; in summer, they breed all the way up to the scree fields on the tops of the highest mountains. Usually, one hears this cryptic bird long before seeing it. The Rock Wren's song is endlessly varied and complex, but always has a telltale repetitious quality. Eventually its incessant droning will give it away, and the persistent observer may be treated to the sight of the parent birds at the nest—the approach to which they often mark with a peculiar "paved" entrance of pebbles (Lowther et al. 2000).

DISTRIBUTION

Rock Wrens proved to be widespread and numerous in Nevada. They were reported from more than three hundred blocks, and multiple pairs were detected in almost 90% of those blocks. More than one hundred pairs were estimated in seventeen of the blocks for which abundance estimates were provided. Breeding was confirmed in about one-third of the blocks where the species occurred. These results fit with previous accounts, which describe the species as a statewide breeder and a year-round resident in the south (Linsdale 1936, Alcorn 1988). Ryser (1985) considered the Rock Wren the most abundant and widespread wren in the Great Basin. During the atlas project, Rock Wrens were reported from almost all atlas habitat types.

The Rock Wren's versatility in nesting is well known: in addition to rocks, it uses any feature with crevices and recesses, including log and brush piles and burrows (Ryser 1985). The predictive map also indicates the Rock Wren's ecological flexibility by predicting a moderate to high chance of finding this species in most of the state. The Rock Wren can be locally uncommon, though, as has been reported for the Lahontan Valley (Chisholm

and Neel 2002) and as is predicted for open playas, wetlands, and urban areas.

Rock Wrens are largely summer residents in the states north of Nevada and year-round residents in those to the south (Lowther et al. 2000). They are common breeders in eastern Oregon, Utah, and Idaho (Behle et al. 1985, Stephens and Sturts 1998, Adamus et al. 2001), and they breed throughout Arizona and much of California (Small 1994, Wise-Gervais in Corman and Wise-Gervais 2005:404–405). They are fairly common even in the lowlands of the lower Colorado River Valley, although not in riparian habitats (Rosenberg et al. 1991).

CONSERVATION AND MANAGEMENT

In the West, Rock Wren populations have declined significantly since the 1960s (Sauer et al. 2005). The causes of the decline are not clear (Lowther et al. 2000). Atlas data indicate that Rock Wrens avoid urban areas, despite reports that they occasionally use human-modified habitats (Lowther et al. 2000). Currently, the Rock Wren is not recognized as a conservation priority in Nevada (Neel 1999a, Rich et al. 2004). The causes of Rock Wren declines in our region are befuddling conservation biologists at the moment, however, and research that can shed light on the situation would be especially beneficial to our understanding of the conservation needs of this species.

HABITAT USE

HABITAT	ATLAS BLOCKS	INCIDENTAL OBSERVATIONS
Agricultural	0 (0%)	1 (1%)
Alpine	21 (5%)	1 (1%)
Barren	80 (19%)	10 (10%)
Grassland	4 (<1%)	0 (0%)
Mesquite	1 (<1%)	0 (0%)
Mojave	39 (9%)	13 (13%)
Montane Forest	11 (3%)	4 (4%)
Montane Parkland	3 (<1%)	1 (1%)
Montane Shrub	45 (10%)	15 (15%)
Pinyon-Juniper	67 (16%)	13 (13%)
Riparian	6 (1%)	15 (15%)
Salt Desert Scrub	26 (6%)	1 (1%)
Sagebrush Scrub	74 (17%)	12 (12%)
Sagebrush Steppe	53 (12%)	13 (13%)
Urban	1 (<1%)	0 (0%)
TOTAL	431	99

Total numbers reported in the Habitat Use, Breeding Status, and Abundance tables may differ from each other (see pp. 22–23 for details). Percentage sums may differ slightly from 100% due to rounding.

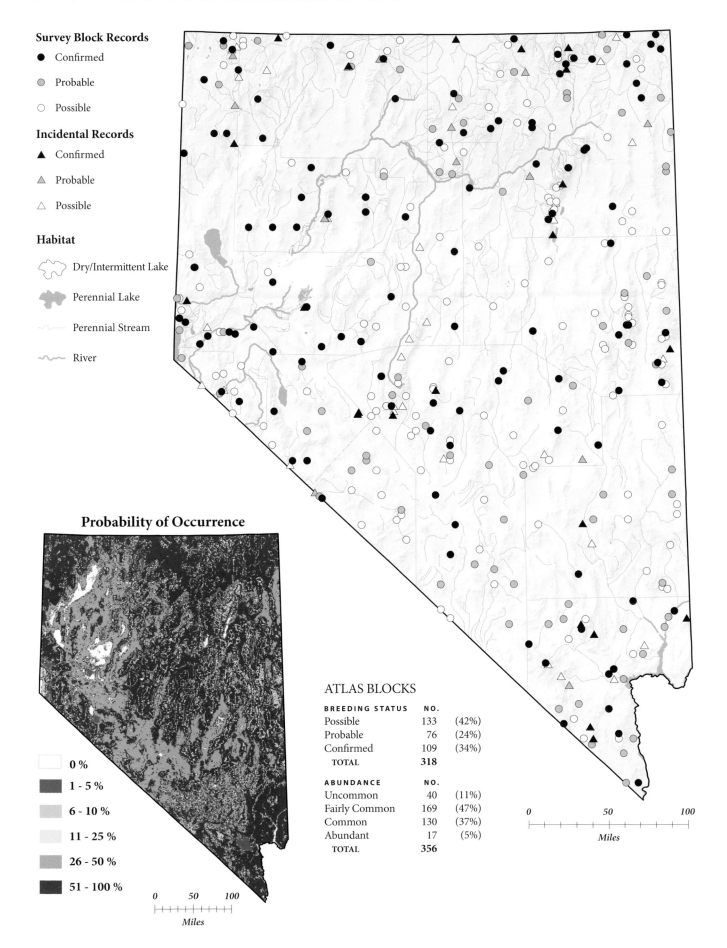

Survey Block Records

- ● Confirmed
- ● Probable
- ○ Possible

Incidental Records

- ▲ Confirmed
- ▲ Probable
- △ Possible

Habitat

- Dry/Intermittent Lake
- Perennial Lake
- Perennial Stream
- River

Probability of Occurrence

- ☐ 0 %
- 1 - 5 %
- 6 - 10 %
- 11 - 25 %
- 26 - 50 %
- 51 - 100 %

0   50   100
Miles

ATLAS BLOCKS

BREEDING STATUS	NO.	
Possible	133	(42%)
Probable	76	(24%)
Confirmed	109	(34%)
TOTAL	**318**	

ABUNDANCE	NO.	
Uncommon	40	(11%)
Fairly Common	169	(47%)
Common	130	(37%)
Abundant	17	(5%)
TOTAL	**356**	

0   50   100
Miles

CANYON WREN
Catherpes mexicanus

The ringing, melodic song of the Canyon Wren is familiar to most who have ventured into the Nevada backcountry. Yet, it is not a particularly well-known bird (Jones and Dieni 1995), nor is it particularly abundant in Nevada. Canyon Wrens inhabit steep canyons and high rocky outcroppings, where they place their nests in inaccessible crevices and crannies. Atlas workers found Canyon Wrens scattered throughout the state, in a variety of habitat types and at a wide variety of elevations. At a microhabitat level, however, the species was always found in intimate association with barren or rocky environments.

DISTRIBUTION

Canyon Wrens were widespread, but not very common, everywhere but in the center of the state. The cliffs or talus patches where they were found were usually small, and thus the habitat associations reported generally reflected the predominant vegetation surrounding the actual site of occurrence. As a consequence, the predictive map shows that the species is most likely to occur in Mojave scrubland and in pinyon-juniper woodland, two habitats that are relatively rich in rock outcroppings. Seemingly suitable rock outcroppings and cliffs occur throughout the whole Nevada landscape, in fact, presumably accounting for the wide variety of other habitats reported. Canyon Wrens do seem to be a bit pickier about their surroundings than the abundant Rock Wrens are, and the chance of finding them in the salt desert lowlands of the Great Basin is generally low. Canyon Wren breeding was confirmed in only 19% of the blocks in which the species was found, likely owing to the inaccessibility of nest sites and the Canyon Wren's secrecy about its whereabouts.

The Canyon Wren's distribution and numbers are closely tied to the availability of suitable cliffs and rock outcroppings in the states surrounding Nevada as well. While present in all of them, the bird is especially common in areas where cliffs abound, such as the Grand Canyon and the Four Corners area (Monson and Phillips 1981, Friederici in Corman and Wise-Gervais 2005:406–407), the lower Snake River corridor in Idaho (Stephens and Sturts 1998), southwestern Utah (Behle et al. 1985), and parts of the lower Colorado River Valley (Rosenberg et al. 1991).

The Canyon Wren's distribution in Nevada appears to have changed little over time. Both Alcorn (1988) and Ryser (1985) characterized the species as widespread but uncommon. Canyon Wrens tend to occur in steeper and rockier situations than Rock Wrens do, but the habitats used by the two species broadly overlap. Canyon Wrens are often reported near temporary or permanent streams, although this probably reflects the frequency of cliffs along waterways rather than a direct affinity for riparian situations (see Jones and Dieni 1995).

The Canyon Wren's use of cliffs and other rocky features makes predicting its occurrence using a model based only on broad habitat classes difficult. Cliffs are probably most common in hilly or mountainous terrain, in desert arroyos of the south, and along streams, and that is where the atlas model generally predicts Canyon Wrens to be. It is likely that the addition of detailed microhabitat data would create much more precise predictions for the distribution of this species.

CONSERVATION AND MANAGEMENT

The Canyon Wren is a poorly understood species (Jones and Dieni 1995). It is a widespread permanent resident throughout the western United States, and it has thus far aroused little conservation concern. Populations are not especially well monitored, but existing data indicate a possible rangewide decline since the 1980s (Sauer et al. 2005). Canyon Wrens are well protected from predators, and their nesting habitat is not typically subject to human disturbance. It is probably safe to say that Canyon Wren populations in Nevada are currently healthy, but it is also important to note that we still know very little about them.

HABITAT USE

HABITAT	ATLAS BLOCKS	INCIDENTAL OBSERVATIONS
Agricultural	1 (2%)	0 (0%)
Alpine	1 (2%)	0 (0%)
Barren	14 (22%)	4 (10%)
Mojave	6 (9%)	6 (15%)
Montane Forest	3 (5%)	4 (10%)
Montane Parkland	1 (2%)	0 (0%)
Montane Shrub	4 (6%)	4 (10%)
Pinyon-Juniper	18 (28%)	7 (18%)
Riparian	10 (15%)	8 (21%)
Salt Desert Scrub	2 (3%)	1 (3%)
Sagebrush Scrub	1 (2%)	2 (5%)
Sagebrush Steppe	4 (6%)	3 (8%)
TOTAL	65	39

Total numbers reported in the Habitat Use, Breeding Status, and Abundance tables may differ from each other (see pp. 22–23 for details). Percentage sums may differ slightly from 100% due to rounding.

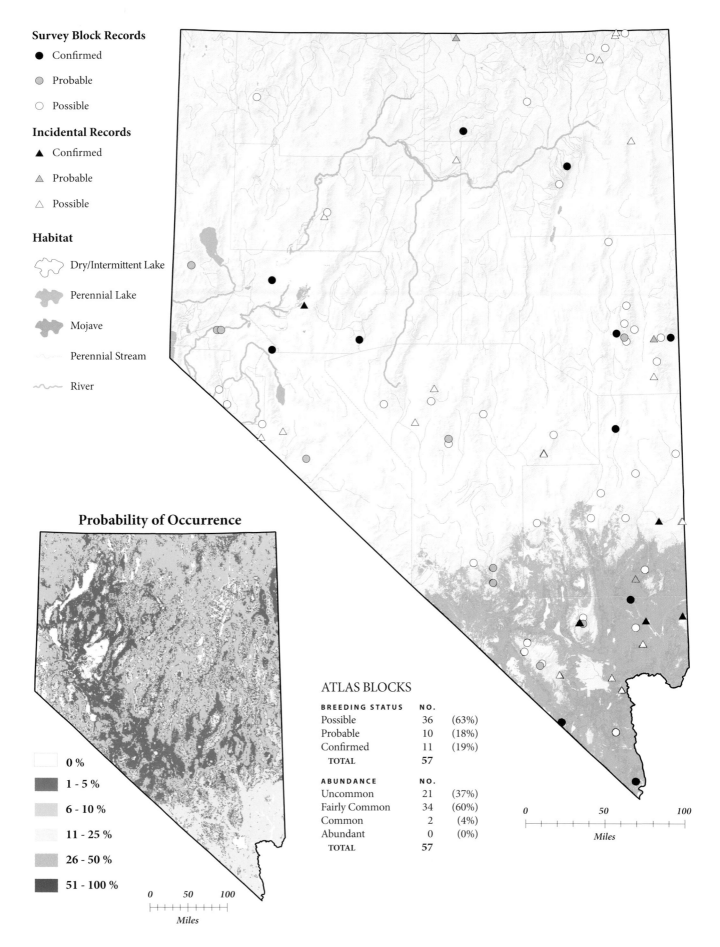

Survey Block Records

● Confirmed

● Probable

○ Possible

Incidental Records

▲ Confirmed

▲ Probable

△ Possible

Habitat

Dry/Intermittent Lake

Perennial Lake

Mojave

Perennial Stream

River

Probability of Occurrence

0 %

1 - 5 %

6 - 10 %

11 - 25 %

26 - 50 %

51 - 100 %

0   50   100

Miles

ATLAS BLOCKS

BREEDING STATUS	NO.	
Possible	36	(63%)
Probable	10	(18%)
Confirmed	11	(19%)
TOTAL	57	

ABUNDANCE	NO.	
Uncommon	21	(37%)
Fairly Common	34	(60%)
Common	2	(4%)
Abundant	0	(0%)
TOTAL	57	

0   50   100

Miles

BEWICK'S WREN
Thryomanes bewickii

There is a joke among Nevada birders: If you can't identify the call or song, it's probably a Bewick's Wren. Each individual seems to have its own inexhaustible lexicon of trills, whistles, chatters, and catcalls, and it does not help that this versatile noisemaker prefers to issue its proclamations from under the cover of dense shrubbery. Fortunately, Bewick's Wrens are not excessively shy, and one can make their acquaintance in suburban parks and gardens. In Nevada, Bewick's Wrens appear to be increasing in numbers in areas where they have long been present and to be expanding into new locales in the northeast.

DISTRIBUTION

Atlas fieldwork greatly clarified the distribution of Bewick's Wrens in Nevada. Records were concentrated throughout the southern and western portions of the state, but there was also a scattering across the northern tier. Linsdale (1936) and Alcorn (1988) regarded this wren as a resident primarily of the southern and western parts of the state, and Ryser (1985) stated that the species was not known to occur in north-central or northeastern Nevada. Where this bird was present, atlas workers also found it to be numerous; abundance estimates exceeded ten pairs in more than 40% of the occupied blocks.

Bewick's Wren is common throughout much of California, Arizona, and southern Utah (Behle et al. 1985, Small 1994, Corman in Corman and Wise-Gervais 2005:408–409). It is rare in eastern Oregon, northern Utah, and nearly all of Idaho (Behle et al. 1985, Stephens and Sturts 1998, Adamus et al. 2001). It is, however, increasing its range in several of the regions where it is currently reported to be rare (Rosenberg et al. 1991, Kennedy and White 1997, Adamus et al. 2001).

Within its Nevada range, Bewick's Wren is most often found in mid-elevation thickets. Favored habitats include riparian shrubbery and woodlands, adjacent pinyon-juniper woodland, and overgrown city parks, yards, and gardens (Ryser 1985), but Bewick's Wrens are also found in various kinds of desert shrublands (Behle et al. 1985, Gaines 1992, Kennedy and White 1997). The predicted occurrence of this species is superficially similar to that of the House Wren, with the highest probabilities in the mountains, but closer inspection reveals that Bewick's Wrens are more likely to be found at lower elevations than House Wrens.

CONSERVATION AND MANAGEMENT

Rangewide, Bewick's Wrens have shown a complex pattern of range expansion and contraction over the past two centuries. The species has been extirpated from much of its eastern range in the United States, yet it is slowly expanding its range in some parts of the West (Kennedy and White 1997). The causes of the population declines are not well understood, but House Wrens, which increased in many of these areas after habitat changes, have been implicated. They can reduce Bewick's Wren nesting success by destroying their nests or removing eggs (Kennedy and White 1997). Bewick's Wrens do not appear to be limited by availability of dead trees, as are many other cavity nesters, because they readily use a wide range of alternative sites, including tin cans, old skulls, and even jacket pockets (Shuford 1993, Gallagher 1997).

Atlas results and other data (Sauer et al. 2005) suggest that the species is expanding in Nevada. In the Lahontan Valley, Bewick's Wrens accept Russian olives and tamarisk as nesting sites (Chisholm and Neel 2002), so prospects may be good even in the many riparian areas of Nevada that have been invaded by exotic vegetation.

HABITAT USE

HABITAT	ATLAS BLOCKS	INCIDENTAL OBSERVATIONS
Agricultural	0 (0%)	2 (3%)
Ash	7 (4%)	0 (0%)
Mesquite	11 (7%)	3 (4%)
Mojave	6 (4%)	10 (14%)
Montane Forest	4 (2%)	2 (3%)
Montane Parkland	1 (<1%)	2 (3%)
Montane Shrub	6 (4%)	5 (7%)
Pinyon-Juniper	48 (29%)	8 (11%)
Riparian	61 (37%)	32 (44%)
Salt Desert Scrub	0 (0%)	1 (1%)
Sagebrush Scrub	12 (7%)	3 (4%)
Urban	9 (5%)	1 (1%)
Wetland	1 (<1%)	3 (4%)
TOTAL	166	72

Total numbers reported in the Habitat Use, Breeding Status, and Abundance tables may differ from each other (see pp. 22–23 for details). Percentage sums may differ slightly from 100% due to rounding.

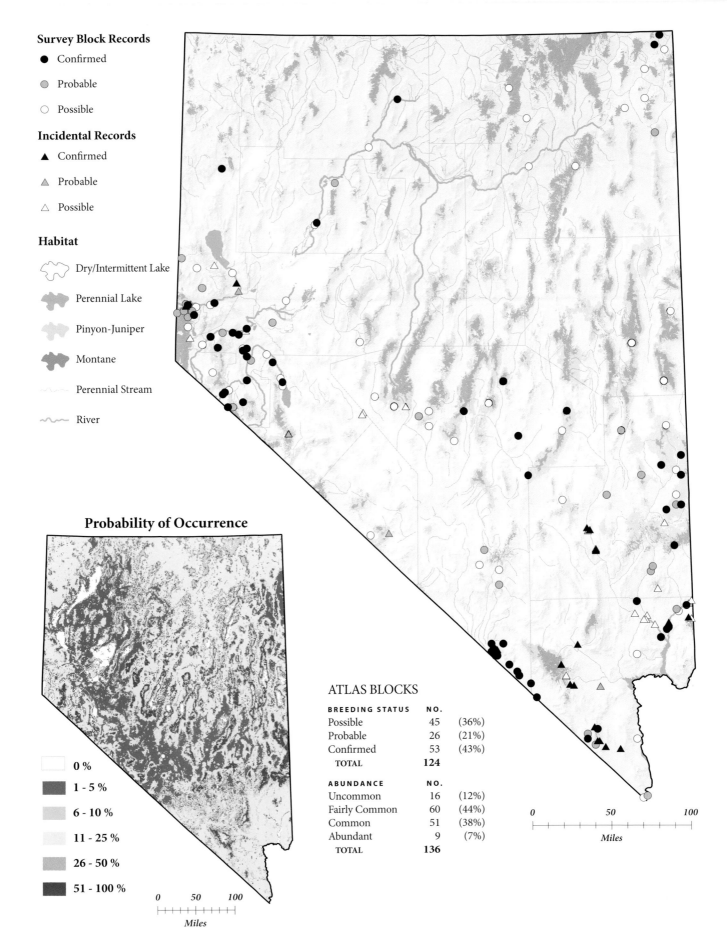

Survey Block Records

● Confirmed

● Probable

○ Possible

Incidental Records

▲ Confirmed

△ Probable

△ Possible

Habitat

Dry/Intermittent Lake

Perennial Lake

Pinyon-Juniper

Montane

Perennial Stream

River

Probability of Occurrence

	0 %
	1 - 5 %
	6 - 10 %
	11 - 25 %
	26 - 50 %
	51 - 100 %

0   50   100
Miles

ATLAS BLOCKS

BREEDING STATUS	NO.	
Possible	45	(36%)
Probable	26	(21%)
Confirmed	53	(43%)
TOTAL	**124**	

ABUNDANCE	NO.	
Uncommon	16	(12%)
Fairly Common	60	(44%)
Common	51	(38%)
Abundant	9	(7%)
TOTAL	**136**	

0   50   100
Miles

HOUSE WREN
Troglodytes aedon

The House Wren may well be one of North America's best-known songbirds. In Nevada, the effervescent singing of the House Wren is a sure sign that forested habitat with an understory of dense brush is nearby. Interestingly, House Wrens seem a bit standoffish here, and they may be less common in residential or agricultural settings than they are in other parts of their range (see Ryser 1985).

DISTRIBUTION

House Wrens were found across most of Nevada, with the greatest concentrations of records in the mountains of the northeastern and western parts of the state, and with almost no records in the dry lowlands of Esmeralda and southern Nye counties. Confirmed breeding in the Spring Mountains of southern Nevada represents a slight extension to the south of the previously recognized limits of the species' breeding range (Johnson 1998), although Linsdale (1936) and Alcorn (1988) did report summer records from this area. Overall, the distribution of the House Wren in Nevada appears to have changed little over the past century (Linsdale 1936, Ryser 1985, Alcorn 1988). The House Wren is also a common and widely distributed breeding bird in the states surrounding Nevada wherever suitable woodlands, riparian areas, or suburban settings exist (Behle et al. 1985, Small 1994, Stephens and Sturts 1998, Adamus et al. 2001, Wise-Gervais in Corman and Wise-Gervais 2005:410–411).

In spite of its name, western House Wrens are not especially likely to be found around houses. Linsdale (1936) and Ryser (1985) considered any location with trees to be acceptable to the species, and in the Lahontan Valley, House Wrens seem to prefer areas with dead and downed cottonwoods (Chisholm and Neel 2002). Dobkin et al. (1995) also reported flexibility in the House Wren's use of natural cavities in montane aspen woodlands of northwestern Nevada.

In Nevada, House Wrens were most often recorded in riparian habitats. Montane habitats were also well represented in the atlas records, and the species occurred at least occasionally in most atlas habitat types. The predictive map closely reflects the observed distribution of House Wrens, with the highest proba-bilities of occurrence in montane and riparian habitats. However, the model overestimates the probability of occurrence in the northern Mojave Desert. In this part of the state the species appears to be even more confined to the mountains than it is farther north.

CONSERVATION AND MANAGEMENT

House Wren populations appear to be in good shape, and recent decades have seen a continent-wide increase (Sauer et al. 2005). Rangewide, House Wrens may have benefited from forest fragmentation (Johnson 1998), and the spread of suburbia has probably also had a positive impact in some areas. In the West, trends are mixed, with overall increases but some local decreases (Johnson 1998). A recent study of House and Bewick's Wrens on the western slope of the Sierra Nevada found no evidence that the two species excluded one another, as studies done in the East have suggested, but did indicate that House Wren populations are sensitive to prolonged drought conditions (Verner and Purcell 1999).

No reliable trend information is available for Nevada. Populations may be limited by the availability of natural cavities (Gaines 1992), and because House Wrens seem less inclined to use human-altered habitats in Nevada than in other areas, the best management strategy is probably to leave intact the understory vegetation and dead timber in native woodlands. It will be interesting to monitor how House Wrens respond to continuing development.

HABITAT USE

HABITAT	ATLAS BLOCKS	INCIDENTAL OBSERVATIONS
Agricultural	6 (3%)	0 (0%)
Alpine	0 (0%)	2 (2%)
Barren	1 (<1%)	0 (0%)
Mojave	0 (0%)	1 (1%)
Montane Forest	39 (17%)	18 (21%)
Montane Parkland	6 (3%)	0 (0%)
Montane Shrub	22 (10%)	4 (5%)
Pinyon-Juniper	15 (7%)	4 (5%)
Riparian	119 (53%)	52 (60%)
Sagebrush Scrub	4 (2%)	0 (0%)
Sagebrush Steppe	2 (1%)	0 (0%)
Urban	8 (4%)	5 (6%)
Wetland	1 (<1%)	1 (1%)
TOTAL	223	87

Total numbers reported in the Habitat Use, Breeding Status, and Abundance tables may differ from each other (see pp. 22–23 for details). Percentage sums may differ slightly from 100% due to rounding.

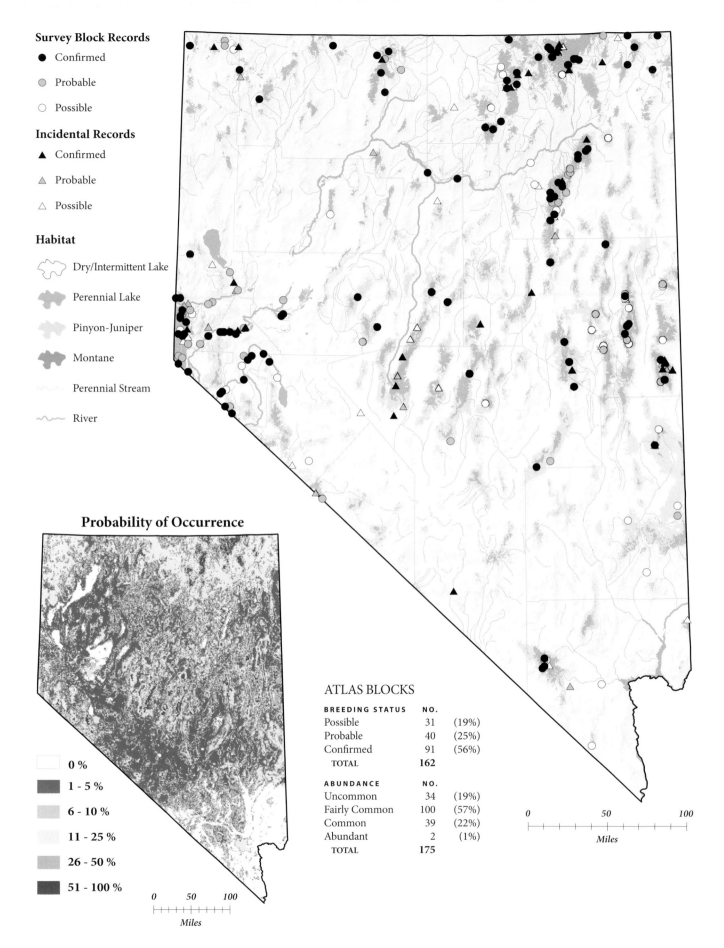

Survey Block Records

● Confirmed

● Probable

○ Possible

Incidental Records

▲ Confirmed

▲ Probable

△ Possible

Habitat

Dry/Intermittent Lake

Perennial Lake

Pinyon-Juniper

Montane

Perennial Stream

River

Probability of Occurrence

☐ 0 %

■ 1 - 5 %

■ 6 - 10 %

☐ 11 - 25 %

☐ 26 - 50 %

■ 51 - 100 %

0   50   100

Miles

ATLAS BLOCKS

BREEDING STATUS	NO.	
Possible	31	(19%)
Probable	40	(25%)
Confirmed	91	(56%)
TOTAL	162	

ABUNDANCE	NO.	
Uncommon	34	(19%)
Fairly Common	100	(57%)
Common	39	(22%)
Abundant	2	(1%)
TOTAL	175	

0   50   100

Miles

WINTER WREN
Troglodytes troglodytes

If you catch a fleeting glimpse of a Winter Wren scurrying through the deadfalls and tangles of the forest floor, you might not even be certain that you have seen a bird. But once you hear this wren's strident and complex song, there is little question about its true identity. Winter Wrens breed mostly to the north and west of Nevada. The atlas produced Nevada's first confirmed breeding records in the Carson Range of western Nevada and a record of probable breeding in the Jarbidge Mountains in the far north.

DISTRIBUTION

The Winter Wren is the only wren that is found outside the Western Hemisphere; its distribution encompasses large portions of Europe, Asia, and northern Africa (Hejl et al. 2002a). In North America, its breeding range reaches across most of southern Canada and dips south into the United States in the Northeast, the upper Midwest, and the Pacific Northwest. In the West, its distribution coincides closely with the presence of extensive tracts of moist coniferous forest such as those found in the Cascade Range and the northern Rocky Mountains (Small 1994, Stephens and Sturts 1998, Adamus et al. 2001, Hejl et al. 2002a). Winter Wrens also breed in the Sierra Nevada and the northern California coastal ranges (Small 1994), and they even have a limited breeding presence in central Arizona (Corman in Corman and Wise-Gervais 2005:412–413). Immediately adjacent to Nevada, Winter Wrens are also established as rare summer residents in Oregon's Malheur Wildlife Refuge (Ryser 1985).

In Nevada, atlas workers reported Winter Wrens from only two atlas blocks: one in the Carson Range, where breeding was confirmed, and another in the Jarbidge Mountains, where breeding was deemed probable. In both blocks, only single pairs were noted. The confirmation of breeding in the Carson Range is apparently Nevada's first. Older reviews provide no breeding records for the state (Linsdale 1936, Ryser 1985, Alcorn 1988), although Ellis and Nelson (1952) considered Winter Wrens to be residents of Nevada's Carson River Basin.

Winter Wrens are unique in their family, not just by virtue of their global distribution, but also because they are the only wrens strongly associated with old-growth coniferous forests over most of their North American range. They prefer shady, moist microhabitats with abundant deadfall material on the forest floor (Hejl et al. 2002a). In the more arid portions of their breeding range, Winter Wrens appear to concentrate locally in moist areas such as densely vegetated ravines and drainages (Small 1994, Corman in Corman and Wise-Gervais 2005:412–413). Although atlas records were limited to Nevada's periphery, there are areas of apparently suitable habitat in the state that appear to be unoccupied, as indicated by the predictive map. Given that the Winter Wren breeds in a highly localized fashion as far south as central Arizona, it seems possible that sporadic breeding could occur in additional Nevada locations.

CONSERVATION AND MANAGEMENT

In recent decades, Winter Wren numbers have held steady or increased moderately throughout most of the breeding range (Sauer et al. 2005). The road-based Breeding Bird Survey, however, may not adequately monitor populations of the forest interior habitats that this species typically uses (Hejl et al. 2002a). Historically, Winter Wren populations may have declined substantially since pre-settlement times because of the loss of old-growth forests. There are some concerns that declines in the Pacific Northwest may continue unless forest managers succeed in preventing further loss and fragmentation of forests (Hejl et al. 2002a). In Nevada, breeding Winter Wrens are still more of a novelty than a conservation issue. Nonetheless, the species presents a good challenge for birders wanting to add to our knowledge of the state's avifauna.

HABITAT USE

HABITAT	ATLAS BLOCKS	INCIDENTAL OBSERVATIONS
Alpine	1 (20%)	0 (0%)
Montane Shrub	1 (20%)	0 (0%)
Riparian	3 (60%)	0 (0%)
TOTAL	5	0

Total numbers reported in the Habitat Use, Breeding Status, and Abundance tables may differ from each other (see pp. 22–23 for details). Percentage sums may differ slightly from 100% due to rounding.

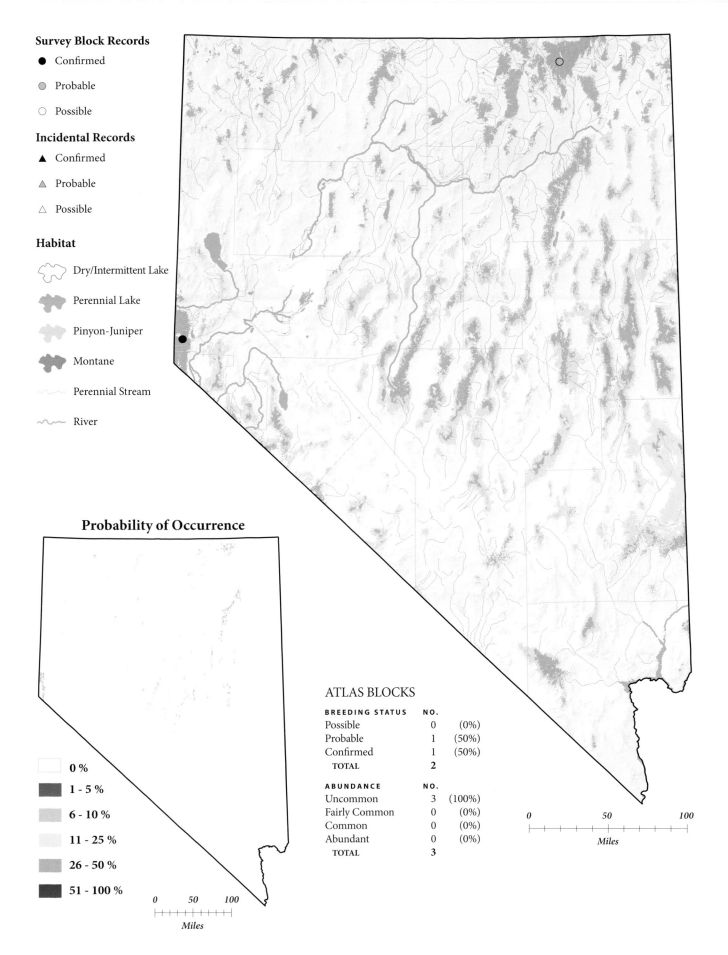

Survey Block Records

● Confirmed

◉ Probable

○ Possible

Incidental Records

▲ Confirmed

△ Probable

△ Possible

Habitat

Dry/Intermittent Lake

Perennial Lake

Pinyon-Juniper

Montane

Perennial Stream

River

Probability of Occurrence

☐ 0 %

■ 1 - 5 %

■ 6 - 10 %

☐ 11 - 25 %

■ 26 - 50 %

■ 51 - 100 %

0   50   100
Miles

ATLAS BLOCKS

BREEDING STATUS	NO.	
Possible	0	(0%)
Probable	1	(50%)
Confirmed	1	(50%)
TOTAL	2	

ABUNDANCE	NO.	
Uncommon	3	(100%)
Fairly Common	0	(0%)
Common	0	(0%)
Abundant	0	(0%)
TOTAL	3	

0   50   100
Miles

MARSH WREN
Cistothorus palustris

The Marsh Wren's eccentric breeding behavior is the stuff of legend. It is a rambunctious and polygamous bundle of energy, inclined to build more nests than it appears to need (Verner and Engelson 1970). It can sing hundreds of different songs (Kroodsma and Canady 1985), and it routinely attacks the nests and nestlings of much larger birds (Bump 1986). It is no wonder that this plucky little bird fascinated atlas workers throughout the state. The Marsh Wren was recorded in marshes of every county in Nevada, from the largest wetlands, where they often reached very high breeding densities, to tiny marshes along the margins of farm ponds and roadside ditches.

DISTRIBUTION

A habitat specialist, the Marsh Wren can be found only in wetlands that provide standing water with dense emergent vegetation (Kroodsma and Verner 1997). Atlas workers found them all over the state, in agreement with earlier reports (Linsdale 1936, Ryser 1985, Alcorn 1988), although observations were few in south-central Nevada. Wetlands or riparian areas, presumably where emergent vegetation was present, were the most frequently recorded habitats. Wetland size was not necessarily a critical factor, and the birds often used small marshes embedded within atlas blocks that were classified in the model by their predominant terrestrial habitat type. This explains why the predictive map shows a chance—albeit a small one—of finding Marsh

Wrens throughout the expansive lowlands of the Great Basin. Higher probabilities are associated with major wetland areas, as expected.

Breeding confirmations were scattered around the state but were notably absent from the Humboldt and Truckee river systems, and from Ash Meadows National Wildlife Refuge in southern Nye County. Confirming breeding is surprisingly complicated in this species, however, because males frequently build "dummy" nests—either to confuse predators or, perhaps, to help attract mates—and the presence of nests alone cannot be taken as firm evidence of breeding. Abundance estimates seemed to be related primarily to marsh size, and were most often in the two–ten-pair range; at a few sites, estimates exceeded one hundred pairs.

Marsh Wrens nest in all of the states adjoining Nevada, but their habitat requirements always make for a disjunct distribution. Of these states, Arizona has the most geographically restricted range, with breeding apparently limited to the lower Colorado River and some of its tributaries (Rosenberg et al. 1991, Corman in Corman and Wise-Gervais 2005:414–415). In southern Utah (Behle et al. 1985) and much of southern California (Small 1994), Marsh Wrens are either absent or uncommon.

CONSERVATION AND MANAGEMENT

The Marsh Wren is something of a conservation paradox: it is widespread and common, yet its specialized wetland habitat is generally thought to be in jeopardy. The secret to its success seems to be its acceptance of even the smallest wetlands, including degraded ones that have been invaded by nonnative species or inundated by agricultural runoff (Kroodsma and Verner 1997, Smith et al. 1997). Perhaps for this reason, populations seem to be generally increasing, both in the West and rangewide (Sauer et al. 2005). Data for Nevada, however, are too limited to allow reliable conclusions.

Despite the apparent health of Marsh Wren populations, the species may be of some interest to conservation biologists. It responds well to wetland mitigation projects (Kroodsma and Verner 1997) and could be a useful indicator species for wetland recovery. Marsh Wrens may, however, affect the nesting success of other species that breed in the same habitat (Leonard and Picman 1986).

HABITAT USE

HABITAT	ATLAS BLOCKS	INCIDENTAL OBSERVATIONS
Agricultural	1 (2%)	0 (0%)
Montane Shrub	1 (2%)	0 (0%)
Open Water	1 (2%)	1 (4%)
Riparian	18 (33%)	5 (18%)
Salt Desert Scrub	0 (0%)	1 (4%)
Sagebrush Scrub	0 (0%)	1 (4%)
Wetland	34 (62%)	20 (71%)
TOTAL	55	28

Total numbers reported in the Habitat Use, Breeding Status, and Abundance tables may differ from each other (see pp. 22–23 for details). Percentage sums may differ slightly from 100% due to rounding.

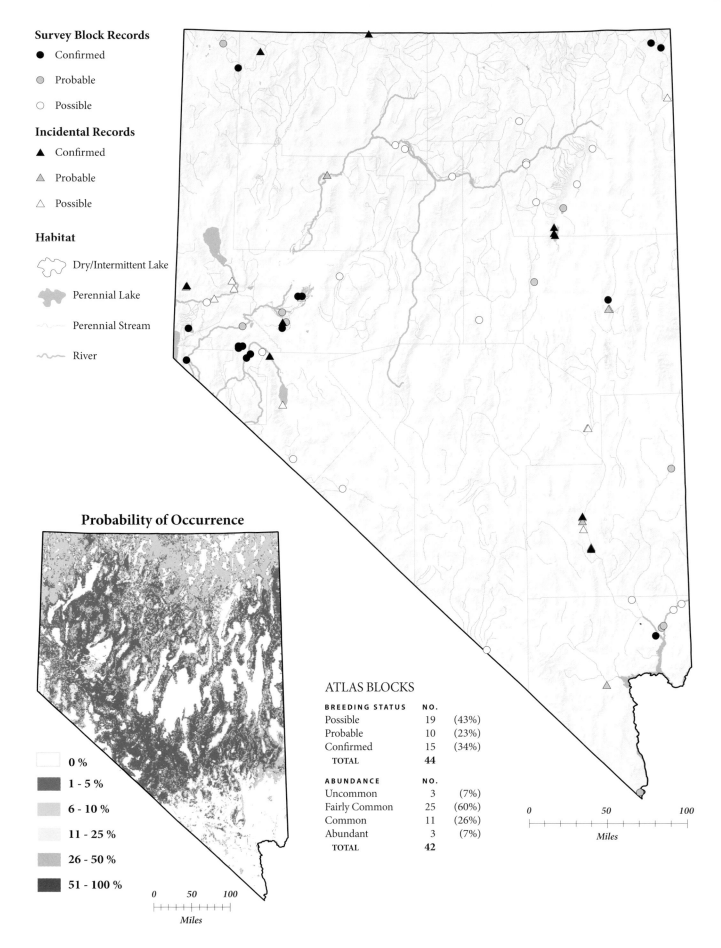

Survey Block Records

● Confirmed

● Probable

○ Possible

Incidental Records

▲ Confirmed

△ Probable

△ Possible

Habitat

Dry/Intermittent Lake

Perennial Lake

Perennial Stream

River

Probability of Occurrence

	0 %
	1 - 5 %
	6 - 10 %
	11 - 25 %
	26 - 50 %
	51 - 100 %

0   50   100
Miles

ATLAS BLOCKS

BREEDING STATUS	NO.	
Possible	19	(43%)
Probable	10	(23%)
Confirmed	15	(34%)
TOTAL	44	

ABUNDANCE	NO.	
Uncommon	3	(7%)
Fairly Common	25	(60%)
Common	11	(26%)
Abundant	3	(7%)
TOTAL	42	

0   50   100
Miles

AMERICAN DIPPER
Cinclus mexicanus

Dippers are among the most popular birds in the western United States. Their unique use of rocky streams and dare-devilish underwater foraging antics can be observed all year long, and they are not shy about displaying their considerable vocal talents. Although the American Dipper is a well-known and common target of birders, atlas workers found them at only a few locations in Nevada, and nowhere was the species particularly common.

DISTRIBUTION

In western Nevada, American Dippers were found only in the Carson Range, and in the south, only in the Spring Mountains. In northeastern Nevada, they were a bit more widespread throughout Elko and White Pine counties. Breeding was not confirmed in all locations, but American Dippers do not migrate far, and most sightings probably were near breeding sites. Abundance estimates were typically low. Linsdale (1936) and Alcorn (1998) reported American Dippers in several areas where they were not found during the atlas project, including northwestern Nevada's Pine Forest Range, the Toiyabe Range, and other locations in Humboldt, Pershing, Esmeralda, Lyon, and Lincoln counties.

Dippers typically use narrow, fast-flowing, perennial streams with clear water and rocky bottoms (Kingery 1996). Rocky stream banks, waterfalls, and the underside of bridges provide protected nest sites. In Nevada, this combination of habitat conditions usually occurs in the upper-elevation headwater sections of streams (Ryser 1985). Unlike other riparian birds, American Dippers seem to care more about the physical attributes of the stream than the streamside vegetation type (Kingery 1996).

The predictive map suggests that American Dippers could occur in most of Nevada's higher mountain ranges, but the chance of finding them is low. Atlas surveys failed to find them in some historical locations, and future surveys will have to determine whether American Dippers are still as widespread in Nevada as Linsdale (1936) and Alcorn (1988) reported them to be, or if they have disappeared from some ranges.

Not surprisingly, the American Dipper's breeding range is noticeably patchier in the desert regions of the Great Basin and the Southwest than it is in many other parts of the species' range (see Kingery 1996, Corman in Corman and Wise-Gervais 2005: 416–417). For instance, American Dippers occur widely in the Sierra Nevada, northern California, the Cascades, the Wasatch Front of Utah, and in other parts of the central and northern Rocky Mountains (Behle et al. 1985, Small 1994, Stephens and Sturts 1998, Adamus et al. 2001).

CONSERVATION AND MANAGEMENT

Breeding Bird Survey data suggest that populations of American Dippers are stable (Sauer et al. 2005), but these records come from road-based surveys, which are not well suited for monitoring montane stream specialists. Dippers do not seem especially susceptible to disturbance or predation. They are, however, sensitive to stream degradation due to pollution, sedimentation, and warming after nearby forest clear-cutting, all of which negatively affect the macroinvertebrates the birds eat (Kingery 1996).

Considerable research on White-throated Dippers in Europe has shown that they are indicators of montane stream health (Ormerod and Tyler 1987). There is also preliminary evidence that American Dippers may be good bioindicators of trout health (Price and Bock 1983). Evidence from Arizona suggests that American Dippers suffer in drainages where introduced crayfish have decimated the aquatic prey base (Corman in Corman and Wise-Gervais 2005:416–417). This possibility is of interest in Nevada, where crayfish flourish in, and may affect the native biota of, many streams.

HABITAT USE

HABITAT	ATLAS BLOCKS	INCIDENTAL OBSERVATIONS
Alpine	2 (12%)	0 (0%)
Montane Forest	1 (6%)	0 (0%)
Montane Shrub	1 (6%)	0 (0%)
Riparian	13 (76%)	7 (100%)
TOTAL	17	7

Total numbers reported in the Habitat Use, Breeding Status, and Abundance tables may differ from each other (see pp. 22–23 for details). Percentage sums may differ slightly from 100% due to rounding.

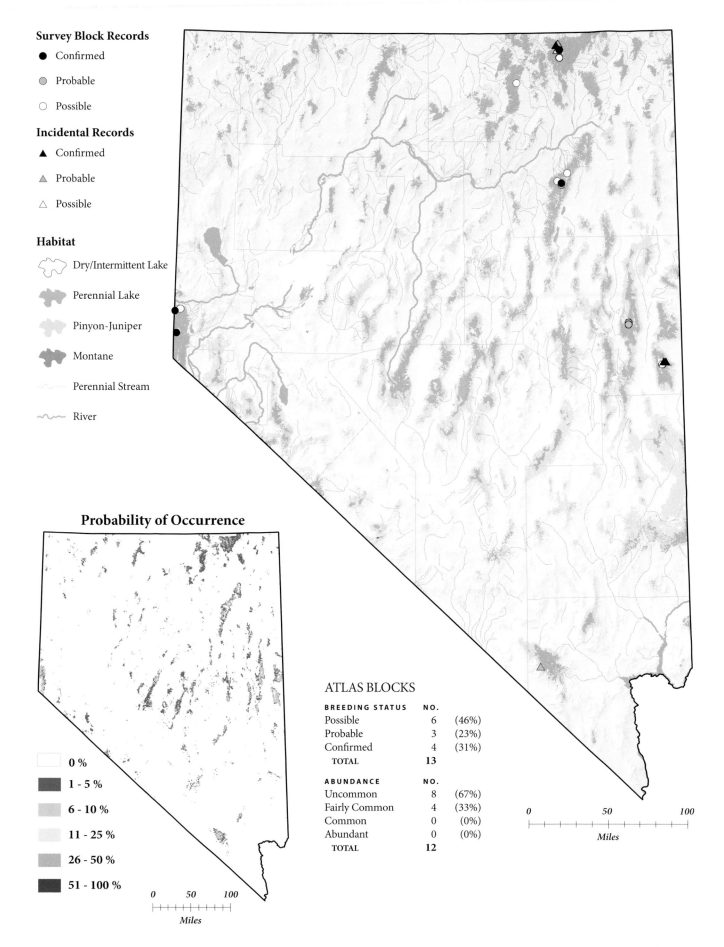

Survey Block Records

● Confirmed

◉ Probable

○ Possible

Incidental Records

▲ Confirmed

△ Probable

△ Possible

Habitat

Dry/Intermittent Lake

Perennial Lake

Pinyon-Juniper

Montane

Perennial Stream

River

Probability of Occurrence

☐ 0 %

■ 1 - 5 %

■ 6 - 10 %

☐ 11 - 25 %

■ 26 - 50 %

■ 51 - 100 %

0   50   100
Miles

ATLAS BLOCKS

BREEDING STATUS	NO.	
Possible	6	(46%)
Probable	3	(23%)
Confirmed	4	(31%)
TOTAL	**13**	
ABUNDANCE	NO.	
Uncommon	8	(67%)
Fairly Common	4	(33%)
Common	0	(0%)
Abundant	0	(0%)
TOTAL	**12**	

0   50   100
Miles

GOLDEN-CROWNED KINGLET
Regulus satrapa

Only a small number of Nevada atlas workers had the pleasure of recording Golden-crowned Kinglets as breeders. The most obvious reason for this bird's apparent scarcity, of course, is that the species is truly uncommon here and throughout the drier portions of the Intermountain West. Golden-crowned Kinglets favor moist old-growth fir and spruce forests, and such forests are limited in extent in Nevada. Beyond that, the Golden-crowned Kinglets that are here can be difficult to find and confirm as breeders. Not only do they inhabit high-elevation conifer forests where access is challenging, but these tiniest of passerines typically build their inconspicuous nests far above the ground in the dense canopy.

DISTRIBUTION

Golden-crowned Kinglets were recorded in only ten atlas blocks and at five other locations. All of the records came from either the Carson Range of far western Nevada, the Jarbidge Mountains of the northeast, or the Snake Range of extreme eastern Nevada. Breeding was confirmed only in the first two of these locations. None of the earlier treatments of the species in Nevada (Linsdale 1936, Ryser 1985, Alcorn 1988) report breeding at any other locations in the state.

The Golden-crowned Kinglet is almost exclusively a bird of tall boreal and subalpine coniferous forests (Ingold and Galati 1997), which are restricted to high elevations in Nevada. The rarity of suitable habitat in Nevada is reflected in the predictive map, which suggests a fair chance of finding this kinglet in the larger mountain ranges, but only in a narrow band of high-elevation forest. The predictive map does indicate that limited

amounts of suitable habitat are present in Nevada's central ranges and in the Ruby Mountains of Elko County, but Golden-crowned Kinglets have never been recorded in these areas. Not only was this kinglet's distribution geographically limited, but its estimated numbers tended to be low, most often just a single pair per block. Of course, the difficulties involved in observing these birds may have caused underestimates of their abundance.

With regard to Golden-crowned Kinglets, Nevada is clearly impoverished in comparison with our neighboring states. Golden-crowneds are fairly common breeders in suitable habitat in California (Raphael et al. 1988, Small 1994), Oregon (Adamus et al. 2001), and Idaho (Stephens and Sturts 1998). While not particularly common in Arizona (Wise-Gervais in Corman and Wise-Gervais 2005:418–419) or Utah (Behle et al. 1985), they are still more widely distributed in those states than they are in Nevada.

CONSERVATION AND MANAGEMENT

The Golden-crowned Kinglet is widespread in North America and is common throughout much of its range, but it is susceptible to cold weather, which can cause local populations to crash (Ingold and Galati 1997). Although local increases have occurred in the East in areas where spruce and fir have been planted (Ingold and Galati 1997), the species on the whole has declined nationwide (Raphael et al. 1988, Sauer et al. 2005).

Golden-crowned Kinglets in the western United States seem to be sensitive to logging (Smith et al. 1997), although they apparently are able to recover after dense forests have been restored (Roth and Potter in Kingery 1998:376–377). In Nevada, the Golden-crowned Kinglet's low numbers are probably the result of the limited extent and disjunct nature of its favored habitat. It is possible, however, that this inconspicuous little bird is more common than we realize. Locally unstable population dynamics have been reported in other parts of the range (Ingold and Galati 1997), and the possibility of short-lived or even permanent populations in central or southern Nevada cannot be ruled out without further surveys. Such work should focus on the potential breeding areas identified on the predictive map.

HABITAT USE

HABITAT	ATLAS BLOCKS	INCIDENTAL OBSERVATIONS
Montane Forest	10 (91%)	3 (75%)
Montane Parkland	1 (9%)	0 (0%)
Riparian	0 (0%)	1 (25%)
TOTAL	11	4

Total numbers reported in the Habitat Use, Breeding Status, and Abundance tables may differ from each other (see pp. 22–23 for details). Percentage sums may differ slightly from 100% due to rounding.

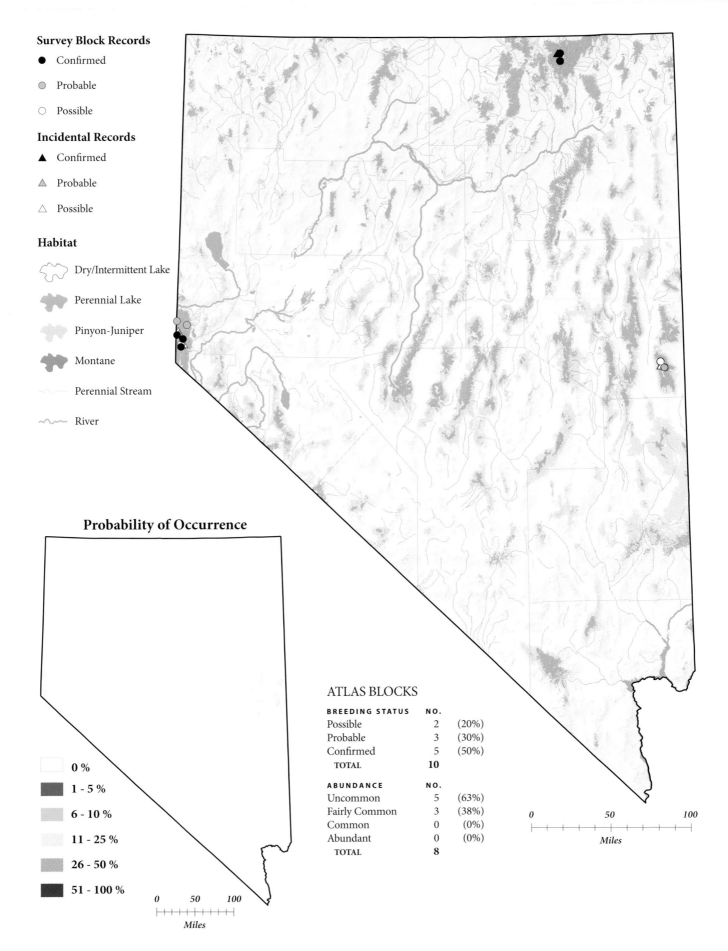

Survey Block Records

● Confirmed

◐ Probable

○ Possible

Incidental Records

▲ Confirmed

△ Probable

△ Possible

Habitat

Dry/Intermittent Lake

Perennial Lake

Pinyon-Juniper

Montane

Perennial Stream

River

Probability of Occurrence

☐ 0 %

▨ 1 - 5 %

▨ 6 - 10 %

▨ 11 - 25 %

▨ 26 - 50 %

▨ 51 - 100 %

0   50   100
Miles

ATLAS BLOCKS

BREEDING STATUS	NO.	
Possible	2	(20%)
Probable	3	(30%)
Confirmed	5	(50%)
TOTAL	10	
ABUNDANCE	NO.	
Uncommon	5	(63%)
Fairly Common	3	(38%)
Common	0	(0%)
Abundant	0	(0%)
TOTAL	8	

0   50   100
Miles

RUBY-CROWNED KINGLET
Regulus calendula

Clearly, what Nevada lacks in Golden-crowned Kinglets it makes up for with Ruby-crowneds (Ryser 1985). Not only do Ruby-crowned Kinglets occur in more locations in Nevada than do Golden-crowneds, their estimated abundances are also considerably higher. A fussbudget without rival among Nevada's breeding birds, the hyperactive Ruby-crowned Kinglet is a conspicuous inhabitant of our montane coniferous forests. While they tend to nest high up in trees, Ruby-crowned Kinglets employ a variety of foraging strategies and can be seen at any height in the canopy, in understory and riparian shrubbery, or even out in the open in clearings and meadows. Whatever the venue, the story is always the same with this avian worrywart: constant wing-flicking, frequent chatter, and endless bustling about in the branches.

DISTRIBUTION

Ruby-crowned Kinglets can be found in most of our higher mountains during the breeding season, but most of the confirmed breeding records came from northeastern and eastern Nevada. Individuals or small flocks are easy to detect, but breeding can be difficult to confirm. In fact, the western half of the state produced no confirmed breeding records, although evidence suggests that Ruby-crowned Kinglets do breed there (Linsdale 1936, Ryser 1985, Alcorn 1988, Ingold and Wallace 1994).

In surrounding states, Ruby-crowned Kinglets appear to breed most commonly in Utah (Behle et al. 1985), where, as in Nevada, they are often residents that migrate only in elevation (Ryser 1985). Ruby-crowned Kinglets are local breeders in parts of

Arizona (Wise-Gervais in Corman and Wise-Gervais 2005:420–421), Oregon (Adamus et al. 2001), and Idaho (Stephens and Sturts 1998). The species can be found at higher elevations in California (Small 1994), although it is found only locally in some of the northern districts (McCaskie et al. 1988) and very locally in the southern part of the state (Garrett and Dunn 1981).

In the West, breeding Ruby-crowned Kinglets are largely restricted to coniferous forests (Ingold and Wallace 1994), especially spruces and firs. Breeding sites of Ruby-crowned Kinglets appear to be somewhat drier than those typical of Golden-crowneds (Smith et al. 1997), perhaps explaining the differences in their ranges in Nevada. According to the predictive map, Ruby-crowned Kinglets could be found in most locations where montane forest occurs. Several mountain ranges in central Nevada are predicted to be suitable for this species, but atlas workers obtained only possible breeding records from the interior of the state, suggesting that the habitat patches might be too small or too isolated to support breeding populations. The areas of high probabilities of occurrence in the Mojave Desert almost certainly are an overestimate by the model, and kinglets undoubtedly do not nest in other extensive low-elevation areas predicted to have a chance of containing them.

CONSERVATION AND MANAGEMENT

Unlike some eastern populations, Ruby-crowned Kinglets in the West have been increasing in numbers in recent decades (Sauer et al. 2005). Determining the local status can be complicated by the presence of migrants, which are commonly noted throughout Nevada and can be mistaken for early breeders. In some areas, breeding populations may also go undetected (Ingold and Wallace 1994, Smith et al. 1997). Timber harvest is usually cited as the main threat to Ruby-crowned Kinglets (see Ingold and Wallace 1994), but the effects of forest management in Nevada are unknown.

The Ruby-crowned Kinglet appears to be faring relatively well in Nevada. In most of our mountain ranges, it greatly outnumbers the Golden-crowned Kinglet—just the opposite of the pattern found as nearby as Yosemite National Park (Gaines 1992). Probably the most important actions to be taken at present on behalf of the Ruby-crowned Kinglet are improved monitoring and additional investigations in western and central Nevada, where the atlas project did not confirm breeding.

HABITAT USE

HABITAT	ATLAS BLOCKS	INCIDENTAL OBSERVATIONS
Alpine	2 (2%)	0 (0%)
Montane Forest	64 (70%)	20 (74%)
Montane Parkland	7 (8%)	0 (0%)
Montane Shrub	2 (2%)	1 (4%)
Pinyon-Juniper	9 (10%)	1 (4%)
Riparian	7 (8%)	5 (19%)
TOTAL	91	27

Total numbers reported in the Habitat Use, Breeding Status, and Abundance tables may differ from each other (see pp. 22–23 for details). Percentage sums may differ slightly from 100% due to rounding.

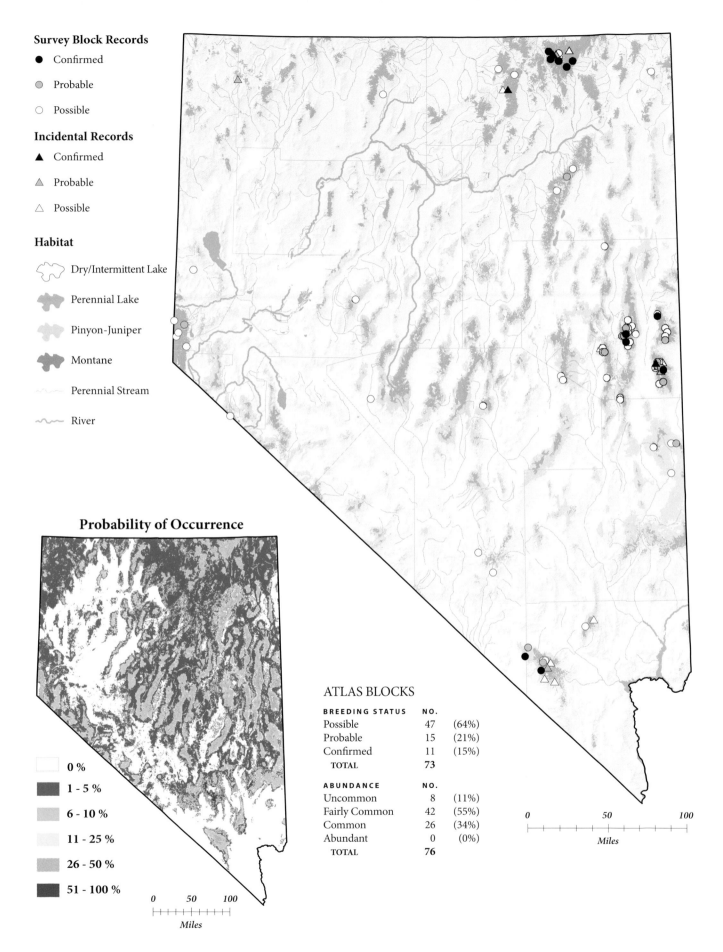

Survey Block Records

● Confirmed

● Probable

○ Possible

Incidental Records

▲ Confirmed

▲ Probable

△ Possible

Habitat

Dry/Intermittent Lake

Perennial Lake

Pinyon-Juniper

Montane

Perennial Stream

River

Probability of Occurrence

☐ 0 %

■ 1 - 5 %

■ 6 - 10 %

■ 11 - 25 %

■ 26 - 50 %

■ 51 - 100 %

0   50   100
Miles

ATLAS BLOCKS

BREEDING STATUS	NO.	
Possible	47	(64%)
Probable	15	(21%)
Confirmed	11	(15%)
TOTAL	73	

ABUNDANCE	NO.	
Uncommon	8	(11%)
Fairly Common	42	(55%)
Common	26	(34%)
Abundant	0	(0%)
TOTAL	76	

0   50   100
Miles

BLUE-GRAY GNATCATCHER
Polioptila caerulea

If it were not for the tail, there wouldn't be much bird at all! Spritely and tiny, the Blue-gray Gnatcatcher is a distinctive inhabitant of dry pinyon-juniper foothills and shrubby habitats throughout much of Nevada. It is often seen jabbing at tiny insects or gathering up strands of cobwebs for its ornate and delicate nest. Whatever it is doing, the Blue-gray Gnatcatcher tends to be vocal: its song is a series of thin, wheezy phrases, and its call note (*wheee!*) has an endearingly giddy quality.

DISTRIBUTION

Atlas workers found Blue-gray Gnatcatchers throughout the state, but there were areas of local scarcity, especially in the lower-elevation portions of the Great Basin. These areas included the Lahontan Valley, Esmeralda County, south-central Nye County north of the Ash Meadows area, and north-central Nevada. Breeding was confirmed in almost a third of the blocks in which the species was found, and abundance estimates were quite variable.

Blue-gray Gnatcatchers approach the northern limit of their western range in Nevada. Linsdale (1936) provided records only as far north as the Toiyabe Range, and Ryser (1985) indicated that numbers decrease with increasing latitudes in the Great Basin. The species has been expanding northward in recent decades (Ellison 1992), however, as evidenced by the breeders fieldworkers found in the far north and by the presence of breeding populations in southern Idaho and Oregon (Stephens and Sturts 1998, Adamus et al. 2001). Still, the atlas data suggest greater densities of birds in the central and southern portions of the state, closer to the core breeding range. In our other adjacent states, Blue-gray Gnatcatchers are common and widespread in southern Utah, most of Arizona, and parts of California (Behle et al. 1985, Small 1994, Wise-Gervais in Corman and Wise-Gervais 2005:422–423).

Throughout its range the Blue-gray Gnatcatcher can be found in a variety of shrubby and forested habitats (Ellison 1992). In Nevada, it is partial to pinyon-juniper woodlands, and predicted probabilities are consequently highest in the foothills of major mountain ranges where this habitat dominates. Willow-cottonwood communities are also used (Ryser 1985), as are areas with buffaloberry and black willow (Chisholm and Neel 2002). In some parts of its range, the species is closely associated with pinyon-juniper stands that have a broadleaved understory (Pavlacky and Anderson 2001), but this type of habitat is rare in Nevada. Even away from the most favored habitats, however, there is a moderately high chance of finding gnatcatchers in much of the state. Only in higher-elevation forest, urban areas, and salt desert scrub does the likelihood of occurrence fall to low levels.

CONSERVATION AND MANAGEMENT

Populations of the Blue-gray Gnatcatcher have increased overall and expanded northward during the past quarter century (Ellison 1992, Sauer et al. 2005). The most dramatic expansions have been in the East, but there have also been modest expansions in the West (Ellison 1992). Local patterns of population change are complex. In California, for example, there have been declines in some lowland areas (Gaines 1992) but increases elsewhere (Shuford 1993). Clearing of woodlands may have depressed gnatcatcher populations in some areas, while fires may have created suitable habitat in others (Shuford 1993). The habitat requirements of Blue-gray Gnatcatchers in Nevada deserve further study, especially in the south, where they share their breeding range with the closely related Black-tailed Gnatcatcher.

HABITAT USE

HABITAT	ATLAS BLOCKS	INCIDENTAL OBSERVATIONS
Agricultural	1 (<1%)	0 (0%)
Ash	5 (3%)	0 (0%)
Barren	0 (0%)	1 (1%)
Mesquite	12 (6%)	1 (1%)
Mojave	13 (7%)	9 (12%)
Montane Forest	2 (1%)	3 (4%)
Montane Parkland	1 (<1%)	2 (3%)
Montane Shrub	18 (9%)	8 (11%)
Pinyon-Juniper	105 (54%)	28 (38%)
Riparian	16 (8%)	13 (18%)
Salt Desert Scrub	6 (3%)	2 (3%)
Sagebrush Scrub	12 (6%)	4 (5%)
Sagebrush Steppe	3 (2%)	2 (3%)
Urban	2 (1%)	0 (0%)
TOTAL	196	73

Total numbers reported in the Habitat Use, Breeding Status, and Abundance tables may differ from each other (see pp. 22–23 for details). Percentage sums may differ slightly from 100% due to rounding.

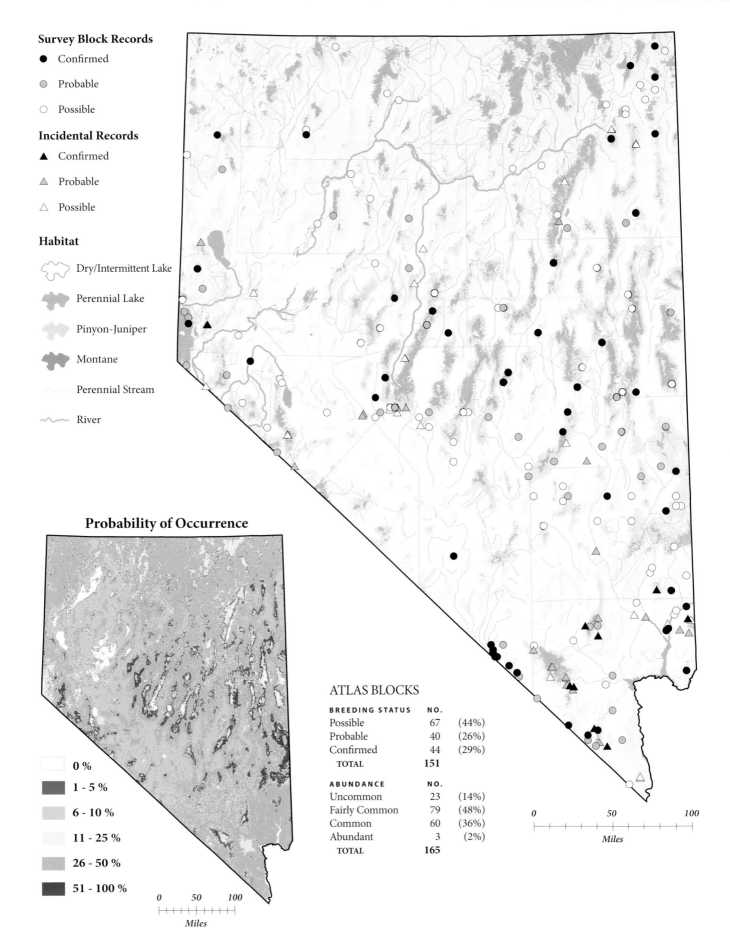

Survey Block Records

- ● Confirmed
- ● Probable
- ○ Possible

Incidental Records

- ▲ Confirmed
- ▲ Probable
- △ Possible

Habitat

- Dry/Intermittent Lake
- Perennial Lake
- Pinyon-Juniper
- Montane
- Perennial Stream
- River

Probability of Occurrence

- 0 %
- 1 - 5 %
- 6 - 10 %
- 11 - 25 %
- 26 - 50 %
- 51 - 100 %

0   50   100

Miles

ATLAS BLOCKS

BREEDING STATUS	NO.	
Possible	67	(44%)
Probable	40	(26%)
Confirmed	44	(29%)
TOTAL	151	

ABUNDANCE	NO.	
Uncommon	23	(14%)
Fairly Common	79	(48%)
Common	60	(36%)
Abundant	3	(2%)
TOTAL	165	

0   50   100

Miles

BLACK-TAILED GNATCATCHER
Polioptila melanura

The Black-tailed Gnatcatcher is a feisty and energetic inhabitant of the warm deserts of southern Nevada. It is found in creosotebush-dominated playas and bajadas and in the dense vegetation of desert washes and oases, where its scolding call is usually heard long before the bird itself is seen. Although the Black-tailed Gnatcatcher is ubiquitous in appropriate habitats in southern Nevada, it is sensitive to habitat alteration, especially that associated with urbanization (Germaine et al. 1998), and its presence thus indicates relatively undisturbed native desert vegetation (Farquhar and Ritchie 2002).

DISTRIBUTION

Black-tailed Gnatcatchers are restricted to the warm deserts of southwestern North America, and they reach their northern limit in southern Nevada (Farquhar and Ritchie 2002). They are common in southern and western Arizona and southeastern California, including the lower Colorado River Valley (Rosenberg et al. 1991, Small 1994, Farquhar and Ritchie 2002, Wise-Gervais in Corman and Wise-Gervais 2005:424–425), and also breed in a small portion of southwestern Utah (Behle et al. 1985).

Atlas fieldworkers found Black-tailed Gnatcatchers in forty-three blocks in the Mojave Desert; more than 80% of these blocks were in Clark County. Breeding was confirmed in almost half of the blocks and in all three southern Nevada counties. The species was not just widespread in the Mojave Desert; it was also common, with more than two pairs estimated in more than 80% of occupied blocks, and more than ten pairs in almost 40% of them. The predicted chance of finding Black-tailed Gnatcatchers in most of southern Nevada is very high, although the species is not expected to occur in montane regions. Its intolerance of houses, roads, and exotic landscaping plants (Emlen 1974, Germaine et al. 1998) explains the lower probability of encountering this species in urban areas.

Early accounts describe much the same distribution for the Black-tailed Gnatcatcher as the atlas project revealed. Alcorn (1988) depicted it as a permanent resident in southern Nevada. Linsdale (1936) did not mention Black-tailed Gnatcatchers in his earlier survey but later listed them as resident in Clark County (Linsdale 1951; see also Gabrielson 1949). Although Alcorn (1988) reported a vagrant individual in Elko County, neither Ryser (1985) nor atlas workers recorded the species in the Great Basin.

Throughout their range, Black-tailed Gnatcatchers are restricted to desert scrub and are most commonly found in riparian areas dominated by creosotebush and other woody species such as mesquites and catclaws (Farquhar and Ritchie 2002). Atlas data revealed the same pattern, with more than 90% of the records coming from Mojave, riparian, and mesquite habitats.

CONSERVATION AND MANAGEMENT

Breeding Bird Survey data indicate declines in Black-tailed Gnatcatcher populations throughout their U.S. range since the 1980s (Sauer et al. 2005). Longer-term or more local trends are more difficult to determine because so few surveys have been conducted in the region where the bird occurs (Farquhar and Ritchie 2002). Conservation issues for the Black-tailed Gnatcatcher include ongoing habitat alteration in dense shrublands and the species' avoidance of heavily developed areas and ornamental landscaping (Emlen 1974, Germaine et al. 1998). These gnatcatchers do use areas that have been invaded by the exotic tamarisk, but to a much more limited extent than they use native mesquite woodlands (Laudenslayer 1981). Due to their restricted distribution, Black-tailed Gnatcatchers are a Partners in Flight Stewardship species for the Southwest (Rich et al. 2004). Given the rapid urbanization of southern Nevada, it will be important to monitor Black-tailed Gnatcatcher populations in this region.

HABITAT USE

HABITAT	ATLAS BLOCKS	INCIDENTAL OBSERVATIONS
Mesquite	15 (23%)	5 (16%)
Mojave	29 (45%)	16 (50%)
Montane Shrub	1 (2%)	0 (0%)
Pinyon-Juniper	1 (2%)	0 (0%)
Riparian	16 (25%)	11 (34%)
Salt Desert Scrub	3 (5%)	0 (0%)
TOTAL	65	32

Total numbers reported in the Habitat Use, Breeding Status, and Abundance tables may differ from each other (see pp. 22–23 for details). Percentage sums may differ slightly from 100% due to rounding.

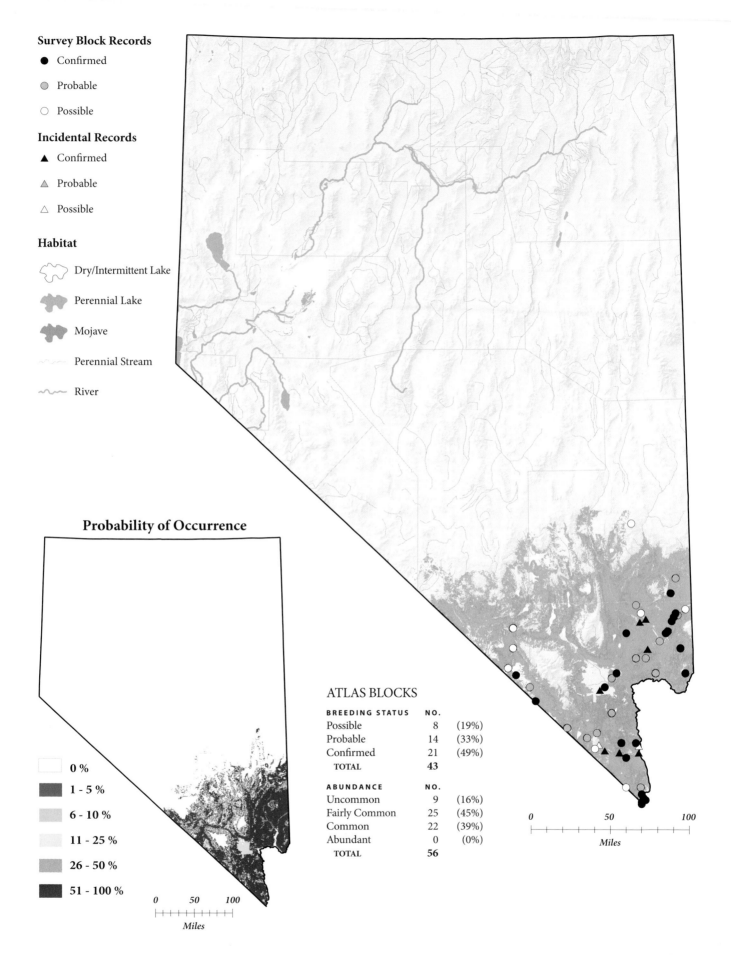

Survey Block Records

● Confirmed

◐ Probable

○ Possible

Incidental Records

▲ Confirmed

△ Probable

△ Possible

Habitat

Dry/Intermittent Lake

Perennial Lake

Mojave

Perennial Stream

River

Probability of Occurrence

	0 %
	1 - 5 %
	6 - 10 %
	11 - 25 %
	26 - 50 %
	51 - 100 %

0   50   100
Miles

ATLAS BLOCKS

BREEDING STATUS	NO.	
Possible	8	(19%)
Probable	14	(33%)
Confirmed	21	(49%)
TOTAL	43	

ABUNDANCE	NO.	
Uncommon	9	(16%)
Fairly Common	25	(45%)
Common	22	(39%)
Abundant	0	(0%)
TOTAL	56	

0   50   100
Miles

WESTERN BLUEBIRD
Sialia mexicana

The Western Bluebird is a bird of contrasts. Clad in bright blue and striking chestnut, its low, whistled calls have a melancholy quality. Western Bluebirds are found most often in tall pine groves, especially those with broken canopies, burned patches, and standing deadwood. As a species of conservation concern, the Western Bluebird can benefit from a number of proactive management activities. These run the gamut from providing artificial nest boxes to conservation-oriented forest management practices.

DISTRIBUTION

Although Western Bluebirds were noted widely throughout Nevada, they were much less common than Mountain Bluebirds, and breeding was confirmed only in the Carson Range and adjoining river drainages of western Nevada and in the woodlands of the south and southeast. Earlier descriptions concur regarding the Western Bluebird's breeding status in western Nevada but are less consistent with regard to its status elsewhere in the state (Linsdale 1936, Ryser 1985, Alcorn 1988). Western Bluebirds were not especially numerous anywhere: abundance estimates rarely exceeded ten pairs per block, and only single pairs were reported for almost one-third of the occupied blocks. They were relatively easy to confirm as breeders because they use relatively conspicuous cavities as nest sites.

The Western Bluebird has a spotty breeding distribution in the states surrounding Nevada. It is fairly widespread in Oregon (Adamus et al. 2001), California (Small 1994), and Arizona (Spence in Corman and Wise-Gervais 2005:430–431), but it is absent from portions of all three states. In Utah, the species is common in the south but rare to absent in the north and west (Behle et al. 1985). In Idaho, it is limited mainly to the western and northern parts of the state (Stephens and Sturts 1998).

On the whole, the Western Bluebird is a woodland species, but one with a clear attraction to forest edges and openings (Guinan et al. 2000). In Nevada, Western Bluebirds are primarily associated with open yellow pine forests in the mountains (Neel 1999a), but they also breed in pinyon-juniper woodlands and in riparian settings away from the mountains (Chisholm and Neel 2002). The predictive map shows a low chance of finding this species throughout the state, even in the mountainous areas where the chances are greatest. Their predicted occurrence in the north is influenced by unconfirmed outlying records, and field observations suggest that the species does not actually occupy many of the interior mountain ranges. Nonetheless, the map indicates that many of these "underbirded" areas may warrant another look.

CONSERVATION AND MANAGEMENT

Rangewide, Western Bluebird population trends appear to be stable, although some data suggest declines in California (Guinan et al. 2000, Sauer et al. 2005). Western Bluebirds are obligate cavity nesters, and they are drawn to burned areas where tree cavities are likely to occur (Neel 1999a). Elsewhere, they have reportedly been harmed by some silvicultural practices such as clearcutting, snag removal, and fire suppression (Guinan et al. 2000).

The Western Bluebird is a Partners in Flight Priority species for Nevada, and a variety of conservation measures have been suggested, including preserving forest remnants in montane parkland habitats, retaining large trees in woodlands with open canopies, and selective thinning in overgrown forests (Neel 1999a). Nest box supplementation in suitable habitat has been beneficial (Tenney in Roberson and Tenney 1993:284–285, Guinan et al. 2000). Competition with European Starlings for nest cavities may be a problem in some areas (Tenney in Roberson and Tenney 1993:284–285, Guinan et al. 2000; but see Koenig 2003) and deserves further study in Nevada.

HABITAT USE

HABITAT	ATLAS BLOCKS	INCIDENTAL OBSERVATIONS
Agricultural	1 (4%)	0 (0%)
Alpine	1 (4%)	0 (0%)
Montane Forest	9 (33%)	17 (68%)
Montane Parkland	5 (19%)	1 (4%)
Montane Shrub	1 (4%)	1 (4%)
Pinyon-Juniper	5 (19%)	1 (4%)
Riparian	3 (11%)	3 (12%)
Sagebrush Scrub	2 (7%)	0 (0%)
Urban	0 (0%)	2 (8%)
TOTAL	27	25

Total numbers reported in the Habitat Use, Breeding Status, and Abundance tables may differ from each other (see pp. 22–23 for details). Percentage sums may differ slightly from 100% due to rounding.

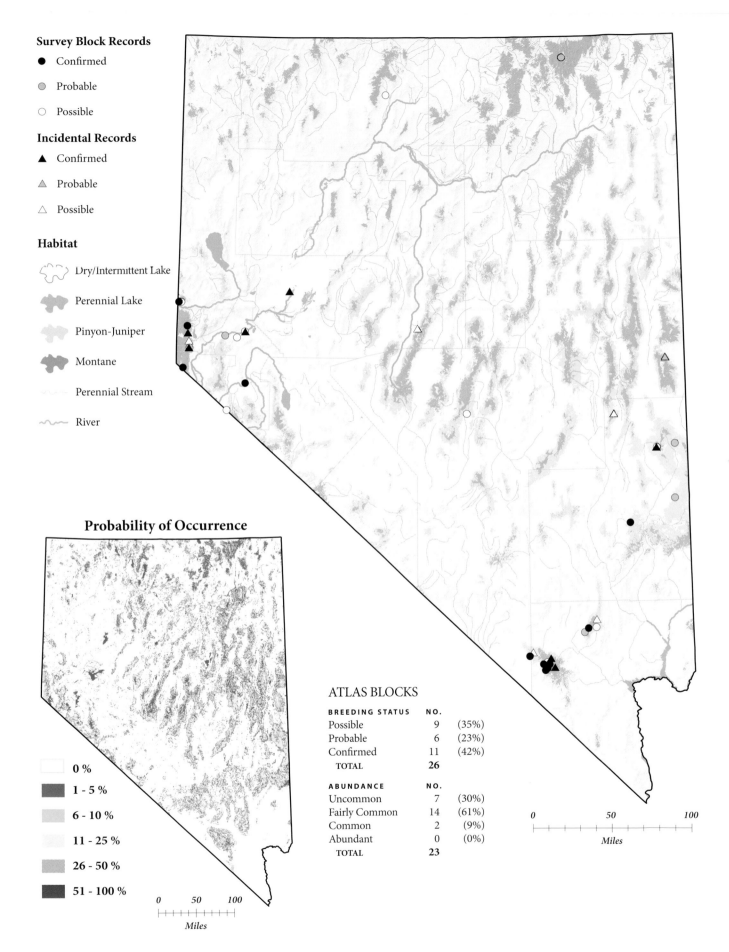

Survey Block Records

● Confirmed

● Probable

○ Possible

Incidental Records

▲ Confirmed

▲ Probable

△ Possible

Habitat

Dry/Intermittent Lake

Perennial Lake

Pinyon-Juniper

Montane

Perennial Stream

River

Probability of Occurrence

	0 %
	1 - 5 %
	6 - 10 %
	11 - 25 %
	26 - 50 %
	51 - 100 %

0   50   100
Miles

ATLAS BLOCKS

BREEDING STATUS	NO.	
Possible	9	(35%)
Probable	6	(23%)
Confirmed	11	(42%)
TOTAL	26	

ABUNDANCE	NO.	
Uncommon	7	(30%)
Fairly Common	14	(61%)
Common	2	(9%)
Abundant	0	(0%)
TOTAL	23	

0   50   100
Miles

MOUNTAIN BLUEBIRD

Sialia currucoides

Nevada's state bird is a thing of beauty—washed all over in azure, as if a bit of sky had fallen to earth. Mountain Bluebirds shun the heavily forested habitats of "typical" thrushes and prefer instead to roam mountain meadows, alpine snowfields, sparse pinyon-juniper woodlands, and sagebrush deserts. They have adapted well to human-altered landscapes and have benefited from nest box supplementation in Nevada and elsewhere. Their future here is probably fairly secure.

DISTRIBUTION

Atlas workers found Mountain Bluebirds throughout the state, with the highest concentration of records in the northeastern and central areas. A high breeding confirmation rate reflected the ease of observing this species at its nest cavities. Mountain Bluebirds have long been considered widespread and common in Nevada (Linsdale 1936, Ryser 1985, Alcorn 1988), although more recent descriptions talk of waxing and waning populations in specific locations (Ryser 1985, Alcorn 1988).

The Mountain Bluebird is generally common and widespread at higher elevations in adjacent states as well (Garrett and Dunn 1981, Monson and Phillips 1981, Behle et al. 1985, Small 1994, Stephens and Sturts 1998, Adamus et al. 2001, Spence in Corman and Wise-Gervais 2005:432–433). It is largely absent as a breeder, however, from coastal and southern California (Small 1994), southern and western Arizona (Spence in Corman and Wise-Gervais 2005:432–433), and western Oregon (Adamus et al. 2001).

Mountain Bluebirds favor semiopen country and seem drawn to transitional areas and habitat edges (Power and Lombardo 1996). In the Great Basin, they are most common at mod-

erate to high elevations (Ryser 1985) in a wide variety of habitats ranging from arid shrublands to montane forests. Farther south in Nevada they are more restricted to montane habitats (Fisher 1893, Linsdale 1936). The predictive map illustrates these relationships well, with high probabilities predicted for areas with pinyon-juniper and montane forests, and progressively lower probabilities as elevation decreases. It seems likely that local habitat features such as the presence of dead trees, cavities, and forest openings influence the occurrence of this species at the local scale.

CONSERVATION AND MANAGEMENT

Overall, Mountain Bluebird populations seem to be increasing, especially north of Nevada (Sauer et al. 2005), but regional population fluctuations were recorded during the twentieth century (Power and Lombardo 1996). For example, some data suggest that populations decreased in Utah and New Mexico during the past century (Behle et al. 1985, Sauer et al. 2005). Conversely, there were range extensions into the Great Plains around 1900 (Power and Lombardo 1996).

Mountain Bluebirds are obligate cavity nesters that are especially partial to tree cavities excavated by Northern Flickers (Dobkin et al. 1995). They are not restricted to tree cavities, however, and nest in sites as diverse as nest boxes, rock crevices above timberline, and crannies in the tufa towers around Mono Lake (Gaines 1992). They have benefited from the indirect effects of forest clearing and livestock grazing, while fire suppression is thought to have negative impacts (Power and Lombardo 1996). Because of its specific habitat needs and restricted distribution, Partners in Flight has designated the Mountain Bluebird a Stewardship species for the Intermountain West (Rich et al. 2004).

In Nevada, there is little cause for immediate concern about this species. Local declines are always a possibility, but fortunately, Mountain Bluebirds respond well to human intervention on their behalf (Weitzel 1988, Power and Lombardo 1996).

HABITAT USE

HABITAT	ATLAS BLOCKS	INCIDENTAL OBSERVATIONS
Agricultural	1 (<1%)	1 (2%)
Alpine	14 (7%)	1 (2%)
Grassland	1 (<1%)	1 (2%)
Montane Forest	38 (18%)	10 (21%)
Montane Parkland	7 (3%)	4 (8%)
Montane Shrub	33 (16%)	3 (6%)
Pinyon-Juniper	77 (36%)	13 (27%)
Riparian	17 (8%)	7 (15%)
Salt Desert Scrub	2 (<1%)	0 (0%)
Sagebrush Scrub	5 (2%)	3 (6%)
Sagebrush Steppe	14 (7%)	3 (6%)
Urban	2 (<1%)	2 (4%)
Wetland	1 (<1%)	0 (0%)
TOTAL	212	48

Total numbers reported in the Habitat Use, Breeding Status, and Abundance tables may differ from each other (see pp. 22–23 for details). Percentage sums may differ slightly from 100% due to rounding.

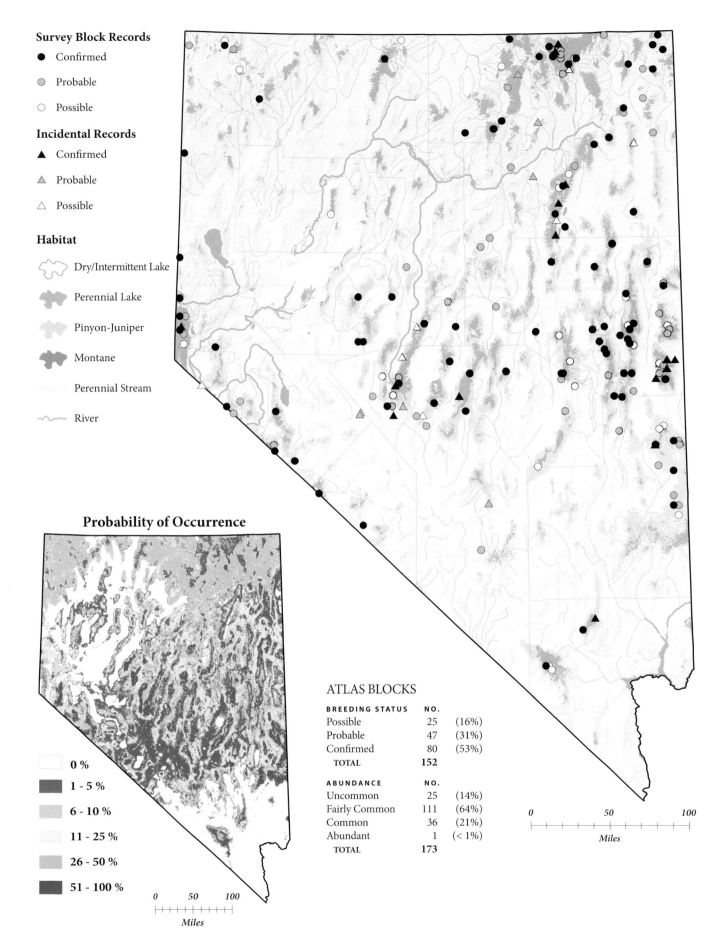

Survey Block Records

- ● Confirmed
- ● Probable
- ○ Possible

Incidental Records

- ▲ Confirmed
- ▲ Probable
- △ Possible

Habitat

- Dry/Intermittent Lake
- Perennial Lake
- Pinyon-Juniper
- Montane
- Perennial Stream
- River

Probability of Occurrence

- ☐ 0 %
- 1 - 5 %
- 6 - 10 %
- 11 - 25 %
- 26 - 50 %
- 51 - 100 %

0   50   100
Miles

ATLAS BLOCKS

BREEDING STATUS	NO.	
Possible	25	(16%)
Probable	47	(31%)
Confirmed	80	(53%)
TOTAL	**152**	

ABUNDANCE	NO.	
Uncommon	25	(14%)
Fairly Common	111	(64%)
Common	36	(21%)
Abundant	1	(< 1%)
TOTAL	**173**	

0   50   100
Miles

TOWNSEND'S SOLITAIRE

Myadestes townsendi

A Townsend's Solitaire perched atop a lonely pine is not much to look at: lean, gray, and nondescript. In flight, the bird is considerably more impressive, twisting wildly as it pursues a flying insect, all the while flashing the white in its tail and the bright buff on its wings. Back at its perch, the solitaire sings madly: a complex series of short, slurred phrases run together in a rich, rambling warble. For the duration of its song, it is the most commanding presence in the forest. Then, song completed, the Townsend's Solitaire resumes its role of quiet observer.

DISTRIBUTION

Townsend's Solitaires occupied most of the major mountain ranges throughout the state, although not in large numbers. The main clusters of records came from the Carson Range; the Spring, Jarbidge, and Ruby mountains; and from eastern White Pine County. Abundance estimates indicated that 87% of the occupied blocks had ten or fewer pairs. The atlas findings are largely consistent with earlier descriptions of the species' distribution and abundance patterns in Nevada (Linsdale 1936, Ryser 1985, Alcorn 1988).

Townsend's Solitaires breed in all of our neighboring states, with the broadest distribution in Utah (Behle et al. 1985). They are fairly common in parts of California (Small 1994) and in the mountains of Arizona (Spence in Corman and Wise-Gervais 2005:434–435), and they are widespread in much of Oregon but

scarce in the region adjoining Nevada (Adamus et al. 2001). In Idaho, they have a northerly distribution and are scarce or absent in the state's southern regions during the breeding season (Stephens and Sturts 1998).

Townsend's Solitaire has a well-known association with juniper woodlands—or, more specifically, with juniper berries—but this is primarily a winter phenomenon. During the nesting season it is mainly a bird of the tall montane forests (Bowen 1997), favoring an open forest structure with a sparse understory (Ryser 1985, Bowen 1997). Nesting Townsend's Solitaires may also be found across a wide elevational gradient, from above timberline down to mid-elevation pinyon-juniper woodlands and riparian zones (Behle et al. 1985, Bowen 1997). As is true for other species that breed primarily in montane forests, the predictive model straightforwardly suggests a fair chance of finding the species in most of Nevada's upper elevations. The predictive model is probably reliable, since Townsend's Solitaires use a range of montane forest types and appear capable of persisting in the state's isolated interior mountain ranges.

CONSERVATION AND MANAGEMENT

Regulation of Townsend's Solitaire populations is not well understood (Bowen 1997), and no clear rangewide trends are evident (Sauer et al. 2005). Populations may respond positively to selective logging and low-intensity burns, but the effects of more intensive forestry practices are unknown (Bowen 1997, Smith et al. 1997). Townsend's Solitaires typically make elevational migrations, and land management practices that affect lower-elevation, mature junipers, where these birds winter, may potentially affect populations.

The population status of Townsend's Solitaire in Nevada is unknown, and the species is often overlooked by observers (Bowen 1997). Although specific conservation actions are not recommended at present, additional monitoring data are needed in Nevada and elsewhere to understand population trends in our region and this species' response to forest management practices (Bowen 1997).

HABITAT USE

HABITAT	ATLAS BLOCKS	INCIDENTAL OBSERVATIONS
Alpine	3 (3%)	1 (3%)
Barren	1 (<1%)	0 (0%)
Montane Forest	67 (61%)	17 (59%)
Montane Parkland	7 (6%)	0 (0%)
Montane Shrub	9 (8%)	3 (10%)
Pinyon-Juniper	10 (9%)	3 (10%)
Riparian	12 (11%)	5 (17%)
TOTAL	109	29

Total numbers reported in the Habitat Use, Breeding Status, and Abundance tables may differ from each other (see pp. 22–23 for details). Percentage sums may differ slightly from 100% due to rounding.

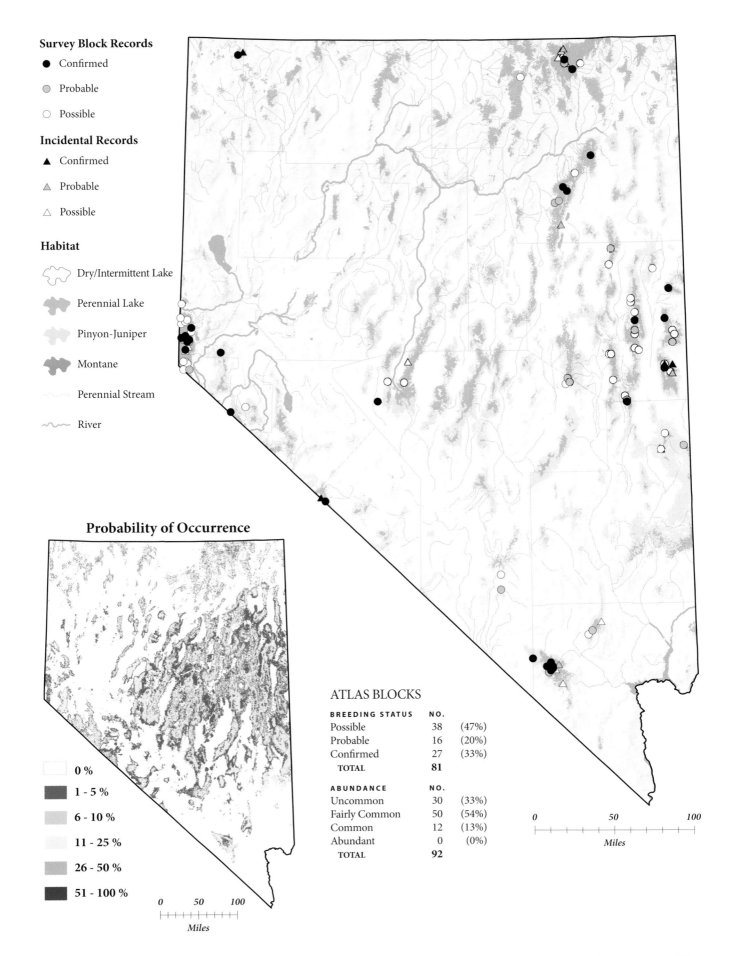

Survey Block Records

● Confirmed

◐ Probable

○ Possible

Incidental Records

▲ Confirmed

△ Probable

△ Possible

Habitat

Dry/Intermittent Lake

Perennial Lake

Pinyon-Juniper

Montane

Perennial Stream

River

Probability of Occurrence

	0 %
	1 - 5 %
	6 - 10 %
	11 - 25 %
	26 - 50 %
	51 - 100 %

0   50   100
Miles

ATLAS BLOCKS

BREEDING STATUS	NO.	
Possible	38	(47%)
Probable	16	(20%)
Confirmed	27	(33%)
TOTAL	81	

ABUNDANCE	NO.	
Uncommon	30	(33%)
Fairly Common	50	(54%)
Common	12	(13%)
Abundant	0	(0%)
TOTAL	92	

0   50   100
Miles

SWAINSON'S THRUSH
Catharus ustulatus

With its restricted geographic range and narrow ecological niche in the Great Basin, Swainson's Thrush is not a bird commonly encountered by most Nevada birders. It can reach fairly high numbers in the mountains of northeastern Nevada, however, and occurs in lower numbers in several other mountain ranges scattered around the state. Were it not for the distinctive and far-reaching song, Swainson's Thrush would surely have been overlooked in many instances. It conceals its nest well; here in Nevada it tends to nest in thick deciduous and riparian forests with dense understories.

DISTRIBUTION

Reports of Swainson's Thrushes came mainly from White Pine and Elko counties. The breeding confirmation rate was 17%, and only one nest was actually discovered. Most records of possible breeding were based only on the presence of singing males, suggesting that some of the southernmost records might involve late migrants. Apart from the core area in the northeast, records were scattered widely about the state, but none of these confirmed breeding. Although the distribution of Swainson's Thrush in Nevada is limited, its numbers tended to be high in the blocks where it was present. Fieldworkers estimated more than ten pairs in nearly one-third of the occupied blocks, which contrasts somewhat with earlier accounts that characterize the species as uncommon (Linsdale 1936, Alcorn 1988).

In terms of distribution, however, the atlas survey's results for Swainson's Thrush are consistent with earlier accounts from Nevada (Linsdale 1936, Ryser 1985, Alcorn 1988). Linsdale (1936) reported the species from the Toiyabe Range in central Nevada and in far western Nevada, which corresponds with several possible and probable atlas records. Given these records, it seems likely that the species actually does breed in low numbers in these locations and perhaps elsewhere in the state. The Nevada breeding population is clearly concentrated in the northeast, however.

Outside Nevada, Swainson's Thrushes are generally common as breeders in forested habitats to the north and east (Behle et al. 1985, Stephens and Sturts 1998, Adamus et al. 2001); locally common in the northern, central, and coastal mountains of California (Small 1994); and rare in Arizona (Corman in Corman and Wise-Gervais 2005:436–437). Throughout most of its breeding range this thrush is strongly associated with coniferous, especially spruce-fir, forests (Evans Mack and Yong 2000). In Nevada and elsewhere along its southern breeding boundary, however, it is more likely to nest in broadleaved montane forests or riparian woodlands that have a well-developed understory (Ryser 1985, Pantle in Kingery 1998:392–393, Evans Mack and Yong 2000, Dobkin and Sauder 2004). The predictive map for this species points to many areas of presumably suitable habitat where additional surveys may find breeding thrushes, although it may overestimate the probability of breeding populations in southern Nevada, where only possible breeding was reported.

CONSERVATION AND MANAGEMENT

In the past two decades, Swainson's Thrush has suffered declines rangewide (Sauer et al. 2005), including in the West (Evans Mack and Yong 2000, Dobkin and Sauder 2004). Suggested causes include cowbird parasitism and loss of riparian habitat (Evans Mack and Yong 2000). At present, however, populations are considered stable in at least some parts of the West (Dobkin and Sauder 2004).

In Nevada, the species is likely limited by the restricted extent of its favored habitat. The species' response to logging varies geographically (Evans Mack and Yong 2000, Dobkin and Sauder 2004), and conservation strategies for this thrush in Nevada, aside from protecting montane broadleaved and riparian woodlands, are unclear.

HABITAT USE

HABITAT	ATLAS BLOCKS	INCIDENTAL OBSERVATIONS
Montane Forest	19 (51%)	1 (14%)
Montane Shrub	3 (8%)	0 (0%)
Pinyon-Juniper	3 (8%)	0 (0%)
Riparian	12 (32%)	6 (86%)
TOTAL	37	7

Total numbers reported in the Habitat Use, Breeding Status, and Abundance tables may differ from each other (see pp. 22–23 for details). Percentage sums may differ slightly from 100% due to rounding.

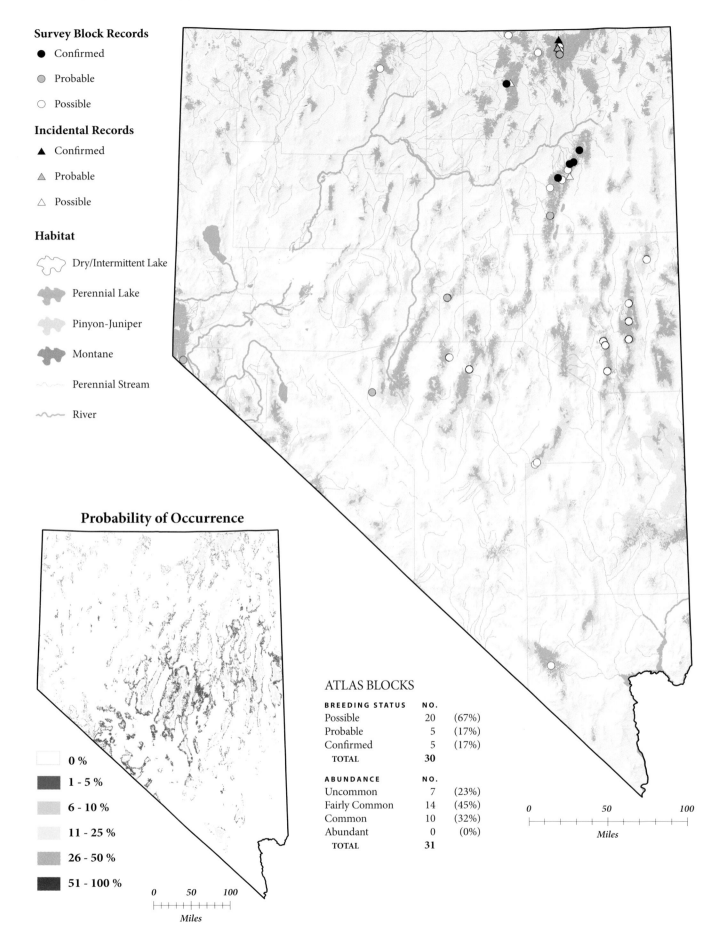

Survey Block Records

● Confirmed

◐ Probable

○ Possible

Incidental Records

▲ Confirmed

△ Probable

△ Possible

Habitat

▱ Dry/Intermittent Lake

Perennial Lake

Pinyon-Juniper

Montane

Perennial Stream

River

Probability of Occurrence

☐ 0 %

■ 1 - 5 %

6 - 10 %

11 - 25 %

26 - 50 %

■ 51 - 100 %

0   50   100
Miles

ATLAS BLOCKS

BREEDING STATUS	NO.	
Possible	20	(67%)
Probable	5	(17%)
Confirmed	5	(17%)
TOTAL	**30**	

ABUNDANCE	NO.	
Uncommon	7	(23%)
Fairly Common	14	(45%)
Common	10	(32%)
Abundant	0	(0%)
TOTAL	**31**	

0   50   100
Miles

HERMIT THRUSH
Catharus guttatus

There is an inscrutable quality to the vocalizations of the Hermit Thrush. The song of a single bird permeates the forest but can be hard to pinpoint in time or space. The sound of the evening chorus is overwhelming: intensely sonorous and yet somehow unobtrusive.

DISTRIBUTION

Hermit Thrushes showed the classic distribution pattern of Nevada's high-mountain avifauna: a broad swath of breeding records running east-to-west across the center of the state intersected by a perpendicular north-to-south column of records through the eastern counties. The confirmation rate for breeding was relatively low, and the proportion of records classified as "possible breeders" was high, no doubt reflecting the difficulty of getting close looks at these unobtrusive inhabitants of shady woods. They were not scarce, though: 85% of the abundance estimates were for multiple pairs, and almost a third of the occupied blocks were estimated to have more than ten pairs.

Older reports describe the Hermit Thrush as widely distributed in Nevada. Linsdale (1936) listed five subspecies that occur in Nevada, three of which breed here. Ryser (1985) and Alcorn (1988) considered the Hermit Thrush to be a common summer resident in montane forests statewide. Beyond Nevada's borders, gaps in suitably forested areas create two noticeable holes in the Hermit Thrush's western breeding distribution (Jones and Donovan 1996): one just to our north, in southeastern Oregon (Adamus et al. 2001) and southwestern Idaho (Stephens and Sturts 1998), and another to our south, in southern California (Garrett and Dunn 1981, Small 1994). Otherwise, the species is common in adjacent areas wherever mid- to upper-elevation forests exist (Monson and Phillips 1981, Behle et al. 1985, Small 1994, Adamus et al. 2001, Spence in Corman and Wise-Gervais 2005:438–439).

Throughout its extensive range the Hermit Thrush uses a broad spectrum of forested habitat types (Jones and Donovan 1996). In the interior West, it is associated primarily with coniferous or mixed coniferous-deciduous forests (Jones and Donovan 1996), including aspen and mountain mahogany stands (Ryser 1985). Populations in the Sierra Nevada seem to favor drier and more open conifer forests than do their coastal counterparts (Small 1994). Ryser (1985), however, noted that Hermit Thrush nests are most frequently located in close proximity to streams in Nevada, a pattern that this species apparently shares with Swainson's Thrush, its close relative. The predictive model clearly reflects the Hermit Thrush's association with montane forests throughout Nevada and predicts little chance of finding it as a breeder anywhere else.

CONSERVATION AND MANAGEMENT

The Hermit Thrush is a taxonomically complex species that exhibits strong geographic variation throughout its range (Jones and Donovan 1996). Monson and Phillips (1981) listed seven subspecies from Arizona alone. Rangewide, numbers are increasing, particularly in the East (Sauer et al. 2005). Western trends are less clear. In California, habitat loss and urbanization may have caused local declines (Roberson in Roberson and Tenney 1993:288–289), but the overall status of the species seems to be stable for now (Sauer et al. 2005).

The effects of human activities on Hermit Thrush populations in Nevada are not known. Changes in forest management can have diverse and contrasting impacts on this species, but specific conservation strategies have not yet been established (Jones and Donovan 1996).

HABITAT USE

HABITAT	ATLAS BLOCKS	INCIDENTAL OBSERVATIONS
Alpine	4 (3%)	0 (0%)
Montane Forest	104 (70%)	26 (58%)
Montane Parkland	6 (4%)	0 (0%)
Montane Shrub	6 (4%)	0 (0%)
Pinyon-Juniper	10 (7%)	5 (11%)
Riparian	17 (11%)	13 (29%)
Sagebrush Scrub	1 (<1%)	0 (0%)
Sagebrush Steppe	0 (0%)	1 (2%)
TOTAL	148	45

Total numbers reported in the Habitat Use, Breeding Status, and Abundance tables may differ from each other (see pp. 22–23 for details). Percentage sums may differ slightly from 100% due to rounding.

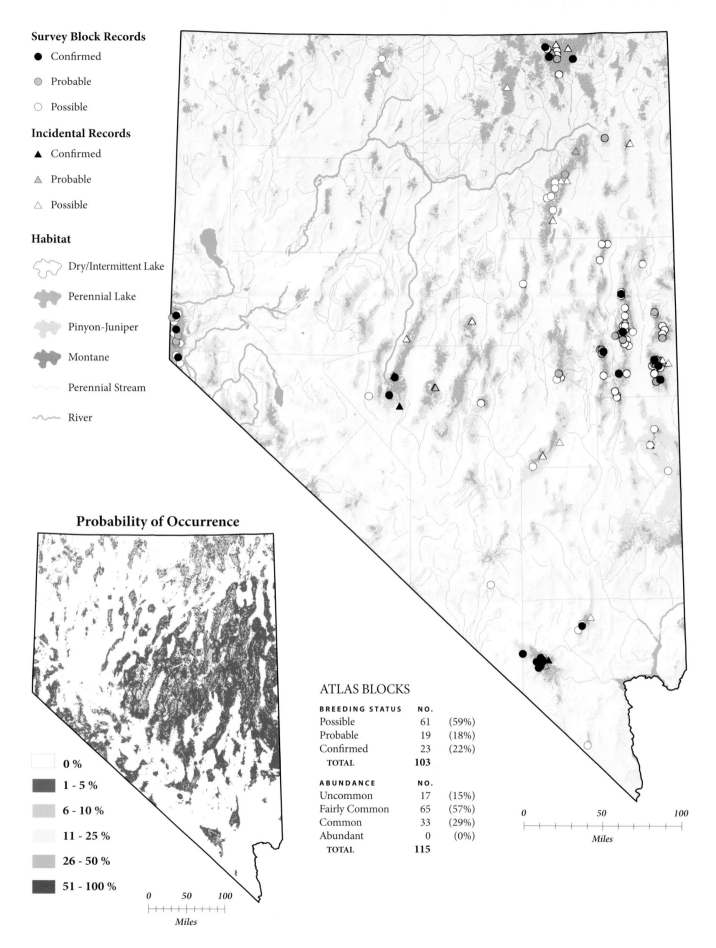

Survey Block Records

● Confirmed

● Probable

○ Possible

Incidental Records

▲ Confirmed

▲ Probable

△ Possible

Habitat

Dry/Intermittent Lake

Perennial Lake

Pinyon-Juniper

Montane

Perennial Stream

River

Probability of Occurrence

0 %

1 - 5 %

6 - 10 %

11 - 25 %

26 - 50 %

51 - 100 %

0   50   100

Miles

ATLAS BLOCKS

BREEDING STATUS	NO.	
Possible	61	(59%)
Probable	19	(18%)
Confirmed	23	(22%)
TOTAL	**103**	

ABUNDANCE	NO.	
Uncommon	17	(15%)
Fairly Common	65	(57%)
Common	33	(29%)
Abundant	0	(0%)
TOTAL	**115**	

0   50   100

Miles

AMERICAN ROBIN
Turdus migratorius

Rn'02

The American Robin is one of the first songbirds to set up its territory each year, and its loud caroling is a sure sign of spring. It is large, conspicuous, easy to identify, and so tame and common that nearly everyone with a lawn is acquainted with its entertaining routine of hunting and capturing earthworms. Because American Robins are such familiar and easily observed birds in urban settings, it is easy to forget that most of them nest in Nevada's remoter woodlands, where they seem to be curiously shy and sensitive to disturbance.

DISTRIBUTION

Atlas workers found American Robins throughout Nevada, and the ease of observing nests and young resulted in a high confirmation rate (61%). American Robins were not uniformly distributed across the state, however, as earlier authors also noted (Linsdale 1936, Alcorn 1998). Observations tended to cluster in the major mountain ranges, and records were more frequent toward the north, a pattern also noted by Alcorn (1988). There were records, but no confirmed breeders, in the hot lowlands of south-central Nevada, just as the Death Valley Expedition reported a hundred years earlier (Fisher 1893). The predictive map reflects both the species' montane association and its latitudinal trend in abundance.

The degree to which American Robins thrive in both human-altered and natural settings is unusual for a native songbird. They are common and widespread throughout the United States and Canada (Behle et al. 1985, Small 1994, Stephens and Sturts 1998, Sallabanks and James 1999, Adamus et al. 2001, Wise-Gervais in Corman and Wise-Gervais 2005:440–441). Despite their ubiquity, however, robins do have some clear habitat associations. In Nevada, atlas workers almost never found them in arid, open

habitats such as salt desert scrub and Mojave, and they were rare in sagebrush habitats. Although many atlas observations came from urban and agricultural habitats, these still constituted a fairly small proportion of all records. The range of native habitats used for nesting is broad (Sallabanks and James 1999), and it includes most woodland types, riparian groves, and even shrublands and grasslands if they have scattered trees.

CONSERVATION AND MANAGEMENT

Most populations of the American Robin are stable or increasing (Sallabanks and James 1999), but moderate declines have been reported for Oregon, California, and Utah (Sauer et al. 2005). Given these declines, it is interesting that monitoring data from Nevada indicate stable populations in recent decades (Sauer et al. 2005).

In Nevada and elsewhere, the American Robin has flourished in spite of human activity (Ryser 1985), and there is a temptation for conservationists to ignore the species. But one should not take the robin, or any common bird, for granted, because even local declines in common species tell us something important about our regional environment. For instance, Sallabanks and James (1999) related several episodes in which declining American Robin populations indicated pesticide contamination. Eiserer (1976) described the perils of taking the familiar for granted in the dedication to his book on robins: "To Robins everywhere, that they may continue to prosper beyond human reckoning of time; and to California Condors of eons now passed, that they may forgive us for what we have done."

HABITAT USE

HABITAT	ATLAS BLOCKS	INCIDENTAL OBSERVATIONS
Agricultural	15 (3%)	0 (0%)
Alpine	3 (<1%)	1 (<1%)
Ash	0 (0%)	1 (<1%)
Grassland	1 (<1%)	0 (0%)
Mesquite	1 (<1%)	1 (<1%)
Montane Forest	120 (24%)	34 (22%)
Montane Parkland	10 (2%)	4 (3%)
Montane Shrub	35 (7%)	4 (3%)
Open Water	0 (0%)	1 (<1%)
Pinyon-Juniper	60 (12%)	8 (5%)
Riparian	187 (37%)	75 (48%)
Sagebrush Scrub	10 (2%)	4 (3%)
Sagebrush Steppe	25 (5%)	1 (<1%)
Urban	39 (8%)	23 (15%)
Wetland	3 (<1%)	0 (0%)
TOTAL	509	157

Total numbers reported in the Habitat Use, Breeding Status, and Abundance tables may differ from each other (see pp. 22–23 for details). Percentage sums may differ slightly from 100% due to rounding.

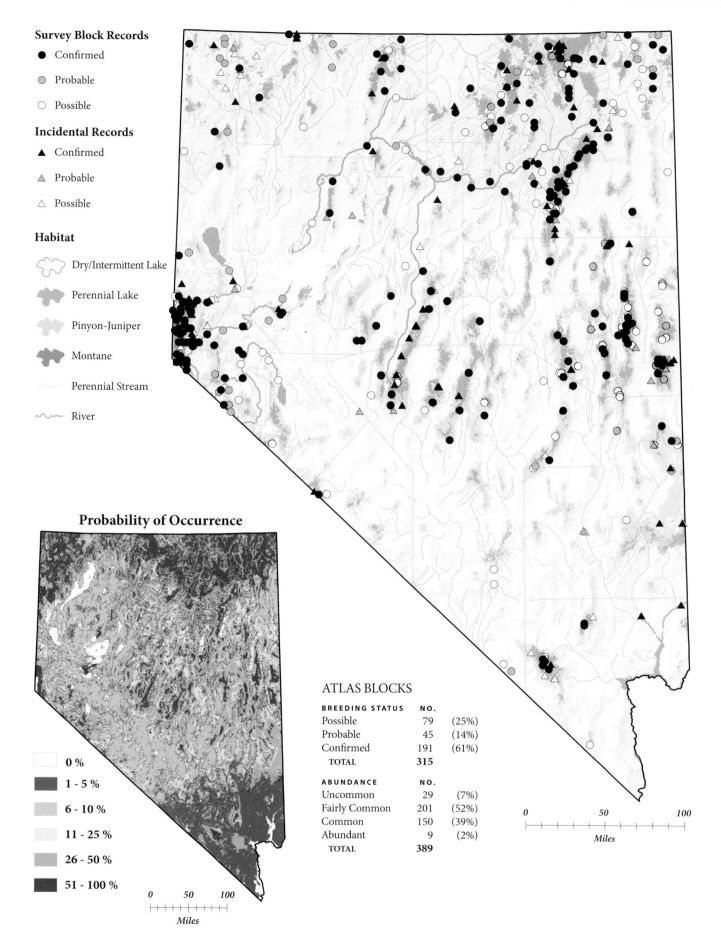

Survey Block Records

● Confirmed

● Probable

○ Possible

Incidental Records

▲ Confirmed

▲ Probable

△ Possible

Habitat

Dry/Intermittent Lake

Perennial Lake

Pinyon-Juniper

Montane

Perennial Stream

River

Probability of Occurrence

0 %

1 - 5 %

6 - 10 %

11 - 25 %

26 - 50 %

51 - 100 %

0   50   100

Miles

ATLAS BLOCKS

BREEDING STATUS	NO.	
Possible	79	(25%)
Probable	45	(14%)
Confirmed	191	(61%)
TOTAL	**315**	

ABUNDANCE	NO.	
Uncommon	29	(7%)
Fairly Common	201	(52%)
Common	150	(39%)
Abundant	9	(2%)
TOTAL	**389**	

0   50   100

Miles

NORTHERN MOCKINGBIRD
Mimus polyglottos

It seems silly to ask which Nevada bird is the best vocalist—not because of the subjective nature of the question, but because of the patently obvious answer. The repertoire of an individual Northern Mockingbird may easily exceed a hundred song elements, each a near-perfect transcription of the song or call of another species. Although the mockingbird is most notable for its vocal abilities, it is interesting for many other reasons as well, including its vigorous territorial defense against all comers and its remarkable adaptability to human-modified landscapes.

DISTRIBUTION

Northern Mockingbirds were found throughout the state, but most of the records came from southern Nevada. Hotspots included the Virgin and Muddy river lowlands, Ash Meadows National Wildlife Refuge, and especially the Las Vegas Valley. Mockingbirds were not especially numerous in these areas; 85% of the occupied blocks were estimated to have ten or fewer pairs. Collectively, the older sources concur with the atlas result that indicates a latitudinal gradient in abundance in Nevada, with the species being common in the south and rare to uncommon farther north (Fisher 1893, Linsdale 1936, Ryser 1985, Alcorn 1988). In the Lahontan Valley, for example, which has seemingly abundant suitable habitat, the mockingbird is a very uncommon summer resident (Chisholm and Neel 2002). The predictive map clearly reflects this trend of decreasing probability of occurrence with increasing latitude within the state, broken up by potential hotspots in urban areas.

The mockingbird's distribution in surrounding states reflects a similar latitudinal gradient. In Arizona, California, and southern Utah, the Northern Mockingbird is a common or fairly common breeder (Behle et al. 1985, Small 1994, Wise-Gervais in Corman and Wise-Gervais 2005:444–445). It is far less common in northern Utah, and in Oregon and Idaho it is still a recent colonist and remains quite rare (Behle et al. 1985, Derrickson and Breitwisch 1992, Stephens and Sturts 1998, Adamus et al. 2001).

Mockingbirds are omnivores, and versatile ones at that. They occur in a wide variety of habitats, but they seem most successful wherever humans have altered the landscape (Derrickson and Breitwisch 1992). They are abundant in developed regions with exotic landscaping (Garrett and Dunn 1981), but they are also found among native shrubs such as greasewood (Chisholm and Neel 2002) and buffaloberry (Gaines 1992).

CONSERVATION AND MANAGEMENT

The mockingbird experienced widespread northerly range extensions in the past century. More recently, it has declined in the Southeast and in other parts of its range (Derrickson and Breitwisch 1992, Sauer et al. 2005). In the West, however, it has generally fared well, with increases in California, Arizona, and apparently in Utah (Behle et al. 1985, Small 1994, Sauer et al. 2005, Wise-Gervais in Corman and Wise-Gervais 2005:444–445). Increases have been greatest in urban regions, but rural landscapes have also proved suitable for mockingbird colonization (Shuford 1993).

Mockingbirds appear to have expanded their range in Nevada as well (Ryser 1985). They are still uncommon in large portions of the state, but Nevada's burgeoning human population may facilitate their continued expansion northward (Ryser 1985). Severe winters regulate some populations (Derrickson and Breitwisch 1992), however, and may serve as a significant impediment to large-scale colonization of Nevada's colder counties.

HABITAT USE

HABITAT	ATLAS BLOCKS	INCIDENTAL OBSERVATIONS
Agricultural	5 (4%)	1 (1%)
Ash	9 (7%)	0 (0%)
Grassland	3 (2%)	0 (0%)
Mesquite	9 (7%)	3 (4%)
Mojave	23 (17%)	14 (19%)
Montane Parkland	1 (<1%)	0 (0%)
Montane Shrub	1 (<1%)	0 (0%)
Open Water	0 (0%)	1 (1%)
Pinyon-Juniper	17 (13%)	1 (1%)
Riparian	9 (7%)	17 (24%)
Salt Desert Scrub	13 (10%)	2 (3%)
Sagebrush Scrub	13 (10%)	4 (6%)
Sagebrush Steppe	6 (4%)	3 (4%)
Urban	26 (19%)	22 (31%)
Wetland	0 (0%)	4 (6%)
TOTAL	135	72

Total numbers reported in the Habitat Use, Breeding Status, and Abundance tables may differ from each other (see pp. 22–23 for details). Percentage sums may differ slightly from 100% due to rounding.

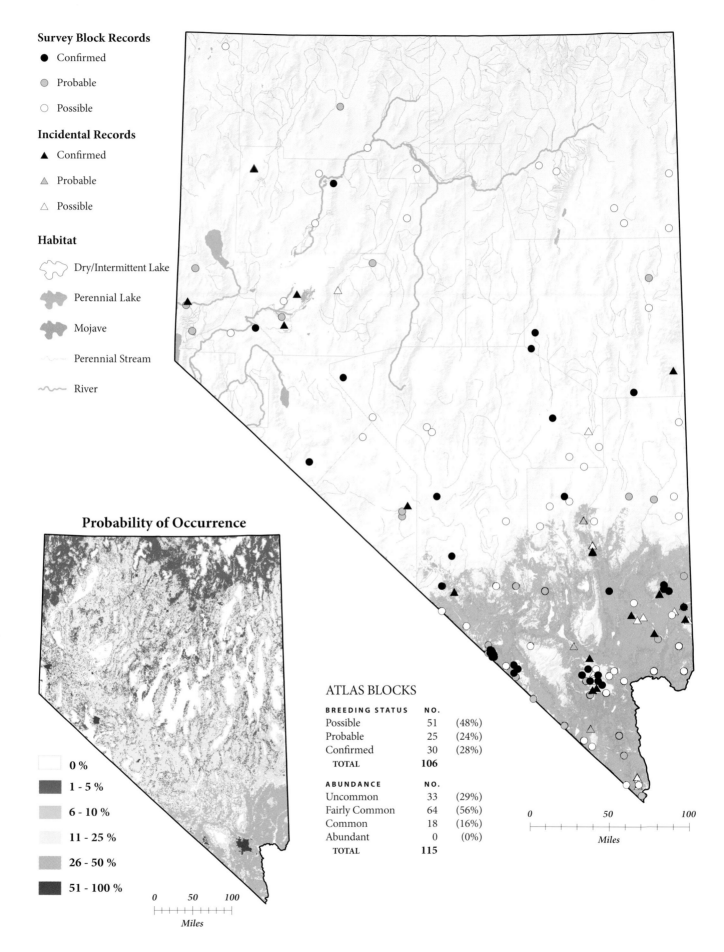

Probability of Occurrence

0 %

1 - 5 %

6 - 10 %

11 - 25 %

26 - 50 %

51 - 100 %

0   50   100

Miles

Survey Block Records

● Confirmed

● Probable

○ Possible

Incidental Records

▲ Confirmed

▲ Probable

△ Possible

Habitat

Dry/Intermittent Lake

Perennial Lake

Mojave

Perennial Stream

River

ATLAS BLOCKS

BREEDING STATUS	NO.	
Possible	51	(48%)
Probable	25	(24%)
Confirmed	30	(28%)
TOTAL	**106**	

ABUNDANCE	NO.	
Uncommon	33	(29%)
Fairly Common	64	(56%)
Common	18	(16%)
Abundant	0	(0%)
TOTAL	**115**	

0   50   100

Miles

SAGE THRASHER

Oreoscoptes montanus

In Nevada, four bird species go by the name of "thrasher." All are shy, shrub-loving songsters, yet they are rarely found together. The three southern species use different habitats, and the northern species—the Sage Thrasher—is generally considered a sagebrush specialist. It favors large expanses of undisturbed, tall sagebrush and is reported to be sensitive to habitat conversion or degradation. Sage Thrasher populations are holding steady where sagebrush remains in good condition, but there is considerable concern about the species' welfare in the face of ongoing rangeland deterioration.

DISTRIBUTION

Sage Thrashers were reported throughout the state, but there were no confirmed nesting records from Clark County. The confirmation rate was fairly high, and Sage Thrashers were numerous where present; more than 40% of block abundance estimates exceeded ten pairs. This pattern of abundant breeding was observed throughout most of the state except in the Mojave Desert, matching earlier accounts of the species (Linsdale 1936, Alcorn 1988, Neel 1999a).

The Sage Thrasher can be found throughout much of the West, but one of its core breeding areas straddles northern Nevada and southeastern Oregon (Adamus et al. 2001, Dobkin and Sauder 2004). Another such core area is in southeastern Idaho and western Wyoming (Stephens and Sturts 1998, Dobkin et al. 2004). Sage Thrashers are also common in parts of Utah (Behle et al. 1985), and they can be locally common in northern Arizona and far eastern California (Garrett and Dunn 1981, Small 1994, Reynolds et al. 1999, Corman in Corman and Wise-Gervais 2005:446–447).

Sage Thrashers are usually associated with intact, fairly dense stands of sagebrush (Holmes and Barton 2003). Big sagebrush is especially favored (Reynolds et al. 1999), but Sage Thrashers

sometimes occur in shrublands dominated by greasewood or bitterbrush (Reynolds et al. 1999). The predictive map reflects these relationships, predicting a high chance of finding the species in sagebrush habitats, a moderate chance in other Great Basin shrublands, and a low chance in montane areas and in the Mojave Desert.

CONSERVATION AND MANAGEMENT

A species of conservation concern throughout much of its range, the Sage Thrasher has been listed or is being considered for listing in several jurisdictions (Reynolds et al. 1999). Recent monitoring data suggest significant declines both rangewide and in Nevada (Sauer et al. 2005). Declines are most likely to occur where sagebrush habitats are degraded or subject to wholesale destruction, conversion, or fragmentation, and populations are likely more stable where large swaths of intact habitat remain (Paige and Ritter 1999, Reynolds et al. 1999, Dobkin and Sauder 2004). Invasion of sagebrush rangelands by cheatgrass is also thought to be a threat (Paige and Ritter 1999, Reynolds et al. 1999). The net effects of livestock grazing, which tends to increase shrub cover but also to facilitate cheatgrass invasion, are unclear (see Reynolds et al. 1999, Dobkin and Sauder 2004).

The Sage Thrasher might benefit greatly from proactive land management. It is a Partners in Flight Priority species in Nevada (Neel 1999a) and a Stewardship species for the Intermountain West (Rich et al. 2004), and it enjoys good publicity for its reputed ability to control Mormon cricket populations (Ryser 1985, Paige and Ritter 1999). Nevada supports a sizable chunk of the species' global breeding population, and the need for quantitative data on population trends and responses to land management is clear (Neel 1999a). Long-term conservation of the Sage Thrasher is likely to require the protection, acquisition, and rehabilitation of suitable habitat (Reynolds et al. 1999).

HABITAT USE

HABITAT	ATLAS BLOCKS	INCIDENTAL OBSERVATIONS
Agricultural	7 (2%)	0 (0%)
Alpine	2 (<1%)	0 (0%)
Mojave	8 (2%)	1 (3%)
Montane Parkland	0 (0%)	1 (3%)
Montane Shrub	12 (3%)	2 (6%)
Pinyon-Juniper	17 (5%)	2 (6%)
Riparian	1 (<1%)	0 (0%)
Salt Desert Scrub	58 (17%)	2 (6%)
Sagebrush Scrub	124 (36%)	13 (41%)
Sagebrush Steppe	120 (34%)	10 (31%)
Wetland	0 (0%)	1 (3%)
TOTAL	349	32

Total numbers reported in the Habitat Use, Breeding Status, and Abundance tables may differ from each other (see pp. 22–23 for details). Percentage sums may differ slightly from 100% due to rounding.

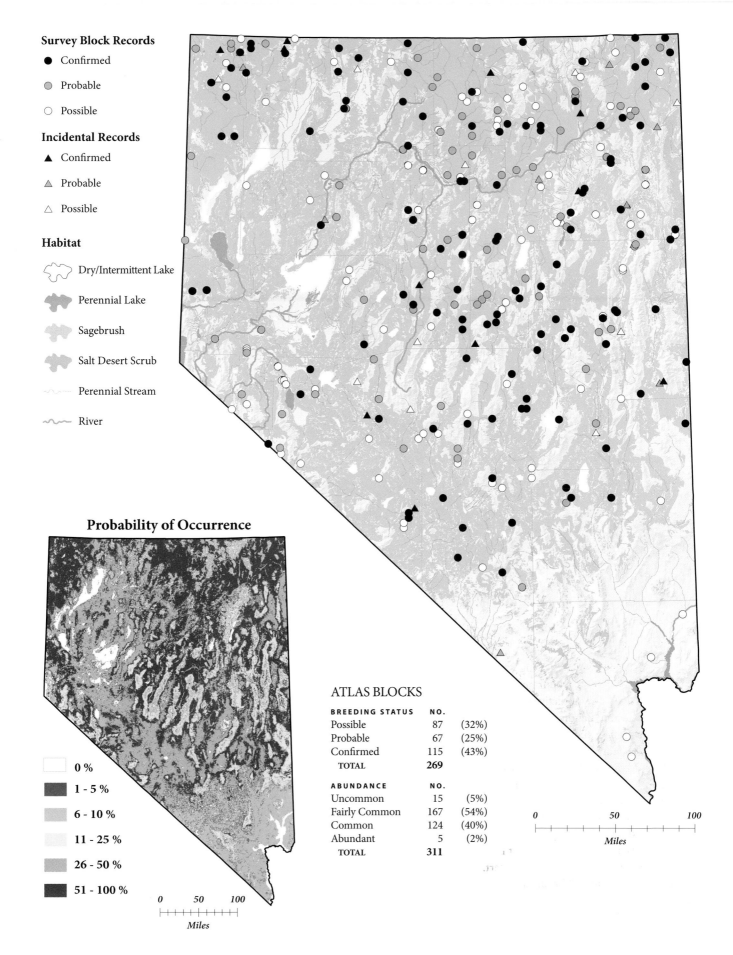

Survey Block Records

- ● Confirmed
- ◐ Probable
- ○ Possible

Incidental Records

- ▲ Confirmed
- ▲ Probable
- △ Possible

Habitat

- Dry/Intermittent Lake
- Perennial Lake
- Sagebrush
- Salt Desert Scrub
- Perennial Stream
- River

Probability of Occurrence

- ▢ 0 %
- ▢ 1 - 5 %
- ▢ 6 - 10 %
- ▢ 11 - 25 %
- ▢ 26 - 50 %
- ▢ 51 - 100 %

0   50   100
Miles

ATLAS BLOCKS

BREEDING STATUS	NO.	
Possible	87	(32%)
Probable	67	(25%)
Confirmed	115	(43%)
TOTAL	**269**	

ABUNDANCE	NO.	
Uncommon	15	(5%)
Fairly Common	167	(54%)
Common	124	(40%)
Abundant	5	(2%)
TOTAL	**311**	

0   50   100
Miles

BENDIRE'S THRASHER
Toxostoma bendirei

Rh '02

Prior to the atlas fieldwork, Bendire's Thrasher was known primarily as a bird that popped up here and there at scattered locales in southern Nevada. Now we know that the species is relatively common in a particular habitat in the southern part of the state: Joshua tree woodlands interspersed with dense grass. In the early spring, this thrasher can be easy to find if one knows the rambling song that it sings while preparing to nest. Later in the year, Bendire's Thrashers can be seen foraging for arthropods on the ground or among Joshua trees.

DISTRIBUTION

Fieldworkers found Bendire's Thrashers in five blocks in southern Nevada: two in Clark County; two in southern Lincoln County, including the Pahranagat Valley; and one at the Nevada Test Site in Nye County. Breeding was not confirmed in any of these blocks, but it was confirmed for two incidental sightings in Clark County. The predictive map reflects the moderate probability of detecting Bendire's Thrashers in suitable habitat in the Mojave Desert. The high-probability zones at the northern edge of this region, however, are almost certainly overestimates.

Older accounts of Nevada birds have little to say about Bendire's Thrasher. Linsdale (1936) did not mention it at all, and the first Nevada record was evidently noted in 1939 (Alcorn 1988). Alcorn (1988:289) ventured to say that "possibly a few are resident in the southern part of the state." England and Laudenslayer (1993), in contrast, stated that Bendire's Thrashers breed, but do not winter, in the Mojave Desert. Farther north, in the Great Basin, Bendire's Thrashers are strictly accidental (Ryser 1985).

Bendire's Thrashers are birds of the hot southwestern deserts,

and most of their range is south of Nevada. They are fairly common but spottily distributed in southern California, more regular in Arizona's Sonoran Desert, and rare in the lower Colorado River Valley (Rosenberg et al. 1991, Small 1994, Corman in Corman and Wise-Gervais 2005:448–449). They are occasionally found in Utah as far north as the Uinta Basin (Behle et al. 1985). The species has not been recorded in Oregon or Idaho (Stephens and Sturts 1998, Adamus et al. 2001).

Bendire's Thrashers prefer low-elevation shrublands and open woodlands, and most atlas records came from Mojave habitat. Typically, they avoid dense vegetation, but they also avoid extremely sparse deserts of the sort used by Le Conte's Thrashers (England and Laudenslayer 1993). This thrasher's partiality for Joshua tree woodlands (Garrett and Dunn 1981) was corroborated by atlas records from the foothills of the McCullough Range. Bendire's Thrashers have also been reported to use agricultural land (Rosenberg et al. 1991) and, occasionally, sagebrush with scattered junipers (England and Laudenslayer 1993).

CONSERVATION AND MANAGEMENT

The scant information available on Bendire's Thrasher is mostly old and anecdotal (England and Laudenslayer 1993), and no reliable population trend data exist (Sauer et al. 2005). Human activities probably have variable effects. Past agricultural activities may have benefited some populations in New Mexico and Arizona (Phillips et al. 1964, Darling 1970), but the species has been locally extirpated by urbanization near Tucson (Emlen 1974).

Bendire's Thrasher is a Watch List species of Partners in Flight, which calls for immediate conservation action due to the restricted continental range and small population size (Rich et al. 2004). It is also an Evaluation species of the Clark County Multiple Species Habitat Conservation Plan (Clark County 2000), which is appropriate given the continuing expansion of Las Vegas into the desert. Bendire's Thrasher deserves further study in Nevada for many reasons. Its current range, status, ecological requirements, and population dynamics are still poorly known, and all of these may have important consequences for preservation.

HABITAT USE

HABITAT	ATLAS BLOCKS	INCIDENTAL OBSERVATIONS
Mesquite	1 (20%)	0 (0%)
Mojave	3 (60%)	3 (60%)
Pinyon-Juniper	1 (20%)	1 (20%)
Salt Desert Scrub	0 (0%)	1 (20%)
TOTAL	5	5

Total numbers reported in the Habitat Use, Breeding Status, and Abundance tables may differ from each other (see pp. 22–23 for details). Percentage sums may differ slightly from 100% due to rounding.

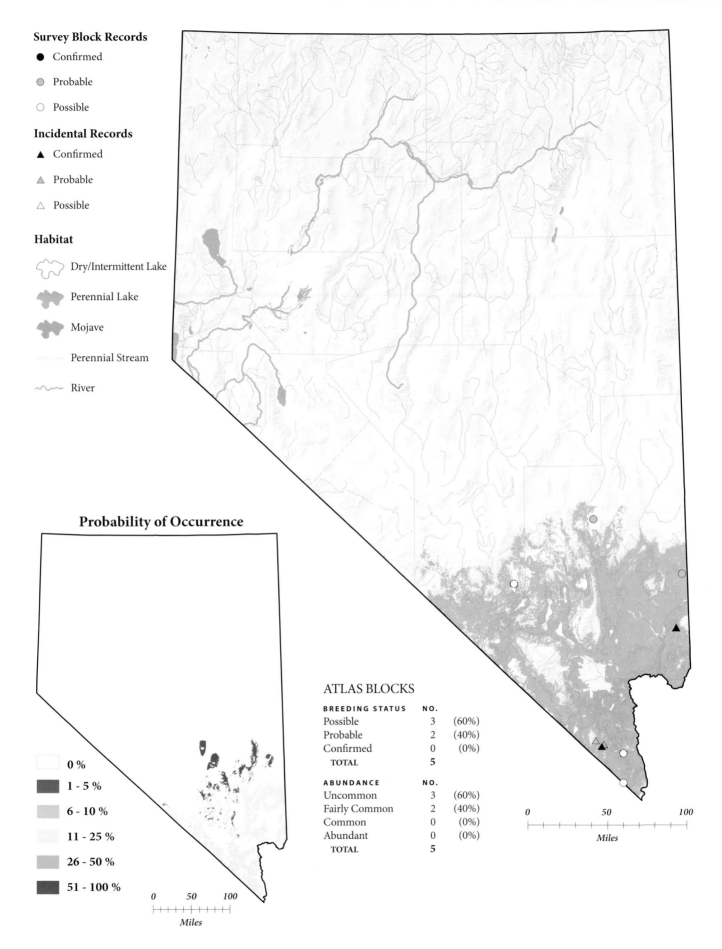

Survey Block Records

● Confirmed

◉ Probable

○ Possible

Incidental Records

▲ Confirmed

△ Probable

△ Possible

Habitat

Dry/Intermittent Lake

Perennial Lake

Mojave

Perennial Stream

River

Probability of Occurrence

☐ 0 %

■ 1 - 5 %

▨ 6 - 10 %

▢ 11 - 25 %

▨ 26 - 50 %

■ 51 - 100 %

0   50   100
Miles

ATLAS BLOCKS

BREEDING STATUS	NO.	
Possible	3	(60%)
Probable	2	(40%)
Confirmed	0	(0%)
TOTAL	**5**	

ABUNDANCE	NO.	
Uncommon	3	(60%)
Fairly Common	2	(40%)
Common	0	(0%)
Abundant	0	(0%)
TOTAL	**5**	

0   50   100
Miles

CRISSAL THRASHER
Toxostoma crissale

Reclusive and agoraphobic, Crissal Thrashers are almost always found in tangled riparian thickets or dense mesquite groves in southern Nevada. The best way to catch a glimpse of this slender, sickle-billed thrasher is to find males at their song posts early in the breeding season. At other times of the year, one must often be content just to listen for the Crissal Thrasher's distinctive *cheer! cheery!* call, which can be heard at all seasons in Mojave Desert lowland and foothill locations that provide luxuriant, impenetrable shrubbery.

DISTRIBUTION

Atlas workers found Crissal Thrashers in much of Clark County as well as in adjacent portions of Lincoln and Nye counties. There were two records, both involving only possible breeding, from as far north as central Lincoln County. Breeding was confirmed in only about one-fifth of the occupied blocks, likely due to the difficulty of observing the behavior of individual birds. Crissal Thrashers were, however, seen more often than the other two southern thrasher species (Bendire's and Le Conte's): at more locations, in more habitats, and in greater numbers. At least two pairs of Crissal Thrashers were estimated to occur in more than 80% of the blocks in which they were detected. The more frequent occurrence of Crissal Thrashers compared with the other Mojave thrashers is also evident when the predictive maps for all three species are compared.

Earlier accounts of Nevada birds portray the Crissal Thrasher as less numerous and less widespread than the atlas data indicate. Linsdale (1936) listed a handful of records, all in Clark County. Alcorn (1988) supplied a few more records but extended the species' range only to southern Nye County. Like all sickle-billed

thrashers within the United States, Crissal Thrashers are restricted to the warm deserts of the Southwest. They are uncommon in southern California (Small 1994) except for the lower Colorado River Valley (Garrett and Dunn 1981, Rosenberg et al. 1991), and in southwestern Utah (Behle et al. 1985), but are common in Arizona except in the northeast (Corman in Corman and Wise-Gervais 2005:452–453).

Crissal Thrashers require very dense shrubbery. Mesquite is especially favored (Mearns 1886, Alcorn 1988, Rosenberg et al. 1991, Small 1994), and this was also the most frequently used habitat type reported from atlas blocks, followed by riparian woodland and Mojave shrubland. The predictive map reflects the high likelihood of encountering Crissal Thrashers in dense habitat along watercourses in the Mojave Desert and the lower probability of detecting them in the open, scrub-dominated desert. The higher probabilities of occurrence predicted for the foothills of the southern mountain ranges and the northern Mojave Desert are not consistent with the atlas field data, however. The overpredictions in the north are probably due to the two northernmost atlas records of possible breeders.

CONSERVATION AND MANAGEMENT

Rangewide, Crissal Thrasher populations seem stable, but the trend estimates are based on few data (Sauer et al. 2005). There have been local population declines in California and Arizona, likely due to loss of breeding habitat (Cody 1999). Because of its restricted global distribution, the species is a Partners in Flight Stewardship species for the southwestern region (Rich et al. 2004).

Crissal Thrashers depend on dense desert vegetation, and they may have even benefited from invasion by tamarisks in some areas (Hunter et al. 1988). Although a few individuals might persist on patches of undeveloped land right in Las Vegas, the Crissal Thrasher is vulnerable to habitat conversion, and the Clark County Multiple Species Habitat Conservation Plan designates it an Evaluation species (Clark County 2000). The situation warrants close attention in Nevada, where the habitats Crissal Thrashers prefer are naturally scarce.

HABITAT USE

HABITAT	ATLAS BLOCKS	INCIDENTAL OBSERVATIONS
Mesquite	10 (32%)	4 (12%)
Mojave	8 (26%)	9 (26%)
Montane Parkland	0 (0%)	1 (3%)
Montane Shrub	1 (3%)	2 (6%)
Pinyon-Juniper	0 (0%)	2 (6%)
Riparian	7 (23%)	15 (44%)
Salt Desert Scrub	2 (6%)	1 (3%)
Sagebrush Scrub	2 (6%)	0 (0%)
Urban	1 (3%)	0 (0%)
TOTAL	31	34

Total numbers reported in the Habitat Use, Breeding Status, and Abundance tables may differ from each other (see pp. 22–23 for details). Percentage sums may differ slightly from 100% due to rounding.

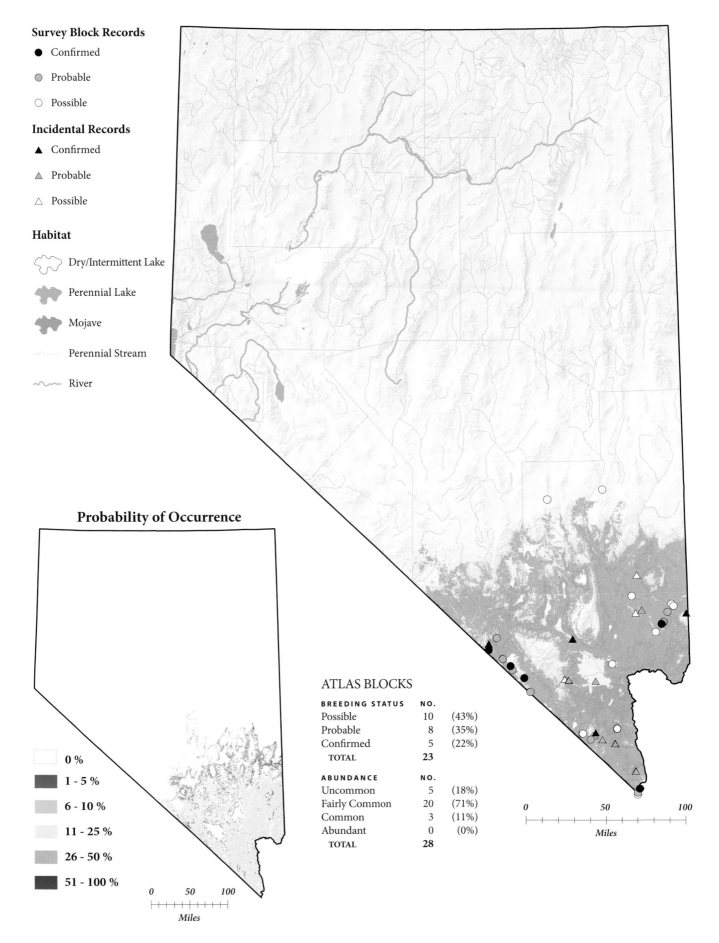

Survey Block Records
● Confirmed
● Probable
○ Possible

Incidental Records
▲ Confirmed
▲ Probable
△ Possible

Habitat
Dry/Intermittent Lake

Perennial Lake

Mojave

Perennial Stream

River

Probability of Occurrence

0 %

1 - 5 %

6 - 10 %

11 - 25 %

26 - 50 %

51 - 100 %

0   50   100
Miles

ATLAS BLOCKS

BREEDING STATUS	NO.	
Possible	10	(43%)
Probable	8	(35%)
Confirmed	5	(22%)
TOTAL	23	

ABUNDANCE	NO.	
Uncommon	5	(18%)
Fairly Common	20	(71%)
Common	3	(11%)
Abundant	0	(0%)
TOTAL	28	

0   50   100
Miles

LE CONTE'S THRASHER

Toxostoma lecontei

Even among desert birds, Le Conte's Thrasher is an extremist. It is usually found in featureless expanses of sun-baked, saltbush-covered valleys and yucca woodlands and seldom ventures into riparian areas or mesquite groves. Except for a wash of apricot across the crissum, Le Conte's Thrasher is the color of desert gravel. Its long bill and sturdy legs are well suited to its ground-feeding habits; it spends most of its time scurrying among low shrubs, pausing now and then to sift through the litter and soil looking for insects. On sunny mornings in late winter, though, singing males on their breeding territories may be the most conspicuous feature in the desert flatlands of southern Nevada.

DISTRIBUTION

Atlas workers found Le Conte's Thrashers throughout Clark County and in southern Lincoln and southeastern Nye counties. Breeding was confirmed in four of ten blocks and two other locations. Le Conte's Thrashers were not numerous, with observers recording only single pairs in half of the blocks and estimating fewer than ten pairs in most others.

Earlier accounts describe a distribution for Le Conte's Thrashers similar to that revealed by the atlas records. Alcorn (1988) considered it uncommon in southern Nevada, and Linsdale (1936) observed that it is restricted to areas south of latitude 37°N. The entire distribution of Le Conte's Thrashers is very limited (Sheppard 1996), and nowhere in its range is it common. It is uncommon to fairly common in southwestern California and the southern San Joaquin Valley (Small 1994, Sheppard 1996), uncommon in extreme western and southwestern Arizona (Corman in Cor-

man and Wise-Gervais 2005:454–455), and rare in southwestern Utah (Behle et al. 1985).

Le Conte's Thrasher is a true xerophile, an inhabitant of the most desolate environments (Sheppard 1996). It favors sparser vegetation than other thrashers (Garrett and Dunn 1981) and often inhabits saltbush-dominated communities. Sandy or pebbly soils, usually covered with some litter, that support arthropods seem to be an essential requirement (Sheppard 1996). The predictive map shows a pattern similar to those for Bendire's and Crissal Thrashers. All three species have a moderate—but not especially high—chance of occurrence in most of the Mojave Desert. Although broadly accurate, this similarity betrays the simplicity of the model, which cannot distinguish subtle habitat differences among species. As it does for several other southern species, the model clearly overestimates the chance of finding Le Conte's Thrashers in the northernmost parts of Nevada's Mojave Desert.

CONSERVATION AND MANAGEMENT

Le Conte's Thrashers are highly intolerant of development and habitat degradation (e.g., due to off-road vehicle use; Sheppard 1996), and their withdrawal from portions of their range in California (McCaskie et al. 1988) and Arizona (Monson and Phillips 1981) is likely due to local habitat degradation (Small 1994). Although Le Conte's Thrasher populations in the Mojave Desert appear stable at present (Sauer et al. 2005), monitoring data are very limited, and this uncommon species may be jeopardized due to its small range and population size (Sheppard 1996).

The outlook for Le Conte's Thrasher in Nevada is unclear but may be poor, given its sensitivity to human disturbance. It is therefore listed as an Evaluation species in the Clark County Multiple Species Habitat Conservation Plan (Clark County 2000), as a Partners in Flight Priority species for Nevada (Neel 1999a), and as a Watch List species for the Southwest (Rich et al. 2004). The desert flats just north of Las Vegas—especially those near Corn Creek—traditionally have been a hotspot for this species (C. Titus pers. comm.), but many of these areas are destined for urbanization. The best hope for both Bendire's and Le Conte's Thrashers is for large expanses of intact upland deserts to be set aside and protected from major habitat alterations (Sheppard 1996).

HABITAT USE

HABITAT	ATLAS BLOCKS	INCIDENTAL OBSERVATIONS
Grassland	2 (18%)	0 (0%)
Mojave	6 (55%)	6 (75%)
Riparian	1 (9%)	0 (0%)
Salt Desert Scrub	1 (9%)	2 (25%)
Urban	1 (9%)	0 (0%)
TOTAL	11	8

Total numbers reported in the Habitat Use, Breeding Status, and Abundance tables may differ from each other (see pp. 22–23 for details). Percentage sums may differ slightly from 100% due to rounding.

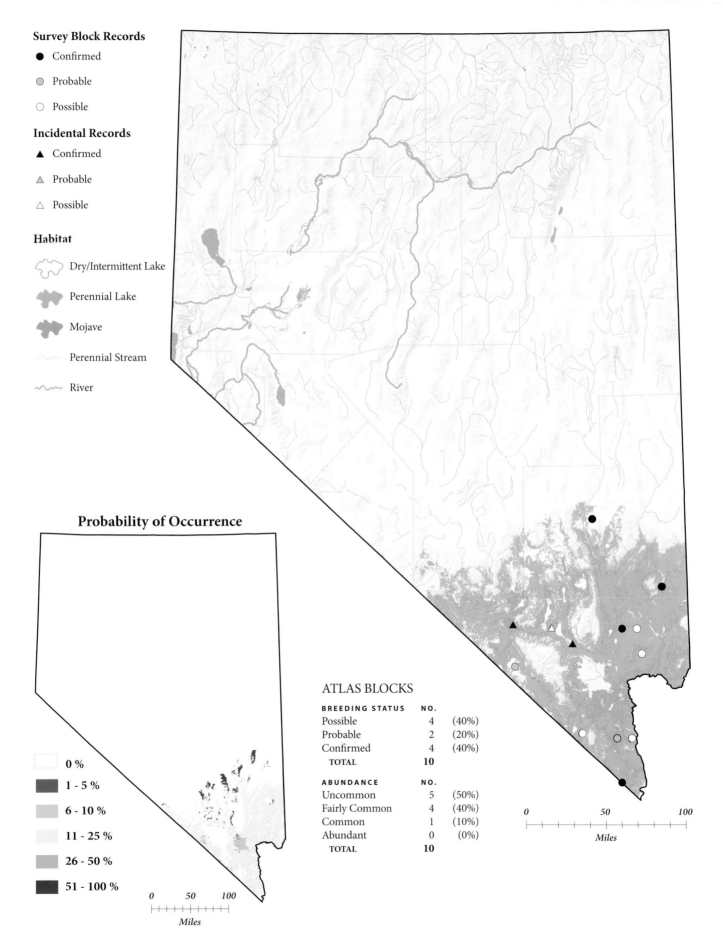

Survey Block Records
● Confirmed
◐ Probable
○ Possible

Incidental Records
▲ Confirmed
◭ Probable
△ Possible

Habitat
⬭ Dry/Intermittent Lake
⬤ Perennial Lake
⬤ Mojave
〜 Perennial Stream
〜 River

Probability of Occurrence

☐ 0 %
■ 1 - 5 %
■ 6 - 10 %
□ 11 - 25 %
■ 26 - 50 %
■ 51 - 100 %

0   50   100
Miles

ATLAS BLOCKS

BREEDING STATUS	NO.	
Possible	4	(40%)
Probable	2	(20%)
Confirmed	4	(40%)
TOTAL	10	

ABUNDANCE	NO.	
Uncommon	5	(50%)
Fairly Common	4	(40%)
Common	1	(10%)
Abundant	0	(0%)
TOTAL	10	

0   50   100
Miles

EUROPEAN STARLING
Sturnus vulgaris

The European Starling is a beautiful bird—suffused with lovely pink and green iridescence during the breeding season, jauntily attired in bright white polka dots in winter. It is also an accomplished vocalist and mimic given to all manner of chattering and whistling. And it is the shining avian exemplar of several qualities humans greatly admire: adaptability, aggressive resourcefulness, and success. Yet, starlings are often held in low esteem, in part because they seem a bit like carpetbaggers, in part because they are often suspected of harming native songbird populations.

DISTRIBUTION

Starlings were found throughout the state, but were much more widespread in the north than in the south, where most records came from the Las Vegas area. Breeding was easy to confirm for this cavity-nesting species that favors human company, with a confirmation rate of 65%. Starling abundance estimates were often high also, as would be expected.

The starling is native neither to Nevada nor to the rest of the United States, and it had not been recorded in the state when Linsdale (1936) published his account of Nevada birds. Ryser (1985) gave a brief chronology of the starling's spread throughout Nevada following the first recorded occurrence in 1938 (Alcorn 1988). It is tempting to think of the starling as uniformly and ubiquitously distributed throughout North America, but in fact it is a relatively recent arrival in the West, and its distribution here is still rather patchy (Cabe 1993). It is abundant in many places, though, and is now considered common to abundant in all of the states adjacent to Nevada (Behle et al. 1985, Small 1994, Stephens and Sturts 1998, Adamus et al. 2001, Wise-Gervais in Corman and Wise-Gervais 2005:456–457).

Although European Starlings can be found in a great variety of environments, they do show clear habitat associations. In Nevada, they are restricted mainly to the valleys and foothills—

especially around towns, cities, ranches, and streams (Ryser 1985)—and most atlas records came from areas with a firmly established human presence. Starlings are among the most common birds in agricultural areas, and they are especially partial to feedlots (Chisholm and Neel 2002). They are generally absent from open desert and unbroken mountain forests (Cabe 1993, Small 1994; but see Garrett and Dunn 1981). A large proportion of the atlas observations were recorded in riparian areas, which are not typically described as the species' favored habitat (Cabe 1993). In Nevada, however, riparian habitat provides both open areas and good nest locations in tree cavities. The predictive map shows two distinct patterns: a geographic, north–south gradient in probability of occurrence, with a spottier pattern of high probability in developed and riparian areas.

CONSERVATION AND MANAGEMENT

A highly successful exotic species like the European Starling is not likely to garner much sympathy from the conservation community; in fact, many despise it for both its actual and presumed aggression toward native cavity-nesting species (Weitzel 1988, Cabe 1993). One Nevada study, for instance, documented the local decline of cavity nesters following the arrival of starlings and their subsequent return after the starlings were exterminated (Weitzel 1988). In contrast, however, a large-scale analysis of population trends found little evidence for a clear link between the abundance of most cavity-nesting species and the occurrence of starlings (Koenig 2003).

Starlings are also considered a nuisance in urban areas because of the large, noisy flocks they form, and in agricultural areas, where starling control is periodically implemented at feedlots and dairies (Chisholm and Neel 2002). The period of dramatically expanding starling populations seems to be over; in fact, recent data show modest declines in several regions (Sauer et al. 2005). Starling numbers in Nevada, however, may increase as the human population continues to grow.

HABITAT USE

HABITAT	ATLAS BLOCKS	INCIDENTAL OBSERVATIONS
Agricultural	26 (13%)	5 (8%)
Grassland	2 (1%)	0 (0%)
Mesquite	1 (<1%)	0 (0%)
Mojave	0 (0%)	1 (2%)
Montane Forest	0 (0%)	1 (2%)
Montane Parkland	0 (0%)	1 (2%)
Montane Shrub	1 (<1%)	0 (0%)
Open Water	1 (<1%)	0 (0%)
Pinyon-Juniper	7 (4%)	0 (0%)
Riparian	64 (32%)	22 (35%)
Salt Desert Scrub	5 (3%)	1 (2%)
Sagebrush Scrub	8 (4%)	2 (3%)
Sagebrush Steppe	9 (5%)	0 (0%)
Urban	73 (37%)	27 (43%)
Wetland	3 (2%)	3 (5%)
TOTAL	200	63

Total numbers reported in the Habitat Use, Breeding Status, and Abundance tables may differ from each other (see pp. 22–23 for details). Percentage sums may differ slightly from 100% due to rounding.

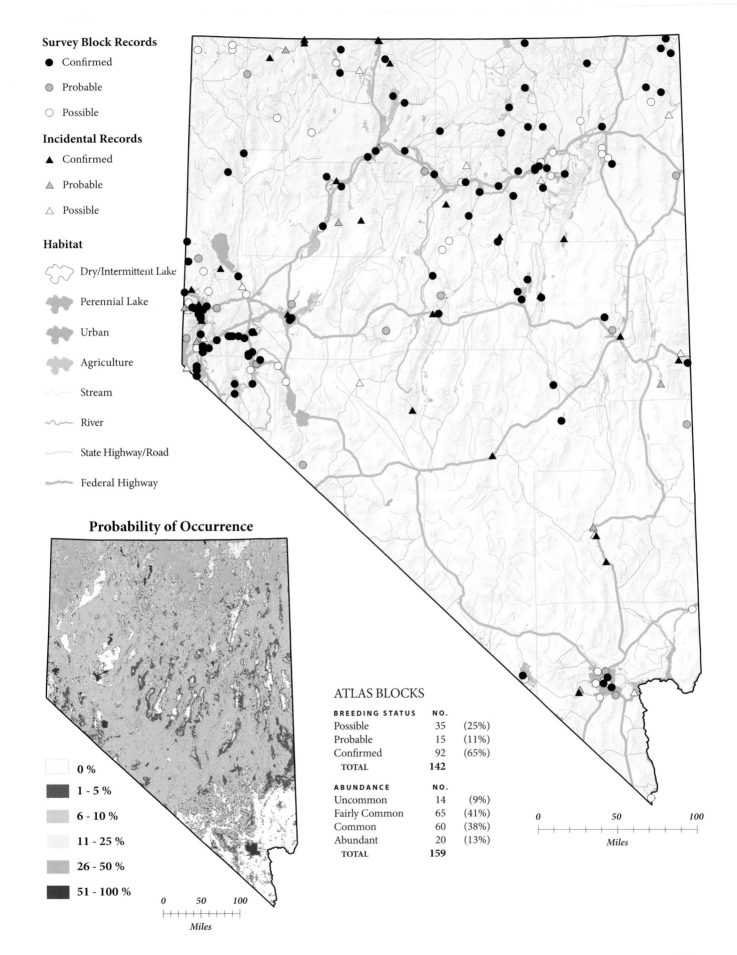

Survey Block Records

- ● Confirmed
- ● Probable
- ○ Possible

Incidental Records

- ▲ Confirmed
- ▲ Probable
- △ Possible

Habitat

Dry/Intermittent Lake

Perennial Lake

Urban

Agriculture

Stream

River

State Highway/Road

Federal Highway

Probability of Occurrence

- 0 %
- 1 - 5 %
- 6 - 10 %
- 11 - 25 %
- 26 - 50 %
- 51 - 100 %

0   50   100

Miles

ATLAS BLOCKS

BREEDING STATUS	NO.	
Possible	35	(25%)
Probable	15	(11%)
Confirmed	92	(65%)
TOTAL	142	

ABUNDANCE	NO.	
Uncommon	14	(9%)
Fairly Common	65	(41%)
Common	60	(38%)
Abundant	20	(13%)
TOTAL	159	

0   50   100

Miles

AMERICAN PIPIT

Anthus rubescens

Few of Nevada's bird species have as disjunct and small a breeding distribution as the American Pipit. Primarily a bird of the high Arctic tundra, this species is restricted as a breeder to alpine areas in Nevada and elsewhere south of the Canadian border. Montane forest clearings that are often mislabeled "alpine" are insufficient for American Pipits. But where there is true tundra they may be among the most commonly encountered birds—perched out in the open on exposed rocks, flying short distances across the snowfields, or tending to their well-concealed nests among tiny flowers and brightly colored lichens.

DISTRIBUTION

Breeding records for American Pipits came only from atop the state's highest mountains, with confirmed breeding in both the Spring Mountains of southern Nevada and the Ruby Mountains in the northeast. Atlas workers could not confirm breeding at two other sites in eastern Nevada where American Pipits were seen—the Jarbidge Mountains and the Snake Range—but the species probably breeds in both ranges (Knorr 2000a).

Knowledge of the American Pipit's Nevada distribution has developed fairly recently. Linsdale (1936) gave no breeding records, but Alcorn (1988) listed breeders in the Snake Range, on Mount Rose in far western Nevada, and on Mount Charleston in the south. Knorr (2000a) offered the most thorough account of the breeding distribution of the species in Nevada, and also reported breeding activity from North Schell Peak in eastern Nevada and Arc Dome in the Toiyabe Range. The predictive map effectively identifies appropriate alpine habitat in eastern Nevada. Because there were no atlas records northwest of the Spring Mountains, however, it does not predict occurrence at the highest elevations in central and western Nevada, where pipits undoubtedly occur.

The American Pipit has a very scattered distribution in the states surrounding Nevada (Verbeek and Hendricks 1994), but it can be locally common. In California, the species is limited for the most part to the southern Sierra Nevada (Garrett and Dunn 1981, Gaines 1992, Small 1994). It is also a known breeder in a few mountain ranges in Oregon (Adamus et al. 2001), Idaho (Stephens and Sturts 1998), Arizona (Corman in Corman and Wise-Gervais 2005:593), and Utah (Behle et al. 1985). The southern stronghold for the species lies to the east, in Colorado and Montana (Versaw in Kingery 1998:410–411).

American Pipits tend to favor fairly moist alpine sites (Smith et al. 1997) and are often found near tarns and alpine lakes with exposed shorelines (Small 1994). In Colorado, they occasionally breed below timberline (Versaw in Kingery 1998:410–411), but this behavior has never been reported in Nevada.

CONSERVATION AND MANAGEMENT

The American Pipit may be expanding its range in the western United States, and it is possible that pipits arrived only recently as breeders in Nevada (cf. Linsdale 1936, Gaines 1992). Pipits may also be recent colonists in the Sierra Nevada (Gaines 1992, Small 1994, Verbeek and Hendricks 1994), where the first confirmed breeders were not documented until 1975 (Gaines 1992) and where additional breeding locations may await discovery (Knorr 2000a).

Pipits are usually thought to be fairly secure on their breeding grounds, simply because of these areas' remoteness and inaccessibility. With its increasing popularity, mountain recreation may become a future threat to the species, though (Versaw in Kingery 1998:410–411). Moreover, improving our knowledge of the current distribution and abundance of American Pipits would provide valuable baseline information about a species that could someday lose its alpine habitat due to global climate changes (see Verbeek and Hendricks 1994).

HABITAT USE

HABITAT	ATLAS BLOCKS	INCIDENTAL OBSERVATIONS
Alpine	4 (67%)	1 (100%)
Barren	1 (17%)	0 (0%)
Riparian	1 (17%)	0 (0%)
TOTAL	6	1

Total numbers reported in the Habitat Use, Breeding Status, and Abundance tables may differ from each other (see pp. 22–23 for details). Percentage sums may differ slightly from 100% due to rounding.

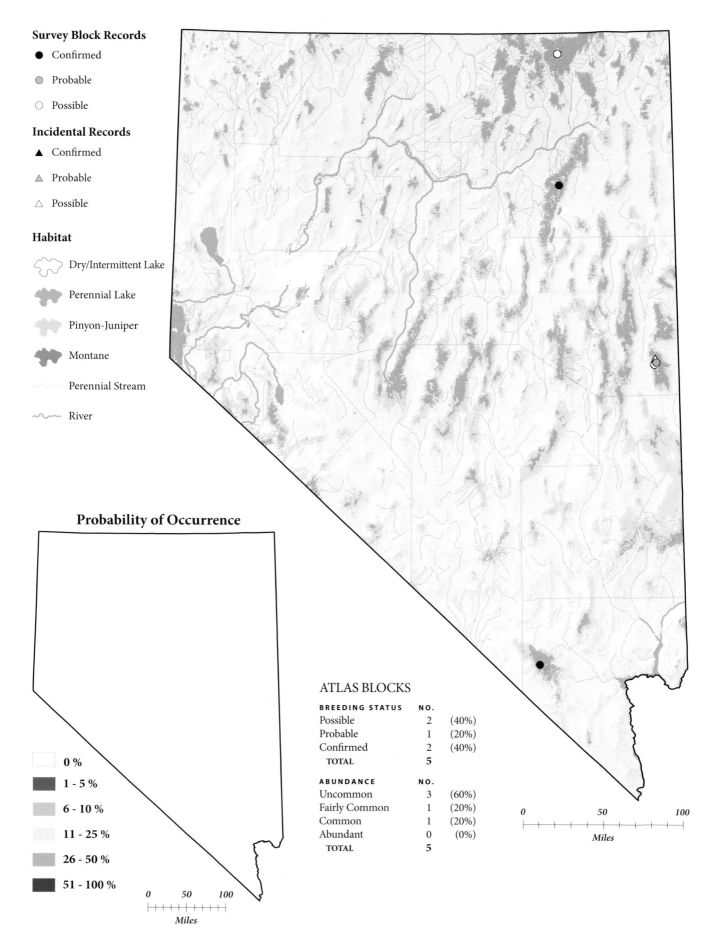

Survey Block Records

● Confirmed

● Probable

○ Possible

Incidental Records

▲ Confirmed

△ Probable

△ Possible

Habitat

Dry/Intermittent Lake

Perennial Lake

Pinyon-Juniper

Montane

Perennial Stream

River

Probability of Occurrence

	0 %
	1 - 5 %
	6 - 10 %
	11 - 25 %
	26 - 50 %
	51 - 100 %

0   50   100

Miles

ATLAS BLOCKS

BREEDING STATUS	NO.	
Possible	2	(40%)
Probable	1	(20%)
Confirmed	2	(40%)
TOTAL	5	

ABUNDANCE	NO.	
Uncommon	3	(60%)
Fairly Common	1	(20%)
Common	1	(20%)
Abundant	0	(0%)
TOTAL	5	

0   50   100

Miles

CEDAR WAXWING
Bombycilla cedrorum

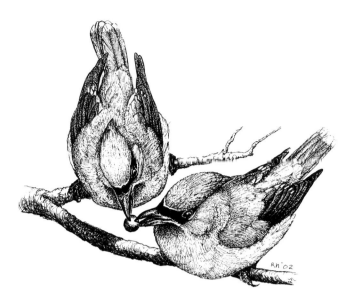

The Cedar Waxwing is the passerine equivalent of the drake Wood Duck: equal parts garish and exquisite, striking and delicate, inevitably destined to capture the attention of artists and the creators of painted porcelain figures. Waxwings are known to most Nevadans as flocking winter itinerants at fruiting junipers and Russian olives. Most of Nevada lies south of their range, but small numbers breed in the northeast and possibly elsewhere.

DISTRIBUTION

During the atlas years, Cedar Waxwings were confirmed as breeders only in the Ruby Mountains and the Lamoille Valley to their west. Possible or probable breeding was also noted at a number of sites throughout northern Nevada. The atlas data are at odds with earlier descriptions of the species' breeding range in Nevada. Linsdale (1936) did not consider the Cedar Waxwing to be a Nevada breeder, whereas both Ryser (1985) and Alcorn (1988) mentioned breeding in Reno and other locations in western Nevada, where atlas data failed to confirm breeding.

Cedar Waxwings are most common in the eastern United States, and their center of abundance in the West lies somewhat to the north of Nevada (Witmer et al. 1997). They are locally common breeders throughout Oregon (Adamus et al. 2001) and Idaho (Stephens and Sturts 1998). In California, the only regular breeding populations are in the extreme northwestern part of the state, although there has been occasional nesting as far south as Orange County (Small 1994). In Utah, breeding is largely limited to the north-central region (Walters et al. 1983). There are no known breeding records from Arizona (Corman and Wise-Gervais 2005).

In the Intermountain West, Cedar Waxwings seem to be most common in wooded stream valleys, city parks and other parklike urban areas, and occasionally in pinyon-juniper groves (Behle et al. 1985, Ryser 1985). They also occur in other forest types, especially in nearby California (Small 1994), Oregon (Adamus et al. 2001), and Washington (Smith et al. 1997). The predictive map indicates a low probability of occurrence in Nevada overall, but it also highlights areas in the foothills of major mountain ranges where breeding birds are more likely to occur.

CONSERVATION AND MANAGEMENT

Cedar Waxwing numbers appear to have increased during the later decades of the twentieth century, and the species may be expanding its range (Witmer et al. 1997, Sauer et al. 2005). Possible reasons for recent population growth vary, but they tend to center on the theme of human modifications to the landscape. In Washington, for example, Cedar Waxwings seem to be increasing in urban areas with ornamental plantings such as fruit orchards or stands of the introduced Russian olive (Smith et al. 1997).

The conservation status of the Cedar Waxwing in Nevada is difficult to gauge. It may be a recent addition to the state's breeding avifauna (cf. Linsdale 1936), but because it can be inconspicuous during nesting, it may simply have been overlooked in the past. Like some other frugivorous birds, the Cedar Waxwing is patchily distributed, difficult to survey, and prone to local colonization and extirpation events (Witmer et al. 1997). It is also a late nester that times its breeding to match berry production, so it is difficult to confirm as a breeder without late-summer visits to possible breeding sites. In addition to its late nesting, the Cedar Waxwing is also a late spring migrant, which makes it difficult to determine the breeding status of early-summer birds.

HABITAT USE

HABITAT	ATLAS BLOCKS	INCIDENTAL OBSERVATIONS
Agricultural	1 (5%)	0 (0%)
Montane Forest	3 (14%)	0 (0%)
Pinyon-Juniper	3 (15%)	0 (0%)
Riparian	10 (48%)	5 (56%)
Urban	4 (19%)	4 (44%)
TOTAL	21	9

Total numbers reported in the Habitat Use, Breeding Status, and Abundance tables may differ from each other (see pp. 22–23 for details). Percentage sums may differ slightly from 100% due to rounding.

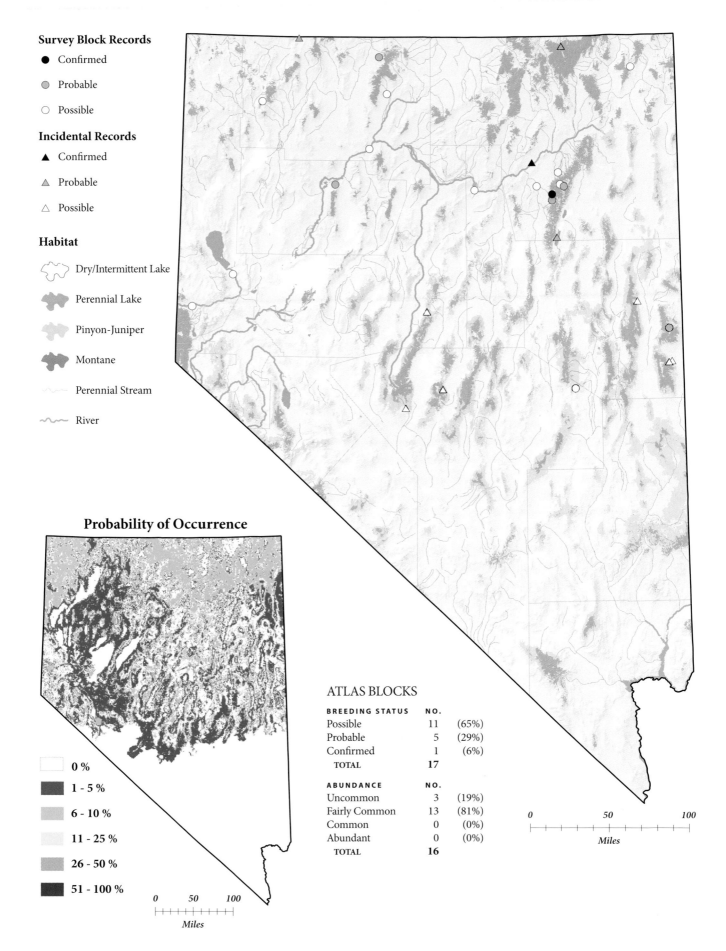

Survey Block Records

● Confirmed

● Probable

○ Possible

Incidental Records

▲ Confirmed

▲ Probable

△ Possible

Habitat

Dry/Intermittent Lake

Perennial Lake

Pinyon-Juniper

Montane

Perennial Stream

River

Probability of Occurrence

0 %

1 - 5 %

6 - 10 %

11 - 25 %

26 - 50 %

51 - 100 %

0   50   100
Miles

ATLAS BLOCKS

BREEDING STATUS	NO.	
Possible	11	(65%)
Probable	5	(29%)
Confirmed	1	(6%)
TOTAL	17	

ABUNDANCE	NO.	
Uncommon	3	(19%)
Fairly Common	13	(81%)
Common	0	(0%)
Abundant	0	(0%)
TOTAL	16	

0   50   100
Miles

PHAINOPEPLA
Phainopepla nitens

Rn'02

Garnet on black velvet. That is the impression left by the male Phainopepla: gleaming black all over with bright red eyes. The Phainopepla is one of the most fascinating birds in Nevada's southern deserts, where it is a widespread and fairly common breeder. Phainopeplas are considered a conservation priority in Nevada, and much remains to be learned about their comings and goings in the region.

DISTRIBUTION

Atlas workers found Phainopeplas throughout the lowlands and foothills of southern Nevada, and almost half of the blocks where the species was present had confirmed breeders. They were also recorded in one block in northeastern Nye County, well beyond their usual Mojave Desert range. Although never abundant, Phainopeplas were often present in moderate numbers, with multiple pairs estimated to be present in almost all of the occupied blocks.

Atlas data approximate the Phainopepla's previously known range in southern Nevada (Linsdale 1936, Alcorn 1988). The species was previously thought to be a year-round resident in Nevada's Mojave region. More recent evidence shows that most birds leave the nesting area after broods are reared, however, and the species' summer distribution in Nevada is still poorly understood (Chu and Walsberg 1999; L. Crampton, pers. comm.). The current thought is that many Phainopeplas breed in desert woodlands dominated by mesquite and acacia between February and June, and then move to other parts of their range, perhaps to breed again in semiarid oak or sycamore woodlands (Chu and Walsberg 1999).

In the United States, Phainopeplas are birds of the arid Southwest (Chu and Walsberg 1999), although they occasionally summer in the more northern parts of California (Small 1994). They are common in southern California and western and southern Arizona, including the lower Colorado River Valley (Rosenberg et al. 1991, Small 1994, Wise-Gervais in Corman and Wise-Gervais 2005:458–459), and uncommon in northern Arizona and southwestern Utah (Behle et al. 1985).

Phainopeplas subsist almost entirely on fruit, and in Nevada breed only in mesquite and acacia stands that produce an abundance of mistletoe berries (Krueger 1998, Crampton 2004). A moderate chance of finding the species exists throughout most of Nevada's Mojave Desert, although field observations suggest that the birds may occur in a localized fashion within this region. Due to a single outlying record in northern Nye County, the predictive model overestimates the probability of detecting Phainopeplas at the northern extent of their range, where they are actually rarer than they are farther south, and in the foothills of southern Nevada's mountains.

CONSERVATION AND MANAGEMENT

Since the 1980s, Phainopepla populations have declined throughout their U.S. range (Sauer et al. 2005). Consequently, the Phainopepla is a Nevada Partners in Flight Priority species (Neel 1999a) and is a Covered species under the Clark County Multiple Species Habitat Conservation Plan (Clark County 2000).

In Nevada, breeding Phainopeplas require areas with abundant mistletoe, and loss of this habitat is probably the greatest threat to the species (Chu and Walsberg 1999, Crampton and Murphy 2006). This knowledge allows a targeted conservation strategy that preserves desert riparian and mesquite woodlands that support mistletoe (Chu and Walsberg 1999, Crampton 2004). Threats to these areas include urbanization, agricultural development, groundwater depletion, firewood cutting, and conversion to tamarisk-dominated woodlands. The Phainopepla's movement patterns and habitat needs following the spring breeding season in the desert remain poorly understood, however, and more research on these topics may provide additional conservation direction.

HABITAT USE

HABITAT	ATLAS BLOCKS	INCIDENTAL OBSERVATIONS
Agricultural	0 (0%)	1 (4%)
Mesquite	16 (37%)	4 (15%)
Mojave	15 (35%)	5 (19%)
Pinyon-Juniper	1 (2%)	1 (4%)
Riparian	10 (23%)	11 (41%)
Salt Desert Scrub	0 (0%)	1 (4%)
Sagebrush Scrub	0 (0%)	2 (7%)
Urban	1 (2%)	2 (7%)
TOTAL	43	27

Total numbers reported in the Habitat Use, Breeding Status, and Abundance tables may differ from each other (see pp. 22–23 for details). Percentage sums may differ slightly from 100% due to rounding.

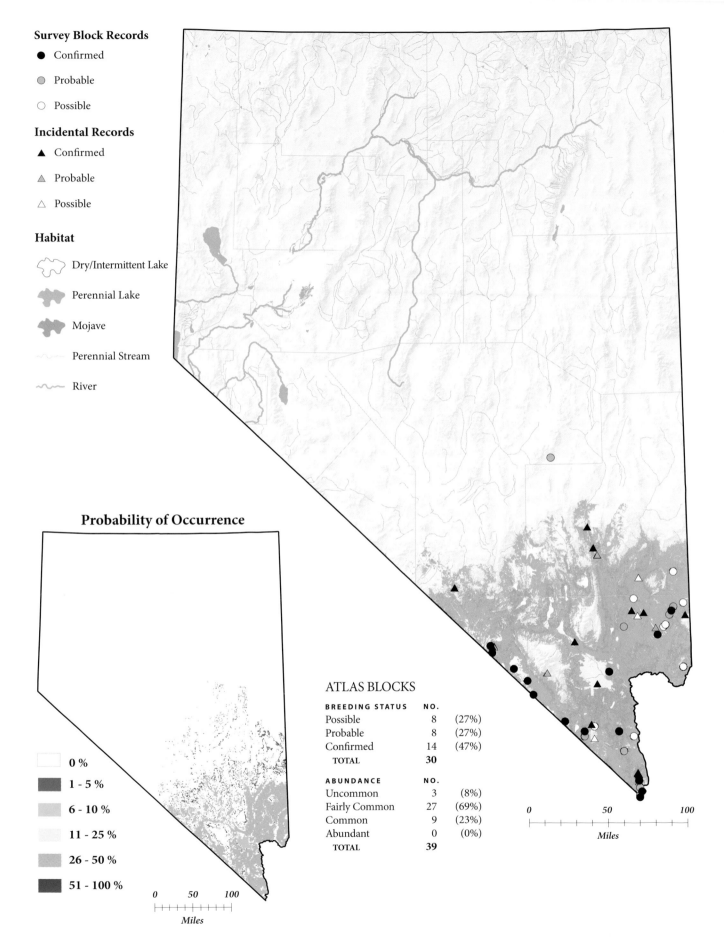

Survey Block Records

● Confirmed

● Probable

○ Possible

Incidental Records

▲ Confirmed

▲ Probable

△ Possible

Habitat

Dry/Intermittent Lake

Perennial Lake

Mojave

Perennial Stream

River

Probability of Occurrence

	0 %
	1 - 5 %
	6 - 10 %
	11 - 25 %
	26 - 50 %
	51 - 100 %

0   50   100
Miles

ATLAS BLOCKS

BREEDING STATUS	NO.	
Possible	8	(27%)
Probable	8	(27%)
Confirmed	14	(47%)
TOTAL	**30**	

ABUNDANCE	NO.	
Uncommon	3	(8%)
Fairly Common	27	(69%)
Common	9	(23%)
Abundant	0	(0%)
TOTAL	**39**	

0   50   100
Miles

ORANGE-CROWNED WARBLER

Vermivora celata

The Orange-crowned Warbler is sometimes noted among birders for its nondescript appearance and colorless song—both of which are in contrast with most other western warblers. While that may be true for some forms, it is not entirely true for most of the Orange-crowned Warblers that breed in Nevada. The *orestera* subspecies that occurs throughout most of the state is almost as distinctively marked as MacGillivray's Warbler—yellow underneath, olive above, and with a gray face. Both *orestera* and the *lutescens* subspecies of far western Nevada are denizens of shady thickets along mountain streams.

DISTRIBUTION

Orange-crowned Warblers were present in mountain ranges throughout the state. Most records came from the north, but some came from as far south as the Spring Mountains of Clark County. Although the species was widespread, breeding was confirmed at only a handful of northern sites: the Carson Range of far western Nevada, the mountains around the Black Rock Desert in northwestern Nevada, and the Jarbidge and Ruby mountains of northeastern Nevada.

Older accounts assume that the subspecies *lutescens* breeds in the Carson Range and *orestera* elsewhere in the state (Linsdale 1936, Alcorn 1988). More recent evidence suggests that *lutescens* breeds on the western slope of the Sierra Nevada and disperses eastward into the Carson Range immediately following the nesting period (Sogge et al. 1994; Dunn and Garrett 1997; J. Steele, pers. comm.). At a bird-banding station operated in the Carson Range for the past ten years, for example, recently hatched birds were often captured around the same time that adults had just arrived from spring migration (J. Eidel, pers. comm.); these juveniles were presumably from early-nesting populations across the Sierra Nevada.

Both subspecies also breed in California (Small 1994) and Oregon (Adamus et al. 2001), with *orestera* to the east and *lutescens* to the west in both states. The subspecies *orestera* is a local breeder in the mountains of Arizona (Corman in Corman and Wise-Gervais 2005:462–463), a common summer resident in Utah (Behle et al. 1985), and widespread in Idaho (Stephens and Sturts 1998).

For nesting, Orange-crowned Warblers seek out fairly dense deciduous woodlands with dense ground cover. This habitat is most often found along streams in Nevada, but it may also occur in upland habitats elsewhere in the species' range (Sogge et al. 1994). Aspen seems to be an especially important habitat component in Nevada (Neel 1999a). The map indicates a moderate chance of finding Orange-crowned Warblers at high elevations throughout Nevada, but chances of seeing them are nowhere especially high, probably reflecting the scarcity of the species' primary breeding habitat. The predictive map also highlights several areas where searches for additional breeding populations are warranted, particularly given that the atlas was able to confirm breeding in only a few mountain ranges.

CONSERVATION AND MANAGEMENT

Population trend analyses suggest recent declines of Orange-crowned Warblers rangewide and in the West (Dobkin and Sauder 2004, Sauer et al. 2005), even though they remain common in many areas (Sogge et al. 1994). They often breed earlier than other warblers (Gaines 1992), and the status of birds in midsummer is difficult to determine. Much remains to be learned, therefore, about the comings and goings of Orange-crowned Warblers in Nevada during the summer months. Despite moderate recent declines, the species is not particularly rare, and no management guidelines have been formulated (Sogge et al. 1994). Nevada Partners in Flight nevertheless ranks the Orange-crowned Warbler as a Priority species because of its association with the state's restricted riparian areas and because its nesting habits and summertime dispersal patterns are still poorly understood in this region (Neel 1999a).

HABITAT USE

HABITAT	ATLAS BLOCKS	INCIDENTAL OBSERVATIONS
Montane Forest	14 (30%)	3 (38%)
Montane Parkland	1 (2%)	0 (0%)
Montane Shrub	6 (13%)	0 (0%)
Pinyon-Juniper	1 (2%)	0 (0%)
Riparian	25 (53%)	5 (63%)
TOTAL	47	8

Total numbers reported in the Habitat Use, Breeding Status, and Abundance tables may differ from each other (see pp. 22–23 for details). Percentage sums may differ slightly from 100% due to rounding.

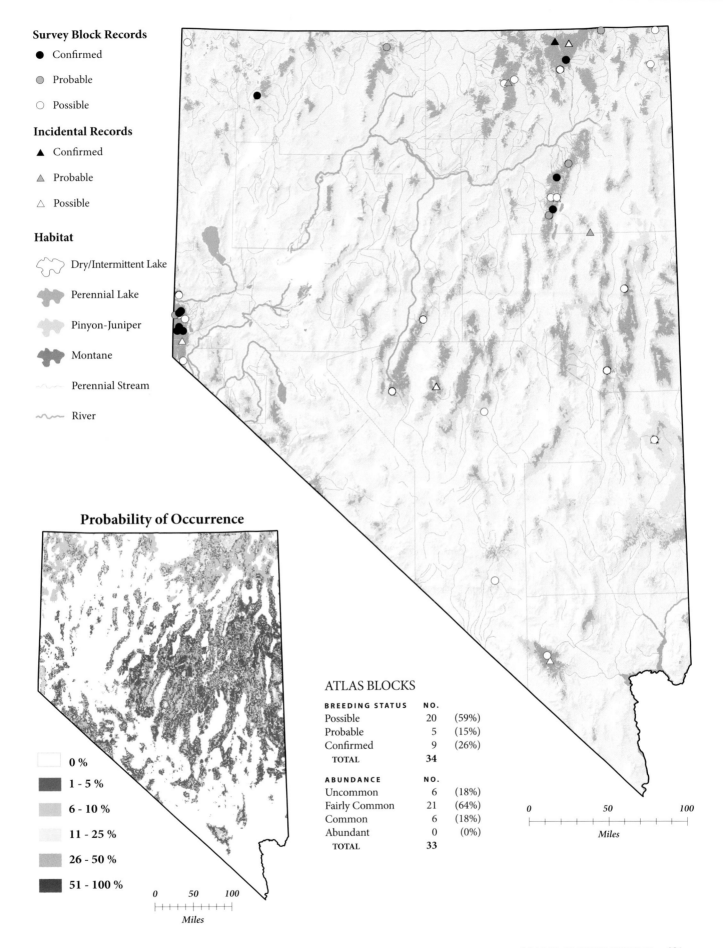

Survey Block Records

- ● Confirmed
- ● Probable
- ○ Possible

Incidental Records

- ▲ Confirmed
- ▲ Probable
- △ Possible

Habitat

- Dry/Intermittent Lake
- Perennial Lake
- Pinyon-Juniper
- Montane
- Perennial Stream
- River

Probability of Occurrence

	0 %
	1 - 5 %
	6 - 10 %
	11 - 25 %
	26 - 50 %
	51 - 100 %

0   50   100

Miles

ATLAS BLOCKS

BREEDING STATUS	NO.	
Possible	20	(59%)
Probable	5	(15%)
Confirmed	9	(26%)
TOTAL	**34**	

ABUNDANCE	NO.	
Uncommon	6	(18%)
Fairly Common	21	(64%)
Common	6	(18%)
Abundant	0	(0%)
TOTAL	**33**	

0   50   100

Miles

NASHVILLE WARBLER
Vermivora ruficapilla

The Nashville Warbler is often thought of as an eastern bird, but it is a regular summer resident within the slim crescent of Sierra Nevada forests that extend into far western Nevada. The population that breeds here comes from the distinctive western subspecies, *ridgwayi,* sometimes also called the "Calaveras Warbler." This subspecies differs in several notable ways from the widespread and familiar eastern subspecies: it sounds different, its coloration is brighter, and it has a habit of continuously bobbing its long tail.

DISTRIBUTION

Breeding Nashville Warblers were reported almost exclusively from the Carson Range. Here and elsewhere in the West, the *ridgwayi* subspecies is found in fairly open forests, often in second growth (Williams 1996). As a ground nester, it also requires a well-developed shrub understory and is commonly found at forest openings and along streams (Williams 1996, Dobkin and Sauder 2004). In Nevada, the Nashville Warbler is most often associated with the deciduous woodlands of riparian zones, and slightly less often with montane forests, where it often nests close to water (Dobkin and Sauder 2004). Dunn and Garrett (1997) provided a good review of the natural history of the *ridgwayi* subspecies.

Nashville Warblers have long been known as summer residents of the Carson Range but are not known to nest elsewhere in the state (Linsdale 1936, Ryser 1985, Alcorn 1988). An intriguing atlas record of a possible breeder from the Spring Mountains may have been a late migrant or postbreeding wanderer. Future surveys in this area would help to solidify our knowledge of the species' breeding range in Nevada. The predictive map shows moderate to high chances of finding Nashville Warblers in the montane forests of western Nevada. This map, however, likely underestimates the true chance of finding them in the Carson Range while overestimating the odds of occurrence elsewhere. These discrepancies result from the Spring Mountain record, which extends the area over which predictions are made well into central Nevada's mountains, where the species is not known to occur.

The western breeding range of the Nashville Warbler is largely limited to the far western and northern mountains of the Intermountain West (Williams 1996). In the states surrounding Nevada, the species is a widespread breeder in the Sierra Nevada and northern parts of California and southwestern Oregon, and to a lesser extent in northeastern Oregon and northwestern Idaho (Williams 1996, Stephens and Sturts 1998, Adamus et al. 2001). It is interesting that this breeding distribution is largely separate from that of Virginia's Warbler, its close relative (Johnson 1976; see also Williams 1996, Olson and Martin 1999).

CONSERVATION AND MANAGEMENT

In the West, Nashville Warblers seem to benefit from recent forest burns and logging, after which they colonize early-successional shrub-dominated habitats (Williams 1996, Dunn and Garrett 1997, Dobkin and Sauder 2004). The Nashville Warbler therefore seems to benefit from forest management practices that may have negative impacts on other western montane species (Ryser 1985). Conversely, outright destruction or heavy grazing of riparian areas almost certainly has negative impacts (Williams 1996, Dobkin and Sauder 2004), and maturation of second-growth vegetation in areas with fire suppression may also cause local declines (Williams 1996, Dobkin and Sauder 2004).

Rangewide and in the West, Nashville Warbler populations are largely stable (Dobkin and Sauder 2004, Sauer et al. 2005). Within its small range in Nevada, the species seems to be fairly common and widespread, but recent population declines in the Sierra Nevada (Sauer et al. 2005) suggest that additional monitoring and research are warranted for this region, which includes Nevada's Carson Range.

HABITAT USE

HABITAT	ATLAS BLOCKS	INCIDENTAL OBSERVATIONS
Montane Forest	5 (38%)	0 (0%)
Montane Shrub	1 (8%)	0 (0%)
Riparian	7 (54%)	0 (0%)
TOTAL	13	0

Total numbers reported in the Habitat Use, Breeding Status, and Abundance tables may differ from each other (see pp. 22–23 for details). Percentage sums may differ slightly from 100% due to rounding.

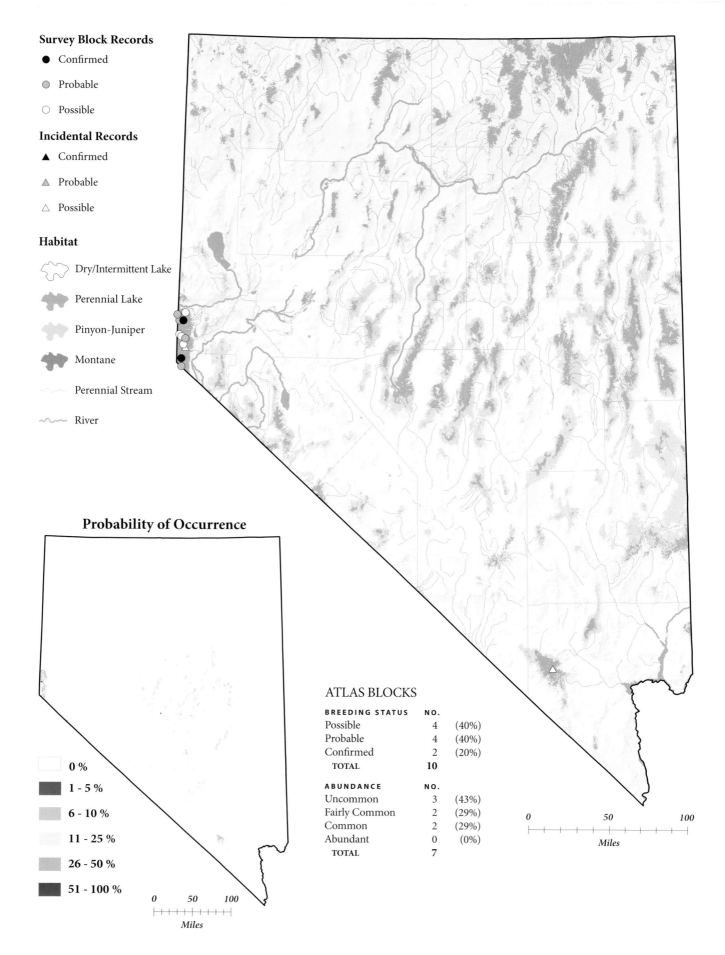

Survey Block Records

● Confirmed

◉ Probable

○ Possible

Incidental Records

▲ Confirmed

◮ Probable

△ Possible

Habitat

Dry/Intermittent Lake

Perennial Lake

Pinyon-Juniper

Montane

Perennial Stream

River

Probability of Occurrence

0 %

1 - 5 %

6 - 10 %

11 - 25 %

26 - 50 %

51 - 100 %

0   50   100
Miles

ATLAS BLOCKS

BREEDING STATUS	NO.	
Possible	4	(40%)
Probable	4	(40%)
Confirmed	2	(20%)
TOTAL	10	

ABUNDANCE	NO.	
Uncommon	3	(43%)
Fairly Common	2	(29%)
Common	2	(29%)
Abundant	0	(0%)
TOTAL	7	

0   50   100
Miles

VIRGINIA'S WARBLER
Vermivora virginiae

Rn'02

Virginia's Warbler has a widespread but patchy range in Nevada, mirroring on a smaller scale its larger breeding distribution in the southern Rocky Mountains and Intermountain West. It is a bird of dry, low-canopy forests such as pinyon-juniper woodlands. Virginia's Warblers are active, easily found birds, their loud, languid singing contributing greatly to the ease of detection. Even so, the species' basic biology and conservation requirements have yet to be studied in detail.

DISTRIBUTION

The center of this warbler's breeding range is the Four Corners region of Utah, Colorado, Arizona, and New Mexico (Behle et al. 1985, Olson and Martin 1999, Wise-Gervais in Corman and Wise-Gervais 2005:464–465). Its Nevada range has therefore been considered somewhat peripheral. Most atlas records came from the state's eastern tier of counties. No records were reported from the Wassuk Range or White Mountains, where Ryser (1985) reported breeding populations. Records of possible breeders from far western Nevada, where the closely related and similar-looking Nashville Warbler occurs, should be viewed with caution, since earlier studies have shown that the breeding ranges of these two species generally do not overlap (Johnson 1976, Dunn and Garrett 1997). No previous breeding records exist for this area (Ryser 1985, Alcorn 1988), but there is anecdotal evidence that Virginia's Warblers have expanded westward into California in recent decades (Johnson 1976, Dunn and Garrett 1997). Thus, the possibility of a small breeding population in western Nevada should not be entirely discounted.

Breeding has never been confirmed in Oregon, but records of possible breeders have been reported from the southeast of that state (Adamus et al. 2001). In California, breeding has been recorded only in small, isolated areas in the east and south (Olson and Martin 1999).

Virginia's Warblers reside in dry, often steep woodlands with substantial brushy cover near ground level (Dunn and Garrett 1997, Olson and Martin 1999). Pinyon-juniper and mountain mahogany stands are often used in Nevada, but other shrubby conifers, oaks, and montane shrublands may also serve as nesting habitat (Olson and Martin 1999). The predictive map could be a valuable tool to focus further investigations of the species' full breeding range in Nevada. Care should be taken, though, because the records of possible breeding from the vicinity of the Carson Range extend the predicted range far to the west and potentially cause overestimates of the species' occurrence in this region. Areas in northern Nevada accorded a high chance of occurrence might be especially worth investigating given the possibility of recent breeding in southeastern Oregon (Adamus et al. 2001).

CONSERVATION AND MANAGEMENT

By all accounts, Virginia's Warbler is a poorly known species. Nests, for example, are notoriously difficult to find (Bent 1953)—a sentiment echoed by Nevada's atlas workers. At this time, no rangewide or regional population declines have been reported (Dobkin and Sauder 2004, Sauer et al. 2005), but monitoring data for this species are quite limited. Evidence for cowbird parasitism is equivocal (Martin 1993, Melcher in Kingery 1998:416–417, Olson and Martin 1999), and the species' response to fire is poorly understood, although potentially negative (Horton 1987, Johnson and Wauer 1996). The effects of land management practices are largely unknown.

Virginia's Warbler could be vulnerable to declines because of its fairly narrow habitat associations, geographically limited range, and patchy distribution (Reed 1992, Olson and Martin 1999, Dobkin and Sauder 2004). For these reasons, Partners in Flight regards it as a Priority species for Nevada (Neel 1999a).

HABITAT USE

HABITAT	ATLAS BLOCKS	INCIDENTAL OBSERVATIONS
Montane Forest	7 (18%)	10 (37%)
Montane Parkland	4 (10%)	0 (0%)
Montane Shrub	11 (28%)	5 (19%)
Pinyon-Juniper	10 (25%)	6 (22%)
Riparian	6 (15%)	5 (19%)
Sagebrush Scrub	1 (3%)	0 (0%)
Sagebrush Steppe	1 (3%)	1 (4%)
TOTAL	40	27

Total numbers reported in the Habitat Use, Breeding Status, and Abundance tables may differ from each other (see pp. 22–23 for details). Percentage sums may differ slightly from 100% due to rounding.

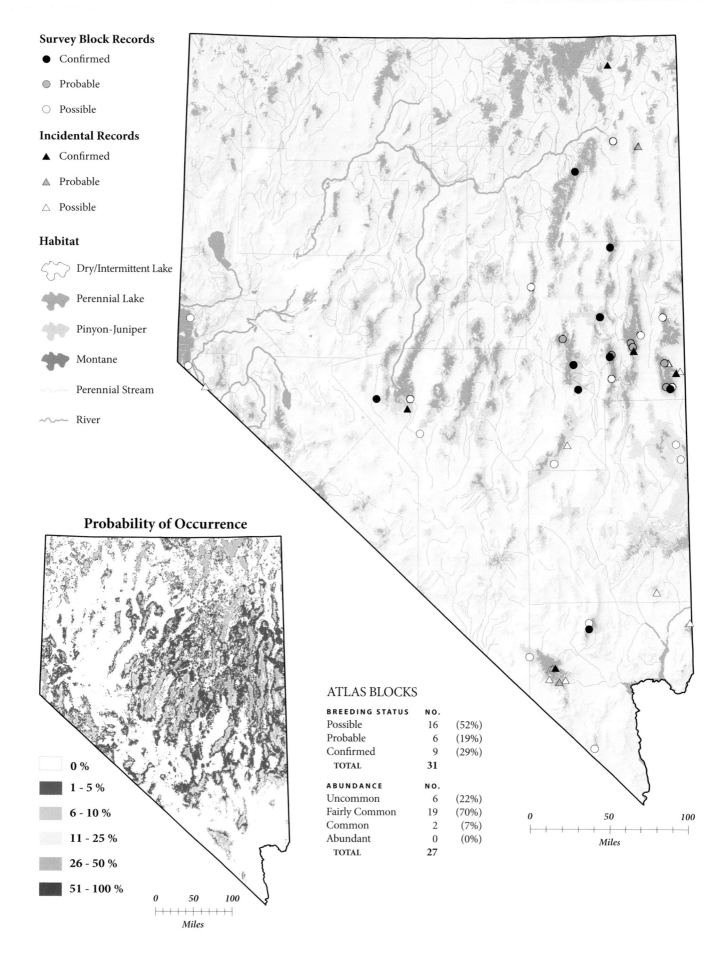

Survey Block Records

● Confirmed

◉ Probable

○ Possible

Incidental Records

▲ Confirmed

△ Probable

△ Possible

Habitat

Dry/Intermittent Lake

Perennial Lake

Pinyon-Juniper

Montane

Perennial Stream

River

Probability of Occurrence

☐ 0 %

■ 1 - 5 %

■ 6 - 10 %

☐ 11 - 25 %

■ 26 - 50 %

■ 51 - 100 %

0   50   100
Miles

ATLAS BLOCKS

BREEDING STATUS	NO.	
Possible	16	(52%)
Probable	6	(19%)
Confirmed	9	(29%)
TOTAL	31	

ABUNDANCE	NO.	
Uncommon	6	(22%)
Fairly Common	19	(70%)
Common	2	(7%)
Abundant	0	(0%)
TOTAL	27	

0   50   100
Miles

LUCY'S WARBLER
Vermivora luciae

Lucy's Warbler is an inconspicuous but often common summer resident in the mesquite bottomlands of southern Nevada. It is one of our smallest warblers, and certainly among the least colorful. It is also one of only two cavity-nesting warblers in North America. Lucy's Warblers are strongly associated with riparian thickets and mesquite woodlands, and although their overall numbers in Nevada seem healthy for now, the threat of degradation and loss of their habitats is causing concern about their long-term prospects.

DISTRIBUTION

Lucy's Warblers reach their northwestern range limits in southern Nevada (Johnson et al. 1997). The rest of their breeding range covers southern Utah, Arizona, southeastern California, southwestern New Mexico, bits of western Colorado, and northern Mexico (Behle et al. 1985, Rosenberg et al. 1991, Small 1994, Johnson et al. 1997, Corman in Corman and Wise-Gervais 2005:466–467).

Breeding was confirmed in all three southern Nevada counties and in more than half of the occupied blocks. Lucy's Warblers were often numerous, with 41% of abundance estimates exceeding ten pairs and only 8% of blocks estimated to have a single pair. The species' geographical distribution in Nevada appears to have changed little since the 1930s (Linsdale 1936, Johnson 1956, Alcorn 1988, Rosenberg et al. 1991).

Lucy's Warblers, once also known as "Mesquite Warblers" (Gilman 1909), are strongly associated with riparian mesquite habitats that have trees large enough to provide cavities or sheltered nooks for nests (Johnson et al. 1997). They also breed in willows, cottonwoods, tamarisks, and the xeroriparian vegetation of desert washes (Johnson et al. 1997). The atlas data match the habitat use reported elsewhere—more than three-quarters of all records were from riparian and mesquite habitats. The best chance of finding this species is along southern Nevada's lowland riparian areas. Unlike some other southern Nevada riparian species, Lucy's Warbler appears to be attracted to small, remote patches of habitat as well as to large floodplains. Use of these small habitat patches, which are often found around springs embedded within Mojave habitat, are the reason the map predicts a moderate chance of finding this species throughout the Mojave Desert lowlands.

CONSERVATION AND MANAGEMENT

Rangewide, Lucy's Warbler populations show no significant trends (Sauer et al. 2005). Historical declines in abundance and range contractions reported for some regions (Small 1994, Johnson et al. 1997) are mostly attributed to loss of riparian habitat. Cowbird parasitism may also reduce reproductive success in some areas (Johnson et al. 1997). In the lower Colorado River system, Lucy's Warblers were historically common but declined substantially in the late 1950s during a period of rapid riparian habitat loss (Rosenberg et al. 1991). Dam projects, altered water flows, and removal of native riparian vegetation allowed tamarisk to invade such areas and proliferate. The species' recovery along the lower Colorado River since the 1960s appears to be due to its increased use of tamarisk thickets for nesting (Rosenberg et al. 1991, Corman in Corman and Wise-Gervais 2005:466–467).

Because of its restricted distribution and habitat, Lucy's Warbler is a Partners in Flight Watch List species for the Southwest (Rich et al. 2004) and a Priority species in Nevada (Neel 1999a). Its frequent use of tamarisk for nesting results in a management dilemma, given the widespread desire to rid riparian areas of this exotic species. Restoration of native riparian woodlands would almost certainly improve the habitat quality for this warbler in the long term, but it is important that revegetation projects be carefully implemented so that some type of suitable nesting habitat is available even during the early stages of restoration.

HABITAT USE

HABITAT	ATLAS BLOCKS	INCIDENTAL OBSERVATIONS
Ash	1 (2%)	0 (0%)
Mesquite	15 (35%)	3 (12%)
Mojave	3 (7%)	6 (23%)
Montane Parkland	0 (0%)	1 (4%)
Riparian	21 (49%)	12 (46%)
Salt Desert Scrub	0 (0%)	1 (4%)
Sagebrush Scrub	0 (0%)	1 (4%)
Urban	2 (5%)	1 (4%)
Wetland	1 (2%)	1 (4%)
TOTAL	43	26

Total numbers reported in the Habitat Use, Breeding Status, and Abundance tables may differ from each other (see pp. 22–23 for details). Percentage sums may differ slightly from 100% due to rounding.

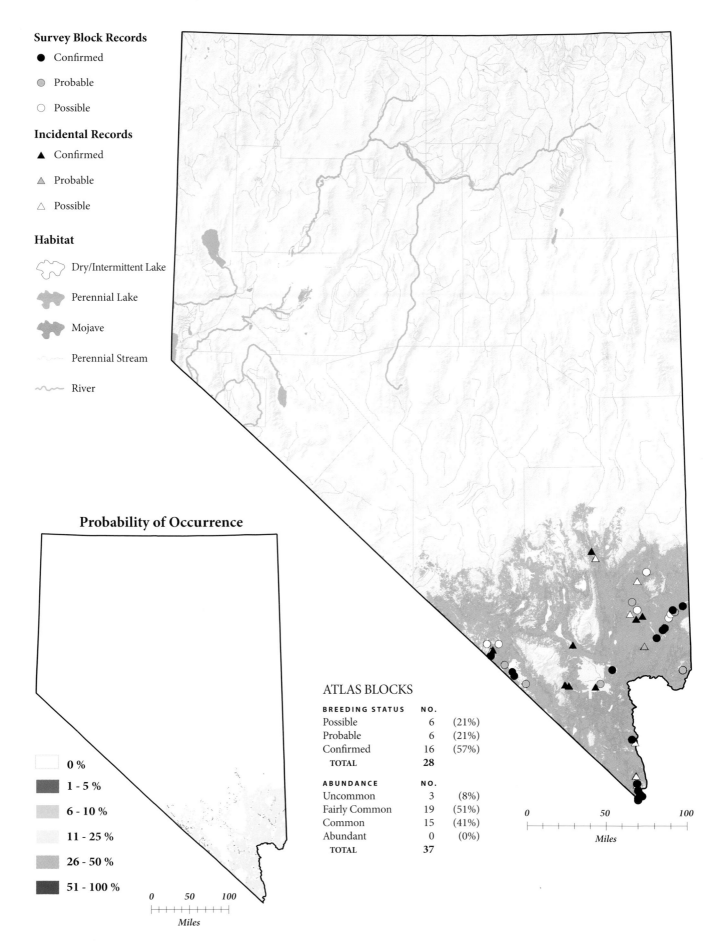

Survey Block Records

● Confirmed

◐ Probable

○ Possible

Incidental Records

▲ Confirmed

▲ Probable

△ Possible

Habitat

⬡ Dry/Intermittent Lake

⬤ Perennial Lake

⬤ Mojave

〜 Perennial Stream

〜 River

Probability of Occurrence

☐	0 %
▨	1 - 5 %
▨	6 - 10 %
▨	11 - 25 %
▨	26 - 50 %
▨	51 - 100 %

0   50   100
Miles

ATLAS BLOCKS

BREEDING STATUS	NO.	
Possible	6	(21%)
Probable	6	(21%)
Confirmed	16	(57%)
TOTAL	**28**	

ABUNDANCE	NO.	
Uncommon	3	(8%)
Fairly Common	19	(51%)
Common	15	(41%)
Abundant	0	(0%)
TOTAL	**37**	

0   50   100
Miles

YELLOW WARBLER
Dendroica petechia

Bright yellow all over, with black eyes and red racing stripes running down its breast, the Yellow Warbler is a splashy streamside bird in Nevada. The species accepts most riparian habitats: the mesquites and ashes in the Virgin and Muddy river drainages, the willows and cottonwoods that line the Truckee and Carson rivers, and the alders and aspens along mountain streams are all satisfactory.

DISTRIBUTION

Just as earlier writers reported (Linsdale 1936, Ryser 1985, Alcorn 1988), Yellow Warblers were found throughout Nevada. Breeding was confirmed as far south as the Colorado River below Hoover Dam, but the highest concentration of confirmed breeders was in Elko County. There were also areas where the species was scarce, such as Esmeralda County and the Black Rock Desert area just north of Pyramid Lake.

For the most part, the Yellow Warbler is common and widespread in the states surrounding Nevada as well, especially Idaho (Stephens and Sturts 1998), Utah (Behle et al. 1985), and Oregon (Adamus et al. 2001). It is common in large parts of California (Small 1994) and Arizona (Wise-Gervais in Corman and Wise-Gervais 2005:468–469), but rare or absent in others. Historically, Yellow Warblers were extremely numerous in the lower Colorado River Valley (Rosenberg et al. 1991). Their subsequent disappearance in the 1950s is often attributed to cowbird parasitism and riparian habitat loss, but recent surveys in the area suggest that numbers are now recovering (Lowther et al. 1999).

Yellow Warblers are usually found along streams or in brushy thickets, and willows seem to be especially suitable (Lowther et al. 1999). They are most often found at lower elevations but may range all the way up to the aspen zones in the Intermountain West (Behle et al. 1985). Yellow Warblers are not always restricted to mesic sites; they are sometimes found in montane chaparral (Gaines 1992, Small 1994) and in urban and rural localities with lush plantings of ornamental trees and shrubs (Ryser 1985). The predictive map shows a moderately high chance of finding the species all across the northern edge of the state and in much of the central part, and a low to moderate chance throughout the rest of Nevada. The Yellow Warbler's willingness to use a range of different habitats and to use even small riparian patches embedded within other habitat types explains why there is at least a low chance of finding it just about anywhere in the state.

CONSERVATION AND MANAGEMENT

Cowbird parasitism has been widely blamed for population declines of Yellow Warblers (e.g., Garrett and Dunn 1981, Monson and Phillips 1981, Gallagher 1997), but it may not always be at fault (Lowther et al. 1999). Habitat destruction and degradation are thought to be the greater current threats in many regions, including the Great Basin (Lowther et al. 1999, Dobkin and Sauder 2004).

In Nevada, the Yellow Warbler has received scant attention in conservation planning. Its status in the state appears to have changed little during the past century (Ryser 1985), and no widespread declines have been reported (Dobkin and Sauder 2004, Sauer et al. 2005). As a riparian shrubland breeder that is fairly common where habitats are suitable, however, the Yellow Warbler may be a good indicator of riparian habitat condition in this region. Perhaps for these reasons, it is a Covered species of the Lower Colorado River Multi-Species Conservation Program (BOR-LCR 2004).

HABITAT USE

HABITAT	ATLAS BLOCKS	INCIDENTAL OBSERVATIONS
Agricultural	6 (3%)	2 (2%)
Alpine	1 (<1%)	0 (0%)
Ash	1 (<1%)	0 (0%)
Mesquite	3 (2%)	2 (2%)
Mojave	0 (0%)	2 (2%)
Montane Forest	11 (6%)	3 (3%)
Montane Parkland	0 (0%)	1 (1%)
Montane Shrub	7 (4%)	1 (1%)
Open Water	0 (0%)	2 (2%)
Pinyon-Juniper	1 (<1%)	1 (1%)
Riparian	146 (75%)	71 (74%)
Salt Desert Scrub	1 (<1%)	0 (0%)
Sagebrush Scrub	3 (2%)	0 (0%)
Urban	10 (5%)	9 (9%)
Wetland	5 (3%)	2 (2%)
TOTAL	195	96

Total numbers reported in the Habitat Use, Breeding Status, and Abundance tables may differ from each other (see pp. 22–23 for details). Percentage sums may differ slightly from 100% due to rounding.

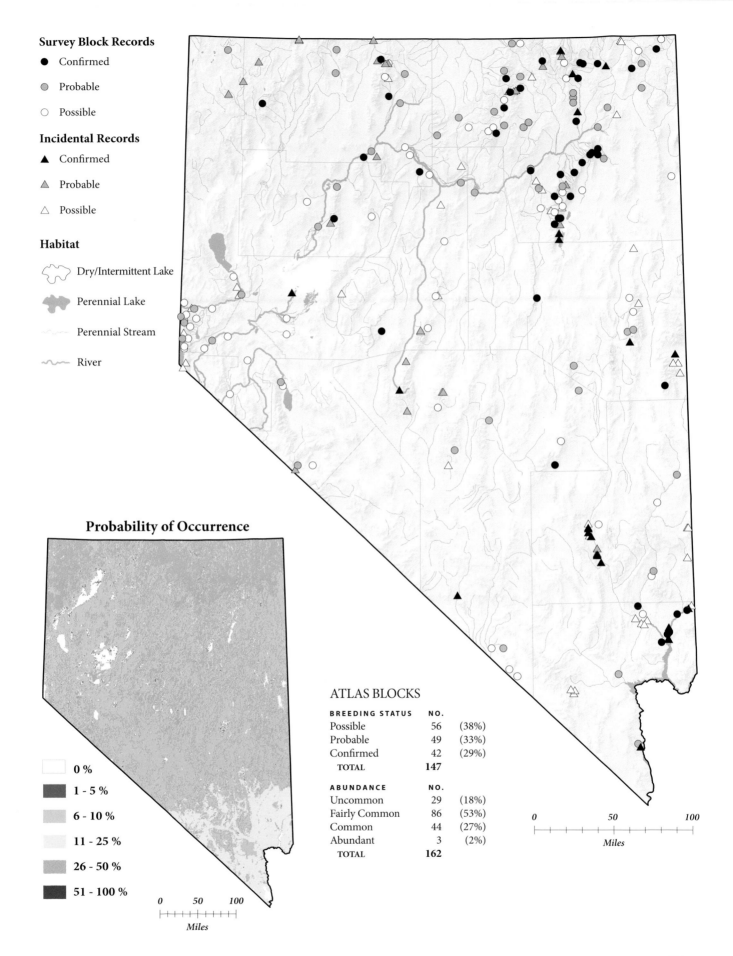

Survey Block Records

● Confirmed

● Probable

○ Possible

Incidental Records

▲ Confirmed

▲ Probable

△ Possible

Habitat

Dry/Intermittent Lake

Perennial Lake

Perennial Stream

River

Probability of Occurrence

	0 %
	1 - 5 %
	6 - 10 %
	11 - 25 %
	26 - 50 %
	51 - 100 %

0   50   100

Miles

ATLAS BLOCKS

BREEDING STATUS	NO.	
Possible	56	(38%)
Probable	49	(33%)
Confirmed	42	(29%)
TOTAL	**147**	

ABUNDANCE	NO.	
Uncommon	29	(18%)
Fairly Common	86	(53%)
Common	44	(27%)
Abundant	3	(2%)
TOTAL	**162**	

0   50   100

Miles

YELLOW-RUMPED WARBLER
Dendroica coronata

In spring, when Yellow-rumped Warblers migrate en masse through the lowlands and foothills of Nevada, there is little doubt about their identity. The males in full breeding colors are distinctively patterned in black, gold, and white. The juveniles in August, almost devoid of useful field marks, are another matter. The species can be found in pine and fir woodlands in much of the state—mixed in with flocks of juncos and other songbirds or in single-species aggregates numbering upward of a hundred individuals. The Yellow-rumped Warbler, an amalgamation of two formerly distinct species ("Myrtle" and "Audubon's Warblers"), is not only common in Nevada; it is one of the most common warblers in North America.

DISTRIBUTION

Yellow-rumped Warblers were found in most of Nevada's major mountain ranges. They were confirmed as breeders north to the Jarbidge Mountains and south to the Spring Mountains and Sheep Range of Clark County. East to west, they were found from the borders of California to Utah. This distribution is consistent with earlier reports by Linsdale (1936), Ryser (1985), and Alcorn (1988). Compared with most of Nevada's other breeding warblers, the Yellow-rumped Warbler appears to be common; multiple pairs were estimated to be present in more than 90% of the occupied blocks.

In the states surrounding Nevada, the Yellow-rumped Warbler is widespread and numerous in coniferous forests. It is common in the mountains of Arizona (Wise-Gervais in Corman and Wise-Gervais 2005:470–471), Utah (Behle et al. 1985), Idaho (Stephens and Sturts 1998), and much of Oregon, though less so in western Oregon (Adamus et al. 2001). Its breeding distribu-

tion has large gaps in California, though it is fairly common in many areas (Small 1994).

Compared with other wood warblers, the Yellow-rumped is something of a generalist (Hunt and Flaspohler 1998). It nests primarily in mature coniferous forest in the West (Gaines 1992, Hunt and Flaspohler 1998) and inhabits a wide elevation range, extending from the highest mountains down to sea level in California (Small 1994). In Nevada as elsewhere, the species is usually associated with conifers, but it may also range into mixed deciduous-coniferous forest and aspen woodlands (Ryser 1985). Occurrence is greatest at middle to high elevations (Ryser 1985), as is evident on the predictive map, which indicates the best chances of finding the species in the high mountains. Away from the mountains, even the low occurrence probabilities that exist in many areas are likely inflated due to a small number of low elevation records that probably involved nonbreeding birds.

CONSERVATION AND MANAGEMENT

The Yellow-rumped Warbler is rarely, if ever, treated as a conservation concern because of its abundance, widespread occurrence, and stable or increasing populations (Hunt and Flaspohler 1998, Sauer et al. 2005). Moreover, it does not seem to have been harmed by human activities on its breeding grounds. On the contrary, Yellow-rumped Warblers may benefit from conifer plantations in the East and from forest fires (Hunt and Flaspohler 1998).

Yellow-rumped Warblers are expected to hold steady in Nevada in the coming years. They are tolerant of low-level human activities, and most of their breeding grounds in Nevada are in areas that currently receive little attention from the resource extraction industry. Actual range expansions, such as have been noted elsewhere in the species' range, however, are not likely.

HABITAT USE

HABITAT	ATLAS BLOCKS	INCIDENTAL OBSERVATIONS
Alpine	7 (3%)	1 (2%)
Montane Forest	125 (55%)	35 (55%)
Montane Parkland	8 (4%)	2 (3%)
Montane Shrub	14 (6%)	1 (2%)
Pinyon-Juniper	25 (11%)	2 (3%)
Riparian	41 (19%)	22 (34%)
Sagebrush Scrub	3 (1%)	0 (0%)
Urban	3 (1%)	1 (2%)
TOTAL	226	64

Total numbers reported in the Habitat Use, Breeding Status, and Abundance tables may differ from each other (see pp. 22–23 for details). Percentage sums may differ slightly from 100% due to rounding.

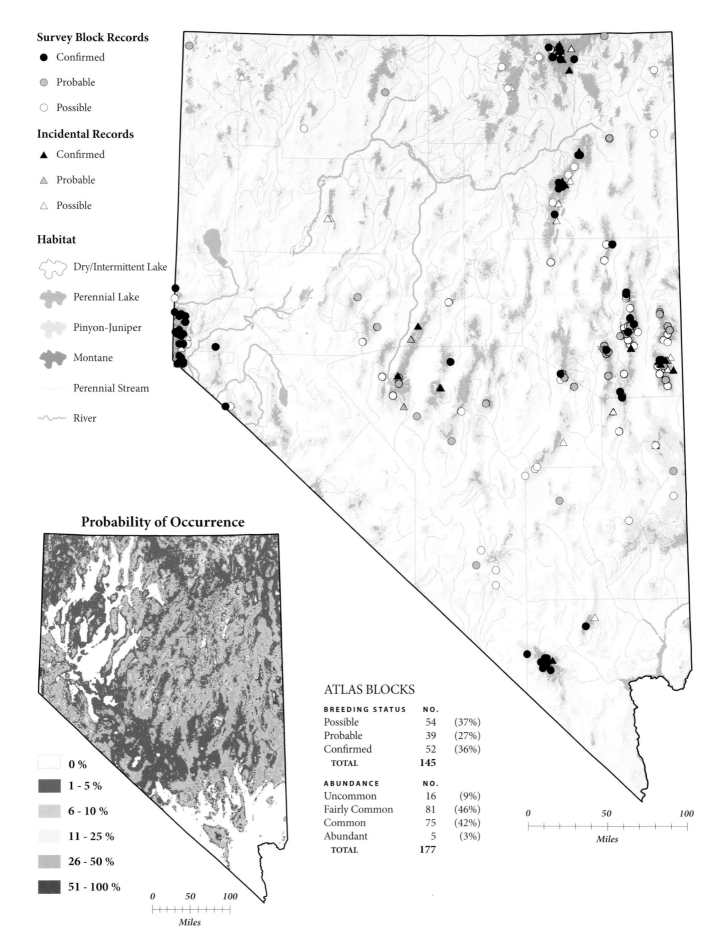

Survey Block Records

- ● Confirmed
- ● Probable
- ○ Possible

Incidental Records

- ▲ Confirmed
- △ Probable
- △ Possible

Habitat

- Dry/Intermittent Lake
- Perennial Lake
- Pinyon-Juniper
- Montane
- Perennial Stream
- River

Probability of Occurrence

- ☐ 0 %
- 1 - 5 %
- 6 - 10 %
- 11 - 25 %
- 26 - 50 %
- 51 - 100 %

0   50   100

Miles

ATLAS BLOCKS

BREEDING STATUS	NO.	
Possible	54	(37%)
Probable	39	(27%)
Confirmed	52	(36%)
TOTAL	**145**	

ABUNDANCE	NO.	
Uncommon	16	(9%)
Fairly Common	81	(46%)
Common	75	(42%)
Abundant	5	(3%)
TOTAL	**177**	

0   50   100

Miles

BLACK-THROATED GRAY WARBLER
Dendroica nigrescens

Rn '02

The gasping song of the Black-throated Gray Warbler is a familiar sound in the pinyon-juniper woodlands of Nevada. This bird is often found in the taller trees within dense stands, and catching a glimpse of one is not always an easy matter. It is worth the effort, though, because the Black-throated Gray Warbler is a dashing member of the pinyon-juniper bird community. Its smartly patterned black, gray, and white plumage adds an element of class to the woodlands. Those lucky enough to get a close look at a male may even be able to make out the distinctive yellow patch in front of the eye.

DISTRIBUTION

Black-throated Gray Warblers were reported throughout the state, with the bulk of the confirmed breeders located along the U.S. Highway 50 corridor between Austin and the Utah border. Breeding was also confirmed in the Spring Mountains, Sheep Range, and Mormon Mountains of Clark County. Abundance estimates provided for occupied blocks suggest that Black-throated Gray Warblers and Yellow-rumped Warblers are perhaps the most numerous of Nevada's breeding warblers, with Yellow and McGillivray's in close competition.

Older accounts portray the Black-throated Gray Warbler as being slightly more widespread in Nevada than the atlas data alone indicate (Linsdale 1936, Alcorn 1988, Guzy and Lowther 1997). Furthermore, Neel's (1999a) account and recent data from the Nevada Bird Count program of the Great Basin Bird Observatory have added records from farther north in Elko County toward the Idaho border and west toward the Carson Range. These observations are consistent with the results shown on our predictive map.

Black-throated Gray Warblers are common and widespread

in Utah (Behle et al. 1985) and Arizona (Wise-Gervais in Corman and Wise-Gervais 2005:472–473). In California, they are fairly common in the Sierra Nevada and a few coastal ranges (Small 1994), and in Oregon they are widespread in the west but scarcer in the east (Adamus et al. 2001). The species is largely absent from all but the far southern part of Idaho (Stephens and Sturts 1998).

The Black-throated Gray Warbler breeds in more arid habitats than most other warblers use (Ryser 1985). The specific forest types used vary regionally (Garrett and Dunn 1981, Small 1994), but in Nevada the species is an inhabitant primarily of mid-elevation pinyon-juniper woodlands (Linsdale 1936, Neel 1999a). The predictive map shows that the best chance of finding the species is at middle elevations in mountain ranges that are skirted by a pinyon-juniper zone. This result includes many local areas where breeding was not documented during atlas fieldwork.

CONSERVATION AND MANAGEMENT

Unlike some other migratory songbirds, the Black-throated Gray Warbler is not known to have suffered widespread declines in recent decades (Guzy and Lowther 1997), and rangewide, populations appear to be stable (Sauer et al. 2005). Population trends from Arizona and the Sierra Nevada, however, suggest regional declines (Sauer et al. 2005) and indicate the need for continued population monitoring. General management strategies have not yet been identified for the Black-throated Gray Warbler (Guzy and Lowther 1997).

Nevada lies squarely in the core of the Black-throated Gray Warbler's breeding range, and a sizable proportion of the world's population occurs here, yet population trends in the state are unknown. The species' primary breeding habitat, pinyon-juniper woodland, is undergoing substantial changes in our region. Altered fire regimes, in particular, have affected the age structure and understory, increased the stand density in many areas, and caused the woodlands to expand into lower elevations (Miller and Rose 1999, Miller and Tausch 2001). For these reasons, Nevada Partners in Flight has designated the Black-throated Gray Warbler a Priority species (Neel 1999a), and additional population monitoring is clearly warranted in Nevada.

HABITAT USE

HABITAT	ATLAS BLOCKS	INCIDENTAL OBSERVATIONS
Montane Forest	7 (6%)	12 (23%)
Montane Parkland	2 (2%)	0 (0%)
Montane Shrub	5 (4%)	8 (15%)
Pinyon-Juniper	95 (76%)	23 (44%)
Riparian	13 (10%)	6 (12%)
Sagebrush Scrub	4 (3%)	1 (2%)
Sagebrush Steppe	0 (0%)	1 (2%)
Urban	0 (0%)	1 (2%)
TOTAL	126	52

Total numbers reported in the Habitat Use, Breeding Status, and Abundance tables may differ from each other (see pp. 22–23 for details). Percentage sums may differ slightly from 100% due to rounding.

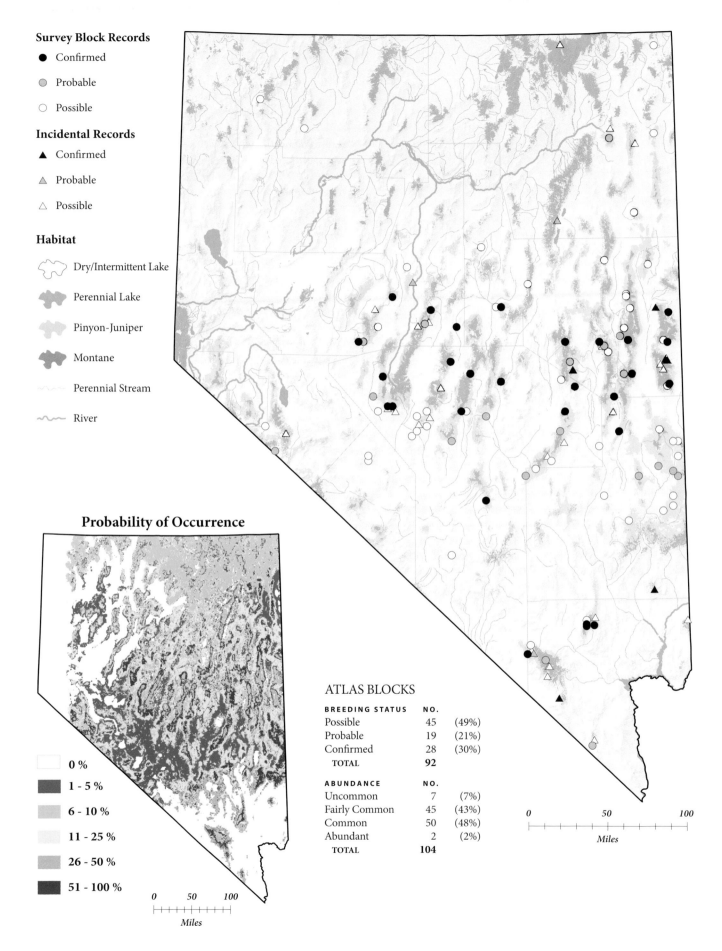

Survey Block Records

● Confirmed

● Probable

○ Possible

Incidental Records

▲ Confirmed

▲ Probable

△ Possible

Habitat

Dry/Intermittent Lake

Perennial Lake

Pinyon-Juniper

Montane

Perennial Stream

River

Probability of Occurrence

0 %

1 - 5 %

6 - 10 %

11 - 25 %

26 - 50 %

51 - 100 %

0   50   100
Miles

ATLAS BLOCKS

BREEDING STATUS	NO.	
Possible	45	(49%)
Probable	19	(21%)
Confirmed	28	(30%)
TOTAL	92	

ABUNDANCE	NO.	
Uncommon	7	(7%)
Fairly Common	45	(43%)
Common	50	(48%)
Abundant	2	(2%)
TOTAL	104	

0   50   100
Miles

GRACE'S WARBLER
Dendroica graciae

Rh'02

Grace's Warbler reaches the extreme northwestern limit of its range in southern Nevada. It is found in the mature, parklike ponderosa pine forests of the Spring Mountains and the Sheep Range, and may also occur in other southern Nevada ranges. This species exhibits a fairly high degree of habitat specialization, and conservation-oriented forest management may hold the key to its continued presence in Nevada.

DISTRIBUTION
Grace's Warblers were recorded only in southern Nevada's Spring Mountains and Sheep Range. Breeding was confirmed in both areas, and these warblers were estimated to be fairly common in the occupied blocks, with six of seven abundance estimates reporting multiple pairs. The predictive map for this species closely reflects the atlas field data, suggesting a high chance of finding it in the higher elevations of its two confirmed mountain ranges and no chance of finding it elsewhere in Nevada. Nonetheless, future surveys in other southern Nevada mountains that have small patches of tall conifers would help to confirm the complete Nevada distribution of this species.

Neither Linsdale (1936, 1951) nor Van Rossem (1936) mentioned the presence of Grace's Warblers in Nevada. Austin (1969), however, reported a Spring Mountains sighting that matched the location of a previous record from 1927. Alcorn (1988) presented several breeding records from the 1960s and 1970s, including one from Mount Irish in Lincoln County and one from Potosi Mountain just south of the Spring Mountains in Clark County. Collectively, these records suggest that Grace's

Warblers could have colonized Nevada sometime during the twentieth century (Alcorn 1988, Stacier and Guzy 2002). Alternatively, they may have simply gone undetected during the early ornithological explorations of the state. Regardless, in the mid-1990s, Johnson (1994) reported Grace's Warblers as breeders in at least five southern Nevada mountain ranges. Stacier and Guzy (2002) tentatively attributed this purported expansion to recent climatic trends.

Beyond Nevada, the U.S. breeding range of Grace's Warbler reaches from the Four Corners area southward, centering on Arizona and New Mexico (Behle et al. 1985, Stacier and Guzy 2002, Wise-Gervais in Corman and Wise-Gervais 2005:474–475). Breeding is suspected, but has not been confirmed, in southeastern California. Within their U.S. range, Grace's Warblers are almost entirely restricted to mature pine, spruce, and fir forests, where they forage primarily in the upper canopy (Levad in Kingery 1998:424–425, Stacier and Guzy 2002). The most typical breeding habitat in the northern parts of the species' range is ponderosa pine, often with an understory of Gambel's oak, a pattern that is consistent with the habitats reported from atlas blocks.

CONSERVATION AND MANAGEMENT
Population trends for Grace's Warbler are not well known, presumably due to insufficient survey coverage in their primary breeding grounds; however, declines may be occurring in the Southwest (Sauer et al. 2005). Widespread threats to the tall conifer forests to which this species is restricted led Partners in Flight to rank Grace's Warbler as a Watch List species for the Southwest, with a conservation goal of increasing populations by 50% (Rich et al. 2004), and as a Priority species for Nevada (Neel 1999a). Recent changes in forest management, such as stand thinning, may benefit Grace's Warblers, but complete regeneration of mature, open stands of tall conifers takes considerable time (Stacier and Guzy 2002). In Nevada, monitoring of known populations and additional surveys in nearby areas such as the Virgin Mountains are warranted, both to provide a better understanding of the species' distribution in the state and to contribute to a better regional estimate of population size and trends for Grace's Warbler.

HABITAT USE

HABITAT	ATLAS BLOCKS	INCIDENTAL OBSERVATIONS
Montane Forest	7 (70%)	7 (100%)
Montane Parkland	1 (10%)	0 (0%)
Pinyon-Juniper	1 (10%)	0 (0%)
Sagebrush Scrub	1 (10%)	0 (0%)
TOTAL	10	7

Total numbers reported in the Habitat Use, Breeding Status, and Abundance tables may differ from each other (see pp. 22–23 for details). Percentage sums may differ slightly from 100% due to rounding.

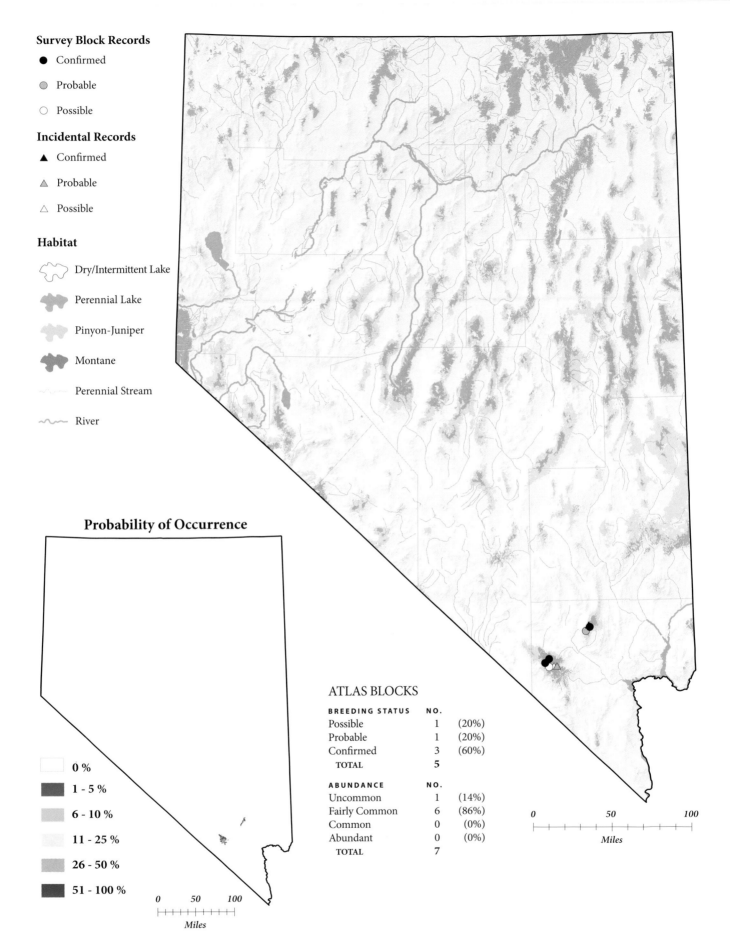

Survey Block Records

● Confirmed

◐ Probable

○ Possible

Incidental Records

▲ Confirmed

△ Probable

△ Possible

Habitat

Dry/Intermittent Lake

Perennial Lake

Pinyon-Juniper

Montane

Perennial Stream

River

Probability of Occurrence

	0 %
	1 - 5 %
	6 - 10 %
	11 - 25 %
	26 - 50 %
	51 - 100 %

0   50   100
Miles

ATLAS BLOCKS

BREEDING STATUS	NO.	
Possible	1	(20%)
Probable	1	(20%)
Confirmed	3	(60%)
TOTAL	5	

ABUNDANCE	NO.	
Uncommon	1	(14%)
Fairly Common	6	(86%)
Common	0	(0%)
Abundant	0	(0%)
TOTAL	7	

0   50   100
Miles

MACGILLIVRAY'S WARBLER

Oporornis tolmiei

In the beautiful aspen woodlands of the Ruby Mountains, it is easy to forget that Nevada is the driest state in the Union. Birds of the dry forests, like Steller's Jay and Pygmy Nuthatch, are nowhere to be found. Swainson's Thrush and Lewis's Woodpecker are more likely to be seen here. Particularly numerous in these woodlands is the wide-eyed MacGillivray's Warbler, which is usually found in brushy areas around streams, at the edges of clearings, and even around campsites. Like other *Oporornis* warblers, MacGillivray's are "skulkers" that are easy to overlook. They are so common, though, that diligent searching in montane riparian areas usually turns up a few singing males.

DISTRIBUTION

MacGillivray's Warblers were found in most of the higher mountains of the state. Population clusters occurred in the northeastern and far western regions, particularly the Santa Rosa, Jarbidge, and Ruby mountains, and the Carson Range. In the south, atlas records came only from the Spring Mountains, and in the far northwest, records were sparse and breeding was not confirmed. This warbler's common presence as a summer resident in central and northern Nevada has long been known (Linsdale 1936, Ryser 1985, Alcorn 1988). With regard to its southern Nevada range, Linsdale (1936) provided no breeding records from the Spring Mountains or from other parts of Clark County. Alcorn (1988) did list some southern Nevada breeding records, however, and reported that nesting was first confirmed in this part of the state in 1965.

MacGillivray's Warbler's distribution lies mainly to the north of Nevada (Pitocchelli 1995) in Oregon (Adamus et al. 2001) and

Idaho (Stephens and Sturts 1998). It is also common and widespread in Utah (Behle et al. 1985), but its range is more restricted in California and Arizona (Small 1994, McCarthey in Corman and Wise-Gervais 2005:476–477).

In the more northerly parts of its range, MacGillivray's Warbler breeds in a variety of brushy habitats, including second growth in recent clear-cuts (Pitocchelli 1995, Dobkin and Sauder 2004). In the south of its breeding range, which includes Nevada, it most often nests in montane riparian zones or in nearby moist, broadleaved woodlands with well-developed shrub understories (Ryser 1985, Pitocchelli 1995, Dobkin and Sauder 2004). The predictive map indicates a strong elevational association, showing the best chance of finding MacGillivray's Warblers at high elevations and decreasing chances with decreasing elevation. Small spots with a high predicted chance of occurrence are also visible in the state's lowland riparian areas, but records from these habitats probably involved migrants; the chances of finding breeding pairs at these low elevations are likely overestimated.

CONSERVATION AND MANAGEMENT

Population trends of MacGillivray's Warbler have been stable rangewide in recent decades (Sauer et al. 2005). Timber harvesting evidently has had a positive impact on this warbler in the northern parts of its range (Pitocchelli 1995, Dobkin and Sauder 2004), even leading to some range expansions (Raphael et al. 1988, Shuford 1993, Pitocchelli 1995). The long-term effects of forest management practices on the species are not known, however. In the south, including in Nevada, the fortunes of the species are more subject to riparian habitat conditions in middle to upper elevations. Threats here include heavy livestock grazing, habitat conversion, and invasive plants (Dobkin and Sauder 2004). Population trends in Nevada are currently unknown, but Partners in Flight ranks MacGillivray's Warbler as a Priority species in the state (Neel 1999a) because of its close ties to riparian areas.

HABITAT USE

HABITAT	ATLAS BLOCKS	INCIDENTAL OBSERVATIONS
Alpine	2 (1%)	0 (0%)
Mojave	0 (0%)	1 (2%)
Montane Forest	17 (11%)	7 (14%)
Montane Parkland	3 (2%)	0 (0%)
Montane Shrub	26 (17%)	2 (4%)
Pinyon-Juniper	10 (7%)	1 (2%)
Riparian	93 (61%)	38 (76%)
Sagebrush Scrub	1 (<1%)	1 (2%)
Sagebrush Steppe	1 (<1%)	0 (0%)
TOTAL	153	50

Total numbers reported in the Habitat Use, Breeding Status, and Abundance tables may differ from each other (see pp. 22–23 for details). Percentage sums may differ slightly from 100% due to rounding.

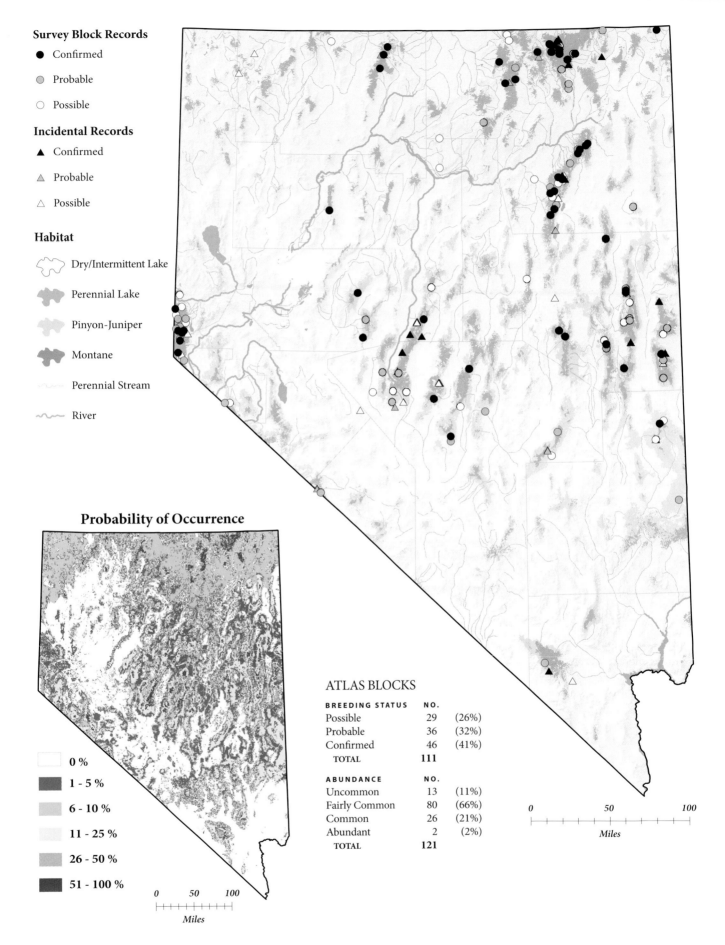

Survey Block Records

- ● Confirmed
- ● Probable
- ○ Possible

Incidental Records

- ▲ Confirmed
- △ Probable
- △ Possible

Habitat

- Dry/Intermittent Lake
- Perennial Lake
- Pinyon-Juniper
- Montane
- Perennial Stream
- River

Probability of Occurrence

- 0 %
- 1 - 5 %
- 6 - 10 %
- 11 - 25 %
- 26 - 50 %
- 51 - 100 %

0   50   100
Miles

ATLAS BLOCKS

BREEDING STATUS	NO.	
Possible	29	(26%)
Probable	36	(32%)
Confirmed	46	(41%)
TOTAL	**111**	

ABUNDANCE	NO.	
Uncommon	13	(11%)
Fairly Common	80	(66%)
Common	26	(21%)
Abundant	2	(2%)
TOTAL	**121**	

0   50   100
Miles

COMMON YELLOWTHROAT
Geothlypis trichas

Rn'02

It is possible to spend a summer day in Nevada's marshlands without seeing a Common Yellowthroat. Failure to hear their slow, grinding *witch-uh-tee witch-uh-tee witch-uh-tee* song, among the most easily learned voices of the marsh, is far less likely. Male Common Yellowthroats typically sing from dense cover, and females and young birds often remain hidden in thick mats of cattails and tules. Their great curiosity, however, makes it easy for a birder to coax them into view with just a small amount of "pishing."

DISTRIBUTION

Common Yellowthroats nest almost exclusively in low, moist, dense vegetation, which they find in a variety of settings throughout their breeding range (Guzy and Ritchison 1999). In a dry state like Nevada, these conditions exist almost exclusively in riparian and wetland areas, which accounted for 94% of the breeding records from atlas blocks. The bird is not invariably present in wetlands, however. Occurrence in the Lahontan Valley is most likely in freshwater wetlands with tule-cattail stands and less likely in drains and areas of closed-stem bulrush (Chisholm and Neel 2002). Atlas breeding records were scattered throughout the state except in the dry regions of Esmeralda County and much of Nye County. This distribution pattern appears to have changed little over the past fifty years (Linsdale 1936, Ryser 1985, Alcorn 1988). The predictive map indicates a widespread, though mostly very low, chance of finding the species throughout the state. This low probability reflects the rare occurrence of small wetlands and riparian areas scattered throughout other habitats across the Nevada landscape. Small areas with high

probabilities of having breeding yellowthroats tend to correspond to the state's larger wetlands, springs, and riparian areas.

Although it is widespread in the states surrounding Nevada, the Common Yellowthroat is often described as localized in these areas, particularly toward the south (Behle et al. 1985, Small 1994, Stephens and Sturts 1998, Adamus et al. 2001, Burger in Corman and Wise-Gervais 2005:478–479). As in Nevada, this pattern coincides with the scattered distribution of wetlands and riparian habitat in much of the West and contrasts with the more contiguous distributional pattern in the eastern part of the species' range (Dunn and Garrett 1997, Guzy and Ritchison 1999).

CONSERVATION AND MANAGEMENT

The Common Yellowthroat is one of the most widespread warblers in North America (Guzy and Ritchison 1999). Continentwide, population trends vary from east to west, with declines reported in several eastern regions. Populations are generally increasing or steady in the West (Sauer et al. 2005), but local declines have also been reported from some areas, such as California's Central Valley (Small 1994). While the species is generally portrayed as tolerant of disturbance (Yahner 1993), it almost certainly suffers when marshes or riparian areas are converted into upland habitats (Guzy and Ritchison 1999).

The status of the Common Yellowthroat in Nevada is thought to have changed little since the nineteenth century (Ryser 1985), and the species is not regarded as a conservation concern at present. While there are some concerns about the possible impacts of cowbird parasitism (Dunn and Garrett 1997), actual negative effects on populations have not been quantified. Isolated populations of the Common Yellowthroat may be vulnerable to ecological impacts, of course. Due to their tight association with wetlands and riparian areas in Nevada, Common Yellowthroat populations may be useful indicators of the condition of riparian areas and the progress of habitat restoration projects.

HABITAT USE

HABITAT	ATLAS BLOCKS	INCIDENTAL OBSERVATIONS
Agricultural	2 (3%)	1 (2%)
Ash	1 (1%)	0 (0%)
Grassland	1 (1%)	0 (0%)
Mojave	0 (0%)	1 (2%)
Open Water	0 (0%)	1 (2%)
Riparian	43 (64%)	23 (55%)
Salt Desert Scrub	0 (0%)	1 (2%)
Sagebrush Scrub	0 (0%)	1 (2%)
Urban	0 (0%)	2 (5%)
Wetland	20 (30%)	12 (29%)
TOTAL	67	42

Total numbers reported in the Habitat Use, Breeding Status, and Abundance tables may differ from each other (see pp. 22–23 for details). Percentage sums may differ slightly from 100% due to rounding.

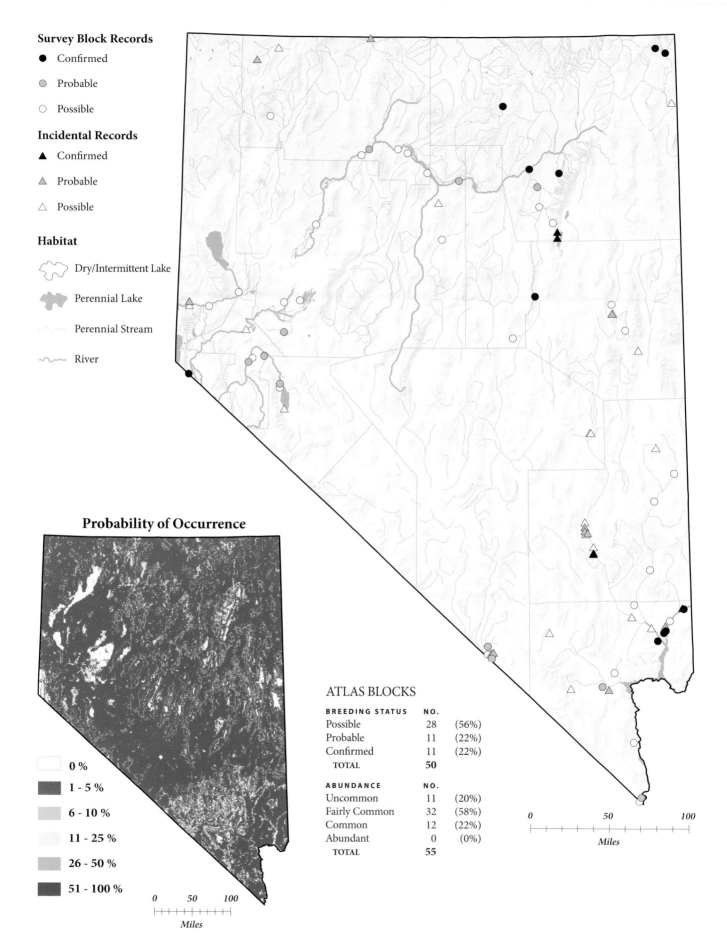

Survey Block Records

● Confirmed

● Probable

○ Possible

Incidental Records

▲ Confirmed

▲ Probable

△ Possible

Habitat

Dry/Intermittent Lake

Perennial Lake

Perennial Stream

River

Probability of Occurrence

☐ 0 %

■ 1 - 5 %

■ 6 - 10 %

■ 11 - 25 %

■ 26 - 50 %

■ 51 - 100 %

0   50   100
Miles

ATLAS BLOCKS

BREEDING STATUS	NO.	
Possible	28	(56%)
Probable	11	(22%)
Confirmed	11	(22%)
TOTAL	50	

ABUNDANCE	NO.	
Uncommon	11	(20%)
Fairly Common	32	(58%)
Common	12	(22%)
Abundant	0	(0%)
TOTAL	55	

0   50   100
Miles

WILSON'S WARBLER
Wilsonia pusilla

Throughout the month of May, Wilson's Warblers can be found in almost every habitat in Nevada. A few linger into June, and then most are gone. The species does breed in the state, but less commonly than many people imagine. Wilson's Warblers were confirmed as breeders only in the Carson Range, although nesting was suspected at scattered high-elevation sites throughout the state, where the birds favor shady thickets of willow and alder.

DISTRIBUTION

During the atlas fieldwork, the only confirmed breeding records of Wilson's Warbler came from the Carson Range of far western Nevada, although unconfirmed records were reported widely throughout the north and east, and south to the Spring Mountains. Both confirmation rates and abundance estimates for Wilson's Warblers were low compared with those for most of Nevada's other warblers. Ryser (1985) considered the species to be an uncommon breeder in Nevada, while Alcorn (1988) considered it to be common at higher elevations. The limited breeding range in Nevada portrayed by Ammon and Gilbert (1999) and Sibley (2003) is consistent with a conservative interpretation of the atlas data, but it could underestimate the actual breeding range in the state.

Wilson's Warbler is for the most part sparsely distributed in the states surrounding Nevada. It is not known to breed in Arizona (Corman and Wise-Gervais 2005), and it is an uncommon summer resident in Utah (Behle et al. 1985). It is common in western Oregon (Adamus et al. 2001) and northern Idaho (Stephens and Sturts 1998), but not in other parts of those states. In California, the species breeds in the north, the Sierra Nevada, and the coastal ranges (Small 1994); local declines and a range restriction have occurred in southern California (Grinnell and Miller 1944, Dunn and Garrett 1997).

During breeding, most Wilson's Warblers in the Intermountain West are restricted to moist, shrubby habitats lacking a dense forest overstory at high elevations, ranging all the way into the treeless alpine zone (Ammon and Gilbert 1999, Dobkin and Sauder 2004). Alcorn (1988) and Ryser (1985) emphasized the importance of willow and alder habitats along mountain streams and lakes. It is likely that most lower-elevation atlas records, such as those from urban and agricultural areas, reflect late migrant sightings. The predictive map shows the same clear elevation pattern seen in several other Nevada warblers, but the chance of finding breeding Wilson's Warblers is seldom high, even at high elevations. The fairly high encounter probabilities predicted for the foothills of the Mojave Desert are almost certainly due to records of late migrants and clearly do not reflect the species' true breeding range in this region. Even in the Great Basin mountains, the chance of finding breeding birds, as opposed to late migrants, is probably lower than the predictive map suggests.

CONSERVATION AND MANAGEMENT

Wilson's Warbler populations have reportedly declined both rangewide and throughout much of the West (Sauer et al. 2005), although some authors argue that western populations are relatively stable (Dunn and Garrett 1997, Dobkin and Sauder 2004). Degradation or loss of riparian areas is a commonly cited source of local declines, particularly in areas away from the Pacific Coast (Ammon and Gilbert 1999, Dobkin and Sauder 2004). Cowbird parasitism has also been cited as a possible reason for population losses in California (Garrett and Dunn 1981, Small 1994), but the net effects of parasitism on productivity elsewhere are unclear (Ammon and Gilbert 1999).

In Nevada, Wilson's Warbler is a Partners in Flight Priority species (Neel 1999a). Significant opportunities still exist to clarify whether breeding actually occurs in the state away from the Carson Range. We also do not yet know whether the declines described elsewhere in the West are occurring in Nevada as well.

HABITAT USE

HABITAT	ATLAS BLOCKS	INCIDENTAL OBSERVATIONS
Agricultural	1 (2%)	0 (0%)
Alpine	1 (2%)	0 (0%)
Montane Forest	10 (21%)	0 (0%)
Montane Shrub	1 (2%)	1 (5%)
Pinyon-Juniper	5 (10%)	2 (10%)
Riparian	28 (58%)	16 (76%)
Sagebrush Steppe	0 (0%)	1 (5%)
Urban	1 (2%)	1 (5%)
Wetland	1 (2%)	0 (0%)
TOTAL	48	21

Total numbers reported in the Habitat Use, Breeding Status, and Abundance tables may differ from each other (see pp. 22–23 for details). Percentage sums may differ slightly from 100% due to rounding.

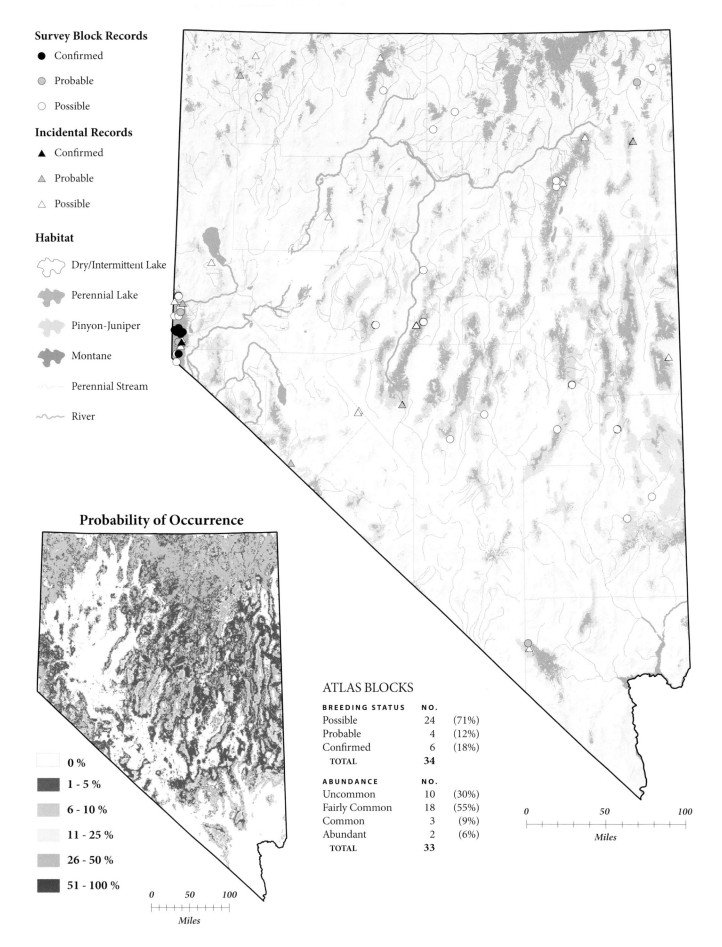

Survey Block Records

● Confirmed

◐ Probable

○ Possible

Incidental Records

▲ Confirmed

△ Probable

△ Possible

Habitat

Dry/Intermittent Lake

Perennial Lake

Pinyon-Juniper

Montane

Perennial Stream

River

Probability of Occurrence

	0 %
	1 - 5 %
	6 - 10 %
	11 - 25 %
	26 - 50 %
	51 - 100 %

0   50   100
Miles

ATLAS BLOCKS

BREEDING STATUS	NO.	
Possible	24	(71%)
Probable	4	(12%)
Confirmed	6	(18%)
TOTAL	**34**	

ABUNDANCE	NO.	
Uncommon	10	(30%)
Fairly Common	18	(55%)
Common	3	(9%)
Abundant	2	(6%)
TOTAL	**33**	

0   50   100
Miles

YELLOW-BREASTED CHAT
Icteria virens

The Yellow-breasted Chat's raucous and penetrating call notes proclaim its dominion over the impenetrable riparian tangles where it hides, yet it is loath to slip into view. This largest and most unusual of our warblers was found throughout the state and was readily detected by its distinctive song. Based on atlas workers' estimates, chats were usually fairly numerous in occupied blocks; nesting takes place in vegetation so dense, though, that breeding was difficult to confirm.

DISTRIBUTION

Ryser (1985) reported that the Yellow-breasted Chat becomes less common as one travels from east to west across the Great Basin. The atlas results are consistent with this pattern, with most records and nearly all breeding confirmations occurring in the eastern half of the state. The low confirmation rate (19%) is similar to that found by other atlas projects (Parmeter in Burridge 1995:153, Jacobs and Wilson 1997).

Throughout their extensive North American breeding range Yellow-breasted Chats can be found in a variety of dense, shrubby habitats. In the East, appropriate thickets may occur far from water, but in the West, the species is partial to lowland and lower-montane riparian settings (Eckerle and Thompson 2001). Gallagher (1997) described mature willow stands as being especially attractive nesting locations, but others have suggested that chats will breed in many kinds of riparian vegetation as long as it is suitably dense (Dunn and Garrett 1997, Eckerle and Thompson 2001). This association with riparian habitat and the correspondingly patchy breeding distribution is also evident in the

states surrounding Nevada (Behle et al. 1985, Small 1994, Stephens and Sturts 1998, Adamus et al. 2001, Averill-Murray and Corman in Corman and Wise-Gervais 2005:484–485), although perhaps less so to the north.

In Nevada, 89% of the reported atlas block habitats involved riparian or similar habitats (e.g., mesquite and ash). The areas where chats are most likely to be found are not easily visible on the predictive map due to its scale, but the birds are most likely to occur along riparian corridors, especially in the southern and eastern parts of the state. The low to moderate chance of finding the species throughout much of the state's lowlands probably reflects the sparse distribution of the chat's primary breeding habitat, dense riparian shrub thickets, on the Nevada landscape.

CONSERVATION AND MANAGEMENT

Rangewide, Yellow-breasted Chat populations seem stable, but there have been significant fluctuations at the periphery of the species' breeding range (Dunn and Garrett 1997, Eckerle and Thompson 2001, Sauer et al. 2005). Local declines in the East are often attributed to the species' frequent use of short-lived early-succession woodlands (Eckerle and Thompson 2001), and those in the West are attributed to degradation or destruction of riparian habitats (Hunter et al. 1988, Eckerle and Thompson 2001), with cowbird parasitism playing a significant contributing role in some areas (Garrett and Dunn 1981, Gallagher 1997).

In Nevada, the Yellow-breasted Chat is considered a Priority species by Partners in Flight (Neel 1999a). Information on population trends is lacking, though, and even some aspects of the biology are still poorly known in the West because of the difficulty of studying this reclusive species. Protection, management, and restoration of riparian areas are likely to be important in managing the Yellow-breasted Chat in Nevada, but better information on its habitat use is needed. Given the species' close association with intact riparian shrublands, it may turn out to be a useful indicator for monitoring riparian restoration projects in the state.

HABITAT USE

HABITAT	ATLAS BLOCKS	INCIDENTAL OBSERVATIONS
Agricultural	1 (2%)	2 (4%)
Ash	5 (8%)	0 (0%)
Mesquite	5 (8%)	1 (2%)
Mojave	1 (2%)	2 (4%)
Montane Shrub	2 (3%)	0 (0%)
Riparian	43 (73%)	41 (84%)
Urban	0 (0%)	1 (2%)
Wetland	2 (3%)	2 (4%)
TOTAL	59	49

Total numbers reported in the Habitat Use, Breeding Status, and Abundance tables may differ from each other (see pp. 22–23 for details). Percentage sums may differ slightly from 100% due to rounding.

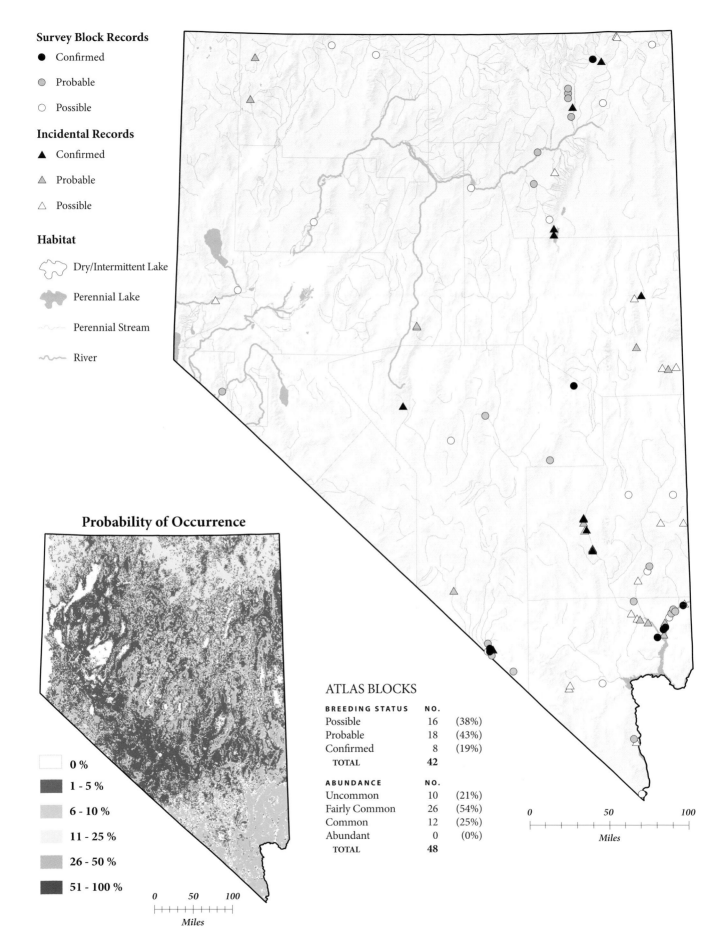

Survey Block Records

- ● Confirmed
- ● Probable
- ○ Possible

Incidental Records

- ▲ Confirmed
- ▲ Probable
- △ Possible

Habitat

Dry/Intermittent Lake

Perennial Lake

Perennial Stream

River

Probability of Occurrence

	0 %
	1 - 5 %
	6 - 10 %
	11 - 25 %
	26 - 50 %
	51 - 100 %

0   50   100
Miles

ATLAS BLOCKS

BREEDING STATUS	NO.	
Possible	16	(38%)
Probable	18	(43%)
Confirmed	8	(19%)
TOTAL	**42**	

ABUNDANCE	NO.	
Uncommon	10	(21%)
Fairly Common	26	(54%)
Common	12	(25%)
Abundant	0	(0%)
TOTAL	**48**	

0   50   100
Miles

SUMMER TANAGER
Piranga rubra

The cottonwood forests of southern Nevada's watercourses are home to a wide variety of colorful passerines—Blue Grosbeaks, Hooded Orioles, and Yellow-breasted Chats among them. And then there is the Summer Tanager, splendidly clad from head to tail in bright rosy red. Summer Tanagers are found in mature riparian forests with extensive canopy cover, and in this regard are reminiscent of Yellow-billed Cuckoos. Like cuckoos, they can be astonishingly difficult to spot, because they spend much of their time in the leafy crowns of tall cottonwoods or in dense tangles of willows and tamarisks.

DISTRIBUTION

Summer Tangers are best known as birds of the southeastern deciduous forests. The distinct *cooperi* subspecies, however, occurs in the southwestern United States and northern Mexico. Southern Nevada and southwestern Utah lie at the northern periphery of *cooperi*'s range, which extends into southeastern California, the river systems of west-central to southeastern Arizona, east into west Texas, and south into northern Mexico (Behle et al. 1985, Small 1994, Robinson 1996, Corman in Corman and Wise-Gervais 2005:488–489).

Summer Tangers were present in only two atlas blocks, both in Clark County, but they were also documented at thirteen other sites in Nevada's three southernmost counties. Earlier reports of the Summer Tanager's Nevada distribution placed them mainly along the Colorado River at the southern tip of Clark County (Linsdale 1936, 1951; Alcorn 1988). Atlas data revealed a wider distribution but, interestingly, failed to document the species along the Colorado River. Because of their low numbers and the scarcity of suitable riparian habitat in Nevada's Mojave Desert, the chances of encountering Summer Tanagers are never high, but—based on the model's predictions—auspicious locations are scattered throughout southern Nevada.

CONSERVATION AND MANAGEMENT

Because it is a neotropical migrant favoring lowland riparian vegetation and having a restricted distribution, one could easily conclude that the Summer Tanager is a conservation concern. Certainly, loss of riparian habitat has been linked to rapid local declines in California and the lower Colorado River Valley (Rosenberg et al. 1991, Robinson 1996). But on a broader scale, Summer Tanager populations appear to be stable both in the Southwest and elsewhere in the United States (Sauer et al. 2005). In fact, Summer Tanagers have expanded their range in several western states, possibly because of climatic trends toward warmer, wetter summers (Johnson 1994, Robinson 1996).

While Summer Tanagers are not a Priority species of Nevada Partners in Flight, they are a Covered species under the Clark County Multiple Species Habitat Conservation Plan (Clark County 2000) and the Lower Colorado River Multi-species Conservation Program (BOR-LCR 2004) because of their specific habitat requirements and low population densities. Both of these plans present a great opportunity to protect Summer Tanagers. A matter of particular interest, and one meriting additional research, is whether Summer Tanagers still occupy their historical breeding grounds at the southern tip of the state (see Linsdale 1936) or whether degradation of riparian habitat in this area has caused local extirpation.

HABITAT USE

HABITAT	ATLAS BLOCKS	INCIDENTAL OBSERVATIONS
Agricultural	0 (0%)	1 (11%)
Mesquite	0 (0%)	1 (11%)
Mojave	0 (0%)	1 (11%)
Riparian	3 (100%)	6 (67%)
TOTAL	3	9

Total numbers reported in the Habitat Use, Breeding Status, and Abundance tables may differ from each other (see pp. 22–23 for details). Percentage sums may differ slightly from 100% due to rounding.

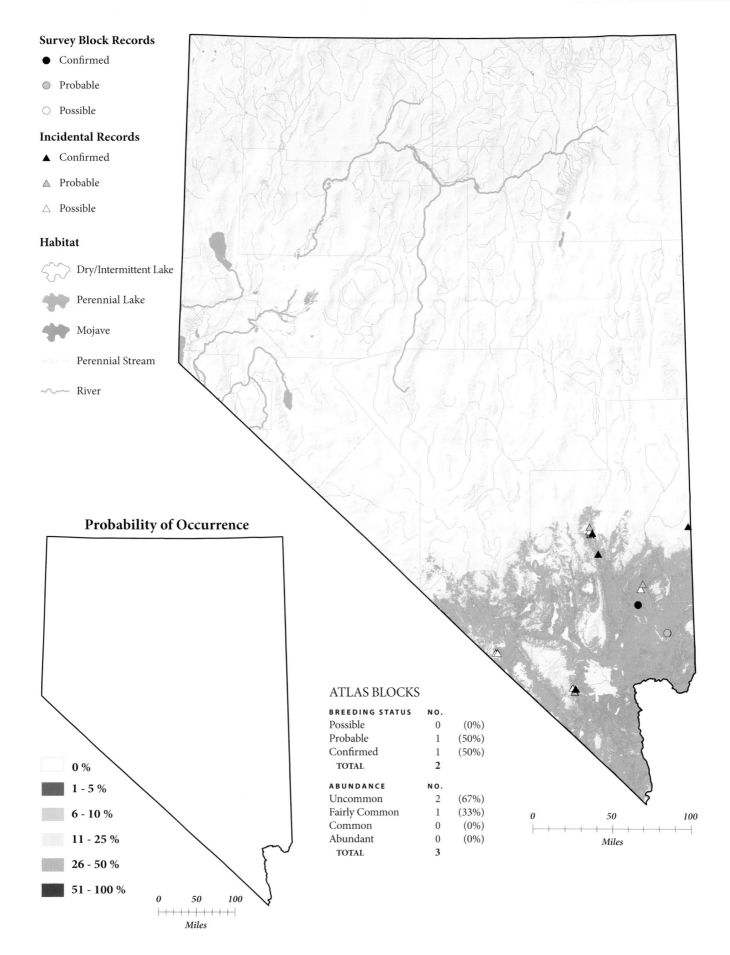

Survey Block Records

● Confirmed

◐ Probable

○ Possible

Incidental Records

▲ Confirmed

△ Probable

△ Possible

Habitat

Dry/Intermittent Lake

Perennial Lake

Mojave

Perennial Stream

River

Probability of Occurrence

0 %

1 - 5 %

6 - 10 %

11 - 25 %

26 - 50 %

51 - 100 %

0   50   100

Miles

ATLAS BLOCKS

BREEDING STATUS	NO.	
Possible	0	(0%)
Probable	1	(50%)
Confirmed	1	(50%)
TOTAL	2	
ABUNDANCE	NO.	
Uncommon	2	(67%)
Fairly Common	1	(33%)
Common	0	(0%)
Abundant	0	(0%)
TOTAL	3	

0   50   100

Miles

WESTERN TANAGER
Piranga ludoviciana

Rh'02

"I never get tired of looking at this one!" seems to be the assessment of most birders, no matter how many times they see a Western Tanager. With its brilliant, strongly contrasting coloration, it resembles an oversized wood warbler. In its overall demeanor, though, it acts more like a vireo: deliberate and perhaps just a bit dopey.

DISTRIBUTION

Atlas workers found Western Tanagers in mountainous regions throughout Nevada, where earlier reports of their distribution had indicated they would be (Linsdale 1936, Ryser 1985, Alcorn 1988). In Nevada and the broader western region, this tanager's breeding distribution closely matches the distribution of montane coniferous forests. Western Tanagers are particularly common and widespread in Oregon (Adamus et al. 2001), Idaho (Stephens and Sturts 1998), and Utah (Behle et al. 1985). Though common in the forests of Arizona and California, they are essentially absent from southwestern Arizona and from the Central Valley and the low deserts of southern California, as well as from the lower Colorado River area (Rosenberg et al. 1991, Small 1994, Moors and Corman in Corman and Wise-Gervais 2005:490–491).

Although usually associated with montane coniferous or mixed deciduous-coniferous forest, the Western Tanager also occurs in riparian woodlands, aspen stands, and pinyon-juniper woodlands (Hudon 1999). Linsdale (1936) and Alcorn (1988) stressed this tanager's association with mountain mahogany woodlands in Nevada, but atlas results show that many reports also came from higher-elevation montane forest.

In some areas, Western Tanagers are known to wander far from their breeding grounds during the summer months (e.g., Small 1994). This tendency for Western Tanagers to show up outside their primary breeding habitat may explain why the model shows a high probability of occurrence in portions of the distinctly unsuitable northern Mojave Desert. Elsewhere, the map shows a high likelihood of finding Western Tanagers throughout the mountain ranges that have extensive montane forest, as expected.

CONSERVATION AND MANAGEMENT

Western Tanager populations have not seen major declines in recent decades; in fact, the species may have increased in some areas since the 1980s (Sauer et al. 2005). There have been some range expansions in California (Small 1994) and population increases in Washington (Smith et al. 1997, Hudon 1999). The species may be a beneficiary of selective logging (see Hudon 1999 for a summary of forestry effects). In Arizona, for example, removal of canopy trees had a positive effect on Western Tanager numbers (Franzreb and Ohmart 1978; see also Moors and Corman in Corman and Wise-Gervais 2005:490–491).

There is no evidence of Western Tanager population declines in Nevada, and the species is not usually viewed as a conservation concern in the state. Ryser (1985) indicated that the status of the Western Tanager has changed little since the nineteenth century. At the present time, the following questions may deserve further investigation: (1) What are the Western Tanager's precise habitat associations in Nevada? (2) How do forest management practices such as fire management affect Western Tanager populations? (3) What are the trends of Western Tanager populations in Nevada?

HABITAT USE

HABITAT	ATLAS BLOCKS	INCIDENTAL OBSERVATIONS
Agricultural	2 (<1%)	0 (0%)
Alpine	2 (<1%)	0 (0%)
Montane Forest	110 (52%)	25 (32%)
Montane Parkland	4 (2%)	0 (0%)
Montane Shrub	7 (3%)	6 (8%)
Pinyon-Juniper	39 (18%)	12 (15%)
Riparian	38 (18%)	27 (35%)
Sagebrush Scrub	2 (<1%)	0 (0%)
Sagebrush Steppe	2 (<1%)	3 (4%)
Urban	5 (2%)	5 (6%)
TOTAL	211	78

Total numbers reported in the Habitat Use, Breeding Status, and Abundance tables may differ from each other (see pp. 22–23 for details). Percentage sums may differ slightly from 100% due to rounding.

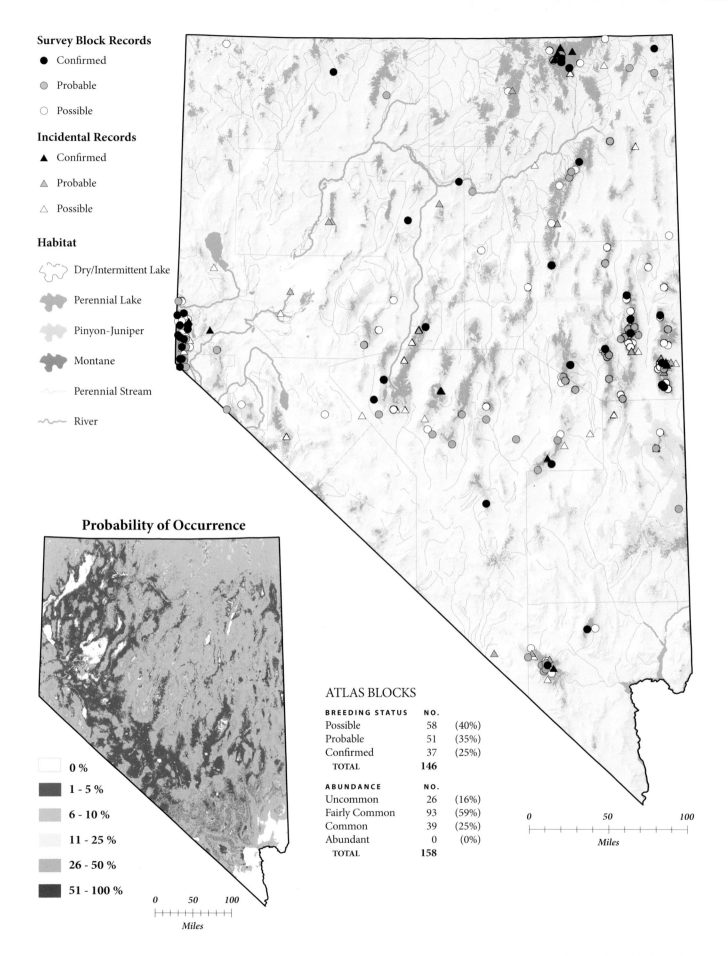

Survey Block Records

● Confirmed

◐ Probable

○ Possible

Incidental Records

▲ Confirmed

◮ Probable

△ Possible

Habitat

Dry/Intermittent Lake

Perennial Lake

Pinyon-Juniper

Montane

Perennial Stream

River

Probability of Occurrence

0 %

1 - 5 %

6 - 10 %

11 - 25 %

26 - 50 %

51 - 100 %

0   50   100
Miles

ATLAS BLOCKS

BREEDING STATUS	NO.	
Possible	58	(40%)
Probable	51	(35%)
Confirmed	37	(25%)
TOTAL	146	

ABUNDANCE	NO.	
Uncommon	26	(16%)
Fairly Common	93	(59%)
Common	39	(25%)
Abundant	0	(0%)
TOTAL	158	

0   50   100
Miles

GREEN-TAILED TOWHEE
Pipilo chlorurus

The Green-tailed Towhee, with its bright rufous crown, white throat and malar stripes, and distinctive greenish wings and tail, is one of our most attractive sparrows. It is also one of our largest, and as competent a songster as any. As it lives in "open" habitats, one might think that it would be easy to see, but the Green-tailed Towhee can be downright furtive. It spends much of its time under the cover of low, dense tangles and ventures into the open only to defend its territory or for short flights between patches of shrubbery.

DISTRIBUTION

The Green-tailed Towhee is very much a bird of the Intermountain West. It ranges no farther east than the Rocky Mountains and barely makes it west of the Sierra Nevada (Dobbs et al. 1998). It is common in suitable habitat throughout Utah (Behle et al. 1985), and the montane areas of California, Oregon, Idaho, and Arizona that are close to Nevada generally have breeding Green-tailed Towhees, while their distant portions do not (Small 1994, Stephens and Sturts 1998, Adamus et al. 2001, Corman in Corman and Wise-Gervais 2005:494–495).

Atlas workers found Green-tailed Towhees throughout Nevada, although the species was rarer in the south, and most observations were clustered around mountain ranges. Green-tailed Towhees tend to avoid dense forests, instead preferring shrubland habitats ranging from mature sagebrush to thickets in the riparian-upland ecotone to higher-elevation forest openings full of manzanita, chokecherry, and serviceberry (Dobbs et al. 1998, Dobkin and Sauder 2004). They are apparently partial to the transition zones between sagebrush and other shrub communities (Knopf et al. 1990) and were found in many atlas habi-

tat types. The predictive map indicates a high likelihood of occurrence in higher-elevation montane habitats and—in contrast with the Spotted Towhee—a lower probability in pinyon-juniper. Away from the high mountains, Green-tailed Towhees are most likely to be found in the sagebrush steppe of northern Nevada. As was true for many montane species, the map's predictions in parts of southern Nevada appear to overestimate the chance of finding breeders there, probably because late migrants were sometimes taken for potential breeders. Fieldworkers estimated high abundance in many atlas blocks, with just over half of the occupied blocks exceeding ten pairs per block.

CONSERVATION AND MANAGEMENT

The Green-tailed Towhee is not a well-studied bird (Dobbs et al. 1998), and its management needs are not fully known. Although some types of logging can create breeding habitat, the healthy shrub communities that it tends to use are likely to be harmed by sagebrush removal, heavy livestock grazing, and changes in fire regimes (Dobbs et al. 1998, Paige and Ritter 1999, Dobkin and Sauder 2004). Such factors may have caused past declines (Dobbs et al. 1998, Dobkin and Sauder 2004). And while it may be true that seventy years ago most Green-tailed Towhees lived in sagebrush (Linsdale 1936), the atlas results suggest that montane shrublands are more often used today. Rangewide population trends from the last few decades are not conclusive (Sauer et al. 2005), and different analyses have produced different interpretations (see Dobbs et al. 1998). Most recently, Dobkin and Sauder (2004) concluded that populations are actually increasing.

While the Green-tailed Towhee is in no danger today, concerns about habitat degradation along with the species' relatively small breeding range have caused Partners in Flight to designate this bird a Stewardship species in the Intermountain West and Southwest (Rich et al. 2004). Management recommendations have emphasized maintenance of sagebrush and associated plant species (Knopf et al. 1990). More work remains to be done before the Green-tailed Towhee's conservation needs are fully understood.

HABITAT USE

HABITAT	ATLAS BLOCKS	INCIDENTAL OBSERVATIONS
Alpine	1 (<1%)	0 (0%)
Grassland	2 (<1%)	0 (0%)
Mojave	1 (<1%)	1 (2%)
Montane Forest	24 (7%)	13 (20%)
Montane Parkland	2 (<1%)	3 (5%)
Montane Shrub	117 (36%)	18 (28%)
Pinyon-Juniper	54 (17%)	7 (11%)
Riparian	43 (13%)	14 (22%)
Salt Desert Scrub	2 (<1%)	0 (0%)
Sagebrush Scrub	35 (11%)	5 (8%)
Sagebrush Steppe	43 (13%)	4 (6%)
Urban	1 (<1%)	0 (0%)
TOTAL	325	65

Total numbers reported in the Habitat Use, Breeding Status, and Abundance tables may differ from each other (see pp. 22–23 for details). Percentage sums may differ slightly from 100% due to rounding.

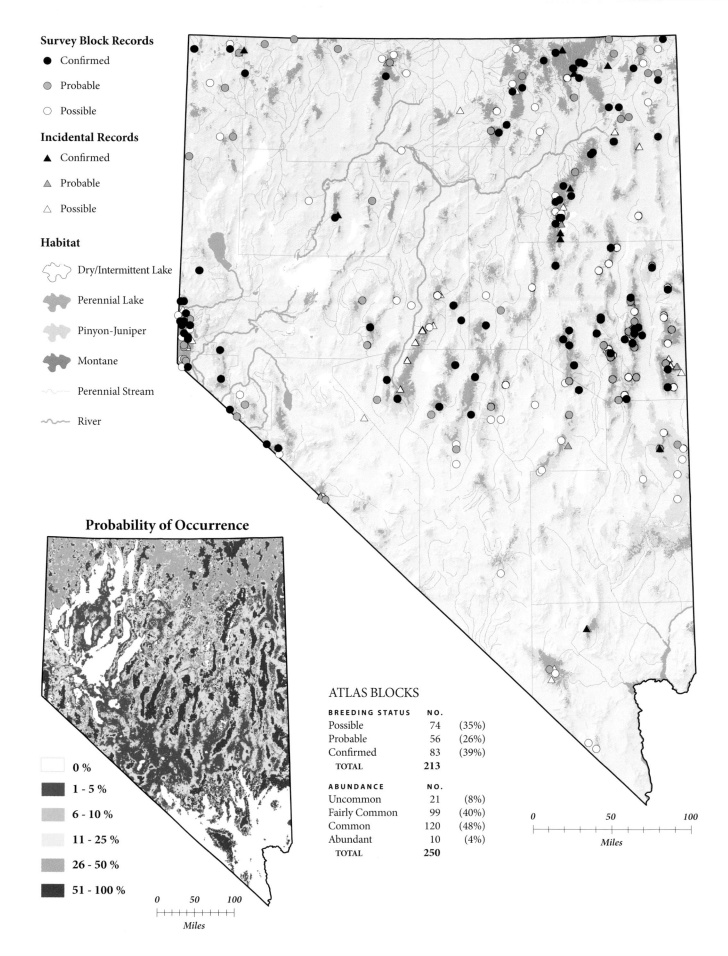

Survey Block Records

- ● Confirmed
- ● Probable
- ○ Possible

Incidental Records

- ▲ Confirmed
- ▲ Probable
- △ Possible

Habitat

- Dry/Intermittent Lake
- Perennial Lake
- Pinyon-Juniper
- Montane
- Perennial Stream
- River

Probability of Occurrence

- ☐ 0 %
- 1 - 5 %
- 6 - 10 %
- 11 - 25 %
- 26 - 50 %
- 51 - 100 %

0   50   100

Miles

ATLAS BLOCKS

BREEDING STATUS	NO.	
Possible	74	(35%)
Probable	56	(26%)
Confirmed	83	(39%)
TOTAL	**213**	

ABUNDANCE	NO.	
Uncommon	21	(8%)
Fairly Common	99	(40%)
Common	120	(48%)
Abundant	10	(4%)
TOTAL	**250**	

0   50   100

Miles

SPOTTED TOWHEE

Pipilo maculatus

If it were not for the Spotted Towhee, summer afternoons in the pinyon-juniper woodlands of Nevada would be much quieter. A flock of descending Pinyon Jays can certainly create a stir, but their appearances are intermittent at best. Bushtits can be vocal, but their whispering calls are audible only at close range. The Spotted Towhee, though, can be relied on to add a sense of exuberance to its surroundings. It has at its command a varied repertoire of loud and distinctive call notes and songs. Even its manner of foraging—kicking around in the litter looking for seeds and soil invertebrates—creates a ruckus.

DISTRIBUTION

Spotted Towhees occur throughout most of the western United States. They were found in every Nevada county, and they were often abundant, with estimates exceeding ten pairs in almost half of the occupied blocks. In the states surrounding Nevada, they are common and widely distributed as breeders except in the low deserts of southern California and southwestern Arizona (Behle et al. 1985, Rosenberg et al. 1991, Small 1994, Greenlaw 1996, Stephens and Sturts 1998, Adamus et al. 2001, Wise-Gervais in Corman and Wise-Gervais 2005:496–497).

Spotted Towhees breed in many vegetation types, especially pinyon-juniper, riparian, montane shrub, sagebrush scrub, and sagebrush steppe. All of these habitats can provide this species' main microhabitat needs: shrubby vegetation in which to nest and a sheltered layer of plant litter in which to forage (Greenlaw 1996). Conditions appropriate for nesting Spotted Towhees are

found throughout much of Nevada, and the estimated chance of occurrence is high in all but the lowest-elevation valleys and playas, and highest in mid-elevation montane areas. The moderate chance of occurrence in urban habitats is undoubtedly the result of atlas records from higher-elevation settlements, especially in western Nevada, and the model's predictions for urban areas in the south are unlikely to be accurate.

Earlier writers portrayed the Spotted Towhee as a primarily montane species in Nevada (Linsdale 1936, Alcorn 1988), but the atlas data indicate a broader elevational range, especially when compared with the Green-tailed Towhee. Though their ranges overlap considerably, Spotted Towhees enter lowland riparian areas such as the Carson and Walker river drainages while Green-tailed Towhees appear to be more restricted to higher elevations. This difference is especially apparent when the predictive maps for the species are compared.

CONSERVATION AND MANAGEMENT

Spotted Towhee populations appear to be healthy rangewide, with no evidence of widespread declines or increases (Greenlaw 1996, Sauer et al. 2005). Human activities appear to help the species in some areas and to harm it in others (Greenlaw 1996). In California, for example, forest clearing and residential plantings have probably benefited the species by increasing the shrub layer, while fire suppression and intensive development may have hurt it (Shuford 1993). Spotted Towhees readily colonize some suburban areas where shrub cover is sufficiently dense (Garrett and Dunn 1981).

The Spotted Towhee is probably currently secure in Nevada, although additional monitoring would help to confirm this assessment. Two subspecies occur in Nevada (Linsdale 1936, Greenlaw 1996), and further study of their range limits would be worthwhile. Ryser (1985) and Gaines (1992) discussed the vocal differences among several of the western subspecies, which can be distinguished by their songs.

HABITAT USE

HABITAT	ATLAS BLOCKS	INCIDENTAL OBSERVATIONS
Agricultural	1 (<1%)	0 (0%)
Grassland	1 (<1%)	0 (0%)
Mojave	0 (0%)	1 (<1%)
Montane Forest	16 (5%)	6 (5%)
Montane Parkland	2 (<1%)	1 (<1%)
Montane Shrub	48 (14%)	19 (17%)
Pinyon-Juniper	128 (37%)	27 (24%)
Riparian	75 (22%)	43 (38%)
Salt Desert Scrub	1 (<1%)	0 (0%)
Sagebrush Scrub	39 (11%)	9 (8%)
Sagebrush Steppe	34 (10%)	4 (4%)
Urban	1 (<1%)	3 (3%)
Wetland	0 (0%)	1 (<1%)
TOTAL	346	114

Total numbers reported in the Habitat Use, Breeding Status, and Abundance tables may differ from each other (see pp. 22–23 for details). Percentage sums may differ slightly from 100% due to rounding.

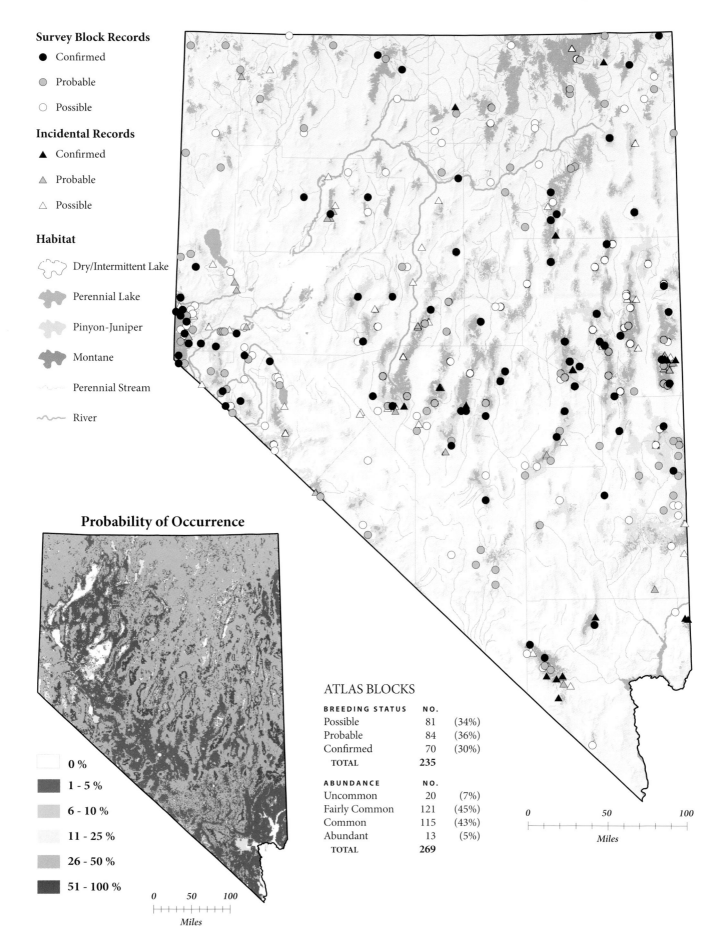

Survey Block Records

● Confirmed

● Probable

○ Possible

Incidental Records

▲ Confirmed

△ Probable

△ Possible

Habitat

Dry/Intermittent Lake

Perennial Lake

Pinyon-Juniper

Montane

Perennial Stream

River

Probability of Occurrence

☐ 0 %

1 - 5 %

6 - 10 %

11 - 25 %

26 - 50 %

51 - 100 %

0   50   100

Miles

ATLAS BLOCKS

BREEDING STATUS	NO.	
Possible	81	(34%)
Probable	84	(36%)
Confirmed	70	(30%)
TOTAL	**235**	

ABUNDANCE	NO.	
Uncommon	20	(7%)
Fairly Common	121	(45%)
Common	115	(43%)
Abundant	13	(5%)
TOTAL	**269**	

0   50   100

Miles

ABERT'S TOWHEE
Pipilo aberti

Abert's Towhees live in year-round territories along the major rivers of the Mojave and Sonoran deserts, where they can be quite secretive. They also occur in the suburbs, where they have an alternate personality and often behave more boldly. In most places, though, this bird is more easily heard than seen. The "reunion duet" produced by both members of a mated pair on reuniting is among the calls most likely to be heard.

DISTRIBUTION

Abert's Towhee is essentially an Arizona bird (Corman in Corman and Wise-Gervais 2005:500–501) that reaches ever so slightly into southern Nevada, southern California, southwestern Utah, and southwestern New Mexico (Behle et al. 1985, Small 1994, Tweit and Finch 1994). Its breeding range barely extends into northern Mexico, unlike the many southwestern birds that have a substantial presence south of the border. Within its limited range Abert's Towhee requires the dense cover of woody vegetation and is most often found in riparian cottonwood-willow and mesquite environs. Shrub-covered desert washes and marsh edges are also used. With increasing habitat conversion and urbanization, this towhee's habitat repertoire has expanded to include exotic tamarisk thickets and suburban plantings (Rosenberg et al. 1991, Tweit and Finch 1994, Corman in Corman and Wise-Gervais 2005:500–501), and in the Las Vegas area it is not hard to find in Sunset Park.

The atlas survey confirmed that Abert's Towhee occurs along the Virgin, Muddy, and lower Colorado rivers, and also in the Las Vegas area, where the species was first reported in the 1960s (Linsdale 1936, Alcorn 1988). The westernmost record was from Ash Meadows National Wildlife Refuge in southern Nye County, and the northernmost was from the Pahranagat Valley in Lincoln County—both areas where the species was poorly documented previously. Breeding was not confirmed in these locations, but the species is sedentary (Tweit and Finch 1994), and the likelihood of nonbreeding transients seems small. Where present, Abert's Towhees were numerous: over two-thirds of the abundance estimates exceeded ten pairs per block.

Most atlas records came from riparian and mesquite habitats, but there were several incidental records from urban areas. Away from rivers, this bird was rarely found in the Mojave uplands. The predictive map suggests that Abert's Towhee is most likely to occur in riparian areas and much less likely to breed throughout the remainder of the Mojave Desert. Whether the species regularly occurs in parts of the northwestern Mojave, as predicted, is unclear, but it seems unlikely because the region provides little riparian habitat.

CONSERVATION AND MANAGEMENT

Abert's Towhees have probably declined over the past 150 years as lowland riparian habitats in their range have been decimated (Rosenberg et al. 1991, Tweit and Finch 1994). No population changes have been noted since the 1960s (Sauer et al. 2005), although monitoring data for this species are limited. Still, because of their restricted geographic range, Abert's Towhees are a Partners in Flight Watch List species (Rich et al. 2004).

Much of the species' preferred cottonwood-willow and mesquite habitat has been destroyed by agriculture, water development, and urbanization, and newly available tamarisk stands and ornamental plantings do not seem to replace it fully (Tweit and Finch 1994). In California, habitat loss has led to relatively recent local declines (Small 1994). Abert's Towhees are also parasitized by cowbirds (Finch 1983), and livestock grazing has been implicated as a threat as well (Tweit and Finch 1994). Fortunately, Abert's Towhees are expected to respond positively to habitat management designed to benefit the Southwestern Willow Flycatcher and other riparian birds (Tweit and Finch 1994), and monitoring the entire riparian avifauna should be a high priority in southern Nevada.

HABITAT USE

HABITAT	ATLAS BLOCKS	INCIDENTAL OBSERVATIONS
Agricultural	0 (0%)	1 (3%)
Barren	1 (3%)	0 (0%)
Mesquite	7 (22%)	5 (14%)
Mojave	3 (9%)	1 (3%)
Riparian	18 (56%)	20 (57%)
Salt Desert Scrub	3 (9%)	1 (3%)
Urban	0 (0%)	5 (14%)
Wetland	0 (0%)	2 (6%)
TOTAL	32	35

Total numbers reported in the Habitat Use, Breeding Status, and Abundance tables may differ from each other (see pp. 22–23 for details). Percentage sums may differ slightly from 100% due to rounding.

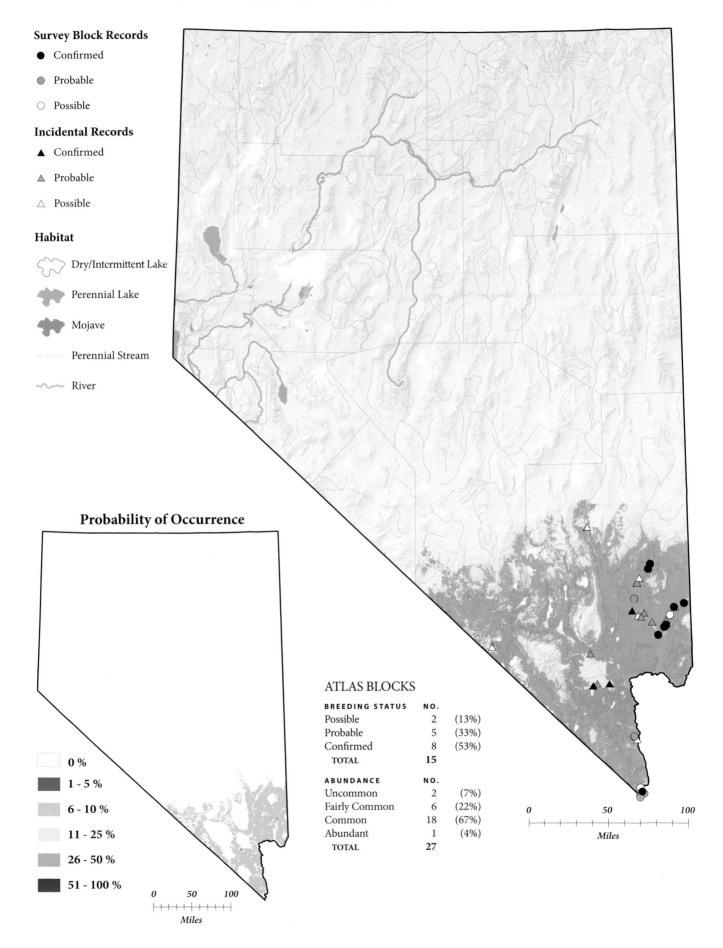

Survey Block Records

● Confirmed

● Probable

○ Possible

Incidental Records

▲ Confirmed

▲ Probable

△ Possible

Habitat

Dry/Intermittent Lake

Perennial Lake

Mojave

Perennial Stream

River

Probability of Occurrence

☐ 0 %

◼ 1 - 5 %

◼ 6 - 10 %

☐ 11 - 25 %

◼ 26 - 50 %

◼ 51 - 100 %

0   50   100
Miles

ATLAS BLOCKS

BREEDING STATUS	NO.	
Possible	2	(13%)
Probable	5	(33%)
Confirmed	8	(53%)
TOTAL	**15**	

ABUNDANCE	NO.	
Uncommon	2	(7%)
Fairly Common	6	(22%)
Common	18	(67%)
Abundant	1	(4%)
TOTAL	**27**	

0   50   100
Miles

RUFOUS-CROWNED SPARROW

Aimophila ruficeps

One can point to a number of reasons why the Rufous-crowned Sparrow is seldom observed in Nevada: it typically inhabits rock- and brush-covered slopes in remote and poorly accessible areas; it is cryptically colored and its song is not terribly conspicuous; and, in sparrowlike fashion, it usually lurks under the cover of vegetation. But the main reason why these little birds are not often seen is that they just barely make it into the southern part of the state. The atlas survey provided the first confirmation that the Rufous-crowned Sparrow is indeed a member of Nevada's breeding avifauna.

DISTRIBUTION

All reports of Rufous-crowned Sparrows during the atlas years came from the McCullough Range in Clark County. This small area produced one block record of possible breeding and five incidental records, one of which involved confirmed breeding. Multiple territorial birds were noted during the incidental observations. These records represent Nevada's first breeding records for a species that Linsdale (1936) did not mention at all and that was previously known in the state only from rare sightings (Alcorn 1988). Most of those old records came from the south, but there are also old observations from the Lehman Caves area of eastern White Pine County (Alcorn 1988), and it seems possible that a breeding population existed in east-central Nevada at one time (Collins 1999).

Rufous-crowned Sparrows generally breed south of Nevada, although their distribution is patchy (Collins 1999). The coastal scrublands of southern and central California and the cactus-covered slopes of central to southeastern Arizona appear to be strongholds (Small 1994, Corman in Corman and Wise-Gervais 2005:508–509). Breeding in Utah has been reported only from Zion National Park in the southwestern part of the state (Behle et

al. 1985). The species' breeding range reaches east to Texas and Oklahoma, and south far into Mexico, but it barely extends north beyond the southern tier of states (Collins 1999).

Rufous-crowned Sparrows are associated with dry, rocky slopes that have good grass and shrub cover. They also use open woodlands, but they generally avoid tall woodlands with much canopy closure and dense, monotypic stands of shrubs or trees (Small 1994, Collins 1999, Corman in Corman and Wise-Gervais 2005:508–509). All of the Nevada records were from pinyon-juniper or montane shrub habitats. The records were few and geographically restricted, however, and the predicted occurrence is therefore limited to essentially the same area where field observations were reported. The historical presence of birds much farther north (Alcorn 1988) suggests that breeding outposts beyond the McCullough Range are a possibility.

CONSERVATION AND MANAGEMENT

Because of the Rufous-crowned Sparrow's habitat preference, habits, and localized distribution, there is little reliable information from which to gauge population trends and conservation status. Breeding Bird Survey data suggest that populations are relatively stable in most areas, but there have been significant declines in Texas (Sauer et al. 2005), Oklahoma, and southern California (Collins 1999). Indeed, three island subspecies off California's coast became extinct during the nineteenth and twentieth centuries (Collins 1999).

Rufous-crowned Sparrows can respond positively to events that create patchy disturbances in maturing shrublands, such as fire and short-term grazing (Collins 1999), but long-term grazing apparently has negative effects by reducing the vegetative cover too much. Fire suppression may also have negative effects by allowing the shrub layer to become overly dense (Shuford 1993). Population declines in California and Texas have been attributed to habitat loss and fragmentation (Collins 1999).

HABITAT USE

HABITAT	ATLAS BLOCKS	INCIDENTAL OBSERVATIONS
Montane Shrub	0 (0%)	2 (33%)
Pinyon-Juniper	1 (100%)	4 (67%)
TOTAL	1	6

Total numbers reported in the Habitat Use, Breeding Status, and Abundance tables may differ from each other (see pp. 22–23 for details). Percentage sums may differ slightly from 100% due to rounding.

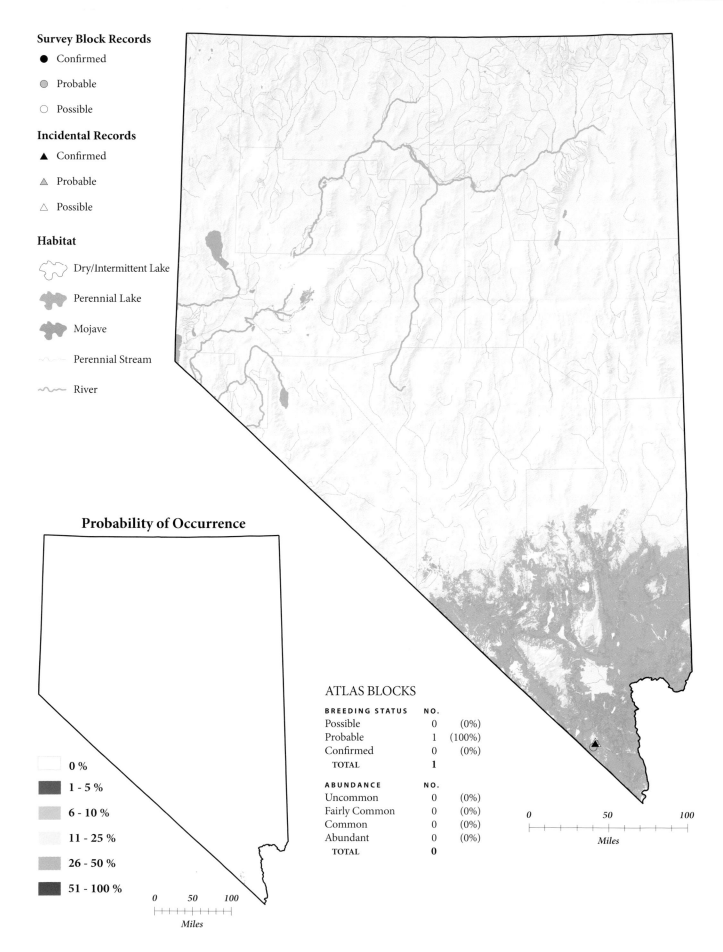

Survey Block Records

● Confirmed

◉ Probable

○ Possible

Incidental Records

▲ Confirmed

△ Probable

△ Possible

Habitat

Dry/Intermittent Lake

Perennial Lake

Mojave

Perennial Stream

River

Probability of Occurrence

☐ 0 %

■ 1 - 5 %

▨ 6 - 10 %

▨ 11 - 25 %

▨ 26 - 50 %

■ 51 - 100 %

0   50   100
Miles

ATLAS BLOCKS

BREEDING STATUS	NO.	
Possible	0	(0%)
Probable	1	(100%)
Confirmed	0	(0%)
TOTAL	**1**	

ABUNDANCE	NO.	
Uncommon	0	(0%)
Fairly Common	0	(0%)
Common	0	(0%)
Abundant	0	(0%)
TOTAL	**0**	

0   50   100
Miles

CHIPPING SPARROW

Spizella passerina

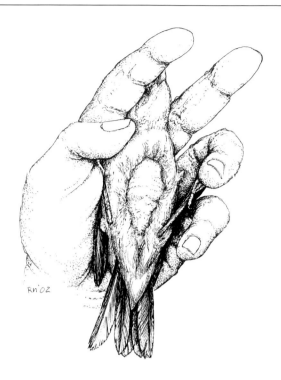

The six American sparrows in the genus *Spizella* sing six very different songs. In the case of the Chipping Sparrow, the song is a long, dry rattle, all on one pitch and somewhat similar to that of a Dark-eyed Junco. It is not an unpleasant sound, though, and it seems to be a perfect fit for the dusty woodlands that the Chipping Sparrow calls home. In many parts of its range, which extends across most of North America, the Chipping Sparrow frequently uses suburban and urban nest sites. But this is not the case in Nevada—at least not yet; here these sparrows appear to be far fonder of open pinyon-juniper and ponderosa pine woodlands.

DISTRIBUTION

Atlas workers found Chipping Sparrows mostly in and around the state's major mountain ranges, with few in the lower valleys (e.g., Chisholm and Neel 2002). The sparrows were often numerous as well, with about half of the abundance estimates exceeding ten pairs per occupied block.

The Chipping Sparrow is common and widely distributed in the states surrounding Nevada, especially to the north and east in Utah, Idaho, and Oregon (Behle et al. 1985, Stephens and Sturts 1998, Adamus et al. 2001). Chipping Sparrows also breed throughout much of California and Arizona, but they are absent from California's Central Valley and southern deserts, and from southwestern Arizona (Small 1994, Wise-Gervais in Corman and Wise-Gervais 2005:512–513).

Unlike many sparrows, this species is primarily a woodland bird in Nevada. Habitat types used for nesting vary considerably across the species' broad summer range, but in the West it is usually some type of open, coniferous forest (Middleton 1998). In Nevada, most habitat records came from dry conifer woodlands—montane forest or pinyon-juniper—and from associated higher-elevation shrublands. This sparrow was not especially common in riparian habitat. The predictive map suggests a moderate to high chance of seeing Chipping Sparrows in all of the state's mountains, with probabilities usually highest at middle elevations. The odds rapidly diminish as one moves away from the mountains. Although the Chipping Sparrow has successfully adapted to urban conditions elsewhere in its range (Middleton 1998), atlas data show little evidence for this in Nevada.

CONSERVATION AND MANAGEMENT

The Chipping Sparrow seems to have benefited from the early European settlement of North America, as forest clearing and tree planting created new habitats, especially in the East (Franzreb and Ohmart 1978, Middleton 1998). These positive impacts are thought to have peaked in the late nineteenth century, after which populations declined, perhaps due to more intensive farming and forestry practices (Middleton 1998). Chipping Sparrow populations in eastern North America remain large, however, and were relatively stable during the latter decades of the twentieth century (Sauer et al. 2005).

In the West, it is a different story. Except for forest edges created by clear-cuts, Chipping Sparrows do not seem to be inclined to use human-altered habitats. Monitoring data for the region indicate population declines of about 1% per year, with even higher rates (4%) in the Sierra Nevada (Dobkin and Sauder 2004, Sauer et al. 2005; see also Tenney in Roberson and Tenney 1993:346–347, Smith et al. 1997). The causes of the declines in the West are not well understood, but they may include cowbird parasitism and other interspecific interactions (Middleton 1998, Dobkin and Sauder 2004). Reliable monitoring data do not exist specifically for Nevada, but Chipping Sparrows remain common in the state despite the ongoing regional declines, and for now the species is not regarded as a conservation concern.

HABITAT USE

HABITAT	ATLAS BLOCKS	INCIDENTAL OBSERVATIONS
Agricultural	1 (<1%)	0 (0%)
Alpine	4 (1%)	0 (0%)
Montane Forest	77 (25%)	25 (34%)
Montane Parkland	8 (3%)	3 (4%)
Montane Shrub	51 (16%)	8 (11%)
Pinyon-Juniper	125 (40%)	21 (28%)
Riparian	17 (5%)	11 (15%)
Salt Desert Scrub	1 (<1%)	0 (0%)
Sagebrush Scrub	15 (5%)	2 (3%)
Sagebrush Steppe	13 (4%)	3 (4%)
Urban	1 (<1%)	0 (0%)
TOTAL	313	73

Total numbers reported in the Habitat Use, Breeding Status, and Abundance tables may differ from each other (see pp. 22–23 for details). Percentage sums may differ slightly from 100% due to rounding.

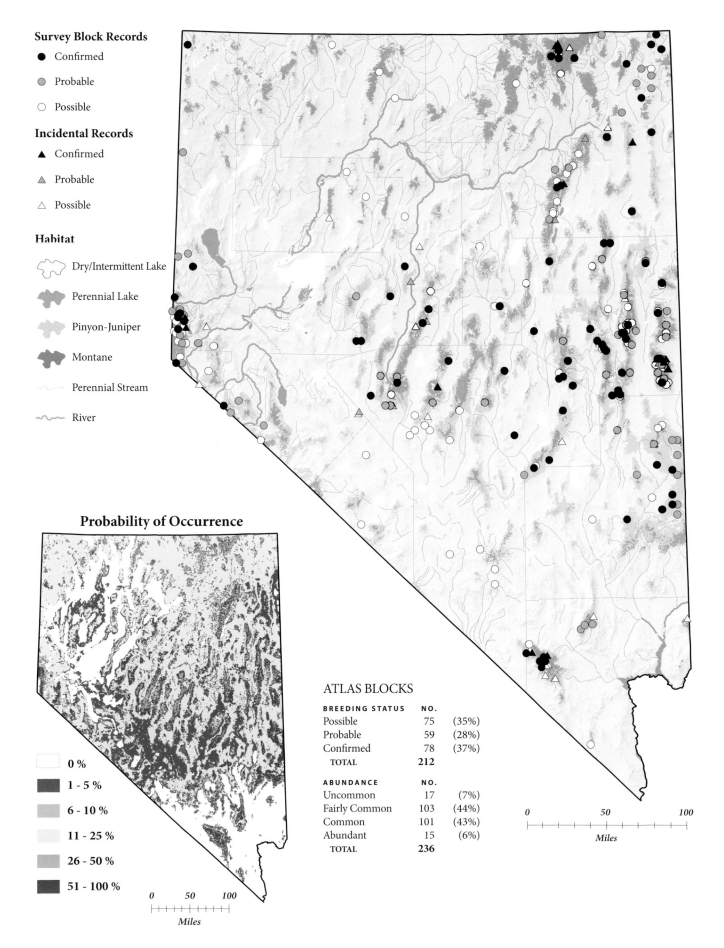

Survey Block Records

- ● Confirmed
- ● Probable
- ○ Possible

Incidental Records

- ▲ Confirmed
- ▲ Probable
- △ Possible

Habitat

- Dry/Intermittent Lake
- Perennial Lake
- Pinyon-Juniper
- Montane
- Perennial Stream
- River

Probability of Occurrence

- 0 %
- 1 - 5 %
- 6 - 10 %
- 11 - 25 %
- 26 - 50 %
- 51 - 100 %

0   50   100
Miles

ATLAS BLOCKS

BREEDING STATUS	NO.	
Possible	75	(35%)
Probable	59	(28%)
Confirmed	78	(37%)
TOTAL	212	

ABUNDANCE	NO.	
Uncommon	17	(7%)
Fairly Common	103	(44%)
Common	101	(43%)
Abundant	15	(6%)
TOTAL	236	

0   50   100
Miles

BREWER'S SPARROW

Spizella breweri

Brewer's Sparrow is one of the most common avian inhabitants of the vast shrublands of northern Nevada. Wherever sagebrush occurs—from basin bottoms to mountain meadows—Brewer's Sparrows seem to be present. Presumably because of the less-than-glamorous habitats it uses, this species tends to attract little interest from most Nevadans. But despite its ubiquity, all is not well with this sandy brown sparrow. It is undergoing a significant rangewide population decline (Sauer et al. 2005), and its breeding habitat is imperiled (Dobkin and Sauder 2004).

DISTRIBUTION

Brewer's Sparrow is a sagebrush specialist (Rotenberry et al. 1999). Its breeding distribution through Utah (Behle et al. 1985), eastern California (Small 1994), northern Arizona (Wise-Gervais in Corman and Wise-Gervais 2005:514–515), southeastern Oregon (Adamus et al. 2001), southern Idaho (Stephens and Sturts 1998), and beyond corresponds closely to the distribution of sagebrush. The species should therefore be common and widespread in Nevada, and indeed, it was recorded in more atlas blocks than any other species—577. As expected, most of the records came from sagebrush habitats or from habitats with a secondary sagebrush component. Brewer's Sparrow was also one of the few species commonly found in salt desert scrub.

Brewer's Sparrows were confirmed breeders in nearly half of the blocks in which they were found, and were generally numerous. One-fifth of the occupied blocks were estimated to contain more than one hundred pairs. Their widespread occurrence and use of several widespread habitat types are also apparent on the predictive map, which indicates very high odds of finding Brewer's Sparrows throughout most of the state. In the Great Basin, only the bleakest deserts and highest mountain peaks are expected to lack Brewer's Sparrows entirely. The species is largely absent from the far south, and the predictive model apparently overestimates occurrence in the northern Mojave Desert.

CONSERVATION AND MANAGEMENT

Brewer's Sparrow could be the poster child for the notion of keeping common birds common. Though numerous throughout much of the Intermountain West, and downright abundant at the center of its range in the Great Basin, Brewer's Sparrow has undergone pronounced declines in recent decades (Rotenberry et al. 1999, Sauer et al. 2005). Dobkin and Sauder (2004) suggested that these declines are concentrated toward the peripheral areas of the species' breeding range, but trends for Nevada also appear negative (Sauer et al. 2005). The primary culprit in the declines appears to be habitat loss and degradation; more specifically, agricultural and urban development, livestock grazing, cheatgrass invasion, altered fire regimes, and herbicides have all been implicated (Kerley and Anderson 1995, Paige and Ritter 1999, Rotenberry et al. 1999, Knick and Rotenberry 2000, Dobkin and Sauder 2004).

Brewer's Sparrow is a Partners in Flight Watch List species, with the conservation goal of doubling its population within thirty years (Rich et al. 2004). Despite its abundance, much remains unknown about this species' ecological requirements and its interactions with the changing high-desert environment (Wiens and Rotenberry 1980, Wiens et al. 1991, Rotenberry et al. 1999). We are fortunate that Brewer's Sparrow is still common in Nevada, but it will require a concerted effort to conserve the sagebrush dominions of this handsome and musical little sparrow.

HABITAT USE

HABITAT	ATLAS BLOCKS	INCIDENTAL OBSERVATIONS
Agricultural	9 (1%)	1 (1%)
Alpine	5 (<1%)	0 (0%)
Grassland	6 (1%)	1 (1%)
Mojave	7 (1%)	1 (1%)
Montane Forest	4 (<1%)	2 (3%)
Montane Parkland	4 (<1%)	0 (0%)
Montane Shrub	67 (11%)	11 (15%)
Open Water	0 (0%)	1 (1%)
Pinyon-Juniper	39 (6%)	8 (11%)
Riparian	6 (<1%)	7 (10%)
Salt Desert Scrub	104 (16%)	3 (4%)
Sagebrush Scrub	199 (31%)	19 (27%)
Sagebrush Steppe	182 (29%)	14 (20%)
Urban	0 (0%)	2 (3%)
Wetland	0 (0%)	1 (1%)
TOTAL	632	71

Total numbers reported in the Habitat Use, Breeding Status, and Abundance tables may differ from each other (see pp. 22–23 for details). Percentage sums may differ slightly from 100% due to rounding.

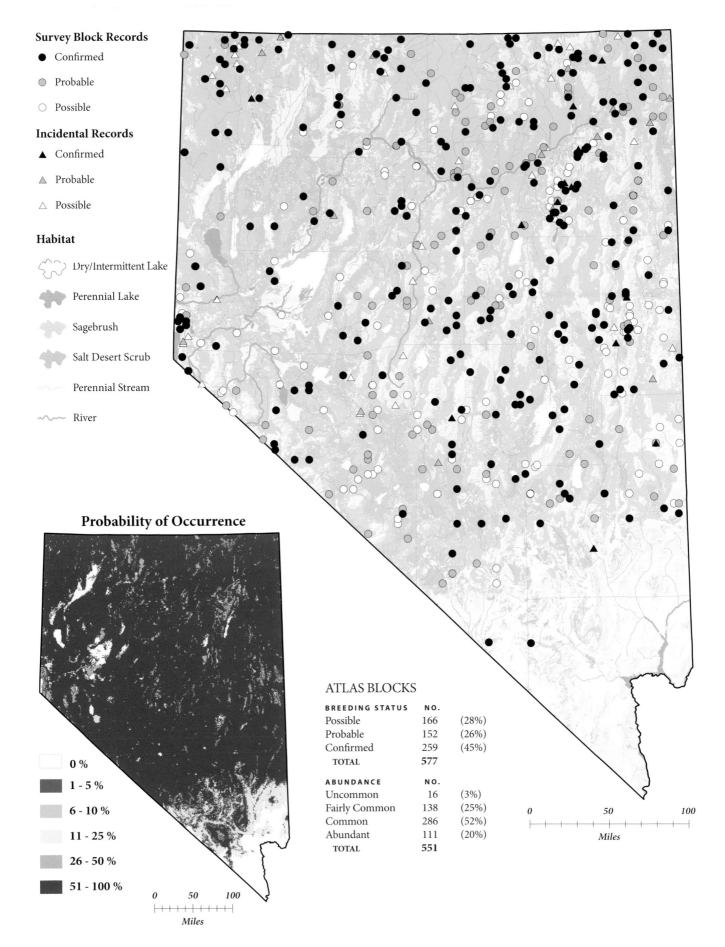

Survey Block Records

- ● Confirmed
- ● Probable
- ○ Possible

Incidental Records

- ▲ Confirmed
- ▲ Probable
- △ Possible

Habitat

- Dry/Intermittent Lake
- Perennial Lake
- Sagebrush
- Salt Desert Scrub
- Perennial Stream
- River

Probability of Occurrence

- 0 %
- 1 - 5 %
- 6 - 10 %
- 11 - 25 %
- 26 - 50 %
- 51 - 100 %

0   50   100
Miles

ATLAS BLOCKS

BREEDING STATUS	NO.	
Possible	166	(28%)
Probable	152	(26%)
Confirmed	259	(45%)
TOTAL	577	

ABUNDANCE	NO.	
Uncommon	16	(3%)
Fairly Common	138	(25%)
Common	286	(52%)
Abundant	111	(20%)
TOTAL	551	

0   50   100
Miles

BLACK-CHINNED SPARROW
Spizella atrogularis

In Nevada, the Black-chinned Sparrow is found on arid, rocky slopes in the Mojave region and seldom ventures far from the cover of thick brush. Black-chinned Sparrows may have a fondness for such terrain, but birders usually do not. Consequently, this southwestern species remains enigmatic, and its natural history is inadequately understood; even the true extent of its breeding range is a matter of speculation. Nevada birders who like a physical challenge should venture into potential Black-chinned Sparrow country and keep their ears open for its unique trilled song.

DISTRIBUTION

Linsdale (1936) and Alcorn (1988) described the Black-chinned Sparrow as a scarce resident of southern Nevada. The atlas data, consisting mostly of incidental observations, also show a sporadic, southerly distribution in the state. The handful of abundance estimates derived from atlas blocks, however, suggests that the species can be fairly numerous in appropriate habitat. A similar pattern was noted in Arizona, where atlas fieldwork turned up more Black-chinned Sparrows than expected (Corman in Corman and Wise-Gervais 2005:516–517). Black-chinned Sparrows also breed widely but sporadically in California, in Utah's southwestern corner, in southwestern New Mexico, and south into Mexico (Behle et al. 1985, Small 1994, Tenney 1997). The northern limits of the breeding range are uncertain, and nesting may occasionally occur in southern Oregon (Adamus et al. 2001). There were no Nevada records very far north of the Mojave Desert, but the possibility of occurrences to the north should not be discounted. The bird's preference for inaccessible, rocky slopes ensures that it will be frequently overlooked (Shuford 1993, Tenney 1997).

Black-chinned Sparrows occur in moderately dense, mixed-species brush on rocky slopes, often on south-facing exposures,

at elevations up to 8,900 feet (2,700 meters). They seem to prefer areas with openings within the brush cover made by larger shrubs, rocks, or trees (Tenney 1997). Atlas fieldworkers found Black-chinned Sparrows primarily in pinyon-juniper and montane shrub habitats within the Mojave Desert—more specifically, in the transitional zone between shrubland and pinyon-juniper habitat, which often spans only a few hundred yards of vertical relief. Because the predictive model uses only block data, the map predictions are based on few records, most of which involved only possible breeding. Nonetheless, the predictive map highlights mid-elevation areas with pinyon-juniper, as expected, and suggests a low chance of finding the species elsewhere in the Mojave Desert. Whether or not Black-chinned Sparrows are actually as common in the northwestern Mojave as the model predicts is not yet known.

CONSERVATION AND MANAGEMENT

Black-chinned Sparrows are not uniformly distributed within suitable habitat, and in certain areas they are prone to local irruptions in favorable years; hence, population trend data can be especially "noisy" (Tenney 1997). Even so, large and significant rangewide declines have been documented since 1966, driven primarily by losses in California (Sauer et al. 2005). Declines have been attributed to heavy livestock grazing on both the breeding and wintering grounds, intensive recreational activities in some areas, and urbanization (Tenney 1997).

Because of its limited distribution and declining trends, the Black-chinned Sparrow is a Partners in Flight Watch List species, with a goal of increasing populations by 50% in the next thirty years (Rich et al. 2004). A better understanding of the species' distribution, northern breeding limits, and habitat requirements would improve our ability to assess its status and develop specific guidelines for its management in Nevada.

HABITAT USE

HABITAT	ATLAS BLOCKS	INCIDENTAL OBSERVATIONS
Mojave	1 (13%)	0 (0%)
Montane Parkland	0 (0%)	1 (6%)
Montane Shrub	3 (38%)	7 (39%)
Pinyon-Juniper	4 (50%)	9 (50%)
Sagebrush Scrub	0 (0%)	1 (6%)
TOTAL	8	18

Total numbers reported in the Habitat Use, Breeding Status, and Abundance tables may differ from each other (see pp. 22–23 for details). Percentage sums may differ slightly from 100% due to rounding.

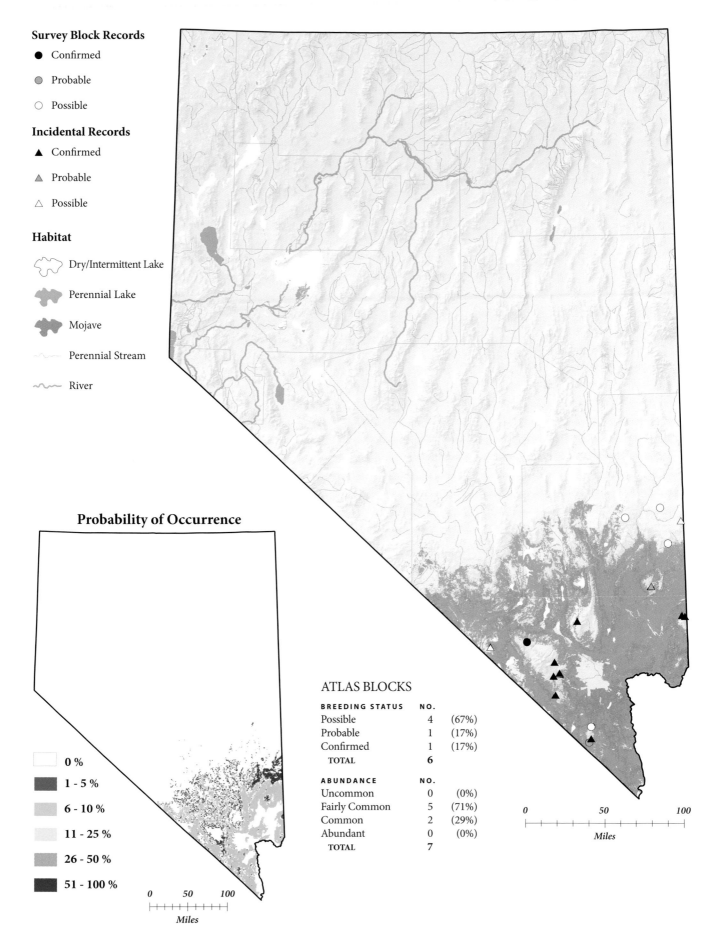

Survey Block Records

● Confirmed

◐ Probable

○ Possible

Incidental Records

▲ Confirmed

△ Probable

△ Possible

Habitat

⬡ Dry/Intermittent Lake

⬢ Perennial Lake

⬢ Mojave

〜 Perennial Stream

〜 River

Probability of Occurrence

☐	**0 %**
■	**1 - 5 %**
☐	**6 - 10 %**
☐	**11 - 25 %**
☐	**26 - 50 %**
■	**51 - 100 %**

0   50   100
├─┼─┼─┼─┼─┼─┼─┼─┼─┤
        Miles

ATLAS BLOCKS

BREEDING STATUS	NO.	
Possible	4	(67%)
Probable	1	(17%)
Confirmed	1	(17%)
TOTAL	**6**	

ABUNDANCE	NO.	
Uncommon	0	(0%)
Fairly Common	5	(71%)
Common	2	(29%)
Abundant	0	(0%)
TOTAL	**7**	

0   50   100
├─────────────┼─────────────┤
                Miles

VESPER SPARROW

Pooecetes gramineus

Some of Nevada's sparrow species are endowed with an obvious beauty; the towhees and juncos, Black-throated and Sage Sparrows, and Lark Sparrow are examples. The Vesper Sparrow is as lovely as any of these, but in an understated way: its thin eye ring, crisp white tail feathers, and muted rufous wing patch have a more subtle splendor.

DISTRIBUTION

Vesper Sparrows have a northerly distribution in Nevada, with most nesting reported from the northernmost counties. Breeding was confirmed from as far south as Churchill and northern Nye counties, and possible breeders also occurred in Mineral and Lincoln counties. Overall, the atlas data match the distributions described in earlier accounts (Linsdale 1936, Alcorn 1988, Neel 1999a). Alcorn (1988) considered the species to be an uncommon breeder in Nevada, but atlas fieldwork suggested that, at least in their strongholds, they can be numerous. In excess of ten pairs were thought to occur in more than half of the blocks for which abundances were estimated.

Beyond Nevada, the Vesper Sparrow's range extends far into Canada and all the way to the Atlantic. In the southwestern portion of the range, the species occurs throughout Utah, Idaho, and eastern Oregon (Behle et al. 1985, Stephens and Sturts 1998, Adamus et al. 2001). In California, Vesper Sparrows are found mainly in the northeast (Small 1994), although breeding has also been documented farther south (Garrett and Dunn 1981). In Arizona, they breed from the Mogollon Plateau northward (Wise-Gervais in Corman and Wise-Gervais 2005:518–519).

Throughout its range, the Vesper Sparrow nests in grasslands and shrublands that feature low, patchy vegetation (Paige and

Ritter 1999, Jones and Cornely 2002). In the Great Basin, it is found primarily in mixed grass and sagebrush habitat where shrub cover is limited and bare ground is often present (Ryser 1985, Jones and Cornely 2002, Holmes and Barton 2003, Dobkin and Sauder 2004). As indicated by the predictive map, these conditions are most widespread in Nevada's northernmost tier, which represents a small sliver of the shrub steppe–covered Columbia Plateau. Within this area, the predicted probability of Vesper Sparrow occurrence is generally high. Vesper Sparrows also use montane shrublands and grasslands, and in central Nevada the highest chance of finding them is predicted to be in the mountains, especially at middle elevations.

CONSERVATION AND MANAGEMENT

Vesper Sparrow populations in the East expanded following European settlement and large-scale forest clearing but then declined as forests recovered (Paige and Ritter 1999, Jones and Cornely 2002). In the West, populations are alternately described as declining (Paige and Ritter 1999), stable (Sauer et al. 2005), or increasing (Dobkin and Sauder 2004), depending on how the data were analyzed and which regions were included in the analysis. Locally, management of Vesper Sparrows is complicated by variation in how quickly the species responds to habitat changes such as burns (Jones and Cornely 2002) and by its complex responses to livestock grazing (Saab et al. 1995, Paige and Ritter 1999) and to nonnative grasslands such as those dominated by crested wheatgrass (Jones and Cornely 2002).

As a sagebrush associate, the Vesper Sparrow faces the same threats as do other shrub steppe birds, and it is therefore a Nevada Partners in Flight Priority species (Neel 1999a). Recommended actions include active species management in northern Nevada. Specific guidelines for prescribed burns and provision of adequate forb and grass cover are outlined in Neel 1999a. Additional research and monitoring of its responses to habitat change will help us to better understand the Vesper Sparrow's status and requirements in Nevada.

HABITAT USE

HABITAT	ATLAS BLOCKS	INCIDENTAL OBSERVATIONS
Agricultural	5 (3%)	1 (5%)
Alpine	1 (<1%)	0 (0%)
Grassland	12 (6%)	1 (5%)
Montane Forest	0 (0%)	1 (5%)
Montane Parkland	1 (<1%)	0 (0%)
Montane Shrub	21 (11%)	5 (25%)
Pinyon-Juniper	10 (5%)	0 (0%)
Riparian	3 (2%)	2 (10%)
Salt Desert Scrub	9 (5%)	0 (0%)
Sagebrush Scrub	35 (19%)	4 (20%)
Sagebrush Steppe	86 (46%)	6 (30%)
Urban	1 (<1%)	0 (0%)
Wetland	1 (<1%)	0 (0%)
TOTAL	185	20

Total numbers reported in the Habitat Use, Breeding Status, and Abundance tables may differ from each other (see pp. 22–23 for details). Percentage sums may differ slightly from 100% due to rounding.

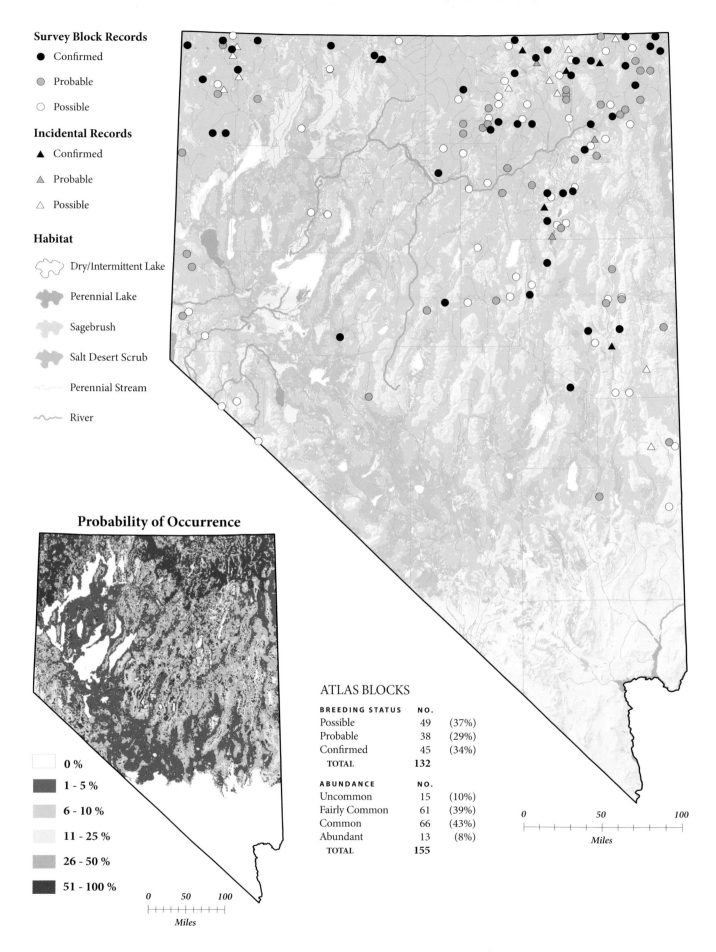

Survey Block Records

- ● Confirmed
- ● Probable
- ○ Possible

Incidental Records

- ▲ Confirmed
- ▲ Probable
- △ Possible

Habitat

- Dry/Intermittent Lake
- Perennial Lake
- Sagebrush
- Salt Desert Scrub
- Perennial Stream
- River

Probability of Occurrence

- 0 %
- 1 - 5 %
- 6 - 10 %
- 11 - 25 %
- 26 - 50 %
- 51 - 100 %

0   50   100

Miles

ATLAS BLOCKS

BREEDING STATUS	NO.	
Possible	49	(37%)
Probable	38	(29%)
Confirmed	45	(34%)
TOTAL	132	

ABUNDANCE	NO.	
Uncommon	15	(10%)
Fairly Common	61	(39%)
Common	66	(43%)
Abundant	13	(8%)
TOTAL	155	

0   50   100

Miles

LARK SPARROW
Chondestes grammacus

Many people view sparrows as the quintessential "little brown jobs." But not the Lark Sparrow. The face and crown of this large sparrow are strikingly patterned with black, white, and chestnut, and when it is flushed from cover, its boldly marked tail flashes black and white. Lark Sparrows are widely distributed in a variety of habitats in Nevada, but they seem to reach their greatest numbers in the northern part of the state, where scattered junipers encroach on the sagebrush steppe.

DISTRIBUTION

Lark Sparrows were found throughout most of Nevada, although sparingly so in the south. Southern Nye County had the most southerly confirmed breeders in the state, and the species was largely absent as a breeder from the Mojave Desert. Atlas records showed a more contiguous distribution throughout the state than Linsdale (1936) reported, and atlas abundance estimates indicate larger numbers than were previously suggested (Alcorn 1988).

The bulk of the Lark Sparrow's distribution lies to the east of Nevada and extends throughout the center of the continent. The species is common as a breeder in most of Utah (Behle et al. 1985), and is fairly common in eastern Oregon (Adamus et al. 2001), southern Idaho (Stephens and Sturts 1998), and northern and eastern Arizona (Wise-Gervais in Corman and Wise-Gervais 2005:520–521). In California, Lark Sparrows breed in the northeast, from the Central Valley west to the coast, and around the lower Colorado River (Rosenberg et al. 1991, Small 1994).

In most of their range, Lark Sparrows occupy open grassland and shrub habitats, including greasewood and sagebrush, or areas with scattered trees (Paige and Ritter 1999, Martin and Parrish 2000). Most atlas habitat data came from lower-elevation shrub habitat, but Lark Sparrows frequently used areas of pinyon-

juniper as well. The highest likelihood of occurrence was predicted in sagebrush and pinyon-juniper habitats, but there is also a moderately good chance of occurrence in salt desert scrub. The species is predicted to be absent from the higher mountains and unlikely to breed in most of the Mojave Desert. In Nevada, Lark Sparrows often occur alongside Vesper Sparrows, but the former are less common at high elevations, while the latter are largely absent from the lower-elevation valleys.

CONSERVATION AND MANAGEMENT

Like Vesper Sparrow populations, Lark Sparrow populations expanded in the East and Midwest after initial European settlement and forest clearing but subsequently declined (Martin and Parrish 2000, Sauer et al. 2005). Trends in the West are less simply characterized. Paige and Ritter (1999) expressed concern about the declines suggested by monitoring data (see Sauer et al. 2005), but others view these declines as localized and feel that the overall pattern implies that western Lark Sparrow populations are stable (Martin and Parrish 2000, Dobkin and Sauder 2004, Sauer et al. 2005).

Although most are found in native habitats, Lark Sparrows show some tolerance for modified habitats. Evidently, the species has recently colonized the lower Colorado River Valley's agricultural areas (Rosenberg et al. 1991), has responded positively to some livestock grazing regimes (Bock and Webb 1984), and copes with physical disturbances such as roads and off-road traffic better than most shrubland birds (Martin and Parrish 2000). Grasshopper control, however, may reduce the species' prey base and negatively affect populations (Paige and Ritter 1999). Altered fire regimes have also had mixed effects on Lark Sparrow populations (Martin and Parrish 2000, Dobkin and Sauder 2004).

At present, Lark Sparrows seem to be doing well in Nevada. Given the bird's complex habitat use and responses to disturbances, however, additional population monitoring and research are needed to elucidate its status and management needs, if any, in Nevada.

HABITAT USE

HABITAT	ATLAS BLOCKS	INCIDENTAL OBSERVATIONS
Agricultural	5 (3%)	4 (14%)
Barren	1 (<1%)	0 (0%)
Grassland	9 (5%)	0 (0%)
Mesquite	1 (<1%)	1 (3%)
Mojave	1 (<1%)	0 (0%)
Montane Shrub	6 (4%)	2 (7%)
Pinyon-Juniper	28 (17%)	1 (3%)
Riparian	6 (4%)	3 (10%)
Salt Desert Scrub	17 (10%)	4 (14%)
Sagebrush Scrub	47 (28%)	7 (24%)
Sagebrush Steppe	41 (25%)	2 (7%)
Urban	5 (3%)	5 (17%)
TOTAL	167	29

Total numbers reported in the Habitat Use, Breeding Status, and Abundance tables may differ from each other (see pp. 22–23 for details). Percentage sums may differ slightly from 100% due to rounding.

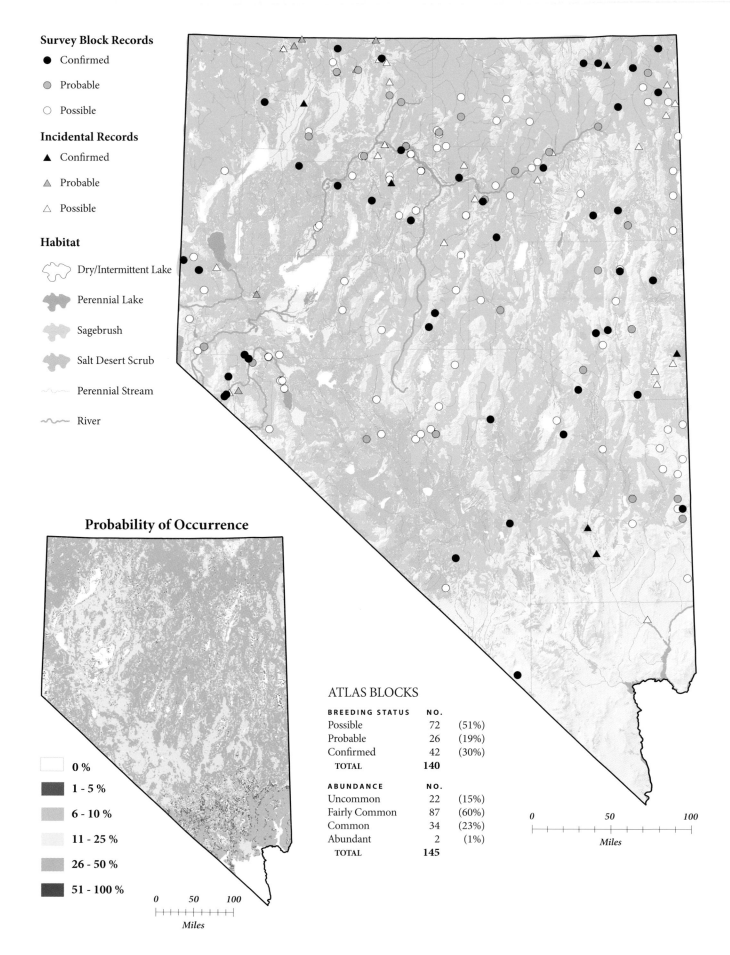

Survey Block Records

● Confirmed

● Probable

○ Possible

Incidental Records

▲ Confirmed

△ Probable

△ Possible

Habitat

Dry/Intermittent Lake

Perennial Lake

Sagebrush

Salt Desert Scrub

Perennial Stream

River

Probability of Occurrence

☐ 0 %

■ 1 - 5 %

6 - 10 %

11 - 25 %

26 - 50 %

■ 51 - 100 %

0   50   100
Miles

ATLAS BLOCKS

BREEDING STATUS	NO.	
Possible	72	(51%)
Probable	26	(19%)
Confirmed	42	(30%)
TOTAL	140	
ABUNDANCE	NO.	
Uncommon	22	(15%)
Fairly Common	87	(60%)
Common	34	(23%)
Abundant	2	(1%)
TOTAL	145	

0   50   100
Miles

BLACK-THROATED SPARROW
Amphispiza bilineata

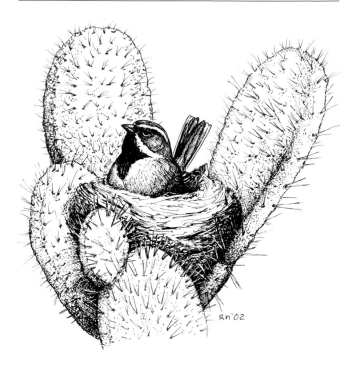

Some places in Nevada are too hot, too dry, and too inhospitable to support even such desert lovers as Greater Roadrunners and Horned Larks. But it is hard to imagine a shrubland too hot and dry for the beautiful Black-throated Sparrow. Its languid song and tinkling call notes add a welcome note of animation to the creosotebush-covered desert flats in the southern part of the state. It is also one of very few birds that thrive in the extensive salt desert scrublands of western Nevada.

DISTRIBUTION

Atlas workers found Black-throated Sparrows to be among the most widely distributed birds in the hot lowlands of Nevada. They were almost impossible to miss in southern Nevada and occurred throughout the central and western portions of the state as well. In the northeast, however, where they approach the limit of their range, they were much more sparsely distributed. The Black-throated Sparrow was numerous in most places where it occurred, and was one of the few species estimated to have more than one hundred pairs in several dozen blocks. These results agree with earlier accounts of the species' distribution and abundance (Fisher 1893, Linsdale 1936, Alcorn 1988).

Looking beyond the state's borders, the Black-throated Sparrow's breeding distribution closely matches a map of arid western shrublands. It breeds throughout much of Utah (Behle et al. 1985) and Arizona (Wise-Gervais in Corman and Wise-Gervais 2005:522–523), but is restricted to dry shrublands in northeastern and southern California (Rosenberg et al. 1991, Small 1994), southeastern Oregon (Adamus et al. 2001), and southern Idaho (Stephens and Sturts 1998).

Breeding Black-throated Sparrows can be found in areas covered with saltbush, shadscale, greasewood, creosotebush, sagebrush, rabbitbrush, sand dune vegetation, and Joshua trees, and they even occur up into montane shrublands and in open stands of pinyon-juniper (Ryser 1985, Paige and Ritter 1999, Chisholm and Neel 2002, Johnson et al. 2002). They generally avoid urban and suburban settings (Mills et al. 1989, Chisholm and Neel 2002) and are not found at the highest elevations. But in bleak and desolate places they may be the most plentiful—and sometimes the only—breeding birds (Gaines 1992). In most of the Mojave Desert, they are the most likely sparrow species to be present. Like Brewer's and Sage Sparrows, they are predicted to occur with high probability in much of the state. Unlike those species, however, Black-throated Sparrows have a more southerly orientation, with lower probabilities of occurrence in the sagebrush steppe habitats of northern Nevada.

CONSERVATION AND MANAGEMENT

The Black-throated Sparrow is well suited to life in extreme desert conditions, but it is not immune to human impacts on the environment (Johnson et al. 2002). Fire suppression causes undesirable effects, and the species does not tolerate urbanization well (Mills et al. 1989, Johnson et al. 2002). The effects of livestock grazing are not well understood and may vary from one location to the next (Paige and Ritter 1999, Dobkin and Sauder 2004).

Monitoring data indicate that Black-throated Sparrows have declined severely throughout the West since the 1960s (Paige and Ritter 1999, Johnson et al. 2002, Sauer et al. 2005), although there is some evidence for increases toward the northern edge of the range (Dobkin and Sauder 2004). The causes of the declines are unclear (Johnson et al. 2002), and Partners in Flight regards this sparrow as a Stewardship species for the southwestern region, with the goal of maintaining its current population size (Rich et al. 2004). In Nevada, Black-throated Sparrow population trends are currently unknown and require monitoring to inform conservation planners.

HABITAT USE

HABITAT	ATLAS BLOCKS	INCIDENTAL OBSERVATIONS
Grassland	4 (1%)	0 (0%)
Mesquite	10 (3%)	0 (0%)
Mojave	83 (21%)	39 (52%)
Montane Forest	0 (0%)	1 (1%)
Montane Parkland	0 (0%)	1 (1%)
Montane Shrub	4 (1%)	5 (7%)
Pinyon-Juniper	22 (5%)	8 (11%)
Riparian	7 (2%)	6 (8%)
Salt Desert Scrub	100 (26%)	4 (5%)
Sagebrush Scrub	109 (28%)	6 (8%)
Sagebrush Steppe	49 (13%)	4 (5%)
Urban	1 (<1%)	0 (0%)
Wetland	0 (0%)	1 (1%)
TOTAL	389	75

Total numbers reported in the Habitat Use, Breeding Status, and Abundance tables may differ from each other (see pp. 22–23 for details). Percentage sums may differ slightly from 100% due to rounding.

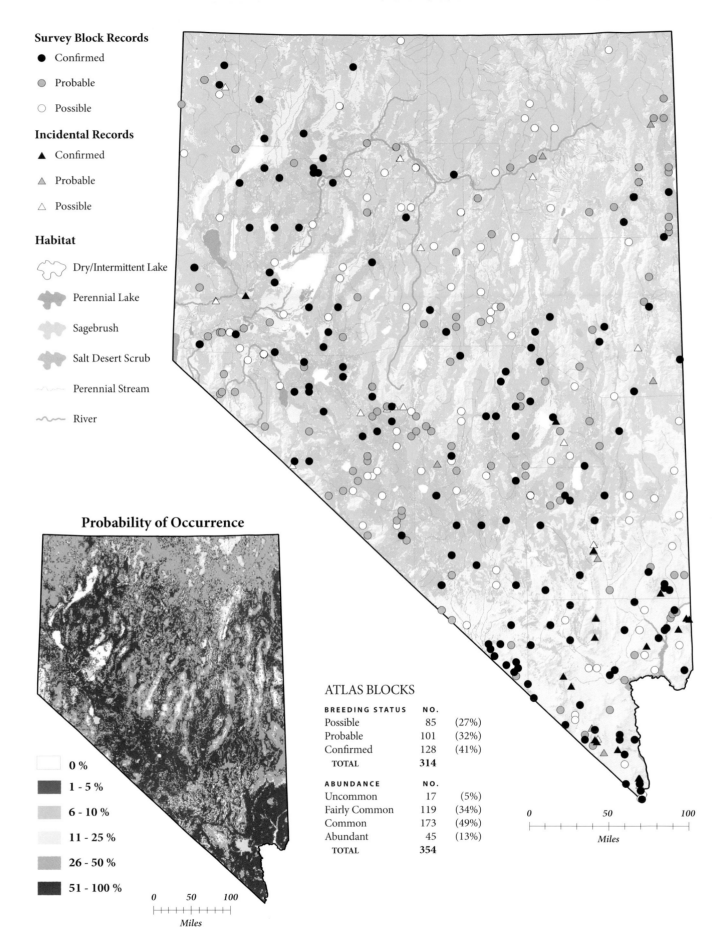

Survey Block Records

● Confirmed

● Probable

○ Possible

Incidental Records

▲ Confirmed

▲ Probable

△ Possible

Habitat

Dry/Intermittent Lake

Perennial Lake

Sagebrush

Salt Desert Scrub

Perennial Stream

River

Probability of Occurrence

	0 %
	1 - 5 %
	6 - 10 %
	11 - 25 %
	26 - 50 %
	51 - 100 %

0   50   100
Miles

ATLAS BLOCKS

BREEDING STATUS	NO.	
Possible	85	(27%)
Probable	101	(32%)
Confirmed	128	(41%)
TOTAL	**314**	

ABUNDANCE	NO.	
Uncommon	17	(5%)
Fairly Common	119	(34%)
Common	173	(49%)
Abundant	45	(13%)
TOTAL	**354**	

0   50   100
Miles

SAGE SPARROW
Amphispiza belli

The Sage Sparrow dashes from shrub to shrub like a miniature roadrunner, and it has a real knack for always being on the other side of a bush or hummock. Birders must often be content with hearing its brief, languid song, one of the distinctive sounds of Nevada's vast sagebrush deserts. At present, the Sage Sparrow is still common, but there is much concern about the species in the face of ongoing sagebrush degradation and conversion.

DISTRIBUTION

In Nevada, Sage Sparrows are widespread north of about latitude 37°N; south of this line, atlas workers confirmed breeding only in Ash Meadows National Wildlife Refuge. Sage Sparrows were numerous, with more than ten pairs in nearly two-thirds of all blocks for which abundance estimates were given. The atlas findings concur with Linsdale's (1936) earlier account but contrast somewhat with that of Alcorn (1988), who stated that Sage Sparrows occur statewide in summer. Neel's (1999a) report that the Sage Sparrow does not occur in Clark County agrees with atlas data and other current range maps (e.g., Martin and Carlson 1998, Sibley 2003).

The Sage Sparrow's summer range lies mostly within Nevada and Utah (Behle et al. 1985, Martin and Carlson 1998), although the species also breeds in southeastern Oregon and southern Idaho, and southward into southern California and northeastern Arizona (Small 1994, Stephens and Sturts 1998, Adamus et al. 2001, Corman in Corman and Wise-Gervais 2005:524–525). In California, a different subspecies of Sage Sparrow resides along the eastern slope of the coastal ranges (Small 1994, Martin and Carlson 1998).

Sage Sparrows in the Intermountain West are usually associated with sagebrush, whether it is the dominant vegetation or part of mixed-shrub communities (Martin and Carlson 1998, Dobkin and Sauder 2004); big sagebrush has been identified as especially important for the species (Ryser 1985, Martin and

Carlson 1998, Neel 1999a). Sage Sparrows are often absent from seemingly suitable habitat, though, and factors other than the presence of sagebrush—such as shrubland patch size, vegetative structure, or the extent of bare ground—may also be important (Martin and Carlson 1998, Paige and Ritter 1999). Sage Sparrows also occur frequently in salt desert scrub, and the predictive map suggests that they are highly likely to occur throughout the Great Basin valleys of Nevada. Atlas field data suggest that Sage Sparrows avoid Mojave habitat and may be even less likely to occur in the south than the predictive map suggests. The chance of finding Sage Sparrows also diminishes at higher elevations, and the species is more restricted to the valleys than either Black-throated or Brewer's Sparrows.

CONSERVATION AND MANAGEMENT

Overall, Sage Sparrow populations appear to be stable (Sauer et al. 2005), although Dobkin and Sauder (2004) suggested declines in some parts of the Intermountain West. In addition to the potential threats from sagebrush degradation caused by such factors as altered fire regimes, heavy grazing, and the spread of introduced plants, fragmentation of intact habitat into smaller patches may harm Sage Sparrows (Martin and Carlson 1998, Paige and Ritter 1999, Dobkin and Sauder 2004).

Sage Sparrows remain common in Nevada, although limited monitoring in the past makes it difficult to assess current population trends. Because the species is something of a habitat specialist and shows declines in some areas, it is a Partners in Flight Stewardship species for the Intermountain West (Rich et al. 2004) and a Priority species in Nevada (Neel 1999a), which is home to a large portion of the bird's global population.

HABITAT USE

HABITAT	ATLAS BLOCKS	INCIDENTAL OBSERVATIONS
Agricultural	0 (0%)	1 (4%)
Barren	1 (<1%)	0 (0%)
Grassland	0 (0%)	1 (4%)
Montane Forest	1 (<1%)	0 (0%)
Montane Shrub	5 (2%)	2 (7%)
Pinyon-Juniper	3 (1%)	3 (11%)
Riparian	1 (<1%)	3 (11%)
Salt Desert Scrub	70 (23%)	3 (11%)
Sagebrush Scrub	134 (45%)	9 (33%)
Sagebrush Steppe	83 (28%)	2 (7%)
Urban	0 (0%)	2 (7%)
Wetland	0 (0%)	1 (4%)
TOTAL	298	27

Total numbers reported in the Habitat Use, Breeding Status, and Abundance tables may differ from each other (see pp. 22–23 for details). Percentage sums may differ slightly from 100% due to rounding.

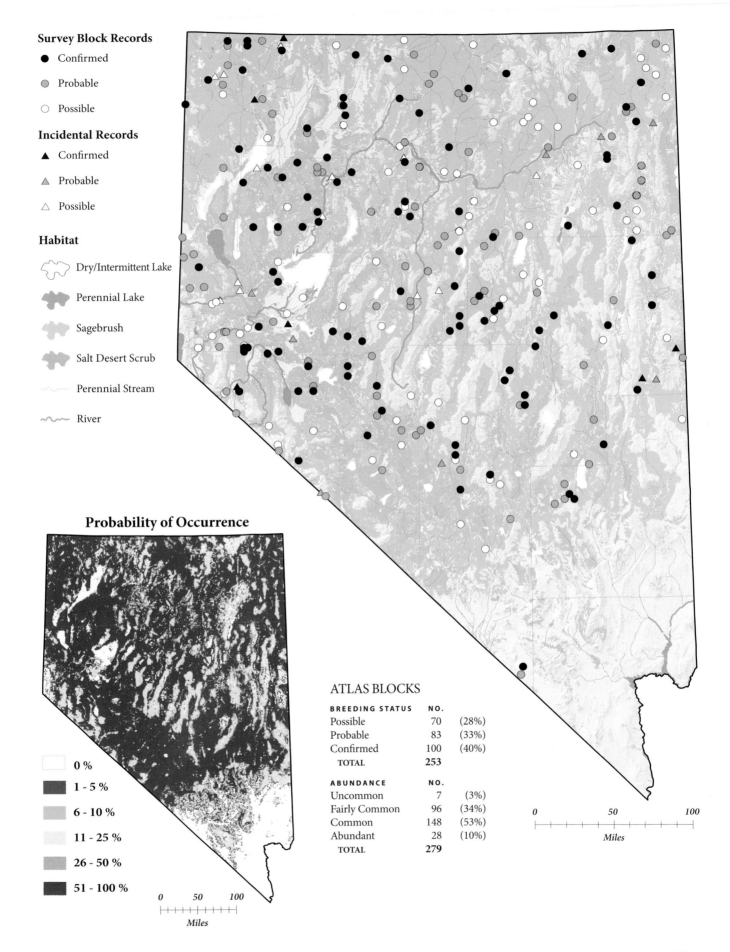

Survey Block Records

- ● Confirmed
- ● Probable
- ○ Possible

Incidental Records

- ▲ Confirmed
- ▲ Probable
- △ Possible

Habitat

- Dry/Intermittent Lake
- Perennial Lake
- Sagebrush
- Salt Desert Scrub
- Perennial Stream
- River

Probability of Occurrence

- ☐ 0 %
- 1 - 5 %
- 6 - 10 %
- 11 - 25 %
- 26 - 50 %
- 51 - 100 %

0   50   100
Miles

ATLAS BLOCKS

BREEDING STATUS	NO.	
Possible	70	(28%)
Probable	83	(33%)
Confirmed	100	(40%)
TOTAL	**253**	

ABUNDANCE	NO.	
Uncommon	7	(3%)
Fairly Common	96	(34%)
Common	148	(53%)
Abundant	28	(10%)
TOTAL	**279**	

0   50   100
Miles

SAVANNAH SPARROW

Passerculus sandwichensis

Compared with some sparrows, the Savannah Sparrow is not much to look at: short-tailed, drab brown, streaky all over, with a little yellow spot in front of the eye providing the only splash of color. Its song is not very striking either, but not for want of trying. In a typical pose, the Savannah Sparrow is seen with its head thrown back and its mouth wide open—giving its all to produce its weak, hissing song. Among our more perplexingly named birds (see Ryser 1985), Savannah Sparrows are present in many types of open, grassy country, including those created by humans. In Nevada, they tend to be associated with moist lowland habitats.

DISTRIBUTION

Savannah Sparrows are fairly widespread in Nevada at latitudes north of about 39°N. Farther south, there were only scattered records, and no confirmed breeders. Fisher (1893) reported nesting in the southern deserts, but the atlas data do not confirm this. Other early accounts suggest a breeding distribution more similar to that described by the atlas data (Linsdale 1936, Alcorn 1988).

The Savannah Sparrow's breeding range extends throughout northern and western North America. Numerous subspecies exist, with the one breeding in the Great Basin aptly named *nevadensis* (Wheelwright and Rising 1993). The Savannah Sparrow occurs throughout the states adjacent to Nevada, except in Arizona, where it is limited to the northeast (Monson and Phillips 1981, Behle et al. 1985, Small 1994, Stephens and Sturts 1998, Adamus et al. 2001, Corman in Corman and Wise-Gervais 2005:526–527).

In Nevada and elsewhere in the arid West, Savannah Sparrows are often associated with wet meadows such as irrigated fields and pastures, wetland margins, and riparian zones with suitably low vegetation (Ryser 1985, Wheelwright and Rising 1993). They tend to be found where there is substantial herbaceous ground cover, and their numbers decline with increasing shrub and tree cover (Wheelwright and Rising 1993). Atlas workers most commonly reported them from wetland and agricultural areas. Somewhat surprisingly, a quarter of the atlas records came from sagebrush or salt desert scrub habitats, although it is possible that there were moist habitats near many of these locations. The predictive map suggests an increasing likelihood of finding Savannah Sparrows as one moves north in the state, but even in the sagebrush steppe of the north they are not especially likely to be found. Within this overall pattern, Savannah Sparrows are predicted to occur with high probability in the agricultural and wetland complexes of northern Nevada and to be absent from the highest elevations and from most of southern Nevada. Las Vegas is predicted to be an area where the species might occur, but this is an error caused by the occasional occurrence of the species in northern urban areas that also contain agricultural or riparian habitat.

CONSERVATION AND MANAGEMENT

Like many grassland birds, the Savannah Sparrow is secretive and easily overlooked (Wheelwright and Rising 1993). Monitoring data suggest a slight overall population decline, most pronounced in the East but partly balanced by increases in the continent's center (Sauer et al. 2005). In many areas, Savannah Sparrows appear to have benefited from human activity (Wheelwright and Rising 1993), and in arid parts of the West they often do well in areas of agricultural development (e.g., Rosenberg et al. 1991). In Nevada, Savannah Sparrows occur widely in irrigated fields and pastures (Ryser 1985, Chisholm and Neel 2002). Reproductive success in some human-created grasslands may be lower than in native grasslands (Wray et al. 1982), though, and studies of nesting productivity are needed to augment descriptions of habitat use.

HABITAT USE

HABITAT	ATLAS BLOCKS	INCIDENTAL OBSERVATIONS
Agricultural	26 (20%)	2 (8%)
Grassland	18 (14%)	4 (15%)
Montane Shrub	1 (1%)	0 (0%)
Open Water	0 (0%)	1 (4%)
Riparian	19 (14%)	5 (19%)
Salt Desert Scrub	15 (11%)	0 (0%)
Sagebrush Scrub	7 (5%)	2 (8%)
Sagebrush Steppe	14 (11%)	3 (12%)
Urban	0 (0%)	1 (4%)
Wetland	32 (24%)	8 (31%)
TOTAL	132	26

Total numbers reported in the Habitat Use, Breeding Status, and Abundance tables may differ from each other (see pp. 22–23 for details). Percentage sums may differ slightly from 100% due to rounding.

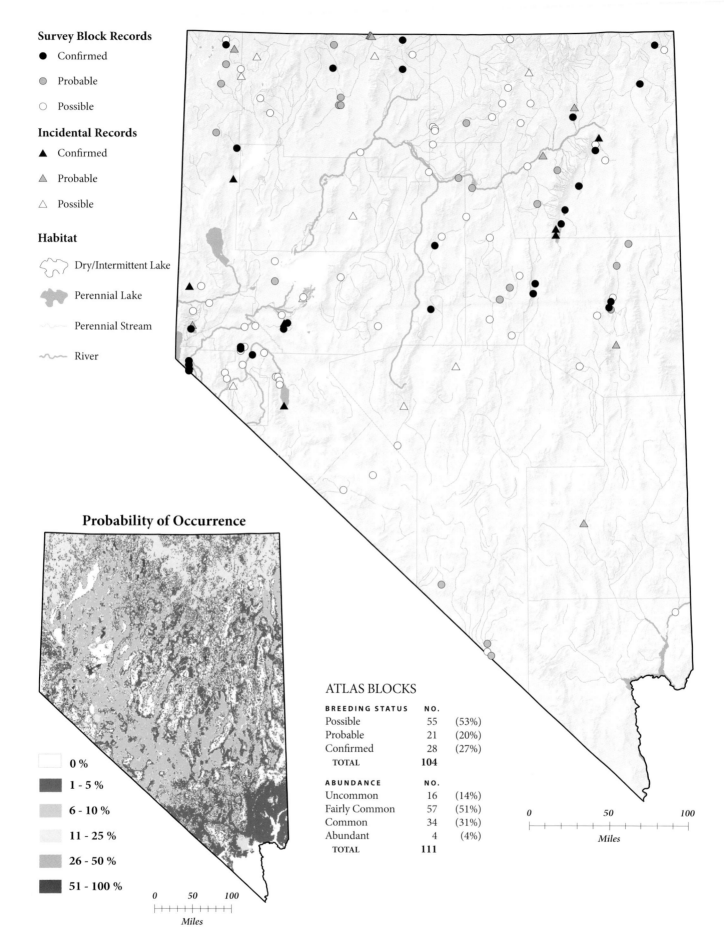

Survey Block Records

- ● Confirmed
- ● Probable
- ○ Possible

Incidental Records

- ▲ Confirmed
- ▲ Probable
- △ Possible

Habitat

Dry/Intermittent Lake

Perennial Lake

Perennial Stream

River

Probability of Occurrence

- ☐ 0 %
- 1 - 5 %
- 6 - 10 %
- 11 - 25 %
- 26 - 50 %
- 51 - 100 %

0   50   100

Miles

ATLAS BLOCKS

BREEDING STATUS	NO.	
Possible	55	(53%)
Probable	21	(20%)
Confirmed	28	(27%)
TOTAL	104	

ABUNDANCE	NO.	
Uncommon	16	(14%)
Fairly Common	57	(51%)
Common	34	(31%)
Abundant	4	(4%)
TOTAL	111	

0   50   100

Miles

FOX SPARROW

Passerella iliaca

The Fox Sparrow is easily recognized. It is large for a sparrow, gray and brown on top, and boldly spotted beneath. But identifying these birds may soon become more difficult, because the current Fox Sparrow could well be reclassified, with a split into as many as four species on the basis of morphological, vocal, and genetic differences. The birds that nest in Nevada belong to four subspecies that fall into two of the groups that constitute putative species. The birds breeding in most of the state are members of the *schistacea* ("Slate-colored") group, while those at the far western border are in the *megarhyncha* group (variously referred to as "Thick-billed" or "Large-billed"). In between, the situation is messier; indeed, one of the few remaining impediments to the four-way split of the Fox Sparrow is the unclear situation in west-central Nevada.

DISTRIBUTION

Fox Sparrows occur at high elevations throughout northern Nevada, from the far northern Santa Rosa and Jarbidge mountains to as far south as the White Mountains of Esmeralda County. A dense cluster of records came from the Carson Range of far western Nevada, but this may be because survey work was especially intense in that area. The breeding ranges of the four Fox Sparrow subspecies listed by Linsdale (1936) match the distribution described by atlas records and by Alcorn (1988).

Most of the Fox Sparrow's breeding range lies in Canada and Alaska, but it does extend south through the northwestern states and the Rocky Mountains, with small populations scattered throughout the Intermountain West and south all the way to San Diego. Geographic variation within the species is substantial. In California, for instance, as many as fifteen subspecies have been described, although breeding occurs mostly in the north and only locally in the south (Garrett and Dunn 1981, Small 1994). The species breeds in much of Oregon and throughout Idaho (Stephens and Sturts 1998, Adamus et al. 2001), but in Utah it nests only in the north, where it is uncommon (Behle et al. 1985). The Fox Sparrow breeds neither in Arizona nor in the Mojave Desert portion of Nevada.

In the Intermountain West, Fox Sparrows generally breed on brushy mountain slopes or in dense riparian thickets of alders, willows, currants, and other montane shrubs (Ryser 1985, Contreras and Kindschy 1996, Weckstein et al. 2002). Most atlas records came from riparian, montane shrub, and forest habitats, and both atlas data and the predictive map indicate the species' predominant use of high elevations in Nevada. The model predicts a moderately high chance of finding Fox Sparrows in the state's major Great Basin ranges; away from the mountains, they are very unlikely to be found.

CONSERVATION AND MANAGEMENT

The Fox Sparrow is not a conservation concern in our region at the present time. Fox Sparrows are among the most numerous birds in their habitats in the southern Sierra Nevada (Gaines 1992), and population declines have been reported only for the Cascades region of the Pacific Northwest (Sauer et al. 2005). Even these declines are occurring in an area that may have been colonized by Fox Sparrows following forest clear-cutting during the mid-twentieth century (Banks 1970).

The status of Nevada's Fox Sparrows is not well established. The boundaries between the "Thick-billed" and "Slate-colored" groups are not well characterized in the state, and it is unclear whether the overall pattern of population stability applies to all subspecies. Populations along the California border and throughout the Great Basin are likely to be small and fairly isolated, potentially increasing their vulnerability. Of particular interest will be any future work that sheds light on the songs and calls of Fox Sparrows in Nye, Mineral, and Esmeralda counties—where the status of the *megarhynca* and *schistacea* groups has not been resolved.

HABITAT USE

HABITAT	ATLAS BLOCKS	INCIDENTAL OBSERVATIONS
Alpine	3 (3%)	0 (0%)
Montane Forest	17 (18%)	5 (19%)
Montane Parkland	3 (3%)	1 (4%)
Montane Shrub	25 (27%)	3 (12%)
Pinyon-Juniper	4 (4%)	2 (8%)
Riparian	37 (39%)	14 (54%)
Sagebrush Scrub	1 (1%)	1 (4%)
Sagebrush Steppe	2 (2%)	0 (0%)
Urban	1 (1%)	0 (0%)
Wetland	1 (1%)	0 (0%)
TOTAL	94	26

Total numbers reported in the Habitat Use, Breeding Status, and Abundance tables may differ from each other (see pp. 22–23 for details). Percentage sums may differ slightly from 100% due to rounding.

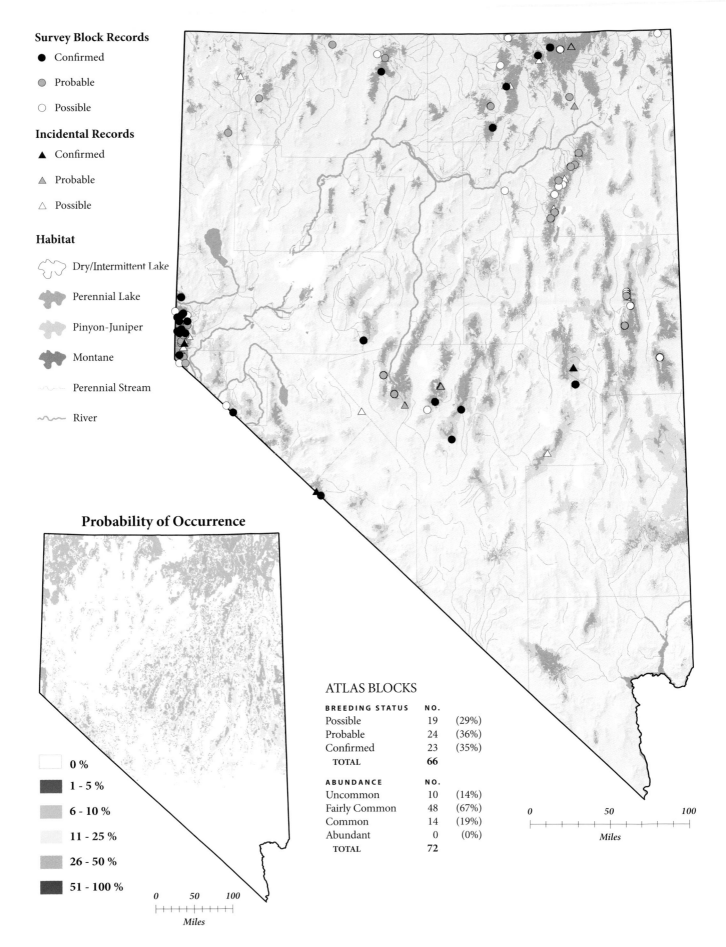

Survey Block Records

● Confirmed

● Probable

○ Possible

Incidental Records

▲ Confirmed

▲ Probable

△ Possible

Habitat

Dry/Intermittent Lake

Perennial Lake

Pinyon-Juniper

Montane

Perennial Stream

River

Probability of Occurrence

☐ 0 %

1 - 5 %

6 - 10 %

11 - 25 %

26 - 50 %

51 - 100 %

0   50   100

Miles

ATLAS BLOCKS

BREEDING STATUS	NO.	
Possible	19	(29%)
Probable	24	(36%)
Confirmed	23	(35%)
TOTAL	66	

ABUNDANCE	NO.	
Uncommon	10	(14%)
Fairly Common	48	(67%)
Common	14	(19%)
Abundant	0	(0%)
TOTAL	72	

0   50   100

Miles

SONG SPARROW

Melospiza melodia

"No two Song Sparrows look alike!" is not just the call of the confused ornithologist. This species exhibits incredible geographic variation, and up to fifty-two subspecies have been described (Arcese et al. 2002). Song Sparrows vary greatly in their vocalizations, too. In fact, it sometimes seems that every drainage in Nevada has its own Song Sparrow song, and even individual birds personalize their repertoire enough to make this one of the more difficult bird songs to memorize.

DISTRIBUTION

Song Sparrows occur widely in Nevada but are rarely found far from wet ground. Atlas records were especially numerous in the northern half of the state and along the eastern border in the south, with most clusters of records coming from major drainages. Overall, the atlas distribution fits with earlier descriptions (Linsdale 1936, Alcorn 1988). No attempt was made to differentiate subspecies during the atlas project, but Linsdale (1936) described the ranges of three that breed in the state.

Song Sparrows are widely distributed—and wildly variable—in the states surrounding Nevada. At least seventeen subspecies have been reported from California, twelve or more of them as breeders, and most of these do not occur outside of California's borders (Shuford 1993, Small 1994). Song Sparrows also occur throughout Utah, Idaho, and Oregon, but are confined to river systems and wetlands in Arizona (Behle et al. 1985, Stephens and Sturts 1998, Adamus et al. 2001, Shrout in Corman and Wise-Gervais 2005:530–531).

Song Sparrows are usually found in wet, low-growing, shrubby thickets, and in the more arid parts of the West they are strictly riparian and wetland birds (Arcese et al. 2002). In the Great Basin, they are usually more common at lower elevations

but do occur into the lower montane zone (Ryser 1985). Song Sparrows are also found in less-than-pristine settings, including agricultural and urban wetlands, and even tamarisk stands (Arcese et al. 2002). Most atlas records came from riparian or wetland habitat, but the species also occurred on occasion in many other habitat types, and was consequently predicted to have a moderately high chance of occurring over much of the state. Presumably, Song Sparrows are largely restricted to wet areas within these habitats. Predictions based on patterns of habitat use can be misleading in some areas, and this appears to be true for Song Sparrows in southern Nevada. Overall, the species is correctly predicted to be largely absent from the Mojave Desert, with occurrence most likely along the major drainages, but the predicted occurrence in the southern mountains and urban areas appears to be an overestimate.

CONSERVATION AND MANAGEMENT

The Song Sparrow appears to be relatively easy to please, as riparian birds go (Gallagher 1997); a trickle of water and a smattering of appropriately dense brush are usually sufficient. It accepts at least some nonnative vegetation and often does well in human-altered landscapes (Arcese et al. 2002). Nonetheless, monitoring data suggest slight population declines rangewide and in the western region. Although monitoring in the Intermountain West has been limited, the available information suggests more stable populations there (Dobkin and Sauder 2004, Sauer et al. 2005). Potential threats to Song Sparrows include heavy livestock grazing, water diversion and habitat conversion in riparian zones, nest predation, and brood parasitism (Arcese et al. 2002, Dobkin and Sauder 2004).

In Nevada, Song Sparrows, like Yellow Warblers, may be useful monitors of riparian habitat conditions. Since they tolerate a fair amount of disturbance, local declines would suggest serious local threats; conversely, local population recovery often indicates the first step toward recovery of riparian habitats.

HABITAT USE

HABITAT	ATLAS BLOCKS	INCIDENTAL OBSERVATIONS
Agricultural	6 (3%)	0 (0%)
Ash	1 (<1%)	0 (0%)
Grassland	3 (1%)	1 (1%)
Mesquite	4 (2%)	1 (1%)
Mojave	0 (0%)	1 (1%)
Montane Forest	6 (3%)	2 (3%)
Montane Shrub	4 (2%)	5 (6%)
Open Water	0 (0%)	1 (1%)
Pinyon-Juniper	4 (2%)	1 (1%)
Riparian	144 (67%)	56 (72%)
Salt Desert Scrub	2 (<1%)	0 (0%)
Sagebrush Scrub	5 (2%)	1 (1%)
Sagebrush Steppe	2 (<1%)	0 (0%)
Urban	7 (3%)	3 (4%)
Wetland	26 (12%)	6 (8%)
TOTAL	214	78

Total numbers reported in the Habitat Use, Breeding Status, and Abundance tables may differ from each other (see pp. 22–23 for details). Percentage sums may differ slightly from 100% due to rounding.

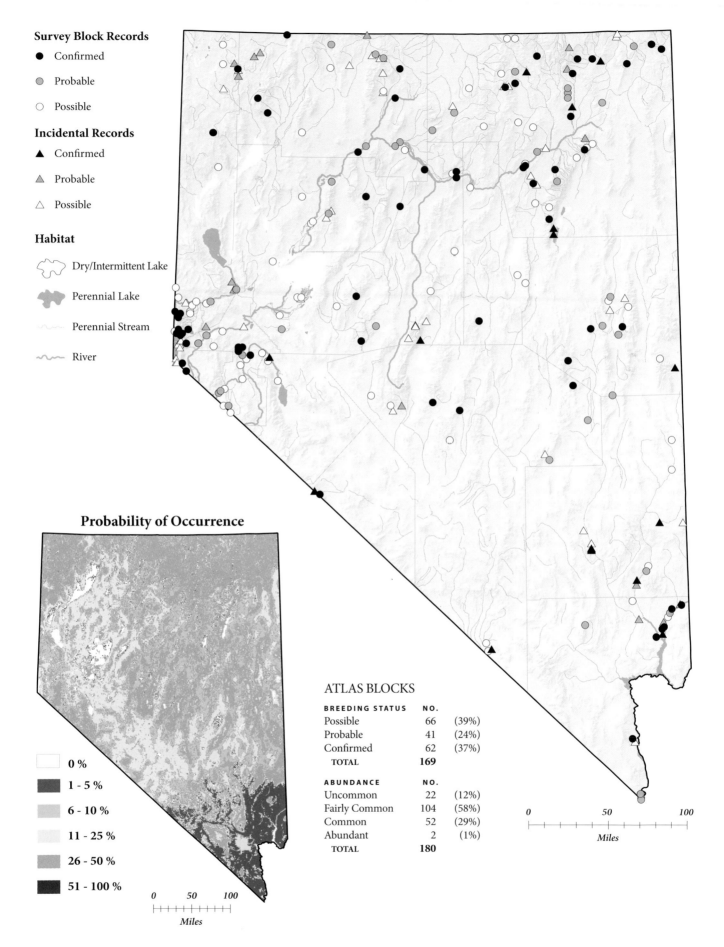

Survey Block Records

● Confirmed

● Probable

○ Possible

Incidental Records

▲ Confirmed

▲ Probable

△ Possible

Habitat

Dry/Intermittent Lake

Perennial Lake

Perennial Stream

River

Probability of Occurrence

0 %

1 - 5 %

6 - 10 %

11 - 25 %

26 - 50 %

51 - 100 %

0   50   100

Miles

ATLAS BLOCKS

BREEDING STATUS	NO.	
Possible	66	(39%)
Probable	41	(24%)
Confirmed	62	(37%)
TOTAL	**169**	

ABUNDANCE	NO.	
Uncommon	22	(12%)
Fairly Common	104	(58%)
Common	52	(29%)
Abundant	2	(1%)
TOTAL	**180**	

0   50   100

Miles

LINCOLN'S SPARROW

Melospiza lincolnii

Lincoln's Sparrow looks a bit like a daintier, more crisply marked version of the familiar Song Sparrow. There can be no mistaking its song, though: a rapid-fire onrush of bubbly, melodic phrases with a distinctive three-part element. Lincoln's Sparrow is commonly seen during migration in Nevada, but as a breeder it is restricted to a few montane wetlands in widely scattered sites. Even more than other members of the genus *Melospiza*, nesting Lincoln's Sparrows are habitual "skulkers" rarely found far from the cover of dense vegetation.

DISTRIBUTION

The atlas confirmed nesting by Lincoln's Sparrows only in the Jarbidge region of far northeastern Nevada and in the Carson Range of far western Nevada. In the year following the atlas fieldwork, several breeding pairs were seen in the Ruby Mountains (Mark Ports, pers. comm.). Neither Linsdale (1936) nor Alcorn (1988) included nesting records for Nevada. Ryser (1985) characterized the species as a summer resident in the Carson Range and noted the apparent lack of breeding in the central Great Basin ranges.

Primarily a boreal-nesting species, Lincoln's Sparrow breeds in most mountains of the western United States. It is more widespread in the surrounding states than in Nevada, except for Arizona, where breeding is limited to a few high-elevation sites (Corman in Corman and Wise-Gervais 2005:532–544). It breeds in parts of Oregon, central and northern Idaho, and eastern and southwestern Utah (Behle et al. 1985, Stephens and Sturts 1998, Adamus et al. 2001). Lincoln's Sparrows are fairly common in parts of northern California (Small 1994) but are very local in the south (Garrett and Dunn 1981). They occur on the western slope of the Sierra Nevada but are largely absent from the eastern slope (Gaines 1992), presumably because it is too arid.

Lincoln's Sparrows require very wet conditions and dense shrub and ground cover with little or no forest canopy (Ammon 1995). In their southern breeding range, they occur primarily in high-elevation, boggy meadows that are dominated by low willows and sedges (Ryser 1985, Small 1994, Ammon 1995). Most atlas data came from riparian sites, and the predictive map identifies potential areas with suitable habitat in the state's mountain ranges. The complete lack of records from central Nevada, however, suggests that the atlas model overpredicts the extent of the species' range.

CONSERVATION AND MANAGEMENT

Lincoln's Sparrows' specific microhabitat requirements for nesting (Ammon 1995) may contribute to their absence from apparently suitable habitats in many parts of Nevada. Assessing trends for this species is difficult due to insufficient monitoring data from much of its range. Declines may be occurring in parts of Canada, but the species seems to be stable or increasing in the western United States (Sauer et al. 2005). Given its tight association with high-elevation boggy habitats, disturbances to those areas, such as livestock grazing or intense recreational activity, are the most likely threats to Lincoln's Sparrow on its breeding grounds (Ammon 1995).

Lincoln's Sparrow is one of our more poorly known breeding birds. For example, a seemingly substantial population in the Ruby Mountains went undetected during atlas fieldwork. Or perhaps this population was absent during the atlas years—the species is subject to periodic fluctuations (Ammon 1995), and local extirpation and colonization events in a naturally fragmented region such as Nevada are thus fairly likely. Ryser (1985) called for additional fieldwork in the central Great Basin, and undetected populations may await discovery in the mountains of central and eastern Nevada.

HABITAT USE

HABITAT	ATLAS BLOCKS	INCIDENTAL OBSERVATIONS
Alpine	1 (20%)	0 (0%)
Montane Forest	0 (0%)	1 (20%)
Montane Parkland	1 (20%)	1 (20%)
Riparian	3 (60%)	3 (60%)
TOTAL	5	5

Total numbers reported in the Habitat Use, Breeding Status, and Abundance tables may differ from each other (see pp. 22–23 for details). Percentage sums may differ slightly from 100% due to rounding.

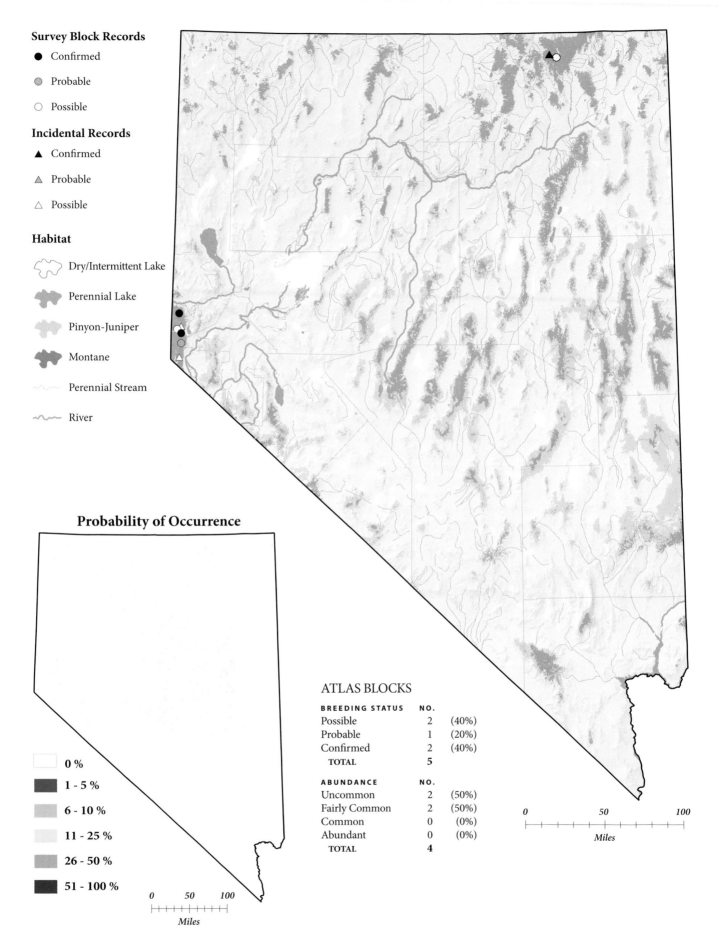

Survey Block Records

● Confirmed

◉ Probable

○ Possible

Incidental Records

▲ Confirmed

△ Probable

△ Possible

Habitat

Dry/Intermittent Lake

Perennial Lake

Pinyon-Juniper

Montane

Perennial Stream

River

Probability of Occurrence

☐ 0 %

■ 1 - 5 %

■ 6 - 10 %

☐ 11 - 25 %

■ 26 - 50 %

■ 51 - 100 %

0   50   100

Miles

ATLAS BLOCKS

BREEDING STATUS	NO.	
Possible	2	(40%)
Probable	1	(20%)
Confirmed	2	(40%)
TOTAL	5	

ABUNDANCE	NO.	
Uncommon	2	(50%)
Fairly Common	2	(50%)
Common	0	(0%)
Abundant	0	(0%)
TOTAL	4	

0   50   100

Miles

WHITE-CROWNED SPARROW
Zonotrichia leucophrys

The smart-looking White-crowned Sparrow is locally common in the open country of Nevada's high mountains. It is found in thickets and tangles in mountain meadows and montane sagebrush, where it tends to be one of the more conspicuous elements of its environment: it sings frequently, it is easy to identify, and it is not at all reclusive in the manner of so many other sparrows.

DISTRIBUTION

The distribution of White-crowned Sparrows in Nevada is scattered. There were concentrations of records in the Jarbidge region, the Santa Rosa and Ruby mountains, and in the Carson Range. Other records were from mountains around the state's margins. The species was essentially absent from central Nevada, a pattern noted previously (Ryser 1985). Breeding was confirmed no farther south than the White Mountains of Esmeralda County and does not appear to occur in the mountains of southern Nevada (Linsdale 1936, Alcorn 1988). The southernmost atlas records likely involved late migrants.

White-crowned Sparrows breed primarily in Canada and Alaska, but their summer range extends south through the high mountains of the western United States. In Nevada's neighboring states, breeding is probably most widespread in Utah (Behle et al. 1985) and Idaho (Stephens and Sturts 1998) and slightly less so in Oregon (Adamus et al. 2001). The species' California distribution lies mainly along the coast and in the Sierra Nevada (Small 1994). Breeding in Arizona is very limited and occurs only in the north (Corman in Corman and Wise-Gervais 2005:594).

Throughout their range, White-crowned Sparrows use a wide variety of habitats for breeding (Chilton et al. 1995). The sub-species that nests in Nevada, *oriantha*, is a bird of high-altitude meadows (Ryser 1985, Chilton et al. 1995), where its breeding habitat is characterized by dense shrubs and a mix of bare ground and grass cover (Dobkin and Sauder 2004). White-crowned Sparrows also nest occasionally in the wide expanses of montane sagebrush of northern Nevada. Atlas data show the species occurring primarily in riparian and montane shrub habitats, with scattered records from a wide range of other habitats. Most records of possible breeders from outside the high mountains may have involved late migrants. Excluding these records would probably increase the accuracy of the predictive map by shifting the southern edge of the range northward and by reducing the estimated probabilities of occurrence at lower elevations. The moderately high chance of finding this species in high-elevation shrublands throughout northern Nevada does reflect its occurrence there.

CONSERVATION AND MANAGEMENT

Long-term declines of White-crowned Sparrow populations have been reported for much of the West (Sauer et al. 2005). Known declines occurred primarily between the 1960s and the early 1980s, however, and populations have since stabilized (Chilton et al. 1995, Dobkin and Sauder 2004). The causes of past declines are poorly understood. Timber harvest seems an unlikely culprit, because White-crowned Sparrows probably benefit from the creation of shrub-dominated habitats (Chilton et al. 1995, Smith et al. 1997, Dobkin and Sauder 2004). Perhaps a more likely explanation involves degradation of riparian areas (Dobkin and Sauder 2004), although these impacts are often more severe at lower elevations than in the montane meadows that White-crowned Sparrows generally occupy.

Although the White-crowned Sparrow has been studied intensively elsewhere in its range, questions remain about Nevada's populations. In particular, we still do not know whether small numbers actually breed in the state's central and southern mountains, and if so, whether these populations are stable (see King and Mewaldt 1987). Also of interest is quantifying just how often the species nests in high-elevation sagebrush habitat.

HABITAT USE

HABITAT	ATLAS BLOCKS	INCIDENTAL OBSERVATIONS
Alpine	5 (6%)	1 (4%)
Mojave	1 (1%)	0 (0%)
Montane Forest	7 (8%)	5 (19%)
Montane Parkland	4 (5%)	3 (12%)
Montane Shrub	24 (27%)	7 (27%)
Pinyon-Juniper	2 (2%)	0 (0%)
Riparian	30 (34%)	8 (31%)
Sagebrush Scrub	5 (6%)	1 (4%)
Sagebrush Steppe	7 (8%)	1 (4%)
Urban	1 (1%)	0 (0%)
Wetland	2 (2%)	0 (0%)
TOTAL	88	26

Total numbers reported in the Habitat Use, Breeding Status, and Abundance tables may differ from each other (see pp. 22–23 for details). Percentage sums may differ slightly from 100% due to rounding.

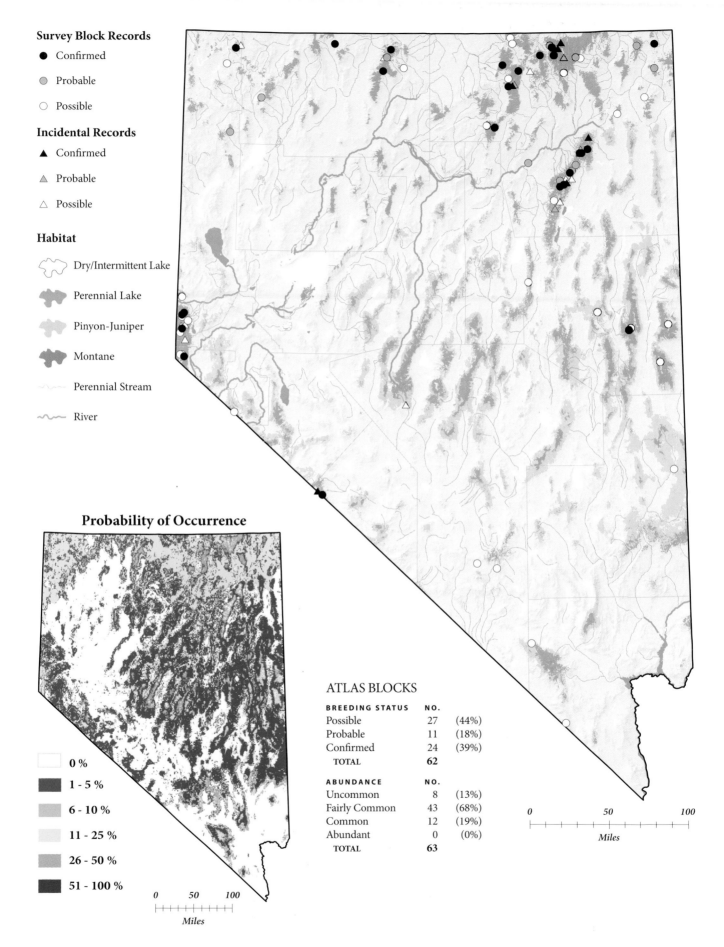

Survey Block Records

● Confirmed

● Probable

○ Possible

Incidental Records

▲ Confirmed

▲ Probable

△ Possible

Habitat

Dry/Intermittent Lake

Perennial Lake

Pinyon-Juniper

Montane

Perennial Stream

River

Probability of Occurrence

☐ 0 %

■ 1 - 5 %

■ 6 - 10 %

☐ 11 - 25 %

■ 26 - 50 %

■ 51 - 100 %

0   50   100

Miles

ATLAS BLOCKS

BREEDING STATUS	NO.	
Possible	27	(44%)
Probable	11	(18%)
Confirmed	24	(39%)
TOTAL	**62**	

ABUNDANCE	NO.	
Uncommon	8	(13%)
Fairly Common	43	(68%)
Common	12	(19%)
Abundant	0	(0%)
TOTAL	**63**	

0   50   100

Miles

DARK-EYED JUNCO
Junco hyemalis

In western Nevada, there is a boldly marked sparrow with a striking black hood and extensive rufous on the flanks. To the east and south, there is another with a bright rufous back, a small black mask, and clean gray flanks. Once considered different species, taxonomists now regard them as two versions of the variable Dark-eyed Junco, with "Oregon Juncos" (*oreganus* group) to the west, and "Gray-headed Juncos" (*caniceps* group) to the east. Atlas fieldworkers identified more than 80% of the atlas records to subspecies group. On the facing page we present data for the two forms combined, with separate information on each form on the following two pages.

DISTRIBUTION

Atlas data for the Dark-eyed Junco were both fascinating and frustrating because the junco's distribution in Nevada was rather more complex than had previously been documented (Linsdale 1936, Ryser 1985, Alcorn 1988, Nolan et al. 2002). "Oregon Juncos" were most concentrated in the far western mountains, especially the Carson Range, but were also confirmed as breeders at scattered sites across northern Nevada. As expected, "Gray-headed Juncos" were more concentrated in eastern Nevada; they accounted for about two-thirds of the records from the state's center and all the confirmed breeding records in the south. In the central, eastern, and northern mountains, where both forms occur, hybrids are also found (C. Wood, pers. comm.). Juncos of both types were abundant wherever they were present, with at least ten pairs estimated in about half of the occupied blocks, and were quite easy to confirm as breeders.

In California, "Oregon Juncos" predominate, although "Gray-headeds" reportedly nest in the White and Inyo mountains near the Nevada border (Small 1994). Atlas fieldworkers found no "Gray-headed Juncos" in nearby portions of Nevada, but the predictive model for "Gray-headed Juncos" does identify ranges close by. The "Gray-headed Junco" is common in northern Arizona but scarce in the south (LaRue in Corman and Wise-Gervais 2005:534–535), and it is a common summer resident in the mountains of Utah. Farther north in Utah, yet another form, the "Pink-sided Junco," occurs (Behle et al. 1985). Appropriately, the "Oregon Junco" breeds statewide in Oregon and in Idaho,

with "Gray-headed Juncos" also nesting locally in both states (Burleigh 1972, Adamus et al. 2001).

In the West, juncos breed in coniferous, montane riparian, and mixed forests (Nolan et al. 2002), and most atlas data came from these habitats. Habitat use by "Gray-headed" and "Oregon Juncos" is similar, although the former tend to occur in more arid woodlands than the latter (Garrett and Dunn 1981, Gaines 1992, Shuford 1993), possibly just reflecting the different habitats present in the regions where each predominates. More than half of the block records came from montane forest, a habitat that generally lies where the model predictions suggest the species is most likely to be found. The high estimated chance of occurrence at lower elevations in the Mojave Desert and the low chance in the valleys of the Great Basin are doubtless overestimates caused by late migrants. The predictive map for the "Gray-headed Junco" closely matches expectations, but that for the "Oregon Junco" does not capture the higher odds of finding the species in the westernmost ranges.

CONSERVATION AND MANAGEMENT

Dark-eyed Juncos are common, widespread, and versatile, and do not generally arouse conservation concern (Nolan et al. 2002). Still, populations have declined steadily in much of the range (Sauer et al. 2005), although perhaps not those of the "Gray-headed Junco" (Nolan et al. 2002). The effects of logging on populations are unclear, since juncos will use brushy clearcuts and regenerating stands as well as mature forests (Smith et al. 1997). In Nevada, Dark-eyed Juncos have not been well monitored, but they may be the most abundant breeding bird in some montane forests (Ryser 1985).

HABITAT USE (ALL FORMS)

HABITAT	ATLAS BLOCKS	INCIDENTAL OBSERVATIONS
Alpine	12 (5%)	0 (0%)
Montane Forest	142 (56%)	35 (63%)
Montane Parkland	10 (4%)	3 (5%)
Montane Shrub	31 (12%)	5 (9%)
Pinyon-Juniper	26 (10%)	4 (7%)
Riparian	23 (9%)	9 (16%)
Sagebrush Scrub	4 (2%)	0 (0%)
Sagebrush Steppe	2 (>1%)	0 (0%)
Urban	2 (>1%)	0 (0%)
TOTAL	252	56

Total numbers reported in the Habitat Use, Breeding Status, and Abundance tables may differ from each other (see pp. 22–23 for details). Percentage sums may differ slightly from 100% due to rounding.

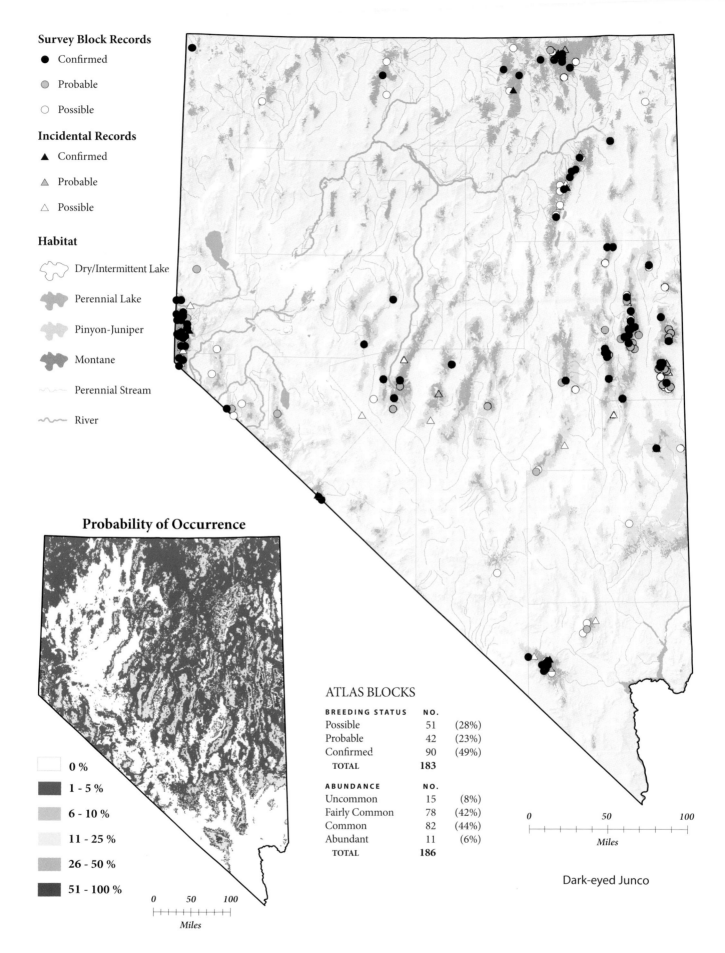

Survey Block Records

● Confirmed

● Probable

○ Possible

Incidental Records

▲ Confirmed

△ Probable

△ Possible

Habitat

Dry/Intermittent Lake

Perennial Lake

Pinyon-Juniper

Montane

Perennial Stream

River

Probability of Occurrence

  0 %

  1 - 5 %

  6 - 10 %

  11 - 25 %

  26 - 50 %

  51 - 100 %

0   50   100
Miles

ATLAS BLOCKS

BREEDING STATUS	NO.	
Possible	51	(28%)
Probable	42	(23%)
Confirmed	90	(49%)
TOTAL	**183**	

ABUNDANCE	NO.	
Uncommon	15	(8%)
Fairly Common	78	(42%)
Common	82	(44%)
Abundant	11	(6%)
TOTAL	**186**	

0   50   100
Miles

Dark-eyed Junco

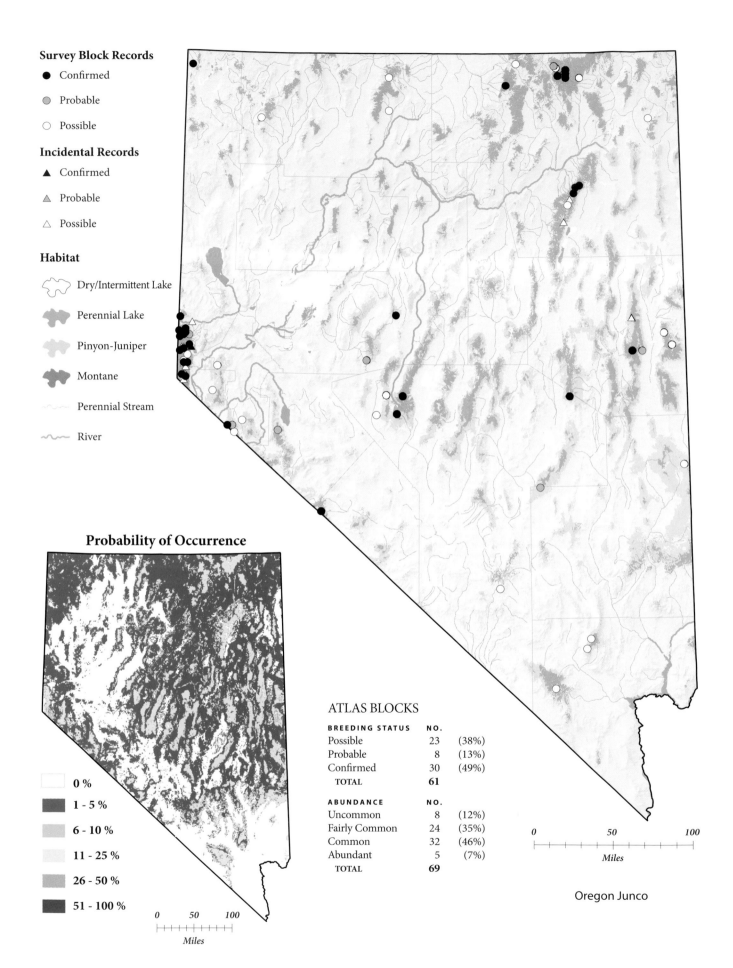

Survey Block Records

● Confirmed

● Probable

○ Possible

Incidental Records

▲ Confirmed

△ Probable

△ Possible

Habitat

Dry/Intermittent Lake

Perennial Lake

Pinyon-Juniper

Montane

Perennial Stream

River

Probability of Occurrence

	0 %
	1 - 5 %
	6 - 10 %
	11 - 25 %
	26 - 50 %
	51 - 100 %

0   50   100

Miles

ATLAS BLOCKS

BREEDING STATUS	NO.	
Possible	23	(38%)
Probable	8	(13%)
Confirmed	30	(49%)
TOTAL	**61**	

ABUNDANCE	NO.	
Uncommon	8	(12%)
Fairly Common	24	(35%)
Common	32	(46%)
Abundant	5	(7%)
TOTAL	**69**	

0   50   100

Miles

Oregon Junco

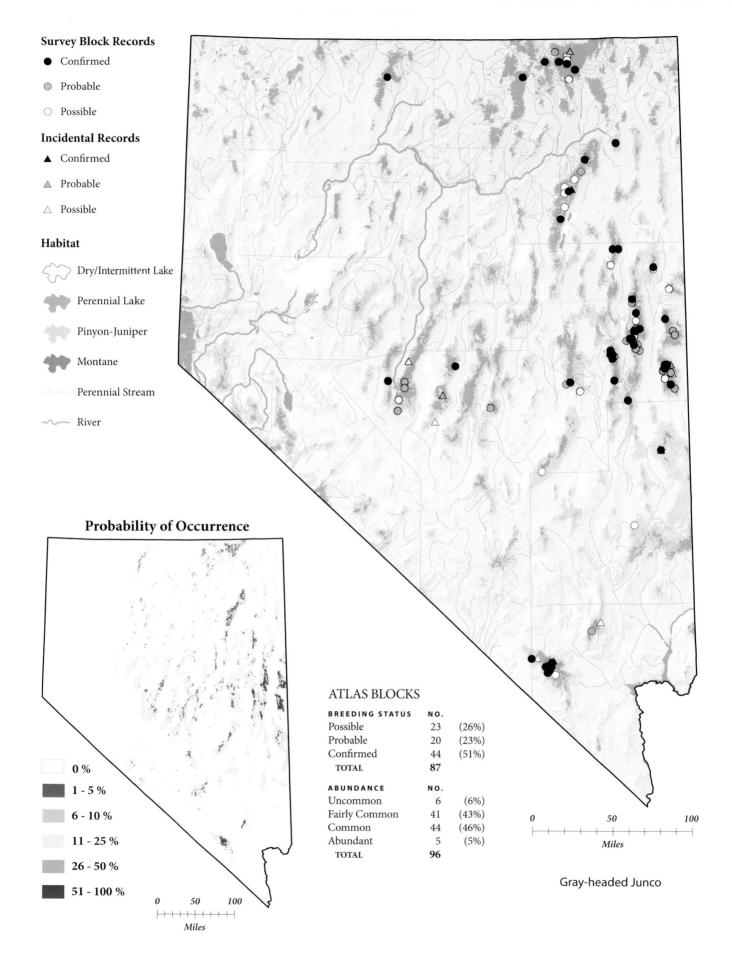

Survey Block Records

- ● Confirmed
- ◉ Probable
- ○ Possible

Incidental Records

- ▲ Confirmed
- △ Probable
- △ Possible

Habitat

- Dry/Intermittent Lake
- Perennial Lake
- Pinyon-Juniper
- Montane
- Perennial Stream
- River

Probability of Occurrence

- ☐ 0 %
- 1 - 5 %
- 6 - 10 %
- 11 - 25 %
- 26 - 50 %
- 51 - 100 %

0   50   100
Miles

ATLAS BLOCKS

BREEDING STATUS	NO.	
Possible	23	(26%)
Probable	20	(23%)
Confirmed	44	(51%)
TOTAL	**87**	

ABUNDANCE	NO.	
Uncommon	6	(6%)
Fairly Common	41	(43%)
Common	44	(46%)
Abundant	5	(5%)
TOTAL	**96**	

0   50   100
Miles

Gray-headed Junco

BLACK-HEADED GROSBEAK
Pheucticus melanocephalus

There is something ebullient and effervescent about Nevada's mountain streams in the early summer. Bright green aspen leaves flutter in the morning sunshine, and icy waters splash on boulders as the last pockets of snow slowly melt from shaded banks. The gurgling of the streams is punctuated by the unrestrained oratory of the male Black-headed Grosbeak, supplemented by the simpler counterpoint of the female's song. Although these birds are most frequently encountered at the middle elevations, they often range down along river drainages to the valley floors, where they may even breed in wooded backyards.

DISTRIBUTION

Atlas workers found Black-headed Grosbeaks commonly in Nevada's mountains and more sporadically in the valleys. Although records were concentrated in the major mountain ranges, Black-headed Grosbeaks were found throughout the state, matching earlier accounts of their distribution (Linsdale 1936, Alcorn 1988). Chisholm and Neel (2002) noted that Black-headed Grosbeaks are also common summer residents in riparian corridors and residential areas of the Lahontan Valley, where there were few atlas records. The Black-headed Grosbeak is widely distributed throughout the West and is a characteristic and common bird of forested habitats in Nevada's neighboring states (Behle et al. 1985, Small 1994, Stephens and Sturts 1998, Adamus et al. 2001, Wise-Gervais in Corman and Wise-Gervais 2005:542–543).

Throughout their range, Black-headed Grosbeaks breed in diverse settings, including lowland cottonwood forests, pinyon-juniper woodlands, aspen groves, backyards, and montane conifers. Nests are located in structurally diverse woodlands, often with a deciduous vegetation component, and are often close to forest openings, edges, and water. This species mostly avoids large tracts of dense, high-elevation forests (Hill 1995), and in Nevada it is most often associated with mid- to high-elevation riparian zones. Breeders can also be common in pinyon-juniper and mountain mahogany (Linsdale 1936). The predictive map highlights these mid-elevation habitats as the areas where the species is most likely to be found. With the exception of a few records along the state's major drainages, most lowland records were assigned only possible breeding codes, suggesting that these reports may have involved migrants, and that the probability of finding the species nesting at lower elevations may be even less than the predictions suggest.

CONSERVATION AND MANAGEMENT

The Black-headed Grosbeak is an adaptable species that seems to be doing well throughout most of its western range (Hill 1995). Populations are slowly increasing overall, and possibly also in Nevada (Sauer et al. 2005). On balance, this species has probably benefited from logging activities that converted conifer forests to second-growth broadleaved woodlands (Smith et al. 1997), the increasing availability of residential developments with suitable ornamental plantings, and the planting and irrigation of trees in agricultural areas (Hill 1995). Still, there is concern for the species in some situations (Hill 1995). It appears to have declined in Orange County, California (Gallagher 1997), and perhaps also in parts of northern California (Shuford 1993).

Overall, the species' Nevada distribution has apparently changed little since the nineteenth century (Ryser 1985). Local declines were reported after decades of severe riparian habitat degradation along the lower Truckee River (Klebenow and Oakleaf 1984). When riparian habitats recovered, however, so did the local Black-headed Grosbeak population, suggesting that the species may be a useful indicator of riparian restoration progress in the region (Ammon 2002).

HABITAT USE

HABITAT	ATLAS BLOCKS	INCIDENTAL OBSERVATIONS
Agricultural	2 (1%)	0 (0%)
Ash	1 (<1%)	0 (0%)
Mojave	0 (0%)	1 (1%)
Montane Forest	31 (21%)	14 (20%)
Montane Parkland	2 (1%)	1 (1%)
Montane Shrub	6 (4%)	9 (13%)
Pinyon-Juniper	28 (19%)	14 (19%)
Riparian	62 (42%)	27 (38%)
Sagebrush Scrub	6 (4%)	0 (0%)
Sagebrush Steppe	1 (<1%)	2 (3%)
Urban	8 (5%)	3 (4%)
TOTAL	147	71

Total numbers reported in the Habitat Use, Breeding Status, and Abundance tables may differ from each other (see pp. 22–23 for details). Percentage sums may differ slightly from 100% due to rounding.

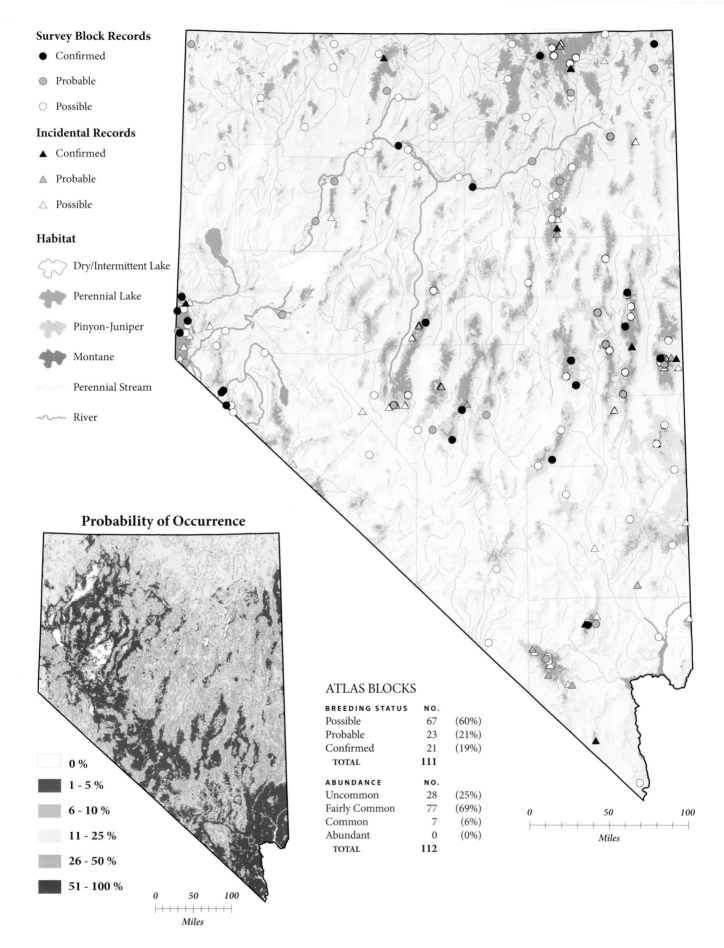

Survey Block Records

● Confirmed

● Probable

○ Possible

Incidental Records

▲ Confirmed

▲ Probable

△ Possible

Habitat

Dry/Intermittent Lake

Perennial Lake

Pinyon-Juniper

Montane

Perennial Stream

River

Probability of Occurrence

	0 %
	1 - 5 %
	6 - 10 %
	11 - 25 %
	26 - 50 %
	51 - 100 %

0   50   100

Miles

ATLAS BLOCKS

BREEDING STATUS	NO.	
Possible	67	(60%)
Probable	23	(21%)
Confirmed	21	(19%)
TOTAL	**111**	

ABUNDANCE	NO.	
Uncommon	28	(25%)
Fairly Common	77	(69%)
Common	7	(6%)
Abundant	0	(0%)
TOTAL	**112**	

0   50   100

Miles

BLUE GROSBEAK
Passerina caerulea

The twangy song and sharp call notes of the beautiful Blue Grosbeak are common summer sounds around oases and river bottomlands of southern Nevada, where nesting pairs can be found in a variety of riparian plant communities, both native and exotic. Farther north the Blue Grosbeak is scarcer, but there are scattered records from the major river drainages of the western Great Basin.

DISTRIBUTION

Blue Grosbeaks were widely noted at lowland riparian sites in southern Nevada, although they were rare along Nevada's portion of the Colorado River. To the north, a small number of records, including confirmed breeders, came from the Mason Valley. The species was widely reported during the Death Valley Expedition (Fisher 1893), while Linsdale (1936) mentioned records only in southern Nevada and noted its rarity in the Colorado River Valley. Alcorn's (1988) records extended the Blue Grosbeak's breeding range into western Nevada, and Chisholm and Neel (2002) discussed its increasing presence along Nevada's western rivers.

Blue Grosbeaks nest across the southern half of the United States and north into the Great Plains. In the West, the bulk of the range lies to the south of Nevada in the southern half of California, southern Utah, and Arizona (Behle et al. 1985, Small 1994, LaRue in Corman and Wise-Gervais 2005:544–545). Interestingly, Blue Grosbeaks are common along the Colorado River both north and south of Nevada (Rosenberg et al. 1991), suggesting that habitat restoration along the Nevada reaches could result in colonization. Small, disjunct breeding populations exist in northern Utah and southern Idaho (Behle et al. 1985, Stephens and Sturts 1998), and Adamus et al. (2001) reported Oregon's first breeding record in the eastern part of the state.

In the East, Blue Grosbeaks nest in many kinds of forest edge and open woodland habitats, but in the West they tend to be found in riparian areas (Ingold 1993). Various riparian settings appear to be acceptable, including tamarisk stands, irrigation ditches, and sloughs, as well as more natural plant communities of willow, mesquite, ash, and cottonwood (Ryser 1985, Rosenberg et al. 1991, Ingold 1993). As expected, most of the atlas records came from riparian habitats, and the model predicts a moderately high chance of finding the species in most riparian areas. Occurrence is predicted to be even more likely in mesquite and ash habitats in southern Nevada, although these habitats occur in patches so small that they are barely discernible on the predictive map. There is also a fairly high chance of finding Blue Grosbeaks in agricultural areas, presumably reflecting their ability to use shrub habitats in irrigated landscapes.

CONSERVATION AND MANAGEMENT

Overall, Blue Grosbeak populations in the United States have steadily increased over the past few decades, and the species appears to be faring well in the West (Sauer et al. 2005). The Blue Grosbeak probably owes at least part of its success to its ability to use disturbed sites and invasive vegetation (Rosenberg et al. 1991). For instance, Garrett and Dunn (1981) noted Blue Grosbeaks breeding in young willow thickets bordering flood control basins and channels. Declines in some California populations, however, have been attributed to agricultural development, cowbird parasitism, and flood control projects (Small 1994, Gallagher 1997).

Detailed information on Blue Grosbeak trends is not available for Nevada, but our populations seem to be healthy. Nonetheless, the species has a restricted distribution in the region and is dependent on riparian habitat. Consequently, Blue Grosbeaks are covered under the Clark County Multiple Species Habitat Conservation Plan (Clark County 2000) and are considered a Priority species by Nevada Partners in Flight (Neel 1999a).

HABITAT USE

HABITAT	ATLAS BLOCKS	INCIDENTAL OBSERVATIONS
Agricultural	1 (3%)	6 (15%)
Ash	7 (18%)	0 (0%)
Mesquite	8 (20%)	1 (2%)
Mojave	1 (3%)	1 (2%)
Open Water	0 (0%)	1 (2%)
Riparian	18 (45%)	25 (61%)
Salt Desert Scrub	1 (3%)	2 (5%)
Sagebrush Scrub	1 (3%)	0 (0%)
Sagebrush Steppe	0 (0%)	1 (2%)
Urban	1 (3%)	2 (5%)
Wetland	2 (5%)	2 (5%)
TOTAL	40	41

Total numbers reported in the Habitat Use, Breeding Status, and Abundance tables may differ from each other (see pp. 22–23 for details). Percentage sums may differ slightly from 100% due to rounding.

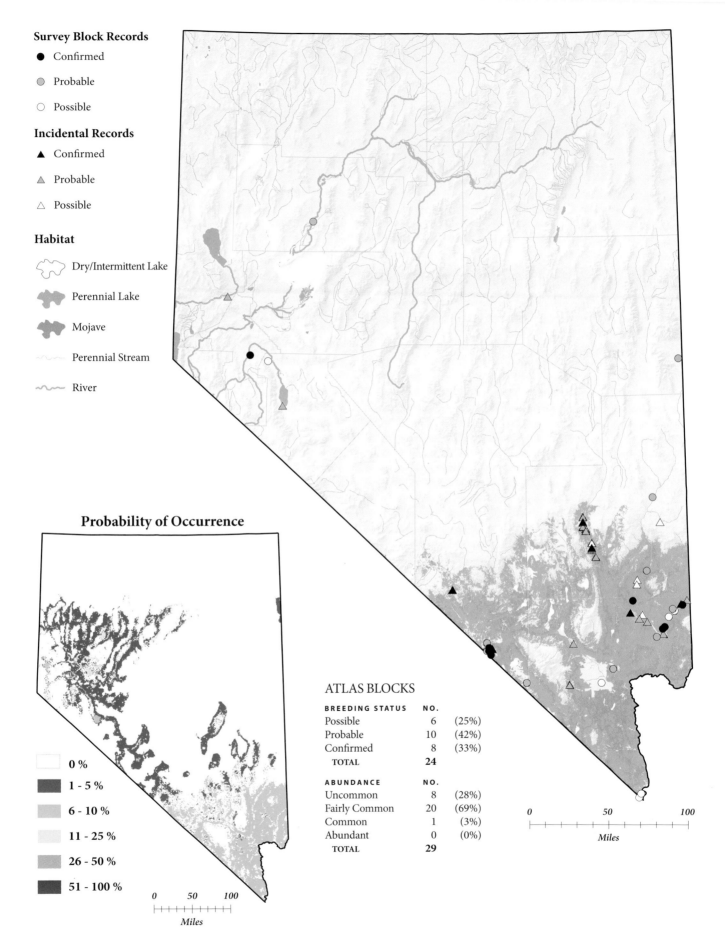

Survey Block Records

● Confirmed

● Probable

○ Possible

Incidental Records

▲ Confirmed

▲ Probable

△ Possible

Habitat

Dry/Intermittent Lake

Perennial Lake

Mojave

Perennial Stream

River

Probability of Occurrence

	0 %
	1 - 5 %
	6 - 10 %
	11 - 25 %
	26 - 50 %
	51 - 100 %

0   50   100
Miles

ATLAS BLOCKS

BREEDING STATUS	NO.	
Possible	6	(25%)
Probable	10	(42%)
Confirmed	8	(33%)
TOTAL	24	

ABUNDANCE	NO.	
Uncommon	8	(28%)
Fairly Common	20	(69%)
Common	1	(3%)
Abundant	0	(0%)
TOTAL	29	

0   50   100
Miles

LAZULI BUNTING
Passerina amoena

Reddish, white, and blue are the colors of the bright little bunting that inhabits riparian woodlands in many of Nevada's mountain ranges. But in this desert state, the Lazuli Bunting seems most characteristic of the willows and tall sagebrush that line streams as they wind their way through the dusty foothills. The Lazuli Bunting's brilliant plumage and lusty song, with each male providing a unique variant, add a certain charm to what can otherwise be a foreboding environment.

DISTRIBUTION

Lazuli Buntings have long been considered widespread and common in Nevada, especially near streams (Fisher 1893, Linsdale 1936, Ryser 1985, Alcorn 1988). Atlas workers found them statewide, although breeding records were sparser in the central and lower-elevation portions of the state.

The bulk of the Lazuli Bunting's breeding range lies north and east of Nevada, and the species occurs in most of Utah, Idaho, and Oregon (Behle et al. 1985, Stephens and Sturts 1998, Adamus et al. 2001). In California, it breeds in the north and along the southern coast into Mexico but is absent from the southeast (Small 1994). Breeding in Arizona is less common, and the species does not nest in the southwestern portion of that state (Wise-Gervais in Corman and Wise-Gervais 2005:546–547).

Some breeding birds in Nevada are more closely tied to riparian habitats than they are in less arid portions of their range, and the Lazuli Bunting provides a fine example of this pattern. Elsewhere, the species is commonly reported from a range of brushy habitats, from coastal areas to high mountains (Greene et al. 1996). Atlas data show that riparian habitats are used far more often than all others in Nevada, and the predictive map highlights riparian patches—often in the foothills of mountain ranges—as the sites where the species is most likely to occur. The predictions also suggest a fairly good chance of finding these buntings in the state's mountains and throughout the sagebrush steppe habitats of the north. Within these areas, Lazuli Buntings are most likely to be in brushy areas near water.

CONSERVATION AND MANAGEMENT

In the United States as a whole, Lazuli Bunting populations appear to be fairly stable, although local declines have certainly occurred in some areas (Sauer et al. 2005). Habitat changes, such as the creation of brushy habitat along irrigation ditches during the past century, may have benefited the species (Dobkin 1994), and some eastward range expansion may still be occurring (Greene et al. 1996). Population losses in some areas have been provisionally attributed to cowbird parasitism, and degradation of riparian habitat is a concern as well (Greene et al. 1996). Overall, though, Lazuli Buntings remain widespread and abundant, and specific management for this species has been limited (Greene et al. 1996).

The status of the Lazuli Bunting in Nevada appears to have changed little since the nineteenth century (Ryser 1985). Detailed information is lacking, though, due to the paucity of long-term monitoring data for the state's riparian habitats. The species makes at least some use of artificial habitats such as willow-lined ditches in the Lahontan Valley (Chisholm and Neel 2002), and fully intact riparian communities seem less critical to the Lazuli Bunting than they are to some other riparian species. This flexibility may have buffered the species against the worst effects of habitat degradation. It will be interesting to monitor future interactions with the Indigo Bunting, which may be expanding its range in Nevada. The two species are closely related, use similar habitats, and hybridize where they co-occur.

HABITAT USE

HABITAT	ATLAS BLOCKS	INCIDENTAL OBSERVATIONS
Agricultural	3 (2%)	3 (7%)
Alpine	1 (1%)	0 (0%)
Ash	2 (1%)	0 (0%)
Grassland	2 (1%)	0 (0%)
Mesquite	1 (1%)	0 (0%)
Mojave	2 (1%)	0 (0%)
Montane Forest	12 (9%)	3 (7%)
Montane Parkland	2 (1%)	0 (0%)
Montane Shrub	10 (7%)	2 (5%)
Pinyon-Juniper	11 (8%)	2 (5%)
Riparian	77 (57%)	29 (67%)
Salt Desert Scrub	0 (0%)	1 (2%)
Sagebrush Scrub	4 (3%)	0 (0%)
Sagebrush Steppe	3 (2%)	0 (0%)
Urban	4 (3%)	1 (2%)
Wetland	1 (1%)	2 (5%)
TOTAL	135	43

Total numbers reported in the Habitat Use, Breeding Status, and Abundance tables may differ from each other (see pp. 22–23 for details). Percentage sums may differ slightly from 100% due to rounding.

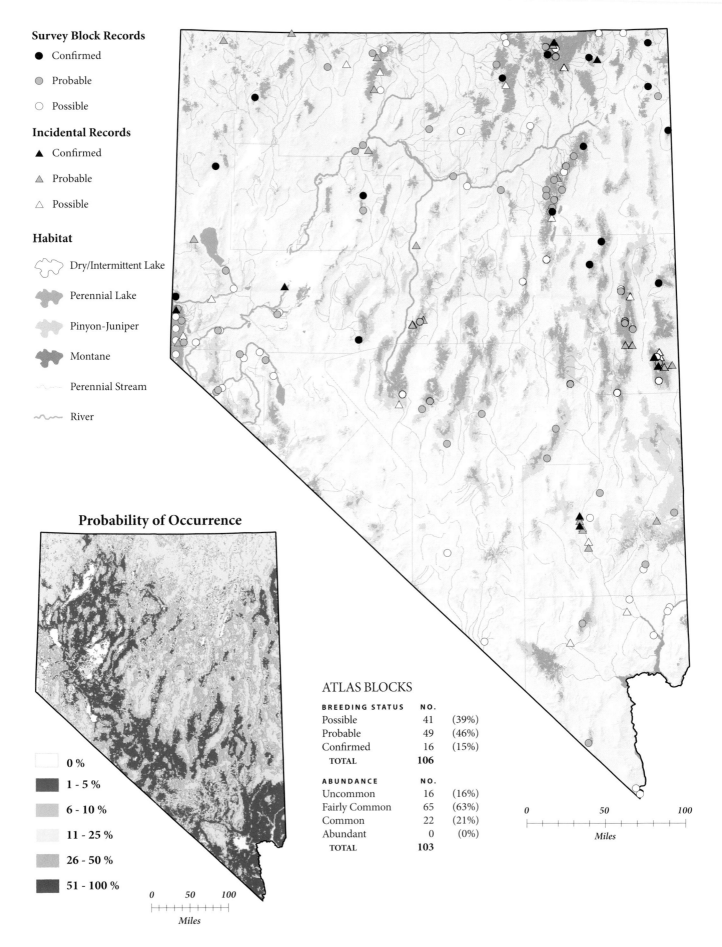

Survey Block Records

● Confirmed

● Probable

○ Possible

Incidental Records

▲ Confirmed

▲ Probable

△ Possible

Habitat

Dry/Intermittent Lake

Perennial Lake

Pinyon-Juniper

Montane

Perennial Stream

River

Probability of Occurrence

☐	0 %
■	1 - 5 %
■	6 - 10 %
☐	11 - 25 %
■	26 - 50 %
■	51 - 100 %

0   50   100
Miles

ATLAS BLOCKS

BREEDING STATUS	NO.	
Possible	41	(39%)
Probable	49	(46%)
Confirmed	16	(15%)
TOTAL	**106**	

ABUNDANCE	NO.	
Uncommon	16	(16%)
Fairly Common	65	(63%)
Common	22	(21%)
Abundant	0	(0%)
TOTAL	**103**	

0   50   100
Miles

INDIGO BUNTING

Passerina cyanea

Like many Easterners, Indigo Buntings seem to be increasingly aware of Nevada's charms, and a growing number of these lovely blue expatriates are becoming established in the southern part of the state. In the West, Indigo Buntings typically occur in deciduous riparian woods, and this is certainly true in Nevada. They even find Lazuli Buntings to be suitable approximations for mates at times, and hybrids between the two species are sometimes noted, including one recorded during the atlas project.

DISTRIBUTION

Indigo Buntings are a fairly recent addition to Nevada's avifauna. Linsdale (1936) did not mention the species at all, and it was not regularly found in the Southwest until about the 1940s (Corman in Corman and Wise-Gervais 2005:548–549). Atlas records were largely limited to southern and east-central Nevada. Breeding was not confirmed at any of the sixteen sites where Indigo Buntings were seen, but four southern records suggested probable breeding. Both Ryser (1985) and Alcorn (1998) suggested that nesting may occur in southern Nevada, though neither reported confirmed cases. Specific areas named by these authors include Ash Meadows National Wildlife Refuge in Nye County, the Spring and Virgin mountains in Clark County, and the Caliente area in Lincoln County; and, indeed, atlas workers reported Indigo Buntings near most of these locations. In Arizona and the lower Colorado River Valley, Indigo Buntings have a better-established presence, but nesting is still a local phenomenon (Rosenberg et al. 1991, Corman in Corman and Wise-Gervais 2005:548–549). Breeding is occasionally noted in Utah, too, especially in the south (Behle 1985).

Several records of possible breeding came from more northern locations, mostly from White Pine County in the east, and there was a single record from the lower Truckee River in the west. These records generally involved single birds that were probably nonbreeding transients. Given the Indigo Bunting's

apparent expansion in Nevada, however, the possibility of breeding in these areas should not be dismissed. Since the species was found in only three blocks, the predictive map provides crude distributional information at best. It does, however, indicate general areas where breeding might occur, while also reflecting the low chance of finding this species anywhere in the state.

In the East, Indigo Buntings tend to nest in edge habitats such as brushy vegetation along roadsides, forest margins, and old fields, and typically avoid closed-canopy forests (Payne 1992). In the West, they are primarily riparian birds, and riparian and wetland habitats accounted for most atlas habitat records, much like the pattern noted in Arizona (Corman in Corman and Wise-Gervais 2005:548–549).

CONSERVATION AND MANAGEMENT

Payne (1992) reported that Indigo Buntings were expanding their range and numbers throughout North America. More recent Breeding Bird Survey data, however, suggest that Indigo Buntings have been declining since the 1980s in the core of their breeding range in the East and Midwest (Sauer et al. 2005). These declines have not been clearly attributed to a single cause, but many shrubland-associated species show similar patterns, and the regeneration of eastern forests and conversion of remaining early successional habitats presumably play a role (Askins 2002).

Data from western states are limited, making trends in this region difficult to determine. Casual observations suggest that the species' spread into the southwestern United States is continuing. In Arizona, Corman (in Corman and Wise-Gervais 2005:548–549) anticipated increased breeding of Indigo Buntings so long as remaining riparian areas are protected from development or overgrazing. The Indigo Bunting's prospects in Nevada are most likely tied to the conservation of riparian woodlands as well.

HABITAT USE

HABITAT	ATLAS BLOCKS	INCIDENTAL OBSERVATIONS
Agricultural	0 (0%)	1 (8%)
Montane Forest	1 (33%)	0 (0%)
Montane Shrub	0 (0%)	1 (8%)
Riparian	2 (67%)	8 (62%)
Salt Desert Scrub	0 (0%)	1 (8%)
Wetland	0 (0%)	2 (15%)
TOTAL	3	13

Total numbers reported in the Habitat Use, Breeding Status, and Abundance tables may differ from each other (see pp. 22–23 for details). Percentage sums may differ slightly from 100% due to rounding.

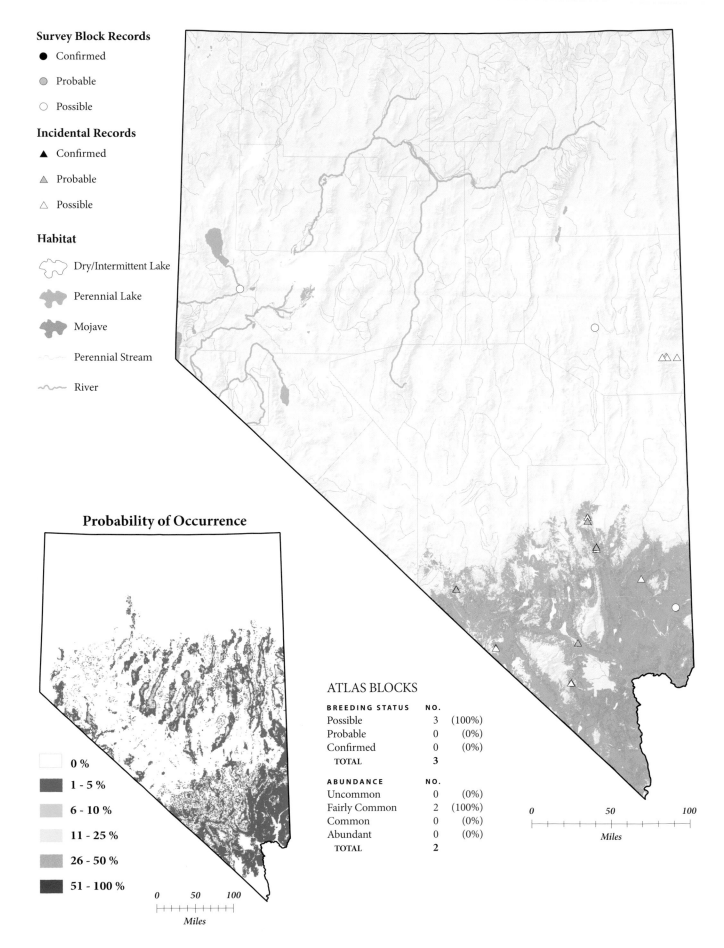

Survey Block Records
● Confirmed
◐ Probable
○ Possible

Incidental Records
▲ Confirmed
△ Probable
△ Possible

Habitat
Dry/Intermittent Lake
Perennial Lake
Mojave
Perennial Stream
River

Probability of Occurrence

0 %
1 - 5 %
6 - 10 %
11 - 25 %
26 - 50 %
51 - 100 %

0   50   100
Miles

ATLAS BLOCKS

BREEDING STATUS	NO.	
Possible	3	(100%)
Probable	0	(0%)
Confirmed	0	(0%)
TOTAL	3	

ABUNDANCE	NO.	
Uncommon	0	(0%)
Fairly Common	2	(100%)
Common	0	(0%)
Abundant	0	(0%)
TOTAL	2	

0   50   100
Miles

BOBOLINK
Dolichonyx oryzivorus

The hayfields of Elko County are home to one of the state's more eye-catching songbirds—the Bobolink—which reaches the southwestern limit of its breeding range in northern Nevada. Small breeding colonies are fairly easily found in and around the Lamoille Valley, and single birds can be seen flying throughout this area. In early summer, "skylarking" males steal the show with their elaborate display flights and exuberant songs. Even after birds have settled down to nest, the distinctive twangy call notes of both sexes are a characteristic sound of the agricultural valleys of northeastern Nevada.

DISTRIBUTION

Atlas workers found Bobolinks at seventeen sites, all but two of which were in Elko County. Linsdale (1936) gave no definite breeding records, but breeding was confirmed at scattered locations in northeastern Nevada by the time Alcorn (1988) published his work on Nevada birds. Neel (1999a) provided a detailed review of the species' known breeding locations.

The heart of the Bobolink's breeding range lies along the Canadian-U.S. border, from the Dakotas to east of the Great Lakes. Populations in the West are patchy and vary from year to year. Breeding is well established in eastern Oregon (Adamus et al. 2001), far northern Utah (Behle et al. 1985), and several parts of Idaho (Stephens and Sturts 1998). Southeastern Arizona has a small, disjunct breeding population (Martin and Gavin 1995, Corman in Corman and Wise-Gervais 2005:599); in California, nesting has been suspected but not confirmed (McCaskie et al. 1988, Small 1994).

The tall-grass and mixed-grass prairies of the north-central United States and south-central Canada were the historical breeding habitats of the ground-nesting Bobolink (Martin and Gavin 1995). In the West, Bobolinks are found mostly in irrigated hayfields, pastures, or riparian meadows; at the landscape level, they favor broad river valleys with little or no topographic relief (Behle et al. 1985, Neel 1999a). Atlas records came from agricultural, grassland, riparian, and wetland habitats. The predictive map highlights the Bobolink's use of agricultural habitats in the state and shows many other areas across northern Nevada's valleys where there is estimated to be a low to moderate chance of finding the species. At first glance, these predictions seem overly extensive and appear to be based on a single atlas record near the California border. Monitoring data collected after the atlas project, however, show that some agricultural areas in north-central Nevada, such as the Paradise Valley, do have small breeding populations and suggest that additional sites may be discovered.

CONSERVATION AND MANAGEMENT

It is generally thought that the Bobolink's range began to spread outward from the northern Great Plains when irrigated pastures were created in the arid West and forests were cleared in the East (Martin and Gavin 1995). More recently, populations have declined severely everywhere except in the core of the species' range (Sauer et al. 2005), probably as a result of loss of meadow habitat, changes in haying schedules, and the rise of mechanized farming (Martin and Gavin 1995). Early-season mowing, in particular, can cause significant nestling mortality (Martin and Gavin 1995). Bad weather, flooding, and predation can also reduce breeding productivity and cause annual fluctuations in abundance (Behle et al. 1985, Martin and Gavin 1995).

Although the Bobolink may have joined the Great Basin avifauna only during the nineteenth century, following the spread of agriculture (Ryser 1985), Nevada Partners in Flight considers it a Priority species because of its rangewide declines and apparent dependence on current land use practices. It is estimated that 95% of Nevada's populations occur on private land (Neel 1999a), and conservation for this species will depend largely on landowners' stewardship efforts—most specifically, the use of Bobolink-friendly agricultural practices (Neel 1999a).

HABITAT USE

HABITAT	ATLAS BLOCKS	INCIDENTAL OBSERVATIONS
Agricultural	3 (20%)	6 (75%)
Grassland	3 (20%)	0 (0%)
Riparian	3 (20%)	0 (0%)
Wetland	6 (40%)	2 (25%)
TOTAL	15	8

Total numbers reported in the Habitat Use, Breeding Status, and Abundance tables may differ from each other (see pp. 22–23 for details). Percentage sums may differ slightly from 100% due to rounding.

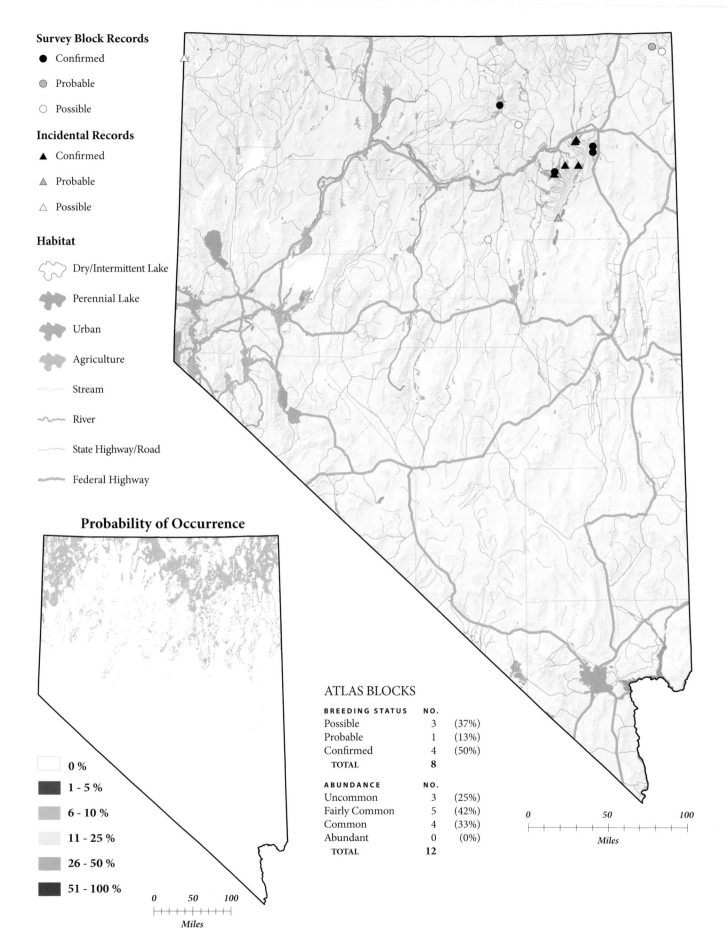

Survey Block Records

● Confirmed

● Probable

○ Possible

Incidental Records

▲ Confirmed

▲ Probable

△ Possible

Habitat

Dry/Intermittent Lake

Perennial Lake

Urban

Agriculture

Stream

River

State Highway/Road

Federal Highway

Probability of Occurrence

☐ 0 %

■ 1 - 5 %

■ 6 - 10 %

■ 11 - 25 %

■ 26 - 50 %

■ 51 - 100 %

0   50   100
Miles

ATLAS BLOCKS

BREEDING STATUS	NO.	
Possible	3	(37%)
Probable	1	(13%)
Confirmed	4	(50%)
TOTAL	**8**	

ABUNDANCE	NO.	
Uncommon	3	(25%)
Fairly Common	5	(42%)
Common	4	(33%)
Abundant	0	(0%)
TOTAL	**12**	

0   50   100
Miles

RED-WINGED BLACKBIRD

Agelaius phoeniceus

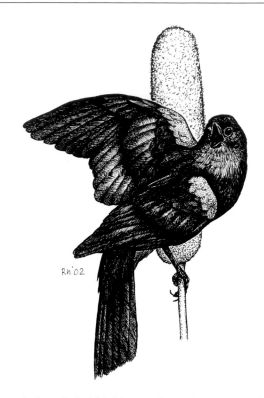

Rh'02

The Red-winged Blackbird is one of America's most widespread and abundant birds. In the summertime, whole marshes come alive with displaying males. And on a dreary afternoon in late winter, few spectacles are more uplifting than the sight—and sound—of a male Red-winged Blackbird singing loudly in a snowy cattail marsh. Although they are wetland associates, Red-winged Blackbirds are not in particular jeopardy in Nevada. They are willing to breed in very small and makeshift wetlands, and they seem to be impervious to most forms of habitat degradation. Populations are declining across North America, but in Nevada the species is one of the few that is adequately monitored and is apparently doing well.

DISTRIBUTION

Atlas workers found Red-winged Blackbirds throughout northern Nevada and more sparsely in the south. The birds were present in all of the state's major wetlands, and as the map reveals, most observations were associated with water features. The areas with few records were primarily those with a dearth of aquatic habitat, such as Esmeralda County and surrounding areas. Breeding was relatively easy to confirm, and abundance was usually high in atlas blocks, with more than half of the estimates exceeding ten pairs, often substantially. As long ago as the nineteenth century the species was recognized to be an abundant breeder in the state's marshes (Fisher 1893), and more recent assessments have reached a similar conclusion (Ryser 1985, Alcorn 1988, Chisholm and Neel 2002). Red-winged Blackbirds are likely to breed in most wetlands in neighboring states as well (Behle et al. 1985, Small 1994, Yasukawa and Searcy 1995, Stephens and Sturts 1998, Adamus et al. 2001, Clark in Corman and Wise-Gervais 2005:552–553).

Throughout its extensive North American range, the Red-winged Blackbird breeds in a variety of wetland and moist upland habitats (Yasukawa and Searcy 1995). In the West, it is essentially an obligate wetland species, albeit one that defines wetlands broadly. Agricultural ditches, borrow pits, and sewage ponds are perfectly acceptable as long as they have some emergent vegetation, as are more natural habitats ranging from tule marshes and wet pastures to mountain meadows and willow thickets (Fisher 1893, Linsdale 1936, Ryser 1985). Atlas data came predominantly from riparian and wetland sites, with agricultural habitats making up the bulk of the remaining observations. Not surprisingly, these habitats are abundant where the species is predicted to be most likely to occur. Elsewhere, the probability of occurrence appears to decline with elevation and to increase with latitude.

CONSERVATION AND MANAGEMENT

The Red-winged Blackbird is one of the most abundant birds in North America (Nero 1984, Yasukawa and Searcy 1995), and it is not a conservation concern in most areas. Its numbers increased greatly in the Midwest in the early part of the twentieth century, but populations have since declined across much of its range (Yasukawa and Searcy 1995, Sauer et al. 2005).

In Nevada, Red-winged Blackbird populations appear healthy, and monitoring data suggest that they have been increasing over the past twenty-five years (Sauer et al. 2005). They have adapted well to agricultural development, which has probably increased the area of suitable habitat in the state. They also fare moderately well in suburban areas, where the fringe vegetation of park ponds and artificial drainages often supports a few pairs. Conflicts with humans occur in agricultural areas elsewhere in the country, but this is primarily an issue where large nonbreeding flocks descend on grain fields and appears not to be a major problem in Nevada.

HABITAT USE

HABITAT	ATLAS BLOCKS	INCIDENTAL OBSERVATIONS
Agricultural	30 (13%)	1 (1%)
Grassland	5 (2%)	1 (1%)
Mojave	0 (0%)	1 (1%)
Montane Forest	0 (0%)	1 (1%)
Montane Shrub	0 (0%)	1 (1%)
Open Water	3 (1%)	2 (2%)
Riparian	100 (43%)	36 (42%)
Salt Desert Scrub	6 (3%)	1 (1%)
Sagebrush Scrub	5 (2%)	3 (4%)
Sagebrush Steppe	3 (1%)	0 (0%)
Urban	8 (3%)	4 (5%)
Wetland	72 (31%)	34 (40%)
TOTAL	232	85

Total numbers reported in the Habitat Use, Breeding Status, and Abundance tables may differ from each other (see pp. 22–23 for details). Percentage sums may differ slightly from 100% due to rounding.

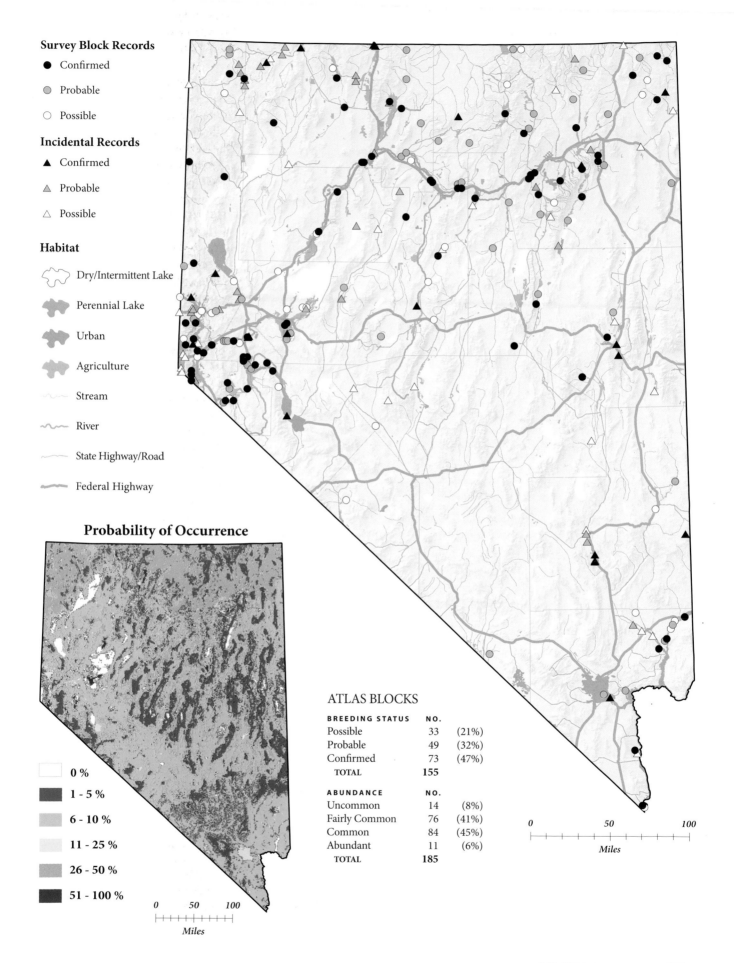

Probability of Occurrence

Survey Block Records
- ● Confirmed
- ● Probable
- ○ Possible

Incidental Records
- ▲ Confirmed
- ▲ Probable
- △ Possible

Habitat
- Dry/Intermittent Lake
- Perennial Lake
- Urban
- Agriculture
- Stream
- River
- State Highway/Road
- Federal Highway

0 %
1 - 5 %
6 - 10 %
11 - 25 %
26 - 50 %
51 - 100 %

0   50   100
Miles

ATLAS BLOCKS

BREEDING STATUS	NO.	
Possible	33	(21%)
Probable	49	(32%)
Confirmed	73	(47%)
TOTAL	**155**	

ABUNDANCE	NO.	
Uncommon	14	(8%)
Fairly Common	76	(41%)
Common	84	(45%)
Abundant	11	(6%)
TOTAL	**185**	

0   50   100
Miles

TRICOLORED BLACKBIRD
Agelaius tricolor

The Carson Valley of far western Nevada offers a taste of the Central Valley of California: sprawling agricultural fields, scattered but diminishing wetlands, and even a small colony of Tricolored Blackbirds. Nearly all of the Tricolored Blackbird's breeding populations are thought to occur in central California. In recent years, though, breeding has also been recorded in Oregon, Washington, and Nevada. The single Nevada colony of Tricolored Blackbirds is probably of little consequence to conservation of the whole species, but it represents an intriguing satellite population that occurs in a slightly different environment than the lower-elevation populations of central California.

DISTRIBUTION

Tricolored Blackbirds were found at only two sites in Nevada, neither in an atlas block. The main colony is in a small freshwater marsh in the Carson Valley, and it was occupied during spring and summer in all atlas years. The colony never seemed to exceed twenty pairs. The other location was a wetland in the Carson Range, but no convincing evidence of breeding was ever detected there.

The Tricolored Blackbird may be a recent breeder in Nevada. Linsdale (1936) did not mention it, and Alcorn (1988) considered it merely a transient in far western Nevada. Ryser (1985) mentioned Honey Lake in Lassen County, California, as the only breeding site in the Great Basin. The first known instance of breeding in Nevada was reported in 1996 (Beedy and Hamilton 1999). The species visually resembles the Red-winged Blackbird, however, and its small Carson Valley colony is located in an area where other blackbirds are abundant, so there is a chance that the colony went undetected for some time.

Almost all of the world's Tricolored Blackbirds breed in California, historically in very large colonies (Beedy and Hamilton 1999). These colonies exhibit substantial annual variation in numbers (Garrett and Dunn 1981, Small 1994), and individuals move frequently among them (Shuford 1993). The latter tendency may explain why there are now scattered breeding populations in Oregon, Washington, and Nevada, well beyond the species' core range (Beedy and Hamilton 1999, Adamus et al. 2001).

Like the Yellow-headed and Red-winged Blackbirds, the Tricolored Blackbird nests in wetlands. Freshwater marshes with dense stands of cattails (such the Carson Valley marshes where the birds nest) or bulrushes are especially favored, but brushy vegetation close to water is sometimes acceptable nesting habitat (Beedy and Hamilton 1999). Foraging individuals may also be encountered in nearby meadows and pastures (Alcorn 1988).

CONSERVATION AND MANAGEMENT

Because of their colonial habits, their need for intact wetlands, and their limited geographic distribution, Tricolored Blackbird populations face significant threats (Beedy and Hamilton 1999). In the main population center in California, populations have declined at a rate of at least 4% per year in recent decades (Sauer et al. 2005), and perhaps even faster, according to other studies summarized in Beedy and Hamilton (1999). The species has several conservation designations; for example, it is a Partners in Flight Watch List species, for which immediate conservation action is recommended (Rich et al. 2004). The main threat is the conversion of wetlands in the Central Valley (Small 1994). It is interesting that some individuals have begun to use alternate habitats such as blackberries, willows, and cash crops (Shuford 1993), and the Tricolored Blackbird may actually be expanding its range beyond California despite the decline in its numbers.

Numerically speaking, the Nevada population of Tricolored Blackbirds is trivial, but it may provide clues about the species' vagrancy and apparent range extension. Birders should be on the lookout for additional breeding sites in the Carson Valley and elsewhere in western Nevada.

HABITAT USE

HABITAT	ATLAS BLOCKS	INCIDENTAL OBSERVATIONS
Wetland	0 (0%)	4 (100%)
TOTAL	0	4

Total numbers reported in the Habitat Use, Breeding Status, and Abundance tables may differ from each other (see pp. 22–23 for details). Percentage sums may differ slightly from 100% due to rounding.

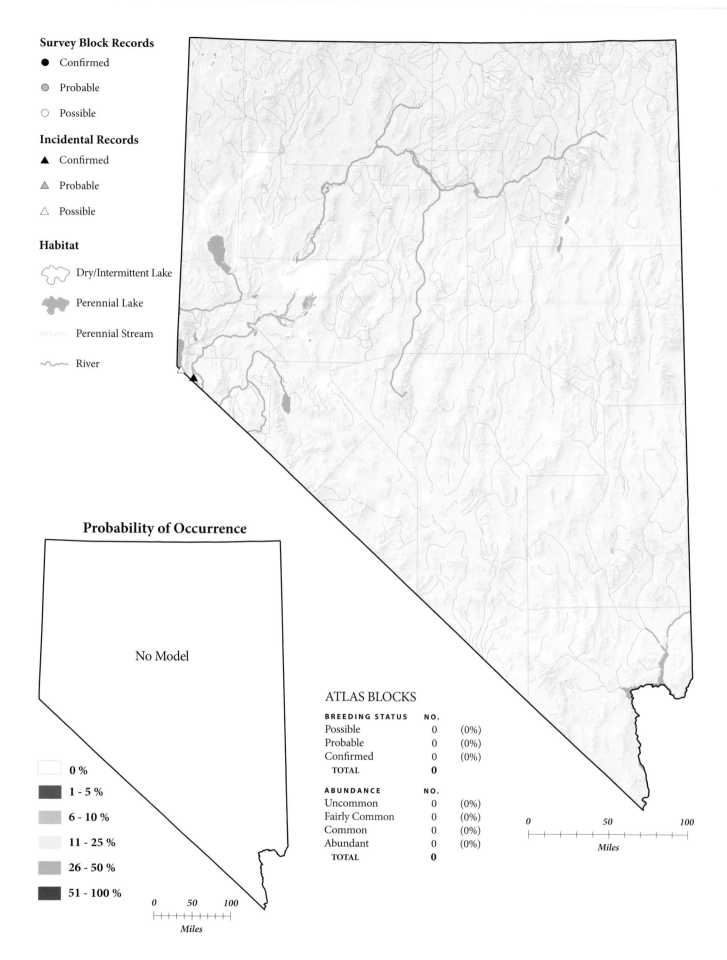

Survey Block Records

● Confirmed

● Probable

○ Possible

Incidental Records

▲ Confirmed

△ Probable

△ Possible

Habitat

Dry/Intermittent Lake

Perennial Lake

Perennial Stream

River

Probability of Occurrence

No Model

☐ 0 %

■ 1 - 5 %

■ 6 - 10 %

11 - 25 %

■ 26 - 50 %

■ 51 - 100 %

0   50   100

Miles

ATLAS BLOCKS

BREEDING STATUS	NO.	
Possible	0	(0%)
Probable	0	(0%)
Confirmed	0	(0%)
TOTAL	0	

ABUNDANCE	NO.	
Uncommon	0	(0%)
Fairly Common	0	(0%)
Common	0	(0%)
Abundant	0	(0%)
TOTAL	0	

0   50   100

Miles

WESTERN MEADOWLARK
Sturnella neglecta

Pure avian bedlam: That is the only way to describe the situation on the Western Meadowlark's breeding grounds in the sagebrush of northern Nevada. Unmated males sing from the ground, from the brush, and from the air. Mated pairs chase each other about. Males with offspring in the nest start courting new females. Birds of both sexes cluck, gurgle, and rattle incessantly. Western Meadowlarks are widespread, abundant, conspicuous, and widely known, and are among Nevada's most characteristic shrub steppe birds.

DISTRIBUTION

Breeding Western Meadowlarks were found primarily north of latitude 39°N, but there were confirmed breeders as far south as Pahrump in southern Nye County. The Western Meadowlark was one of the most frequently encountered birds during the atlas project, matching earlier accounts of the species' occurrence in the state (Linsdale 1936, Ryser 1985, Alcorn 1988), and abundance estimates exceeded ten pairs in more than half of the blocks for which estimates were provided.

The Western Meadowlark breeds throughout the western United States and adjacent portions of Canada and Mexico. Densities are at their highest in the western Great Plains, in parts of northern Nevada, and in nearby portions of Utah, Oregon, and Idaho (Lanyon 1994, Sauer et al. 2005). The species has expanded its breeding range into the Great Lakes region over the past century, but has apparently not moved eastward in the more southerly parts of its range (Lanyon 1994). Western Meadowlarks breed throughout the states surrounding Nevada, although

they are less common in the drier deserts of southern Arizona and California (Behle et al. 1985, Small 1994, Stephens and Sturts 1998, Adamus et al. 2001, Wise-Gervais in Corman and Wise-Gervais 2005:556–557).

From a rangewide perspective, Western Meadowlarks are primarily grassland breeders (Lanyon 1994). They will use pastures and alfalfa fields, although breeding densities in agricultural settings are often lower than in native grasslands (Ryser 1985, Lanyon 1994). In Nevada, where pure grasslands are uncommon, Western Meadowlarks commonly breed in shrublands that have a prominent grass component (Holmes and Barton 2003). Almost half of the atlas records came from sagebrush habitats, and the chance of finding the species in these areas is predicted to be high. Salt desert scrublands and other lower-elevation Great Basin habitats are less likely to be occupied, but there is still a good chance of finding Western Meadowlarks in these areas. In contrast, there is little chance of them breeding in more forested areas or in the southern deserts.

CONSERVATION AND MANAGEMENT

Although a common species, the Western Meadowlark is declining throughout much of its range (Sauer et al. 2005; but see Dobkin and Sauder 2004). Nevada is one of the few areas where populations appear to be stable (Sauer et al. 2005). Factors that can have negative impacts on the species include heavy grazing that reduces grass cover; conversion of native habitats to agriculture, which can reduce breeding productivity due to harvesting activities and increased predator densities; and disturbance that causes nest abandonment (Lanyon 1994, Dobkin and Sauder 2004). The Western Meadowlark's sensitivity to human disturbance is particularly noteworthy and should be considered by anyone studying its breeding biology. Nonetheless, there is a clear need for research on Western Meadowlarks, in particular to investigate the reasons why the species appears to fare better in Nevada than in some other parts of its range.

HABITAT USE

HABITAT	ATLAS BLOCKS	INCIDENTAL OBSERVATIONS
Agricultural	42 (8%)	7 (9%)
Ash	2 (<1%)	0 (0%)
Barren	2 (<1%)	0 (0%)
Grassland	51 (10%)	7 (9%)
Mesquite	1 (<1%)	0 (0%)
Mojave	3 (<1%)	2 (3%)
Montane Forest	1 (<1%)	0 (0%)
Montane Shrub	7 (1%)	1 (1%)
Pinyon-Juniper	25 (5%)	1 (1%)
Riparian	34 (7%)	10 (13%)
Salt Desert Scrub	63 (13%)	6 (8%)
Sagebrush Scrub	101 (20%)	18 (23%)
Sagebrush Steppe	122 (25%)	13 (16%)
Urban	12 (2%)	6 (8%)
Wetland	30 (6%)	8 (10%)
TOTAL	496	79

Total numbers reported in the Habitat Use, Breeding Status, and Abundance tables may differ from each other (see pp. 22–23 for details). Percentage sums may differ slightly from 100% due to rounding.

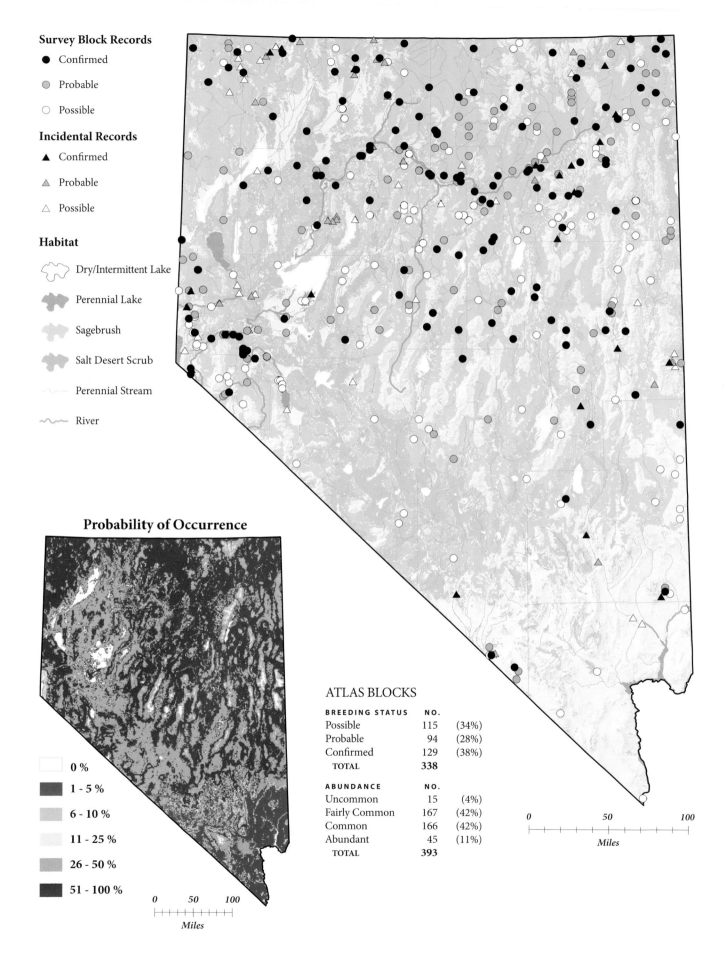

Survey Block Records

● Confirmed

● Probable

○ Possible

Incidental Records

▲ Confirmed

▲ Probable

△ Possible

Habitat

Dry/Intermittent Lake

Perennial Lake

Sagebrush

Salt Desert Scrub

Perennial Stream

River

Probability of Occurrence

☐	**0 %**
■	**1 - 5 %**
■	**6 - 10 %**
☐	**11 - 25 %**
■	**26 - 50 %**
■	**51 - 100 %**

0   50   100

Miles

ATLAS BLOCKS

BREEDING STATUS	NO.	
Possible	115	(34%)
Probable	94	(28%)
Confirmed	129	(38%)
TOTAL	**338**	

ABUNDANCE	NO.	
Uncommon	15	(4%)
Fairly Common	167	(42%)
Common	166	(42%)
Abundant	45	(11%)
TOTAL	**393**	

0   50   100

Miles

YELLOW-HEADED BLACKBIRD
Xanthocephalus xanthocephalus

The sight of a Yellow-headed Blackbird just doesn't seem to match up with the sound of a Yellow-headed Blackbird. The adult male is visually spectacular—a big, regal-looking blackbird with a gleaming yellow hood. But his vocalizations are decidedly vulgar. His loud, shrill, drawn-out shrieking and his piercing, unmusical, distressed-sounding song drown out all the other sounds of the marsh. The racket is audible even through the closed windows of passing cars with the air-conditioning and radio on.

DISTRIBUTION

Atlas workers found Yellow-headed Blackbird colonies at scattered locations throughout Nevada. The records were concentrated along major rivers and wetlands; for example, in the Lahontan Valley. Most of the breeding sites were in the northern half of the state, but breeding was confirmed as far south as the Las Vegas area. Because this species is colonial, confirmation rates and estimated abundances were relatively high. The atlas data match earlier accounts, which note the Yellow-headed Blackbird's occurrence in marshy areas throughout the entire state (Linsdale 1936, Ryser 1985, Alcorn 1988).

The greatest breeding densities of Yellow-headed Blackbirds occur in the prairie pothole country of the northern Great Plains and Canada (Twedt and Crawford 1995). Nevada is near the western edge of the species' breeding range, although localized breeding extends into California (McCaskie et al. 1988, Small 1994). It is a widespread breeder in eastern Oregon, Idaho, and Utah, but occurs more sporadically in Arizona (Behle et al. 1985, Stephens and Sturts 1998, Adamus et al. 2001, Corman in Corman and Wise-Gervais 2005:558–559).

For breeding habitat, Yellow-headed Blackbirds opt for freshwater marshes with tall emergent vegetation into which they weave their nests. They generally use areas of deeper water than do Red-winged Blackbirds, but the two species often share the same wetlands. Yellow-headed Blackbirds are usually found in valley-bottom wetlands in areas where they can forage in nearby grasslands, agricultural fields, and shrublands (Twedt and Crawford 1995). Chances of finding breeding populations in Nevada are high where suitable wetlands are concentrated and low in the rest of the state, where wetlands are few and far between.

CONSERVATION AND MANAGEMENT

Overall, Yellow-headed Blackbird populations appear to be in good health (Twedt and Crawford 1995), although declines in some of the Canadian provinces, Minnesota, and Wisconsin merit further investigation (Sauer et al. 2005). Populations have apparently also declined in California's San Joaquin Valley because of irrigation projects, wetland destruction, and habitat conversion to agriculture (Small 1994). The species is also subject to local population booms and busts. Populations fluctuate with wetland condition, and declines may be noted during extended drought periods, with recovery during periods of higher precipitation (Twedt and Crawford 1995). The Yellow-headed Blackbird's habit of foraging in agricultural fields exposes it to pesticides and persecution as a suspected pest species, but neither is currently considered a major threat. Nevertheless, because a large portion of the global population resides in the Southwest, the Yellow-headed Blackbird is ranked as a Partners in Flight Stewardship species for that region (Rich et al. 2004).

Management for the Yellow-headed Blackbird in Nevada and elsewhere should emphasize protection of wetlands. The species is also amenable to using artificial habitats, such as the agricultural drains that are found throughout the Lahontan Valley (Chisholm and Neel 2002). Thus, it may respond well to wetland restoration projects and may serve as a useful indicator of wetland recovery.

HABITAT USE

HABITAT	ATLAS BLOCKS	INCIDENTAL OBSERVATIONS
Agricultural	8 (7%)	0 (0%)
Grassland	2 (2%)	0 (0%)
Mesquite	1 (1%)	0 (0%)
Open Water	4 (4%)	5 (8%)
Pinyon-Juniper	1 (<1%)	0 (0%)
Riparian	30 (28%)	13 (21%)
Salt Desert Scrub	4 (4%)	0 (0%)
Sagebrush Scrub	4 (4%)	0 (0%)
Sagebrush Steppe	2 (2%)	1 (2%)
Urban	5 (5%)	4 (6%)
Wetland	46 (43%)	40 (63%)
TOTAL	107	63

Total numbers reported in the Habitat Use, Breeding Status, and Abundance tables may differ from each other (see pp. 22–23 for details). Percentage sums may differ slightly from 100% due to rounding.

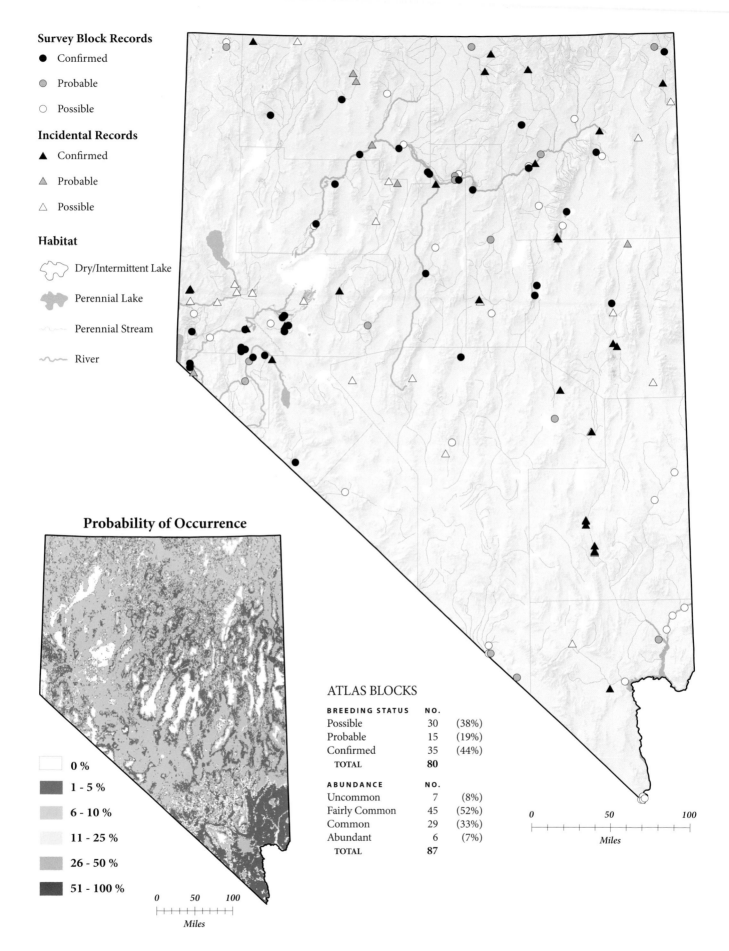

Survey Block Records

● Confirmed

● Probable

○ Possible

Incidental Records

▲ Confirmed

▲ Probable

△ Possible

Habitat

Dry/Intermittent Lake

Perennial Lake

Perennial Stream

River

Probability of Occurrence

	0 %
	1 - 5 %
	6 - 10 %
	11 - 25 %
	26 - 50 %
	51 - 100 %

0   50   100
Miles

ATLAS BLOCKS

BREEDING STATUS	NO.	
Possible	30	(38%)
Probable	15	(19%)
Confirmed	35	(44%)
TOTAL	**80**	

ABUNDANCE	NO.	
Uncommon	7	(8%)
Fairly Common	45	(52%)
Common	29	(33%)
Abundant	6	(7%)
TOTAL	**87**	

0   50   100
Miles

BREWER'S BLACKBIRD

Euphagus cyanocephalus

Rh'02

A highly adaptable species, Brewer's Blackbird seems to flourish wherever there are humans, but it also occurs in areas remote from human presence. It can be found around farmhouses and irrigation systems in rural lands, but it is just as likely to show up in parking lots and landscaped gardens in the city.

DISTRIBUTION

Atlas workers found Brewer's Blackbirds throughout Nevada, but records were concentrated in the northern two-thirds of the state. The birds are conspicuous and easily observed, and breeding was confirmed in more than half of the occupied blocks. Breeding confirmations were few in the south, however, with none in Clark County.

Older accounts describe Brewer's Blackbirds as widespread and abundant in Nevada (Linsdale 1936, Ryser 1985, Alcorn 1988), and this description also applies in all the surrounding states save Arizona, where they are absent in the south and along the lower Colorado River (Behle et al. 1985, Small 1994, Stephens and Sturts 1998, Adamus et al. 2001, Martin 2002, Corman in Corman and Wise-Gervais 2005:560–561). Numbers are especially high in California, northwestern Nevada, and southeastern Oregon (Martin 2002, Dobkin and Sauder 2004, Sauer et al. 2005).

Historically, Brewer's Blackbirds bred in a number of naturally open habitat types, frequently near water. For example, Linsdale (1936) listed streams in high valleys and mountain meadows as the species' primary breeding habitat, and Small (1994) cited wet meadows, stream margins, and rivers. With human settlement, though, Brewer's Blackbirds have been increasingly won over by human-modified habitats such as farms, roadsides, parks, cemeteries, and suburban neighborhoods. This pattern is now so pronounced that Brewer's Blackbirds are commonly thought of as preferring human-modified habitats to their original habitats (Martin 2002, Dobkin and Sauder 2004). Atlas records came from a wide range of habitat types, with riparian areas leading the way. The predictive map shows the increasing likelihood of finding this species in the northern reaches of the state, in wetland, riparian, and urban habitats.

CONSERVATION AND MANAGEMENT

Throughout the first part of the twentieth century, Brewer's Blackbird's acceptance of human-modified habitats allowed it both to spread outward beyond its historic breeding range and to increase locally within its original western range. In more recent decades, however, the species has declined in the West (Dobkin and Sauder 2004, Sauer et al. 2005). The causes of this decline are not known, although pesticides, livestock grazing, predation, and brood parasitism have been proposed as possible explanations (Martin 2002, Dobkin and Sauder 2004). A study conducted in the Toiyabe Range failed to detect any effect of grazing in riparian habitats on blackbird numbers (Warkentin and Reed 1999), but most of the other proposed explanations have not been tested. Brewer's Blackbirds also flock with species that are crop pests and can be killed along with them, even though they are unlikely to inflict much crop damage themselves (Martin 2002).

In Nevada, Brewer's Blackbird numbers are probably higher now than they were in the nineteenth century (Ryser 1985), and the species is clearly not at serious risk. Identifying the causes of population declines and determining whether they extend into Nevada are valuable goals, however, because they might indicate broader environmental changes. It will also be interesting to learn more about the southern limits of the species' breeding range in Nevada, and especially to determine whether the prediction of nesting in the Las Vegas area is upheld.

HABITAT USE

HABITAT	ATLAS BLOCKS	INCIDENTAL OBSERVATIONS
Agricultural	36 (9%)	3 (4%)
Grassland	15 (4%)	3 (4%)
Mojave	1 (<1%)	1 (1%)
Montane Forest	7 (2%)	2 (2%)
Montane Parkland	1 (<1%)	1 (1%)
Montane Shrub	17 (4%)	2 (2%)
Open Water	0 (0%)	1 (1%)
Pinyon-Juniper	13 (3%)	2 (2%)
Riparian	116 (30%)	24 (29%)
Salt Desert Scrub	8 (2%)	3 (4%)
Sagebrush Scrub	42 (11%)	10 (12%)
Sagebrush Steppe	50 (13%)	5 (6%)
Urban	37 (10%)	14 (17%)
Wetland	40 (10%)	12 (14%)
TOTAL	383	83

Total numbers reported in the Habitat Use, Breeding Status, and Abundance tables may differ from each other (see pp. 22–23 for details). Percentage sums may differ slightly from 100% due to rounding.

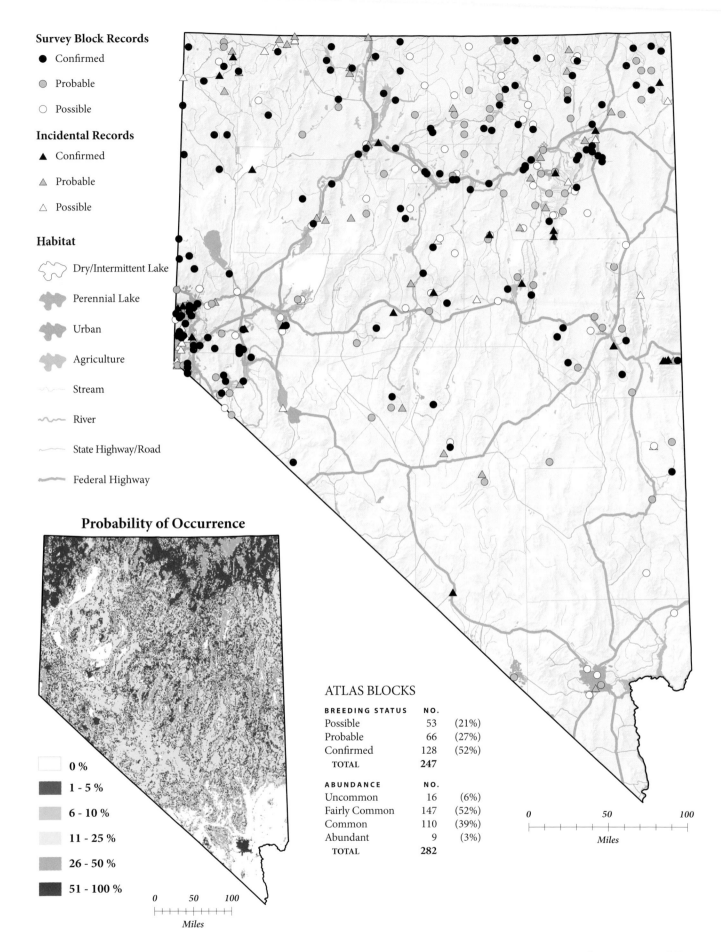

Survey Block Records

- ● Confirmed
- ● Probable
- ○ Possible

Incidental Records

- ▲ Confirmed
- ▲ Probable
- △ Possible

Habitat

- Dry/Intermittent Lake
- Perennial Lake
- Urban
- Agriculture
- Stream
- River
- State Highway/Road
- Federal Highway

Probability of Occurrence

	0 %
	1 - 5 %
	6 - 10 %
	11 - 25 %
	26 - 50 %
	51 - 100 %

0   50   100
Miles

ATLAS BLOCKS

BREEDING STATUS	NO.	
Possible	53	(21%)
Probable	66	(27%)
Confirmed	128	(52%)
TOTAL	247	

ABUNDANCE	NO.	
Uncommon	16	(6%)
Fairly Common	147	(52%)
Common	110	(39%)
Abundant	9	(3%)
TOTAL	282	

0   50   100
Miles

GREAT-TAILED GRACKLE

Quiscalus mexicanus

It is hard to believe that the Great-tailed Grackle arrived in Nevada just three decades ago, for it is now abundant in many areas. The species first appeared in the state along the Colorado River and then proceeded quickly northward all the way to Elko County. More recently, these birds have begun to establish a presence in western Nevada. Great-tailed Grackles need water nearby and readily use a variety of human-modified landscapes. As long as people continue to spread in Nevada, this spectacular and resourceful bird will likely flourish and continue to expand its range.

DISTRIBUTION

Great-tailed Grackles were first seen in southern Nevada in the early 1970s (Ryser 1985, Alcorn 1988) and began breeding shortly thereafter (Holmes et al. 1985). Their arrival was part of a larger pattern of northward expansion from Mexico into the American Southwest and southern Great Plains states that started in the late nineteenth century and continues to this day (Johnson and Peer 2001, Wehtje 2003). Arizona's first grackles arrived in 1935, and the species now occurs statewide (Wise-Gervais in Corman and Wise-Gervais 2005:562–563). The grackles reached southern California and southern Utah at approximately the same time that they reached Nevada (Johnson and Peer 2001), and the species had reached as far north as Oregon by the 1990s (Contreras 1999, Adamus et al. 2001).

Atlas records were concentrated in southern Nevada, where Great-tailed Grackles are year-round residents. Farther north, records were more sparsely distributed, but breeding was confirmed in a number of locations. The northern breeders are generally migratory, as is true for more northerly populations elsewhere (e.g., Versaw in Kingery 1998:512–513). Breeding in far western Nevada is a fairly recent phenomenon, and the species has taken hold there a bit more slowly than expected (Chisholm and Neel 2002).

Great-tailed Grackles generally nest in open habitats with scattered trees and accessible water. They have a pronounced association with human-altered landscapes (Johnson and Peer 2001), which may include orchards, open stream banks, degraded wetlands, less-than-pristine riparian zones, irrigated pastures, sewage treatment plants, lawns, golf courses, and even fountains and palm trees along the Las Vegas Strip. Most atlas habitat records were from urban, riparian, and wetland habitats, and the odds of finding grackles are predicted to be highest in areas where these habitats predominate. With the species' range continuing to spread, the predictive map may act as a useful indicator of potential new breeding sites.

CONSERVATION AND MANAGEMENT

The Great-tailed Grackle is currently undergoing a spectacular range expansion, and populations in the West are rapidly increasing (Marzluff et al. 1994, Wehtje 2003, Sauer et al. 2005). This expansion is likely due to anthropogenic changes to the landscape, such as cattle grazing, agricultural irrigation, and urbanization (Versaw in Kingery 1998:512–513, Johnson and Peer 2001). Reduced rates of nest predation in human-modified areas may also be working to the advantage of the grackle (Wehtje 2003). Few monitoring data exist for Nevada, but clearly the increase has occurred here, too, with grackles doing especially well in the major urban and riparian areas in the south.

Great-tailed Grackles are colonial and omnivorous. They are often regarded as agricultural pests, although this is unlikely to be a problem in Nevada. Concerns have also been raised about their predation on other bird species, but evidence for negative impacts of such predation on prey populations seems to be lacking or equivocal (Johnson and Peer 2001). Nevada birders should be on the lookout for new nesting locations and should ensure that any continued spread of the species is well documented. It also would be helpful to clarify whether subspecies other than the *nelsoni* form occur here.

HABITAT USE

HABITAT	ATLAS BLOCKS	INCIDENTAL OBSERVATIONS
Agricultural	2 (4%)	0 (0%)
Ash	2 (4%)	0 (0%)
Mesquite	2 (4%)	1 (2%)
Mojave	0 (0%)	2 (4%)
Montane Parkland	0 (0%)	1 (2%)
Open Water	2 (4%)	1 (2%)
Riparian	19 (38%)	13 (29%)
Salt Desert Scrub	2 (4%)	0 (0%)
Sagebrush Steppe	0 (0%)	1 (2%)
Urban	19 (38%)	18 (40%)
Wetland	2 (4%)	8 (18%)
TOTAL	50	45

Total numbers reported in the Habitat Use, Breeding Status, and Abundance tables may differ from each other (see pp. 22–23 for details). Percentage sums may differ slightly from 100% due to rounding.

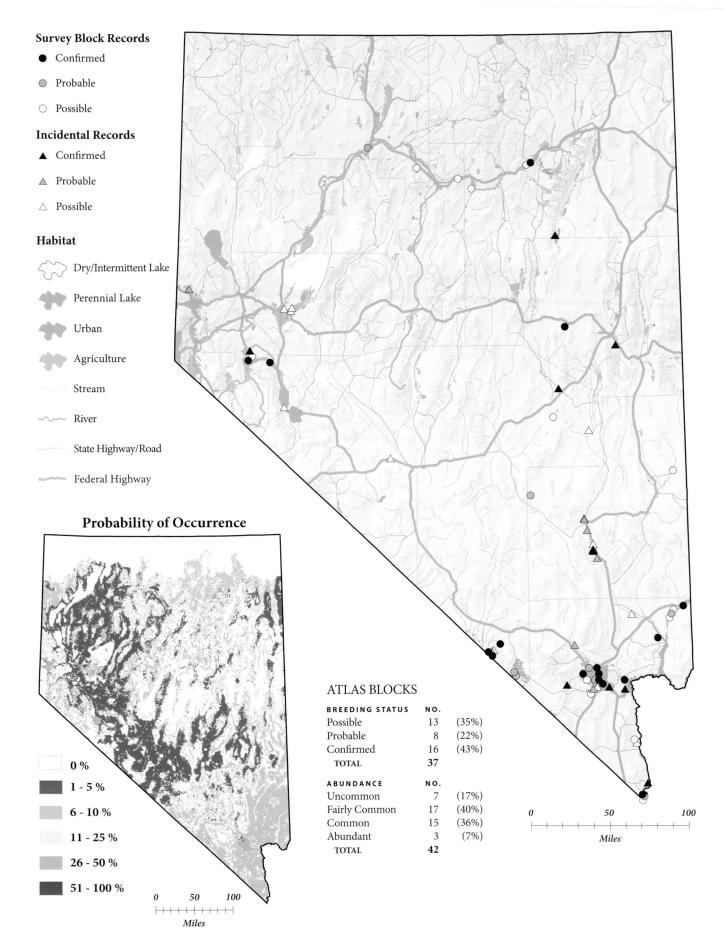

Survey Block Records

- ● Confirmed
- ● Probable
- ○ Possible

Incidental Records

- ▲ Confirmed
- ▲ Probable
- △ Possible

Habitat

- Dry/Intermittent Lake
- Perennial Lake
- Urban
- Agriculture
- Stream
- River
- State Highway/Road
- Federal Highway

Probability of Occurrence

- ☐ 0 %
- 1 - 5 %
- 6 - 10 %
- 11 - 25 %
- 26 - 50 %
- 51 - 100 %

0   50   100
Miles

ATLAS BLOCKS

BREEDING STATUS	NO.	
Possible	13	(35%)
Probable	8	(22%)
Confirmed	16	(43%)
TOTAL	**37**	

ABUNDANCE	NO.	
Uncommon	7	(17%)
Fairly Common	17	(40%)
Common	15	(36%)
Abundant	3	(7%)
TOTAL	**42**	

0   50   100
Miles

BROWN-HEADED COWBIRD

Molothrus ater

The fascinating Brown-headed Cowbird is the most widely distributed North American bird that always lays its eggs in the nests of other species. The male is a colorful little bronze-and-black bird with a lovely tinkling song and a charming courtship display. Cowbirds have expanded their range tremendously in the wake of logging, agriculture, and residential development, and the pool of host species that they parasitize has grown accordingly. Cowbird parasitism is often invoked as a cause of declines in many songbirds, but this impact is often secondary to the effects of habitat loss and degradation.

DISTRIBUTION

Brown-headed Cowbirds were recorded all over Nevada, but breeding was difficult to confirm because cowbirds do not build their own nests. Confirmed breeding records tended to come from mesic or riparian areas such as the Carson Range, Virgin River, and Humboldt River system. Breeding was not confirmed in most of the state's central region, but the survey effort was less strong in this area than elsewhere.

Originally, cowbirds resided primarily in the prairies of the central and western Great Plains (Lowther 1993). Forest clearing brought them eastward in the 1800s, and agriculture, irrigation, and ranching opened the West to them several decades later. Now, cowbirds breed almost everywhere in the continental United States (Lowther 1993) and are abundant in all of Nevada's neighboring states (Behle et al. 1985, Small 1994, Stephens and Sturts 1998, Adamus et al. 2001, Averill-Murray in Corman and Wise-Gervais 2005:566–567). The cowbird's expansion into Nevada apparently began around the early nineteenth century and was well under way by the time Linsdale published on Nevada birds (1936).

Brown-headed Cowbirds typically occur in grassland environments with trees nearby, and they especially favor forest–field ecotones (Lowther 1993). In the West, the grassland component may be provided by pastures, fields, lawns, and rangelands; and the trees are often provided by agricultural and suburban plantings and by riparian woodlands, which are typically quite rich in potential avian hosts. Suitable conditions occur throughout Nevada, and the chance of finding the species is high in the Great Basin portion of the state. Occurrence is predicted to be somewhat lower at lower elevations, especially in the Mojave Desert, and in dense montane forests and alpine areas.

CONSERVATION AND MANAGEMENT

The Brown-headed Cowbird's range has expanded so dramatically that it seems there is virtually no place left for the bird to go. In the last few decades, though, its numbers have declined significantly in many areas (Lowther 1993, Sauer et al. 2005). Nevada appears to be an exception to this general trend, with limited monitoring suggesting that the species is still increasing in the state (Sauer et al. 2005). These increases may represent, at least in part, continued spread of the western subspecies *obscurus*, which has moved into the western Great Basin from California and is now interbreeding with the *artemisiae* subspecies that first occupied the region (Fleischer and Rothstein 1988, Fleischer et al. 1991).

The most important conservation question regarding cowbirds involves their impacts on rare or imperiled host species. Proposed management strategies for host species at risk often focus on cowbird control (e.g., Laymon and Halterman 1989), but such approaches have become increasingly controversial because of concerns that control efforts might deflect attention from more fundamental problems such as habitat loss (Ortega et al. 2005). The challenge in Nevada, as elsewhere, is to determine the extent of cowbird impacts and to identify management strategies that balance different conservation objectives.

HABITAT USE

HABITAT	ATLAS BLOCKS	INCIDENTAL OBSERVATIONS
Agricultural	24 (6%)	2 (2%)
Alpine	1 (<1%)	0 (0%)
Ash	5 (1%)	0 (0%)
Grassland	12 (3%)	1 (<1%)
Mesquite	12 (3%)	3 (3%)
Mojave	5 (1%)	7 (6%)
Montane Forest	11 (3%)	5 (5%)
Montane Parkland	6 (1%)	1 (1%)
Montane Shrub	9 (2%)	7 (6%)
Open Water	0 (0%)	2 (2%)
Pinyon-Juniper	53 (13%)	8 (7%)
Riparian	113 (27%)	47 (44%)
Salt Desert Scrub	33 (8%)	2 (2%)
Sagebrush Scrub	33 (8%)	7 (6%)
Sagebrush Steppe	47 (11%)	3 (3%)
Urban	41 (10%)	7 (6%)
Wetland	10 (2%)	6 (6%)
TOTAL	415	108

Total numbers reported in the Habitat Use, Breeding Status, and Abundance tables may differ from each other (see pp. 22–23 for details). Percentage sums may differ slightly from 100% due to rounding.

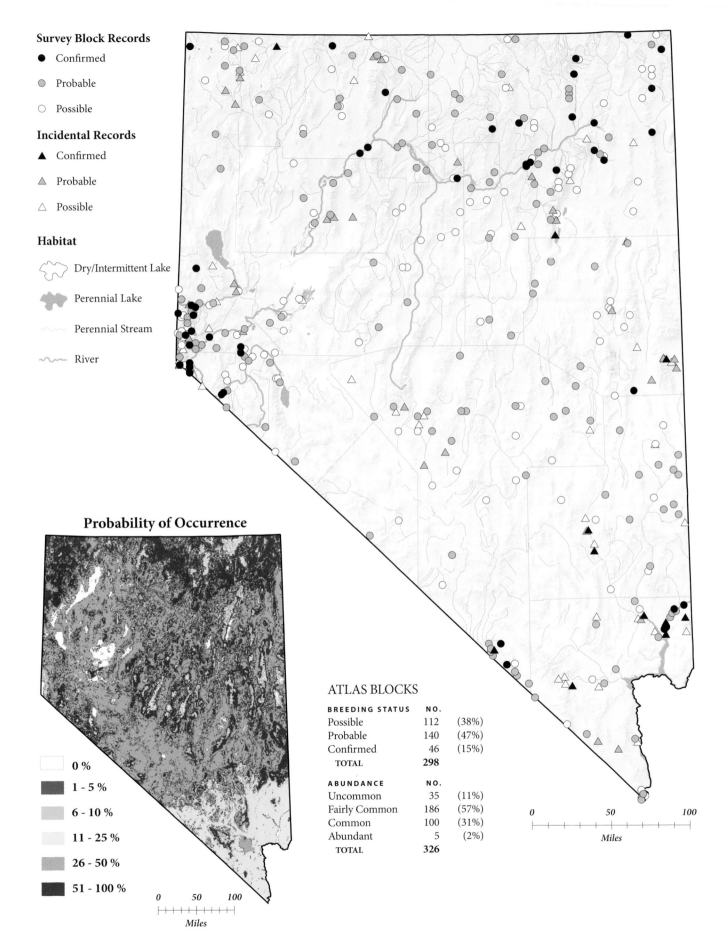

Survey Block Records

- ● Confirmed
- ● Probable
- ○ Possible

Incidental Records

- ▲ Confirmed
- ▲ Probable
- △ Possible

Habitat

Dry/Intermittent Lake

Perennial Lake

Perennial Stream

River

Probability of Occurrence

☐	0 %
■	1 - 5 %
■	6 - 10 %
☐	11 - 25 %
■	26 - 50 %
■	51 - 100 %

0   50   100

Miles

ATLAS BLOCKS

BREEDING STATUS	NO.	
Possible	112	(38%)
Probable	140	(47%)
Confirmed	46	(15%)
TOTAL	**298**	

ABUNDANCE	NO.	
Uncommon	35	(11%)
Fairly Common	186	(57%)
Common	100	(31%)
Abundant	5	(2%)
TOTAL	**326**	

0   50   100

Miles

HOODED ORIOLE

Icterus cucullatus

The brightly colored Hooded Oriole is an uncommon and widely scattered breeder in southern Nevada. Within this region its basic breeding requirement is tall, woody vegetation from which to suspend its distinctive hanging nest. Appropriate sites include wooded streams, palm oases, city trees, and riparian thickets. Although populations have declined in certain areas, Hooded Orioles, like several other members of the blackbird family, seem to be expanding their range in the Southwest in areas where suitable human-altered habitats are available.

DISTRIBUTION

Hooded Orioles were noted in four atlas blocks and at several additional sites in Clark and Lincoln counties. Confirmed breeding was observed in Las Vegas, along the Virgin River, in Pahranagat National Wildlife Refuge, and in the southeastern portion of Nellis Air Force Range. Probable breeding was also noted in the foothills of the Spring Mountains and in the Meadow Valley Wash. This pattern matches previous assessments (Alcorn 1988) that the Hooded Oriole is essentially limited to the Mojave Desert portion of the state, and is thinly distributed even there. Alcorn (1988) did, however, note two extralimital records in northern Nevada: one in Reno and another at Ruby Lake. Hooded Orioles, along with Scott's Orioles, a closely related southern species, are known for vagrant movements northward (Alcorn 1988, Flood 2002). Thus, the possibility of northern breeding outposts should not be dismissed, even though our model predicts no occurrences outside the Mojave Desert.

Outside Nevada, Hooded Orioles breed in the coastal ranges, the Central Valley, and the southern interior of California; in extreme southwestern Utah; in central and southern Arizona; and farther south and east in parts of Texas, New Mexico, and Mexico (Behle et al. 1985, Small 1994, Pleasants and Albano 2001, Corman in Corman and Wise-Gervais 2005:568–569). There are at least thirteen records from Oregon (Contreras 1999), although the Oregon atlas project did not confirm breeding there (Adamus et al. 2001).

Throughout their range Hooded Orioles nest in a variety of woodland habitats, and it is not unusual for them to occur alongside Bullock's Orioles (Pleasants and Albano 2001). Most of the atlas records came from riparian areas, and tall deciduous trees are often favored nest sites. Willows, mesquite-lined washes, and ornamental plantings are also readily used, and Hooded Orioles even nest in palm trees in cities (Pleasants and Albano 2001, Corman in Corman and Wise-Gervais 2005:568–569). Because the predictive map is based on so few occupied blocks, it probably should not be used to infer much more than a good chance of finding this species throughout the lower-elevation portions of the Mojave Desert wherever suitable trees are present.

CONSERVATION AND MANAGEMENT

Hooded Orioles appear to be faring well throughout most of their U.S. range (Sauer et al. 2005), especially along the West Coast, where the species has apparently benefited from the widespread planting of palms in urban areas (Pleasants and Albano 2001). Little information is available from Nevada, but there are no signs of any major shifts in range or abundance.

Not much is known about the conservation status of the species, and few threats have been identified. Local declines in Texas have been associated with increased numbers of Brown-headed Cowbirds, but there is little evidence for a direct link (Pleasants and Albano 2001). With accelerating development likely to bring more palm trees and other exotic plantings to southern Nevada, it seems likely that the species has a good future in the state.

HABITAT USE

HABITAT	ATLAS BLOCKS	INCIDENTAL OBSERVATIONS
Mojave	2 (50%)	1 (8%)
Montane Parkland	0 (0%)	1 (8%)
Pinyon-Juniper	0 (0%)	1 (8%)
Riparian	1 (25%)	6 (50%)
Sagebrush Scrub	0 (0%)	1 (8%)
Urban	1 (25%)	2 (17%)
TOTAL	4	12

Total numbers reported in the Habitat Use, Breeding Status, and Abundance tables may differ from each other (see pp. 22–23 for details). Percentage sums may differ slightly from 100% due to rounding.

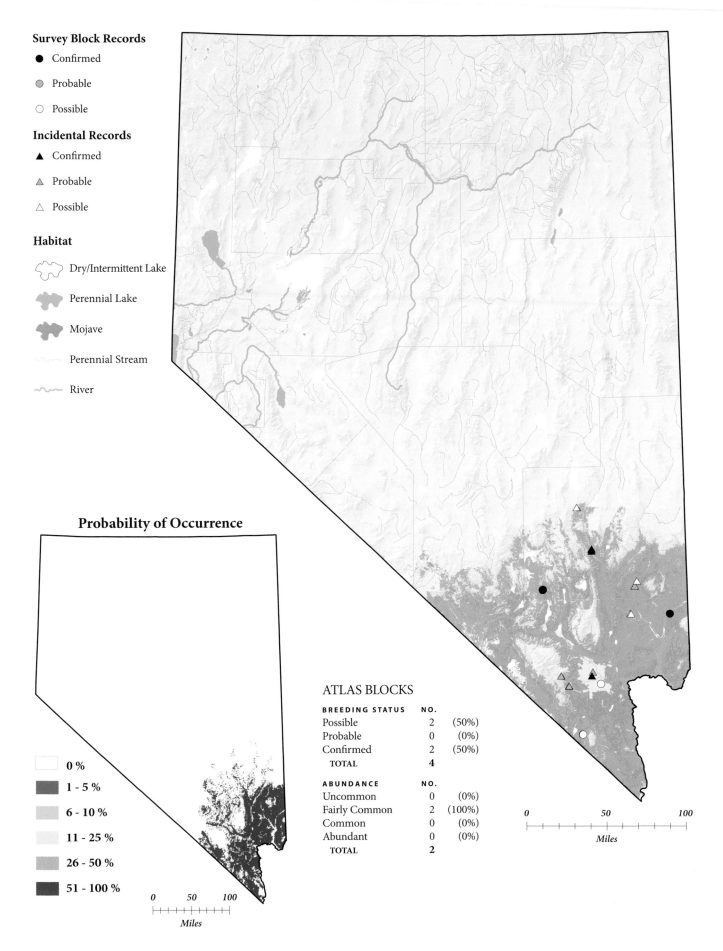

Survey Block Records

● Confirmed

◐ Probable

○ Possible

Incidental Records

▲ Confirmed

◮ Probable

△ Possible

Habitat

Dry/Intermittent Lake

Perennial Lake

Mojave

Perennial Stream

River

Probability of Occurrence

☐ 0 %

▨ 1 - 5 %

▨ 6 - 10 %

▨ 11 - 25 %

▨ 26 - 50 %

▨ 51 - 100 %

0   50   100

Miles

ATLAS BLOCKS

BREEDING STATUS	NO.	
Possible	2	(50%)
Probable	0	(0%)
Confirmed	2	(50%)
TOTAL	**4**	

ABUNDANCE	NO.	
Uncommon	0	(0%)
Fairly Common	2	(100%)
Common	0	(0%)
Abundant	0	(0%)
TOTAL	**2**	

0   50   100

Miles

BULLOCK'S ORIOLE

Icterus bullockii

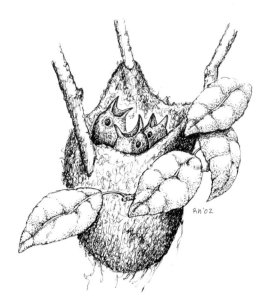

After spending more than two decades as a subspecies of the "Northern Oriole," Bullock's Oriole has once again been accorded full species status. With its bright orange–and-black plumage, the male is one of the splashier-looking inhabitants of Nevada's riparian woodlands. Even when this oriole is hidden high in a cottonwood, its chattering call notes and rich piping song render it one of the most conspicuous birds in the woodlands, as does its distinctive pendant nest.

DISTRIBUTION

Atlas workers found Bullock's Orioles to be widely distributed in Nevada. The largest concentrations of records came from the rivers and urban areas of the far western, northern, and southern parts of state. Records were notably sparse along the Colorado River, however. Comparatively few records came from the central portions of the state, and none from Esmeralda County. Breeding was confirmed in a relatively high proportion of blocks, presumably in large part because the long, pendant nests are so distinctive and visible. Earlier Nevada accounts also indicate the species' widespread occurrence in the state and its association with riparian zones in the valleys and foothills (Linsdale 1936, Ryser 1985, Alcorn 1988, Chisholm and Neel 2002).

Bullock's Oriole is a widespread species in the West. It is common in most areas immediately adjacent to Nevada but is sparse or absent as a breeder in Oregon's Cascade Range and coastal region, in northeastern Idaho, at higher elevations in the Sierra Nevada, and in far southwestern Arizona (Behle et al. 1985, Small 1994, Stephens and Sturts 1998, Rising and Williams 1999, Adamus et al. 2001, Friederici in Corman and Wise-Gervais 2005:572–573). Interestingly, Bullock's Orioles breed more commonly in Arizona's portions of the lower Colorado River than in Nevada's (Rosenberg et al. 1991).

Bullock's Orioles tend to nest in large riparian trees; these are often cottonwoods, but willows, mesquite, and others are also acceptable (Rising and Williams 1999). They readily nest in the scattered trees found in urban and agricultural settings as well, especially where water is available. The probability of occurrence is estimated to be moderately high in riparian, agricultural, and urban areas. Away from human habitats the chance of seeing this bird is relatively low, except in the lower foothills, where well-developed riparian habitat is most likely to occur. The model does predict at least some chance of finding the species in a range of habitats across the state, however, reflecting this oriole's ability to take advantage of scattered trees wherever they occur.

CONSERVATION AND MANAGEMENT

Overall, Bullock's Orioles appear to be declining in the West, although changes in California are largely responsible for this pattern; populations appear to be more stable in the other western states (Dobkin and Sauder 2004, Sauer et al. 2005). The limited data from Nevada hint at possible declines, but numbers are highly variable and are based on a small number of survey sites, making it difficult to draw firm conclusions (Sauer et al. 2005). While Bullock's Oriole remains widespread and common generally, the California declines are a concern because this state has among the highest breeding densities known (Rising and Williams 1999).

Dobkin and Sauder (2004) suggested degradation of riparian habitat to be the main threat to the species. Rosenberg et al. (1991), though, found that Bullock's Oriole has not declined substantially in the lower Colorado River Valley—unlike most riparian nesting migrants—a fact they attributed to its adaptability in the face of habitat loss and resistance to cowbird parasitism. Ammon (2002) reported a similar resilience in the face of habitat change along the lower Truckee River, and the species' use of artificial landscaping has allowed local expansions in some areas (Rising and Williams 1999).

HABITAT USE

HABITAT	ATLAS BLOCKS	INCIDENTAL OBSERVATIONS
Agricultural	11 (6%)	2 (2%)
Ash	6 (3%)	1 (1%)
Mesquite	7 (4%)	3 (3%)
Mojave	2 (1%)	3 (3%)
Montane Forest	2 (1%)	3 (3%)
Montane Shrub	3 (2%)	1 (1%)
Open Water	0 (0%)	1 (1%)
Pinyon-Juniper	9 (5%)	1 (1%)
Riparian	84 (49%)	48 (55%)
Salt Desert Scrub	5 (3%)	1 (1%)
Sagebrush Scrub	11 (6%)	1 (1%)
Sagebrush Steppe	5 (3%)	0 (0%)
Urban	28 (16%)	19 (22%)
Wetland	0 (0%)	3 (3%)
TOTAL	173	87

Total numbers reported in the Habitat Use, Breeding Status, and Abundance tables may differ from each other (see pp. 22–23 for details). Percentage sums may differ slightly from 100% due to rounding.

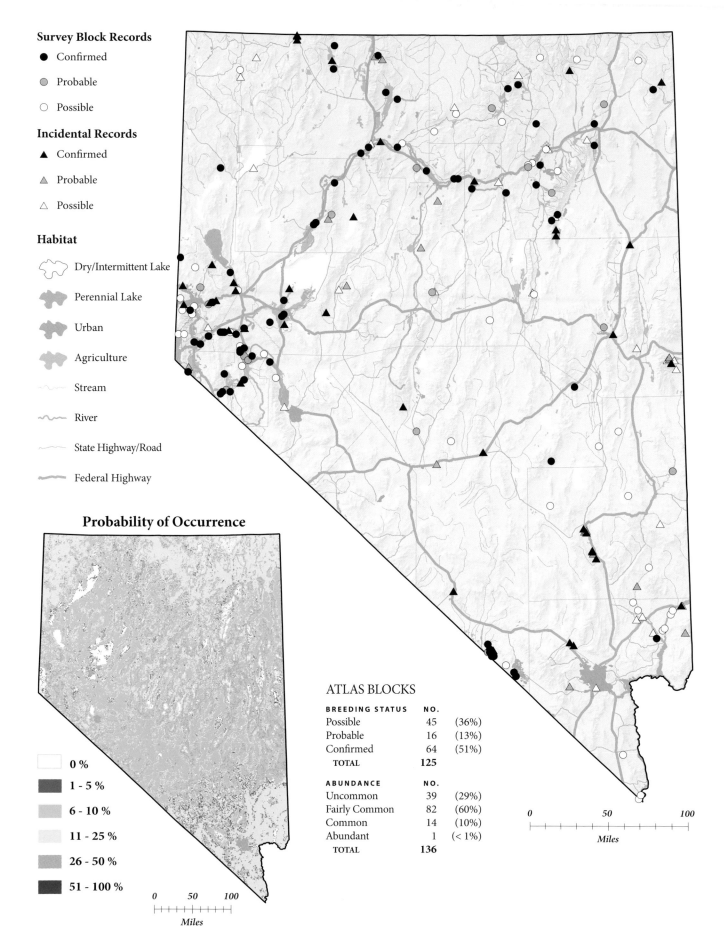

Survey Block Records

- ● Confirmed
- ● Probable
- ○ Possible

Incidental Records

- ▲ Confirmed
- ▲ Probable
- △ Possible

Habitat

- ◌ Dry/Intermittent Lake
- Perennial Lake
- Urban
- Agriculture
- Stream
- River
- State Highway/Road
- Federal Highway

Probability of Occurrence

- 0 %
- 1 - 5 %
- 6 - 10 %
- 11 - 25 %
- 26 - 50 %
- 51 - 100 %

0   50   100
Miles

ATLAS BLOCKS

BREEDING STATUS	NO.	
Possible	45	(36%)
Probable	16	(13%)
Confirmed	64	(51%)
TOTAL	**125**	

ABUNDANCE	NO.	
Uncommon	39	(29%)
Fairly Common	82	(60%)
Common	14	(10%)
Abundant	1	(< 1%)
TOTAL	**136**	

0   50   100
Miles

SCOTT'S ORIOLE
Icterus parisorum

In Nevada, Scott's Orioles occur primarily in the south, where they nest in the Joshua trees and yuccas of the Mojave Desert. Perhaps surprisingly for a species associated with the southern deserts, they are also widely distributed, though uncommon, in pinyon-juniper woodlands as far north as Elko County. Scott's Orioles were not recorded in western Nevada during the atlas years, but there are older records from the northwest, and breeding probably should not be discounted anywhere in the state.

DISTRIBUTION

Most Scott's Oriole records came from within or slightly north of the Mojave Desert. Other records were scattered through the eastern tier of Nevada counties, although confirmed breeding was limited to the south. Similar distribution patterns have been observed elsewhere in the species' southwestern U.S. range: while the bird is most common in the southern deserts, breeding outposts occur far to the north of the core range (see Dexter in Kingery 1998:520–521, Flood 2002).

Linsdale (1936), Ryser (1985), and Alcorn (1998) all regarded Scott's Oriole as primarily a southern Nevada bird, but collectively they provided definite or likely breeding records from several northern locations, including Pershing, Churchill, and Eureka counties. In the 1970s, nesting was also confirmed in the Granite Range in Washoe County and in the Toiyabe Range of central Nevada (A. Wallace, pers. comm.). Neel (1999a) speculated that Scott's Oriole may once have been more widespread in northern Nevada than it is now, and the contrast between historical and atlas records lends some support to this notion.

Scott's Orioles occur primarily in Mexico, and Nevada is at the northern limit of their range. They do not breed in Oregon (Adamus et al. 2001) and nest only rarely in far southern Idaho (Stephens and Sturts 1998). Elsewhere in the region, Scott's Orioles breed in disjunct portions of southern California, through

southern Utah, and in most of Arizona (Behle et al. 1985, Small 1994, Flood 2002, Corman in Corman and Wise-Gervais 2005:574–575).

Most atlas records came from Mojave habitat, and the chance of finding this species is clearly highest there. Within the Mojave, however, Scott's Orioles are not found just anywhere. They are strongly associated with Joshua trees, yuccas, and foothill areas where there are suitable nest trees. North of the Mojave, there is quite a large area over which the species might occur, perhaps even more extensive than the predictive map suggests. Within the Great Basin, the probability of occurrence is predicted to be greatest at middle elevations in areas dominated by sagebrush. Actual habitat reports, though, indicate a strong pinyon-juniper association, perhaps suggesting that the species occurs primarily in areas where these habitats intergrade. Unlike other orioles in Nevada, Scott's Orioles are not often found in riparian zones or in the scattered trees of agricultural and urban areas.

CONSERVATION AND MANAGEMENT

Little is known about the natural history and conservation needs of this species (Flood 2002). Because Scott's Oriole is so strongly associated with the limited Joshua tree and yucca woodlands in southern Nevada, Partners in Flight recognizes it as a Priority species in the state (Neel 1999a) and as a Stewardship species for the Southwest (Rich et al. 2004). The northern populations in Nevada, Utah, and Idaho, which are probably derived from a range expansion that occurred in the second half of the twentieth century (Flood 2002, Dobkin and Sauder 2004), may have different conservation issues than southern populations due to their different habitat use patterns. Breeding Bird Survey data suggest recent declines in California (Sauer et al. 2005) but are insufficient to determine broader-scale patterns (Dobkin and Sauder 2004). In Nevada, additional monitoring is warranted to clarify both the bird's conservation status and its distribution in the northern part of the state.

HABITAT USE

HABITAT	ATLAS BLOCKS	INCIDENTAL OBSERVATIONS
Mojave	14 (52%)	13 (45%)
Montane Forest	1 (4%)	0 (0%)
Montane Shrub	1 (4%)	1 (3%)
Pinyon-Juniper	10 (37%)	9 (31%)
Riparian	0 (0%)	1 (3%)
Salt Desert Scrub	0 (0%)	1 (3%)
Sagebrush Scrub	1 (4%)	2 (7%)
Urban	0 (0%)	2 (7%)
TOTAL	27	29

Total numbers reported in the Habitat Use, Breeding Status, and Abundance tables may differ from each other (see pp. 22–23 for details). Percentage sums may differ slightly from 100% due to rounding.

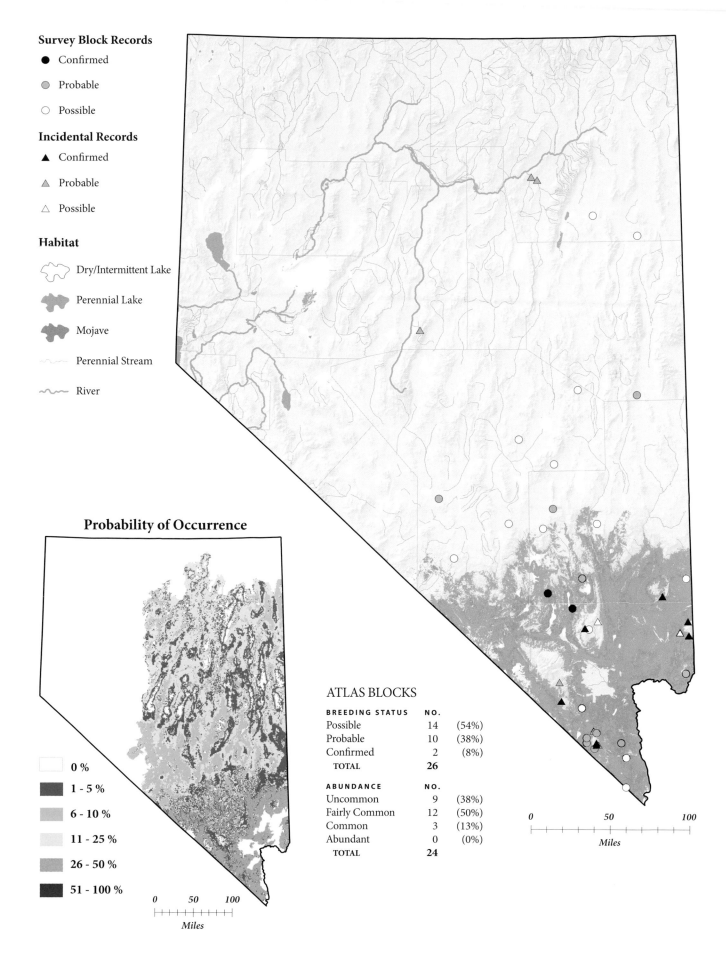

Survey Block Records

● Confirmed

● Probable

○ Possible

Incidental Records

▲ Confirmed

▲ Probable

△ Possible

Habitat

Dry/Intermittent Lake

Perennial Lake

Mojave

Perennial Stream

River

Probability of Occurrence

	0 %
	1 - 5 %
	6 - 10 %
	11 - 25 %
	26 - 50 %
	51 - 100 %

0   50   100
Miles

ATLAS BLOCKS

BREEDING STATUS	NO.	
Possible	14	(54%)
Probable	10	(38%)
Confirmed	2	(8%)
TOTAL	26	

ABUNDANCE	NO.	
Uncommon	9	(38%)
Fairly Common	12	(50%)
Common	3	(13%)
Abundant	0	(0%)
TOTAL	24	

0   50   100
Miles

BLACK ROSY-FINCH
Leucosticte atrata

The possibility of encountering a breeding Black Rosy-Finch may be a mere afterthought for birders in pursuit of the Himalayan Snowcock in Nevada's Ruby Mountains, but perhaps it shouldn't be. Rosy-finches are altitudinal migrants and are much more readily seen during the nonbreeding season than in summer. The species' breeding biology is not well known, as their alpine breeding grounds occupy a small area and are generally difficult to reach. Thomas Cirque, above Island Lake in the Rubies, is not hard to get to, though, and the sight of this small finch nesting amongst the rocks can nicely complement a search for the state's rarest game bird.

DISTRIBUTION

Atlas workers found Black Rosy-Finches in only two areas: the tundra above Island Lake in the Ruby Mountains and in similar habitat near the summit of Wheeler Peak in the Snake Mountains. The status of the species in the alpine zones of other high mountains in Nevada is not well known, but the possibility that it occurs elsewhere should not be discounted.

Black Rosy-Finches had not been recorded in Nevada at the time Linsdale (1936) wrote, but Alcorn (1988) claimed that either Black or Gray-crowned Rosy-Finches bred in most of the mountain ranges in the northern two-thirds of Nevada. He offered no confirmed breeding locations to support this statement, however, and most of the records he included were from winter or migration periods (Alcorn 1988). Ryser (1985), in contrast, mentioned only the Ruby and Snake Mountains as breeding sites in Nevada. Nesting has also been reported in the Santa Rosa Range of far northern Nevada, and birds have been observed in the Independence, Jarbidge, and East Humboldt

ranges; meanwhile, surveys in the Toiyabe, Toquima, White Pine, and Schell Creek ranges have failed to document nesting birds (Johnson 2002).

The bulk of the Black Rosy-Finch's breeding range lies north and east of Nevada. The species is a summer resident at the highest elevations in Utah, especially in the Uinta Mountains (Behle et al. 1985), and it also breeds in scattered alpine areas in Idaho, Montana, and Wyoming (Stephens and Sturts 1998, Johnson 2002). Southeastern Oregon has a single confirmed breeding location at Steens Mountain (Adamus et al. 2001), not too far from Johnson's (2002) record from Nevada's Santa Rosa Range.

Black Rosy-Finches breed only in alpine tundra atop high mountains, where they nest in crevices in cliff sides and talus fields (Neel 1999a, Johnson 2002). Seemingly suitable habitat is present in several mountain ranges of central and eastern Nevada (though difficult to see on the predictive map due to its limited extent), and further exploration of these areas might produce additional breeding locations.

CONSERVATION AND MANAGEMENT

The Black Rosy-Finch is probably the least known of the three North American rosy-finches. The global breeding range of the species is quite small, and the overall population size cannot be very large. The Gray-crowned Rosy-Finch has a much broader breeding distribution, and breeding populations of the Brown-capped Rosy-Finch are often found in areas more accessible to researchers. Data adequate to estimate Black Rosy-Finch population trends are lacking; however, winter counts—although inconclusive—suggest local declines in some regions (Johnson 2002).

Because of its small range and habitat specialization, the Black Rosy-Finch is recognized as a Partners in Flight Watch List species in the Intermountain West (Rich et al. 2004) and as a Priority species in Nevada, which probably hosts a sizable fraction of the world's population (Neel 1999a). The primary risk to breeding habitat is probably climate change, which threatens alpine habitats in the temperate zone. The species may also be vulnerable to habitat degradation on its lower-elevation wintering grounds (Johnson 2002).

HABITAT USE

HABITAT	ATLAS BLOCKS	INCIDENTAL OBSERVATIONS
Alpine	11 (92%)	1 (100%)
Barren	1 (8%)	0 (0%)
TOTAL	12	1

Total numbers reported in the Habitat Use, Breeding Status, and Abundance tables may differ from each other (see pp. 22–23 for details). Percentage sums may differ slightly from 100% due to rounding.

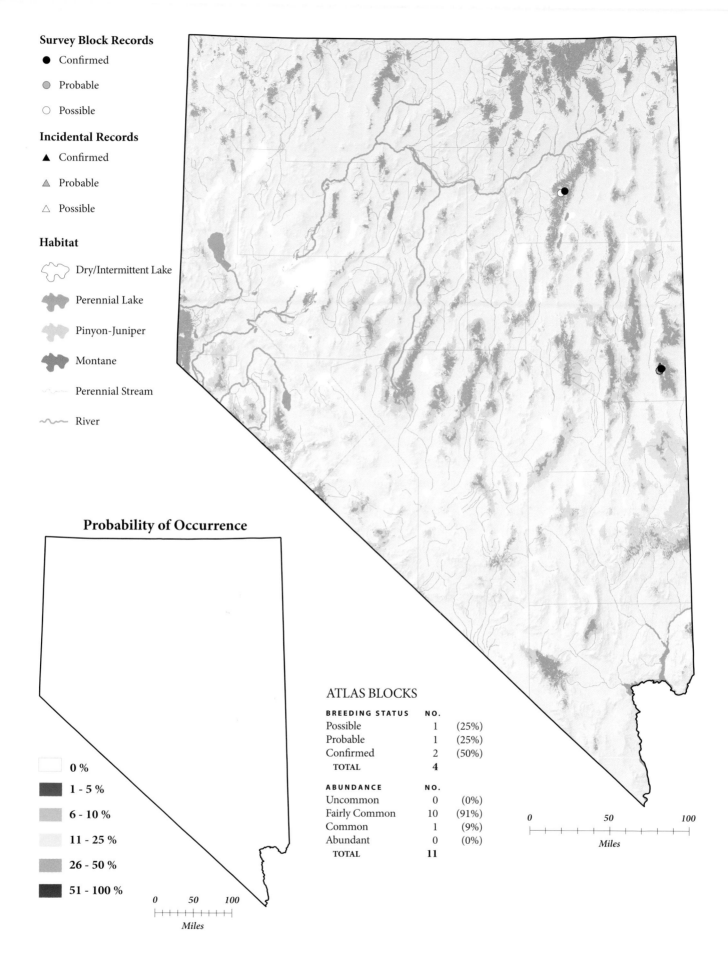

Survey Block Records

- ● Confirmed
- ◉ Probable
- ○ Possible

Incidental Records

- ▲ Confirmed
- △ Probable
- △ Possible

Habitat

Dry/Intermittent Lake

Perennial Lake

Pinyon-Juniper

Montane

Perennial Stream

River

Probability of Occurrence

- ☐ 0 %
- 1 - 5 %
- 6 - 10 %
- 11 - 25 %
- 26 - 50 %
- 51 - 100 %

0   50   100
Miles

ATLAS BLOCKS

BREEDING STATUS	NO.	
Possible	1	(25%)
Probable	1	(25%)
Confirmed	2	(50%)
TOTAL	4	

ABUNDANCE	NO.	
Uncommon	0	(0%)
Fairly Common	10	(91%)
Common	1	(9%)
Abundant	0	(0%)
TOTAL	11	

0   50   100
Miles

PINE GROSBEAK
Pinicola enucleator

Of all the reasons to visit the Tahoe Meadows above Lake Tahoe, the prospect of finding a Pine Grosbeak is among the best. This oversized finch with its wild warbling song is typically quite tame and easily approached, and males sing on conspicuous perches atop medium-sized conifers. Although the Tahoe Meadows is the most accessible site in Nevada for viewing the Pine Grosbeak, the species probably is present throughout the Carson Range wherever there are high-elevation clearings or meadows. Every once in a while there are sightings in other parts of the state, but breeding has never been confirmed away from the Carson Range.

DISTRIBUTION

Pine Grosbeaks were found in five atlas blocks and at two other locations, all within the Carson Range of far western Nevada. Breeding was confirmed at only two of these sites. Pine Grosbeaks were easy to observe but never numerous; fewer than ten pairs were estimated to occur in all of the occupied blocks. The earlier literature provides relatively little information on Nevada breeding sites. Ryser (1985) and Alcorn (1988) recounted scattered records across northern Nevada but provided no evidence of breeding at any of those locations. Both authors regarded Pine Grosbeaks as uncommon within the Great Basin and noted that they were concentrated in the Carson Range.

The Pine Grosbeak's overall breeding range is rather extensive, but nearly all of it lies well to our north in the taiga forests of Canada (as well as in northern Asia and Europe). In the western United States, Pine Grosbeak populations are scattered and disjunct. Probably the largest contiguous breeding grounds in our neighboring states lie along the Idaho-Montana border (Stephens and Sturts 1998, Adkisson 1999). Utah's montane forests also have relatively extensive breeding areas (Behle et al. 1985). There are suspected or confirmed breeding populations in Oregon, California, and Arizona, but these are quite local, isolated, and of limited extent (Small 1994, Adkisson 1999, Adamus et al. 2001, Corman in Corman and Wise-Gervais 2005:600).

Pine Grosbeaks are birds of open coniferous forests. They are typically found near forest clearings, often in close proximity to water (Adkisson 1999). The population on the eastern slope of the Sierra Nevada occurs mostly at elevations between 8,000 and 10,000 feet (ca. 2,400–3,000 meters; Small 1994). All descriptions of Pine Grosbeak habitat for the Sierra Nevada population emphasize the importance of meadows, streams, and lakes (Ryser 1985, Gaines 1992, Small 1994).

CONSERVATION AND MANAGEMENT

Adkisson (1999) summarized the little that has been written about Pine Grosbeak conservation. There is no solid information about the species' population trends and conservation status in North America. Limited and largely anecdotal information, such as the reduced frequency of southward irruptions, suggests that populations may have declined somewhat over the past several decades. The clearest potential threat to the species is deforestation.

The Pine Grosbeaks that breed in western Nevada belong to the subspecies *californicus* (see Gaines 1992, Adkisson 1999). This population is virtually sedentary and is not prone to irruptions, but even so, its numbers can vary greatly from year to year (Gaines 1992). It seems safe to say that nothing is known about the Pine Grosbeak's conservation status in Nevada at the present time, and Ryser's (1985) call for further research still needs to be heeded. It would be especially interesting to determine whether breeding occurs in any of the Great Basin ranges for which historic records exist.

HABITAT USE

HABITAT	ATLAS BLOCKS	INCIDENTAL OBSERVATIONS
Montane Forest	7 (100%)	1 (100%)
TOTAL	7	1

Total numbers reported in the Habitat Use, Breeding Status, and Abundance tables may differ from each other (see pp. 22–23 for details). Percentage sums may differ slightly from 100% due to rounding.

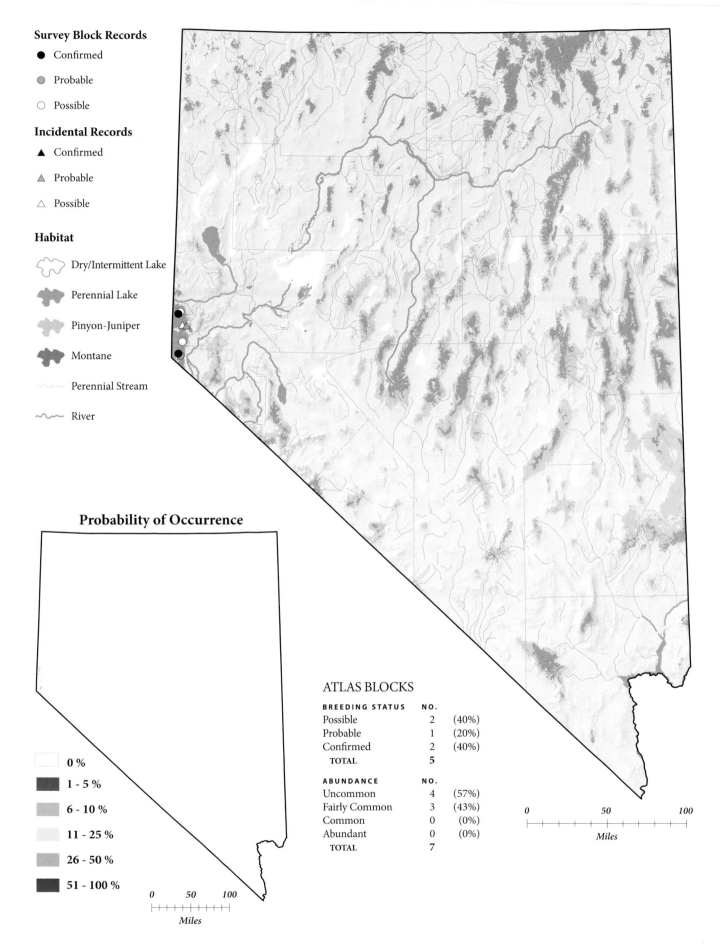

Survey Block Records

● Confirmed

◉ Probable

○ Possible

Incidental Records

▲ Confirmed

△ Probable

△ Possible

Habitat

Dry/Intermittent Lake

Perennial Lake

Pinyon-Juniper

Montane

Perennial Stream

River

Probability of Occurrence

	0 %
	1 - 5 %
	6 - 10 %
	11 - 25 %
	26 - 50 %
	51 - 100 %

0   50   100
Miles

ATLAS BLOCKS

BREEDING STATUS	NO.	
Possible	2	(40%)
Probable	1	(20%)
Confirmed	2	(40%)
TOTAL	**5**	

ABUNDANCE	NO.	
Uncommon	4	(57%)
Fairly Common	3	(43%)
Common	0	(0%)
Abundant	0	(0%)
TOTAL	**7**	

0   50   100
Miles

CASSIN'S FINCH
Carpodacus cassinii

The thought of avian mimics usually brings to mind birds like the Northern Mockingbird and European Starling. Most Nevada birders are aware that Steller's Jays and American Crows are also among the more talented mimics. But the capabilities of Cassin's Finch in that regard—which are impressive—are less well known. The next time you are in the montane forests of Nevada and hear a complex song that sounds just a little too "busy," chances are you have stumbled on a Cassin's Finch.

DISTRIBUTION

After the closely related House Finch, Cassin's Finch is the most widely distributed member of its family in Nevada and the finch most likely to be found in montane forests. Atlas workers found Cassin's Finches in most of the state's higher mountains, where observations were noticeably clustered and estimated densities were relatively high. There were also scattered observations, including confirmed breeders, in the foothills regions and several lower ranges, especially in northwestern and west-central Nevada. The atlas findings correspond closely to earlier accounts (Linsdale 1936, Ryser 1985, Alcorn 1988) that describe Cassin's Finch as a widespread breeder in nearly all of Nevada's higher mountain ranges. The predictive map reveals the same pattern, showing a very good chance of finding this species in higher-elevation forests, reduced chances in the foothills, and—at best—a low chance elsewhere in the state.

Cassin's Finch is mainly a bird of the Intermountain West; its breeding range just reaches into southern Canada and northern Baja California (Hahn 1996). It breeds throughout the mountains of Idaho (Stephens and Sturts 1998), Utah (Behle et al. 1985), and eastern Oregon (Adamus et al. 2001); in the major in-

terior mountains of northern California (Small 1994); and into far northern Arizona (Corman in Corman and Wise-Gervais 2005:576–577). Breeding also occurs locally in southern California, but the species is absent from the coastal areas of the state (Small 1994, Hahn 1996).

Cassin's Finches usually nest in coniferous forests, but they are not associated with particular conifer species (Hahn 1996). In the Great Basin, they sometimes include aspen stands in their repertoire of habitats (Behle et al. 1985, Ryser 1985), and there are even reports of nesting in sagebrush (Sullivan et al. 1986). Atlas workers generally found Cassin's Finches in montane forest, but a good number of records came from pinyon-juniper and riparian habitats. Cassin's Finches apparently exhibit little site fidelity, and breeding concentrations may shift geographically from year to year (Ryser 1985, Hahn 1996).

CONSERVATION AND MANAGEMENT

Although Cassin's Finches are widespread and abundant, monitoring data reveal significant long-term declines throughout most of the range (Sauer et al. 2005). There is little information on specific threats, but forestry practices probably play a role. Good data on recent population trends in Nevada are lacking (Sauer et al. 2005), and this is a species that will benefit greatly from increased statewide monitoring. The species' abundance and broad distribution in the state suggest that it is likely to be secure in the near future, but declines reported for the larger region warrant attention in Nevada as well. It is easy to make the mistake of overlooking declines in common species until they have proceeded so far that recovery becomes difficult. For this reason, Cassin's Finch is a Partners in Flight Stewardship species for the Intermountain West. Investigating the causes of population declines in this species and how they relate to other birds in Nevada is an important area for future research.

HABITAT USE

HABITAT	ATLAS BLOCKS	INCIDENTAL OBSERVATIONS
Agricultural	1 (<1%)	0 (0%)
Alpine	7 (3%)	0 (0%)
Montane Forest	108 (50%)	30 (39%)
Montane Parkland	12 (6%)	6 (8%)
Montane Shrub	17 (8%)	5 (6%)
Open Water	0 (0%)	1 (1%)
Pinyon-Juniper	37 (17%)	7 (9%)
Riparian	29 (13%)	25 (32%)
Sagebrush Scrub	2 (<1%)	1 (1%)
Sagebrush Steppe	4 (2%)	0 (0%)
Urban	1 (<1%)	3 (4%)
TOTAL	218	78

Total numbers reported in the Habitat Use, Breeding Status, and Abundance tables may differ from each other (see pp. 22–23 for details). Percentage sums may differ slightly from 100% due to rounding.

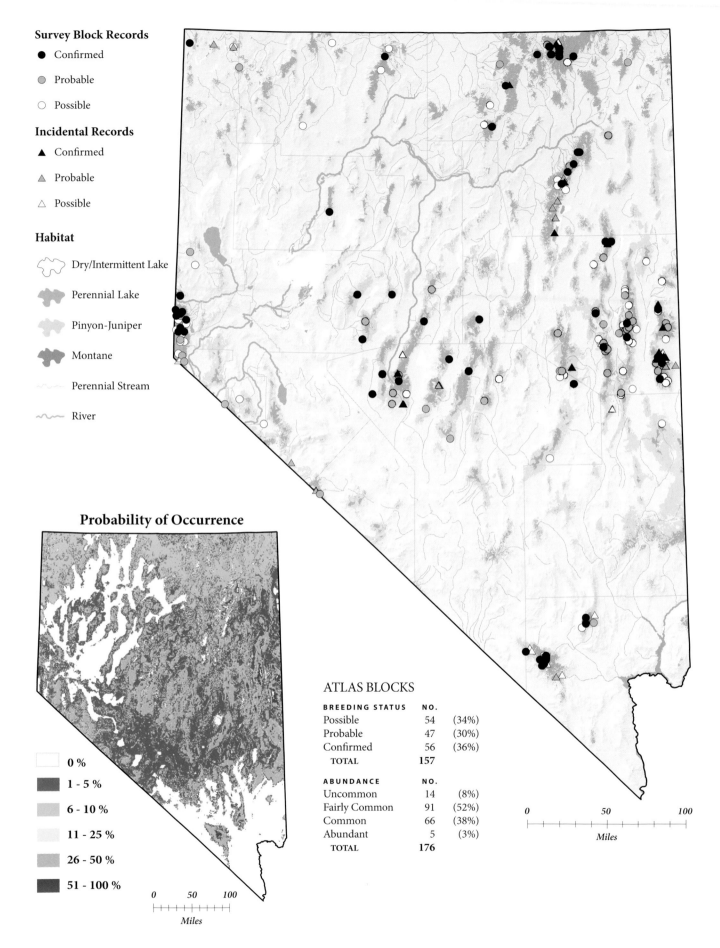

Survey Block Records

- ● Confirmed
- ● Probable
- ○ Possible

Incidental Records

- ▲ Confirmed
- ▲ Probable
- △ Possible

Habitat

- Dry/Intermittent Lake
- Perennial Lake
- Pinyon-Juniper
- Montane
- Perennial Stream
- River

Probability of Occurrence

- 0 %
- 1 - 5 %
- 6 - 10 %
- 11 - 25 %
- 26 - 50 %
- 51 - 100 %

0   50   100
Miles

ATLAS BLOCKS

BREEDING STATUS	NO.	
Possible	54	(34%)
Probable	47	(30%)
Confirmed	56	(36%)
TOTAL	157	

ABUNDANCE	NO.	
Uncommon	14	(8%)
Fairly Common	91	(52%)
Common	66	(38%)
Abundant	5	(3%)
TOTAL	176	

0   50   100
Miles

HOUSE FINCH
Carpodacus mexicanus

The pronounced plumage variation in males, complex behavioral repertoire, and bright cheery song are but a few of the reasons why the House Finch is so interesting and appealing. The species is highly adaptable, and it is one of the most common birds around human habitations as diverse as lonely ranch houses, highway rest stops, and the Las Vegas Strip. In the West, House Finches can also be found far from human abodes—in riparian groves, Joshua tree woodlands, and even in caves and cliff sides. Nesting House Finches have one basic requirement, though: proximity to water.

DISTRIBUTION

Atlas workers confirmed House Finches as breeders in all of Nevada's counties—a feat accomplished for very few species. There were dense clusters of atlas records in the Reno and Las Vegas areas, with many other sightings widely scattered throughout the state. The distribution and abundance of nonurban populations seem to have changed little over the past century in Nevada; earlier accounts all characterize House Finches as widespread and common in the state (Linsdale 1936, Ryser 1985, Alcorn 1988). House Finches also breed widely in all the states surrounding Nevada (Behle et al. 1985, Hill 1993, Small 1994, Adamus et al. 2001, Wise-Gervais in Corman and Wise-Gervais 2005:578–579) except Idaho, where breeding is mostly confined to the state's southern regions (Stephens and Sturts 1998).

It sometimes seems that House Finches can occur virtually anywhere. Gaines (1992) noted that their nests can even be found in tufa groves at Mono Lake. It is more accurate to point out that House Finches are generally not found in areas where

water is completely unavailable, in extensive and dense forests, at very high altitudes, or in places that completely lack elevated nesting sites (Hill 1993). But they can be found virtually everywhere else, and human activity, including bird-feeding, has greatly expanded the availability of suitable habitat. From a broader-scale point of view, the predictive map also indicates that House Finches are especially likely breeders in southern Nevada.

CONSERVATION AND MANAGEMENT

The House Finch's dramatic range expansion in the eastern United States stems from the release of captive birds in New York and perhaps elsewhere. These eastern populations are almost exclusively associated with human habitations. The native western populations have also undergone local expansions into human-modified areas and now thrive in major urban centers and residential backyards (Hill 1993). At present, the expanding eastern populations appear poised to merge into the species' original breeding range in the West, which extends into the western Great Plains. Interestingly, while eastern populations are rapidly expanding westward, western populations are also expanding northward and eastward, albeit at a slower rate. Trend data indicate substantial population increases in the eastern and central United States, especially the Great Plains states that are currently being invaded from the east (Sauer et al. 2005).

Population trends in the West are more mixed (Sauer et al. 2005). Some states, such as Montana and Washington, show increases. California, where population densities are especially high, has a decreasing trend (Hill 1993, Sauer et al. 2005). Breeding Bird Survey data hint at increases in Nevada, but these data are not conclusive. In any case, House Finches are very common here, and they certainly seem to be secure for the time being. It would be interesting to compare population trends and demographic patterns among House Finches using primarily native habitats and those that depend heavily on human-modified habitat.

HABITAT USE

HABITAT	ATLAS BLOCKS	INCIDENTAL OBSERVATIONS
Agricultural	9 (3%)	3 (3%)
Mesquite	9 (3%)	1 (1%)
Mojave	30 (11%)	16 (16%)
Montane Forest	5 (2%)	0 (0%)
Montane Parkland	1 (<1%)	1 (1%)
Montane Shrub	5 (2%)	3 (3%)
Pinyon-Juniper	50 (18%)	8 (8%)
Riparian	42 (15%)	31 (31%)
Salt Desert Scrub	11 (4%)	0 (0%)
Sagebrush Scrub	28 (10%)	4 (4%)
Sagebrush Steppe	12 (4%)	0 (0%)
Urban	71 (26%)	33 (33%)
Wetland	4 (1%)	1 (1%)
TOTAL	277	101

Total numbers reported in the Habitat Use, Breeding Status, and Abundance tables may differ from each other (see pp. 22–23 for details). Percentage sums may differ slightly from 100% due to rounding.

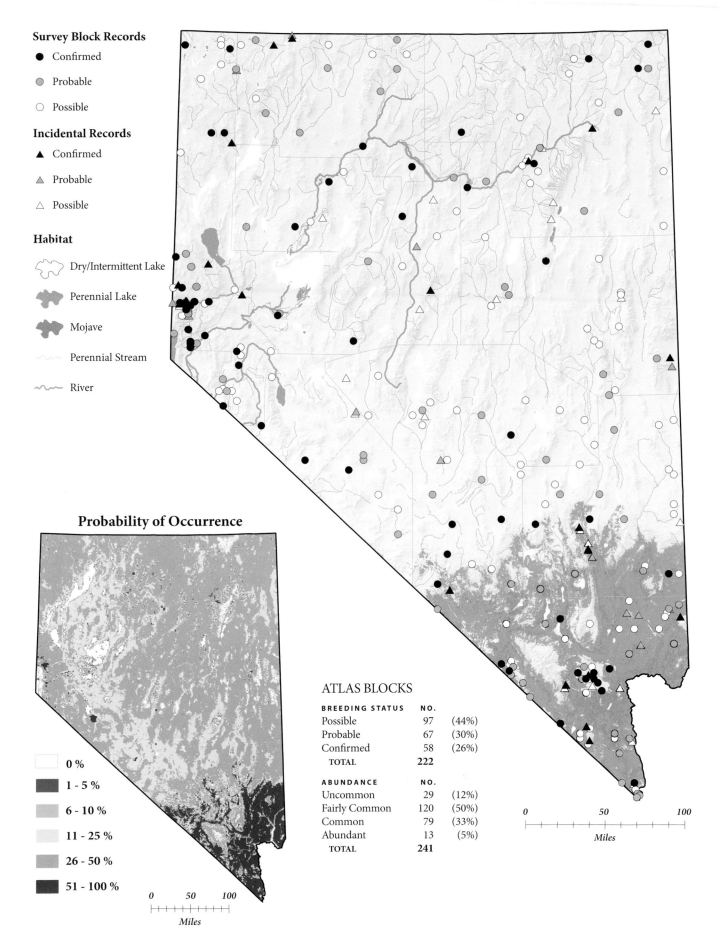

Survey Block Records
- ● Confirmed
- ● Probable
- ○ Possible

Incidental Records
- ▲ Confirmed
- ▲ Probable
- △ Possible

Habitat
- Dry/Intermittent Lake
- Perennial Lake
- Mojave
- Perennial Stream
- River

Probability of Occurrence

- 0 %
- 1 - 5 %
- 6 - 10 %
- 11 - 25 %
- 26 - 50 %
- 51 - 100 %

0   50   100
Miles

ATLAS BLOCKS

BREEDING STATUS	NO.	
Possible	97	(44%)
Probable	67	(30%)
Confirmed	58	(26%)
TOTAL	**222**	

ABUNDANCE	NO.	
Uncommon	29	(12%)
Fairly Common	120	(50%)
Common	79	(33%)
Abundant	13	(5%)
TOTAL	**241**	

0   50   100
Miles

RED CROSSBILL

Loxia curvirostra

Few Nevada birds are more erratic or unpredictable than the Red Crossbill. Flocks travel widely, and during some years populations may explode, or "irrupt," and then later return to more typical numbers. This propensity makes it difficult to know whether a particular mountain range will harbor dozens of breeding Red Crossbills—or none at all. Furthermore, Red Crossbills can breed almost anytime of the year except during late fall. This behavioral flexibility is related to the variable availability of conifer seeds, the Red Crossbill's primary food. And as if the situation were not complex enough, it turns out that Red Crossbills in Nevada may represent several subspecies with different vocalizations that reinforce their reproductive isolation (see Adkisson 1996).

DISTRIBUTION

Atlas workers confirmed Red Crossbills as breeders in the Desatoya Mountains, Sheep Range, Snake Range, and Schell Creek Range. There were clusters of unconfirmed breeding records from other areas, including the Jarbidge, Ruby, and Spring mountains and the Carson Range. The earlier literature confirms that at one time or another, Red Crossbills have been seen in most of Nevada's mountains, and occasionally in the valleys, but offers limited specific breeding records (Linsdale 1936, Ryser 1985, Alcorn 1988). It is probably safe to assume that whenever and wherever conditions are right (i.e., an abundant cone crop), there is a good chance that this nomadic species will establish a breeding presence. The same opportunistic behavior characterizes Red Crossbill breeding activity in all of the states surrounding Nevada. Some parts of the region are deficient in potential breeding habitat, however, including the lower Colorado River

region, southeastern Oregon, and the environs of Idaho's Snake River (Behle et al. 1985, Small 1994, Stephens and Sturts 1998, Adamus et al. 2001, Corman in Corman and Wise-Gervais 2005:580–581).

Throughout its extensive range, the Red Crossbill's distribution is very closely tied to the seed production of coniferous trees (Benkman 1993a, 1993b, Adkisson 1996). This phenomenon has also been noted in Nevada (Ryser 1985), and the predictive map generated from atlas data corresponds closely to the distribution of cone-bearing forests and woodlands. But in keeping with the spirit that little about the Red Crossbill is predictable, Gaines (1992) listed a variety of broadleaved trees in which Red Crossbills may also be seen, probably primarily during irruptive years, when bird numbers outpace the capacity of the species' main habitats to support them (Adkisson 1996).

CONSERVATION AND MANAGEMENT

The Red Crossbill's year-round breeding, wandering, and periodic irruptive cycles (Ryser 1985, Adkisson 1996) render atlas methods insufficient for determining the full extent of breeding in a particular region (Shuford 1993). These factors also make it difficult to effectively monitor populations. With this caveat in mind, it still appears that Red Crossbills have declined somewhat in the western United States (Sauer et al. 2005). Extensive logging and decline of older-aged forest stands, which produce the largest seed crops, are likely the most important threats to a species as specialized as the Red Crossbill (Benkman 1993a, Adkisson 1996), but detailed information is not available.

Atlas data almost certainly provide an incomplete picture of the Red Crossbill's status in Nevada and elsewhere. At the right time, Red Crossbills could nest just about anywhere in our mountains, sometimes even in the valleys. Future observers may be able to expand the list of the Red Crossbill's confirmed breeding sites, and continued monitoring may provide usable estimates of abundance and population trends. The fascinating biology of the Red Crossbill may also spawn interest by researchers in Nevada, where it is still an understudied species.

HABITAT USE

HABITAT	ATLAS BLOCKS	INCIDENTAL OBSERVATIONS
Montane Forest	13 (65%)	8 (62%)
Montane Parkland	0 (0%)	1 (8%)
Montane Shrub	1 (5%)	2 (15%)
Pinyon-Juniper	3 (15%)	2 (15%)
Riparian	2 (10%)	0 (0%)
Urban	1 (5%)	0 (0%)
TOTAL	20	13

Total numbers reported in the Habitat Use, Breeding Status, and Abundance tables may differ from each other (see pp. 22–23 for details). Percentage sums may differ slightly from 100% due to rounding.

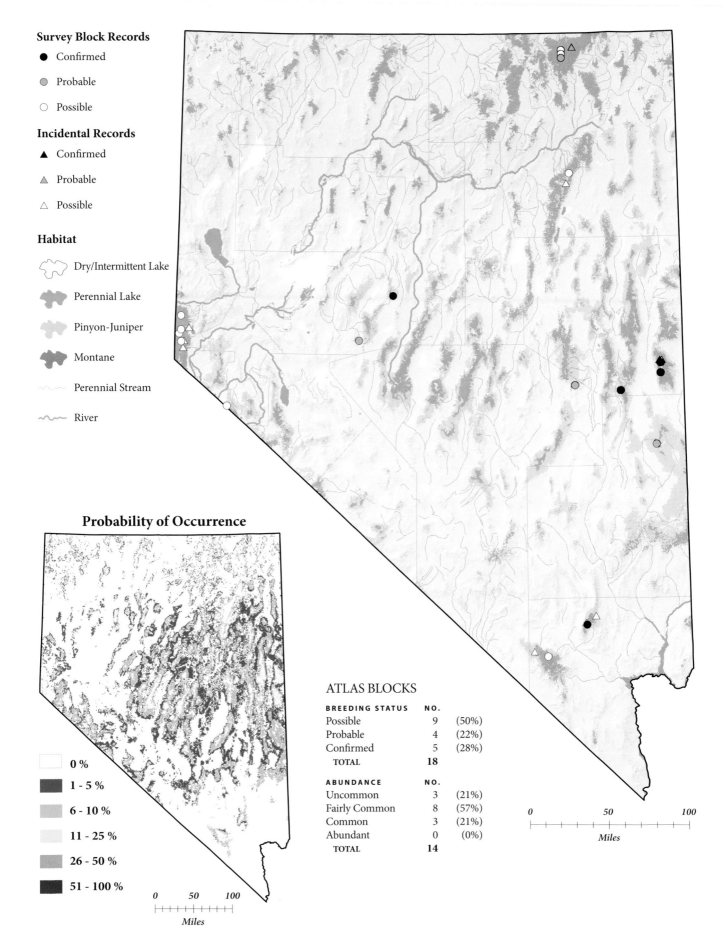

Survey Block Records

● Confirmed

● Probable

○ Possible

Incidental Records

▲ Confirmed

▲ Probable

△ Possible

Habitat

Dry/Intermittent Lake

Perennial Lake

Pinyon-Juniper

Montane

Perennial Stream

River

Probability of Occurrence

☐ 0 %

■ 1 - 5 %

■ 6 - 10 %

☐ 11 - 25 %

■ 26 - 50 %

■ 51 - 100 %

0   50   100
Miles

ATLAS BLOCKS

BREEDING STATUS	NO.	
Possible	9	(50%)
Probable	4	(22%)
Confirmed	5	(28%)
TOTAL	18	

ABUNDANCE	NO.	
Uncommon	3	(21%)
Fairly Common	8	(57%)
Common	3	(21%)
Abundant	0	(0%)
TOTAL	14	

0   50   100
Miles

PINE SISKIN
Carduelis pinus

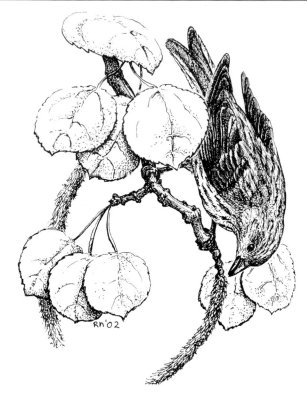

Of all our finches, the Pine Siskin is the drabbest. The male is tinged with a little bit of yellow on the wings and tail, but the female is brownish everywhere. Many of the Pine Siskin's call notes are soft, twittering, and easily overlooked. There is one call, though, that is surprisingly loud, long, and distinctive: a husky, rising, *sshreeeEE!!!* Pine Siskins are also known for outbursts of a different sort, namely, their periodic population irruptions, which drive wandering birds well beyond their usual haunts and habitats (see Bock and Lepthien 1976).

DISTRIBUTION

Atlas workers found Pine Siskins in most, but not all, of the major mountain ranges in Nevada. They were relatively scarce in the northwestern part of the state, and in the south they were limited to the Spring Mountains. Breeding was confirmed in only about one-fifth of the blocks in which it was suspected, with no confirmed breeding records south of latitude 38°30′N. The atlas findings were generally consistent with earlier accounts of the Pine Siskin's distribution in Nevada (Linsdale 1936, Alcorn 1988) but also provided records from some additional locations. As might be expected with these gregarious birds, Pine Siskins tended to be common where present. Observers estimated a single pair to be present in only one-fifth of the occupied blocks, while more than one-third of the blocks had more than ten pairs.

Pine Siskins breed throughout most of the western United States and across much of Canada. They seem to be somewhat more widespread in Oregon, Idaho, and Utah than they are in Nevada (Burleigh 1972, Behle et al. 1985, Stephens and Sturts 1998, Adamus et al. 2001). In California, they occur in most of the mountain ranges, but populations are much more scattered in the south (Small 1994). Breeding is limited to the northeastern half of Arizona (Wise-Gervais in Corman and Wise-Gervais 2005:582–583).

Throughout their breeding range, Pine Siskins are found primarily in open coniferous forests (Dawson 1997). Like Red Crossbills, their presence or absence in a given patch of ostensibly suitable habitat is influenced by the current availability of conifer seeds. Pine Siskins seem to have a broader diet than Red Crossbills, and they occur in more habitat types, including ornamental conifers, mixed coniferous-deciduous forests, and pure deciduous forest stands (Dawson 1997). There were also a few records from urban areas in Nevada, but perhaps even more interesting was the fact that riparian habitats accounted for 20% of the block records.

CONSERVATION AND MANAGEMENT

Like many finches, Pine Siskins are nomadic and difficult to survey accurately (Gaines 1992, Dawson 1997). Additionally, their pronounced annual fluctuations in abundance can obfuscate underlying trends. Breeding Bird Survey data, however, leave little doubt that the species is declining, sometimes substantially, in the western United States and in Canada. In the West, California (−10.4% per year), Oregon (−11.2% per year), and Idaho (−6% per year) show the most notable declines since 1980 (Sauer et al. 2005). Data from Nevada are insufficient to calculate trends, but likely match the broader pattern.

Because it is associated year-round with conifer forests, there is an understandable tendency to focus on the potential impacts of silviculture on the Pine Siskin. Available research, however, suggests that logging has mixed effects that depend on its scale and timing (Dawson 1997). Pine Siskins in California and Utah also readily accept exotic pine plantings of many types (Behle et al. 1985, Small 1994), which may partly mitigate logging impacts elsewhere. Clearly, additional research is needed if we are to understand the causes of apparent declines and identify management strategies that could reverse recent trends. In Nevada, the first priority is to continue to monitor the species and clarify its habitat use.

HABITAT USE

HABITAT	ATLAS BLOCKS	INCIDENTAL OBSERVATIONS
Alpine	5 (3%)	1 (3%)
Montane Forest	89 (59%)	14 (41%)
Montane Parkland	4 (3%)	3 (9%)
Montane Shrub	10 (7%)	2 (6%)
Pinyon-Juniper	7 (5%)	2 (6%)
Riparian	30 (20%)	10 (29%)
Sagebrush Steppe	1 (<1%)	0 (0%)
Urban	4 (3%)	2 (6%)
TOTAL	150	34

Total numbers reported in the Habitat Use, Breeding Status, and Abundance tables may differ from each other (see pp. 22–23 for details). Percentage sums may differ slightly from 100% due to rounding.

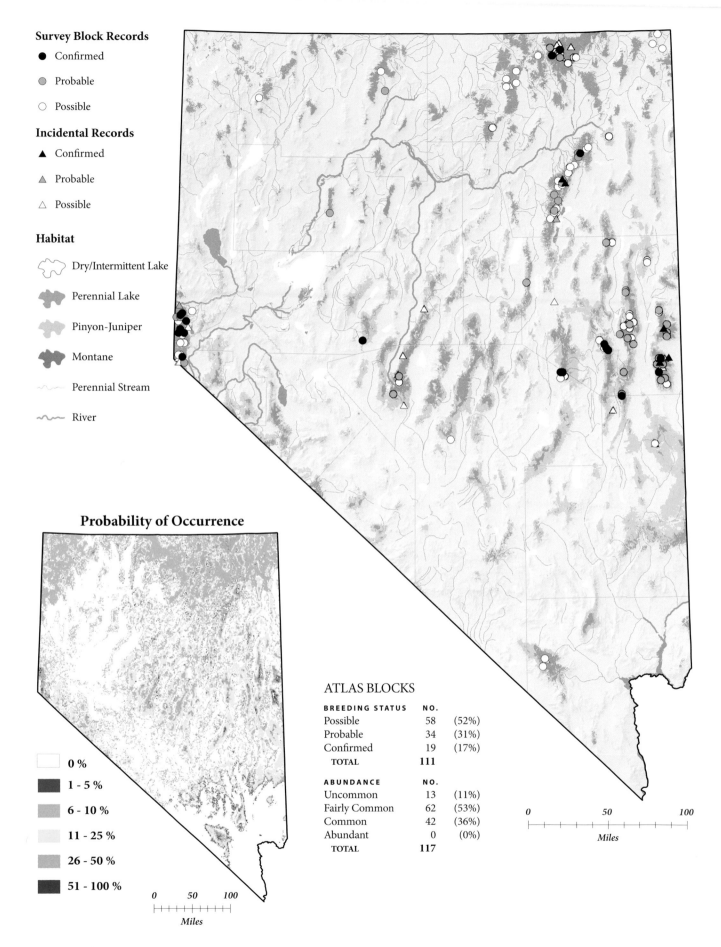

Survey Block Records

- ● Confirmed
- ● Probable
- ○ Possible

Incidental Records

- ▲ Confirmed
- ▲ Probable
- △ Possible

Habitat

- Dry/Intermittent Lake
- Perennial Lake
- Pinyon-Juniper
- Montane
- Perennial Stream
- River

Probability of Occurrence

- ☐ 0 %
- ■ 1 - 5 %
- ■ 6 - 10 %
- ☐ 11 - 25 %
- ■ 26 - 50 %
- ■ 51 - 100 %

0   50   100
Miles

ATLAS BLOCKS

BREEDING STATUS	NO.	
Possible	58	(52%)
Probable	34	(31%)
Confirmed	19	(17%)
TOTAL	**111**	

ABUNDANCE	NO.	
Uncommon	13	(11%)
Fairly Common	62	(53%)
Common	42	(36%)
Abundant	0	(0%)
TOTAL	**117**	

0   50   100
Miles

LESSER GOLDFINCH
Carduelis psaltria

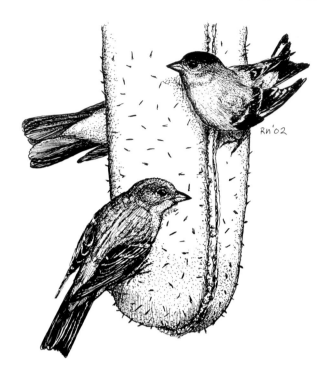

The tiny Lesser Goldfinch has become one of the most visible inhabitants of the residential neighborhoods of Reno and Carson City. Even among finches it is a midget, weighing less than 4 ounces (10 grams). This goldfinch is a feisty little sprite, though, and small flocks sometimes go into a frenzy of plaintive mews and rapid-fire chatter. Although there is no definitive information on breeding numbers in past years, many longtime observers are convinced that the species has been undergoing a recent and dramatic range expansion and population increase in Nevada.

DISTRIBUTION

Most atlas records for the Lesser Goldfinch came from western and southern Nevada, but there were widely scattered records from much of the rest of the state. Breeding was confirmed in many places in western Nevada, but also more sporadically in the east, from the Elko area down to the far south. The Lesser Goldfinch was previously considered widespread but not particularly numerous in Nevada (Linsdale 1936, Ryser 1985, Alcorn 1988). The heart of the species' distribution runs roughly from California, where abundances are particularly high, through Arizona and New Mexico to west Texas, and southward into Mexico (Small 1994, Watt and Willoughby 1999, Wise-Gervais in Corman and Wise-Gervais 2005:584–585). Breeding is also fairly widespread in Oregon and Utah, but these states, like Nevada, lie at or near the Lesser Goldfinch's northern range limit (Behle et al. 1985, Adamus et al. 2001). The species also breeds locally in far southeastern Idaho (Stephens and Sturts 1998).

Nesting Lesser Goldfinches use a wide variety of habitats, including woodlands, brushlands, ranches, and residential areas. Riparian or semiriparian woodlands are especially favored (Watt and Willoughby 1999), and atlas data confirm this to be the case in Nevada. In fact, nearby open water may be a good general indicator of the species' presence (Small 1994). Of particular interest in Nevada is the Lesser Goldfinch's association with residential settings, which accounted for more than one-third of the block records. This pattern seemed especially pronounced around Reno and Carson City, and occurred to a lesser extent in other areas, too. The predictive map indicates the likelihood of finding the species to be fairly low overall but higher in the south, and notably higher in urbanized regions and major riparian areas.

CONSERVATION AND MANAGEMENT

Deciphering the population trends of the Lesser Goldfinch is fascinating from a Nevada perspective. Breeding Bird Survey data and the account by Watt and Willoughby (1999) suggest that the species is stable or perhaps slightly declining in its traditional U.S. range (Sauer et al. 2005). Trend data for Nevada, the Sierra Nevada, the Basin and Range region, and the Great Basin, however, indicate a substantial increase in numbers (Sauer et al. 2005). These trends are not definitive due to limited survey coverage, but they do mesh with the impressions of local observers who have noted a burgeoning of Lesser Goldfinch populations in the state's population centers. Certainly, dramatic increases have been documented for wintering populations in Nevada and Utah, as reported by Versaw (2001). He suggested that winter bird-feeding plays a primary role in this phenomenon, and also noted that the Lesser Goldfinch's breeding range is largely concurrent with the winter range. Other factors that may contribute to the Lesser Goldfinch's apparent expansion in Nevada are increased availability of irrigated landscaping, exotic plantings, and certain weed species whose seeds provide food (Bradley 1980, Watt and Willoughby 1999).

HABITAT USE

HABITAT	ATLAS BLOCKS	INCIDENTAL OBSERVATIONS
Agricultural	1 (2%)	1 (3%)
Grassland	0 (0%)	1 (3%)
Mojave	1 (2%)	2 (6%)
Montane Forest	3 (6%)	5 (14%)
Montane Parkland	1 (2%)	0 (0%)
Pinyon-Juniper	7 (13%)	4 (11%)
Riparian	18 (34%)	14 (39%)
Sagebrush Scrub	3 (6%)	0 (0%)
Urban	19 (36%)	9 (25%)
TOTAL	53	36

Total numbers reported in the Habitat Use, Breeding Status, and Abundance tables may differ from each other (see pp. 22–23 for details). Percentage sums may differ slightly from 100% due to rounding.

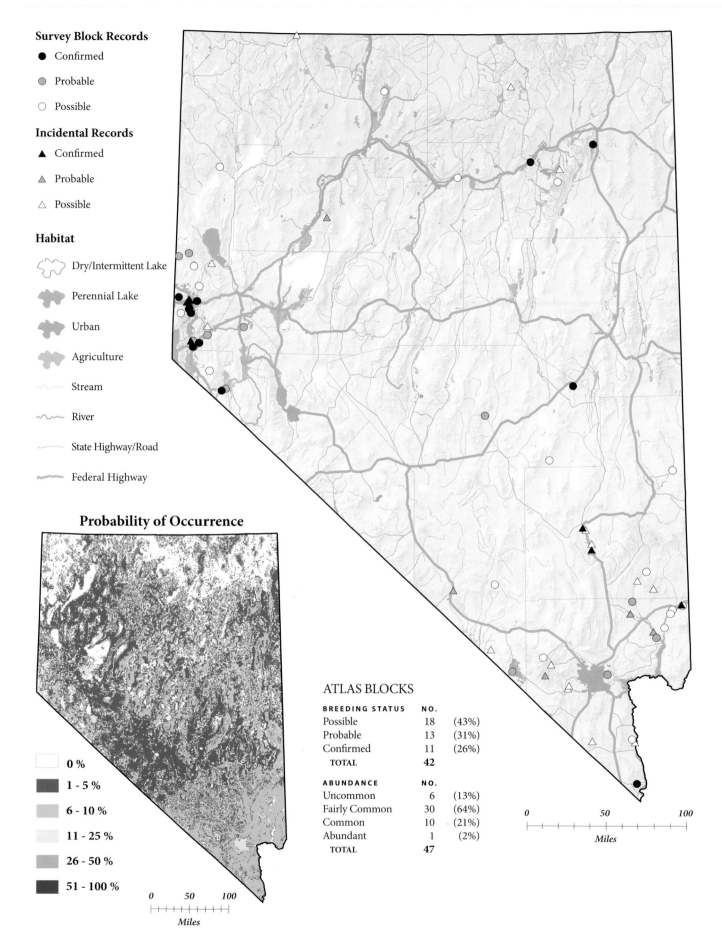

Survey Block Records

● Confirmed

● Probable

○ Possible

Incidental Records

▲ Confirmed

▲ Probable

△ Possible

Habitat

Dry/Intermittent Lake

Perennial Lake

Urban

Agriculture

Stream

River

State Highway/Road

Federal Highway

Probability of Occurrence

☐ 0 %

■ 1 - 5 %

6 - 10 %

11 - 25 %

26 - 50 %

■ 51 - 100 %

0   50   100
Miles

ATLAS BLOCKS

BREEDING STATUS	NO.	
Possible	18	(43%)
Probable	13	(31%)
Confirmed	11	(26%)
TOTAL	**42**	

ABUNDANCE	NO.	
Uncommon	6	(13%)
Fairly Common	30	(64%)
Common	10	(21%)
Abundant	1	(2%)
TOTAL	**47**	

0   50   100
Miles

AMERICAN GOLDFINCH

Carduelis tristis

Rh'02

It seems as though the American Goldfinch should be among the best-known birds wherever it occurs. The breeding male is distinctive, attractive, and easily identified, and the distinctive flight call (*potato chip!*) can be heard all year long. A fairly tame bird, it is a frequent visitor to residential gardens and is addicted to thistle feeders. Yet the American Goldfinch is a scarce breeder in Nevada, and its range in the state remains somewhat uncertain. The fact that it breeds late in the summer, after most birders have stopped looking for nesting activity, makes clarifying its status especially challenging.

DISTRIBUTION

Atlas workers found American Goldfinches at widely scattered locations across the state. Probable breeding was noted at only five sites, however, and the only two confirmed breeding records came from northern Elko County and the Sheep Range of Clark County. These goldfinches were usually numerous where present, with multiple pairs estimated to occur in most of the occupied blocks, and more than ten pairs in one-fifth of the blocks.

Although the early literature describes records from across Nevada, it provides little information about the species' breeding status (Linsdale 1936, Ryser 1985, Alcorn 1988). Given its sporadic distribution in Nevada, it is a bit of a surprise that the American Goldfinch is a common breeder statewide in Oregon, Idaho, and Utah (Behle et al. 1985, Stephens and Sturts 1998, Adamus et al. 2001). It nests in the lowlands of California but is absent as a breeder from the eastern parts of the state that adjoin Nevada (Small 1994). The American Goldfinch also is largely absent as a breeder from Arizona, although occasional nesting has occurred there (Corman in Corman and Wise-Gervais 2005:594).

American Goldfinches tend to breed near disturbed habitats such as weedy fields and floodplains. Scattered trees and tall shrubs provide nest sites, and breeding often coincides with the availability of thistle seeds (Middleton 1993). In the West, nesting birds are often found in lower-elevation riparian zones and canyons (Ryser 1985, Small 1994), and sometimes in city parks and residential areas, and that is where atlas workers found them. The predictive map reflects the low probabilities of finding this species throughout the state, with a slightly higher chance in the north. Lesser Goldfinches, in contrast, are somewhat more likely to be found in the south.

CONSERVATION AND MANAGEMENT

American Goldfinches appear to be doing well in North America as a whole, but western populations have declined in the past few decades (Sauer et al. 2005). Declines have been most noticeable in Oregon, Idaho, Colorado, and British Columbia. Middleton (1993) also reported declines in the East and Midwest, but these appear to have reversed recently (Sauer et al. 2005). No monitoring data are available for Nevada, making it difficult to assess the species' status in this state.

Much of Nevada seems to be marginally suitable, at best, for the American Goldfinch. This species is generally considered to be difficult to monitor because it often nests late in the summer. Late breeding, however, might not be as common in western North America as it is in the East, perhaps because the timing of seed availability differs (Middleton 1993). Improved monitoring in Nevada would be valuable given the declines noted elsewhere in the West. Further study would also allow comparisons with the ecologically similar Lesser Goldfinch, which appears to be doing much better in Nevada. Overall, though, American Goldfinches remain common in much of their breeding range, probably having benefited from human settlement, and they are not a species of conservation concern.

HABITAT USE

HABITAT	ATLAS BLOCKS	INCIDENTAL OBSERVATIONS
Montane Forest	2 (13%)	0 (0%)
Montane Shrub	1 (6%)	0 (0%)
Riparian	6 (38%)	5 (83%)
Sagebrush Steppe	1 (6%)	1 (17%)
Urban	5 (31%)	0 (0%)
Wetland	1 (6%)	0 (0%)
TOTAL	16	6

Total numbers reported in the Habitat Use, Breeding Status, and Abundance tables may differ from each other (see pp. 22–23 for details). Percentage sums may differ slightly from 100% due to rounding.

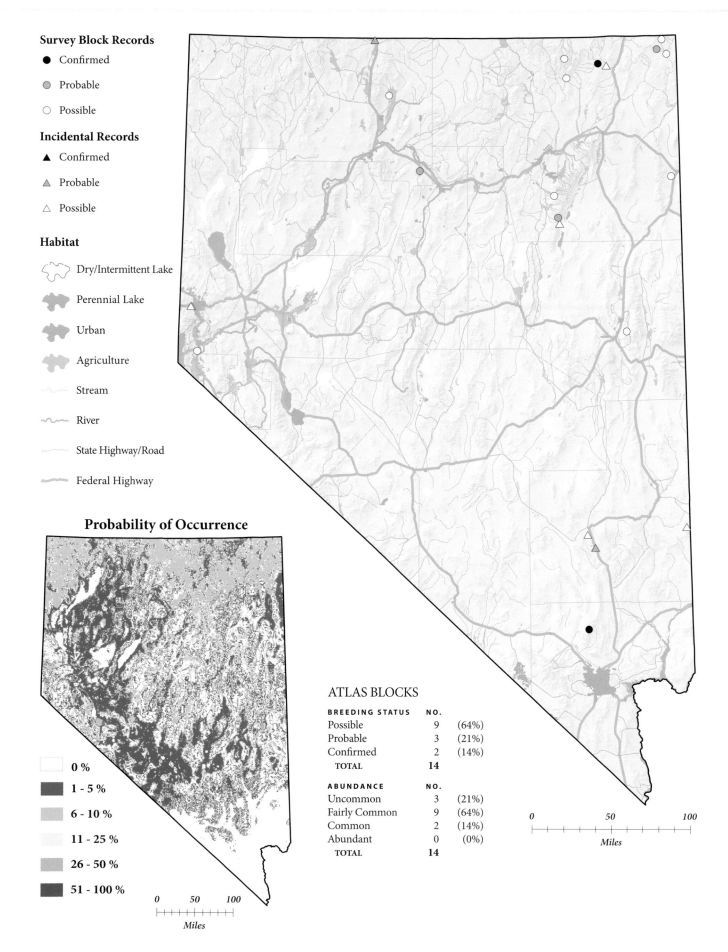

Survey Block Records

● Confirmed

● Probable

○ Possible

Incidental Records

▲ Confirmed

▲ Probable

△ Possible

Habitat

Dry/Intermittent Lake

Perennial Lake

Urban

Agriculture

Stream

River

State Highway/Road

Federal Highway

Probability of Occurrence

	0 %
	1 - 5 %
	6 - 10 %
	11 - 25 %
	26 - 50 %
	51 - 100 %

0   50   100
Miles

ATLAS BLOCKS

BREEDING STATUS	NO.	
Possible	9	(64%)
Probable	3	(21%)
Confirmed	2	(14%)
TOTAL	**14**	

ABUNDANCE	NO.	
Uncommon	3	(21%)
Fairly Common	9	(64%)
Common	2	(14%)
Abundant	0	(0%)
TOTAL	**14**	

0   50   100
Miles

EVENING GROSBEAK
Coccothraustes vespertinus

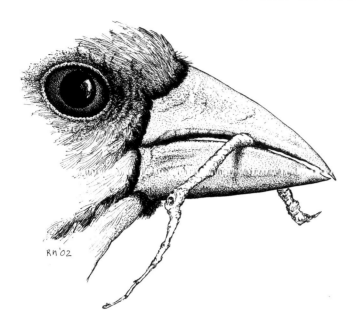

RN'02

Nobody would ever accuse the Evening Grosbeak of being shy. Feeding flocks are spirited, noisy, and spectacular to observe. Males have bright golden bodies, a bold yellow eyestripe, and gleaming white wing patches, all contrasting with their otherwise black flight feathers and dusky head. Females are a little duller but are nonetheless striking with their gigantic, pale olive bills.

DISTRIBUTION

Very few Evening Grosbeaks were recorded in Nevada. Most atlas records came from the Carson Range, including two records of confirmed breeders. There were also isolated records of unconfirmed breeders from the Sweetwater Range of southern Douglas County and from the Sheep Range of northern Clark County. These results correspond well with the account of Linsdale (1936), who noted the species mostly in the western mountains but also provided a likely breeding record in the White Mountains of Esmeralda County. Alcorn (1988) presented a somewhat broader range of records, including confirmed breeding at Mount Wilson in Clark County. Ryser (1985) also made passing reference to an additional breeding record in Nevada's eastern mountains.

The Evening Grosbeak's breeding range extends across southern Canada and south through the mountains of the American West into central Mexico (Gillihan and Byers 2001). North of Nevada, Evening Grosbeaks breed widely in Idaho and Oregon (Stephens and Sturts 1998, Adamus et al. 2001). In California, Utah, and Arizona, however, breeding is much more restricted, and the species is most likely to be found in the higher-elevation portions of all three states (Small 1994, Gillihan and Byers 2001, Corman in Corman and Wise-Gervais 2005:586–587).

Evening Grosbeaks are very much birds of coniferous and mixed coniferous-deciduous forests throughout their range (Gillihan and Byers 2001). In Nevada, breeding Evening Grosbeaks tend to be found on the higher mountain slopes (Linsdale 1936, Ryser 1985). The Carson Range is clearly the heart of their Nevada range, and this is where Evening Grosbeaks are predicted to be easiest to find. The predictive map also suggests the potential for breeding in several ranges in central and eastern Nevada. Realizing this potential requires that dispersing birds reach these ranges in numbers large enough to avoid rapid extirpation, and it is likely that remote populations occur unpredictably and are short-lived. An interesting feature of the atlas habitat records was the Evening Grosbeak's summer presence in some urban areas. A similar phenomenon has also been noted in Arizona (Corman in Corman and Wise-Gervais 2005:586–587).

CONSERVATION AND MANAGEMENT

In the West, Evening Grosbeak populations have apparently declined since 1980, particularly in the Rocky Mountain and Sierra Nevada regions (Sauer et al. 2005). Populations tend to fluctuate quite a lot, though, often in response to insect outbreaks (Bekoff et al. 1987, Gillihan and Beyers 2001), and it is difficult to definitively separate these recent declining trends from normal fluctuations. For the present, Evening Grosbeaks are widespread and numerous in many parts of their range, and there have been few, if any, clear demonstrations of serious impacts resulting from human activities (Pantle in Kingery 1998:540–541, Gillihan and Byers 2001).

The Evening Grosbeak made a rather disappointing showing in Nevada during the atlas years, but it is not clear if this is cause for concern. The current challenge is to better document the full breeding range of the species in the state and to obtain monitoring data sufficient to determine whether the apparent declining trends in the Sierra Nevada extend farther into the state.

HABITAT USE

HABITAT	ATLAS BLOCKS	INCIDENTAL OBSERVATIONS
Montane Forest	4 (40%)	1 (100%)
Pinyon-Juniper	2 (20%)	0 (0%)
Urban	4 (40%)	0 (0%)
TOTAL	10	1

Total numbers reported in the Habitat Use, Breeding Status, and Abundance tables may differ from each other (see pp. 22–23 for details). Percentage sums may differ slightly from 100% due to rounding.

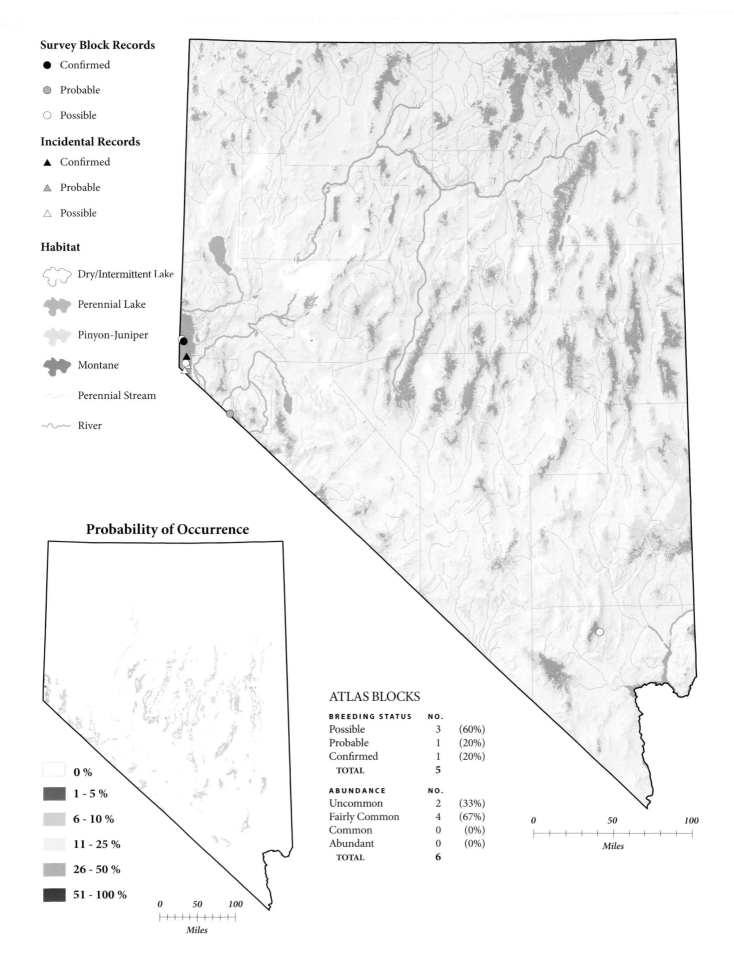

Survey Block Records
● Confirmed
◐ Probable
○ Possible

Incidental Records
▲ Confirmed
△ Probable
△ Possible

Habitat
⬜ Dry/Intermittent Lake
⬛ Perennial Lake
⬛ Pinyon-Juniper
⬛ Montane
∿ Perennial Stream
∿ River

Probability of Occurrence

⬜ 0 %
⬛ 1 - 5 %
⬛ 6 - 10 %
⬛ 11 - 25 %
⬛ 26 - 50 %
⬛ 51 - 100 %

0   50   100
Miles

ATLAS BLOCKS

BREEDING STATUS	NO.	
Possible	3	(60%)
Probable	1	(20%)
Confirmed	1	(20%)
TOTAL	5	

ABUNDANCE	NO.	
Uncommon	2	(33%)
Fairly Common	4	(67%)
Common	0	(0%)
Abundant	0	(0%)
TOTAL	6	

0   50   100
Miles

HOUSE SPARROW

Passer domesticus

No Nevada bird is more dependent on human company than the House Sparrow. European Starlings are often found in riparian groves that have old, dead trees, and even Rock Pigeons sometimes break away from life in the city to breed on high cliffs. In contrast, a House Sparrow away from human habitation is a rare sight. This introduced species flourishes everywhere that humans do, from the downtowns of large cities to tiny ranch houses and outbuildings, and from desert regions to the Arctic.

DISTRIBUTION

House Sparrows were found at widely scattered locales throughout the state. Nearly four in five habitat records came from either urban or agricultural sites, and many others came from riparian areas that were probably near ranches or other human dwellings. Where present, the House Sparrow was easily confirmed as a breeder and numerous: more than two-thirds of the occupied blocks definitely had breeders, and more than two-thirds of the estimated abundances exceeded ten pairs per block.

The House Sparrows that occur in Nevada today are not necessarily descendants of the famous population that was introduced in New York City in the mid-nineteenth century. Our House Sparrows could also be derived from either of two introductions in San Francisco and Salt Lake City in the 1870s (Lowther and Cink 1992). The species was already broadly distributed here by the time Linsdale (1936) was writing, and it is now widespread and locally abundant, especially in the greater Reno and Las Vegas areas and at settlements along the Interstate 80 corridor.

In California, Oregon, Utah, and Arizona, too, the species is common to abundant wherever there are humans (Behle et al. 1985, Small 1994, Adamus et al. 2001, Corman in Corman and

Wise-Gervais 2005:588–589). The House Sparrow is absent from parts of central and northern Idaho (Stephens and Sturts 1998), but this is an area where the human presence is quite limited. Similar patterns of human-influenced distribution occur from northern Canada to Central America.

House Sparrows basically occur everywhere there are houses and will nest in crevices in buildings, in nearby shrubs and trees, and in nest boxes (Lowther and Cink 1992). They can be found in the largest urban centers as well as the smallest desert communities (Small 1994). Not surprisingly, the predictive map closely approximates a map of human-modified areas, with the highest chance of House Sparrow occurrence in towns and cities, and a lesser chance in agricultural areas. The prediction for the Hawthorne Army Depot in Mineral County, however, is probably an error, because even though this site was classified as urban on the vegetation map, it lacks many of the features that attract urban birds.

CONSERVATION AND MANAGEMENT

Although it may be difficult to believe, the House Sparrow has long been in decline throughout most of North America (Sauer et al. 2005) and in its native range in western Europe (Hole et al. 2002). Changes in human behavior and technology are likely responsible for these declines, which probably began when horse-based transportation, and the food that it provided, gave way to motor vehicles. More sophisticated agricultural practices and the resulting reduction in available grain are considered to be an important factor in explaining declines in rural areas (Hole et al. 2002). The causes of declines in urban centers are less clear.

Nevada is one of the few parts of the continent where House Sparrows appear to be increasing (Sauer et al. 2005), presumably because of the state's rapid urbanization. Even if House Sparrows were to decline in the state, however, they would be unlikely to receive much sympathy from birders. The irony in this disdain is that the House Sparrow is perhaps the only bird species that would simply disappear from Nevada if humans were to do the same.

HABITAT USE

HABITAT	ATLAS BLOCKS	INCIDENTAL OBSERVATIONS
Agricultural	18 (14%)	0 (0%)
Mesquite	2 (2%)	0 (0%)
Mojave	0 (0%)	4 (7%)
Montane Parkland	0 (0%)	1 (2%)
Riparian	22 (17%)	13 (22%)
Salt Desert Scrub	2 (2%)	0 (0%)
Sagebrush Scrub	1 (<1%)	0 (0%)
Urban	85 (65%)	40 (67%)
Wetland	1 (<1%)	2 (3%)
TOTAL	131	60

Total numbers reported in the Habitat Use, Breeding Status, and Abundance tables may differ from each other (see pp. 22–23 for details). Percentage sums may differ slightly from 100% due to rounding.

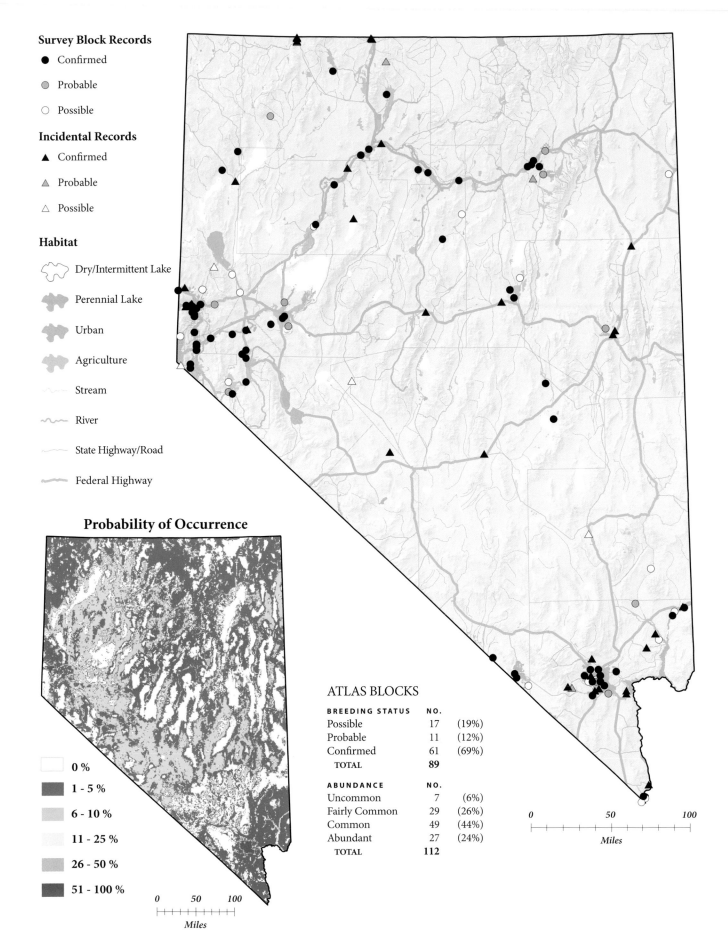

Survey Block Records

- ● Confirmed
- ● Probable
- ○ Possible

Incidental Records

- ▲ Confirmed
- ▲ Probable
- △ Possible

Habitat

- Dry/Intermittent Lake
- Perennial Lake
- Urban
- Agriculture
- Stream
- River
- State Highway/Road
- Federal Highway

Probability of Occurrence

- ☐ 0 %
- ☐ 1 - 5 %
- ☐ 6 - 10 %
- ☐ 11 - 25 %
- ☐ 26 - 50 %
- ☐ 51 - 100 %

0   50   100
Miles

ATLAS BLOCKS

BREEDING STATUS	NO.	
Possible	17	(19%)
Probable	11	(12%)
Confirmed	61	(69%)
TOTAL	89	

ABUNDANCE	NO.	
Uncommon	7	(6%)
Fairly Common	29	(26%)
Common	49	(44%)
Abundant	27	(24%)
TOTAL	112	

0   50   100
Miles

PART III

SUPPLEMENTAL SPECIES ACCOUNTS

This section treats twenty-eight species that, for a variety of reasons, are not included in the main atlas text. Some, such as the Elf Owl and Gila Woodpecker, formerly occurred in the state, presumably as breeders. Others, such as the Clapper Rail and Eurasian Collared-Dove, were not confirmed as Nevada breeders until after the atlas fieldwork was completed. And yet others fall into the intriguing "hypothetical" category, with breeding suspected, but not confirmed, during the atlas years.

All of the species in this section should be of interest to birders and field ornithologists in Nevada because it is their diligent efforts that will likely confirm breeding for many—perhaps most—in the years to come. The breeding avifauna of Nevada is dynamic, and the prepared birder should expect to document range expansions and new breeding records well into the future. The accounts below will provide some guidance, but expect surprises. Curve-billed Thrasher? Veery? Purple Finch?

MUTE SWAN
Cygnus olor

The exotic Mute Swan, admired by many for its grace and elegance, was not known to nest in Nevada until recently. Alcorn (1988:38) described its status as limited to a "few domesticated birds in parks," and Ryser (1985) did not mention the species at all. During the atlas project, however, an introduced, wing-clipped pair bred successfully on a small lake on the University of Nevada campus in Reno, and additional birds were present on other small lakes in the city. The pair continued to breed at the university at least until 2003, although most of the cygnets either died or were removed each year. By 2003, two of the young birds born on the pond had joined their parents, increasing the uni-

versity population to four (A. Wallace, pers. comm.). The potential for subsequent releases, and eventually for free-living birds capable of flight to become established in the area, seems quite real. Two main conservation issues surround introduced Mute Swan populations: (1) the species is extremely aggressive and sometimes displaces other species (Ciaranca et al. 1997); and (2) Mute Swans eat large amounts of aquatic vegetation, which can directly affect plant populations and potentially affect other species that depend on those plants (Willey and Halla 1972, Allin et al. 1987, Allin and Husband 2003). Mute Swans have a tenuous toehold in Nevada at present, but the experience of recent years suggests that they could eventually breed in urban parts of the state.

BUFFLEHEAD
Bucephala albeola

The diminutive Bufflehead has not been confirmed as a breeder in Nevada, but there were three observations during the course of atlas fieldwork. Catnip Reservoir in northwestern Nevada's Sheldon National Wildlife Refuge produced two records (one possible in 2000, one probable in 1998), and the Lahontan Valley produced another (possible). Earlier authors described the Bufflehead as strictly a nonbreeding visitor to the state (Linsdale 1936, Ryser 1985, Alcorn 1988). With its distinctly northern breeding distribution, centered on the boreal forests of Canada, this species is unlikely ever to be more than a rare breeder in Nevada. There are disjunct populations in adjacent states, however. Indeed, there is an isolated breeding population not far from the Nevada border in northern California, and more distant ones are present in western Oregon, the area around Yellow-

RN'04

stone National Park, and northern Colorado (Gauthier 1993). With this pattern of occasional breeding at fairly southerly latitudes, it is possible that Buffleheads also occasionally nest in Nevada. If Buffleheads do breed in Nevada, ponds located in older forest stands with large aspens are probably where they will be found. Buffleheads readily take to nest boxes, and it is possible that Nevada's first breeding records might arise from efforts to provide nest sites for Wood Ducks.

HOODED MERGANSER
Lophodytes cucullatus

With its striking plumage and distinctive profile, the Hooded Merganser is hard to miss on the quiet ponds where it spends its winters in Nevada. But has it been overlooked as a Nevada breeder? Although atlas workers found no Hooded Mergansers, the species apparently has bred here in the past. Ridgway (1877) cited breeding in the Truckee and Carson river valleys, and Ryser (1985) and Alcorn (1988) regarded the species as a possible breeder in western Nevada. In their treatment of western Nevada's Lahontan Valley, however, Chisholm and Neel (2002) described the species as a migrant and winterer only. Hooded Merganser numbers are increasing in various regions of North America (Dugger et al. 1994), and Stallcup (2002) discussed its range expansion in California. Of particular interest is a 2001 report of two broods in Plumas County, California (Stallcup 2002), on the eastern slope of the Sierra Nevada and just 31 miles (50 kilometers) from Nevada. Hooded Mergansers are cavity nesters, and they have shown a proclivity, especially in recent years, to ac-

cept human-made nest boxes. Potential Nevada colonists could come from either California or Oregon, and the species should be looked for wherever Wood Ducks already nest.

SHARP-TAILED GROUSE
Tympanuchus phasianellus

Formerly common in northern Nevada, the Sharp-tailed Grouse was extirpated here in the twentieth century but has recently been reintroduced. The exact details of its historical distribution are unknown, but the birds were evidently numerous (Linsdale 1936). By the time of Ryser's (1985) review, the species was restricted in Nevada to a few remnant populations in the northeast, and before the commencement of atlas fieldwork, native populations had evidently been extirpated there as well. The race that occurred in Nevada—and that is currently being reintroduced here—is *columbianus,* which has suffered widely from habitat loss and hunting pressure (Ryser 1985). Walters (2004) included details of recent reintroductions overseen (see also Coates 2001, Coates and Delehanty 2006). The reintroduction effort has targeted the mountains of northern Elko County, which is relatively close to extant populations of the Sharp-tailed Grouse from which the transplanted birds are derived. Introduced birds have been lekking here for several years, and there has been some dispersal away from the release sites; however, breeding has not yet been confirmed (D. Delehanty, pers. comm.). Atlas workers did not detect or record any introduced birds.

CALIFORNIA CONDOR
Gymnogyps californianus

It is a near certainty that the California Condor—the colossus of the North American avifauna—does not breed in Nevada at the present time. There have been recent sightings in southern Nevada, however, of individuals presumed to be from the Grand Canyon release population. The first known Nevada sighting was in April 2002 (D. Blake, pers. comm.), and sightings continue sparingly to this day (D. McIvor, pers. comm.). The home range of an individual California Condor is huge (Snyder and Schmitt 2002), and conservation strategies for the Arizona (and possibly other) release populations should factor in the use of Nevada habitats by breeding condors. Could California Condors actually breed in Nevada someday? There are no plans to establish release populations in Nevada, but the species occurred in Nevada in the past (Linsdale 1936, Ryser 1985, Alcorn 1988), and it is not inconceivable that offshoots of one of the current release populations might establish themselves as breeders in the state. In the years to come, flyovers should be expected almost anywhere in southern Nevada, as California Condors forage far from their nest sites. Released birds, in particular, encounter many hazards while foraging (Moir 2005), and thus managers in Nevada share responsibility for their well-being. Eventual breeding might be expected among the high cliffs above Lake Mead or in the rugged and arid mountains of Lincoln and Clark Counties.

WHITE-TAILED KITE
Elanus leucurus

Although it is a frequent sight in a variety of open habitats west of the Sierra Nevada, the White-tailed Kite is only occasionally found in Nevada. It is possible that the species now wanders into Nevada more regularly than it used to. Linsdale (1936) gave no records in his summary for the state, but Ryser (1985) documented a flurry of records after 1971, and Titus (2003) considered the species to be an accidental visitor. There was one incidental record of possible breeding at Pahranagat National Wildlife Refuge during the atlas years. Additionally, several observers saw an adult near the town of Pahrump, in southern Nye County, during the summer of 2001. More recently, an adult was seen in Churchill County in 2005 (J. Lytle, pers. comm.). The White-tailed Kite is a conspicuous and easily identified species, and it is unlikely that regular breeding in Nevada has gone undetected. It is also a resilient species, and it continues to recover from earlier population losses in California (Ryser 1985) and to spread through Arizona, where it began breeding only recently (Corman in Corman and Wise-Gervais 2005:601). Observers in Nevada should be on the lookout for breeders here, especially in and around the wetlands and associated agricultural districts of western and southern Nevada.

COMMON BLACK-HAWK
Buteogallus anthracinus

The Common Black-Hawk, a frog-eating denizen of lowland riparian habitats, has not yet been confirmed as a breeder in Nevada. The species breeds just over the border in both Utah (Fridell 2004) and Arizona (Corman in Corman and Wise-Gervais 2005:138–139), however, and future breeding in Nevada is possible. In particular, the Muddy and Virgin river drainages of southern Lincoln County—where there have been numerous observations in the early twenty-first century (Floyd 2002, Walters 2004)—are prime candidates for breeding sites. It is possible that Common Black-Hawks bred in Meadow Valley Wash, an extensive north–south drainage in central Lincoln County, in 2001 (Walters 2004). Larger wetlands in southern Nevada should be checked for breeders, including Pahranagat National Wildlife Refuge, the Warm Springs Ranch near Moapa, and the riparian woodlands along the Virgin River near the Utah border. The breeding avifauna of southern Nevada's riparian communities is still in the process of being documented, and it is probably dynamic and unstable due to habitat changes. Birders should be on the lookout for evidence of breeding by Common Black-Hawks and other riparian-dependent species in an ecosystem that continues to offer many surprises.

ZONE-TAILED HAWK
Buteo albonotatus

Sightings of the Zone-tailed Hawk have increased in recent years in southern Nevada, and several birders have raised the possibility of local breeding. Linsdale (1936) listed no records for the state, but by the time of Alcorn's (1988) review there had been observations from Pahranagat National Wildlife Refuge, Corn Creek, Mesquite, and Las Vegas. In recent years, one or more individuals have summered in the Pahranagat Valley and have been noted by multiple observers. The two atlas records for the species (both of possible breeders) were from this region. The most convincing evidence of breeding was noted in June 2005 by Jim Boone (pers. comm.), who observed a pair of Zone-tailed Hawks on a stick nest, exhibiting nest defense behaviors, in the Bridge Canyon Wilderness northwest of Searchlight. However, successful breeding was not confirmed. Zone-tailed Hawks have been confirmed as breeders in the Lower Colorado River valley (Rosenberg et al. 1991), and riparian groves associated with this drainage in Nevada would seem to be promising places for future attempts to confirm breeding. Observers on the lookout for Zone-tailed Hawks in southern Nevada should be aware of two look-alike species: the widespread and unrelated Turkey Vulture, which shares many behavioral, plumage, and superficial structural similarities with the Zone-tailed Hawk; and the similarly plumaged Common Black-Hawk, which differs significantly from the Zone-tailed Hawk in overall structure and proportions.

BLACK RAIL
Laterallus jamaicensis

Most of the species treated in these short accounts were anticipated, to varying degrees, to be found breeding in Nevada during the atlas years. But the secretive Black Rail was on the radar screens of few, if any, Nevada birders. There were no atlas records for the species, and none of the older accounts of Nevada birds mentions the species. After atlas fieldwork was completed, however, in July 2003, one or more Black Rails were reported along the Virgin River by a research team with the San Bernardino County Museum (Braden et al. 2004). Birders alerted to the possibility of finding the species in Nevada also reported Black Rails at the Henderson Bird Viewing Preserve in 2003, and possibly at Peterson Reservoir in Ash Meadows National Wildlife Refuge in 2004 (C. Titus, pers. comm.). To date, none of these reports has been accompanied by physical evidence that confirms the species' occurrence in the state; however, this flurry of possible sightings suggests that Nevada birders should take the species' potential occurrence in the state seriously. In Arizona, Black Rails have been confirmed as breeders along the Colorado River in the southern part of the state. Additional Arizona records occurred north to Lake Havasu City (Corman in Corman and Wise-Gervais 2005:160–161), fairly close to Nevada, but breeding was not confirmed for these records. This suggests that any birds sighted in Nevada might be nonbreeding wanderers far from the established breeding range well to the south. Then again, the Black Rail is a poorly known and difficult-to-observe species that was first recorded in Arizona less than forty years ago (Corman in Corman and Wise-Gervais 2005:160–161).

CLAPPER RAIL
Rallus longirostrus

The Clapper Rail—like an oversized version of the widespread Virginia Rail—was thought until recently to be accidental in Nevada. Alcorn (1988) treated the species as hypothetical, and earlier writers did not even mention it. By the end of the atlas project, however, sightings had been reported throughout the Muddy and Virgin river valleys, with three incidental observations of possible breeders reported by atlas workers. In the years since, the Clapper Rail has proved to be more widespread and common than previously anticipated in suitable habitat in southern Nevada, though it is difficult to detect. Nesting by multiple pairs was confirmed at Big Marsh, Clark County, in 2001 (McKernan and Braden 2001), and the Clapper Rail is now a Covered species under both the Clark County Multiple Species Habitat Conservation Plan (Clark County 2000) and the Lower Colorado River Multi-Species Conservation Program (BOR-LCR 2004). Nevada's Clapper Rails are of the genetically distinct Yuma Clapper Rail (*yumanensis*) subspecies—now known to be fairly common along the Colorado River and its associated wetlands in Arizona and California (Rosenberg et al. 1991, Garnett et al. 2004, Wise-Gervais in Corman and Wise-Gervais 2005: 162–163). Has the Yuma Clapper Rail been in Nevada, undetected, for a long time, or is it a recent colonist? And which management strategies need to be implemented now to ensure its future in Nevada?

HEERMANN'S GULL
Larus heermanni

The unusual-looking and handsome Heermann's Gull has never been confirmed as a breeder in the United States (Islam 2002). There have, however, been a few unsuccessful nesting attempts along the California coast and also, surprisingly, in Nevada. Chisholm and Neel (2002) discussed an adult Heermann's Gull that paired with a California Gull in Churchill County in 1990, and Alcorn (1988) described an earlier report of a Heermann's Gull in Washoe County that exhibited breeding behavior, although no mate was observed. In addition to these unsuccessful breeders, there have been at least ten records of strays in Nevada, with a concentration of sightings in the late 1990s (Chisholm and Neel 2002, Walters 2004). The most recent were well-documented birds in Reno in April 2004 (F. Petersen, pers. comm.) and at Pyramid Lake in October 2004 (G. Scyphers, pers. comm.). The saga of Heermann's Gull in Nevada is strange in every way. It does not seem out of the question that the first successful breeding event for the species in the United States might take place in Nevada, perhaps at Pyramid Lake or Lahontan Reservoir, or maybe even at the California Gull colony at Virginia Lake near downtown Reno.

EURASIAN COLLARED-DOVE
Streptopelia decaocto

Quite large and easily approached, the introduced Eurasian Collared-Dove is found in several places in Nevada. Yet it was not seen during the atlas years. How can that be? The answer is that this rapidly expanding species did not arrive in Nevada until 2001. Breeding has not yet been confirmed in Nevada, but it is suspected to have occurred already (Walters 2004), and in any event appears inevitable. The speed with which Eurasian Collared-Doves are invading western North America is startling. For example, the species was completely absent from Colorado during its atlas project in the 1980s (Kingery 1998), but less than fifteen years later it was a widely established and locally abundant breeder. Eurasian Collared-Doves apparently entered Nevada in the south, with the first records coming from a survey crew doing work along the Muddy River in May 2001 (N. McDonal, pers. comm.). Since that time there have been numerous records within about 50 miles (80 kilometers) of Las Vegas, and the species has also colonized western Nevada in the past few years (Fridell 2004). It is hoped that Nevada's birders and field ornithologists will carefully document the advances by the Eurasian Collared-Dove in the years to come. The invasion by this species presents a variety of fascinating questions.

ELF OWL
Micrathene whitneyi

Tiny but easily detected by its yelping song, the Elf Owl formerly occurred in Nevada but is now scarce or absent. Linsdale (1936) considered the species' occurrence in the state to be hypothetical, whereas Alcorn (1988) considered it uncommon in far southern Nevada, mainly in and around cottonwood groves in the Fort Mojave area. Walters (2004) reported searching for Elf Owls in this very region, but he failed to find any, noting that the habitat there had been destroyed. Atlas workers, too, failed to find the species at this and other plausible locales in southern Nevada, such as riparian woodlands along the Colorado River and its tributaries and old-growth Joshua tree stands near Searchlight. Should we thus conclude that the species is extirpated here? Not necessarily; there are several recent records from the Virgin River drainage in southwestern Utah (Fridell 2004), a region that was presumably reached via Nevada. If Elf Owls should be discovered in suitable habitat in Nevada, monitoring and recovery efforts should be quickly established. The species faces a host of threats farther south in the Colorado River drainage (Rosenberg et al. 1991), and it is a Covered species under the Lower Colorado River Multi-Species Conservation Program (BOR-LCR 2004).

SPOTTED OWL
Strix occidentalis

Spotted Owls are rarely reported from Nevada, but the species may be an occasional breeder here. The most likely region for breeding in Nevada is the Carson Range, which is adjacent to known breeding populations in California (Small 1994). The atlas yielded no Nevada records for the species, but USDA Forest Service surveys in 2001–2002 and subsequently produced several sightings from the Carson Range (G. Wilson, pers. comm.), including confirmed breeding just over the border in California in 2005 (M. Easton, pers. comm.). All Spotted Owls reported to date in and around the Carson Range are presumed to have been of the *occidentalis* subspecies, the California Spotted Owl—not the same as the threatened Northern Spotted Owl (*caurina*) of the Pacific Northwest. A more remote possibility is that the Mexican Spotted Owl (*lucida*) of the Southwest might enter the mountains of southern Nevada from Arizona or Utah. The Spotted Owl is considered an indicator of forest health (Gutiérrez et al. 1995; see also Verner et al. 1992), and its occurrence in any forested site in Nevada should warrant further monitoring and, ideally, protection of that site. Those in a position to better document occurrences of the Spotted Owl in Nevada should also keep in mind the complication that the Barred Owl, a rapidly spreading species that is similar to and hybridizes with the Spotted Owl, may well arrive in Nevada soon.

BARRED OWL
Strix varia

During the twentieth century, the Barred Owl expanded its range westward, eventually reaching all the way to the mountains of northern California (Mazur and James 2000). The range expansion appears to be continuing, with individuals having been recently detected in California as far south as San Francisco. Invading Barred Owls frequently hybridize with Spotted Owls (Hamer et al. 1994, Haig et al. 2004) and a Barred × Spotted Owl hybrid was found in 2003 in Placer County, California, not far from Nevada (Seamans et al. 2004). In addition to invading from the north, Barred Owls also appear to be coming in from the east, with recent records from Colorado (C. Wood, pers. comm.) and Utah (R. Fridell, pers. comm.). Thus, atlas worker Michael Janik's report of hearing the distinctive "eight-hooter" song of the Barred Owl in a forested stretch of the Humboldt River drainage in the summer of 2001 was not completely unexpected. Janik treated the record as uncertain, as he never did see the bird. The Barred Owl's hooting is usually unmistakable, though, and the habitat and location seem ideal for a Barred Owl. At present, the Barred Owl is not on the official Nevada state list, but additional sightings may be expected, especially in western and northern Nevada.

BLACK SWIFT
Cypseloides niger

The little-known Black Swift is not a confirmed Nevada breeder, but breeding Black Swifts apparently forage in Nevada airspace. Long ago, Owen A. Knorr, an expert on the Black Swift, documented breeding by Black Swifts in eastern Alpine County, Cali-

fornia, just across the Nevada state line. Breeding presumably still occurs there, and the far-ranging adults may well reach nearby Douglas County on their foraging bouts away from the nest site. The best chance of seeing Black Swifts probably would be on the Douglas County–Alpine County border during inclement weather, when they descend to lower altitudes, or in the early evening as the birds are returning to roost. Undetected nesting is possible in Nevada, especially since breeding is notoriously difficult to confirm for this species. Owen Knorr (pers. comm.) considered the Humboldt and Ruby Mountains to be the best bet for Nevada breeders but judged much of state to be unsuitable (Knorr 2000b).

VAUX'S SWIFT
Chaetura vauxi

The diminutive and twittery Vaux's Swift is seen annually, usually in small numbers, in western Nevada. Most sightings are in spring, presumably of individuals en route to their breeding grounds in the Sierra Nevada, the Cascades, and farther north. Do some of them ever stay to breed in Nevada? That question was raised by Sochi (2001), who documented an unusually strong incursion of Vaux's Swifts into Nevada in the spring of 2000. During the atlas years, two pairs were seen engaged in courtship display (designated probable breeding) over a wooded pond in Reno as late as May 30, but they were not seen beyond June 4, leaving the question of local nesting unresolved. The regular breeding range of Vaux's Swift almost grazes the Nevada border, and observers should be on the lookout for nesting in forested, or even developed, areas around Lake Tahoe or Reno. An interesting complication is that breeding *Chaetura* swifts in Nevada should not automatically be assumed to be Vaux's Swifts. The Chimney Swift of eastern North America—similar in appearance to Vaux's, although usually discernible in the field with careful study—has bred at various locations in southern California (Small 1994), and its appearance in southern Nevada is not out of the question.

RUFOUS HUMMINGBIRD
Selasphorus rufus

The belligerent Rufous Hummingbird has never been confirmed as a breeder in Nevada, but it is treated here for several reasons. First, the species is widely—although erroneously—assumed to be a common breeder here. Males on their southward migration are widespread in the state by the beginning of July, and the species is thus very much a part of our summer avifauna. Second, Rufous Hummingbirds breed very close to Nevada, ranging south into northern California, southeastern Oregon, and southwestern Idaho (Calder 1993). It would not be much of a range extension for the species to show up as a breeder in, for example, the Santa Rosa Mountains of northern Humboldt County. Third, and perhaps most intriguing, atlas workers reported several instances of behaviors that were consistent with breeding;

for example, females seen carrying bits of spider web. Should breeding occur in Nevada, it would likely take place from mid-April to mid-July (Calder 1993). Rufous Hummingbirds are rarely noted during spring migration in Nevada, and any sightings before mid-June would be of particular interest.

ACORN WOODPECKER
Melanerpes formicivorus

Four seasons of atlas fieldwork produced no records of the charismatic Acorn Woodpecker. Subsequently, however, in 2003–2004, there was a mini-invasion by this distinctive species in Nevada (Fridell 2004), highlighted by a long-staying pair at Verdi, Washoe County, in the summer of 2004 first discovered by R. Bruno. The birds appeared to be territorial and were seen entering a tree cavity. Acorn Woodpeckers are common residents of the western slope of the Sierra Nevada, but breeders spill over to the eastern side near Susanville, California (Chisholm and Neel 2002), a little north of where the Verdi pair was observed. The species appears to be in the process of invading the southern reaches of the eastern Sierra Nevada (Kirk and Kirk 2004), too, and the Verdi record would seem to be consistent with a general range expansion by the Acorn Woodpecker. Could Acorn Woodpeckers become established in western Nevada? The determining factor appears to be the presence of native oaks (Kirk and Kirk 2004), which are present, but sparsely so, in the foothills and middle elevations of the Carson Range (J. Nachlinger, pers. comm.).

GILA WOODPECKER
Melanerpes uropygialis

The Gila Woodpecker—a rather plain but noisy member of the genus *Melanerpes*—was not recorded in Nevada during the atlas years. It has occurred here in the past, however, although breeding was not confirmed. Linsdale (1936) considered the Gila Woodpecker to be common in the wooded stretches along the Colorado River in the extreme southern reaches of the state, and Alcorn (1988) called it a permanent resident in the same region. These areas have subsequently been flooded and logged (e.g., Walters 2004), and are currently not inhabited by Gila Woodpeckers. The extirpation of the Gila Woodpecker in Nevada seems to have followed the same pattern as that of the Elf Owl. But whereas the Elf Owl's best chance of recovering may be restoration of woodlands along the Colorado River (as part of the U.S. Bureau of Reclamation's Lower Colorado River Multi-Species Conservation Program), it is conceivable that the Gila Woodpecker might go in a very different direction. The Gila Woodpecker flourishes in large desert cities with golf courses and planted trees, such as Phoenix and Tucson, and Las Vegas would seem to have a lot to offer the species. For better or for worse, urban centers in the hot deserts of the United States have proven attractive to a surprising number of species (Mills et al. 1989), and dispersal barriers might be the only impediment to the colonization of Las Vegas by the Gila Woodpecker.

PACIFIC-SLOPE FLYCATCHER
Empidonax difficilis

The status of *E. difficilis* in Nevada is indeed vexing (see Lowther 2000 and the Cordilleran Flycatcher account for details). The species is presumably a common migrant throughout the lowlands of the state, yet it is almost never definitively identified in the field. In most of Nevada, the Pacific-slope Flycatcher is not the breeding "Western Flycatcher"; the Cordilleran Flycatcher is, even in the isolated mountain ranges of southern Nevada. But what about "Western Flycatchers" in the Carson Range and in the forested drainages of far northwestern Nevada? Puzzlingly, the atlas produced no records of either "Western Flycatcher," despite an apparent surfeit of suitable habitat, save for a record of one bird—judged by the observer who heard its call to be a Pacific-slope—from Slumgullion Creek in far northwestern Humboldt County. A complication with this record is that it comes from the periphery of a region notorious in recent years as a hotbed for "Western Flycatchers" that do not entirely match the field characters of either Cordillerans or Pacific-slopes. Careful work by birders in Oregon has also called into question the entire matter of field separation of Pacific-Slope and Cordilleran Flycatchers in those areas where their ranges overlap (Canterbury 2003). The situation in Nevada is simply unknown at present.

GRAY CATBIRD
Dumetella carolinensis

This sleek skulker is probably an annual visitor to the broad-leaved thickets of Elko County, and many Nevada birders expected confirmation of breeding in the northwestern part of the state during the atlas survey. This did not occur, however; there was only one possible breeding record from all of Elko County. Eastern Kingbirds, Bobolinks, and other eastern species reach the limits of their range in Elko County, and it is conceivable that the catbird does, too. Titus (2003) called the species a summer visitor in northeastern Nevada, and Behle et al. (1985) described it as a rare and localized summer resident in Utah. Surprisingly, the best evidence for breeding during the atlas years, involving a probable breeding record, came from the southern Toiyabe Range of central Nevada. A careful observer (Owen Knorr) saw the species in the same drainage for several years prior to and during the atlas project and suspected breeding there. Both Ryser (1985) and Alcorn (1988) implied that catbirds definitely breed in the Toiyabe Mountains, but the details were a little unclear. The Gray Catbird has probably bred in Nevada; whether it still does and with what frequency remain to be determined. Finally, in June 2006, John Woodyard (pers. comm.) confirmed a bird with nest material near Baker in White Pine County.

HERMIT WARBLER
Dendroica occidentalis

The handsome Hermit Warbler is one of a large suite of Sierra Nevada species that barely enter Nevada in the Carson Range.

Has the species bred here, and does it currently? The answers to both questions are unclear. Linsdale (1936) considered the Hermit Warbler to be a transient in Nevada, whereas Ryser (1985), Alcorn (1988), and Chisholm and Neel (2002) rated it as either a summer resident or a rare breeder in the Carson Range—but without elaborating. In recent years, individuals have been seen throughout the summer months in the Carson Range, where atlas fieldwork produced three possible breeding records. Between 1995 and 2005, twenty-eight Hermit Warblers, including one hatch-year bird, were captured during the summer months at a breeding bird banding station in Little Valley, on the eastern slope of the Carson Range above Carson City (J. Eidel, pers. comm.). Although several of these captured Hermit Warblers exhibited brood patches, this is not sufficient to definitively demonstrate local breeding (Cringan et al. 1992; J. Eidel, pers. comm.). Currently, the Hermit Warbler is best viewed as a likely breeder in Nevada, but confirmation awaits the fieldwork of intrepid birders in the mountains south and west of Reno.

PAINTED REDSTART
Myioborus pictus

Many Nevada birders anticipated recording the Painted Redstart during the course of atlas fieldwork. Alcorn (1988), citing several authorities, had reported the species to be a rare to uncommon breeder in the Spring Mountains, and Titus (2003) treated it as a rare vagrant in southern Nevada. No Painted Redstarts were documented in Nevada during the atlas survey, however, even though migrants are noted, albeit rarely, in the lowlands of southern Nevada. The core of the Painted Redstart's U.S. range lies in Arizona, but the known breeding distribution extends northwestward toward southern Nevada (Corman in Corman and Wise-Gervais 2005:482–483). There have been multiple records from Zion National Park in southwestern Utah, and breeding has been confirmed in Utah's Beaver Dam Mountains, just across the Nevada border (Wauer 1997). Given that the Painted Redstart is easily identified, it is probably safe to say that it is currently absent from the more popular campgrounds and roadside stops on the eastern and northern slopes of the Spring Mountains. It is quite possible, though, that the species breeds in small numbers in the less-well-covered parts of the range, and possibly in the Sheep Range or Virgin Mountains, too.

HEPATIC TANAGER
Piranga flava

As the atlas project began, the brick-colored Hepatic Tanager seemed like a good bet in the Spring Mountains and maybe the Sheep Range of southern Nevada. The conventional wisdom was that the species was present in these Vegas-area sky islands, along with other southwestern specialties such as Grace's Warbler and the Whip-poor-will. Not only was the Hepatic Tanager not confirmed as a breeder, it went totally unrecorded during the project. In the years since the atlas, however, there have been at least four records of migrants in lowlands throughout the state. The

species occurred in the state in the past, too, although earlier writers were reluctant to ascribe confirmed breeding status to it. Alcorn (1988) defined the Hepatic Tanager as a summer resident, without assigning breeding status, and Titus (2003) regarded it as accidental in the south. Ryser (1985) considered the Hepatic Tanager to be a breeder to the south of the Great Basin portion of Nevada, presumably referring to the mountains in the southern part of the state. Relatively speaking, the Spring Mountains were fairly well covered during the atlas project, so it is tempting to conclude that the Hepatic Tanager was absent from Nevada, at least during this period. But the Spring Mountains are extensive, and populations of low-density summer visitors such as the Hepatic Tanager are variable and unstable. The lesson of the Hepatic Tanager is that even Nevada's better-known birding locales are deserving of additional fieldwork.

GRASSHOPPER SPARROW
Ammodramus savannarum

This buzzy-voiced inhabitant of grasslands was unexpectedly hard to find during the atlas years. Extensive grasslands in Elko County and in northern Humboldt County produced no records of confirmed breeding during the course of the atlas fieldwork. In fact, there were only two records from the atlas years (both from northeastern Nevada, as expected): a possible breeder in Lander County and a probable breeder in White Pine County. The year after atlas fieldwork was completed, Walter Wehtje of Riverside, California, observed an adult carrying an object—perhaps a fecal sac—in Lincoln County in July 2001. Earlier writers described the Grasshopper Sparrow as a summer resident in northern Nevada (Ridgway 1877, Linsdale 1936, Ryser 1985, Alcorn 1988, Titus 2003), but confirmation of breeding, although assumed, was not formally reported by these writers. If it is present in Nevada, this generally declining species is rare here. The Grasshopper Sparrow may be of interest from a conservation perspective, as its presence is often indicative of high-quality rangeland and native grasslands (Dobkin and Sauder 2004). Observers should be aware of superficial similarities between the songs of Grasshopper and Savannah Sparrows; during the atlas years, singing Savannah Sparrows were incorrectly reported as Grasshopper Sparrows on several occasions.

COMMON GRACKLE
Quiscalus quiscula

The Common Grackle is one of the most characteristic breeding birds of eastern North America, but it is found only occasionally in Nevada. Vagrants are seen from time to time, the most recent being an October 2001 adult female reported by Jon Dunn in Nye County. Prior to the atlas project there were scattered reports from across Nevada, with the sightings tending to aggregate in the northern and eastern parts of the state (Linsdale 1936, Ryser 1985, Alcorn 1988). There were three Elko County reports of Common Grackles during the atlas years, but it was felt that all three could have pertained to Great-tailed Grackles, a species on the increase in the county. Alcorn (1988) cited two breeding records, both in 1987. One was from an oasis near Dyer, Esmeralda County; the other was from Stillwater National Wildlife Refuge, Churchill County. Chisholm and Neel (2002), however, questioned the Stillwater record. To our east, Common Grackles breed in Utah, with recent breeding as far west as the Salt Lake City area. Although the Common Grackle should not be considered to be regular in Nevada at the present time, it may establish itself—possibly as a breeder—in the not-too-distant future. Colonists are probably most likely to arrive from the east, via the Interstate 80 and U.S. Highway 50 corridors.

GRAY-CROWNED ROSY-FINCH
Leucosticte tephrocotis

Scenic Mount Rose, looming over the city of Reno, is a popular destination for day hikers. And the hike to this Carson Range peak offers special appeal to birders, for Mount Rose is known to have harbored breeding American Pipits (Knorr 2000a) and Gray-crowned Rosy-Finches (Ryser 1985), with sightings as late as the mid-1990s (C. Elphick, pers. comm.). Somewhat surprisingly, however, Gray-crowned Rosy-Finches were not observed on Mount Rose during the atlas years. Does that mean that the species is now absent from this well-birded locality? Not necessarily, as the tundra habitat atop Mount Rose is fairly extensive, and very few birders wander far from the main summit trail. Similarly, in Rocky Mountain National Park, Colorado, which receives intensive coverage by birders, Brown-capped Rosy-Finches, although common, are missed more often than they are found. Interestingly, there *was* an atlas record of a Gray-crowned Rosy-Finch, although not from Mount Rose. Larry Allen, a visiting birder from Los Angeles, reported, with good details, an apparent Gray-crowned Rosy Finch from the mountains of White Pine County in eastern Nevada. Separation of Gray-crowned and Black Rosy-Finches can be surprisingly difficult, however, and this single record of a possible breeder is best treated as hypothetical. At the same time, observers are encouraged to sort through Nevada's breeding Black Rosy-Finches for possible Gray-crowneds—and, of course, to continue to visit the historic Gray-crowned stronghold of Mount Rose.

CONCLUSIONS

Steve Mlodinow, the prolific western North American field ornithologist, quipped that "a truism regarding status-and-distribution works is that they are out-of-date before they are published." And that is not a bad thing: It reflects the fundamentally dynamic and healthy nature of the status-and-distribution literature (and it keeps authors, editors, and publishers forever employed).

To be sure, the *Atlas of the Breeding Birds of Nevada* was out of date—and incomplete—even before it was published. That is partly a reflection of the inevitable time lag between the conclusion of fieldwork and publication, and partly due to the fact that any atlas project can sample only a fraction of the entire avifauna of a state or province. In this regard, the reader is encouraged to study the Supplemental Species Accounts (pp. 537–44), which demonstrate just how quickly our knowledge of the Nevada avifauna is changing.

There is something else at play here. Nevada's bird populations are themselves changing. Even if we documented every single breeding bird in the state during the period 1997–2000, the atlas would still be out-of-date at the moment of publication. Far from being the final say on the breeding avifauna of Nevada, the atlas should instead be viewed as a springboard for future discovery. That significant ornithological discoveries may be made anywhere in the state is a lesson we learned again and again during the atlas years. Still, there are certain regions and habitats that might be especially good starting points for the birder or field ornithologist hoping to push forward the frontiers of ornithological knowledge in the years to come. A few are listed below.

Las Vegas. Yes, the city itself. Las Vegas presents a fast-changing and not entirely undesirable suite of habitats, and many bird species are making good use of this new resource. Anna's Hummingbirds and Great-tailed Grackles are now widespread in the metropolitan area, and species such as the Peregrine Falcon, Eurasian Collared-Dove, and Hooded Oriole may be increasing. Exotics—especially parrots—are increasingly detected in and around the city, too. Other species, such as Gambel's Quail, are declining and have been extirpated from many regions of the city. Without question, the avifauna of Las Vegas is dynamic and fascinating—and deserving of serious study by anybody with interests in either the "pure biology" questions of adaptation and population dynamics or the "applied biology" matters of managing vertebrate populations in rapidly transforming landscapes.

The Northern Mojave Desert. If counting Anna's Hummingbirds in suburban Las Vegas just isn't your thing, there is still plenty to be discovered in the varied habitats a little farther afield in southern Nevada. Two major mountain ranges—the Spring Mountains and the Sheep Range—may well support as-yet-undocumented populations of Painted Redstarts, Hepatic Tanagers, and perhaps even Magnificent Hummingbirds. Smaller mountain ranges—among them the Virgin, Mormon, and Lucy Grey mountains—may also harbor surprises. For example, California Condors, should they become established in Nevada, might first colonize these lower, drier ranges. Lowland riparian areas of southern Nevada's Mojave Desert are practically guaranteed to provide significant new records in the years to come. A short list of species in need of documentation or discovery includes the Common Black-Hawk, Zone-tailed Hawk, Clapper and Black Rails, Elf Owl, Gila Woodpecker, Gilded Flicker, Bendire's and Curve-billed Thrashers, Indigo Bunting, and Rufous-crowned Sparrow.

Reno. In the years to come, the city of Reno itself, like Las Vegas, is likely to be of considerable interest to the student of avian status and distribution, as will the areas surrounding it. Will introduced populations of such species as the Mute Swan and Eurasian Collared-Dove become established, or will they wink out? Might Hooded Mergansers or even Heermann's Gulls establish an outpost at one of the city's many waterways? What are the Reno-area population trends of the distinctive *californica* and *woodhouseii* races of the species currently called the Western Scrub-Jay? Will Canada Geese continue to expand as breeders in the city and continue to exhibit novel adaptations to urban life? Have Western Screech-Owls been extirpated from greater Reno? Will Anna's Hummingbirds and Red-shouldered Hawks, currently on the increase, become established breeders in Reno and its surroundings? Reno is forecast to grow and to "green up" through the foreseeable future, and major avifaunal changes should be expected.

The Carson Range. Even though the foothills of the Carson Range are within fifteen minutes of Reno, getting into some of the more rugged terrain of this eastern fringe of the Sierra Nevada requires many hours of hiking. How many breeding pairs of diminutive Winter Wrens and strapping Pileated Woodpeckers could a dedicated team of backcountry birders tally in the Carson Range? We know that Vaux's Swifts and Spotted Owls are present, but are they breeding? Strong evidence for breeding in the latter species became available just as the atlas went to press, but further research is needed. Have we perhaps overlooked Purple Finches and maybe Purple Martins, which breed just a few miles west of Lake Tahoe on the California side? Will invading species such as the Barred Owl and Acorn Woodpecker become established? Are they here already?

Elko and Environs. From an ornithological perspective, the agricultural valleys and high mountains around Elko provide Nevadans with a sample of the avifauna primarily associated with more northern and eastern regions. However, basic population data for such species as the Sandhill Crane, Eastern Kingbird, Bobolink, and Black Rosy-Finch are still needed. The breeding status of low-density summer visitors such as the Gray Catbird and Grasshopper Sparrow remains undetermined. And what about even more "exotic" fare? Might a few Upland Sandpipers breed in agricultural settings in Elko County? Could there be small populations of Northern Waterthrushes or American Redstarts or maybe Veerys?—perhaps in the Jarbidge Mountains? Is even the Boreal Owl a possibility?

The Northern Tier. Although most Nevada birders are aware of the excitement that awaits in the Carson Range and along the Jarbidge–Rubies axis, there is comparatively little current interest in the region that arcs east-northeastward from the Reno area to the Jarbidge area. Here, near the Oregon border, one can find some of the most extensive sagebrush country in the state, along with not unimposing mountains and not insignificant lowland riparian habitats. Do Rufous Hummingbirds extend into this region of Nevada? Could a small population of Yellow Rails be holed up in one of the numerous wet meadows in the area? And what about the birds of the "sagebrush sea"? Nevada's healthiest

populations of Greater Sage-Grouse occur here, but they are in need of extensive monitoring. What are the ranges and microhabitat limits of the many species of sparrows that occur in the sagebrush country of far northern Nevada?

Wetlands and Riparian Zones. The marshes and river systems of the Intermountain West are naturally dynamic, and human agency—for better or for worse—has made the situation all the more interesting. Happily, populations of many wetland species in Nevada have been increasing in recent years, and field ornithologists in the state will want to continue aquatic monitoring programs and be on the lookout for range expansions and continuing population increases. Water-loving Black Phoebes, for example, are spreading northward through the state, but their movement has received little quantitative study here. Many piscivorous birds—cormorants, herons, raptors, terns, and others—are generally recovering in northern Nevada, but their future status in the state bears careful monitoring. The situation is especially dynamic in the south, where some species are apparently colonizing our riparian and wetland habitats and others are in danger of extirpation.

Cryptic Species and the Great Middle. The Great Basin, with its complex mosaic of mountains and valleys, is a natural laboratory for exploring the speciation process. In all regions of the state—very much including its center—speciation is unfolding right now. Many fascinating and fundamental questions are unanswered at the present time. Where is the breakpoint between the *schistacea* and *megarhynca* groups of the Fox Sparrow? Between Red-breasted and Red-naped Sapsuckers? Cassin's and Plumbeous Vireos? Virginia's and *ridgwayi* Nashville Warblers? Even more exciting, could Nevada harbor unexpected or even undescribed species? Do Timberline (Brewer's) Sparrows (*Spizella breweri taverneri*) breed in the Ruby Mountains? Do sage-grouse populations in the highlands of Mineral County constitute a distinct species, the as-yet-unnamed "Bodie Hills" Sage-Grouse?

Climate Change. It is difficult enough to model climate change per se, let alone the effects of climate change on regional avifaunas. Nonetheless, many ornithologists believe that climate change is starting to have measurable impacts on bird populations that will intensify in the decades to come. What might the impacts look like in Nevada? With warming temperatures, look for long-term movements of populations northward—potentially of Mojave Desert species northward into Nevada and of sage steppe and spruce-fir species northward out of Nevada. And expect the unexpected, as many models predict the landscape-level impacts of climate change to be disruptive, "bumpy," and possibly extreme.

Epilogue: 2057. Fifty years isn't all that long in the grand scheme of things. Yet who in 1957 could have predicted the changes that would sweep across the Nevada landscape? Few biologists would have guessed that the Greater Sage-Grouse would plunge into precipitous decline, and few would have imagined that fish-eating predators—from Ospreys to Bald Eagles—would respond so well to pesticide abatement, habitat conservation and restoration, and protection from persecution. Few biologists raised alarm bells for Gila Woodpeckers and Elf Owls, and few anticipated the rapid advances of the Great-tailed Grackle and Anna's Hummingbird. And not even Steve Wynne, in his wildest fantasies, could have foreseen the growth of Las Vegas.

There is every reason to believe that the next fifty years will see just as much change. Things are different now, though. We have long-term data sets and computational power that our forebears in 1957 did not have. We have a better grasp of ecological theory, and we have the new discipline of conservation biology. We live in a society that is generally more concerned about the environment and more attuned to environmental degradation than was the case in 1957. In the specific case of Nevada, we have generated predictive models that can be put to use, right now, by birders and biologists who care about the state's dynamic avifauna. We have the opportunity to manage and conserve bird populations into the next half-century and beyond.

ACKNOWLEDGMENTS

A generation from now, when historians of science assess the whole concept of the breeding bird atlas, they will arrive at a peculiar and paradoxical conclusion. Breeding bird atlases are complex, large-scale, analytically sophisticated undertakings of the sort that are usually associated with university or government research; in short, they are "Big Science." Yet they are carried out largely by citizen scientists who sometimes have little formal training and typically have no expectation of reward beyond the satisfaction of getting out into the field to study birds.

One of the biggest challenges that we faced early in the atlas project was finding fieldworkers. We asked ourselves: "Can we really find a hundred skilled and reliable birders in Nevada? Maybe fifty or seventy-five would be more realistic." Well, we exceeded everybody's expectations: by the end of the project, more than four hundred people from Nevada and surrounding regions had contributed to atlas fieldwork. Here are their names:

Gail Abend	Cristi Baldino
Marc Adamus	Laura Baldwin
Paul Adamus	Jennifer Ballard
Nelsene Alford	John Bare
Larry Allen	Jenny Barnett
Dawn Alvarez	Martin Barr
John Anderson	Curt Baughman
David Arsenault	Jason Beason
Michele Attaway	Maury Beck
Millie Ayers	Bill Belli
Serena Ayers	Marla Bennett
Ted Bacike	M. Berggren
Shirley Badame	Michelle Berkowitz
Aaron Baker	Teri Bernston

Kathy Berry
John Bialecki
Michael Bish
Nancy Bish
Joy Black
Wayne Bliss
Jay Block
John Blow
Sunya Bolinger
Lynda Booth
Nikki Bottum
Mike Boyles
Jed Bradley
Peter Bradley
Gary Branzell
Wendy Broadhead
Richard Brune
Jessica Bulloch
Gerrit Buma
Rob Bundy
Michelle Buonopane
Gretchen Burris
Suzanne Cardinal
Loretta Cartner
Dwight Carver
Terry Carver
Ali Chaney
Chris Cheney
Candy Chicon
Graham Chisholm
Beth Clark
Bill Clark
Lucie Clark
Joan Clarke
Jim Clement
Greg Clune
Jerry Coe
Jeanie Cole
Marti Collins
Terri Corlett
Gary Cottle
Sean Cottle
Bob Cox
Mike Cox
Nancy Cox
Cliff Creger
James Cressman
Marian Cressman
Beth Cristobal
Chris Crookshanks
L. Crow
Dorothy Crowe
Hillie Crowfoot
David Daily
Karen Danner

Neil Darby
Aaron Dawson
Justin Dean
Merry Lynn Denny
Mike Denny
Sally Denton
Kevin DesRoberts
Garry Dierks
Linda Dierks
Danny Dixon
Joe Doucette
Cheryl Doyle
Barney Drake
Richard Duncan
Kate Dunlap
Jon Dunn
Mike Eichelberger
Jim Eidel
John Elliot
Mary Jo Elpers
Chris Elphick
Dennis Elphick
Duane Erickson
Becky Estes
Sue Farley
Tim Farrell
Joselyn Fenstermacher
Cy Fernandez
Robert Flores
Ted Floyd
Loraine Ford
Steve Foree
J. Forest
Phoebe Fowler
Sue Fox
Jay Frederick
Maggie Freese
Mike Freese
Larry French
Bob Furtek
Travis Gallagher
Joel Geier
Rebecca Geier
Wil Geier
Dennis Ghiglieri
Bill Goggin
Dori Goldman
Sue Golish
Bob Goodman
Bruce Gordon
Peter Grant
Ken Gray
Blair Green
Jennifer Green
Paul Greger

Mike Gregg
Janet Grigg
Aaron Gross
Pam Groves
Alan Gubanich
Anne Haglund
Marvin Hall
Nancy Hall
Steve Hampton
J. Hanselmann
K. Hanselmann
Corbett Harrison
Karla Harrison
Jim Harvey
Hugh Hawkins
Mary Haworth
Jim Healey
Don Helling
Jocelyn Helling
Max Helling
James Henry
William Henry
Sue Herrera
Gary Herron
Carol Heward
Larry Heward
Hermi Hiatt
John Hiatt
Holly Hicks
George Hill
Margaret Hill
Jack Hilsabeck
Brian Hobbs
Nancy Hoffman
David Holway
Eric Hopson
Heather Hundt
Marcus Hurd
Linda Hussa
Ester Hutchison
Mary Hutton
Larry Hyslop
Marshall Iliff
Shonna Ingram
Francesca Innocenti
Michael Janik
Stephanie Jentsch
Ken Johns
Gary Johnson
Greg Jones
Sally Jones
Hugh Judd
Joe Kahl
Richard Katschke
Terry Katzer

Virginia Katzer
Donna Ketner
Keith Ketner
Virgil Ketner
Mike King
Hugh Kingery
Urling Kingery
Karen Kish
David Klauberg
Anna Knipps
Marjorie Knorr
Owen A. Knorr
Tom Kobylarz
Lawrence Krenzien
Marjorie Krenzien
Jeri Krueger
Edward Kuklinski
Wendy Kuntz
Sherry Lacker
Jessica Langsam
John Lanning
Kit Larsen
James Lawrence
Anastasia Leigh
Tim Lenz
Steve Leslie
Lynn Lindsay
N. Little
Kelly Long
Bruce Lund
Floice Lund
Sheree Luttrell
Jim Lytle
Marsha Lytle
Kevin Mack
Laura Mack
Jeff Mackay
Paul Mackesey
Carrie Marten
Jon Martin
Leslie Martin
Joe Maslach
Sara Mattan
Frank Maxwell
Mike Mayberry
Jeff McColcum
John McCormick
Keiker McCormick
Bart McDermitt
Karen McDonal
Neil McDonal
Birnie McGavin
Millie McGavin
Kristian McIntyre
Martin Meyers

Kerstan Micone
Jim Mills
Rebecca Mills
Sally Mills
Carolyn Molica
Barbara Moore
Jim Moore
Brittani Morse
Callie Morse
Chantey Morse
Cole Morse
Kit Morse
Kris Morse
Shane Morse
Tom Myers
Jan Nachlinger
Jennifer Napier
Larry Neel
Ray Nelson
Helen Neville
Pam Nickels
Chris Niemela
Kathy Oakes
Martin Osborn
Aggie Owens
Felix Owens
Elaine Oxborrow
Phyllis Paxton
Bill Pelle
Tom Peppersack
Mike Perkins
Ann Peterka
Al Pfister
John Poole
Lois Ports
Mark Ports
Adam Quinn
Jim Ramakka
Donnie Ratcliff
Floyd Rathbun
Robert Rauch
Barbara Raulston
John Reagan
Carol Reed
Michael Reed
Melissa Renfro
Pamela Repp
Jennifer Richard
Chuck Ritter
Deenie Ritter
Andrea Robb
Ben Roberts
Audrey Rollo
Shane Romos
Margaret Rubega

Chuck Rumsey
Ronnie Ryno
Mike San Miguel
Rick Saval
Burt Sawade
Joe Sawyer
Joe Schlageter
Rita Schlageter
Matt Schlenker
Matthew Schlesinger
Stephen Schmidt
Steve Schmidt
Katja Schott
Erin Schuldheiss
Monica Schwalbach
Greg Scyphers
Jeff Sedy
Betty Seebeck
M. Seebeck
Shirley Sekarajasingham
Dennis Serdehely
Rebecca Serdehely
Catina Sevidal
Amanda Seymour
Ken Seymour
Pam Shandrick
Pat Shanley
Jenn Shinnick
Huston Shoopman
Trudy Sharpe
Sophie Sheppard
Gayle Shields
Seth Shteir
Steven Siegel
Grace Simeone
Stephanie Sims
Kevin Sloan
Eric Skov
Heather Smith
Jerry Smith
Randy Smith
Rob Smith
Deb Snyder
Kei Sochi
Mark Sogge
John Spence
Mark Stackhouse
Beth St. George
David St. George
Dave Straley
Gwen Straley
Pam Straley
Rose Strickland
Je Anne Strott-Branca
Cindy Suitor

Brian Sullivan
Jane Sunday
John Swett
Michael Taft
Judy Taggard
Liz Terry
Pat Terry
Larry Teske
Mei Thai
Jane Thompson
Jeanne Tinsman
Carolyn Titus
Richard Titus
Cris Tomlinson
Kathleen Trever
John Trijonis
Dennis Trousdale
Monica Turi
Kent Undlin
Cathi Unruh
Alan Versaw
Ken Voget
Martha Voget
Sue Wainscott
Alan Wallace

Jack Walters
John Warder
Nathan Welch
Therese Werst
Cameron Whaitman
Melanie Wilhelm
Brian Williams
Jill Williams
Robert Williams
Phil Wilson
Abby Wines
Mary Winter
Robin Wolcott
Diane Wong
John Woodyard
Tasha Woodyard
David Worley
JoLynn Worley
Tom Wurster
Eileen Wynkoop
Brad Young
Craig Young
Desna Young
Sandy Young
Randy Yuen

In addition to on-the-ground support from atlas workers in the field, we are very grateful for financial, logistical, and in-kind support from a large number of government, university, tribal, nonprofit, and corporate sponsors. We owe a great debt of gratitude to each of the following organizations and institutions:

Barrick Goldstrike Mines
Barrick Museum of Natural History, University of Nevada, Las Vegas
Biological Resources Research Center, University of Nevada, Reno
Clark County Desert Conservation Program: Multiple Species Habitat Conservation Plan
Confederated Tribes of the Goshute Reservation, Nevada and Utah
Duckwater Shoshone Tribe of the Duckwater Reservation, Nevada
Fallon Naval Air Station, U.S. Department of Defense
Fort Mojave Indian Tribe of Arizona, California, and Nevada
Great Basin National Park, U.S. National Park Service
Hawthorne Army Depot, U.S. Department of Defense
Independence Mining Company
Lahontan Audubon Society
Lake Mead National Recreation Area, U.S. National Park Service
Moapa Band of the Paiute Indians of the Moapa River Indian Reservation, Nevada
National Fish and Wildlife Foundation
Natural Resources Conservation Service
The Nature Conservancy of Nevada
Nellis Air Force Range, U.S. Department of Defense

Nevada Bell
Nevada Department of Wildlife
Nevada Land and Resources Company
Nevada Mining Association
Nevada Natural Heritage Program
Nevada Partners in Flight
Nevada Power
Norcross Wildlife Foundation
Patagonia, Inc.
Pyramid Lake Paiute Tribe of the Pyramid Lake Reservation, Nevada
Red Rock Audubon Society
Sierra Pacific Resources
Shoshone-Paiute Tribe of the Duck Valley Reservation, Nevada
Southern Nevada Water Authority
U.S. Bureau of Land Management
U.S. Bureau of Reclamation
U.S. Department of Defense Partners in Flight
U.S. Department of Energy
U.S. Fish and Wildlife Service
USDA Forest Service
Walker River Paiute Tribe of the Walker River Reservation, Nevada
Yerington Paiute Tribe of the Yerington Colony and Campbell Ranch, Nevada

The atlas project received guidance and counsel from its original inception through the conclusion of fieldwork from a steering committee whose insights and energy were essential for a project of this magnitude. Robert Flores (U.S. Fish and Wildlife Service), Jim Ramakka (U.S. Bureau of Land Management), and John Swett (U.S. Bureau of Reclamation) deserve special thanks for their key roles and dedication in this function, as well as for numerous other contributions to the atlas project. Other individuals participating in the steering committee were: Don Baepler (Barrick Museum of Natural History, University of Nevada, Las Vegas), Janet Bair (The Nature Conservancy), Mike Boyles (U.S. National Park Service, Lake Mead NRA), Peter Bradley (Nevada Department of Wildlife), Peter Brussard (Biological Resources Research Center, University of Nevada, Reno), Karl Burke (Barrick Goldstrike Mines), Eric Campbell (U.S. Bureau of Land Management), Lucie Clark (Lahontan Audubon Society), Garry Cottle (Fallon Naval Air Station, U.S. Department of Defense), Deb Couche (USDA Forest Service), Walt DeVaurs (U.S. Bureau of Land Management), Kerwin Dewberry (USDA Forest Service), Chris Eberly (U.S. Department of Defense, Partners in Flight), Brad Edwards (U.S. Bureau of Reclamation), Jim Eidel (Great Basin Bird Observatory), Mary Jo Elpers (U.S. Fish and Wildlife Service), Alan Gubanich (Lahontan Audubon Society, Great Basin Bird Observatory), Meg Jensen (U.S. Bureau of Land Management), Jeri Krueger (U.S. Fish and Wildlife Service), Gayle Marrs-Smith (U.S. Bureau of Land Management), Zane Marshall (Southern Nevada Water Authority, Great Basin Bird Observatory), Larry Neel (Nevada Department of Wildlife), Jennifer Newmark (Nevada Natural Heritage Program), Robert

Schmidt (Natural Resources Conservation Service), Steve Siegel (Sierra Pacific Resources), Carolyn Titus (Red Rock Audubon Society), Cris Tomlinson (Nevada Department of Wildlife), Richard Tracy (Biological Resources Research Center, University of Nevada, Reno), Vicki Tripoli (Nevada Power), Ken Voget (U.S. Fish and Wildlife Service), Alan Wallace, Eric Watkins (Nellis Air Force Range, U.S. Department of Defense), and Genny Wilson (USDA Forest Service).

It is furthermore a pleasure to acknowledge the contributions of the following individuals for their direct assistance in the areas of fundraising, scientific advice, information about Nevada's birds, logistical support, review of the draft manuscript, or some combination thereof:

Larry Allen
John Anderson (University of Nevada, Reno)
Sheila Anderson
Don Baepler (Barrick Museum University of Nevada, Las Vegas)
John Bare (Nevada Power)
Janet Bair (The Nature Conservancy)
Donn Blake (Red Rock Audubon Society)
Jim Boone
Douglas Booth (USDA Forest Service)
Mike Boyles (National Park Service Lake Mead NRA)
Pete Bradley (Nevada Department of Wildlife)
Richard Brune
Peter Brussard (Biological Resources Research Center, University of Nevada, Reno)
Karl Burke (Barrick Goldstrike Mines)
Carrie Carreño
David Catalano
Lucie Clark (Lahontan Audubon Society)
Glenn Clemmer (Nevada Natural Heritage Program)
Troy Corman (Arizona Fish and Game Department)
Gary Cottle (Fallon Naval Air Station)
Lisa Crampton (Biological Resources Research Center, University of Nevada, Reno)
Dorothy Crowe (University of Nevada, Las Vegas)
David Delehanty (Idaho State University)
Walt DeVaurs (U.S. Bureau of Land Management)
S. Dunham (Oregon State University)
Jon Dunn
Maureen Easton (USDA Forest Service)
Chris Eberly (U.S. Department of Defense, Partners in Flight)
Brad Edwards (U.S. Bureau of Reclamation)
Jim Eidel (Great Basin Bird Observatory)
Mary Jo Elpers (U.S. Fish and Wildlife Service)
Robert Flores (U.S. Fish and Wildlife Service)
Rick Fridell (Utah Division of Wildlife Resources)
Robert Furlow (U.S. Department of Energy)
Joel Geier
Christina Gibson (Clark County Desert Conservation Program)
Bob Goodman (Desert Images)
Alan Gubanich (Lahontan Audubon Society, Great Basin Bird Observatory)

Jeanene Hafen
Ross Haley (National Park Service, Lake Mead NRA)
Dana Hartley (Great Basin Bird Observatory)
Ame Hellman (The Nature Conservancy)
Larry Hillerman (Biological Resources Research Center, University of Nevada, Reno)
Carrie House
Michael Janik
Joe Kahl (U.S. Bureau of Reclamation)
Anna Lou Kelso (Great Basin Bird Observatory)
D. M. Keppie
Hugh Kingery (Colorado Breeding Bird Atlas)
Owen Knorr (Institute of Alpine Ecology)
Frank Kratzer
Jeri Krueger (U.S. Fish and Wildlife Service)
Paul Lehman
Bruce Lund (The Nature Conservancy)
Jim Lytle
Jeff Mackay (U.S. Fish and Wildlife Service)
Gayle Marrs-Smith (U.S. Bureau of Reclamation)
Zane Marshall (Southern Nevada Water Authority, Great Basin Bird Observatory)
Neil McDonal
Don McIvor (Lahontan Audubon Society)
Brian McMenamy (Biological Resources Research Center, University of Nevada, Reno)
Martin Meyers
Willie Molini (Great Basin Bird Observatory)
Milton Moody (Utah Ornithological Society)
Craig Mortimore (Nevada Department of Wildlife)
Jan Nachlinger (The Nature Conservancy)
Larry Neel (Nevada Department of Wildlife)
Ray Nelson (Lahontan Audubon Society)
Jennifer Newmark (Nevada Natural Heritage Program)
Chris Nicolai
Joanne O'Hare (University of Nevada Press)
Colleen O'Laughlin (U.S. Department of Energy)
Felix Owens
David Parmelee
Bill Pelle (National Park Service, Lake Mead NRA)
Fred Petersen
Dave Pulliam (U.S. Bureau of Reclamation)

Jim Ramakka (U.S. Bureau of Land Management)
Will Richardson (Biological Resources Research Center, University of Nevada, Reno)
R. Rovansek
Margret Rubega (University of Connecticut)
Mike San Miguel
Rick Saval
Rob Scanland (The Nature Conservancy)
Rita Schlageter (Red Rock Audubon Society)
Robert Schmidt (Natural Resources Conservation Service)
Monica Schwalbach (USDA Forest Service)
Greg Scyphers
Steve Siegel (Sierra Pacific Resources)
Kei Sochi (The Nature Conservancy)
Jim Steele
Je Anne Strott-Branca
John Swett (U.S. Bureau of Reclamation)
Kathleen Szawiola (University of Nevada Press)
Jeanne Tinsman (U.S. Fish and Wildlife Service)
Carolyn Titus (Red Rock Audubon Society)
Cris Tomlinson (Nevada Department of Wildlife)
Richard Tracy (Biological Resources Research Center, University of Nevada, Reno)
Sara Vélez Mallea (University of Nevada Press)
Ken Voget (U.S. Fish and Wildlife Service)
Tim Wade (Biological Resources Research Center, University of Nevada, Reno)
Jon Walker (Biological Resources Research Center, University of Nevada, Reno)
Alan Wallace
Eric Watkins (U.S. Department of Defense)
Walter Wehtje
Genny Wilson (USDA Forest Service)
Christopher L. Wood (Cornell Laboratory of Ornithology)
John Woodyard
Tom Wurster
Tara Zimmerman (U.S. Fish and Wildlife Service)

Finally, our sincere thanks to two anonymous reviewers of the atlas manuscript, whose detailed and extensive comments helped to greatly improve the book. Any remaining errors or omissions are, of course, the responsibility of the authors.

Southern Nevada
Water Authority

University of Nevada, Reno

*B*iological
*R*esources
*R*esearch
*C*enter

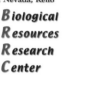

APPENDIX: SCIENTIFIC NAMES

BIRDS NOT TREATED IN SPECIES ACCOUNTS

COMMON NAME	SCIENTIFIC NAME
American Redstart	*Setophaga ruticilla*
Boreal Owl	*Aegolius funereus*
Brown-capped Rosy-Finch	*Leucosticte australis*
Curve-billed Thrasher	*Toxostoma curvirostre*
Magnificent Hummingbird	*Eugenes fulgens*
Northern Waterthrush	*Seiurus noveboracensis*
Oak Titmouse	*Baeolophus inornatus*
Purple Finch	*Carpodacus purpureus*
Purple Martin	*Progne subis*
Upland Sandpiper	*Bartramia longicauda*
Veery	*Catharus fuscescens*
White-throated Dipper	*Cinclus cinclus*
White-throated Sparrow	*Zonotrichia albicollis*
Yellow Rail	*Coturnicops noveboracensis*

OTHER VERTEBRATES

COMMON NAME	SCIENTIFIC NAME
Black bear	*Ursus americanus*
Bullfrog	*Rana catesbeiana*
Coyote	*Canis latrans*
Cui-ui	*Chasmistes cujus*
Desert tortoise	*Gopherus agassizii*
Domestic cat	*Felis silvestris*
Ground squirrel	*Spermophilus* spp.
Kangaroo rat	*Dipodomys* spp.
Mountain lion	*Puma concolor*
Muskrat	*Ondatra zibethica*
Pika	*Ochotona princeps*
Prairie dog	*Cynomys* spp.
Rattlesnake	in Nevada, *Crotalus viridis, C. cerastes, C. scutulatus, C. mitchellii,* or *C. atrox*
Vole	in Nevada, *Microtus* spp. or *Lemmiscus curtatus*

INVERTEBRATES

COMMON NAME	SCIENTIFIC NAME
Brine fly	*Ephydra* spp.
Brine shrimp	*Artemia* spp.
Crayfish	*Pacifastacus* spp.

555

COMMON NAME	SCIENTIFIC NAME	COMMON NAME	SCIENTIFIC NAME
Earthworm	*Lumbricus* spp.	Hackberry	*Celtis occidentalis*
Mormon cricket	*Anabrus simplex*	Hardstem bulrush	*Scirpus acutus*
Tarantula	*Aphonopelma* spp.	Jeffrey pine	*Pinus jeffreyi*
Zebra mussel	*Dreissena polymorpha*	Joshua tree	*Yucca brevifolia*
		Juniper	*Juniper* spp.

PLANTS

COMMON NAME	SCIENTIFIC NAME		
		Limber pine	*Pinus flexilis*
		Lodgepole pine	*Pinus contorta*
Acacia = catclaw	*Acacia* spp.	Manzanita	*Arctostaphylos* spp.
Alder	*Alnus* spp.	Mesquite	*Prosopis* spp.
Ash = velvet ash	*Fraxinus velutina*	Mistletoe	*Phoradendron* spp.
Aspen = quaking aspen	*Populus tremuloides*	Mountain mahogany	*Cercocarpus* spp.
Bailey's greasewood	*Sarcobatus baileyi*	Oak	*Quercus* spp.
Big sagebrush	*Artemisia tridentata*	Ponderosa pine	*Pinus ponderosa*
Bitterbrush	*Purshia* spp.	Pinyon-juniper—see pinyon pine	
Blackberry	*Rubus* spp.	and juniper	
Black willow	*Salix laevigata*	Pinyon pine	*Pinus monophylla*
Bristlecone pine	*Pinus longaeva*	Pine	*Pinus* spp.
Buffaloberry	*Shepherdia argentea*	Quaking aspen	*Populus tremuloides*
Bulrush	*Scirpus* spp.	Rabbitbrush	*Chrysothamnus* spp.
Catclaw	*Acacia* spp.	Russian olive	*Eleagnus angustifolius*
Cattail	*Typha* spp.	Rush	*Juncus* spp.
Cheatgrass	*Bromus tectorum*	Sagebrush	*Artemisia* spp.
Chokecherry	*Prunus virginiana*	Saguaro cactus	*Carnegia gigantea*
Cholla cactus	*Opuntia* spp.	Saltbush	*Atriplex* spp.
Corn	*Zea mays*	Screwbean mesquite	*Prosopis pubescens*
Cottonwood	*Populus fremontii* and other	Sedge	*Carex* spp.
	Populus species	Serviceberry	*Amelanchier* spp.
Creosotebush	*Larrea tridentata*	Shadscale	*Atriplex confertifolia*
Crested wheatgrass	*Agropyron cristatum*	Snowberry	*Symphoricarpos* spp.
Currant	*Ribes* spp.	Spruce	*Picea* spp.
Cypress	*Cupressus sempervirens*	Subalpine fir	*Abies lasiocarpa*
Desert willow	*Chilopsis linearis*	Sycamore	*Platanus* spp.
Elephant tree	*Bursera microphylla*	Tamarisk	*Tamarix chinensis*
Eucalyptus	*Eucalyptus* spp.	Torrey saltbush	*Atriplex torreyi*
Fir	*Abies* spp. and *Pseudotsuga* spp.	Utah juniper	*Juniperus osteosperma*
Fourwing saltbush	*Atriplex canescens*	White fir	*Abies concolor*
Gambel's oak	*Quercus gambelii*	Willow	*Salix* spp.
Giant cactus = saguaro cactus	*Carnegia gigantea*	Yellow pine	*Pinus* spp., primarily *ponderosa*
Goldenrod	*Solidago* spp.		and/or *jeffreyi* in Nevada
Greasewood	*Sarcobatus* spp.	Yucca	*Yucca baccata*

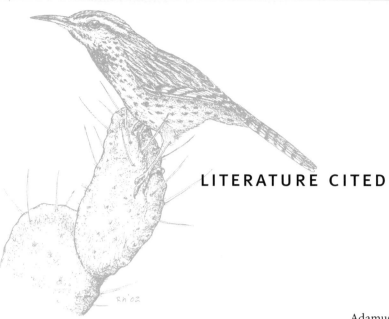

LITERATURE CITED

Adamus PR, Larsen K, Gillson G, and Miller CR. 2001. *Oregon Breeding Bird Atlas.* Eugene: Oregon Field Ornithologists.

Adkisson CS. 1996. Red Crossbill (*Loxia curvirostra*). *In* Poole A and Gill F, eds., *The Birds of North America* no. 256. Philadelphia: Academy of Natural Sciences, and Washington: American Ornithologists' Union.

———. 1999. Pine Grosbeak (*Pinicola enucleator*). *In* Poole A and Gill F, eds., *The Birds of North America* no. 456. Philadelphia: The Birds of North America Inc.

Alcorn JR. 1988. *The Birds of Nevada.* Fallon: Fairview West.

Allin CC, Chasko GG, and Husband TP. 1987. Mute Swans in the Atlantic flyway: A review of the history, population growth, and management needs. *Transactions of the Northeastern Section of the Wildlife Society* 44:32–47.

Allin CC and Husband TP. 2003. Mute Swan (*Cygnus olor*) impact on submerged aquatic vegetation and macroinvertebrates in a Rhode Island coastal pond. *Northeastern Naturalist* 10:305–318.

Altman B and Sallabanks R. 2000. Olive-sided Flycatcher (*Contopus cooperi*). *In* Poole A and Gill F, eds., *The Birds of North America* no. 502. Philadelphia: The Birds of North America Inc.

Ammon EM. 1995. Lincoln's Sparrow (*Melospiza lincolnii*). *In* Poole A and Gill F, eds., *The Birds of North America* no. 191. Philadelphia: Academy of Natural Sciences, and Washington: American Ornithologists' Union.

———. 2002. Changes in the bird community of the lower Truckee River, Nevada, 1868–2001. *Great Basin Birds* 5:13–20.

Ammon EM and Gilbert WM. 1999. Wilson's Warbler (*Wilsonia pusilla*). *In* Poole A and Gill F, eds., *The Birds of North America* no. 478. Philadelphia: The Birds of North America Inc.

Andrews R and Righter R. 1992. *Colorado Birds: A Reference to Their Distribution and Habitat.* Denver: Denver Museum of Natural History.

Andrle RF and Carroll JR, eds. 1988. *The Atlas of Breeding Birds in New York State.* Ithaca: Cornell University Press.

AOU [American Ornithologists' Union]. 1985. Thirty-fifth supplement to the American Ornithologists' Union *Check-list of North American Birds. Auk* 102:680–686.

———. 1995. Fortieth supplement to the American Ornithologists' Union *Check-list of North American Birds. Auk* 112:819–830.

———. 1998. *Check-list of North American Birds.* 7th ed. Washington: American Ornithologists' Union.

———. 2003. Forty-fourth supplement to the American Ornithologists' Union *Check-list of North American Birds. Auk* 120:923–931.

Arcese P, Sogge MK, Marr AB, and Patten MA. 2002. Song Sparrow (*Melospiza melodia*). *In* Poole A and Gill F, eds., *The Birds of North America* no. 704. Philadelphia: The Birds of North America Inc.

Arsenault DP, Wilson GE, and Neel L. 2003. Flammulated Owls in the Spring Mountains, Nevada. *Great Basin Birds* 6:45–51.

Askins RA. 2002. *Restoring North America's Birds: Lessons from Landscape Ecology, Second Edition.* New Haven: Yale University Press.

Austin GT. 1969. New and additional records of some passerine birds in southern Nevada. *Condor* 71:75–76.

———. 1976. Behavioral adaptations of the Verdin to the desert. *Auk* 93:245–262.

Austin GT and Bradley WG. 1971. The avifauna of Clark County, Nevada. *Journal of the Arizona Academy of Science* 6:283–303.

Austin JE, Afton AD, Anderson MG, Clark RG, Custer CM, Lawrence JS, Pollard JB, and Ringelman JK. 2000. Declining scaup populations: Issues, hypotheses, and research needs. *Wildlife Society Bulletin* 28:254–263.

Austin JE, Custer CM, and Afton AD. 1998. Lesser Scaup (*Aythya affinis*). *In* Poole A and Gill F, eds., *The Birds of North America* no. 338. Philadelphia: The Birds of North America Inc.

Austin JE and Miller MR. 1995. Northern Pintail (*Anas acuta*). *In* Poole A and Gill F, eds., *The Birds of North America* no. 163. Philadelphia: Academy of Natural Sciences, and Washington: American Ornithologists' Union.

Balda RP. 2002. Pinyon Jay (*Gymnorhinus cyanocephalus*). *In* Poole A and Gill F, eds., *The Birds of North America* no. 605. Philadelphia: The Birds of North America Inc.

Baltosser WH and Russell SM. 2000. Black-chinned Hummingbird (*Archilochus alexandri*). *In* Poole A and Gill F, eds., *The Birds of North America* no. 495. Philadelphia: The Birds of North America Inc.

Baltosser WH and Scott PE. 1996. Costa's Hummingbird (*Calypte costae*). *In* Poole A and Gill F, eds., *The Birds of North America* no. 251. Philadelphia: Academy of Natural Sciences, and Washington: American Ornithologists' Union.

Banko WE. 1960. *The Trumpeter Swan: Its History, Habits and Population in the United States.* Washington: U.S. Department of the Interior.

Banks RC. 1970. The Fox Sparrow on the west slope of the Oregon Cascades. *Condor* 72:369–370.

Bannor BK and Kiviat E. 2002. Common Moorhen (*Gallinula chloropus*). *In* Poole A and Gill F, eds., *The Birds of North America* no. 685. Philadelphia: The Birds of North America Inc.

Barlow JC. 1980. Patterns of ecological interactions among migrant and resident vireos on the wintering grounds. *In* Keast A and Morton ES, eds., *Migrant Birds in the Neotropics: Ecology, Distribution, and Conservation,* pp. 79–107. Washington: Smithsonian Institution Press.

Barlow JC, Leckie SN, and Baril CT. 1999. Gray Vireo (*Vireo vicinior*). *In* Poole A and Gill F, eds., *The Birds of North America* no. 447. Philadelphia: Academy of Natural Sciences, and Washington: American Ornithologists' Union.

Bates JM. 1992. Frugivory on *Bursera microphylla* (Burseraceae) by wintering Gray Vireos (*Vireo vicinior,* Vireonidae) in the coastal deserts of Sonora, Mexico. *Southwestern Naturalist* 37:252–258.

Beason RC. 1995. Horned Lark (*Eremophila alpestris*). *In* Poole A and Gill F, eds., *The Birds of North America* no. 195. Philadelphia: Academy of Natural Sciences, and Washington: American Ornithologists' Union.

Bechard MJ and Schmutz JK. 1995. Ferruginous Hawk (*Buteo regalis*). *In* Poole A and Gill F, eds., *The Birds of North America* no. 172. Philadelphia: Academy of Natural Sciences, and Washington: American Ornithologists' Union.

Beedy EC and Hamilton WJ. 1999. Tricolored Blackbird (*Agelaius tricolor*). *In* Poole A and Gill F, eds., *The Birds of North America* no. 423. Philadelphia: The Birds of North America Inc.

Behle WH. 1985. *Utah Birds: Geographic Distribution and Systematics.* Utah Museum of Natural History Occasional Publication no. 5. Salt Lake City: Utah Museum of Natural History.

Behle WH, Sorensen ED, and White CM. 1985. *Utah Birds: A Revised Checklist.* Utah Museum of Natural History Occasional Publication no. 4. Salt Lake City: Utah Museum of Natural History.

Bekoff M, Scott AC, and Conner DA. 1987. Nonrandom nest site selection in Evening Grosbeaks. *Condor* 89:819–829.

Bellrose FC. 1959. Lead poisoning as a mortality factor in waterfowl populations. *Illinois Natural History Survey Bulletin* 27:235–288.

———. 1980. *Ducks, Geese, and Swans of North America.* 3rd ed. Harrisburg, Pa.: Stackpole Books.

Bellrose FC and Holm DJ. 1994. *Ecology and Management of the Wood Duck.* Harrisburg, Pa.: Stackpole Books.

Bemis C and Rising JD. 1999. Western Wood-Pewee (*Contopus sordidulus*). *In* Poole A and Gill F, eds., *The Birds of North America* no. 451. Philadelphia: The Birds of North America Inc.

Benkman CW. 1993a. Adaptation to single resources and the evolution of crossbill (*Loxia*) diversity. *Ecological Monographs* 63:305–325.

———. 1993b. Logging, conifers, and the conservation of crossbills. *Conservation Biology* 7:473–479.

Bent AC. 1953. *Life Histories of North American Wood Warblers. United States National Museum Bulletin* no. 203.

Bergstrom E. 1998. Willow Flycatchers on the western edge of the Great Basin. *Great Basin Birds* 1:10–11.

Bildstein KL and Meyer K. 2000. Sharp-shinned Hawk (*Accipiter striatus*). *In* Poole A and Gill F, eds., *The Birds of North America* no. 482. Philadelphia: The Birds of North America Inc.

Blancher PJ and Robertson RJ. 1984. Resource use by sympatric kingbirds. *Condor* 86:305–313.

Bloom PH. 1994. The biology and current status of the Long-eared Owl in coastal southern California. *Bulletin of the Southern California Academy of Science* 93:1–12.

Blus LJ, Stroud RK, Reiswig B, and McEneaney T. 1989. Lead poisoning and other mortality factors in Trumpeter Swans. *Environmental Toxicology and Chemistry* 8:263–271.

Boarman WI. 2002. *Reducing Predation by Common Ravens on Desert Tortoises in the Mojave and Colorado Deserts.* Report to Bureau of Land Management. San Diego: U.S. Bureau of Land Management.

Boarman WI and Heinrich B. 1999. Common Raven (*Corvus corax*). *In* Poole A and Gill F, eds., *The Birds of North America* no. 476. Philadelphia: The Birds of North America Inc.

Bock CE. 1970. The ecology and behavior of the Lewis' Woodpecker (*Asyndesmus lewis*). *University of California Publications in Zoology* 92:1–100.

Bock CE and Bock JH. 1974. On the geographical ecology and evolution of the Three-toed Woodpeckers, *Picoides tridactylus* and *P. arcticus*. *American Midland Naturalist* 92:397–405.

Bock CE and Fleck DE. 1995. Avian response to nest box addition in two forests of the Colorado Front Range. *Journal of Field Ornithology* 3:352–362.

Bock CE and Lepthien LW. 1976. Synchronous eruptions of boreal seed-eating birds. *American Naturalist* 110:559–579.

Bock CE and Webb B. 1984. Birds as grazing indicator species in southeastern Arizona. *Journal of Wildlife Management* 48:1045–1049.

BOR-LCR [Bureau of Reclamation, Lower Colorado Region]. 2004. Lower Colorado River Multi-species Conservation Program: Final Environmental Impact Report/Environmental Impact Statement. Boulder City, Nev.: U.S. Bureau of Reclamation, Lower Colorado Region.

Bowen RV. 1997. Townsend's Solitaire (*Myadestes townsendi*). *In* Poole A and Gill F, eds., *The Birds of North America* no. 269. Philadelphia: Academy of Natural Sciences, and Washington: American Ornithologists' Union.

Boyd WS and Jehl JR Jr. 1998. Air-photo censuses of Eared Grebes on Mono Lake, California. *Colonial Waterbirds* 21:236–241.

Braden GT, Carter K, and Rathbun MR. 2004. The status of Yuma Clapper Rail and Yellow-billed Cuckoo along portions of the Virgin River and Muddy River in southern Nevada: 2003. Report to the Southern Nevada Water Authority. Las Vegas: U.S. Bureau of Reclamation.

Bradley RA. 1980. Avifauna of the Palos Verdes Peninsula, California. *Western Birds* 11:1–24.

Brauning DW, ed. 1992. *Atlas of Breeding Birds in Pennsylvania.* Pittsburgh: University of Pittsburgh Press.

Brisbin IL and Mowbray TB. 2002. American Coot (*Fulica americana*). *In* Poole A and Gill F, eds., *The Birds of North America* no. 697. Philadelphia: The Birds of North America Inc.

Brown BT. 1993. Bell's Vireo (*Vireo bellii*). *In* Poole A, Stettenheim P, and Gill F, eds., *The Birds of North America* no. 35. Philadelphia: Academy of Natural Sciences.

Brown BT, Carothers SW, and Johnson RR. 1983. Breeding range expansion of Bell's Vireo in Grand Canyon, Arizona. *Condor* 85:499–500.

Brown BT and Trosset MW. 1989. Nesting-habitat relationships of riparian birds along the Colorado River in Grand Canyon, Arizona. *Southwestern Naturalist* 34:260–270.

Brown CR and Brown MB. 1995. Cliff Swallow (*Hirundo pyrrhonota*). *In* Poole A and Gill F, eds., *The Birds of North America* no. 149. Philadelphia: Academy of Natural Sciences, and Washington: American Ornithologists' Union.

———. 1999. Barn Swallow (*Hirundo rustica*). *In* Poole A and Gill F, eds., *The Birds of North America* no. 452. Philadelphia: The Birds of North America Inc.

Brown CR, Knott AM, and Damrose EJ. 1992. Violet-green Swallow (*Tachycineta thalassina*). *In* Poole A, Stettenheim P, and Gill F, eds., *The Birds of North America* no. 14. Philadelphia: Academy of Natural Sciences, and Washington: American Ornithologists' Union.

Brown DE, Hagelin JC, Taylor M, and Galloway J. 1998. Gambel's Quail (*Callipepla gambeli*). *In* Poole A and Gill F, eds., *The Birds of North America* no. 321. Philadelphia: The Birds of North America Inc.

Brown S, Hickey C, Harrington B, and Gill R, eds. 2001. The U.S. Shorebird Conservation Plan. 2nd ed. Manomet, Mass.: Manomet Center for Conservation Sciences. For more information and updates, see: http://www.fws.gov/shorebirdplan/USShorebird.htm

Brua RB. 2001. Ruddy Duck (*Oxyura jamaicensis*). *In* Poole A and Gill F, eds., *The Birds of North America* no. 696. Philadelphia: The Birds of North America Inc.

Buckelew AR and Hall GA. 1994. *The West Virginia Breeding Bird Atlas.* Pittsburgh: University of Pittsburgh Press.

Buckland ST, Anderson DR, Burnham KP, and Laake JL. 1993. *Distance Sampling: Estimating Abundance of Biological Populations.* London: Chapman and Hall.

Budnik JM, Thompson FR, and Ryan MR. 2002. Effect of habitat characteristics on the probability of parasitism and predation of Bell's Vireo nests. *Journal of Wildlife Management* 66:232–239.

Buehler DA. 2000. Bald Eagle (*Haliaeetus leucocephalus*). *In* Poole A and Gill F, eds., *The Birds of North America* no. 506. Philadelphia: The Birds of North America Inc.

Bull EL and Jackson JA. 1995. Pileated Woodpecker (*Dryocopus pileatus*). *In* Poole A, and Gill F, eds., *The Birds of North America* no. 148. Philadelphia: Academy of Natural Sciences.

Bump SR. 1986. Yellow-headed Blackbird (*Xanthocephalus xanthocephalus*) nest defense: Aggressive responses to Marsh Wrens (*Cistothorus palustris*). *Condor* 88:328–335.

Burger J and Gochfeld M. 1994. Franklin's Gull (*Larus pipixcan*). *In* Poole A and Gill F, eds., *The Birds of North America* no. 116. Philadelphia: Academy of Natural Sciences, and Washington: American Ornithologists' Union.

Burleigh TD. 1972. *Birds of Idaho.* Caldwell, Ida.: Caxton Printers.

Burridge B, ed. 1995. *Sonoma County Breeding Bird Atlas.* Santa Rosa: Madrone Audubon Society.

Butler RW. 1992. Great Blue Heron (*Ardea herodias*). *In* Poole A, Stettenheim P, and Gill F, eds., *The Birds of North America* no. 25. Philadelphia: Academy of Natural Sciences, and Washington: American Ornithologists' Union.

Cabe PR. 1993. European Starling (*Sturnus vulgaris*). *In* Poole A and Gill F, eds., *The Birds of North America* no. 48. Philadelphia: Academy of Natural Sciences, and Washington: American Ornithologists' Union.

Calder WA. 1993. Rufous Hummingbird (*Selasphorus rufus*). *In* Poole A and Gill F, eds., *The Birds of North America* no. 53. Philadelphia: Academy of Natural Sciences, and Washington: American Ornithologists' Union.

Calder WA and Calder LL. 1992. Broad-tailed Hummingbird (*Selasphorus platycercus*). *In* Poole A, Stettenheim P, and Gill F, eds., *The Birds of North America* no. 16. Philadelphia: Academy of Natural Sciences, and Washington: American Ornithologists' Union.

———. 1994. Calliope Hummingbird (*Stellula calliope*). *In* Poole A and Gill F, eds., *The Birds of North America* no. 135. Philadelphia: Academy of Natural Sciences, and Washington: American Ornithologists' Union.

Calkins JD, Hagelin JC, and Lott DF. 1999. California Quail (*Callipepla californica*). *In* Poole A and Gill F, eds., *The Birds of North America* no. 473. Philadelphia: The Birds of North America Inc.

Cannings RJ. 1993. Northern Saw-whet Owl (*Aegolius acadicus*). *In* Poole A and Gill F, eds., *The Birds of North America* no. 42. Philadelphia: Academy of Natural Sciences, and Washington: American Ornithologists' Union.

Cannings RJ and Angell T. 2001. Western Screech-Owl (*Otus kennicottii*). *In* Poole A and Gill F, eds., *The Birds of North America* no. 597. Philadelphia: The Birds of North America Inc.

Canterbury GE. 2003. Pacific-slope and Cordilleran Flycatchers. *In* Marshall DB, Hunter MG, and Contreras AL, eds., *Birds of Oregon: A General Reference*, pp. 386–389. Corvallis: Oregon State University Press.

Cardiff SW and Dittmann DL. 2000. Brown-crested Flycatcher (*Myiarchus tyrannulus*). *In* Poole A and Gill F, eds., *The Birds of North America* no. 496. Philadelphia: The Birds of North America Inc.

———. 2002. Ash-throated Flycatcher (*Myiarchus cinerascens*). *In* Poole A and Gill F, eds., *The Birds of North America* no. 664. Philadelphia: The Birds of North America Inc.

Carratello R. 1993. Cliff Swallow (*Hirundo pyrrhonota*). *In* Roberson D and Tenney C, eds., *Atlas of the Breeding Birds of Monterey County*, pp. 236–237. Carmel: Monterey Peninsula Audubon Society.

Carroll JP. 1993. Gray Partridge (*Perdix perdix*). *In* Poole A and Gill F, eds., *The Birds of North America* no. 58. Philadelphia: Academy of Natural Sciences, and Washington: American Ornithologists' Union.

Carter M, Fenwick G, Hunter C, Pashley D, Petit D, Price J, and Trapp J. 1996. For the future. *American Birds* 50:238–240.

Chace JF and Cruz A. 1996. Knowledge of the Colorado host relations of the parasitic Brown-headed Cowbird (*Molothrus ater*). *Journal of the Colorado Field Ornithologists* 30:67–81.

Charlet DA. 1996. *Atlas of Nevada Conifers: A Phytogeographic Reference*. Reno: University of Nevada Press.

Chilton G, Baker MC, Barrentine CD, and Cunningham MA. 1995. White-crowned Sparrow (*Zonotrichia leucophrys*). *In* Poole A and Gill F, eds., *The Birds of North America* no. 183. Philadelphia: Academy of Natural Sciences, and Washington: American Ornithologists' Union.

Chisholm G and Neel L. 2002. *Birds of the Lahontan Valley, Nevada*. Reno: University of Nevada Press.

Christensen GC. 1954. *The Chukar Partridge in Nevada*. Biological Bulletin 1. Carson City: Nevada Fish and Game Commission.

———. 1996. Chukar (*Alectoris chukar*). *In* Poole A and Gill F, eds., *The Birds of North America* no. 258. Philadelphia: Academy of Natural Sciences, and Washington: American Ornithologists' Union.

———. 1998. Himalayan Snowcock (*Tetraogallus himalayensis*). *In* Poole A and Gill F, eds., *The Birds of North America* no. 328. Philadelphia: The Birds of North America Inc.

Chu M and Walsberg G. 1999. Phainopepla (*Phainopepla nitens*). *In* Poole A and Gill F, eds., *The Birds of North America* no. 415. Philadelphia: The Birds of North America Inc.

Ciaranca MA, Allin CC, and Jones GS. 1977. Mute Swan (*Cygnus olor*). *In* Poole A and Gill F, eds., *The Birds of North America* no. 273. Philadelphia: The Birds of North America Inc.

Cicero C. 1996. Sibling species of titmice in the *Parus inornatus* complex (Aves: Paridae). *University of California Publications in Zoology* 128: 1–217.

———. 2000. Oak Titmouse (*Baeolophus inornatus*) and Juniper Titmouse (*Baeolophus ridgwayi*). *In* Poole A and Gill F, eds., *The Birds of North America* no. 485. Philadelphia: The Birds of North America Inc.

Cink CL. 2002. Whip-poor-will (*Caprimulgus vociferus*). *In* Poole A and Gill F, eds., *The Birds of North America* no. 620. Philadelphia: The Birds of North America Inc.

Clark County. 2000. Final Clark County Multiple Species Habitat Conservation Plan and Environmental Impact Statement for Issuance of a Permit to Allow Incidental Take of 79 Species in Clark County, Nevada. Las Vegas, Clark County, Nev.

Coates PS. 2001. Movement, Survivorship, and Reproductive Behavior of Columbian Sharp-tailed Grouse Translocated to Nevada. M.S. thesis, University of Nevada, Reno.

Coates PS and Delehanty DJ. 2006. Effect of capture date on nest-attempt rate of translocated Sharp-tailed Grouse. *Wildlife Biology* 12:277–283.

Cody ML. 1999. Crissal Thrasher (*Toxostoma crissale*). *In* Poole A and Gill F, eds., *The Birds of North America* no. 419. Philadelphia: The Birds of North America Inc.

Collins PW. 1999. Rufous-crowned Sparrow (*Aimophila ruficeps*). *In* Poole A and Gill F, eds., *The Birds of North America* no. 472. Philadelphia: The Birds of North America Inc.

Colwell MA and Jehl JR Jr. 1994. Wilson's Phalarope (*Phalaropus tricolor*). *In* Poole A and Gill F, eds., *The Birds of North America* no. 83. Philadelphia: The Birds of North America Inc.

Connelly, JW, Knick ST, Schroeder MA, and Stiver SJ. 2004. Conservation assessment of Greater Sage-Grouse and sagebrush habitats. Cheyenne, Wyo.: Western Association of Fish and Wildlife Agencies.

Conover MR. 1983. Recent changes in Ring-billed Gull and California Gull populations in the western United States. *Wilson Bulletin* 95:362–383.

Contreras A. 1999. *A Pocket Guide to Oregon Birds.* 2nd ed. Eugene: Oregon Field Ornithologists.

Contreras A and Kindschy RR. 1996. *Birds of Malheur County, Oregon.* Eugene: Oregon Field Ornithologists.

Conway CJ. 1995. Virginia Rail (*Rallus limicola*). *In* Poole A and Gill F, eds., *The Birds of North America* no. 173. Philadelphia: Academy of Natural Sciences, and Washington: American Ornithologists' Union.

Conway CJ and Gibbs JP. 2005. Effectiveness of call-broadcast surveys for monitoring marsh birds. *Auk* 122:26–35.

Corman TE and Magill RT. *Western Yellow-billed Cuckoo in Arizona: 1998 and 1999 Survey Report.* Technical Report 150. Pheonix: Arizona Game and Fish Department.

Corman TE and Wise-Gervais C, eds. 2005. *Arizona Breeding Bird Atlas.* Albuquerque: University of New Mexico Press.

Crampton LH. 2004. Ecological determinants of the distribution, abundance, and breeding success of Phainopeplas (*Phainopepla nitens,* Cl. Aves) at the northern edge of their range. Ph.D. diss., University of Nevada, Reno.

Crampton LH and Murphy DD. 2006. Ecology and conservation of Phainopeplas in southern Nevada: The challenges of managing a moving target. *Great Basin Birds* 8:21–31.

Cringan AT, Kingery H, and Wills D. 1992. *Proceedings of the Fourth Breeding Bird Atlas Conference, 13–15 September 1991.* Denver: North American Ornithological Atlas Committee.

Crockett AB and Hadow HH. 1975. Nest site selection by Williamson's and Red-naped Sapsuckers. *Condor* 77:365–368.

Crocoll ST. 1994. Red-shouldered Hawk (*Buteo lineatus*). *In* Poole A and Gill F, eds., *The Birds of North America* no. 107. Philadelphia: Academy of Natural Sciences, and Washington: American Ornithologists' Union.

Cronquist A, Holmgren AH, Holmgren NH, Reveal JL, and Holmgren PK. 1994. *Intermountain Flora.* Vol. 5. New York: New York Botanical Garden.

Cullen SA, Jehl JR Jr., and Nuechterlein GL. 1999. Eared Grebe (*Podiceps nigricollis*). *In* Poole A and Gill F, eds., *The Birds of North America* no. 433. Philadelphia: Academy of Natural Sciences, and Washington: American Ornithologists' Union.

Curry RL, Peterson AT, and Langen TA. 2002. Western Scrub-Jay (*Aphelocoma californica*). *In* Poole A and Gill F, eds., *The Birds of North America* no. 712. Philadelphia: The Birds of North America Inc.

Curson DR and Goguen CB. 1998. Plumbeous Vireo (*Vireo plumbeus*). *In* Poole A and Gill F, eds., *The Birds of North America* no. 366. Philadelphia: Academy of Natural Sciences, and Washington: American Ornithologists' Union.

Custer CM and Custer TW. 1996. Food habits of diving ducks in the Great Lakes after the zebra mussel invasion. *Journal of Field Ornithology* 67:86–99.

Custer TW, Hensler GL, and Kaiser TE. 1983. Clutch size, reproductive success, and organochlorine contaminants in Atlantic Coast Black-crowned Night-Herons. *Auk* 100:699–710.

Cuthbert FJ and Wires LR. 1999. Caspian Tern (*Sterna caspia*). *In* Poole A and Gill F, eds., *The Birds of North America* no. 403. Philadelphia: The Birds of North America Inc.

Darling JL. 1970. New breeding records of *Toxostoma curvirostre* and *T. bendirei* in New Mexico. *Condor* 72:366–367.

Davis WE. 1993. Black-crowned Night-Heron (*Nycticorax nycticorax*). *In* Poole A and Gill F, eds., *The Birds of North America* no. 74. Philadelphia: Academy of Natural Sciences, and Washington: American Ornithologists' Union.

Davis WE and Kushlan JA. 1994. Green Heron (*Butorides virescens*). *In* Poole A and Gill F, eds., *The Birds of North America* no. 129. Philadelphia: Academy of Natural Sciences, and Washington: American Ornithologists' Union.

Dawson WR. 1997. Pine Siskin (*Carduelis pinus*). *In* Poole A and Gill F, eds., *The Birds of North America* no. 280. Philadelphia: Academy of Natural Sciences, and Washington: American Ornithologists' Union.

DeJong MJ. 1996. Northern Rough-winged Swallow (*Stelgidopteryx serripennis*). *In* Poole A and Gill F, eds., *The Birds of North America* no. 234. Philadelphia: Academy of Natural Sciences, and Washington: American Ornithologists' Union.

Derrickson KC and Breitwisch R. 1992. Northern Mockingbird (*Mimus polyglottos*). *In* Poole A, Stettenheim P, and Gill F, eds., *The Birds of North America* no. 7. Philadelphia: Academy of Natural Sciences, and Washington: American Ornithologists' Union.

DeSante DF and George TL. 1994. Population trends in the landbirds of western North America. *Studies in Avian Biology* 15:173–190.

Dixon RD and Saab VA. 2000. Black-backed Woodpecker (*Picoides arcticus*). *In* Poole A and Gill F, eds., *The Birds of North America* no. 509. Philadelphia: The Birds of North America Inc.

Dobbs RC, Martin PR, and Martin TE. 1998. Green-tailed Towhee (*Pipilo chlorurus*). *In* Poole A and Gill F, eds., *The Birds of North America* no. 368. Philadelphia: The Birds of North America Inc.

Dobbs RC, Martin TE, and Conway CJ. 1997. Williamson's Sapsucker (*Sphyrapicus thyroideus*). *In* Poole A and Gill F, eds., *The Birds of North America* no. 285. Philadelphia: Academy of Natural Sciences, and Washington: American Ornithologists' Union.

Dobkin DS. 1994. *Conservation and Management of Neotropical Migrant Landbirds in the Northern Rockies and Great Plains.* Moscow: University of Idaho Press.

Dobkin DS, Rich AC, Pretare JA, and Pyle WH. 1995. Nest-site relationships among cavity-nesting birds of riparian and snowpocket aspen woodlands in the northwestern Great Basin. *Condor* 97:694–707.

Dobkin DS and Sauder JD. 2004. *Shrubsteppe Landscapes in Jeopardy: Distributions, Abundances and the Uncertain Future of*

Birds and Small Mammals in the Intermountain West. Bend, Ore.: High Desert Ecological Research Institute.

Dobkin DS and Wilcox BA. 1986. Analysis of natural forest fragments: Riparian birds in the Toiyabe Mountains, Nevada. *In* Verneer J, Morrison ML, and Ralph CJ, eds., *Wildlife 2000: Modeling Habitat Relationships of Terrestrial Vertebrates,* pp. 293–299. Madison: University of Wisconsin Press.

Drilling N, Titman R, and McKinney F. 2002. Mallard (*Anas platyrhynchos*). *In* Poole A and Gill F, eds., *The Birds of North America* no. 658. Philadelphia: The Birds of North America Inc.

DuBowy PJ. 1996. Northern Shoveler (*Anas clypeata*). *In* Poole A and Gill F, eds., *The Birds of North America* no. 217. Philadelphia: Academy of Natural Sciences, and Washington: American Ornithologists' Union.

Dugger BD and Dugger KM. 2002. Long-billed Curlew (*Numenius americanus*). *In* Poole A and Gill F, eds., *The Birds of North America* no. 628. Philadelphia: Academy of Natural Sciences, and Washington: American Ornithologists' Union.

Dugger BD, Dugger KM, and Fredrickson LH. 1994. Hooded Merganser (*Lophodytes cucullatus*). *In* Poole A and Gill F, eds., *The Birds of North America* no. 98. Philadelphia: Academy of Natural Sciences, and Washington: American Ornithologists' Union.

Dunham S, Butcher L, Charlet DA, and Reed JM. 1996. Breeding range and conservation of Flammulated Owls (*Otus flammeolus*) in Nevada. *Journal of Raptor Research* 30:189–193.

Dunn EH and Agro DJ. 1995. Black Tern (*Chlidonias niger*). *In* Poole A and Gill F, eds., *The Birds of North America* no. 147. Philadelphia: Academy of Natural Sciences, and Washington: American Ornithologists' Union.

Dunn JL and Garrett KL. 1997. *A Field Guide to the Warblers of North America.* Boston: Houghton Mifflin.

Eadie J, Sherman P, and Semel B. 1998. Conspecific brood parasitism, population dynamics, and the conservation of cavity-nesting birds. *In* Caro T, ed., *Behavioral Ecology and Conservation Biology,* pp. 306–340. Oxford: Oxford University Press.

Earnst SL, Neel L, Ivey GL, and Zimmerman T. 1998. Status of the White-faced Ibis: Breeding colony dynamics of the Great Basin population, 1985–1997. *Colonial Waterbirds* 20:301–313.

Eaton SW. 1992. Wild Turkey (*Meleagris gallopavo*). *In* Poole A, Stettenheim P, and Gill F, eds., *The Birds of North America* no. 22. Philadelphia: Academy of Natural Sciences, and Washington: American Ornithologists' Union.

Eckerle KP and Thompson CF. 2001. Yellow-breasted Chat (*Icteria virens*). *In* Poole A and Gill F, eds., *The Birds of North America* no. 575. Philadelphia: Academy of Natural Sciences, and Washington: American Ornithologists' Union.

Ehrlich PR and Daily GC. 1988. Red-naped Sapsuckers feeding at willows: Possible keystone herbivores. *American Birds* 47:357–365.

Ehrlich PR, Dobkin DS, and Wheye D. 1988. *The Birder's Handbook.* New York: Simon and Schuster.

Eiserer L. 1976. *The American Robin.* Chicago: Nelson-Hall.

Ellis E and Nelson J, eds. 1952. *Birds of the Carson River Basin.* Minden: Record-Courier Press.

Ellison WG. 1992. Blue-gray Gnatcatcher (*Polioptila caerulea*). *In* Poole A, Stettenheim P, and Gill F, eds., *The Birds of North America* no. 23. Philadelphia: Academy of Natural Sciences, and Washington: American Ornithologists' Union.

Elphick C. 2001. A comprehensive revision of the Nevada Bird Records Committee review list. *Great Basin Birds* 4:13–18.

Elphick C, Dunning JB, and Sibley DA. 2001. *The Sibley Guide to Bird Life and Behavior.* New York: Knopf.

Emlen JT. 1974. An urban bird community in Tucson, AZ: derivation, structure, regulation. *Condor* 76:184–197.

England AS, Bechard MJ, and Houston CS. 1997. Swainson's Hawk (*Buteo swainsoni*). *In* Poole A and Gill F, eds., *The Birds of North America* no. 265. Philadelphia: Academy of Natural Sciences, and Washington: American Ornithologists' Union.

England AS and Laudenslayer WF. 1993. Bendire's Thrasher (*Toxostoma bendirei*). *In* Poole A and Gill F, eds., *The Birds of North America* no. 71. Philadelphia: Academy of Natural Sciences, and Washington: American Ornithologists' Union.

Espinosa S. 2004. Greater Sage-Grouse conservation plan for Nevada and eastern California completed. *Great Basin Birds* 7:40–45.

Evans RM and Knopf FL. 1993. American White Pelican (*Pelecanus erythrorhynchos*). *In* Poole A and Gill F, eds., *The Birds of North America* no. 57. Philadelphia: Academy of Natural Sciences, and Washington: American Ornithologists' Union.

Evans Mack D and Yong W. 2000. Swainson's Thrush (*Catharus ustulatus*). *In* Poole A and Gill F, eds., *The Birds of North America* no. 540. Philadelphia: The Birds of North America Inc.

Ewell H and Cruz A. 1998. Foraging behavior of the Pygmy Nuthatch (*Sitta pygmaea*). *Western Birds* 29:169–173.

Farquhar CC and Ritchie KL. 2002. Black-tailed Gnatcatcher (*Polioptila melanura*). *In* Poole A and Gill F, eds., *The Birds of North America* no. 690. Philadelphia: The Birds of North America Inc.

Faulkner D, Dexter C, Levad R, and Leukering T. 2005. Black Phoebe breeding range expansion into Colorado. *Western Birds* 36:114–120.

Federal Register. 2001. Endangered and Threatened wildlife and plants; 12-month finding for a petition to list the Yellow-billed Cuckoo (*Coccyzus americanus*) in the western continental United States. *Federal Register* 66:38611–38626.

Field CA. 1997. Estimating abundance for a breeding bird atlas. *In* Brillinger DR, Fernholz LT, and Morgenthaler S, eds., *The Practice of Data Analysis: Essays in Honor of John W. Tukey,* pp. 93–99. Princeton: Princeton University Press.

Fiero W. 1986. *Geology of the Great Basin.* Reno: University of Nevada Press.

Finch DM. 1983. Brood parasitism of the Abert's Towhee: Timing, frequency, and effects. *Condor* 85:355–359.

Fisher AK. 1893. Report on the ornithology of the Death Valley Expedition of 1891, comprising notes on the birds observed in southern California, southern Nevada, and parts of Arizona and Utah. *North American Fauna* 7:7–158.

Fleischer RC and Rothstein SI. 1988. Known secondary contact and rapid gene flow among subspecies and dialects in the Brown-headed Cowbird. *Evolution* 42:1146–1158.

Fleischer RC, Rothstein SI, and Miller L. 1991. Mitochondrial-DNA variation indicates gene flow between subspecies of the Brown-headed Cowbird. *Condor* 93:185–189.

Fleury SA. 1998. Studies on the population biology and habitat use of the Red-naped Sapsucker (*Sphyrapicus nuchalis*) and its role as a keystone species in central Nevada. *Great Basin Birds* 1:60–61.

———. 2000. Population and community dynamics in western riparian avifauna: The role of the Red-naped Sapsucker (*Sphyrapicus nuchalis*). Ph.D. diss., University of Nevada, Reno.

Flood NJ. 2002. Scott's Oriole (*Icterus parisorum*). *In* Poole A and Gill F, eds., *The Birds of North America* no. 608. Philadelphia: The Birds of North America Inc.

Floyd T. 2001. Great Basin regional report. *North American Birds* 55:330–332.

———. 2002. Summary of significant sightings in the Great Basin: 2001. *Great Basin Birds* 5:68–72.

———. 2005. Great Basin: State of the region. *North American Birds* 58:574–575.

Floyd T and Stackhouse M. 2001. Fall 2000 regional report: Great Basin. *North American Birds* 55:79–82.

Forbes MR, Barkhouse HP, and Smith PC. 1989. Nest-site selection by Pied-billed Grebes, *Podilymbus podiceps. Ornis Scandinavica* 20:211–218.

Forbush EH. 1929. *Birds of Massachussetts and Other New England States.* Norwood, Mass.: Norwood Press.

Franzreb KE and Ohmart RD. 1978. The effects of timber harvesting on breeding birds in a mixed-coniferous forest. *Condor* 80:431–441.

Freer VM. 1979. Factors affecting site tenacity in New York Bank Swallows. *Bird Banding* 50:349–357.

Fridell R. 2004. Summary of significant sightings in the Great Basin: 2003. *Great Basin Birds* 7:63–70.

Gabrielson IN. 1949. Bird notes from Nevada. *Condor* 51:179–187.

Gaines D. 1992. *Birds of Yosemite and the East Slope.* 2nd ed. Lee Vining, Calif.: Artemisia Press.

Gallagher SR. 1997. *Atlas of Breeding Birds of Orange County.* Irvine, Calif.: Sea and Sage Audubon Press.

Gamble LR and Bergin TM. 1996. Western Kingbird (*Tyrannus verticalis*). *In* Poole A and Gill F, eds., *The Birds of North America* no. 227. Philadelphia: Academy of Natural Sciences, and Washington: American Ornithologists' Union.

Gammonley JH. 1996. Cinnamon Teal (*Anas cyanoptera*). *In* Poole A and Gill F, eds., *The Birds of North America* no. 209. Philadelphia: Academy of Natural Sciences, and Washington: American Ornithologists' Union.

Gardali T and Ballard G. 2000. Warbling Vireo (*Vireo gilvus*). *In* Poole A and Gill F, eds., *The Birds of North America* no. 551. Philadelphia: Academy of Natural Sciences, and Washington: American Ornithologists' Union.

Garnett MC, Kahl J, Swett J, and Ammon EM. 2004. Status of the Yuma Clapper Rail (*Rallus longirostris yumanensis*) in the northern Mojave Desert compared with other parts of its range. *Great Basin Birds* 7:6–15.

Garrett KL and Dunn JL. 1981. *Birds of Southern California.* Los Angeles: Los Angeles Audubon Society.

Garrett KL, Raphael MG, and Dixon RD. 1996. White-headed Woodpecker (*Picoides albolarvatus*). *In* Poole A and Gill F, eds., *The Birds of North America* no. 252. Philadelphia: Academy of Natural Sciences, and Washington: American Ornithologists' Union.

Garrison BA. 1999. Bank Swallow (*Riparia riparia*). *In* Poole A and Gill F, eds., *The Birds of North America* no. 414. Philadelphia: The Birds of North America Inc.

Gauthier G. 1993. Bufflehead (*Bucephala albeola*). *In* Poole A and Gill F, eds., *The Birds of North America* no. 67. Philadelphia: Academy of Natural Sciences, and Washington: American Ornithologists' Union.

Germaine SS, Rosenstock SS, Schweinsburg RE, and Richardson WS. 1988. Relationships among breeding birds, habitat, and residential development in Greater Tucson, AZ. *Ecological Applications* 8:680–691.

Ghalambor CK and Martin TE. 1999. Red-breasted Nuthatch (*Sitta candensis*). *In* Poole A and Gill F, eds., *The Birds of North America* no. 459. Philadelphia: The Birds of North America Inc.

Gibbs JP, Melvin SM, and Reid FA. 1992a. American Bittern (*Botaurus lentiginosus*). *In* Poole A, Stettenheim P, and Gill F, eds., *The Birds of North America* no. 18. Philadelphia: Academy of Natural Sciences, and Washington: American Ornithologists' Union.

Gibbs JP, Reid FA, and Melvin SM. 1992b. Least Bittern (*Ixobrychus exilis*). *In* Poole A, Stettenheim P, and Gill F, eds., *The Birds of North America* no. 17. Philadelphia: Academy of Natural Sciences, and Washington: American Ornithologists' Union.

Gillihan SW and Byers B. 2001. Evening Grosbeak (*Coccothraustes vespertinus*). *In* Poole A and Gill F, eds., *The Birds of North America* no. 599. Philadelphia: The Birds of North America Inc.

Gilman MF. 1909. Nesting notes on the Lucy Warbler. *Condor* 11:166–168.

Giudice JH and Ratti JT. 2001. Ring-necked Pheasant (*Phasianus colchicus*). *In* Poole A and Gill F, eds., *The Birds of North America* no. 572. Philadelphia: The Birds of North America Inc.

Goguen CB and Curson DR. 2002. Cassin's Vireo (*Vireo cassinii*). *In* Poole A and Gill F, eds., *The Birds of North America* no. 615. Philadelphia: Academy of Natural Sciences, and Washington: American Ornithologists' Union.

Goldstein DL. 1984. The thermal environment and its constraint on activity of desert quail in summer. *Auk* 101:542–550.

Greene E, Davison W, and Muehter VR. 1998. Steller's Jay (*Cyanocitta stelleri*). *In* Poole A and Gill F, eds., *The Birds of North America* no. 343. Philadelphia: Academy of Natural Sciences, and Washington: American Ornithologists' Union.

Greene E, Muehter VR, and Davison W. 1996. Lazuli Bunting (*Passerina amoena*). *In* Poole A and Gill F, eds., *The Birds of North America* no. 232. Philadelphia: Academy of Natural Sciences, and Washington: American Ornithologists' Union.

Greenlaw JS. 1996. Spotted Towhee (*Pipilo maculatus*). *In* Poole A and Gill F, eds., *The Birds of North America* no. 263. Philadelphia: Academy of Natural Sciences, and Washington: American Ornithologists' Union.

Griese HJ, Ryder RA, and Braun CE. 1980. Spatial and temporal distribution of rails in Colorado. *Wilson Bulletin* 92:96–102.

Grinnell J and Miller AH. 1944. *The Distribution of the Birds of California.* Berkeley: Artemisia Press.

Grudzien TA, Moore WS, Cook JR, and Tagle D. 1987. Genic population structure and gene flow in the Northern Flicker (*Colaptes auratus*) hybrid zone. *Auk* 104:654–664.

Guinan JA, Gowaty PA, and Eltzroth EK. 2000. Western Bluebird (*Sialia mexicana*). *In* Poole A and Gill F, eds., *The Birds of North America* no. 510. Philadelphia: The Birds of North America Inc.

Gullion GW, Pulich WM, and Evenden FG. 1959. Notes on the occurrence of birds in southern Nevada. *Condor* 61:278–297.

Gutiérrez RJ and Delehanty DJ. 1999. Mountain Quail (*Oreortyx pictus*). *In* Poole A and Gill F, eds., *The Birds of North America* no. 457. Philadelphia: The Birds of North America Inc.

Gutiérrez RJ, Franklin AB, and LaHaye WS. 1995. Spotted Owl (*Strix occidentalis*). *In* Poole A and Gill F, eds., *The Birds of North America* no. 179. Philadelphia: The Birds of North America Inc.

Guzy MJ and Lowther PE. 1997. Black-throated Gray Warbler (*Dendroica nigrescens*). *In* Poole A and Gill F, eds., *The Birds of North America* no. 319. Philadelphia: Academy of Natural Sciences, and Washington: American Ornithologists' Union.

Guzy MJ and Ritchison G. 1999. Common Yellowthroat (*Geothlypis trichas*). *In* Poole A and Gill F, eds., *The Birds of North America* no. 448. Philadelphia: The Birds of North America Inc.

Hagemeijer JM and Blair MJ, eds. 1997. *The EBCC Atlas of European Breeding Birds.* London: T & AD Poyser.

Hahn TP. 1996. Cassin's Finch (*Carpodacus cassinii*). *In* Poole A and Gill F, eds., *The Birds of North America* no. 240. Philadelphia: Academy of Natural Sciences, and Washington: American Ornithologists' Union.

Haig SM, Mullins TD, Forsman ED, Tail PW, and Wennerberg L. 2004. Genetic identification of Spotted Owls, Barred Owls, and their hybrids: Legal implications of hybrid identity. *Conservation Biology* 18:1347–1357.

Haig SM, Oring LW, Sanzenbacher PM, and Taft OW. 2002. Space use, migratory connectivity, and population segregation among Willets breeding in the western Great Basin. *Condor* 104:620–630.

Halterman MD. 1991. Distribution and habitat use of the Yellow-billed Cuckoo (*Coccyzus americanus*) on the Sacramento River, 1987–1990. M.S. thesis, California State University, Chico.

———. 2002. Surveys and life history studies of the Yellow-billed Cuckoo: Summer 2001. Unpublished report, Bureau of Reclamation Lower Colorado Region, Boulder City, Nev.

———. 2003. Surveys and life history studies of the Yellow-billed Cuckoo: Summer 2002. Unpublished report, U.S. Bureau of Reclamation, Lower Colorado Region, Boulder City, Nev.

Hamas MJ. 1994. Belted Kingfisher (*Ceryle alcyon*). *In* Poole A and Gill F, eds., *The Birds of North America* no. 84. Philadelphia: Academy of Natural Sciences, and Washington: American Ornithologists' Union.

Hamer TE, Forsman ED, Fuchs AD, and Walters ML. 1994. Hybridization between Barred and Spotted Owls. *Auk* 111: 487–492.

Hart J. 1996. *Storm over Mono: The Mono Lake Battle and the California Water Future.* Berkeley: University of California Press.

Hartman CA and Oring LW. 2006. Hayfields in Nevada: Critical habitat for the imperiled Long-billed Curlew. *Great Basin Birds* 8:11–15.

Hatch JJ. 1995. Changing populations of Double-crested Cormorants. *Colonial Waterbirds* 18:8–24.

Hatch JJ and Weseloh DV. 1999. Double-crested Cormorant (*Phalacrocorax auritus*). *In* Poole A and Gill F, eds., *The Birds of North America* no. 441. Philadelphia: The Birds of North America Inc.

Haug EA, Millsap BA, and Martell MS. 1993. Burrowing Owl (*Speotyto cunicularia*). *In* Poole A and Gill F, eds., *The Birds of North America* no. 61. Philadelphia: Academy of Natural Sciences, and Washington: American Ornithologists' Union.

Hejl SJ, Holmes JA, and Kroodsma DE. 2002a. Winter Wren (*Troglodytes troglodytes*). *In* Poole A and Gill F, eds., *The Birds of North America* no. 623. Philadelphia: The Birds of North America Inc.

Hejl SJ, Newlon KR, McFadzen ME, Young JS, and Ghalambor CK. 2002b. Brown Creeper (*Certhia americana*). *In* Poole A and Gill F, eds., *The Birds of North America* no. 669. Philadelphia: The Birds of North America Inc.

Hepp GR and Bellrose FC. 1995. Wood Duck (*Aix sponsa*). *In* Poole A and Gill F, eds., *The Birds of North America* no. 169. Philadelphia: Academy of Natural Sciences, and Washington: American Ornithologists' Union.

Herlan PJ. 1965. *The Nevada Highway Bird Watcher.* Carson City: Nevada State Museum.

Herron GB. 1999. Status of the Northern Goshawk in the Great Basin. *Great Basin Birds* 2:9–12.

Herron GB, Mortimore CA, and Rawlings MS. 1985. *Nevada Raptors: Their Biology and Management.* Reno: Nevada Division of Wildlife.

Hill GE. 1993. House Finch (*Carpodacus mexicanus*). *In* Poole A and Gill F, eds., *The Birds of North America* no. 46. Philadelphia: Academy of Natural Sciences, and Washington: American Ornithologists' Union.

———. 1995. Black-headed Grosbeak (*Pheucticus melanocephalus*). *In* Poole A and Gill F, eds., *The Birds of North America* no. 143. Philadelphia: Academy of Natural Sciences, and Washington: American Ornithologists' Union.

Hill RL. 2002. Observations of Three-toed Woodpeckers (*Picoides tridactylus*) in the breeding season. *Great Basin Birds* 5:21–25.

Hochachka WM, Dhondt AA, McGowan KJ, and Kramer LD. 2004. Impact of West Nile virus on American Crows in the northeastern United States, and its relevance to existing monitoring programs. *EcoHealth* 1:60–68.

Hofmann D. 1985. Hairy Woodpecker (*Picoides villosus*). *In* Burridge B, ed., *Sonoma County Breeding Bird Atlas*, p. 95. Santa Rosa, Calif.: Madrone Audubon Society.

Hohman WL and Eberhardt RT. 1998. Ring-necked Duck (*Aythya collaris*). *In* Poole A and Gill F, eds., *The Birds of North America* no. 329. Philadelphia: The Birds of North America Inc.

Hole DG, Whittingham MJ, Bradbury RB, Anderson GQA, Lee PLM, Wilson JD, and Krebs JR. 2002. Widespread local extinctions of house sparrows. *Nature* 418:931.

Holmes AL and Barton DC. 2003. *Determinants of Songbird Abundance and Distribution in Sagebrush Habitats of Eastern Oregon and Washington*. Point Reyes Bird Observatory Contribution No. 1094. Stinson Beach, Calif.

Holmes JA, Dobkin DS, and Wilcox BA. 1985. Second nesting record and northward advance of the Great-tailed Grackle (*Quiscalus mexicanus*) in Nevada. *Great Basin Naturalist* 45:483–484.

Holt DW and Leasure SM. 1993. Short-eared Owl (*Asio flammeus*). *In* Poole A and Gill F, eds., *The Birds of North America* no. 62. Philadelphia: Academy of Natural Sciences, and Washington: American Ornithologists' Union.

Holt DW and Petersen JL. 2000. Northern Pygmy-Owl (*Glaucidium gnoma*). *In* Poole A and Gill F, eds., *The Birds of North America* no. 494. Philadelphia: The Birds of North America Inc.

Homer CG. 1997. *Nevada Landcover Classification*. Logan: Utah State University. Online at http://earth.gis.usu.edu/

Hoogland JL and Sherman PW. 1976. Advantages and disadvantages of Bank Swallow (*Riparia riparia*) coloniality. *Ecological Monographs* 46:33–58.

Horton SP. 1987. Effects of prescribed burning on breeding birds in a ponderosa pine forest, southeastern Arizona. M.S. thesis, University of Arizona, Tucson.

Houston CS, Smith DG, and Rohner C. 1998. Great Horned Owl (*Bubo virginianus*). *In* Poole A and Gill F, eds., *The Birds of North America* no. 372. Philadelphia: The Birds of North America Inc.

Hudon J. 1999. Western Tanager (*Piranga ludoviciana*). *In* Poole A and Gill F, eds., *The Birds of North America* no. 432. Philadelphia: The Birds of North America Inc.

Hughes JM. 1996. Greater Roadrunner (*Geococcyx californianus*). *In* Poole A and Gill F, eds., *The Birds of North America* no. 244. Philadelphia: Academy of Natural Sciences, and Washington: American Ornithologists' Union.

———. 1999. Yellow-billed Cuckoo (*Coccyzus americanus*). *In* Poole A and Gill F, eds., *The Birds of North America* no. 418. Philadelphia: The Birds of North America Inc.

Hunt PD and Flaspohler DJ. 1998. Yellow-rumped Warbler (*Dendroica coronata*). *In* Poole A and Gill F, eds., *The Birds of North America* no. 376. Philadelphia: The Birds of North America Inc.

Hunter WC, Ohmart RD, and Anderson BW. 1988. Use of exotic saltcedar (*Tamarix chinensis*) by birds in arid riparian systems. *Condor* 90:113–123.

Ingold JL. 1993. Blue Grosbeak (*Guiraca caerulea*). *In* Poole A and Gill F, eds., *The Birds of North America* no. 79. Philadelphia: Academy of Natural Sciences, and Washington: American Ornithologists' Union.

Ingold JL and Galati R. 1997. Golden-crowned Kinglet (*Regulus satrapa*). *In* Poole A and Gill F, eds., *The Birds of North America* no. 301. Philadelphia: Academy of Natural Sciences, and Washington: American Ornithologists' Union.

Ingold JL and Wallace GE. 1994. Ruby-crowned Kinglet (*Regulus calendula*). *In* Poole A and Gill F, eds., *The Birds of North America* no. 119. Philadelphia: Academy of Natural Sciences, and Washington: American Ornithologists' Union.

Islam K. 2002. Heermann's Gull (*Larus heermanni*). *In* Poole A and Gill F, eds., *The Birds of North America* no. 643. Philadelphia: Academy of Natural Sciences, and Washington: American Ornithologists' Union.

Ivey GL and Herziger CP, coords. 2005. *Intermountain West Waterbird Conservation Plan—A Plan Associated with the Waterbird Conservation for the Americas Initiative*. Version 1.0. U.S. Fish and Wildlife Service Pacific Region, Portland, Ore. Online at http://birds.fws.gov/waterbirds/Intermountainwest/

Jackson BJS and Jackson JA. 2000. Killdeer (*Charadrius vociferus*). *In* Poole A and Gill F, eds., *The Birds of North America* no. 517. Philadelphia: The Birds of North America Inc.

Jackson JA, Ouellet HR, and Jackson BJS. 2002. Hairy Woodpecker (*Picoides villosus*). *In* Poole A and Gill F, eds., *The Birds of North America* no. 702. Philadelphia: The Birds of North America Inc.

Jackson LS, Thompson CA, and Dinsmore JJ, eds. 1996. *The Iowa Breeding Bird Atlas*. Iowa City: University of Iowa.

Jacobs B and Wilson JD. 1997. *The Missouri Breeding Bird Atlas*. Jefferson City: Conservation Commission of the State of Missouri.

James FC, McCulloch CE, and Wiedenfeld DA. 1996. New approaches to the analysis of population trends in land birds. *Ecology* 77:13–27.

Janssen RB. 1987. *Birds in Minnesota*. Minneapolis: University of Minnesota Press.

Jehl JR Jr. 1993. Observations on the fall migration of Eared Grebes, based on evidence from a mass downing in Utah. *Condor* 95:470–473.

———. 1994. Changes in saline and alkaline lake avifaunas in western North America in the past 150 years. *Studies in Avian Biology* 15:258–272.

———. 1996. Mass mortality events of Eared Grebes in North America. *Journal of Field Ornithology* 67:471–476.

Johnson K. 1995. Green-winged Teal (*Anas crecca*). *In* Poole A and Gill F, eds., *The Birds of North America* no. 193. Philadelphia: Academy of Natural Sciences, and Washington: American Ornithologists' Union.

Johnson K and Peer BD. 2001. Great-tailed Grackle (*Quiscalus mexicanus*). *In* Poole A and Gill F, eds., *The Birds of North America* no. 576. Philadelphia: The Birds of North America Inc.

Johnson LS. 1998. House Wren (*Troglodytes aedon*). *In* Poole A and Gill F, eds., *The Birds of North America* no. 380. Philadelphia: The Birds of North America Inc.

Johnson M, Beckman JP, and Oring LW. 2003. Diurnal and nocturnal behavior of breeding American Avocets. *Wilson Bulletin* 115:176–185.

Johnson MJ, van Riper C III, and Pearson KM. 2002. Black-throated Sparrow (*Amphispiza bilineata*). *In* Poole A and Gill F, eds., *The Birds of North America* no. 637. Philadelphia: The Birds of North America Inc.

Johnson NK. 1956. Recent bird records for Nevada. *Condor* 58: 449–452.

———. 1963. Biosystematics of sibling species of flycatchers in the *Empidonax hammondii-oberholseri-wrightii* complex. *University of California Publications in Zoology* 66:79–238.

———. 1965. The breeding avifaunas of the Sheep and Spring ranges in southern Nevada. *Condor* 67:93–124.

———. 1966. Bill size and the question of competition in allopatric and sympatric populations of Dusky and Gray Flycatchers. *Systematic Zoology* 15:70–87.

———. 1972. Breeding distribution and habitat preference of the Gray Vireo in Nevada. *California Birds* 3:73–78.

———. 1973. The distribution of boreal avifaunas in southern Nevada. *Biological Society of Nevada Occasional Papers* no. 36.

———. 1975. Controls of number of bird species on montane islands in the Great Basin. *Evolution* 29:545–567.

———. 1976. Breeding distribution of Nashville and Virginia's Warblers. *Auk* 93:219–230.

———. 1994. Pioneering and natural expansion of breeding distributions in western North American birds. *Studies in Avian Biology* 15:27–44.

Johnson RE. 2002. Black Rosy-Finch (*Leucosticte atrata*). *In* Poole A and Gill F, eds., *The Birds of North America* no. 678. Philadelphia: The Birds of North America Inc.

Johnson RF. 1992. Rock Dove (*Columba livia*). *In* Poole A and Gill F, eds., *The Birds of North America* no. 13. Philadelphia: Academy of Natural Sciences.

Johnson RR and Dinsmore JJ. 1986. Habitat use by breeding Virginia Rails and Soras. *Journal of Wildlife Management* 50:387–392.

Johnson RR, Yard HK, and Brown BT. 1997. Lucy's Warbler (*Vermivora luciae*). *In* Poole A and Gill F, eds., *The Birds of North America* no. 318. Philadelphia: Academy of Natural Sciences, and Washington: American Ornithologists' Union.

Johnson T and Wauer RH. 1996. Avifaunal response to the 1977 La Mesa fire. *In* Allen CD, ed., *Fire Effects in Southwestern Forests: Proceedings of the Second La Mesa Fire Symposium*, pp. 70–94. USDA Forest Service, General Technical Report RM-GTR-286. Rocky Mountain Forest and Range Experiment Station, Fort Collins, Colo.

Jones PW and Donovan TM. 1996. Hermit Thrush (*Catharus guttatus*). *In* Poole A and Gill F, eds., *The Birds of North America* no. 261. Philadelphia: Academy of Natural Sciences, and Washington: American Ornithologists' Union.

Jones RM and Rosenberg GH. 2001. Fall 2000 regional report: Arizona. *North American Birds* 55:82–85.

Jones SL and Cornely JE. 2002. Vesper Sparrow (*Pooecetes gramineus*). *In* Poole A and Gill F, eds., *The Birds of North America* no. 624. Philadelphia: The Birds of North America Inc.

Jones SL and Dieni JS. 1995. Canyon Wren (*Catherpes mexicanus*). *In* Poole A and Gill F, eds., *The Birds of North America* no. 197. Philadelphia: Academy of Natural Sciences.

Joyner DE. 1983. Parasitic egg laying in Redheads and Ruddy Ducks in Utah: Incidence and success. *Auk* 100:717–725.

Kaminski RM and Prince HH. 1981. Dabbling duck and aquatic macroinvertebrate responses to manipulated wetland habitat. *Journal of Wildlife Management* 45:1–15.

Kaufman, K. 1996. *Lives of North American Birds.* New York: Houghton Mifflin.

Kennedy ED and White DW. 1997. Bewick's Wren (*Thryomanes bewickii*). *In* Poole A and Gill F, eds., *The Birds of North America* no. 315. Philadelphia: Academy of Natural Sciences, and Washington: American Ornithologists' Union.

Keppie DM and Braun CE. 2000. Band-tailed Pigeon (*Columba fasciata*). *In* Poole A and Gill F, eds., *The Birds of North America* no. 530. Philadelphia: The Birds of North America Inc.

Kerley LL and Anderson SH. 1995. Songbird responses to sagebrush removal in a high elevation sagebrush steppe ecosystem. *Prairie Naturalist* 27:129–146.

King JR and Mewaldt LR. 1987. The summer biology of an unstable insular population of White-crowned Sparrows in Oregon. *Condor* 89:549–565.

Kingery HE. 1996. American Dipper (*Cinclus mexicanus*). *In* Poole A and Gill F, eds., *The Birds of North America* no. 229. Philadelphia: Academy of Natural Sciences.

———, ed. 1998. *Colorado Breeding Bird Atlas.* Denver: Colorado Bird Atlas Partnership.

Kingery HE and Ghalambor CK. 2001. Pygmy Nuthatch (*Sitta pygmaea*). *In* Poole A and Gill F, eds., *The Birds of North America* no. 567. Philadelphia: The Birds of North America Inc.

Kirk A and Kirk L. 2004. Expansion of the breeding range of the Acorn Woodpecker east of the Sierra Nevada, California. *Western Birds* 35:221–223.

Kirk DA and Mossman MJ. 1998. Turkey Vulture (*Cathartes aura*). *In* Poole A and Gill F, eds., *The Birds of North America* no. 339. Philadelphia: The Birds of North America Inc.

Klebenow DA and Oakleaf RJ. 1984. Historical avifaunal changes in the riparian zone of the Truckee River, Nevada. *In* Warner RE and Hendrix KM, eds., *California Riparian Systems: Ecology, Conservation, and Productive Management*, pp. 203–209. Berkeley: University of California Press.

Klute DS, Ayres LW, Green MT, Howe WH, Jones SJ, Shaffer JA, Sheffield SR, and Zimmerman TS. 2003. *Status Assessment and Conservation Plan for the Western Burrowing Owl in the United States.* U.S. Fish and Wildlife Service Biological Technical Publication FWS/BTR-R6001-2003. Washington: U.S. Fish and Wildlife Service.

Knick ST, Holmes AL, and Miller RF. 2003. The role of fire in structuring sagebrush habitats and bird communities. *Studies in Avian Biology* 30:63–75.

Knick ST and Rotenberry JT. 2000. Ghosts of habitats past: Contribution of landscape to current habitats used by shrubland birds. *Ecology* 81:220–227.

Knopf FL, Sedgwick JA, and Inkley DB. 1990. Regional correspondence among shrubsteppe bird habitats. *Condor* 92:45–53.

Knorr OA. 1957. Communal roosting of the Pygmy Nuthatch. *Condor* 59:398.

———. 2000a. Breeding of the American Pipit (*Anthus rubescens*) in Nevada. *Great Basin Birds* 3:7–9.

———. 2000b. Robert Ridgway and the Black Swift (*Cypseloides niger*) in the Great Basin. *Great Basin Birds* 3:13–16.

Kochert MN, Steenhof K, McIntyre CL, and Craig EH. 2002. Golden Eagle (*Aquila chrysaetos*). *In* Poole A and Gill F, eds., *The Birds of North America* no. 684. Philadelphia: The Birds of North America Inc.

Koenig WD. 2003. European Starlings and their effect on native cavity-nesting birds. *Conservation Biology* 17:1134–1140.

Kroodsma DE and Canady RA. 1985. Differences in repertoire size, singing behavior, and associated neuroanatomy among Marsh Wren populations have a genetic basis. *Auk* 102:439–446.

Kroodsma DE and Verner J. 1997. Marsh Wren (*Cistothorus palustris*). *In* Poole A and Gill F, eds., *The Birds of North America* no. 308. Philadelphia: Academy of Natural Sciences, and Washington: American Ornithologists' Union.

Krueger J. 1998. Use of mesquite woodlands in southern Nevada as breeding habitat for Phainopepla (*Phainopepla nitens*). *Great Basin Birds* 1:59–60.

Kruse AD and Bowen BS. 1996. Effects of grazing and burning on densities and habitats of breeding ducks in North Dakota. *Journal of Wildlife Management* 60:233–246.

Kruse KL, Lovvorn JR, Takekawa JY, and Mackay J. 2003. Long-term productivity of Canvasback (*Aythya valisineria*) in a snowpack-driven desert marsh. *Auk* 120:107–119.

Kus BE. 1998. Use of restored riparian habitat by the endangered Least Bell's Vireo (*Vireo bellii pusillus*). *Restoration Ecology* 6:75–82.

Kushlan JA, Steinkamp MJ, Parsons KC, Capp J, Cruz AC, Coulter M, Davidson I, Dickson L, Edelson N, Elliot R, Erwin RM, Hatch S, Kress S, Milko R, Miller S, Mills K, Paul R, Phillips R, Saliva JE, Sydeman B, Trapp J, Wheeler J, and Wohl K. 2002. *Waterbird Conservation for the Americas: The North American Waterbird Conservation Plan.* Version 1. Washington: Waterbird Conservation for the Americas.

Lanyon WE. 1994. Western Meadowlark (*Sturnella neglecta*). *In* Poole A and Gill F, eds., *The Birds of North America* no. 104. Philadelphia: Academy of Natural Sciences, and Washington: American Ornithologists' Union.

Latta SC and Baltz ME. 1997. Lesser Nighthawk (*Chordeiles acutipennis*). *In* Poole A and Gill F, eds., *The Birds of North America* no. 314. Philadelphia: Academy of Natural Sciences, and Washington: American Ornithologists' Union.

Laudenslayer WF Jr. 1981. Habitat utilization by birds of three desert riparian communities. Ph.D. diss., Arizona State University, Tempe.

Laymon SA. 1998. Yellow-billed Cuckoo (*Coccycus americanus*). *In* California Partners in Flight, ed., *The Riparian Bird Conservation Plan: A Strategy for Reversing the Decline of Riparian-Associated Birds in California.* Online at http://www.prbo.org/calpif/htmldocs/riparian_v-2.html

Laymon SA and Halterman MD. 1989. A proposed habitat management plan for Yellow-billed Cuckoos in California. *In* Abell DL, tech. coord., *Proceedings of the California Riparian Systems Conference: Protection, Management, and Restoration for the 1990s,* pp. 272–277. USDA Forest Service General Technical Report PSW-110. Pacific Southwest Forest and Range Experiment Station, Berkeley.

———. 1998. The effects of Brown-headed Cowbirds on neotropical migrants in the Great Basin National Park and Lake Mead National Recreational Area: An overview report for 1995–1996. *Great Basin Birds* 1:42–45.

Leonard ML and Picman J. 1986. Why are nesting Marsh Wrens and Yellow-headed Blackbirds spatially segregated? *Auk* 103:135–140.

LeSchack CR, McKnight SK, and Hepp GR. 1997. Gadwall (*Anas strepera*). *In* Poole A and Gill F, eds., *The Birds of North America* no. 283. Philadelphia: Academy of Natural Sciences, and Washington: American Ornithologists' Union.

Liebezeit JR and George TL. 2002. Nest predators, nest-site selection, and nesting success of the Dusky Flycatcher in a managed ponderosa pine forest. *Condor* 104:507–517.

Ligon JD. 1973. Foraging behavior of the White-headed Woodpecker in Idaho. *Auk* 90:862–869.

Ligon JS. 1961. *New Mexico Birds and Where to Find Them.* Albuquerque: University of New Mexico Press.

Linsdale JM. 1936. *The Birds of Nevada.* Berkeley: Cooper Ornithological Club.

———. 1951. A list of the birds in Nevada. *Condor* 53:228–249.

Littlefield CD. 1990. *Birds of Malheur National Wildlife Refuge, Oregon.* Corvallis: Oregon State University Press.

Lowther PE. 1993. Brown-headed Cowbird (*Molothrus ater*). *In* Poole A and Gill F, eds., *The Birds of North America* no. 47. Philadelphia: Academy of Natural Sciences, and Washington: American Ornithologists' Union.

———. 2000. Pacific-slope Flycatcher (*Empidonax difficilis*) and Cordilleran Flycatcher (*Empidonax occidentalis*). *In* Poole A and Gill F, eds., *The Birds of North America* no. 556. Philadelphia: The Birds of North America Inc.

———. 2001. Ladder-backed Woodpecker (*Picoides scalaris*). *In* Poole A and Gill F, eds., *The Birds of North America* no. 565. Philadelphia: Academy of Natural Sciences, and Washington: American Ornithologists' Union.

Lowther PE, Celada C, Klein NK, Rimmer CC, and Spector DA. 1999. Yellow Warbler (*Dendroica petechia*). *In* Poole A and Gill F, eds., *The Birds of North America* no. 454. Philadelphia: The Birds of North America Inc.

Lowther PE and Cink CL. 1992. House Sparrow (*Passer domesti-*

cus). *In* Poole A, Stettenheim P, and Gill F, eds., *The Birds of North America* no. 12. Philadelphia: Academy of Natural Sciences, and Washington: American Ornithologists' Union.

Lowther PE, Douglas HD III, and Gratto-Trevor CL. 2001. Willet (*Catoptrophorus semipalmatus*). *In* Poole A and Gill F, eds., *The Birds of North America* no. 579. Philadelphia: The Birds of North America Inc.

Lowther PE, Kroodsma DE, and Farley GH. 2000. Rock Wren (*Salpinctes obsoletus*). *In* Poole A and Gill F, eds., *The Birds of North America* no. 486. Philadelphia: The Birds of North America Inc.

Mack K. 2000. 1999 Pyramid Lake water bird census. *Great Basin Birds* 3:37.

———. 2001. 2000 Pyramid Lake water bird census. *Great Basin Birds* 4:44–45.

MacWhirter RB and Bildstein KL. 1996. Northern Harrier (*Circus cyaneus*). *In* Poole A and Gill F, eds., *The Birds of North America* no. 210. Philadelphia: Academy of Natural Sciences, and Washington: American Ornithologists' Union.

Mahoney SA and Jehl JR Jr. 1985. Adaptations of migratory shorebirds to highly saline and alkaline lakes: Wilson's Phalarope and American Avocet. *Condor* 87:520–527.

Mallory M and Metz K. 1999. Common Merganser (*Mergus merganser*). *In* Poole A and Gill F, eds., *The Birds of North America* no. 442. Philadelphia: The Birds of North America Inc.

Marks JS, Evans DL, and Holt DW. 1994. Long-eared Owl (*Asio otus*). *In* Poole A and Gill F, eds., *The Birds of North America* no. 133. Philadelphia: Academy of Natural Sciences, and Washington: American Ornithologists' Union.

Marshall DB, Hunter MG, and Contreras AL, eds. 2003. *Birds of Oregon: A General Reference*. Corvallis: Oregon State University Press.

Marti CD. 1992. Barn Owl (*Tyto alba*). *In* Poole A, Stettenheim P, and Gill F, eds., *The Birds of North America* no. 1. Philadelphia: Academy of Natural Sciences, and Washington: American Ornithologists' Union.

Martin JW and Carlson BA. 1998. Sage Sparrow (*Amphispiza belli*). *In* Poole A and Gill F, eds., *The Birds of North America* no. 326. Philadelphia: The Birds of North America Inc.

Martin JW and Parrish JR. 2000. Lark Sparrow (*Chondestes grammacus*). *In* Poole A and Gill F, eds., *The Birds of North America* no. 488. Philadelphia: The Birds of North America Inc.

Martin SG. 2002. Brewer's Blackbird (*Euphagus cyanocephalus*). *In* Poole A and Gill F, eds., *The Birds of North America* no. 616. Philadelphia: The Birds of North America Inc.

Martin SG and Gavin TA. 1995. Bobolink (*Dolichonyx oryzivorus*). *In* Poole A and Gill F, eds., *The Birds of North America* no. 176. Philadelphia: Academy of Natural Sciences, and Washington: American Ornithologists' Union.

Martin TE. 1992. Breeding productivity considerations: What are the appropriate habitat considerations for management? *In* Hagan JM and Johnston DW, eds., *Ecology and Conservation of Neotropical Migrant Landbirds*, pp. 455–473. Washington: Smithsonian Institution Press.

———. 1993. Nest predation among vegetation layers and habitat types: Revising the dogmas. *American Naturalist* 141:897–913.

Marzluff JM, Boone RB, and Cox GW. 1994. Historical changes in populations and perceptions of native pest bird species in the West. *Studies in Avian Biology* 15:202–220.

Mazur KM and James PC. 2000. Barred Owl (*Strix varia*). *In* Poole A and Gill F, eds., *The Birds of North America* no. 508. Philadelphia: The Birds of North America Inc.

McCallum DA. 1994. Flammulated Owl (*Otus flammeolus*). *In* Poole A and Gill F, eds., *The Birds of North America* no. 94. Philadelphia: Academy of Natural Sciences, and Washington: American Ornithologists' Union.

McCallum DA, Grundel R, and Dahlsten DL. 1999. Mountain Chickadee (*Poecile gambeli*). *In* Poole A and Gill F, eds., *The Birds of North America* no. 453. Philadelphia: The Birds of North America Inc.

McCaskie G, DeBenedictis P, Erickson R, and Morlan J. 1988. *Birds of Northern California*. 2nd ed. Berkeley: Golden Gate Audubon Society.

McCrimmon DA Jr., Ogden JC, and Bancroft G. 2001. Great Egret (*Area alba*). *In* Poole A and Gill F, eds., *The Birds of North America* no. 570. Philadelphia: The Birds of North America Inc.

McIvor D. 2003. Walker Lake Important Bird Area at risk. *Great Basin Birds* 6:68–70.

McKernan RL and Braden GT. 2001. *The Status of Yuma Clapper Rail and Yellow-billed Cuckoo along Portions of the Virgin River, Muddy River, and Las Vegas Wash, Southern Nevada: 2000*. Las Vegas: U.S. Fish and Wildlife Service.

———. 2002. Status, distribution, and habitat affinities of the Southwestern Willow Flycatcher along the lower Colorado River, year 6. Unpublished report, U.S. Bureau of Reclamation, Lower Colorado Region, Boulder City, Nev.

McLeod MA, Koronkiewicz TJ, Brown BT, and Carothers SW. 2005. Southwestern Willow Flycatcher surveys, demography, and ecology along the lower Colorado River and its tributaries, 2004. Unpublished report, U.S. Bureau of Reclamation, Lower Colorado Region, Boulder City, Nev.

McNicholl MK, Lowther PE, and Jall JA. 2001. Forster's Tern (*Sterna forsteri*). *In* Poole A and Gill F, eds., *The Birds of North America* no. 595. Philadelphia: The Birds of North America Inc.

Mearns EA. 1886. Some birds of Arizona. *Auk* 3:289–307.

Meehan TD and George TL. 2003. Short-term effects of moderate- to high-severity wildfire on a disturbance-dependent flycatcher in northwest California. *Auk* 120:1102–1113.

Melvin SM and Gibbs JP. 1996. Sora (*Porzana carolina*). *In* Poole A and Gill F, eds., *The Birds of North America* no. 250. Philadelphia: Academy of Natural Sciences, and Washington: American Ornithologists' Union.

Mensing SA, Elston RG, Raines GL, Tausch RJ, and Nowak CL. 2000. A GIS model to predict the location of fossil packrat (*Neotoma*) middens in central Nevada. *Western North American Naturalist* 60:111–120.

Middleton ALA. 1993. American Goldfinch (*Carduelis tristis*). *In* Poole A and Gill F, eds., *The Birds of North America* no. 280.

Philadelphia: Academy of Natural Sciences, and Washington: American Ornithologists' Union.

———. 1998. Chipping Sparrow (*Spizella passerina*). *In* Poole A and Gill F, eds., *The Birds of North America* no. 334. Philadelphia: The Birds of North America Inc.

Miller RF and Rose JA. 1999. Fire history and western juniper encroachment in sagebrush steppe. *Journal of Range Management* 52:550–559

Miller RF and Tausch RJ. 2001. The role of fire in juniper and pinyon woodlands: A descriptive analysis. *Proceedings of the First National Congress on Fire, Ecology, Prevention, and Management*, pp. 15–30. November 27–December 1, 2000, San Diego, California. Tall Timbers Research Station Miscellaneous Publication 11. Tallahassee, Fla.

Mills GS, Dunning JB, and Mates JM. 1989. Effects of urbanization on breeding bird community structure in southwestern desert habitats. *Condor* 91:416–428.

Mirarchi RE and Baskett TS. 1994. Mourning Dove (*Zenaida macroura*). *In* Poole A and Gill F, eds., *The Birds of North America* no. 117. Philadelphia: Academy of Natural Sciences, and Washington: American Ornithologists' Union.

Mitchell CD. 1994. Trumpeter Swan (*Cygnus buccinator*). *In* Poole A and Gill F, eds., *The Birds of North America* no. 105. Philadelphia: Academy of Natural Sciences, and Washington: American Ornithologists' Union.

Mitchell RM. 1975. The current status of the Double-crested Cormorant in Utah: A plea for protection. *American Birds* 29:927–930.

Moir J. 2005. Bringing back the condor: Adaptive management guides the recovery effort. *Birding* 37:44–50.

Monson G and Phillips AR. 1981. *Annotated Checklist of the Birds of Arizona.* 2nd ed. Tucson: University of Arizona Press.

Moore WS. 1995. Northern Flicker (*Colaptes auratus*). *In* Poole A and Gill F, eds., *The Birds of North America* no. 166. Philadelphia: Academy of Natural Sciences, and Washington: American Ornithologists' Union.

Morrison ML and With KA. 1987. Interseasonal and intersexual resource partitioning in Hairy and White-headed Woodpeckers. *Auk* 104:225–233.

Morrison RIG, Gill RE, Harrington BA, Skagen S, Page GW, Gratto-Trevor CL, and Haig SM. 2000. Population estimates of Nearctic shorebirds. *Waterbirds* 23:337–352.

Morse B, Peterson T, and Smith M. 1995. Locating the Himalayan Snowcock. *Winging It* 7(2):1–6.

Mowbray T. 1999. American Wigeon (*Anas americana*). *In* Poole A and Gill F, eds., *The Birds of North America* no. 401. Philadelphia: The Birds of North America Inc.

Mowbray TB, Ely CR, Sedinger JS, and Trost RE. 2002. Canada Goose (*Branta canadensis*). *In* Poole A and Gill F, eds., *The Birds of North America* no. 682. Philadelphia: The Birds of North America Inc.

Mueller AJ. 1992. Inca Dove (*Columbina inca*). *In* Poole A, Stettenheim P, and Gill F, eds., *The Birds of North America* no. 28. Philadelphia: Academy of Natural Sciences, and Washington: American Ornithologists' Union.

Mueller H. 1999. Common Snipe (*Gallinago gallinago*). *In* Poole A and Gill F, eds. *The Birds of North America* no. 417. Philadelphia: The Birds of North America Inc.

Muller MJ and Storer RW. 1999. Pied-billed Grebe (*Podilymbus podiceps*). *In* Poole A and Gill F, eds., *The Birds of North America* no. 410. Philadelphia: The Birds of North America Inc.

Murphy MT. 1996. Eastern Kingbird (*Tyrannus tyrannus*). *In* Poole A and Gill F, eds., *The Birds of North America* no. 253. Philadelphia: Academy of Natural Sciences, and Washington: American Ornithologists' Union.

Nachlinger JL, Sochi K, Comer P, Kittel G, and Dorfman D. 2001. *The Great Basin: An Eco-region Based Conservation Blueprint.* Reno: The Nature Conservancy.

Neel LA. 1998. Survey of colony-nesting birds in northwestern Nevada–1997. *Great Basin Birds* 1:30–34.

———. 1999a. *Nevada Bird Conservation Plan.* Reno: Nevada Partners in Flight.

———. 1999b. Summary of Nevada Division of Wildlife 1998 non-game activities. *Great Basin Birds* 2:30–38.

———. 2001. Northern Nevada bird surveys. *Great Basin Birds* 4:59–64.

Neel LA and Henry WG. 1997. Shorebirds of the Lahontan Valley, Nevada, USA: a case history of western Great Basin shorebirds. *International Wader Studies* 9:15–19.

Nero RW. 1984. *Redwings.* Washington: Smithsonian Institution Press.

NGS [National Geographic Society]. 1999. *Field Guide to the Birds of North America.* Washington: National Geographic Society.

Nish DH. 1973. *Guidelines for Managing the Habitat of Merriam's Turkey.* Publication no. 73-5, Utah State Department of Natural Resources, Division of Wildlife Resources, Salt Lake City.

Nolan V Jr., Ketterson ED, Cristol DA, Rogers CM, Clotfelter ED, Titus RC, Schoech SJ, and Snajdr E. 2002. Dark-eyed Junco (*Junco hyemalis*). *In* Poole A and Gill F, eds., *The Birds of North America* no. 716. Philadelphia: The Birds of North America Inc.

NORAC [North American Ornithological Atlas Committee]. 1990. *Handbook for Atlasing American Breeding Birds.* Woodstock: Vermont Institute of Natural Science.

Noyes JH and Jarvis RL. 1985. Diet and nutrition of breeding female Redhead and Canvasback ducks in Nevada. *Journal of Wildlife Management* 49:203–211.

Nuechterlein GL. 1981. Courtship behavior and reproductive isolation between Western Grebe color morphs. *Auk* 98:335–349.

NWF [Nevada Wildlife Federation]. 2002. *Enhancing Sage Grouse Habitat: A Nevada Landowner's Guide.* Reno: Nevada Wildlife Federation.

Olson CR and Martin TE. 1999. Virginia's Warbler (*Vermivora virginiae*). *In* Poole A and Gill F, eds., *The Birds of North America* no. 477. Philadelphia: The Birds of North America Inc.

Omland KE, Tarr CL, Boarman WI, Marzluff JM, and Fleischer RC. 2000. Cryptic genetic variation and paraphyly in ravens. *Proceedings of the Royal Society of London B* 267:2475–2482.

Oring LW, Gray EM, and Reed JM. 1997. Spotted Sandpiper (*Actitis macularia*). *In* Poole A and Gill F, eds., *The Birds of North*

America no. 289. Philadelphia: Academy of Natural Sciences, and Washington: American Ornithologists' Union.

Oring LW, Neel LA, and Oring KE. 2005. *Intermountain West Regional Shorebird Plan.* Version 1.0. U.S. Shorebird Conservation Plan Initiative.

Ormerod SJ and Tyler SJ. 1987. Dippers *Cinclus cinclus* and Grey Wagtails *Motacilla cinerea* as indicators of stream acidity in upland Wales. *International Council for Bird Preservation Technical Publication* 6:191–208.

Ortega CP, Cruz A, and Mermoz ME. 2005. Issues and controversies of cowbird (*Molothrus* spp.) management. *In* Ortega CP, Chace JF, and Peer BD, eds., *Management of Cowbirds and Their Hosts: Balancing Science, Ethics, and Mandates.* Ornithological Monographs no. 57, pp. 6–15. Washington: American Ornithologists' Union.

Page GW and Stenzel LE. 1981. The breeding status of the Snowy Plover in California. *Western Birds* 12:1–40.

Page GW, Warriner JS, Warriner JC, and Paton PWC. 1995. Snowy Plover (*Charadrius alexandrinus*). *In* Poole A and Gill F, eds., *The Birds of North America* no. 52. Philadelphia: Academy of Natural Sciences, and Washington: American Ornithologists' Union.

Paige C and Ritter SA. 1999. *Birds in a Sagebrush Sea.* Boise: Partners in Flight Western Working Group.

Palmer RS. 1976. *Handbook of North American Birds.* New Haven: Yale University Press.

Palmer-Ball BL. 1996. *The Kentucky Breeding Bird Atlas.* Lexington: University Press of Kentucky.

Parsons KC and Master TL. 2000. Snowy Egret (*Egretta thula*). *In* Poole A and Gill F, eds., *The Birds of North America* no. 489. Philadelphia: Academy of Natural Sciences, and Washington: American Ornithologists' Union.

Paton PWC and Bachman VC. 1997. Impoundment drawdown and artificial nest structures as management strategies for Snowy Plovers. *In* Reed JM, Warnock N, and Oring LW, eds., *Conservation and Management of Shorebirds in the Western Great Basin. International Wader Studies* 9:64–70.

Paton PWC, Kneedy C, and Sorensen E. 1992. Chronology of shorebird and ibis use of selected marshes at Great Salt Lake. *Utah Birds* 8:1–19.

Pavlacky DC Jr. and Anderson SH. 2001. Habitat preferences of pinyon-juniper specialists near the limit of their geographic range. *Condor* 103:322–331.

Payne RB. 1992. Indigo Bunting (*Passerina cyanea*). *In* Poole A, Stettenheim P, and Gill F, eds., *The Birds of North America* no. 4. Philadelphia: Academy of Natural Sciences, and Washington: American Ornithologists' Union.

Peterson JMC. 1988. Black-backed Woodpecker (*Picoides arcticus*). *In* Andrle RF and Carroll JR, eds., *The Atlas of Breeding Birds in New York State*, pp. 238–239. Ithaca: Cornell University Press.

Peterson RA. 1995. *South Dakota Breeding Bird Atlas.* South Dakota Ornithologists' Union.

Peterson RT. 1990. *A Field Guide to Western Birds.* Boston: Houghton Mifflin.

Phillips AR, Marshall J, and Monson G. 1964. *The Birds of Arizona.* Tucson: University of Arizona Press.

Pinkowski B. 1981. High density of avian cavity nesters in aspen. *Southwestern Naturalist* 25:560–562.

Pitocchelli J. 1995. MacGillivray's Warbler (*Oporornis tolmiei*). *In* Poole A and Gill F, eds., *The Birds of North America* no. 159. Philadelphia: Academy of Natural Sciences, and Washington: American Ornithologists' Union.

Pleasants BY and Albano DJ. 2001. Hooded Oriole (*Icterus cucullatus*). *In* Poole A and Gill F, eds., *The Birds of North America* no. 568. Philadelphia: The Birds of North America Inc.

Plissner JH, Haig SM, and Oring LW. 1999. Within and among year movements of American Avocets in the western Great Basin. *Wilson Bulletin* 111:314–320.

——— 2000. Post-breeding movements among American Avocets and wetland connectivity in the U.S. western Great Basin. *Auk* 117:290–298.

Poole AF, Bierregaard RO, and Martell MS. 2002. Osprey (*Pandion haliaetus*). *In* Poole A and Gill F, eds., *The Birds of North America* no. 683. Philadelphia: The Birds of North America Inc.

Potts GR. 1986. *The Partridge: Pesticides, Predation, and Conservation.* London: Collins.

Poulin RG, Grindal SD, and Brigham RM. 1996. Common Nighthawk (*Chordeiles minor*). *In* Poole A and Gill F, eds., *The Birds of North America* no. 213. Philadelphia: Academy of Natural Sciences, and Washington: American Ornithologists' Union.

Power HW and Lombardo MP. 1996. Mountain Bluebird (*Sialia currucoides*). *In* Poole A and Gill F, eds., *The Birds of North America* no. 222. Philadelphia: Academy of Natural Sciences, and Washington: American Ornithologists' Union.

Pratt HM and Winkler DW. 1985. Clutch size, timing of laying, and reproductive success in a colony of Great Blue Herons and Great Egrets. *Auk* 102:49–63.

Pravosudov VV and Grubb TC Jr. 1993. White-breasted Nuthatch (*Sitta carolinensis*). *In* Poole A and Gill F, eds., *The Birds of North America* no. 54. Philadelphia: Academy of Natural Sciences, and Washington: American Ornithologists' Union.

Preston CR and Beane RD. 1993. Red-tailed Hawk (*Buteo jamaicensis*). *In* Poole A and Gill F, eds., *The Birds of North America* no. 52. Philadelphia: Academy of Natural Sciences, and Washington: American Ornithologists' Union.

Price FE and Bock CE. 1983. Population ecology of the Dipper (*Cinclus mexicanus*) in the Front Range of Colorado. *Studies in Avian Biology* no. 7. Los Angeles: Cooper Ornithological Society.

Proudfoot GA, Sherry DA, and Johnson S. 2000. Cactus Wren (*Campylorhynchus brunneicapillus*). *In* Poole A and Gill F, eds., *The Birds of North America* no. 558. Philadelphia: The Birds of North America Inc.

Raphael MG, Rosenberg KV, and Marcot G. 1988. Large-scale changes in bird populations of Douglas fir forests, northwestern California. *Bird Conservation* 3:63–83.

Ratti JT. 1981. Identification and distribution of Clark's Grebe. *Western Birds* 12:41–46.

Reed JM. 1992. A system for ranking conservation priorities for Neotropical migrant birds based on relative susceptibility to extinction. *In* Hagan JM and Johnston DW, eds., *Ecology and Conservation of Neotropical Migrant Landbirds*, pp. 524–536. Washington: Smithsonian Institution Press.

Reed JM, Warnock N, and Oring LW. 1997. Censusing shorebirds in the western Great Basin of North America. *International Wader Studies* 9:29–36.

Reynolds RT, Graham RT, Reiser MH, Bassett RL, Kennedy PL, Boyce DA Jr., Goodwin G, Smith R, and Fisher EL. 1992. *Management Recommendations for the Northern Goshawk in the Southwestern United States.* General Technical Report RM-217. USDA Forest Service, Rocky Mountain Forest and Range Experiment Station, Fort Collins, Colo.

Reynolds TD, Rick TD, and Stephens DA. 1999. Sage Thrasher (*Oreoscoptes montanus*). *In* Poole A and Gill F, eds., *The Birds of North America* no. 463. Philadelphia: The Birds of North America Inc.

Rich TD, Beardmore CJ, Berlanga H, Blancher PJ, Bradstreet MSW, Butcher GS, Demarest DW, Dunn EH, Hunter WC, Inigo-Elias EE, Kennedy JA, Martell AM, Panjabi AO, Pashley DN, Rosenberg KV, Rustay CM, Wendt JS, and Will TC. 2004. *Partners in Flight: North American Landbird Conservation Plan.* Ithaca: Cornell Laboratory of Ornithology.

Richardson TW. 2003. First records of Black-backed Woodpecker (*Picoides arcticus*) nesting in Nevada. *Great Basin Birds* 6:52–55.

Ridgway R. 1877. *United States Geological Exploration of the Fortieth Parallel: Ornithology.* Washington: U.S. Government Printing Office.

Rising JD and Williams PL. 1999. Bullock's Oriole (*Icterus bullockii*). *In* Poole A and Gill F, eds., *The Birds of North America* no. 416. Philadelphia: The Birds of North America Inc.

Robbins C, Bystrak D, and Geisler P. 1986. The breeding bird survey: Its first fifteen years, 1965–1979. Research Publication no. 157. Washington: U.S. Fish and Wildlife Service.

Roberson D and Tenney C, eds. 1993. *Atlas of the Breeding Birds of Monterey County.* Carmel, Calif.: Monterey Peninsula Audubon Society.

Robertson RJ, Stutchbury BJ, and Cohen RR. 1992. Tree Swallow (*Tachycineta bicolor*). *In* Poole A, Stettenheim P, and Gill F, eds., *The Birds of North America* no. 11. Philadelphia: Academy of Natural Sciences, and Washington: American Ornithologists' Union.

Robinson JA and Oring LW. 1997. Natal and breeding dispersal in American Avocets. *Auk* 114:416–430.

Robinson JA, Oring LW, Skorupa JP, and Boettcher R. 1997. American Avocet (*Recurvirostra americana*). *In* Poole A and Gill F, eds., *The Birds of North America* no. 275. Philadelphia: The Birds of North America Inc.

Robinson JA, Reed JM, Skorupa JP, and Oring LW. 1999. Black-necked Stilt (*Himantopus mexicanus*). *In* Poole A and Gill F, eds., *The Birds of North America* no. 449. Philadelphia: The Birds of North America Inc.

Robinson SK, Rothstein SI, Brittingham MC, Petit LJ, and Grzy-

bowski JA. 1995. Ecology and behavior of cowbirds and their impact on host populations. *In* Martin TE and Finch DM, eds., *Ecology and Management of Neotropical Migratory Birds*, pp. 428–460. New York: Oxford University Press.

Robinson WD. 1996. Summer Tanager (*Piranga rubra*). *In* Poole A and Gill F, eds., *The Birds of North America* no. 248. Philadelphia: Academy of Natural Sciences, and Washington: American Ornithologists' Union.

Rosenberg KV, Ohmart RD, Hunter WC, and Anderson BW. 1991. *Birds of the Lower Colorado River Valley.* Tucson: University of Arizona Press.

Rosenfield RN and Bielefeldt J. 1993. Cooper's Hawk (*Accipiter cooperii*). *In* Poole A and Gill F, eds., *The Birds of North America* no. 75. Philadelphia: Academy of Natural Sciences, and Washington: American Ornithologists' Union.

Rotenberry JT, Patten MA, and Preston KL. 1999. Brewer's Sparrow (*Spizella breweri*). *In* Poole A and Gill F, eds., *The Birds of North America* no. 390. Philadelphia: Academy of Natural Sciences, and Washington: American Ornithologists' Union.

Rowher FC, Johnson WP, and Loos ER. 2002. Blue-winged Teal (*Anas discors*). *In* Poole A and Gill F, eds., *The Birds of North America* no. 625. Philadelphia: Academy of Natural Sciences, and Washington: American Ornithologists' Union.

Rubega MA and Robinson JA. 1996. Water salinization and shorebirds: Emerging issues. *International Wader Studies* 9:45–54.

Rusch DH, DeStefano S, Reynolds MC, and Lauten D. 2000. Ruffed Grouse (*Bonasa umbellus*). *In* Poole A and Gill F, eds., *The Birds of North America* no. 515. Philadelphia: The Birds of North America Inc.

Russell SM. 1996. Anna's Hummingbird (*Calypte anna*). *In* Poole A and Gill F, eds., *The Birds of North America* no. 226. Philadelphia: Academy of Natural Sciences, and Washington: American Ornithologists' Union.

Ryan MR and Renken RB. 1987. Habitat use by breeding Willets in the northern Great Plains. *Wilson Bulletin* 99:175–189.

Ryan TP and Collins CT. 2000. White-throated Swift (*Aeronautes saxatalis*). *In* Poole A and Gill F, eds., *The Birds of North America* no. 526. Philadelphia: The Birds of North America Inc.

Ryder JP. 1993. Ring-billed Gull (*Larus delawarensis*). *In* Poole A, Stettenheim P, and Gill F, eds., *The Birds of North America* no. 33. Philadelphia: Academy of Natural Sciences, and Washington: American Ornithologists' Union.

Ryder RA and Manry DE. 1994. White-faced Ibis (*Plegadis chihi*). *In* Poole A and Gill F, eds., *The Birds of North America* no. 130. Philadelphia: Academy of Natural Sciences, and Washington: American Ornithologists' Union.

Ryser FA. 1985. *Birds of the Great Basin.* Reno: University of Nevada Press.

Saab VA, Bock CE, Rich TD, and Dobkin DS. 1995. Livestock grazing effects in western North America. *In* Martin TE and Finch DM, eds., *Ecology and Management of Neotropical Migratory Birds*, pp. 311–353. New York: Oxford University Press.

Saab VA and Vierling KT. 2001. Reproductive success of Lewis's Woodpecker in burned pine and cottonwood riparian forests. *Condor* 103:491–501.

Sakai HF and Noon BR. 1991. Nest site characteristics of Hammond's and Pacific-slope Flycatchers in northwestern California. *Condor* 93:563–574.

Sallabanks R and James FC. 1999. American Robin (*Turdus migratorius*). *In* Poole A and Gill F, eds., *The Birds of North America* no. 462. Philadelphia: The Birds of North America Inc.

Sanzenbacher PM and Haig SM. 2001. Killdeer population trends in North America. *Journal of Field Ornithology* 72:160–169.

Sauer JR, Hines JE, and Fallon J. 2005. *The North American Breeding Bird Survey, Results and Analysis 1966–2004*. Version 2005.2. USGS Patuxent Wildlife Research Center, Laurel, Md.

Schroeder MA, Young JR, and Braun CE. 1999. Sage Grouse (*Centrocercus urophasianus*). *In* Poole A and Gill F, eds., *The Birds of North America* no. 425. Philadelphia: The Birds of North America Inc.

Schukman JM and Wolf BO. 1998. Say's Phoebe (*Sayornis saya*). *In* Poole A and Gill F, eds., *The Birds of North America* no. 374. Philadelphia: The Birds of North America Inc.

Scott JM, Heglund PJ, and Morrison ML, eds. 2002. *Predicting Species Occurrences: Issues of Accuracy and Scale*. Covello, Calif.: Island Press.

Seamans ME, Corcoran J, and Rex A. 2004. Southernmost record of a Spotted Owl ¥ Barred Owl hybrid in the Sierra Nevada. *Western Birds* 35:173–174.

Sedgwick JA. 1993. Dusky Flycatcher (*Empidonax oberholseri*). *In* Poole A and Gill F, eds., *The Birds of North America* no. 78. Philadelphia: Academy of Natural Sciences, and Washington: American Ornithologists' Union.

———. 1994. Hammond's Flycatcher (*Empidonax hammondii*). *In* Poole A and Gill F, eds., *The Birds of North America* no. 109. Philadelphia: Academy of Natural Sciences, and Washington: American Ornithologists' Union.

———. 2000. Willow Flycatcher (*Empidonax traillii*). *In* Poole A and Gill F, eds., *The Birds of North America* no. 533. Philadelphia: The Birds of North America Inc.

Sheppard JM. 1996. Le Conte's Thrasher (*Toxostoma lecontei*). *In* Poole A and Gill F, eds., *The Birds of North America* no. 230. Philadelphia: Academy of Natural Sciences, and Washington: American Ornithologists' Union.

Shuford WD. 1993. *The Marin County Breeding Bird Atlas*. Bolinas, Calif.: Bushtit Books.

Shuford WD, Humphrey MB, and Nur N. 2001. Breeding status of the Black Tern in California. *Western Birds* 32:189–217.

Sibley DA. 2003. *The Sibley Field Guide to Birds of Western North America*. New York: Knopf.

Sidle JG, Koonz WH, and Roney K. 1985. Status of the American White Pelican: An update. *American Birds* 39:859–864.

Sloane SA. 2001. Bushtit (*Psaltriparus minimus*). *In* Poole A and Gill F, eds., *The Birds of North America* no. 598. Philadelphia: The Birds of North America Inc.

Small A. 1994. *California Birds: Their Status and Distribution*. Vista, Calif.: Ibis Publishing.

Smallwood JF and Bird DM. 2002. American Kestrel (*Falco sparverius*). *In* Poole A and Gill F, eds., *The Birds of North America* no. 602. Philadelphia: The Birds of North America Inc.

Smith AR. 1996. *Atlas of Saskatchewan Birds*. Regina: Saskatchewan Natural History Society.

Smith MR, Mattocks PW, and Cassidy KM. 1997. *Breeding Birds of Washington State*. Seattle: Seattle Audubon Society.

Smith SM. 1993. Black-capped Chickadee (*Poecile atricapillus*). *In* Poole A, Stettenheim P, and Gill F, eds., *The Birds of North America* no. 39. Philadelphia: Academy of Natural Sciences, and Washington: American Ornithologists' Union.

Snyder NFR and Schmitt NJ. 2002. California Condor (*Gymnogyps californianus*). *In* Poole A and Gill F, eds., *The Birds of North America* no. 610. Philadelphia: The Birds of North America Inc.

Sochi, K. 2001. Spring 2000 flight of Vaux's Swift (*Chaetura vauxi*) in the Great Basin. *Great Basin Birds* 4:31–32.

Sogge MK, Gilbert WM, and van Riper C. 1994. Orange-crowned Warbler (*Vermivora celata*). *In* Poole A and Gill F, eds., *The Birds of North America* no. 101. Philadelphia: Academy of Natural Sciences, and Washington: American Ornithologists' Union.

Sorenson MD. 1998. Patterns of parasitic egg laying and typical nesting in Redhead and Canvasback ducks. *In* Rothstein SI and Robinson SK, eds., *Parasitic Birds and Their Hosts: Studies in Coevolution*, 357–375. Oxford: Oxford University Press.

Squires JR and Reynolds RT. 1997. Northern Goshawk (*Accipiter gentilis*). *In* Poole A and Gill F, eds., *The Birds of North America* no. 298. Philadelphia: Academy of Natural Sciences, and Washington: American Ornithologists' Union.

Stacier CA and Guzy MJ. 2002. Grace's Warbler (*Dendroica graciae*). *In* Poole A and Gill F, eds., *The Birds of North America* no. 677. Philadelphia: The Birds of North America Inc.

Stallcup R. 2002. Hooded Merganser (*Lophodytes cucullatus*): Recent breeding range expansions in California and potential for breeding in Nevada. *Great Basin Birds* 5:45–46.

Steenhof K. 1998. Prairie Falcon (*Falco mexicanus*). *In* Poole A and Gill F, eds., *The Birds of North America* no. 346. Philadelphia: The Birds of North America Inc.

Stephens DA and Sturts SH. 1998. *Idaho Bird Distribution*. 2nd ed. Special Publication no. 13. Boise: Idaho Museum of Natural History.

Sterling JC. 1999. Gray Flycatcher (*Empidonax wrightii*). *In* Poole A and Gill F, eds., *The Birds of North America* no. 458. Philadelphia: The Birds of North America Inc.

Stiver S. 2001. Nevada's Greater Sage-Grouse (*Centrocercus urophasianus*) conservation strategy. *Great Basin Birds* 4:27–30.

Storer RW. 1965. The color phases of the Western Grebe. *Living Bird* 4:59–63.

Storer RW and Nuechterlein GL. 1992. Western and Clark's Grebes (*Aechmophorus occidentalis* and *A. clarkii*). *In* Poole A, Stettenheim P, and Gill F, eds., *The Birds of North America* no. 26. Philadelphia: Academy of Natural Sciences, and Washington: American Ornithologists' Union.

Sullivan P and Titus C. 2001. The Vermilion Flycatcher (*Pyrocephalus rubinus*) in southern Nevada. *Great Basin Birds* 4:19–24.

Sullivan SL, Pyle WH, and Herman WG. 1986. Cassin's Finch nesting in big sagebrush. *Condor* 88:378–379.

SWCA Environmental Consultants. 2005. Survey for Yuma Clapper Rail, Yellow-billed Cuckoos and Southwestern Willow Flycatcher along Las Vegas Wash, Clark County, Nevada. Unpublished report, Southern Nevada Water Authority, Las Vegas.

Swett J. 1999. The Southwestern Willow Flycatcher: An update. *Great Basin Birds* 2:12–15.

Tacha TC, Nesbitt SA, and Vohs PA. 1992. Sandhill Crane (*Grus canadensis*). *In* Poole A, Stettenheim P, and Gill F, eds., *The Birds of North America* no. 31. Philadelphia: Academy of Natural Sciences, and Washington: American Ornithologists' Union.

Tausch RJ. 1999. Transitions and thresholds: Influences and implications for management in pinyon and Utah juniper woodlands. *In* Monsen SB, Stevens R, Tausch RJ, Miller R, and Goodrich S, *Ecology and Management of Pinyon-Juniper Communities within the Interior West*, pp. 361–365. Proceedings of a symposium held September 15–18, 1997, Provo, Utah. Proceedings RMRS-P-9. Ogden, Utah: USDA Forest Service, Rocky Mountain Research Station.

Telfair RC. 1994. Cattle Egret (*Bubulcus ibis*). *In* Poole A and Gill F, eds., *The Birds of North America* no. 113. Philadelphia: Academy of Natural Sciences, and Washington: American Ornithologists' Union.

Tenney CR. 1997. Black-chinned Sparrow (*Spizella atrogularis*). *In* Poole A and Gill F, eds., *The Birds of North America* no. 270. Philadelphia: Academy of Natural Sciences, and Washington: American Ornithologists' Union.

Thelander CG and Crabtree M. 1994. *Life on the Edge: A Guide to California's Endangered Natural Resources*. Santa Cruz, Calif.: Biosystems Books.

Titus CK. 2003. *Field List of the Birds of Nevada*. Las Vegas: Red Rock Audubon Society.

Tobalske BW. 1997. Lewis' Woodpecker (*Melanerpes lewis*). *In* Poole A and Gill F, eds., *The Birds of North America* no. 284. Philadelphia: Academy of Natural Sciences, and Washington: American Ornithologists' Union.

Tomback DF. 1998. Clark's Nutcracker (*Nucifraga columbiana*). *In* Poole A and Gill F, eds., *The Birds of North America* no. 331. Philadelphia: The Birds of North America Inc.

Tomlinson C. 2001. Southern Nevada bird surveys. *Great Basin Birds* 4:57–58.

Trimble S. 1989. *The Sagebrush Ocean: A Natural History of the Great Basin*. Reno: University of Nevada Press.

Trost CH. 1999. Black-billed Magpie (*Pica pica*). *In* Poole A and Gill F, eds., *The Birds of North America* no. 389. Philadelphia: The Birds of North America Inc.

Twedt DJ and Crawford RD. 1995. Yellow-headed Blackbird (*Xanthocephalus xanthocephalus*). *In* Poole A and Gill F, eds., *The Birds of North America* no. 192. Philadelphia: Academy of Natural Sciences, and Washington: American Ornithologists' Union.

Tweit RC and Finch DM. 1994. Abert's Towhee (*Pipilo aberti*). *In* Poole A and Gill F, eds., *The Birds of North America* no. 111. Philadelphia: Academy of Natural Sciences, and Washington: American Ornithologists' Union.

Tweit RC and Tweit JC. 2000. Cassin's Kingbird (*Tyrannus vociferans*). *In* Poole A and Gill F, eds., *The Birds of North America* no. 534. Philadelphia: The Birds of North America Inc.

UOSBRC [Utah Ornithological Society Bird Records Committee]. 2005. *Field Checklist of the Birds of Utah*. Cedar City: Utah Ornithological Society.

USFWS [U.S. Fish and Wildlife Service]. 1986. Determination of Endangered Status for the Least Bell's Vireo. *Federal Register* 51:16474–16481.

———. 1995. Final rule determining endangered status for the Southwestern Willow Flycatcher. *Federal Register* 60:10694–10715.

———. 2001. Ruby Lake National Wildlife Refuge annual narrative report. Unpublished report, pp. 18–26.

———. 2003. *Waterfowl Population Status, 2003*. Washington: U.S. Department of the Interior.

USGS [U.S. Geological Survey]. 1998. Distribution and status of avifauna utilizing riparian habitat as in Clark County, Nevada. Unpublished report, U.S. Bureau of Reclamation, Lower Colorado River Region, Boulder City, Nev.

———. 2000. Distribution and status of avifauna utilizing riparian habitat as in Clark County, Nevada. Unpublished report, U.S. Bureau of Reclamation, Lower Colorado River Region, Boulder City, Nev.

Vander Wall SB and Balda RP. 1977. Coadaptations of the Clark's Nutcracker and pinyon pine for efficient seed harvest and dispersal. *Ecological Monographs* 47:89–111.

Van Rossem AJ. 1936. Birds of the Charleston Mountains, Nevada. *Pacific Coast Avifauna* 24:5–67.

Verbeek NAM and Caffrey C. 2002. American Crow (*Corvus brachyrhynchos*). *In* Poole A and Gill F, eds., *The Birds of North America* no. 647. Philadelphia: The Birds of North America Inc.

Verbeek NAM and Hendricks P. 1994. American Pipit (*Anthus rubescens*). *In* Poole A and Gill F, eds., *The Birds of North America* no. 95. Philadelphia: Academy of Natural Sciences, and Washington: American Ornithologists' Union.

Verner J and Engelsen GH. 1970. Territories, multiple nest building, and polygyny in the Long-billed Marsh Wren. *Auk* 87:557–567.

Verner J, McKelvey KS, Noon BR, Gutiérrez RJ, Gould GI Jr., and Beck TW, tech. coords. 1992. *The California Spotted Owl: A Technical Assessment of Its Current Status*. USDA Forest Service Pacific Southwest Research Station, SW-GTR-133. Berkeley: USDA Forest Service.

Verner J and Purcell KL. 1999. Fluctuating populations of House Wrens and Bewick's Wrens in foothills of the western Sierra Nevada of California. *Condor* 101:219–229.

Versaw AE. 2001. Winter range expansion and identification of the Lesser Goldfinch (*Carduelis psaltria*). *Great Basin Birds* 4:6–11.

Viverrette CB, Struve S, Goodrich LJ, and Bildstein KL. 1996. Decreases in migrating Sharp-shinned Hawks at traditional raptor-migration watchsites in eastern North America. *Auk* 113:32–40.

Voget K. 1998. Status of the Southwestern Willow Flycatcher (*Empidonax traillii extimus*) in Nevada. *Great Basin Birds* 1: 7–10.

Walsh J, Elia V, Kane R, and Halliwell T. 1999. *Birds of New Jersey*. Bernardsville: New Jersey Audubon Society.

Walter H. 2002. Natural history and ecology of the Chukar (*Alectoris chukar*) in the northern Great Basin. *Great Basin Birds* 5:28–37.

Walters EL, Miller EH, and Lowther PE. 2002. Red-breasted Sapsucker (*Sphyrapicus ruber*) and Red-naped Sapsucker (*Sphyrapicus nuchalis*). *In* Poole A and Gill F, eds., *The Birds of North America* no. 663. Philadelphia: Academy of Natural Sciences, and Washington: American Ornithologists' Union.

Walters JC. 2004. *Bird Stories and Sightings in Nevada*. Vol. 1. Carson City: Walters Publishing.

Walters RE, Sorensen E, and Casjens S, eds. 1983. *Utah Bird Distribution: Latilong Study 1983*. Salt Lake City: Utah Division of Wildlife Resources.

Warkentin IG and Reed JM. 1999. Effects of habitat type and degradation on avian species richness in Great Basin riparian habitats. *Great Basin Naturalist* 59:205–212.

Watt DJ and Willoughby EJ. 1999. Lesser Goldfinch (*Carduelis psaltria*). *In* Poole A and Gill F, eds., *The Birds of North America* no. 392. Philadelphia: The Birds of North America Inc.

Wauer RH. 1977. Changes in the breeding avifauna within the Chisos Mountain system. *In* Wauer RH and Riskind DH, eds., *Transactions of a Symposium on the Biological Resources of the Chihuahua Desert Region, United States and Mexico*, pp. 597–608. National Park Service Transaction Proceedings Series 3.

———. 1997. *Birds of Zion National Park and Vicinity*. Logan: Utah State University Press.

Webster MD. 1999. Verdin (*Auriparus flaviceps*). *In* Poole A and Gill F, eds., *The Birds of North America* no. 470. Philadelphia: The Birds of North America Inc.

Weckstein JD, Droodsma DE, and Faucett RC. 2002. Fox Sparrow (*Passerella iliaca*). *In* Poole A and Gill F, eds., *The Birds of North America* no. 715. Philadelphia: The Birds of North America Inc.

Weed BJ. 1993. Brown Creeper (*Certhia americana*). *In* Roberson D and Tenney C, eds., *Atlas of the Breeding Birds of Monterey County*, pp. 266–267. Carmel, Calif.: Monterey Peninsula Audubon Society.

Wehtje W. 2003. The range expansion of the Great-tailed Grackle (*Quiscalus mexicanus* Gmelin) in North America since 1880. *Journal of Biogeography* 30:1593–1607.

Weitzel NH. 1988. Nest-site competition between the European Starling and native breeding birds in northwestern Nevada. *Condor* 90:515–517.

Wheelwright NT and Rising JD. 1993. Savannah Sparrow (*Passerculus sandwichensis*). *In* Poole A and Gill F, eds., *The Birds of North America* no. 45. Philadelphia: Academy of Natural Sciences, and Washington: American Ornithologists' Union.

White CM, Clum NJ, Cade TJ, and Hunt WG. 2002. Peregrine Falcon (*Falco peregrinus*). *In* Poole A and Gill F, eds., *The Birds of North America* no. 660. Philadelphia: The Birds of North America Inc.

Wiens JA, Cates RG, Rotenberry JT, Cobb N, Van Horne B, and Redak RA. 1991. Arthropod dynamics on sagebrush (*Artemisia tridentata*): Effects of plant chemistry and avian predation. *Ecological Monographs* 61:229–321.

Wiens JA and Rotenberry JT. 1980. Patterns of morphology and ecology in grassland and shrubsteppe bird populations. *Ecological Monographs* 50:287–308.

———. 1985. Response of breeding passerine birds to rangeland alteration in a North American shrubsteppe locality. *Journal of Applied Ecology* 22:655–668.

Willey CH, and Halla BF. 1972. *Mute Swans of Rhode Island*. Providence: Rhode Island Department of Natural Resources.

Williams JM. 1996. Nashville Warbler (*Vermivora ruficapilla*). *In* Poole A and Gill F, eds., *The Birds of North America* no. 205. Philadelphia: Academy of Natural Sciences, and Washington: American Ornithologists' Union.

Wilson UW. 1991. Responses of three seabird species to the El Niño events and other warm episodes on the Washington coast, 1979–1990. *Condor* 93:853–858.

Winkler DW. 1996. California Gull (*Larus californicus*). *In* Poole A and Gill F, eds., *The Birds of North America* no. 259. Philadelphia: Academy of Natural Sciences, and Washington: American Ornithologists' Union.

Winkler DW and Shuford WD. 1988. Changes in the numbers and locations of California Gulls nesting at Mono Lake, California, in the period 1863–1986. *Colonial Waterbirds* 11:263–274.

Witmer MC, Mountjoy DJ, and Elliot L. 1997. Cedar Waxwing (*Bombycilla cedrorum*). *In* Poole A and Gill F, eds., *The Birds of North America* no. 309. Philadelphia: Academy of Natural Sciences, and Washington: American Ornithologists' Union.

Wolf BO. 1997. Black Phoebe (*Sayornis nigricans*). *In* Poole A and Gill F, eds., *The Birds of North America* no. 268. Philadelphia: Academy of Natural Sciences, and Washington: American Ornithologists' Union.

Wolf BO and Jones SL. 2000. Vermilion Flycatcher (*Pyrocephalus rubinus*). *In* Poole A and Gill F, eds., *The Birds of North America* no. 484. Philadelphia: The Birds of North America Inc.

Wolf BO and Walsberg GE. 1996. Thermal effects of radiation and wind on a small bird and implications for microsite selection. *Ecology* 77: 2228–2236.

Wolfe LM. 1979. *John of the Mountains: The Unpublished Journals of John Muir*. Madison: University of Wisconsin Press.

Woods CP, Csada RD, and Brigham RM. 2005. Common Poorwill (*Phalaenoptilus nuttallii*). *In* Poole A, ed., *The Birds of North America Online*. Ithaca: Cornell Laboratory of Ornithology; retrieved from The Birds of North American Online database: http://bna.birds.cornell.edu/BNA/account/Common Poorwill/

Wray T, Strait KA, and Whitmore RC. 1982. Reproductive success of grassland sparrows on a reclaimed surface mine in West Virginia. *Auk* 99:157–164.

Yahner RH. 1993. Effects of long-term forest clear-cutting on wintering and breeding birds. *Wilson Bulletin* 105:239–255.

Yasukawa K and Searcy WA. 1995. Red-winged Blackbird (*Agelaius phoeniceus*). *In* Poole A and Gill F, eds., *The Birds of North America* no. 184. Philadelphia: Academy of Natural Sciences, and Washington: American Ornithologists' Union.

Yates M. 1999. Satellite and conventional telemetry study of American White Pelicans in northern Nevada. *Great Basin Birds* 2:4–9.

Yosef R. 1996. Loggerhead Shrike (*Lanius ludovicianus*). *In* Poole A and Gill F, eds., *The Birds of North America* no. 231. Philadelphia: Academy of Natural Sciences, and Washington: American Ornithologists' Union.

Young GM and Oring LW. 2006. Long-billed Curlew nesting at Carson Lake, Nevada. *Great Basin Birds 8:16–20.*

Younk JV and Bechard MJ. 1994. Breeding ecology of the Northern Goshawk in high-elevation aspen forests of northern Nevada. *Studies in Avian Biology* 16:83–87.

Yunick RP. 1985. A review of recent irruptions of the Black-backed Woodpecker and Three-toed Woodpecker in eastern North America. *Journal of Field Ornithology* 56:138–152.

Zwickel FC. 1992. Blue Grouse (*Dendragapus obscurus*). *In* Poole A, Stettenheim P, and Gill F, eds., *The Birds of North America* no. 15. Philadelphia: Academy of Natural Sciences, and Washington: American Ornithologists' Union.

ABOUT THE AUTHORS

TED FLOYD was the project coordinator for the Nevada Breeding Bird Atlas from 1999 to 2002. He was responsible for all aspects of day-to-day oversight of the atlas project, including fieldwork, data analysis, and fundraising. During his stint in Nevada, Ted was also coeditor of the journal *Great Basin Birds,* Great Basin regional editor for the journal *North American Birds,* and list owner of the popular NVBIRDS listserver. He is currently the editor of *Birding,* the flagship publication of the American Birding Association. Ted is broadly involved in the birding community in his current home state of Colorado and has recently been appointed to the steering committee for the second-generation *Colorado Breeding Bird Atlas.*

CHRIS ELPHICK is a conservation biologist at the University of Connecticut. He earned his Ph.D. from the Program in Ecology, Evolution and Conservation Biology at the University of Nevada, Reno, for his work on the conservation implications of different methods of managing rice fields in California. His current research focuses primarily on the ecology and conservation of birds in wetlands and agricultural settings. Along with David Sibley and Barny Dunning, he wrote the syndicated column *Sibley on Birds* and coedited the *Sibley Guide to Bird Life and Behavior.*

In his role as cofounder and first director for the Great Basin Bird Observatory (GBBO), **GRAHAM CHISHOLM** helped build the partnerships to get the Nevada breeding bird atlas under way. Previously Graham served as an aide in the U.S. Senate and worked extensively on land and water conservation issues in the West while serving as director of both the Nevada and California programs for The Nature Conservancy. He coauthored with Larry Neel *The Birds of the Lahontan Valley.* A birder since childhood, Graham now lives in Berkeley, California, with his family and is the director of conservation for Audubon California.

KEVIN MACK grew up in Queens, New York, where birds were the one accessible part of the natural world. After spending summers in the Catskills and along the coast, he headed to Orono, Maine, where his college years found him birding as a complement to mountain biking, cross-country skiing, and rock climbing. On his arrival in Reno in 1996, he sought out opportunities to engage in various community conservation projects as an employee of Patagonia, Inc. Kevin began working with GBBO as part of an internship in 1998 and was soon hired to help coordinate various aspects of the atlas project with Ted Floyd. At the same time, Kevin helped found the Nevada Wilderness Project (NWP), a nonprofit group whose aim is to increase the amount of congressionally designated wilderness on Nevada's federal public lands. He now works on behalf of NWP in Washington, D.C., and has found time to volunteer (although not enough!) as a fieldworker for the second generation of the *Maryland Breeding Bird Atlas* project. He finds the humidity a challenge but has a much easier time finding Yellow-billed Cuckoos.

ROBERT G. ELSTON is a GIS (Geographic Information Systems) analyst and cartographer for the Biological Resources Research Center at the University of Nevada, Reno. In that role he supports ecologists and conservation biologists working in Nevada and the greater Intermountain West. His M.S. thesis examined the concept of island biography as applied to the high mountain ranges of the Great Basin.

ELISABETH M. AMMON completed her dissertation on habitat relationships and predictors of nest success in Lincoln's Sparrow and Wilson's Warbler at the University of Colorado, Boulder. At the University of Nevada, Reno, she later pursued research on riparian restoration projects in order to facilitate restoration of wildlife habitats and monitored bird and amphibian populations as the projects were implemented. Since 2002 she has been working for the Great Basin Bird Observatory as its bird monitoring coordinator and science director.

JOHN D. BOONE received his Ph.D. from the University of Colorado, Boulder, in 1995. His research there focused on habitat restoration for mammals and birds at the Rocky Mountain Arsenal, an EPA superfund site and national wildlife refuge near Denver. He came to the University of Nevada, Reno as a postdoctoral fellow, and later an assistant research professor, to work with the research group of Dr. Stephen St. Jeor, which specializes in the ecological aspects of zoonotic host-pathogen systems. He now focuses on conservation-oriented projects in the Intermountain West, and enjoyed the opportunity provided by the Nevada breeding bird atlas project to indulge his long-standing interests in wildlife management issues and in the generation and assembly of hefty manuscripts.

INDEX